Perspectives on Animal Behavior

Judith Goodenough
BIOLOGY DEPARTMENT
UNIVERSITY OF MASSACHUSETTS, AMHERST

Betty McGuire
PSYCHOLOGY DEPARTMENT
UNIVERSITY OF MASSACHUSETTS, AMHERST

Robert A. Wallace
ZOOLOGY DEPARTMENT
UNIVERSITY OF FLORIDA, GAINESVILLE

JOHN WILEY & SONS, INC.
NEW YORK • **CHICHESTER** • **BRISBANE** • **TORONTO** • **SINGAPORE**

Cover Photo by Jonathan Blair/Woodfin Camp Assoc.

Acquisitions Editor	Sally Cheney
Production Manager	Pam Kennedy
Marketing Manager	Catherine Faduska
Production Supervisor	Sandra Russell
Designer	Laura Nicholls
Manufacturing Manager	Andrea Price
Copy Editing Supervisor	Elizabeth Swain
Photo Research Director	Stella Kupferberg
Photo Researchers	Jennifer Atkins/Eloise Marion/Pat Cadley/Elyse Rieder/Ray Segal/Linda Sykes
Photo Research Assistant	Lisa Passmore
Illustration Coordinator	Jaime Perea
Illustrations	Precision Graphics
Scientific Illustrations	Howard S. Friedman

This book was set in 10/12 Garamond Light by Ruttle, Shaw & Wetherill, Inc. and printed and bound by R.R. Donnelley (Crawfordsville). The cover was printed by Phoenix Color Corporation. Color insert printed by Lehigh Press Colortronics.

Recognizing the importance of preserving what has been written, it is a policy of John Wiley & Sons, Inc. to have books of enduring value published in the United States printed on acid-free paper, and we exert our best efforts to that end.

Library of Congress Cataloging-in-Publication Data:

Goodenough, Judith.
    Perspectives on animal behavior / Judith Goodenough, Betty McGuire, Robert Wallace.
    Includes bibliographical references and index.
    ISBN 0-471-53623-7
    1. Animal behavior.   I. McGuire, Betty.   II. Wallace, Robert A.
  III. Title.
  QL751.G59   1993
  591.51—dc20   92-42640
                                                        CIP

Printed in the United States of America

10 9 8 7 6 5

Dedication

To Stephen, my husband, best friend, personal hero and the funniest person I know.

To Aimee and Heather, my daughters, the colors in my rainbow and the sparkle in my life. You fill me with love, wonder and amazement.

To Betty and Raymond Levrat, my parents, who instilled in me their love of learning and who may have planted the seed of my interest in animal behavior when I was a child by patiently reading *The Tawny Scrawny Lion* at each of my requests.

J.E.G.

To Dora and James P. McGuire and Kate and Willy Bemis.

B.M.

To Fulton Crews, fearless wilderness companion, first-rate scientist, bon vivant, and true friend.

R.A.W.

Preface

It is not true that each of us, the authors, had a sign over our desks that said:

Make it
Current!
Balanced!
Integrated!

But we may as well have had. These are the three things we talked most about, the things we worried about and aimed for. When we began to think about this book, we all agreed that there are several other fine texts in animal behavior, and that writing a new one would be entirely too time consuming to take lightly. If we couldn't bring something special to the book, we shouldn't even begin. As time went on, however, we found that our thoughts were merging into the single idea that we could bring something new and useful to the study of animal behavior.

The field has had a long and interesting (some would say rambunctious) history, marked at times by very special views that have, in some cases, led to discoveries. Those days are mostly behind us now, and the disagreements that exist currently are those one would expect in any mature discipline. There are still very pronounced ways of approaching animal behavior, however. Some people focus on its evolution, others on its ecology or physiology or history. We finally decided to write a textbook that began with the history of animal behavior studies and, once that perspective was established, moved to the cutting edge of modern research in all the major areas of animal behavior. It was at times upsetting (even maddening) to find that the arguments were continuing even as we were writing about some topic. So, in a few cases, we had to wait until the dust settled before concluding a discussion.

We were keenly aware that animal behavior is often taught as an upper level course in both biology and psychology departments, so not much could be taken for granted regarding the students' backgrounds. We decided, therefore, to provide the necessary background, for example in genetics, enabling the student to follow the finer points of any discussion without first having to bone up the hard way, through independent study.

We chose not only to develop a balanced view of animal behavior, including all the major approaches to the field, but also to integrate the information, where we could, to present the bottom line of current thinking. The field is a vigorous one, however, so in some cases we had to report there is yet no bottom line, no definitive answer. In those cases, we could only give a status report. But, even then we tried to include a brief summary of the arguments and the evidence so the reader could get the flavor of the battle. We hope such efforts stand as an invitation for young scientists to join the fray in the spirit of scientific adventure.

The road from those first discussions with our editor Sally Cheney to the book you hold in your hand has been longer and more arduous than any of us expected. And, even though we're glad it's done, we're already reading about new findings, new ideas, and unexpected observations, and eyeing each other apprehensively.

Judith Goodenough
Biology Department,
University of Massachusetts,
Amherst, MA

Betty McGuire
Psychology Department,
University of Massachusetts,
Amherst, MA

Robert A. Wallace
Zoology Department,
University of Florida,
Gainesville, FL.

Acknowledgments

We would not have even considered a project such as this without an editor like ours. Sally Cheney has brought an intelligence, energy, and enthusiasm that encouraged us every step of the way. As a team we have decades of experience in publishing, and Sally Cheney is, we think, hands down, the best editor that any of us has ever worked with. We would not hesitate to begin a new project, even one as daunting as this one, with her. We are indeed grateful to her.

There are many others at John Wiley & Sons who were involved with the production of this text, each one a professional and a pleasure to work with. Sandra Russell, our production supervisor, and Pam Kennedy, our production manager, somehow managed to keep track of everything. We'd like to thank Laura Nicholls for the design of the text & the insert. We feel that the art program is an asset of this text and thank the illustration coordinator, Jaime Perea, and illustration manager, Ishaya Monokoff. The superb drawings of Howard Friedman, resulted from his talent and also the hours he spent doing his own research on the appearance and behavior of many of the species illustrated. Even a quick flip through the pages of this book reveals the high standards of the Photo Research Manager, Stella Kupferberg. The photo researchers Jennifer Atkins and Eloise Marion were diligent in their pursuit of striking and pedagogically essential photographs. (Eloise Marion even managed to make this part of the production process fun.) The copyeditor supervisor was Elizabeth Swain. Our copyeditor, Linda Pawelchak was excellent. Her attention to detail greatly improved the manuscript. We would like to thank Bruce Winn and Victoria Ingalls for their contributions to chapters 6 and 12 respectively.

The preparation of this text proved to be a staggering and complex task for the two coauthors of this team who had never before worked on a project of this size. We could never have navigated our way through what seemed to be insurmountable obstacles without the help of our friend and colleague, Robert Wallace. He massaged the manuscript and our spirits.

We would also like to thank those who played indirect roles in the development of this book. The staff of the Biological Sciences library at UMass, especially James Craig, were an enormous help to us, as were our colleagues in the Departments of Biology and Psychology.

For Judith Goodenough: I would like to thank the friends, family and colleagues who read and criticized (usually gently) various drafts of my chapters: Diane Rossiter, Ellen Zagajeski, Karen Morse, Carol Goodenough, and John D. Palmer. I would also like to thank my friends in "the group" for their support and encouragement.

My husband Steve kept the home fires burning while I was working on this book, provided endless encouragement, and held my hand when I needed confidence. Laughter is the best reliever of stress and his wit kept me sane. (Do you have to make a joke out of everything?) My daughters, Aimee and Heather, are the joy in my life and my toughest critics. They were patient and understanding each time I was late picking them up at the dance studio. They lovingly helped me keep my priorities straight with reminders such as, "Have some fun," or "Get a life." The willingness of my parents, Betty and Ray Levrat, to help in any way—driving the girls to the doctor or dance lessons, painting, wallpapering, and even cleaning—allowed me to focus on writing.

Kate Bemis, the new daughter of Betty McGuire, could not have timed her birth at a more opportune time relative to the production of this book. Thanks to Kate's sense of timing and the willingness to babysit of her grandmothers, Dora McGuire and Jane Bemis, we somehow made it through galleys and pages without a delay. (Betty, I still don't know how you did it.)

For Betty McGuire: I relied on many institutions during this project, including the Department of Psychology at UMass and Scripps Institute of Oceanography, surely one of the world's most inspirational environments for biology. Melinda Novak, Lowell Getz, Glenn and Mary Sue Northcutt, and Willy Bemis remain my patient and supportive friends (at least for this edition). My style and examples improved thanks to reviews by Dora McGuire, George Strieter, Donald Dewsbury, and Willy Bemis.

Contents

CHAPTER
1

Introduction

Four Questions About Animal Behavior

Animal Behavior as an Interdisciplinary
  Study

*I*n one way or another, people have been studying animal behavior for thousands of years. The behavior of animals was important in earlier ages for some of the same reasons it is important now. The most skillful hunters and fishermen are usually those who can make predictions regarding the behavior of their prey (Figure 1.1). It was important to know that when salmon are spawning, they will not respond to a fisherman's bait, that many rodents escape toward the dark while most birds escape toward the light, and that many kinds of animals will fight, some ferociously, if they are trapped.

The study of animal behavior may have occupied the fringes of human consciousness for centuries on just such a practical basis. Later,

Figure 1.1 People have been studying animal behavior for centuries, sometimes for very practical reasons. Knowledge of the behavior of game species may make it easier to put food on the table.

when animals were domesticated and put to work, it was necessary to learn new things about them. Horses could be trained for riding or for pulling wagons or tools. Dogs could be trained to track prey or to protect individual humans; cats could not.

In time, the study of animal behavior took on new dimensions. The goals, as well as the techniques, changed. Animals are no longer studied simply so that we can exploit them better, although this may still be one reason for our attention. Now, though, we have become aware that increased knowledge of the behavior of specific species in their natural habitats may help us save some endangered groups from extinction. In addition, information on their normal behavior may help us ensure their welfare, not just in the wild but also in laboratories or zoos. We may be interested in behavior as an example of a broader intellectual concern, such as evolutionary theory. And, sometimes we study animal behavior simply because our curiosity prompts us to ask questions.

Four Questions about Animal Behavior

As casual observations of animal behavior crystallized into a field of scientific study, Niko Tinbergen (1963) identified four types of questions that should be asked about behavior: What are the mechanisms that cause it? How does it develop? What is its survival value? How did it evolve? Tinbergen believed that the biological study of behavior, ethology, should "give equal attention to each of them and to their integration." It was later suggested that consideration of the mechanisms of behavior should, at least in some cases, include both cognitive and emotional mental processes (Sherman, 1988).

To better appreciate the types of questions we may ask about animal behavior, consider those that may be raised as a group of us watch in amazement as moonlight seems to flicker as a mass of thousands of tiny songbirds fly through it during migration. We may each become curious about migration, but depending on our personal interests, we may each ask different types of questions. How do they "know" it is time? How do they find their way? Such questions focus on the mechanisms underlying the behavior. Must those making this journey for the first time learn the route from experienced travelers? Do they inherit a directional tendency from their parents? Questions such as these concern development. Why do they do it? What do they gain that outweighs the risks and demands of such a journey? These are questions regarding the survival value, or adaptiveness, of migration. Finally, how did it all begin? Were the advancing glaciers responsible? Were the migratory paths modified during the thousands of years each species has been migrating? These questions center on the evolution of the behavior. So we see that when we ask *why* an animal behaves a certain way, some of us may be asking about immediate causes (the machinery underlying the response) and others may be asking about the evolutionary causes.

No one type of question is better than the others. Answers to all types are necessary as we weave the fabric of our understanding. These are not competing avenues of investigation. Rather, they are complementary. Each may feed back on the others, deepening our understanding and broadening our avenues of investigation (Armstrong, 1991; Halpin, 1991; Stamps, 1991).

Animal Behavior as an Interdisciplinary Study

Marion Stamp Dawkins (1989) has drawn an analogy between Tinbergen's four aspects of investigation of behavior and the four legs of an animal. An animal lacking one of its legs can only hobble along. Similarly, progress in the study of animal behavior is hampered by a lack of information in any one of these areas of study. This is not to imply that each investigator must ask all types of questions. Often we find that individuals are more excited by one type of question than by others. However, each investigator will be more successful in finding the answer to the question of personal interest if he or she is armed with information and techniques from all four areas of study.

The progress of scientists whose primary interest is in the mechanisms of behavior will be slowed if they fail to take note of the evolutionary history of the species they are studying. For example, at one time, psychologists interested in learning theory strove to find the "general laws of learning." They found, however, that many species seemed to break the general laws. Although animals learn some things easily, they learn other things very slowly or not at all (see Chapters 2 and 5). In the laboratory, it is often difficult to understand why these predispositions or difficulties in learning should exist. Why, for example, does a hummingbird quickly learn to leave an artificial flower at which it just fed to feed at a new one, but have difficulty learning to return to the same artificial flower to feed several times (Cole et al., 1982)? Such biases often begin to make more sense when they are considered in light of the animal's natural environment. In nature, hummingbirds feed from small flowers (Figure 1.2). A flower's nectar supply is quickly exhausted and slowly replenished. Therefore, the most efficient feeding strategy for a hummingbird is to avoid revisiting flowers it has drained.

As a result of this and many similar observations, it was gradually realized that evolution shapes the learning "styles" of different species to suit their ecological demands. Individuals may learn certain things more easily because, in the natural ecological setting, those that do have a better chance of surviving and leaving offspring than those that do not.

Figure 1.2 A hummingbird gathering nectar. A hummingbird learns to leave a flower at which it fed more easily than it learns to revisit a flower that it has drained. Such learning baises may be seen as specializations suited to the natural environment in which the organism learns.

Scientists who find evolutionary questions most fascinating may find their research hampered by a lack of information on the mechanisms of behavior. One of the evolutionary questions that has intrigued some investigators is the function of kin discrimination. Throughout the animal kingdom we find many examples of species in which individuals treat their relatives differently from others. Although the phenomenon is well documented, its evolutionary value has been debated. For example, some have suggested that it functions to reduce inbreeding among relatives, a practice that might allow damaging genetic traits to surface. Others suggest that it is necessary for nepotism—that is, favoritism shown to relatives. However, if more were known about the mechanisms underlying kin recognition, behavioral ecologists might be better able to predict the circumstances in which relatives are treated differently, and, in turn, this information might provide some hints as to the function of kin discrimination (Waldman, Frumhoff, and Sherman, 1988).

There are many examples of techniques borrowed from one area of study that have been useful in answering other types of questions. For instance, some of the techniques developed to study animal learning have been used to study the evolutionary aspects of foraging behavior (see Chapter 12). A species can be viewed as having evolved such that individuals obtain the greatest amount of energy during the time spent foraging. The forager is, therefore, seen as making decisions about where and how to find food using information about the environment, information that, in many cases, must be learned. Psychologists have developed techniques for studying learning in the laboratory that allow behavioral ecologists to ask the following questions: How does an optimal forager decide where to look for food? What rules govern its decision to leave its current feeding area in search of a new place where food could be gathered more efficiently? Do animals form "search images" that help them find camouflaged prey more easily? Indeed, the behavioral ecologist interested in studying foraging might "optimize" research gains by taking advantage of the techniques already developed by psychologists (Kamil, 1983).

Likewise, techniques for assessing genetic variation in populations have been applied to

Figure 1.3 Florida scrub jay helpers at the nest. The helpers are nonbreeding individuals that live on the territory of a breeding pair and help care for the young by feeding them and defending them against snakes. The Florida scrub jay helpers are, therefore, altruists.

studies of kinship and paternity, issues that are related to the evolution of behavior (see Chapters 14 and 19). Kinship, the degree to which individuals are related, is important in understanding how altruism may have evolved. Altruism, the act of helping another individual, involves some cost to the altruist. Therefore, according to the logic of evolution, it should be selected against relative to more selfish behaviors. But there are many examples in which animals appear to be doing things that assist others (Figure 1.3). When we look at the situation more closely, we often find that the costs of the good deed are outweighed by its benefits. Evolution measures benefits in terms of the number of copies of one's genes that are passed into the next generation. Relatives are likely to share a proportion of one's genes. Therefore, if a good deed benefits close relatives, the altruist may gain more than it loses. This explanation for the

Figure 1.4 A mountain bluebird at its nest with young. Experiments designed to determine whether this species was altruistic had to be interpreted in a manner that was consistent with the known mechanisms for the altruistic actions.

evolution of altruism, called kin selection, predicts that the altruist would assist relatives. To test this prediction, it is often necessary to determine the degree of kinship.

There are instances in which information regarding the underlying mechanisms of behavior may be helpful in designing experiments to test hypotheses about the evolution of behavior. As an example, consider the dialog between Harry Power and Stephen Emlen regarding Power's experiment to determine whether the mountain bluebird (*Sialia currucoides*) exhibits altruism (Figure 1.4). Power (1975) reasoned that "One way to determine the frequency of true altruism is to give the animals the choice of behaving altruistically or selfishly." The choice he presented to adult bluebirds was whether to care for young to which they were not related. He did this by removing one member of a mated pair that was caring for nestlings. A replacement moved into the nest from the nonbreeding surplus in the surrounding area and established a relationship with the widowed parent. Would

the replacement care for the young? It turns out that replacement males did not. They did not feed the young or sound an alarm when a predator approached. Power concluded that they were not altruistic.

However, Emlen (1976) pointed out that the conclusion was not justified because the design of the experiment failed to consider the immediate causes of behavior. He maintained that the birds did not have the choice of behaving altruistically or selfishly because they were not in the proper hormonal condition to be responsive to the young nestlings. To become hormonally prepared, a bird must experience the sequence of events in a normal breeding cycle—courting, nest building, egg laying, and incubation. Since the replacement birds did not experience the entire breeding cycle, they were physiologically incapable of responding to the cues that would cause them to feed the young.

Power (1976) then argued that whether or not a bird must be in the proper hormonal condition to feed nestlings, it is capable of giving alarm calls in the presence of a predator at any time of the year. Indeed, the replacement males in this study did, on occasion, give alarm calls. However, they did so only when the female became excited about apparent danger to her young. Nonetheless, they were *capable* of sounding an alarm and, therefore, did have the choice of whether or not to defend the young. So we see that even if one's interests are far from the immediate mechanisms of a behavior, these mechanisms bear on the conclusions that can be drawn from an experiment.

It is apparent that our understanding of animal behavior will be fuller if both the immediate causes and the evolutionary causes are considered. As Marian Stamp Dawkins (1989) has said,

... genes operate through making bodies do things. These bodies have to develop and they need machinery (sense organs, decision centers, and means of executing action) to be able to pass their genes on to the next generation. To understand this process fully, we need a science that is not only aware of the evolutionary ebb and flow of genotypes over evolutionary time, but can look at the bridge between generations, at the bodies that grow and move and court and find food and pass their genetic cargo on through time with the frailest and most marvelous of flesh-and-blood machinery.

In the pages that follow, we will consider aspects of behavior related to all four of Tinbergen's questions. We will treat each aspect as interesting in its own right and, in some discussions, we will consider how the various aspects may act synergistically to broaden our understanding of animal behavior.

Summary

Animal behavior is studied for many reasons, both practical and intellectual. A full understanding of behavior requires answers to four types of questions, those about (1) immediate mechanisms, (2) development, (3) survival value, and (4) evolution. Our progress in understanding the behavior of animals will be enhanced by considering all four types of questions.

PART
One

Approaches to the Study of Animal Behavior

CHAPTER 2

History of the Study of Animal Behavior

In order to understand a field of study today, we must know something about its past. In this chapter we consider the history of the study of animal behavior. Our focus is on the development of key concepts in the field. We begin with some of the earliest ideas concerning the minds and emotions of animals and describe the conceptual framework provided to the study of behavior by Darwin's theory of evolution. We then describe the development during the early part of the twentieth century of ethology and comparative psychology, two approaches to animal behavior. Despite initial hostility between proponents of the two approaches, hostility that eventually culminated in the nature/nurture controversy, interactions between the two camps modified in positive ways the types of questions that we ask about animal behavior today. Indeed, many of the contemporary approaches to the study of animal behavior that are discussed in subsequent chapters were inspired by the arguments of ethologists and comparative psychologists. We conclude the chapter with a brief examination of sociobiology, an approach to the study of behavior that achieved prominence in the mid-1970s.

The Beginnings

INTELLECTUAL CONTINUITY IN THE ANIMAL WORLD

It is difficult, perhaps impossible, to pinpoint the precise beginnings of the study of animal behavior. Rather than attempting this feat, we will simply consider some of the highlights in the development of the discipline. One idea, that of intellectual continuity among animals, was important in shaping some of the earliest views of animal behavior. Although the idea of intellectual continuity was summarized in 1855 by Herbert Spencer in his book entitled *Principles of Psychology*, its roots can be traced back to the ideas of the ancient Greek philosophers. The idea focused on a continuity in mental states between "lower" and "higher" animals and was based on a picture of evolution similar to Aristotle's scala naturae, the Great Chain of Being. In the scala naturae, the evolution of species was viewed as linear and continuous. This classification system was hierarchical with animals ranked according to their degree of relationship, and with the highest point of evolution being humans (Hodos and Campbell, 1969). At the bottom of the scale would be creatures such as sponges; then moving up the scale, there would be insects, fish, amphibians, reptiles, birds, nonhuman mammals, and finally, humans. Because evolution was seen as a linear process with each higher species evolving from a lower one until, finally, humans emerged, it was thought that the animal mind and the human mind were simply points on a continuum.

DARWIN'S EVOLUTIONARY FRAMEWORK

A few years after the publication of Spencer's book, Charles Darwin (Figure 2.1) published his thoughts on evolution by natural selection in *The Origin of Species* (1859). Although Darwin's focus in this book was not on animal behavior, his ideas provided a conceptual framework within which the field of animal behavior could develop. His ideas, discussed in more detail in Chapter 4, can be briefly summarized as follows:

1. Within a species, there is usually variation among individuals and some of this variation is inherited.

2. Most of the offspring produced by animals do not survive to reproduce.

3. Some individuals survive longer and produce more offspring than others, as a consequence of their particular inherited characteristics.

Figure 2.1 Charles Darwin. His ideas on evolution by natural selection provided an evolutionary framework for the study of animal behavior.

4. Natural selection is the differential survival and reproduction of individuals that results from genetically-based variation in their behavior, morphology, physiology, and so on.

5. Evolutionary change occurs as the heritable traits of successful individuals (i.e., those that survive and reproduce) are spread throughout the population, while those traits of less successful individuals are lost.

In two later books, *The Descent of Man, and Selection in Relation to Sex* (1871) and *Expression of the Emotions in Man and Animals* (1873), Darwin applied his evolutionary theory to behavior. In these volumes Darwin recorded his careful and thorough observations on animal behavior, but his records were anecdotal and often anthropomorphic. This was not, however, sloppy science. In the tradition of his day, he believed that careful observations were useless

unless they were connected to a general theory. Darwin's general theory was evolution by natural selection. Because humans evolved from other animals, he considered the minds of humans and animals to be similar in kind and to differ only in complexity. Because of his belief, he described the behavior of animals using terms that denote human emotions and feelings: Ants despaired, and dogs expressed pleasure, shame, and love. Darwin's opinion was influential and it is not surprising that others interested in animal behavior followed his lead. Both ethologists and comparative psychologists trace the beginnings of their respective fields to the ideas of Darwin.

For about a decade after the publication of *Expression of Emotions in Man and Animals*, descriptions of animal behavior were usually stories about the accomplishments of individual animals that were believed to think and experience emotions as humans do. For example, on the basis of his subjective interpretation of what he observed, George J. Romanes, a protégé of Darwin's, constructed a table of emotions that charted the evolution of the mind and listed the emotions in order of their historical or evolutionary appearance (Figure 2.2). In his books, *Animal Intelligence* (1882), *Mental Evolution in Animals* (1884), and *Mental Evolution in Man* (1889), Romanes examined the implications of Darwinian thinking about the continuity of species for the behavior of nonhuman and human animals.

In addition to Romanes, several other scientists made notable contributions to the study of animal behavior at the turn of the century. Jacques Loeb believed that all patterns of behavior were simply "forced movements" or tropisms, physiochemical reactions toward or away from stimuli. Herbert Spencer Jennings, perhaps best known for his book *Behavior of the Lower Organisms* (1906), disagreed with Loeb's ideas and emphasized instead the variability and modifiability of behavior. Douglas A. Spalding, another scientist, is best known for his experimen-

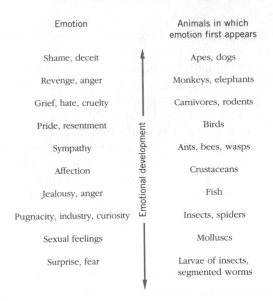

Emotion	Animals in which emotion first appears
Shame, deceit	Apes, dogs
Revenge, anger	Monkeys, elephants
Grief, hate, cruelty	Carnivores, rodents
Pride, resentment	Birds
Sympathy	Ants, bees, wasps
Affection	Crustaceans
Jealousy, anger	Fish
Pugnacity, industry, curiosity	Insects, spiders
Sexual feelings	Molluscs
Surprise, fear	Larvae of insects, segmented worms

Figure 2.2 The ideas of George J. Romanes on the evolutionary appearance of emotions in animals. (Modified from Romanes, 1889.)

tal work on the behavioral development of chicks. Of course there were many other pioneers in the study of animal behavior, but we will move on to the turmoil that developed in the discipline during the twentieth century.

As interest in animal behavior grew during that time, differences in opinion developed. These differences eventually led to the development of two major approaches to the study of animal behavior, ethology and comparative psychology, centered in Europe and the United States, respectively. The split that developed between the two approaches seemed at times quite severe. Indeed, one must wonder why two groups of scientists, each striving for a greater understanding of the marvels of animal behavior, could have stood worlds apart. The European ethologists and the American comparative psychologists were separated by more than the Atlantic Ocean. In fact, the basic questions they asked about animal behavior were different (re-

call from Chapter 1 that the four questions outlined by Niko Tinbergen in 1963 concerned the mechanism, function, development, and evolution of behavior). Whereas the ethologists focused their attention on the evolution and function of behavior, the comparative psychologists concentrated on the mechanism and development of behavior. Because they asked different questions, the types of behavior they studied, and even their experimental organisms, differed. Whereas ethologists, by and large, studied innate behavior in birds, fish, and insects, comparative psychologists (particularly those of the behaviorist school, see later discussion) emphasized learned behavior in mammals such as the Norway rat. Furthermore, in order to determine the normal function of a behavior, ethologists often attempted to observe the animal in its natural habitat or in environments designed to simulate that habitat. On the other hand, comparative psychologists believed that learning was best studied in the laboratory, where variables could be controlled. Finally, ethologists were interested in species differences while comparative psychologists searched for general "laws" of behavior.

Of course there were many exceptions to this characterization of ethologists and comparative psychologists. Some ethologists made remarkable discoveries indoors in their homes and laboratories and explored the influence of experience on behavioral development (Barlow, 1989; Bateson and Klopfer, 1989), and some comparative psychologists studied a wide range of species and patterns of behavior and conducted their studies in the field (Dewsbury, 1989). Indeed, within each field there were individuals interested in all four questions of animal behavior. Although traditionally ethology and comparative psychology have been portrayed as very different approaches, some recent accounts of the history of animal behavior have tended to downplay the differences between them (e.g., Dewsbury, 1984, 1989).

Classical Ethology

THE APPROACH: EVOLUTIONARY, COMPARATIVE, DESCRIPTIVE, FIELD-ORIENTED

"Why is that animal doing that?" is perhaps the fundamental question of ethology, the approach to the study of behavior developed largely by Konrad Lorenz, Niko Tinbergen, and Karl von Frisch, European zoologists who shared the Nobel prize in medicine and psychology in 1973 (Figure 2.3). Traditionally, ethology concentrated on the evolutionary basis of animal behavior. Because natural selection can only act on traits that are genetically determined, it seemed

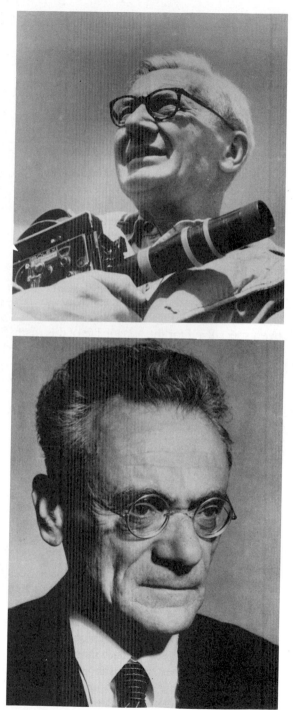

Figure 2.3 Konrad Lorenz (above), Niko Tinbergen (top, right), and Karl von Frisch (right), three ethologists who shared the Nobel Prize in 1973.

a logical outcome of the ethologists' basic interest in the evolution of behavior that they should focus their attention on those behavior patterns that are inherited. An emphasis on phylogeny (the evolutionary history of a species) was particularly characteristic of the work of Konrad Lorenz.

The studies of ethologists often involved comparisons among closely related species. In the words of Lorenz (1958), "Every time a biologist seeks to know why an organism looks and acts as it does, he must resort to the comparative method." The method Lorenz was referring to is that employed by comparative anatomists when they ask the same question about morphology. If a comparative anatomist were to ask why a whale's flipper is structured as it is, the skeleton of the flipper might be compared with that of the forelimb of other vertebrates. It could then be seen that the typical vertebrate forelimb has been specialized for the aquatic life of this mammal. Likewise, if one were to wonder why a male fly of the species *Hilara sartor* spins an elaborate silken balloon to present to a female before mating (Figure 2.4), the significance of the behavior would become apparent after comparing it with the mating behavior of the other species of flies in the family Empididae. Let us consider the gift-giving behavior of *Hilara sartor* in more detail in order to illustrate the ethologist's comparative method.

Among the empid flies there are species that display almost every imaginable evolutionary step progressing toward the balloon display. By observing the manner in which the male empid flies approach the female for mating, one sees that at the heart of the problem is the fact that the male may be a meal, rather than a mate, for the predacious female. In one species, *Empis trigramma*, the male approaches the female while she is eating. Because her mouth is already full, his well-timed approach increases his chances of surviving the encounter. In another species, *Empis poplitea*, the male captures a prey,

perhaps a fly, and gives it to the female, providing her with dinner before attempting to copulate. Males of the species *Hilara quadrivittata* gift wrap the meal in a silky cocoon before offering it to the female. In another species, *Hilara thoracica*, the cocoon, or case, is large and elaborate, while the food inside is small and of little value. In yet another species, *Hilara maura*, only some males place food inside cocoons, others enclose something meaningless, like a daisy petal. Finally, the males of *Hilara sartor* present the females with an empty gauze case that turns off the predatory behavior of the female, thereby allowing them to mate (Kessel, 1955). Without a comparison of the behavior with that of other species, an observer would be hard pressed to explain why males offer silken balloons to females.

In addition to utilizing the comparative method, ethologists often worked in the field rather than in the laboratory. After all, they reasoned, it is in the natural setting that the normal context in which the behavior is displayed is apparent. From this, the function of the behavior may be deduced, and knowledge of the function may allow us to understand why the behavior has been shaped to its present form by natural selection. Tinbergen and his students conducted much of their research in the field. Lorenz and his followers, on the other hand, studied captive animals, but they often attempted to simulate in captivity some characteristics of the animal's natural habitat (Barlow, 1991).

CLASSICAL ETHOLOGICAL CONCEPTS

The Fixed Action Pattern

At the turn of the twentieth century, Charles Otis Whitman of the University of Chicago and Oskar Heinroth of the Berlin Aquarium were pioneering the field of ethology (Lorenz, 1981). Both scientists were interested in the behavior of birds and each independently came to the conclusion that the displays of different species are

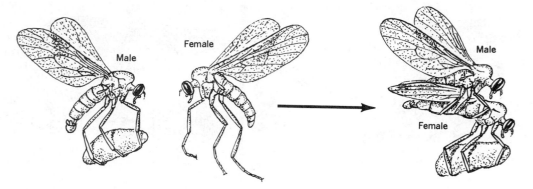

Figure 2.4   Male flies of the species *Hilara sartor* present females with an empty silken balloon before mating. This behavior can be understood by comparing the behavior of closely related species, that is, by using the comparative method characteristic of the ethological approach. (Drawn from descriptions in Kessel, 1955.)

often exceptionally constant. In fact, they considered these patterns of movement to be just as reliable as morphological characters in defining a particular group.

Wallace Craig, a student of Whitman's, pursued the study of avian displays and examined the behavioral sequences of ring doves. He is perhaps best known for making the distinction between appetitive and consummatory components of behavior. He noted that in the beginning of a behavioral sequence, the acts are often variable and serve to bring the animal into the proper stimulus situation for a simpler, more stereotyped response. Impressed by the plasticity and apparent purposefulness of the initial motor patterns, Craig referred to these actions as appetitive behavior. As the name implies, appetitive behavior is thought of as a striving for the correct stimulus situation that will release the more rigid pattern of behavior that concludes the sequence, as if the animal had an appetite for that specific stimulus. Continuing with the epicurean designation, the stereotyped behavior that completes the sequence is called the consummatory act. Just as the consumption

of a meal "satisfies" the appetite, the consummatory act brings about a sudden drop in the animal's motivation to perform those behaviors. Craig published his ideas in 1918 in a paper entitled "Appetites and aversions as constituents of instincts."

The stereotyped patterns of behavior that intrigued ethologists such as Whitman, Heinroth, and Craig were named fixed action patterns by Lorenz. By definition, a fixed action pattern (FAP) is a motor response that may be initiated by some environmental stimulus, but can continue to completion without the influence of external stimuli. For example, Lorenz and Tinbergen (1938) showed that a female greylag goose (*Anser anser*) will retrieve an egg that has rolled just outside her nest by reaching beyond it with her bill and rolling it toward her with the underside of the bill (Figure 2.5). If the egg is experimentally removed once the rolling behavior has begun, the goose will continue the retrieval response until the now imaginary egg is safely returned to the nest. We have emphasized the fact that once initiated, fixed action patterns continue to completion. There is little consen-

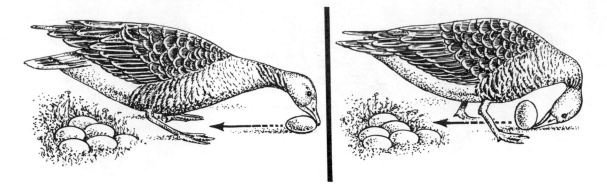

Figure 2.5 The egg retrieval response of the greylag goose. The chin-tucking movements used by the female as she rolls the egg back to the nest are highly stereotyped and are an example of a fixed action pattern. The side-to-side movements that correct for any deviations in the path of the egg are termed the taxis component of the response. If the egg is removed, the female will continue to roll an imaginary egg back to the nest. One defining characteristic of a fixed action pattern is that it will continue to completion even in the absence of guiding stimuli. (Drawn from photograph in Lorenz and Tinbergen, 1938.)

sus, however, regarding the defining attributes of FAPs. Other characteristics that have been used to describe them include (1) the sequence of component acts of an FAP is unalterable; (2) an FAP is not learned; (3) an FAP may be triggered under inappropriate circumstances; and (4) an FAP is performed by all appropriate members of a species (Dewsbury, 1978).

Although it is true that a specific FAP is executed in much the same way by each individual of a species, there is some variation because, in some cases, the FAP must be oriented according to the conditions under which it is performed. These orientation movements are called taxes. As the greylag goose retrieves an egg, for example, the egg might roll slightly to one side or the other. The female must adjust her retrieval motions according to which way the egg rolls. The chin-tucking movements constitute the FAP and the orientation movements, the taxis component.

The concept of a fixed action pattern has been

questioned in the years following Lorenz's first introduction of the term. George Barlow (1968) suggested that in reality, most patterns of behavior are not as stereotyped as suggested by the notion of the FAP, and furthermore, most cannot easily be separated into fixed and orientation components. He suggested replacing the term fixed action pattern with modal action pattern (MAP). In specific cases, however, the term FAP may be appropriate. Finley and colleagues (1983) examined the courtship displays of mallard ducks (*Anas platyrhyncos*) and concluded that the patterns of behavior were indeed as highly stereotyped as suggested by the notion of FAP. We will continue to use the traditional term here, keeping in mind that there is some debate over the appropriateness of its use.

Sign Stimuli and Releasers as Triggers

A fixed action pattern is obviously produced in response to something in the environment. Let's consider the nature of the stimulus that might trigger the behavior. Ethologists called such a stimulus a sign stimulus. If the sign stimulus is emitted by a member of the same species, it is termed a social releaser, or simply a releaser. Releasers are important in communication among animals, as we will see in Chapter 17. Although releasers are technically a type of sign stimulus, the terms are often used interchangeably.

Sign stimuli may be only a small part of any environmental situation. For example, a male European robin (*Erithacus rubecula*) will attack another male robin that enters its territory. Experiments have shown, however, that a tuft of red feathers is attacked as vigorously as an intruding male (Lack, 1943). The attack is not stimulated by the sight of another bird, but only by the sight of red feathers. Of course, in the world of male robins, red feathers usually appear on the breast of a competitor.

Sign stimuli, simple cues that may be indicative of very complex situations, get through to the animal's nervous system where they release patterns of behavior that may consist, in large part, of fixed action patterns. For example, the attack of the male European robin may be composed of FAPs involving pecking, clutching, and wing fluttering. The end result is that when an intruding male robin appears, it is immediately identified and effectively attacked.

Types of Sign Stimuli   Any of the traits possessed by an animal or an object may serve as a sign stimulus. It may have a certain color, a special shape, or a pattern of movement. At the same time it may produce a sound or have an odor. Perhaps it even touches the responding animal in a certain way. Any one of these aspects may be the one that releases the response—that is, that serves as the sign stimulus. How, then, are we to know which character serves as the sign stimulus? One way in which ethologists can tease out the sign stimuli from the barrage of

information reaching the responding animal is with models of the signaling animal with only one trait (or a few) presented at a time.

Many sign stimuli and releasers are visual. The male common flicker (*Colaptes auratus*), a type of woodpecker found throughout North America, has a black stripe or moustache extending from its bill along the sides of its head (Figure 2.6). It is the presence of this moustache on a male entering another's territory that is the specific releaser for aggressive behavior. If such a moustache is painted on a female of a mated pair, domestic harmony disappears to the point at which the male actually attacks his mate. When the moustache is removed, his aggression ceases and she is once more allowed to enter the territory (Noble, 1936).

In some cases, color is a sign stimulus. During the early spring, a male three-spined stickleback,

Figure 2.6   A male (bottom) and female flicker. It is the black moustache, present only on the male, that is the visual releaser for aggression among males of this species.

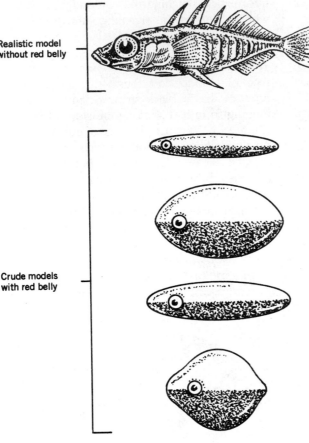

Realistic model
without red belly

Crude models
with red belly

Figure 2.7 Models of male three-spined sticklebacks presented to other males of the species to test their effectiveness as releasers. The crude models with red bellies were more effective in eliciting aggression from males than was the more realistic replica that lacked the characteristic red belly. Experiments by Tinbergen and colleagues using models such as these demonstrated that it is the red color on the underside of the male that releases aggression from other males. More recent experiments, however, suggest substantial individual variation in the responses of males to red bellies. (After Tinbergen, 1948.)

Gasterosteus aculeatus, concentrates his efforts on reproduction. After constructing a nest in shallow fresh waters, he begins to defend his territory from any intruding males. At this time the underside of this fish turns a brilliant red. Tinbergen and his coworkers at the University of Leiden, Netherlands, demonstrated by constructing models of sticklebacks of varying degrees of likeness to the real male and painting them red, pale silver, or green, that it is the red tint of the undersurface of the trespasser that releases an aggressive response from the territory defender (Pelkwijk and Tinbergen, 1937). A very realistic replica lacking the red color was not attacked, but models barely resembling a fish, on which the underside was painted red, provoked assault (Figure 2.7). Red models were always more effective than those of other colors. More recent studies, however, suggest substantial individual variation in the response of male sticklebacks to models with red undersides (e.g., Rowland and Sevenster, 1985), and this has led some researchers to suggest that the motivational state of the test fish may play an important role in determining its response (Baerends, 1985). Thus, the issue of releasers may not be as simple as the early experiments of ethologists suggested.

Supernormal Stimuli When one constructs a model that isolates the effective characteristic of the stimulus, it is sometimes possible to exaggerate that characteristic and create an unrealistic model that elicits the behavior more strongly than the natural stimulus. This model serves as a supernormal stimulus. A well-known example comes from the incubation behavior of the oystercatcher (*Haematopus ostralegus*). Given a choice of one of two eggs of the appropriate shape, a female oystercatcher will choose the larger (Tinbergen, 1951). Although she may topple off, she will persistently attempt to incubate an "egg" 20 times the normal size, even if there is an egg of a more appropriate size available. Similar behavior has been noted in other shore birds (Figure 2.8).

Sign Stimuli and Mimicry Some animals exploit the reliability of the response to a sign stimulus for their own advantage. Brood parasites are birds that lay their eggs in the nest of other species, leaving the care of the offspring to the foster parents. In some such cases, the eggs of the brood parasite must have the appropriate sign stimuli to elicit incubation behavior by the female of the host species. Otherwise, she will eject them from the nest. The eggs of several species of brood parasitic cuckoos mimic the eggs of their host species, being similar in size, color, and markings (e.g., Southern, 1954).

Not only must the host accept the eggs of the brood parasite, it must also feed the nestlings. The releasers for feeding include markings on the inside of the mouth of the young that are apparent as they gape. The markings and gaping behavior of the brood parasites may mimic those of the host, ensuring that the offspring will be cared for. The similarity of these markings in one brood parasite, the long-tailed paradise

Figure 2.8 A gull attempting to incubate a super-egg instead of her own egg (foreground). A model, such as this monstrous egg, that emphasizes the aspect of the natural stimulus that triggers a behavior may be more effective than the natural stimulus in eliciting the response. Such a model is called a supernormal stimulus.

widow bird *Steganura paradisaea*, and its host, the finch *Pytilia melba*, are shown in Figure 2.9 (Nicolai, 1964).

Chain of Reactions

So far we have considered only relatively simple behaviors, but a great deal of complexity can be added to the behavioral repertoire by building sequences of FAPs. The final product is an intricate pattern called a chain of reactions by ethologists. Here, each component FAP brings the animal into the situation that evokes the next FAP.

One of the earliest analyses of a chain of reactions was conducted by Tinbergen (1951) on the courtship ritual of the three-spined stickleback. This complex sequence of behaviors culminates with the synchronization of gamete

Figure 2.9 The finch (top left) is the host of the brood parasitic long-tailed paradise widow bird (top right). The gaping signal and markings on the inside of the mouth of the host species (bottom left) are mimicked by nestlings of the brood parasite (bottom right). These are the key stimuli that release feeding behavior of the parents. (Modified from Wickler, 1968.)

release, an event of obvious adaptive value in an aquatic environment. Each female behavior is triggered by the preceding male behavior which in turn was triggered by the preceding female behavior (Figure 2.10).

A male stickleback in reproductive condition may sometimes attack a female entering his ter-

ritory. If the female does not flee and begins to display the appropriate head-up posture in which she hangs obliquely in the water exposing her egg-swollen abdomen, the male will begin his courtship with a zigzag dance. He repeatedly alternates a quick movement toward her with a sideways turn away. This dance releases ap-

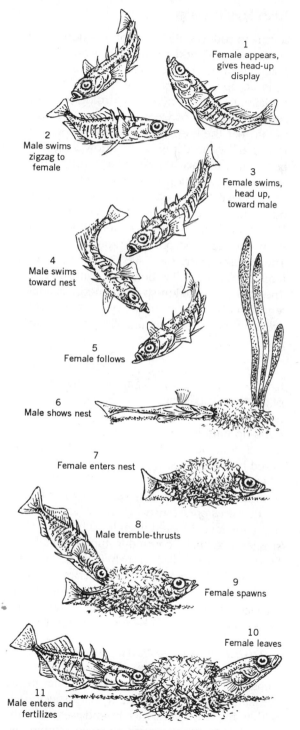

1
Female appears,
gives head-up
display

2
Male swims
zigzag to
female

3
Female swims,
head up,
toward male

4
Male swims
toward nest

5
Female follows

6
Male shows nest

7
Female enters nest

8
Male tremble-thrusts

9
Female spawns

10
Female leaves

11
Male enters and
fertilizes

proach behavior of the female. Her movement induces the male to turn and swim rapidly toward the nest, an action that entices the female to follow. At the nest he lies on his side and makes a series of rapid thrusts with his snout into the entrance, while raising his dorsal spines toward his mate. This behavior is the releaser for the female to enter the nest. The presence of the female in the nest is the releaser for the male to begin to rhythmically prod the base of her tail with his snout, causing the female to release her eggs. She then swims out of the nest, making room for the male to enter and fertilize the eggs. At the completion of this ritual, the male chases the female away and continues to defend his territory against other males until another female can be enticed into the courtship routine. The male mates with three to five females and then cares for the developing eggs by guarding them from predators and fanning water over the eggs to aerate them. We see, then, that this complex sequence is largely a chain of FAPs, each triggered by its own sign stimulus, or releaser.

The chain of reactions is not as rigid as the above description of courtship in the three-spined stickleback implies. There are actually many deviations in the precise order of the events in the ritual (Figure 2.11), and some actions must be repeated several times if one partner is less motivated than the other (Morris, 1958). Such flexibility begins to make sense when the function for which the ritual evolved is considered. For the stickleback, courtship is important to time the release of the gametes, and thus males and females seem to adjust their activities so that they are physiologically ready for gamete release at the same time. Despite some flexibility, however, the component behaviors do not occur randomly. In the display, a particular behavior is more likely to be followed by certain actions than by others.

Figure 2.10   Courtship behavior in the three-spined stickleback. (From Tinbergen, 1989.)

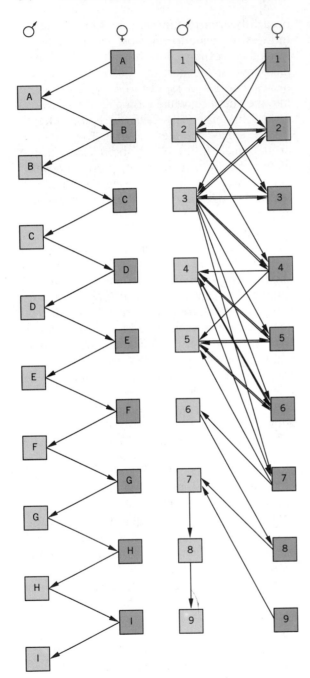

Figure 2.11 The ideal courtship sequence of the three-spined stickleback (left) and the observed sequence (right). Although there is a general pattern to the courtship ritual, there is some variability in the specific reactions to the behavior of the partner. (From Morris, 1958.)

Action Specific Energy

Lorenz introduced the concept of action specific energy in 1937 and renamed it action specific potential in 1981. The concept has not gained wide acceptance because of the difficulties of explaining it in concrete neurological terms. Lorenz used the concept to explain his observation that the ease with which a stimulus released an action depended on how long it had been since the act was previously released. He postulated that action specific energy, or energy for a specific action, is constantly being produced in the animal's nervous system. The energy is held in check by an inhibitory mechanism until the appropriate stimulus releases it. The released energy activates certain muscle systems and a particular behavior pattern results. The concept of action specific energy is consistent with observations of two types of behavioral activity, vacuum activity and displacement activity.

Vacuum Activity   In vacuum activity, a fixed action pattern is displayed in the absence of a stimulus. Basically, if the proper stimulus is not found, action specific energy continues to build up until it can be contained no more, and then it goes off on its own, without being triggered by a stimulus.

One example of vacuum activity was described in the reproductive behavior of ring doves (*Streptopelia risoria*). In an experiment, the male dove, isolated from females of his own species, eventually courted a stuffed model of a female that he had previously ignored. As time went on, he became even less discriminating in his choice of a mate and would bow and coo to a rolled up cloth. Finally, even the corner of his cage elicited courtship behavior (Craig, 1918).

Displacement Activity   During times of conflict, animals are often observed performing behaviors that seem to be totally irrelevant to the situation. For example, a male stickleback within his own territorial boundaries will unfailingly attack any intruding male. Within his neighbor's territory, however, the same male will typically flee from the territory owner. So what happens if two neighboring males meet just at the borders of their territories? Strangely enough, they may both start digging a pit as if for a nest. Why would they do this—an act that appears to have nothing to do with the situation at hand? According to the ideas of Lorenz and Tinbergen, because each fish cannot express its urge to attack nor its desire to flee, it is driven by its tension to find an outlet in another activity. In other words, within each of the male sticklebacks at the territorial boundary, there is a high level of energy driving them to attack their rival and an equally high level of energy driving them to flee. As the nervous energy for the two incompatible behaviors peaks within each male, it spills over into a third system, that controlling nesting behavior. Behavior patterns such as the nest-digging behavior in this example are called displacement activities because they are caused by the displacement of nervous energy from one system to another.

Models for the Organization of Instinctive Behavior

Models are hypothetical constructs that help us understand complex phenomena by framing their essential relationships in simpler terms, with simpler structure. Good models force us to organize our observations and our thinking about the natural phenomena on which they are based. Ideally, they should lead to experiments.

Lorenz's Psycho-hydraulic Model Lorenz (1950) organized his observations on animal behavior into the psycho-hydraulic model, also called the flush toilet model (Figure 2.12). Vac-

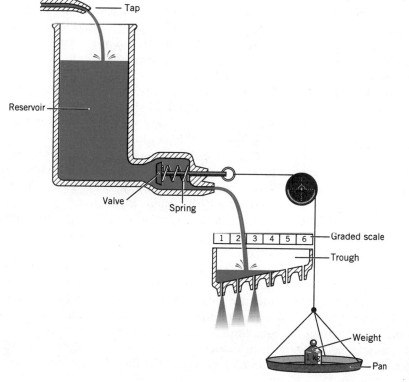

Figure 2.12 Lorenz's psycho-hydraulic model for instinctive behavior. (After Lorenz, 1950.)

uum behavior and displacement activity are among the observations that he hoped to explain in this model. The continuous production and storage of action specific energy during the interim when an animal is not performing some FAP is represented by the damming of water within the reservoir as it pours in from the tap. The idea that action specific energy accumulates until the particular FAP is performed is depicted as the filling of the reservoir while a valve is held shut by the spring. The valve can be opened to permit the outpouring of water. In the real world, FAPs are generally released by the appropriate sign stimuli, but sometimes when the animal has not had an opportunity to perform the FAP in a long time, the behavior seems to appear spontaneously. The model is consistent with the observation of vacuum activity, because it was designed with two ways in which the valve can be opened: (1) weights of various strengths, symbolizing stimuli of varying degrees of effectiveness, can be added to the pan connected to the valve; or (2) the pressure of the water in the reservoir, representing the action specific energy, may become intense enough to compress the spring, thus forcing open the valve. The higher the water level, the smaller the weight required, and eventually hydraulic pressure alone pushes the valve open, which is analogous to a vacuum activity. This model emphasizes that the appearance of an FAP will depend on both the level of action specific energy and the strength of the releaser.

The motor activity itself is represented by the water spouting from the spigot. The graded trough was added to account for the observation that there are often a number of components to a motor pattern, some of which are only displayed at the highest levels of motivation. If the valve is only slightly open, a small trickle of water enters the trough and will drip through the first and lowest hole in the trough, symbolizing the motor activity with the lowest threshold. As the valve opens wider, an increasingly greater amount of water is discharged, allowing it to reach higher trough holes, which represent activities with higher thresholds for elicitation. If the valve is fully open and the reservoir is full, water gushes forth and pours from all openings in the trough. This is analogous to the full-blown response of the animal. In this way, the model accounts for observations of behaviors such as the aggressive display of the cichlid fish, *Astatotilipia strigigena*, that includes five successive behavior patterns of increasing threshold: display coloration, fin spreading, distending the covering of gills, beating the caudal fin sideways, and ramming the adversary. Ramming, the final and most difficult component of the total display to elicit, would be represented by the highest hole in the trough (Seitz, 1940).

According to the hydraulic model, one might describe displacement activity as motor patterns that occur because the trough representing one set of fixed action patterns, such as those for aggression, is blocked while the valve is open. This would cause water to spill over into another trough, one symbolizing a different set of fixed action patterns, such as grooming.

The model also accounts for the decrease in the probability of an FAP's being performed after it has been elicited many times in succession. In hydraulic terms, the reservoir is empty. As time passes, and the reservoir fills, the likelihood of getting an FAP increases until the repeated performance of the behavior pattern drains the tank again. This is consistent with the observation of cyclical changes in responsiveness.

Lorenz's hydraulic model has been criticized, in part, because it does not fit with what we know about the neurophysiology of the brain (Lehrman, 1953). There is no physiological correlate of action specific energy, and the brain does not control behavior in the same way that a flush toilet operates. The hydraulic model has been useful in organizing observations on behavior into a conceptual framework, but now that the ability to determine the actual physiological mechanisms controlling behavior is beginning to be technologically possible, less schematic models will be more useful. Lorenz (1950) himself recognized the limitations of his model.

In his words, "This contraption is, of course, still a very crude simplification of the real processes it is symbolizing, but experience has taught us that even the crudest simplisms often prove a valuable stimulus to investigation."

Tinbergen's Hierarchical Model Tinbergen observed the same regularities of behavior that Lorenz did and was particularly struck by certain features that are not addressed in the hydraulic model. He developed a hierarchical model that emphasizes some different features of behavior and is easier to relate to the nervous structures that control behavior (Tinbergen, 1951). The model makes use of the ideas of appetitive and consummatory components of behavior and includes the notion of the innate releasing mechanism (IRM), an idea developed jointly by Lorenz and Tinbergen, but elaborated upon largely by Tinbergen (Lorenz, 1981). According to their scheme, when a sufficiently motivated animal encounters a certain key stimulus, the stimulus is identified by an innate releasing mechanism. The IRM then removes the inhibition from a particular brain center and allows it to direct the performance of a stereotyped behavior pattern, the fixed action pattern (FAP). As defined by Tinbergen (1951), an IRM is a "special neurosensory mechanism that releases the reaction and is responsible for its (the reaction's) selective susceptibility to a very special combination of sign stimuli." Thus, at least part of its job is to function as a stimulus filter, identifying the sign stimulus and ignoring the irrelevant information.

According to Tinbergen's model, each general instinct is hierarchically organized (Figure 2.13), and as such it is susceptible to certain priming, releasing, and directing influences. These stimuli may be of internal as well as external origin. In other words, the general instinct can be broken down into several levels of consummatory acts. Each level has its own appetitive behaviors and is comprised of a number of specific FAPs, and each is considered to be controlled by a different neural center. When ap-

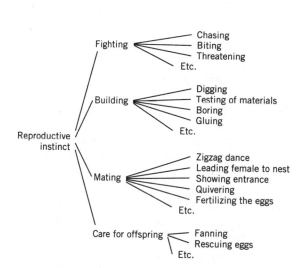

Figure 2.13 The hierarchical organization of an instinct as visualized by Tinbergen. (From Tinbergen, 1942.)

petitive behavior brings the animal into the correct stimulus situation, a neural block is removed so that the "energy" of the general drive can flow to the next hierarchical level, allowing that consummatory act to be performed. The part of the nervous system that identifies the correct releasing stimulus and removes the neural block is the IRM. As the energy flows to lower levels, it progresses along a path determined by the particular stimuli actually encountered that can remove the blocks for specific behaviors. If the appropriate stimulus is not encountered, the energy activates the centers controlling the appetitive behavior for that level of consummatory behavior, and the animal searches for the specific stimuli that will remove the block and allow the behaviors to be performed.

Although the behavior patterns involved are not FAPs, a simple analogy of this hierarchical model would be a man walking along a path until he faces a closed door. If he looks around (appetitive behavior), he will find the key (releaser) to open the door. When the key is turned, the lock mechanism (IRM) allows the door to swing open. The man walks through the portal (the first level consummatory response). Once he is on the other side, he faces a choice of several doors. The specific key he finds will determine which door he proceeds through. Having passed through this door (the second level behavior), he confronts another series of doors, and so on.

Although Tinbergen's hierarchical model has been criticized because it, like Lorenz's hydraulic model, incorporates the concept of "energy" that develops within the nervous system and drives certain behaviors, it is easier to make the transition from hypothetical construct to real neurophysiology with the hierarchical model. Nevertheless, there are problems associated with the notion of innate releasing mechanisms. Historically the IRM has been associated with central filtering in the minds of ethologists, and this has led to certain problems. It has given rise to the notion that there is a special device in the central nervous system of animals that lies waiting to be switched on or off. Such a notion is undoubtedly simplistic. The concept of IRM is useful, however, if we keep in mind that we do not know how central filtering is accomplished, that it may (and probably does) involve a multitude of complex neuronal interactions at various levels (from peripheral to central), and that the systems may differ in various kinds of animals. The concept of IRM also has value if it stimulates research into the nature of central filters.

Comparative Psychology

THE APPROACH: PHYSIOLOGICAL, DEVELOPMENTAL, QUANTITATIVE, LABORATORY ORIENTED

The comparative psychologists' emphasis on laboratory studies of observable, quantifiable patterns of behavior distinguished them from the European ethologists during the first half of the twentieth century. Recall that at this time, many ethologists preferred to study animal behavior under natural conditions. This meant that they went into the field and observed behavior. The problem was that in the field, the unexpected is expected; one cannot control all the variables. The comparative psychologists argued that good, experimental science cannot be done under such uncontrolled conditions. The ethologists were further criticized because, although they described changes in behavior, they often neglected to quantify their results and rarely analyzed the data with statistical procedures. Given the psychologists' penchant for laboratory studies that produce quantifiable results, it is not too surprising that much of their early research focused on learning and the physiological basis of behavior. Again, however, we wish to emphasize that although learning and physiology were popular areas of study among comparative psychologists, the evolution and function of behavior were also examined by some of comparative psychology's practitioners. Next we consider some of the major conceptual developments in the field of comparative psychology.

EARLY CONCEPTS OF COMPARATIVE PSYCHOLOGY

Morgan's Canon

Recall from our previous discussion of the ideas and writings of Darwin and Romanes that the early descriptions of animal behavior were often subjective, anthropomorphic accounts. C. Lloyd Morgan helped stop the anecdotal tradition, thereby helping send comparative psychology toward the objective science it is today. He argued that behavior must be explained in the simplest way that is consistent with the evidence and without the assumption that human emotions or mental abilities are involved. This idea was crystallized in Morgan's Canon (1894): "In no case may we interpret an action as the outcome of the exercise of a higher psychical faculty if it can be interpreted as the outcome of the exercise of one which stands lower in the psychological scale." In other words, when two explanations for a behavior appear equally valid, the simpler is preferred. People were urged to offer explanations of an animal's behavior without referring to the animal's presumed feelings or thought processes.

Learning and Reinforcement

We have already mentioned that many of the early comparative psychologists focused their research efforts on learning. The early days of these studies were exciting times, indeed, and some of the most important work was done by scientists in America. E. L. Thorndike (1898), for example, devised experimental techniques used to study learning in the laboratory. Thorndike

was a pioneer in research on what was called trial-and-error learning, now usually called operant conditioning. In operant conditioning the animal is required to perform a behavior in order to receive a reward. In one series of experiments, Thorndike invented boxes that presented different problems to animals. For instance, one problem box was a crate with a trap door at the top through which a cat might be dropped to the inside. A hungry cat was left in the box until it accidentally operated a mechanism, perhaps pulling a loop or pressing a lever, that opened an escape door on the side of the box, allowing it access to food that had been placed nearby. The length of time it took for each escape provided an objective quantifiable measure of learning progress (Figure 2.14). During repeated trials the animal became more ef-

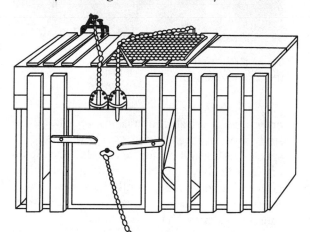

a

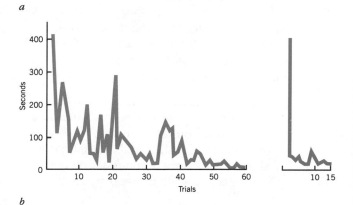

b

Figure 2.14   (*a*) A problem box. Thorndike invented many "problem boxes" to measure the learning ability of animals. An animal would be placed inside the box and would have to learn how to operate an escape mechanism. (*b*) The time required for escape on successive trials was a measure of how quickly the animals mastered the task. (From Thorndike, 1911.)

ficient and required less time to hit the escape latch. Thorndike's studies led him to develop the Law of Effect, a cornerstone of operant conditioning. His basic notion was that responses that are rewarded, that is, followed by a "satisfying" state of affairs, will tend to be repeated (this idea was also described by C. Lloyd Morgan and other investigators of animal behavior). Thorndike began publishing studies on animal intellect and behavior in the late 1800s, and in 1911 published a collection of his writings in a book entitled *Animal Intelligence: Experimental Studies*.

Just a few years after Thorndike introduced the idea of trial-and-error learning, Ivan Pavlov (Figure 2.15), a Russian physiologist, described the conditioned reflex. Pavlov noticed that a dog begins to salivate at the sight of food and reasoned that the sight of food must have come to signal the appearance of food. (In science, the key observations that trigger great ideas are often quite commonplace, as in this case. It is not what you observe, it is what you make of it.) In his well-known experiment, Pavlov rang a bell immediately before feeding a dog and found that, in time, the dog came to salivate at the sound of the bell alone (Pavlov, 1927).

Pavlov's dog learning to salivate at the sound of a bell is just one example of a more general phenomenon called classical conditioning, or establishing a conditioned reflex. In general terms, a conditioned reflex begins with a regular and measurable unlearned response, called the unconditioned response (UR—here, salivation), that is normally elicited by a stimulus called the unconditioned stimulus (UCS—here, food). During conditioning, a neutral stimulus, one that does not ordinarily cause the unconditioned response, is presented just before the UCS is offered. This is the conditioned stimulus (CS—here, the bell). Once the association is made, the conditioned stimulus alone is sufficient to elicit the response. In this case, the sound of the bell is sufficient to cause the dog to salivate.

At first, comparative psychologists used op-

Figure 2.15   Ivan Pavlov described a conditioned reflex in the dog.

erant and classical conditioning techniques to study the learning abilities of a wide variety of species. Thorndike, for example, examined learning in fish, chickens, cats, dogs, and monkeys and noted striking similarities in the learning processes of these animals. His results were, thus, consistent with the idea of intellectual continuity. Thorndike concluded that although animals might differ in what they learned or in how rapidly they learned it, the process of learning must be the same in all species. In his 1911 collection of papers he summarized his idea of intellectual continuity as follows (pp. 294):

[Intellect's] general law is that when in a certain situation an animal acts so that pleasure results, that act is selected from all those performed and associated with that situation, so that, when that situation recurs, the act will be more likely to follow than it was before. . . . The intellectual evolution of the race consists of an increase in the number, delicacy, complexity, permanence and speed of formation of such associations. In man this increase reaches such a point that an apparently new type of mind results, which conceals the real continuity of the process. . . . Amongst the minds of animals that of man leads, not as a demigod from another planet, but as a king from the same race.

Behaviorism

Another important event that steered comparative psychology toward objectivity and laboratory analysis was the birth of behaviorism, a school of psychology that restricts the study of behavior to events that can be seen—a description of the stimulus and the response it elicits. Before behaviorism, psychologists described animal behavior with anecdotes, and attributed human emotions to animals. Behaviorists, however, sought to eliminate subjectivity from their studies by concentrating their research efforts on identifying the stimuli that elicit responses and the rewards and punishments that maintain them. This was, indeed, a step toward better science. They designed experiments that would yield quantifiable data, invented apparatus to measure and record responses, and developed statistical techniques that could be used to analyze behavioral data. The assumptions that an animal's mental capacity could not be measured directly, but its ability to solve a problem could, again focused attention on learning ability as a popular research subject. A learned response could be described objectively and experiments could be conducted under the controlled conditions of the laboratory.

B. F. Skinner, one of the most famous behaviorists, devised an apparatus that was similar to Thorndike's problem box but lacked the Hou-

dini quality. Instead of learning to operate a contrivance that provides a means of escape, a hungry animal placed in a "Skinner box" must manipulate a mechanism that provides a small food reward (Figure 2.16). A rat may learn to press a lever and a pigeon may learn to peck at a key. Patterns of behavior that are rewarded tend to be repeated, or to increase in frequency, so learning can be measured as the number of responses over time. Skinner believed that the control of behavior was basically a matter of reinforcement.

Figure 2.16 B.F. Skinner and his apparatus, the Skinner box. Animals placed in the box learned to operate a mechanism to obtain a food reward.

Behaviorists began to see basic principles underlying learning that were common to all species. They expected to see similarities in the learning process because at that time, the minds of all species were considered similar in kind. Thus, according to the traditional view of learning held by the followers of behaviorism, the minds of humans and animals were similar in kind and differed only in complexity. In short, there were general laws of learning that transcended all species and problems. If this was true, then it was reasonable to study the laws of acquisition, extinction, delay of reinforcement, or any other aspect of the learning process in a simple and convenient animal, such as the domesticated form of the Norway rat (*Rattus norvegicus*), and the results could then be broadly applied to all other species.

The outcome of the beliefs in general laws of learning and intellectual continuity was a narrowing of the number of different animals employed in learning experiments. Frank Beach was among the first to point out this trend. He analyzed the contents of the odd-numbered volumes of the *Journal of Animal Behavior* and its successor, the *Journal of Comparative Physiological Psychology*, from the first volume in 1911 through 1948. He reported, in an article entitled "The snark was a boojum" (1950), that although the number of articles published was increasing, the number of species studied was decreasing (Figure 2.17). Beach's analysis indicated that the laboratory rat had become the favored study animal of comparative psychologists, in fact, it was the subject of 50 percent of all the papers published in these journals (see Dewsbury, 1984, for a critique of this and later analyses based on frequency counts of journal contents). Although it may seem obvious now that studying learning in only one species is not a comparative approach, the psychologists of that time were not bothered by this bias because they were convinced that there were universal laws of learning. Even today, despite a general increase

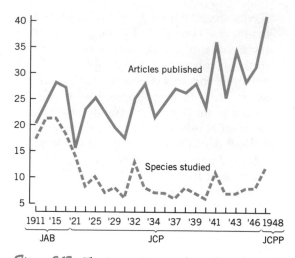

Figure 2.17   The increasing number of articles published by comparative psychologists between 1911 and 1948 concerned a gradually decreasing number of species. (From Beach, 1950.)

in interest among psychologists in intellectual and behavioral differences among species (Kalat, 1983), the historical bias in learning theory against the existence of phyletic differences in learning and intelligence is still evident (e.g., Macphail, 1987).

THE ROOTS OF PHYSIOLOGICAL PSYCHOLOGY

Although learning was a dominant focus of research during the middle of this century, it was not the only research interest of comparative psychologists. Another research topic was the physiological basis of behavior. Part of the psychological foundation of behavior is, of course, the nervous system. The comparative psychologists' interest in the neurological mechanisms of behavior can be traced back to Pierre Flourens (Jaynes, 1969), a protégé of Baron Cuvier, a famous scientist of nineteenth-century France who stressed the importance of laboratory research. Flourens's reputation was earned for his studies of the relationship between behavior and brain structure. For example, he did experiments in which parts of the brain were removed, such as the cerebral hemispheres from a pigeon, to look for the effect on the animal's behavior.

Karl Lashley was one comparative psychologist who maintained an interest in physiology as well as a comparative base of study during the years when learning in the laboratory rat dominated the field. His attempts to localize learning in the cerebral cortex resulted in the rejection of some hypotheses that were widely accepted at the time. For example, Pavlov's ideas assumed that learning depended on the growth or strengthening of neural connections between one part of the cerebral cortex and another. To test this idea, Lashley (1950) trained rats on a variety of mazes and discrimination tests and then tried to disrupt the memory by making a cut into the cerebral cortex in a different place in each animal. After destroying varying amounts of brain tissue, he would then retest the animals to see if their behavior changed. In general, he found that when it came to complex problem solving, the entire cerebral cortex was involved, and any particular area was just as important as any other. He also experimentally addressed questions such as whether a learned response depends on the stimulation of the same receptors that were activated during learning, and whether the learned response depended on a fixed pattern of muscle movements. Contrary to expectations, he found that they do not. But Lashley was not solely concerned with the neurological basis of learning in the rat. He also examined the role of the brain in emotion and in vision.

The comparative psychologist Frank Beach began his career using brain surgery to determine the effects of lesions on the maternal behavior of the rat, but he later went on to study the effects of hormones on behavior. He analyzed the roles of nerves, hormones, and experience in the sexual behavior of fishes, amphibians, reptiles, birds, and mammals. We will discuss some of his work in the field of behavioral endocrinology in more detail in Chapter 7.

Recognizing that animal behavior is concerned with the activities of groups of animals as well as of individuals, some comparative psychologists studied social behavior. Robert Yerkes, for example, established a research facility (later named the Yerkes Laboratory of Primate Biology) at Orange Park, Florida, to study a wide range of behaviors in primates. Some researchers also began to see that although it is often easier to make measurements in the laboratory, it is not impossible to get good measurements in the field. C. R. Carpenter studied a variety of primate species, each in its natural setting: howler monkeys in Panama, spider monkeys in Central America, and gibbons in Thailand, to name a few. T. C. Schneirla used both field observation and laboratory experimentation to investigate the biological basis of social behavior in army ants. In doing so, he applied the rigorous methodology that was used by laboratory researchers to his field studies on the social behavior of army ants. Such pioneering studies began to help weave the two independent sciences of ethology and comparative psychology together.

The Nature/Nurture Controversy and the Modern Synthesis

Differences in beliefs about the importance of inheritance or experience in behavior were one cause for heated debate between the European ethologists and the American comparative psychologists. The early ethologists, concerned with the evolution of behavior, concentrated on the adaptiveness of behavior. They believed that behavior is inherited in a form that is suited to the life circumstances of the organism because it is shaped by natural selection. Their experiments were designed and their results were interpreted within that theoretical framework. Because the ethologists perceived behavior as having evolved to suit the environment, it seemed best to study behavior in as natural a setting as possible. The American psychologists were more interested in how an individual's behavior develops than in explaining the adaptiveness of that behavior. Furthermore, the majority of psychologists believed that given the uncontrolled conditions of the natural environment, the laboratory was the best place to assess the effects of various experiences on behavioral development.

The result of these differences was that a battle, often called the nature/nurture controversy, raged for about 20 years (the nature/nurture controversy actually has a much longer history than the argument between ethologists and comparative psychologists, but we will concentrate on their debate here). Behavior must be inherited the antagonists said, or it must be learned. Behavior is shaped by natural selection, or it is shaped by the experience of the individual. The classic quote representing the behaviorist's position comes from John B. Watson (1930; pp. 104, Figure 2.18):

Figure 2.18 John B. Watson, a prominent behaviorist.

"Give me a dozen healthy infants, well-formed, and my own specified world to bring them up in and I'll guarantee to take any one at random and train him to become any type of specialist I might select—doctor, lawyer, merchant, chief and yes, even beggar man and thief, regardless of his talents, peculiarities, tendencies, abilities, vocations, and race of his ancestors."

The behaviorist position was clear—learning is the ultimate basis of behavior, so instinct and innate tendencies are irrelevant and unnecessary. (It is interesting to note that although the extreme statements made by Watson in his later years came to symbolize the behaviorist's position and resulted in his becoming perhaps the most unpopular of all psychologists among the ethologists, his work as a young man—comparative studies of the behavior of two species of shorebirds on islands off the coast of Florida—was quite compatible with the ethologist's approach [Dewsbury, 1989]. It should also be noted that most of his extreme statements were made with specific reference to human behavior.)

In contrast to the attitudes of behaviorists, the ethologists' position made learning subordinate to instinct. This feeling is reflected in the words of Niko Tinbergen (1952): "...learning and many other higher processes are secondary modifications of innate mechanisms, and. . . therefore a study of learning processes has to be preceded by a study of the innate foundations of behavior." Tinbergen saw learning as "superimposed on the innate patterns and their mechanisms."

One of the most striking events of this time period was the publication in 1953 of Daniel Lehrman's article entitled "A critique of Konrad Lorenz's theory of instinctive behavior." In this paper Lehrman harshly criticized many of the ideas that had formed the foundation of ethology. He argued against Lorenz's belief that all behavior could be divided into the two categories of innate and learned. According to Lehrman, the innate–learned dichotomy was totally inappropriate for an analysis of behavioral development, and he proposed instead an interactionist approach. All behavior, Lehrman argued, arises from interactions within and between a developing animal and its environment. Lehrman was not the first comparative psychologist to criticize the fundamental concepts of ethology. Indeed, it was his mentor, T. C. Schneirla, who initially led the attack. However, because Schneirla's style of writing was somewhat difficult and his critiques of ethological ideas were often published in journals that were not commonly read by the European ethologists (Hinde, 1982), the 1953 article published by his pupil Lehrman in the Quarterly Review of Biology caused the greatest stir.

In the debate that followed, each camp, with its own dogma as a shield, attacked the other with the vehemence that springs from seeing issues as black or white. Even if the primary proponents of each position did not see nature and nurture as a dichotomy, others were perceiving their views in that light. In 1970, as tempers were cooling, Daniel Lehrman said, "I have heard it said so often that I believed that all species-specific behavior develops through individual learning that I almost came to believe that I *had* said it."

Persuaded by cogent arguments and experimental data, each side began to see some merit in the opposing view and reformulated key concepts. Although some differences in emphasis still exist, and accusations of dichotomous thinking about behavioral development are still being made in regard to certain research areas (e.g., Johnston, 1988, and accompanying commentary), compromises have been reached. In the modern synthesis of classical ethology and comparative psychology, both environment and inheritance are seen as important in shaping behavior. But before discussing how the ideas about nature and nurture were modified, we will focus on the issues that fired the controversy.

Difficulties with the Concept of Innate

Problems with the Definition   At the heart of the nature/nurture controversy was the word innate. The first problem was in defining the term. An innate behavior came to be regarded as a behavior that appears as the animal develops, apparently without having been learned. It is often stereotyped (always performed the same way by members of a particular species) and complete (performed in its entirety). The problem was that other factors could also account for such behaviors. Two specific objections were raised.

First, stereotypy may result from similarities in environment. Daniel Lehrman (1953) pointed out that the constancy in the form of a behavior does not mean that it is genetically programmed. Instead, a behavior may be performed in the same way by all the members of a species because all the members of that species developed within the same environment. He cited the example of the Ichneuman fly. This parasitic fly provides food for its offspring by laying its eggs on the larvae of a moth. In nature, flour moth larvae are the victims. For this reason, the parasite's preference could easily be interpreted as a response to stimuli that were determined by its genes. However, when the larvae were raised on the larvae of moths other than the flour moth, these parasites, when adult, preferred to lay their eggs on the larvae of the moth species that had nourished them during development. This was one of the first demonstrations that a species-specific stereotyped behavior—the choice of flour moth larvae for an oviposition site—was influenced by experience (Thorpe and Jones, 1937).

Another characteristic of innate behavior traditionally included in the definition is that it appears in a complete form without practice the first time it is performed. The second objection concerns this characteristic. Numerous studies, including the one just described, have shown that learning or environmental effects can take place so early, before hatching or birth, that the completeness of a new behavior does not demonstrate that learning was unimportant. Prenatal learning has been demonstrated, for example, in humans (Kolata, 1984) and rats (Smotherman, 1982).

Problems Experimentally Demonstrating Innateness   Because observable characteristics such as stereotypy and completeness of form on the first occurrence are not reliable indicators that a behavior is innate, one might wonder, what would demonstrate innateness? Here we consider problems associated with two types of experiments that were traditionally used to demonstrate innateness, the isolation experiment and the breeding experiment.

The isolation experiment has also been called the deprivation experiment or the Kaspar Hauser experiment. (Kaspar Hauser lived in Germany in the early 1800s. He claimed to have lived his life in a dark room.) The isolation experiment is designed to demonstrate that a particular behavior is acquired without experience. Immediately after an animal is born, it is isolated and raised without contact with members of its own species or other environmental stimuli

deemed to be important in the development of the behavior in question. In other words, the basic rationale of an isolation experiment is that if you suspect that something in the environment might be important, eliminate it and see whether the behavior develops without it. For example, if you wanted to determine whether the members of a particular species choose the food items they do because of their genetic makeup or their experience, you would raise an animal without contact with others who might display the preference and on a diet that excludes all the normal foods. Later, offer the animal a variety of foods, including those of the customary diet, and see whether it prefers the normal foods of its group or those it has previously eaten.

There are problems in interpreting this type of experiment. One is that it is impossible to be sure that all the important environmental stimuli have been eliminated. Daniel Lehrman (1953) stated emphatically: "It must be realized that an animal raised in isolation from fellow-members of his species is not necessarily isolated from the effect of processes and events which contribute to the development of any particular behavior pattern. The important question is not "Is the animal isolated?", but "From what is the animal isolated?". Important, but unsuspected factors may remain uncontrolled.

Furthermore, even if an animal is deprived of what is considered to be essential environmental stimulation, it may display normal behavior because it provides the missing stimulation itself. Gilbert Gottlieb has observed this in Peking ducklings, a domestic form of the mallard duck (*Anas platyrhynchos*). Upon hatching, the ducklings recognize and prefer their species-specific maternal call (Gottlieb, 1971). This recognition occurs only if the ducklings hear the contentment calls of their species while they are still inside their shell. If ducklings are reared in isolation during the embryonic period, however, they still prefer their species' call upon hatching. These isolated ducklings apparently learn the characteristics of their species call by listening to their own sounds prior to hatching. It is only if they are isolated and devocalized while still in the egg (and thus cannot hear others or call to themselves) that they do not recognize the maternal call of their species after hatching (Gottlieb, 1976). These experiments show that a behavior may develop normally, even if the animal is apparently isolated from the important environmental stimuli because an individual may provide the stimuli itself. We will discuss the behavioral development of mallard ducklings in more detail in Chapter 8.

A second approach to demonstrating that a particular behavior is innate involves breeding experiments. These may include (1) cross breeding, in which individuals who express a particular behavior in different ways are mated and the behavior in the offspring is examined, (2) inbreeding, in which siblings are mated for many generations to look for behavioral differences between inbred lines, and (3) selective breeding, in which individuals exhibiting a certain behavioral trait are mated in an attempt to select for, or increase the frequency of, that behavioral characteristic. Examples of experiments using these breeding techniques will be discussed in the next chapter. Such breeding experiments can indicate the degree to which genetic differences result in differences in behavior.

It should be apparent, however, that simply demonstrating a genetic basis for a behavioral characteristic does not mean that the behavior develops without being influenced by the environment. In fact, there is a great deal of good information indicating that both genes and environment can be important in the development of behavior. Phenylketonuria is a disease caused by a single recessive gene. If an individual lacks the dominant normal allele for this trait (that is, is homozygous recessive), he or she lacks the enzyme necessary to convert the amino acid phenylalanine to another amino acid, tyrosine. As a result of this metabolic deficiency, phenylalanine and an intermediate product, phenylpy-

ruvic acid, accumulate in the blood, move into the cerebrospinal fluid, and ultimately cause brain damage and mental retardation. Fortunately, the disease can be detected by a urine and blood test administered in the first few weeks of life. If the diet of the afflicted child is modified so that phenylalanine is restricted, mental deficiency can be prevented (Klug and Cummings, 1983). This, then, is an example of the expression of an inherited trait being greatly influenced by an element in the environment, the diet.

Modifying Innate Behavior by Experience   Jack Hailman's (1967, 1969) investigation of the feeding behavior of gull chicks demonstrated that an innate behavior can be modified by learning. Before Hailman's study, Tinbergen and Perdeck (1950) had described the behavior of herring gull chicks (*Larus argentatus argentatus*) and had described their food begging and feeding actions as innate, or genetically based. The chicks peck at their parent's bill causing the parent to regurgitate food. The behavior was called innate because it was performed in the same stereotyped way by all herring gull chicks soon after hatching.

However, Hailman (1967, 1969) showed that the innate feeding responses of gull chicks are modified by learning. Laughing gull chicks (*Larus atricilla*) that are at least one week old beg for food from their parent in much the same manner as do herring gull chicks. The pattern involves the three separate actions, the last of which is shown in Figure 2.19: (*a*) The chick opens and closes its bill; (*b*) the chick moves toward the parent; and (*c*) the chick rotates its head so that it can grasp the bill of the parent. But the behavior of newly hatched chicks is slightly different from that of one-week-old chicks; they do not rotate their heads and are therefore unable to grasp the parent's bill. Instead, they end up just pecking at the bill, if they are lucky enough to actually hit it. Hailman demonstrated that the chicks learn to rotate their

Figure 2.19   An example of an innate behavior pattern that is modified by experience. Right after hatching, a chick begs by striking at the parent's bill and, with luck, hitting it. By the time the chick is 3 days old, its begging is nearly perfected. The begging response involves the correct timing of opening and closing the mandibles around the parent's bill, the movement of the head toward the parent, and the rotation of the head so that the chick can grasp the parent's bill. This last component of the response requires the experience of begging from a parent. (Redrawn from Hailman, 1969.)

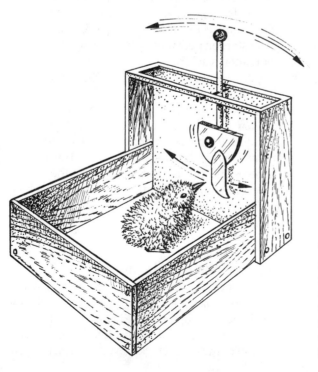

Figure 2.20   The laboratory apparatus used to test the effectiveness of various models in eliciting pecking (the begging response) from gull chicks. The model was attached to a pivot arm and moved in time with a metronome in front of the chick. The number of pecks per unit of time, usually 30 seconds, was recorded. (Redrawn from a photograph in Hailman, 1967.)

heads and grasp the parent's bill. He raised some chicks in the dark so that their visual experience would be limited and force-fed them so that they would not have experience in begging from a parent. At an age when wild chicks showed the rotational component, the hand-reared chicks did not. Obviously, that component of the response does not develop without practice begging.

In addition, the chick's experience during its first week of life refines its image of a parent, the stimulus for the begging response. Laughing gull chicks that had been kept in the dark for the first 24 hours after hatching were presented with a set of five models of gull heads one at a time. The models were attached to a metronome and moved with the same rate and pattern in front of a gull chick (Figure 2.20). The number of pecks aimed at the model in a given amount of time, usually 30 seconds, was recorded. As you can see in Figure 2.21, newly hatched chicks peck with almost the same frequency at anything that vaguely resembles a bill. However, as the

chicks matured, the rate at which they pecked at a model was greater the more realistic the image of an adult laughing gull. At one week old, chicks pecked only at models that closely resembled a parent. We can see, then, that the chicks responded to more lifelike models as they learned the characteristics of their parents.

Again, just because a behavior is innate does not mean that learning plays no role in its development. Conversely, we will see next that the involvement of learning in the shaping of a behavior is not an indication that the behavior has no genetic basis.

Problems with Early Learning Theory

If it was difficult for the early ethologists to accept that what they considered to be innate behavior could be modified by experience, it was equally difficult for the comparative psychologists (particularly the behaviorists) to accept that learning may be influenced by, or constrained by, genetic factors. Let's see how ideas pro-

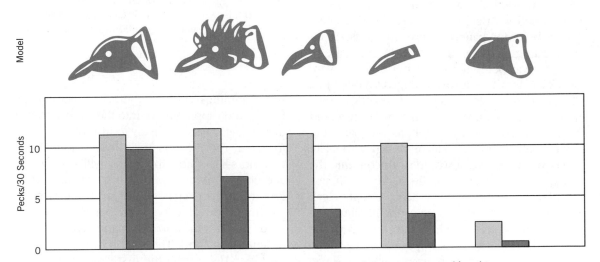

Figure 2.21 Representative models of parent heads and the responses of laughing gull chicks to them. The white bars indicate the responses of newly hatched chicks and the black bars the responses of older chicks. The newly hatched chicks peck at almost anything that resembles a bill, but older chicks preferred a more realistic model of a laughing gull head. (Modified from Hailman, 1969.)

gressed in understanding the development of behavior by taking a look at the position of the behaviorists and some of the problems they encountered.

Interference of Innate Behavior with a Conditioned Response According to learning theory, a response that produces a rewarding outcome becomes more frequent or more probable over time. This idea, which dates back to Thorndike's Law of Effect, is the basis of operant conditioning, the means of training described earlier, in which a behavior is encouraged through reward. The reward is said to reinforce the appropriate response.

The idea that reinforced responses become more frequent was questioned by Keller and Marion Breland (1961), students of B. F. Skinner who used operant conditioning techniques to train animals for commercial purposes. The Brelands discovered that instinctive behavior patterns would sometimes disrupt a trained animal's performance by appearing without reinforcement in place of desired actions that were rewarded. Because their animals were often trained with a food reward, the disruptive innate behavior patterns were those associated with eating or obtaining food. In one case, a raccoon was being trained to perform in a window display where it was to deposit money in a "bank," thereby encouraging observers to save their money (Figure 2.22). The raccoon would pick up a coin and bring it to the bank but occasionally it appeared to be reluctant to make the deposit. Instead, it would dip a coin into the bank, retrieve it, and fondle it. The coin was handled in much the way a raccoon might clean a crayfish in a stream before eating it. After playing with the coin for awhile, the raccoon would eventually deposit the money and receive a food reward. To make matters worse, when the raccoon was required to deposit two coins before being rewarded, it had even greater difficulty parting with its money. The masked miser would

Figure 2.22 A raccoon trained by the Brelands to deposit money in a bank often handled the coin as it would a prey item. In this instance, an innate behavior interfered with a conditioned response.

spend minutes rubbing the coins between its fingers and dipping them into the bank. Because this behavior was not likely to inspire spectators to invest, the raccoon soon lost its job. These behaviors—rubbing the coins with the hands and dipping them into the bank—are instinctive actions for the preparation of food. The raccoon had made the link between performance and food, but it was born with the link between cleaning and food. The researchers were often unable to disconnect the first from the second.

In this example, nonreinforced innate behaviors were interfering with those conditioned by operant techniques. The Brelands (1961) concluded that "the behavior of any species cannot be adequately understood, predicted or controlled without knowledge of its instinctive patterns, evolutionary history, and ecological niche."

Learning Predispositions   Another important principle in classical learning theory has been poetically called the premise of equivalence of associability—through the process of reward and punishment, any stimulus can come to be associated with any response. In classical conditioning, this would mean that any conditioned stimulus could be associated with any unconditioned stimulus with equal ease. The premise is also relevant in operant conditioning; in that case it means that any reinforcer should be equally effective in increasing the frequency of any response.

Martin Seligman (1970) was among the first in the twentieth century to reevaluate this idea (ideas contrary to the premise of equivalence of associability can be found at least as far back as the writings of C. L. Morgan, 1894). Seligman argued that natural selection not only shapes the sensory and motor abilities of the species to suit its ecological niche, it also influences the organism's "associative apparatus." As a result, individuals of different species may differ in their abilities to associate different events because evolution prepared them to make certain associations more easily than others. The number of trials or pairings of stimuli necessary before a response occurs routinely can serve as an operational measure of preparedness. A comparison of the abilities of different species to learn certain associations reveals a continuum of preparedness. At one end of the preparedness spectrum are associations that individuals of a particular species are prepared to learn (associations that they learn easily). At the other end of the continuum are those associations that an animal seems unable to learn. The animal is said to be contraprepared to learn that association. In between are the associations that the animal is prepared to learn, but which are not part of its normal repertoire.

There are some clear lines of evidence of a genetic predisposition to make certain associations. One of the more interesting experiments demonstrating such preparedness was done by Garcia and Koelling (1966) and involved "bright noisy water" and rats. The experimenters had two groups of rats. The first group was trained to drink bright noisy water created by circuitry that caused a flash of light and a click when a rat's tongue touched the drinking spout in its cage. The second group of rats drank water sweetened with saccharin. Both groups were then divided in half and one-half was given an electric shock following drinking, while the other was made sick by X-ray irradiation. The animals were then offered the type of water they had been accustomed to drinking, either bright noisy water or sweet water. The results are summarized in Table 2.1. Notice that the animals that had been given a shock (external pain) developed an aversion to the bright noisy water (external cues), but not to the sweet water (internal cue). Conversely, a distinctive saccharin taste (internal cue) could be associated with nausea (internal discomfort), but not with the external pain of electric shock. Such results clearly support the concept that not every stimulus can be associated with every response.

Experiments with avoidance learning provide additional evidence of genetic restrictions on learning. In avoidance learning, an animal must learn to perform some action to prevent the occurrence of an unpleasant stimulus. These experiments showed that an animal may be able to learn one response to a stimulus, but not another. For example, Bolles (1969) found that rats would readily learn to run to avoid an electric shock but would not learn to rear up and stand on their hind legs to avoid the same stimulus.

So we can see that behind the concept of a

Table 2.1

ASSOCIATIONS BETWEEN CUE AND CONSEQUENCE IN THE DOMESTIC STRAIN OF THE NORWAY RAT: EVIDENCE FOR SELECTIVE LEARNING (Garcia and Koelling, 1966)

CUES	CONSEQUENCES	
	Nausea	Electric shock
Taste (saccharin sweet)	Drinking inhibited	No effect
Sound and light (click and light flash)	No effect	Drinking inhibited

preparedness continuum is the idea that each species evolved so that it is exquisitely designed for its specific niche, and that this specialization process may result in predispositions to learn some responses and not others. This means that it should be possible to interpret species differences in learning ability as adaptations to particular habitats or life-styles. For example, the fact that parent–offspring recognition is better developed in bank swallows (*Riparia riparia*) and cliff swallows (*Hirundo pyrrhonata*) than in northern rough-winged swallows (*Stelgidopteryx serripennis*) and barn swallows (*H. rustica*) makes sense when we consider that the first two species are colonial and the latter two noncolonial (Beecher, 1990). In noncolonial species, the chances that parents will provide care to young other than their own are fairly slim (there are not several other swallow nests nearby), and thus the ability to recognize their own young is less critical. (The often cited example that parental recognition of young is well developed in herring gulls, a colonial species in which the young can mix at an early age, and poorly developed in kittiwakes, a colonial cliff-nesting species in which the young cannot safely wander from their nests, is incorrect. Although this generalization began with the writings of Cullen [1957], more recent studies have shown that gull parents fail to recognize their young during the first few weeks [e.g., Holley, 1984], although chicks appear to recognize their parents.)

CONTEMPORARY VIEWS ON NATURE AND NURTURE

It is important to understand that the terms "innate" and "learned" are not invalidated by examples of the modification of innate behavior by learning, or by the constraints placed on learning by genetic factors. There are, indeed, species-specific behaviors that are expressed in a highly stereotyped manner in a complete form the first time the situation is appropriate. A male fruit fly that has been isolated from the larval stage to adulthood still exhibits species-specific courtship behavior. It is also true that some behaviors are learned. No one would argue that a surgeon was born with the motor patterns for performing a coronary bypass. The point is, the terms innate and learned can be appropriate if certain cautions are kept in mind. First, labeling a behavior as innate or learned should never be interpreted as an explanation of its development, the mechanisms that allow its expression, or the forces active in its selection. If a label is accepted as an explanation, the research required for a complete understanding of the behavior is stifled. Furthermore, innate and learned should not be considered as mutually exclusive terms; just as innate behaviors are not purely genetic, learned behaviors are not entirely environmental. Genes and environment interact in the development of every behavior.

We have learned that there is an intimate

relationship between the genetic and learned contributions to behavior. One way to get at the question of the contribution of genes versus environment is to learn when genes and experience have their greatest relative effects. Although genes and experience work together throughout the organism's development to produce most behaviors, as the organism changes over time, the nature of the interaction will vary. For example, as a result of its genetic makeup, an animal may be more or less responsive to certain environmental stimuli at different times of its life. Conversely, the presence or absence of certain stimuli at critical times during development may alter the expression of the genes.

As the previous discussion indicates, perhaps the most important result of the nature/nurture controversy was the increasing focus on behavioral development in both ethology and comparative psychology. Indeed, although the two approaches to the study of animal behavior moved through a period of extreme divergence and debate, the distinction between them has blurred in recent years.

Sociobiology

The field of animal behavior has grown enormously since the time of reconciliation between the ethologists and comparative psychologists, and there have been several developments in the years that followed. In the 1960s and early 1970s, for example, field researchers such as John Crook (1964, 1970) and John Eisenberg and colleagues (1972) suggested that ecological context was sometimes a better correlate of social behavior than was phylogeny (remember that ethologists often focused on phylogenetic analyses of behavior). Another dramatic development was the birth of a new discipline that focused on the application of evolutionary theory to social behavior. This new discipline was called sociobiology. Many scientists use the terms so-

ciobiology and behavioral ecology interchangeably, although narrower definitions might confine sociobiology to the study of social interactions and behavioral ecology to the study of behavior involved in obtaining and using resources (Barlow, 1989; Krebs, 1985).

In addition to some debate over how narrow or broad the definition of sociobiology should be, there has also been some question concerning the uniqueness of its approach. Some scientists, for example, questioned whether sociobiology is really a new discipline or simply part of contemporary ethology (e.g., M. S. Dawkins, 1989). In contrast, others believe that at least early on, ethology and sociobiology could be separated in several ways (Barlow, 1989). For example, whereas classical ethologists tended to derive hypotheses from detailed observations (i.e., through induction), sociobiologists tended to be more deductive, typically deriving hypotheses from larger theoretical frameworks. Whereas classical ethologists were interested in species differences, sociobiologists began to investigate individual differences, examining the costs and benefits of a particular act. Although ethologists often did not focus their research efforts on the mechanisms of behavior, when they did seek reductionist explanations, they typically considered physiological factors (recall, for example, Tinbergen's hierarchical model for the organization of instinctive behavior). In contrast, sociobiologists sought reductionist explanations in the area of genetics. Having mentioned some of the questions concerning the precise relationship between sociobiology and other fields of animal behavior, let us consider the relatively recent "history" of sociobiology.

During the late 1960s and early 1970s, most scientists were quite comfortable with the idea that natural selection acted primarily on individuals. Despite the existence of widespread agreement, however, some nagging issues that seemed inconsistent with selection at the level of the individual remained (Hinde, 1982). For example, how could one explain the evolution of sterile castes in species of ants, bees, and wasps? How could the evolution of nonreproducing individuals be consistent with Darwinian selection? Similarly, how was one to explain the evolution of certain patterns of behavior, called altruistic behavior, that seemed to benefit others, yet were costly (with respect to survival and reproduction) to the performer? Why, for example, do some animals give alarm calls when they spot a predator, when calling may actually increase their own chances of being detected by that predator? The answer to these questions came in 1964 when W. D. Hamilton published his seminal paper entitled "The genetical evolution of social behaviour, I, II." In this paper Hamilton showed that evolutionary success (contribution of genes to subsequent generations) should be measured not only by the number of surviving offspring produced by an individual, but also by the effects of that individual's actions on nondescendant kin (e.g., siblings, nieces, and nephews). He coined the term "inclusive fitness" to describe an individual's collective genetic success, that is, a combination of direct fitness (own reproduction) and indirect fitness (effects on reproduction by nondescendant kin). When quantifying an individual's inclusive fitness, we count, to varying degree depending on genetic relatedness, all the offspring—personal or of relatives—that are alive due to the actions of that individual. This concept of inclusive fitness paved the way toward an understanding of the evolution of sterile castes and altruistic acts (these topics are discussed in more detail in Chapter 19).

In the years following Hamilton's paper, many studies were conducted in which the idea of inclusive fitness was used to interpret social behavior. However, it was not until 1975, when E. O. Wilson published his landmark text *Sociobiology*, that the true impact of sociobiological ideas was felt. The text, an engaging integration of ideas from fields such as ethology, ecology, and population biology, gained almost instant notoriety from both within and outside the sci-

entific community (see later discussion). In his book, Wilson (1975) defined sociobiology as the "systematic study of the biological basis of all social behavior" and proposed that a knowledge of demography (e.g., information on population growth and age structure) and the genetic structure of populations was essential to understanding the evolution of social behavior.

Although sociobiological ideas had been developing for several years prior to the publication of Wilson's book, the text crystallized many of the relevant issues and soon became the focal point for proponents and critics alike. Criticism arose from both the scientific and political arenas. First, in attempting to establish sociobiology, Wilson attacked fields such as ethology and comparative psychology and made the bold prediction that in due time sociobiology would engulf these disciplines (Figure 2.23). Specifically, he predicted that ethology and comparative psychology would be "cannibalized by neurophysiology and sensory physiology from one end and sociobiology and behavioral ecology from the other" (Wilson, 1975). Another area of great concern, this time from the political arena, was the extension of sociobiological thinking, in the absence of sound evidence, to human social behavior (Cooper, 1985). Opponents of sociobiology claimed that Wilson advocated biological determinism, the idea that the present conditions of human societies are simply the result of the biology of the human species. Although only the final chapter of the 27 in his text was devoted specifically to humans, heated debate over the social and political implications of sociobiological theories ensued (e.g., see collection of papers in Caplan, 1978).

Time and extensive discussion have somewhat eased the arrival of sociobiology, and probably more of each is still necessary to achieve a proper perspective on the full impact of sociobiological ideas. Certainly, Wilson's prediction that sociobiology would swallow up ethology and comparative psychology has not come to pass. Although some researchers feel that socio-

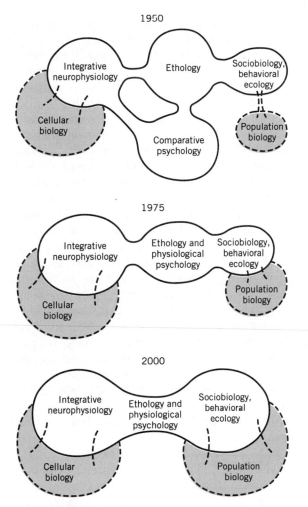

1950

Integrative neurophysiology

Ethology

Sociobiology, behavioral ecology

Cellular biology

Comparative psychology

Population biology

1975

Integrative neurophysiology

Ethology and physiological psychology

Sociobiology, behavioral ecology

Cellular biology

Population biology

2000

Integrative neurophysiology

Ethology and physiological psychology

Sociobiology, behavioral ecology

Cellular biology

Population biology

Figure 2.23 E. O. Wilson's dire predictions for the fields of ethology and comparative psychology. (From Wilson, 1975.)

biological concepts are being integrated into mainstream animal behavior (Dewsbury, 1989), others may see evidence of increasing polarities between sociobiology and other disciplines in animal behavior. Whichever view is most accurate, it is hoped that the contacts between the different disciplines will stimulate research and advance the study of animal behavior.

In our next unit we explore many of the contemporary approaches to the study of animal behavior that have grown out of the early ideas of ethology and comparative psychology, the nature/nurture controversy, and sociobiology.

Summary

One of the earliest ideas to influence the study of animal behavior was that of intellectual continuity. According to this concept, there was a continuity in mental states between "lower" and "higher" animals. Although based on a linear picture of evolution similar to that of Aristotle's scala naturae, the idea of intellectual continuity was first summarized by Herbert Spencer in his book *Principles of Psychology* (1855). Perhaps the most important concept to influence the study of animal behavior came a few years later when Darwin published *The Origin of Species* (1859), in which he outlined his ideas on evolution through natural selection. Darwin's ideas provided the evolutionary framework necessary for the development of animal behavior.

As interest in animal behavior grew, so too did differences in opinion. These differences eventually led to a split in the discipline that began in the early 1900s. The two approaches that resulted were ethology, centered in Europe, and comparative psychology, focused in America. Although there is some debate among historians as to the extent of dissimilarities in the two approaches, ethologists and comparative psychologists often differed in the types of questions they asked, the types of behavior they studied, and the methodologies and experimental animals they used.

Ethologists focused primarily on the function and evolution of behavior. Because the context in which a behavior is displayed is sometimes a clue to its function, ethologists often studied behavior under field conditions. It followed from their interest in evolution that ethologists used a comparative approach and studied primarily innate behaviors.

Early ethologists such as Charles Whitman, Oskar Heinroth, and Wallace Craig were interested in stereotyped patterns of behavior and considered such patterns to be just as reliable as morphological characters in defining a particular group. These stereotyped behaviors were called fixed action patterns (FAPs) by Konrad Lorenz. An FAP is triggered by a very specific stimulus. That portion of the total stimulus that releases the FAP is called the sign stimulus or releaser. The concept of FAP has been questioned in the years following Lorenz's introduction of the term. George Barlow suggested that most behaviors are not as stereotyped as implied by the notion of FAP, and he suggested using the term modal action pattern (MAP) instead.

Action specific energy was a hypothetical construct proposed by Lorenz to describe the specific motivation or drive for a particular action. A buildup of action specific energy can be used to explain vacuum activity (the appearance of an FAP in the absence of a releaser) and displacement activity (behavior that is seemingly inappropriate in a particular context).

Two models were proposed by ethologists to explain the organization of instinctive behavior. Lorenz formulated a psycho-hydraulic model and Tinbergen proposed a hierarchical model that incorporated the notion of the innate releasing mechanism (IRM), a mechanism believed to remove the inhibition from a particular brain center, thereby allowing an FAP to occur. Both models have been criticized for incorporating concepts of energy developing within the nervous system and because of the difficulty in reconciling the models with what is known about neurophysiology.

In contrast to the early ethologists, comparative psychologists emphasized laboratory studies of observable, quantifiable patterns of behavior. In general they asked questions that concerned the development or causation of behavior.

Learning and the physiological bases of behavior were the focus of much research.

The writings of early comparative psychologists were anecdotal and anthropomorphic. However, in what came to be known as Morgan's Canon (1894), C. Lloyd Morgan argued that behavior should be explained in the simplest way possible. This meant that animal behavior should be explained without assuming human emotion or mental ability. In the years following Morgan's proposal for parsimony, many exciting advances were made in the study of learning. Thorndike developed the techniques for studying trial-and-error learning, and Pavlov provided the methodology for classical conditioning.

The birth of behaviorism was another event that steered comparative psychology to become more of an objective science. This school of psychology proposed limiting the study of behavior to actions that can be observed. B. F. Skinner, a prominent behaviorist, found that patterns of behavior that are rewarded tend to be repeated or increase in frequency, and he concluded that the control of behavior was largely a matter of reinforcement. Another important member of the behaviorist school was J. B. Watson.

The physiological basis of behavior was another traditional subject investigated by comparative psychologists. Karl Lashley made many contributions to the study of the neurobiology of behavior, and Frank Beach pioneered the study of the endocrine basis of behavior. Despite their emphasis on learning and physiology in the laboratory, some comparative psychologists, such as C. R. Carpenter and T. C. Schneirla, studied the social behavior of animals in the field.

The ethologists' emphasis on instinctive behavior and the comparative psychologists' focus on learning soon became cause for heated debate between the two groups over the importance of inheritance versus experience in behavior. The result of these differences was the nature/nurture controversy, a battle that peaked in 1953 with Daniel Lehrman's stinging attack on ethological concepts.

Central to the nature/nurture controversy was the concept of innate. Part of the problem was defining the term. Ethologists had applied the word to behaviors that were stereotyped and appeared in a functional form the first time the appropriate stimulus was encountered. It was argued, however, that because members of a given species develop within essentially the same environment, the similarity in the form of behavior could be a product of the uniformity of experience. Furthermore, even though an animal shows the appropriate response on the first suitable occasion, it does not mean that learning was unimportant. Learning can take place prenatally.

Another difficulty with the concept of innate was interpreting experiments designed to demonstrate a genetic basis to a behavior. One such experiment, the isolation experiment, came under fire. In this experiment, the animal is raised without contact with members of its species and without the environmental stimuli thought to be important. One problem is that it is impossible to control all stimuli. A stimulus that was unsuspected, and therefore uncontrolled, may later be shown to be important. Additionally, an animal may provide the stimulation necessary for the development of the behavior itself.

Breeding experiments were also used to demonstrate that genetic differences between individuals can result in behavioral differences. It cannot be concluded, however, that a behavior with a genetic basis is immune to environmental influences. For example, although the begging responses of young gulls have a genetic basis and are present in a functional form soon after hatching, Jack Hailman (1967, 1969) demonstrated that they are modified by the experience of begging from a parent.

Just as it was difficult for the early ethologists to accept that what they considered to be innate behavior could be modified by experience, it was also difficult for comparative psychologists to accept that learning could be constrained by genetic factors. Initially it was believed that any rewarded response would persist. However, Keller and Marion Breland (1961) noticed that some of the animals they trained using operant conditioning methods interrupted their performances with innate behaviors used for obtaining or eating food.

A second principle of classical learning theory is the premise of equivalence of associability: Any stimulus can be associated with any response through the process of reward and punishment. However, it soon became apparent that members of different species may vary in their abilities to associate different events because, through evolution, different species have adapted to different ecological niches. A given species may be prepared to make some associations, unprepared but capable of making other associations, and unable to associate some stimuli. This preparedness continuum is consistent with taste aversion studies and avoidance learning experiments.

In spite of these difficulties, it has proven useful to retain the terms innate and learned. There are differences in the degree of flexibility of behaviors—some are stereotyped and others are obviously acquired or easily modified. The term innate, however, should never be used as an explanation of how the action developed, or to imply that experience is unimportant. Today it is recognized that the categories of innate and learned are not mutually exclusive. Nature and nurture are both important to the development of every behavior. Although ethology and comparative psychology passed through a stage of extreme disagreement, today the distinction between the two fields is blurring. Indeed, perhaps the best result of the nature/nurture controversy has been the increasing study of behavioral development in both fields.

At about the time that ethologists and comparative psychologists were reconciling their differences, a new discipline emerged in the study of animal behavior. The discipline was called sociobiology (a term sometimes used interchangeably with behavioral ecology), and it focused on the application of evolutionary theory to social behavior. One of its central concepts, that of inclusive fitness, was articulated in 1964 by W. D. Hamilton. According to Hamilton, individuals behave in such a manner so as to maximize their inclusive fitness (i.e., their own survival and reproduction plus that of their relatives), rather than acting simply to maximize their own fitness. Suddenly, certain issues that seemed inconsistent with selection at the level of the individual, such as the evolution of sterile castes in insects and altruistic behavior (behav-

ior that benefits others at the expense of the performer), were explainable.

Approximately 10 years after Hamilton's paper, E. O. Wilson crystallized sociobiological ideas in his landmark text *Sociobiology* (1975). The book became a focal point for both critics and proponents of the new discipline. Whereas some critics from within the scientific community were irritated by Wilson's prediction that sociobiology would swallow up ethology and comparative psychology, critics from both within and outside the scientific community attacked the extension of sociobiological ideas to human behavior. Despite its controversial beginnings, some believe that sociobiology has blended with mainstream animal behavior.

CHAPTER
3

Genetic Analysis of Behavior

*O*h what a tangled web we'd weave! And yet, a spider with its minuscule brain fashions a lacy masterpiece (Figure 3.1). Furthermore, it must create this orb without parental guidance on its first attempt since, right after hatching, it floats away from its parent on a gossamer thread. How can it create such a complex and delicate thing on its first try? Obviously, because its behavior is genetically programmed.

The Relationship Between Genes and Behavior

Does that mean that the spider inherited a gene containing a cram course in web weaving? The answer is no, and the reason is that genes do not contain instructions for performing specific behaviors. Nor are genes blueprints for behavior or even for the circuitry of the nervous system.

So then, if a gene is not an instruction manual or a blueprint, what does it mean to say that a behavior has a genetic basis? It simply means that the probability that a behavior will be performed is due to the presence of a gene. Put another way, if the animal has a certain gene it will be able to perform a behavior, but if it lacks that gene it may be unable to perform the behavior, or the form of the behavior may be altered, or the frequency of the behavior may change.

What do genes really do, then? How do they work? As you may already know, genes simply direct the synthesis of proteins. Each protein is specified by a different gene or, if the protein consists of more than one chain of amino acids, a gene specifies one of those chains. The protein may be structural, used as a building block of the organism, or it may be an enzyme that facilitates chemical reactions. In other words, genes code for specific proteins and the proteins affect the composition and organization of the animal and influence the way it behaves. An animal's sensory receptors detect stimuli and send information to the nervous system where it is interpreted and analyzed. The nervous system may then initiate a response by effectors, the muscles and glands, that results in behavior. The structure and function of all these cells—receptors, nerves, muscles, and glands—require information from the genes. Alterations in genes change the proteins they code for with the result that anatomy or physiology may be altered in a manner that changes behavior. We will look at a few examples of the links between genes and behavior later in this chapter.

To understand the relationship between genes and proteins, it is helpful to know a little biochemistry. Genes are composed of DNA (deoxyribonucleic acid). The structure of DNA is somewhat like a long ladder twisted about itself like a spiral staircase. The DNA ladder is composed of two long strings of smaller molecules called nucleotides. Each nucleotide chain makes up one side of the ladder and half of each rung. A nucleotide consists of a phosphate, a nitrogenous base, and a sugar called deoxyribose. There are only four different nitrogenous bases used in DNA: adenine (A), thymine (T), cytosine (C), and guanine (G). Although DNA has only four different nucleotides, a DNA molecule is very long and has thousands of nucleotides. When forming a rung of the ladder, adenine must pair with thymine and cytosine must pair with guanine (Figure 3.2). The specificity of these base pairs is important, not only for the accurate production of new DNA molecules, but also for converting the information in the gene into a protein.

The instructions for each protein are written as the sequence of bases in the DNA molecule. The first step is to transcribe the information in DNA into messenger RNA (mRNA). The DNA unzips for part of its length so that an mRNA molecule can be formed. The mRNA has a structure similar to DNA except for three differences: It is single stranded, its sugar is ribose, and the base uracil substitutes for thymine. Since adenine must pair with uracil and cytosine must pair

Figure 3.1 A spider's web. The ability to construct a web is a genetically programmed behavior.

with guanine, the sequence of bases on RNA is specified by the sequence of bases on the DNA. Now you see why the specificity of base pairing is important in getting the information in a gene translated into a protein. Because the DNA molecule is so long, there is a multitude of ways that the four bases can be ordered. Therefore, with only four bases, DNA can direct the synthesis of a myriad of different proteins.

The next step is to translate the order of bases on the mRNA molecule into a protein. Proteins are long chains of amino acids. Each different protein has a unique order of amino acids. A group of three bases on the mRNA molecule is translated, three bases at a time, into a protein. The order of bases on DNA specifies the sequence of bases in mRNA that can be translated into only one array of amino acids. Thus, the

information in the gene is its sequence of bases that codes for a specific protein.

It should be kept in mind that possession of "a gene for some behavior" in the sense we have just described the relationship is a necessary, but not always sufficient, condition for the expression of that behavior. There are at least two reasons for this: gene regulation and interaction with the environment.

An animal may have all the genes that are needed to perform a behavior and still not perform it because an essential gene is not active. Although all the cells of an animal's body have the same genes, some of those genes are turned off in the normal process of things and do not produce a protein. This should not be unexpected precisely *because* all cells have the same genes. If all the genes functioned all the time,

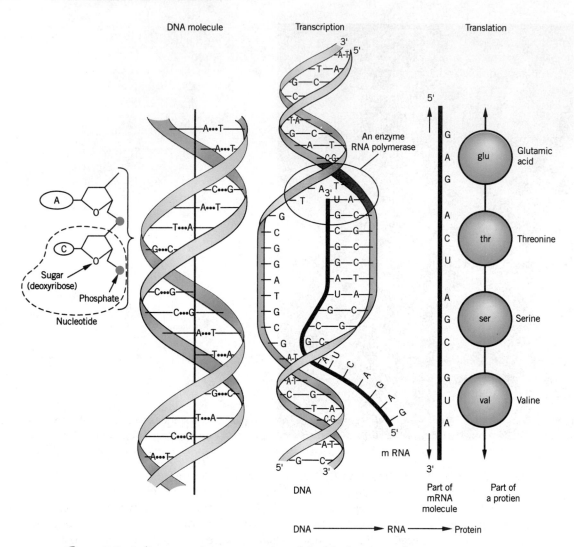

Figure 3.2 A diagrammatic representation of the biochemistry of gene expression. A gene is a region of a DNA molecule that has the information needed to make a specific protein. A DNA molecule is composed of nucleotides. A nucleotide, shown on the extreme left of the figure, is a nitrogenous base, a sugar (deoxyribose), and a phosphate. In the ladderlike DNA molecule, the two uprights are composed of alternating sugar and phosphate groups and the rungs are paired nitrogenous bases. The pairing of bases is specific: adenine with thymine and cytosine with guanine. During transcription, the synthesis of mRNA, the sequence of bases on the DNA molecule is converted to the complementary sequence of bases on the mRNA molecule. Each unit of three bases on the mRNA molecule signifies a particular amino acid. The message of messenger RNA is, therefore, its sequence of bases that determines the order and kinds of amino acids in the protein product.

all cells would have the same appearance and function, and that, clearly, would not do.

So then, how is gene activity regulated? The answer is complex and incomplete, but we do know that in some bacteria, the protein produced from one gene affects the activity of other genes. Gene regulation in higher organisms, those whose behavior is the subject of this book, is not understood as well, but it is suspected that these genes might be influenced in similar ways.

Steroid hormones are known to be a means of gene regulation in higher organisms. One example of hormonal regulation of gene activity may be familiar to you: puberty. Puberty is the time when, if you are male, your voice deepened and you had to start shaving or when, if you are female, your breasts developed and you started menstruating. These physical changes were accompanied by behavioral changes, such as beginning to associate with members of the opposite sex. Studies show that human males have the most interest in sex between the ages of 15 to 25 years when the level of their steroid sex hormones, the androgens, is peaking (Davidson, Camargo, and Smith, 1979). The physiological and behavioral changes experienced at puberty are triggered by the steroid sex hormones that your body began to produce. These hormones were produced when certain genes began to produce the enzymes needed to manufacture them. Only certain body parts change at puberty because a steroid hormone affects the protein production of only the cells that have receptors that can bind to it. In other words, the steroid sex hormones that bring on the anatomical, physiological, and behavioral changes signifying sexual maturity in vertebrates work by stimulating the activity of specific genes, but only in those cells that have receptors that can bind to the hormones.

An example of the environment altering gene expression with dramatic effects on behavior is the determination of castes in some of the social insects such as ants. A caste is a group of indi-

viduals in a society that is anatomically and behaviorally specialized to perform specific functions. A female ant can be in any one of three castes. If she is in the reproductive caste, she is the queen and is specialized for egg laying. Members of the worker caste are specialized for foraging and hive maintenance. Finally, those in the soldier caste are specialized for defending the hive. You can see the difference in appearance of members of each of these castes in Figure 3.3. It is interesting that members of these castes have the genes for becoming a member of any other caste. Environmental factors, the most important of which is diet during the larval stage, determine the developmental fate of each female in the society (Wilson, 1971).

Patrick Bateson (1983) pointed out that when two or more factors are interacting, as genes and environment do during the development of behavior, it is impossible to isolate one of them as the only important factor, or even as the most important factor. He emphasized the importance of gene–environment interactions by likening the development of a behavior to the making of a cake. When a cake is prepared, both the recipe and the cooking temperature are important. Bateson's analogy may hit home for those of you who are not known as gourmet cooks. Most of us have had firsthand experience with ruining a would-be culinary delight by altering "genetic input," or recipe, on some occasions and the "environment," or cooking temperature, on others. If a particular ingredient is lacking, perhaps you simply left it out or substituted another. Inevitably this changes the product in noticeable ways. However, even if the recipe is followed exactly, the product can still be ruined. With dinner guests at the door, it is always tempting to cook foods for a shorter time than is called for and compensate by using a higher temperature than is recommended. This practice can change the consistency of rice pudding to that of concrete. However, even if your only interest in baking is consuming the final product, you must be aware that no matter how it is prepared,

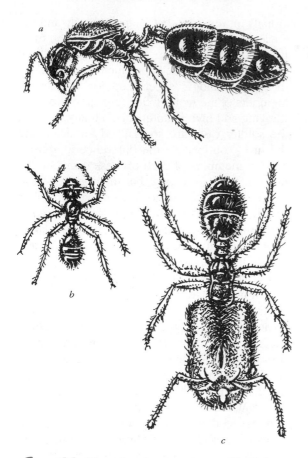

Figure 3.3   Three female castes in ants *Pheidole kingi*, an example of the effect of environment on gene expression: (*a*) a queen, a member of the reproductive caste; (*b*) a worker; (*c*) a soldier. These three females could be genetically identical. Environmental factors, such as diet, determine which caste a larva will develop into. (Redrawn from a figure in Wilson, 1971.)

a cake differs from its ingredients. All analogies are weak in some respect but may be helpful in other ways, so this one should not be pushed too far. The important points to be learned from this digression are that the expression of a behavior depends on both its genetic foundations and environment and that the many factors contributing to its development cannot be teased apart.

Experimental Methods Demonstrating a Genetic Basis to Behavior

While we turn our attention to the experimental methods used to show that genes can influence behavior, keep in mind that genes and environment are actually interacting to produce behavior.

INBREEDING

Inbred lines, strains created by mating close family members with one another, are useful in behavioral genetics because they provide a way to hold constant the genetic input. By using inbred strains, therefore, it is possible to separate the effects of genes from those of the environment. To show the effects of genes, the behavior of members of two inbred strains is compared in the same environment. In this case, any observed difference in behavior must be caused by a difference in the genes of the strain. However, the influence of the environment on behavior can also be shown using inbred strains. If members of the same inbred strain, individuals who are almost genetically identical, behave differently when they are raised under different conditions, then the variation must be caused by environmental effects.

Before considering how inbreeding reduces genetic diversity, we should think about why a population of unrelated individuals is dissimilar. Usually a genetically controlled characteristic or trait can be expressed in more than one way. For example, coat color in mice might be black or brown. In other words, this gene for coat color can be expressed in two different ways; one produces intense black pigment granules and the other produces chocolate brown pigment granules. These alternative forms of a gene are called alleles. Sometimes, as in the eye color of fruit flies, there are many possible alleles. The wild type, or most common eye color of fruit flies, is red, but other alleles of this gene can result in white, vermilion, or white-eosin eyes.

Furthermore, most of the animal species used in genetic studies are diploid, meaning that an individual possesses two alleles of each gene, one from each parent. If the two alleles of a gene are identical, the individual is said to be homozygous for the trait. However, an individual may inherit different alleles for a trait from its mother and father. Such an individual is heterozygous for the trait. The genetic diversity among unrelated individuals results from the variety of alleles they possess and heterozygosity.

Inbreeding reduces genetic diversity because it creates a population that is homozygous for almost all of its genes. How does it do this? In the laboratory, an inbred strain is usually created by mating brothers with sisters for many generations. If the original parents of an inbred line lacked certain alleles found in the general population, those alleles would also be missing in their offspring. So when close family members mate, the alleles present in the parents will be retained and will increase in frequency relative to the general population and those not present in the founders will be lost. In addition, each parent passes only one of its two alleles for each gene on to the offspring. Therefore, if a parent is heterozygous for a trait, one of its alleles will be missing in the offspring unless it is supplied by the other parent. With each successive generation of inbreeding, genetic variability is lost. Each generation of brother–sister matings reduces heterozygosity by 25 percent. After 20 generations of sibling matings, all individuals are expected to have identical alleles for 98 percent of their genes. As a result, no matter which member of each chromosome pair is transmitted from parent to offspring, the same alleles will be passed on. Inbred lines, therefore, are virtually genetically identical (Plomin, DeFries, and McClearn, 1980).

Comparison of Inbred Strains To Show the Role of Genes

As previously mentioned, inbred lines can be used to demonstrate that differences in genetic makeup may result in differences in behavior. According to the scientific method, when you want to explore the effect of one variable, you hold the others constant. Inbred lines provide a way to do this. Inbreeding minimizes genetic diversity within strains but maintains diversity between strains and may even enhance it. If individuals of different inbred strains are raised in the same environment, then any difference in their behavior is due to a difference in their genes. So the comparison of inbred strains raised under the same conditions is a way to demonstrate that a particular behavior has a genetic basis.

Historically, strain comparisons have been used to demonstrate that genes play a role in behavior. There were nearly 1700 studies of this type on inbred mouse strains alone between 1922 and 1978 (Sprott and Staats, 1975, 1978, 1979). Almost all behaviors compared, by the way, showed strain differences.

One of the most consistent strain differences has been in learning ability. We can draw an example from studies on avoidance learning in mice. These studies usually involve a shuttle box, a two-compartment apparatus with a floor that can be electrified (Figure 3.4). The mouse is placed in one compartment. An electric light signals that a mild electric shock is imminent. When the light comes on, the mouse must run through a door into the other compartment to avoid a shock. As you can see in Figure 3.5, not all mice learn the trick. Each dot on the graph represents the percentage of the trials in which a given mouse avoided the shock during five sessions. Notice the responses of the Swiss mice. They are interesting because the mice are not inbred. Since their breeding is random, genetic variability has been maintained. Notice that their responses vary tremendously: Some almost never avoid the shock and others avoid it about 60 percent of the time. Such a range is expected from a genetically heterogeneous stock. Now notice the responses of the other strains of mice, all of which were inbred. The responses of individual mice within each inbred strain are

Figure 3.4 A shuttle box used to study avoidance learning. An animal such as a mouse is placed on one side of the apparatus. When a light comes on, it signifies that the grid on the floor will soon be electrified. The animal learns to avoid the shock by running through the door to the other compartment.

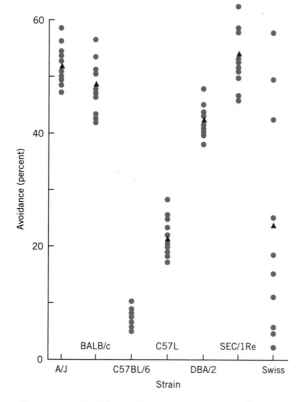

Figure 3.5 Avoidance learning in a strain of genetically heterogeneous mice (Swiss) and six strains of inbred mice. A dot represents the average number of times a mouse of the indicated strain successfully avoided the shock during five sessions. The mean response of all mice of a strain is indicated with a triangle. (From Bovet, 1977.)

much more similar than are those of the Swiss mice. Inbreeding has reduced the variability in the genetic composition and this decreased the range of responses. Furthermore, it is obvious that there are substantial differences in the ability of mice of different inbred strains to learn this task. Four of the strains are quite good and learn to run to safety on more than 40 percent of the trials. However, mice of the C57L strain avoid the shock only about 20 percent of the time and those of the C57BL/6 strain only about 10 percent of the time. For these latter strains, one might say, avoidance learning must be a shocking experience. Since the environment is the same for all strains of mice, but the genetic composition (genotype) of each strain is different as a result of inbreeding, the differences between strains indicate a genetic influence on active avoidance learning (Bovet, 1977).

A brief mention of a few other strain comparisons will give you a feeling for the diverse behaviors that have been shown to have a genetic basis. For example, white strain domestic fowl spend an excessive amount of time pacing about before laying eggs and do not return to incubating the eggs as quickly as do the brown strain (Wood-Gush, 1972). And then in the rock-pool-breeding mosquito *Aedes atropalus*, when females of two strains were compared, it was found that those of the GP strain did not need

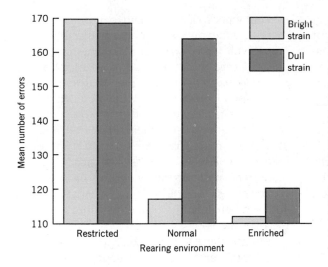

Figure 3.6 The mean number of errors made in maze running by maze-bright and maze-dull rats raised in three different environments. Notice that the effect of the environment depends on the genetic composition of the strain. The restricted rearing environment impaired the maze-learning ability of both strains. Being raised in a normal environment rather than a restricted one had a great effect on the learning ability of maze-bright rats, but only a modest effect on the learning ability of maze-dull rats. In contrast, the enriched rearing environment greatly reduced the number of errors made by maze-dull rats and had a much smaller effect on the errors made by maze-bright rats. (Data from Cooper and Zubek, 1958.)

to gorge on a protein-rich meal such as blood for their eggs to mature and they mated at a much younger age than did the females of the TEX strain. In laboratory tests, 50 percent of the GP females that were 38 hours old were inseminated after spending 24 hours in a culture jar with a male. The TEX females had to be 120 hours (5 days) old before half of them would breed after a day's exposure to a male (Gwadz, 1970). Comparison of inbred strains has also revealed differences in the olfactory ability of both mice (Wysocki, Whitney, and Tucker, 1977) and fruit flies (Hay, 1976) and in the pattern of development of seizures in mice (Deckard et al., 1976).

Comparisons of Inbred Strains to Show the Role of Environment

Because individuals of inbred strains are identical at 98 percent of their genes, such strains, as we mentioned, can provide a way to hold the genetic input constant while varying the environment. Thus, if differences in behavior are found, they must be due to the environment.

In one of the best known experiments of this type, researchers first developed two strains of rats that had distinctly different levels of success

with mazes. To create these strains, "maze-bright" rats were mated with other bright rats and "maze-dull" rats with other dull ones. Rats from each strain were raised under similar conditions until they were weaned at 25 days of age. Then some from each strain were raised in an enriched environment, one that was brightly colored and contained many movable toys, and others were moved to a restricted environment, a cage with gray walls and no movable objects. The rats were kept under these conditions for 40 days and then tested in mazes. Their scores were compared to those of maze-bright and maze-dull rats that in a previous experiment, were raised in a normal environment.

The combined results, shown in Figure 3.6, indicate that the environment influenced the maze-running performance of both strains. Although the enriched environment improved the performance of both strains, the amount of improvement depended on the genes. The maze-bright rats made many more errors when raised in a restricted environment than when raised in a normal one. In contrast, the impoverished environment had only a slightly negative effect on the performance of maze-dull rats. A comparison of the maze-running performance of rats reared in an enriched environment to that of rats raised in normal conditions shows little difference in the error rate of the bright rats, but a great reduction in errors made by the dull rats. Manipulating the rearing environment altered the ability of the rats of each strain to learn to run mazes. This experiment indicates that the degree of the effect of the environment may depend on the genetic composition of the animal (Cooper and Zubek, 1958).

So far we have seen that genetic differences between inbred strains may be accompanied by differences in behavior, but we have also noted that different genotypes may respond differently to specific environments. Of course, behavior is also influenced by early experience.

Eliminating the effects of early learning is tricky, but it can be done. The simplest way is by a reciprocal cross, one in which males of inbred strain A are mated with females of inbred strain B and males of strain B are mated with females of strain A. Since the individuals of each strain are homozygous for almost every gene, the hybrid offspring of both of these crosses have the same genotype even though their mothers are from different strains. So if the behavior of the hybrid offspring of reciprocal crosses differs, it must be an effect of the parental environment. If the young are raised solely by their mothers, as they often are in the laboratory, then the difference must be an effect of the maternal environment.

Theoretically it is even possible to determine whether the maternal environment had its greatest influence on the offspring before or after birth. Cross-fostering, transferring the offspring shortly after birth to a mother of a different strain, is a technique for detecting maternal influences that occur after birth. If offspring that were transferred to a foster mother immediately after birth behave more like individuals of the foster mother's strain than like those of their own strain, then postnatal maternal influences are implicated.

A cross-breeding experiment involving two different species of voles helped in separating the influences of genes and parental environment in the expression of a particular behavior. Prairie voles, *Microtus ochrogaster*, show more parental care than do meadow voles, *Microtus pennsylvanicus*. For example, prairie vole females spend more time in the nest with their young, contacting their offspring by huddling over them and nursing them. Male prairie voles also show more parental care than do male meadow voles. In contrast to meadow vole males who nest separately and rarely enter the female's nest, prairie vole males share a nest with the female and frequently groom and huddle over the young.

To determine whether the species difference in parental care was due to genes or early experience, Betty McGuire (1988) fostered meadow vole pups to prairie vole parents, and as a control, she fostered meadow vole pups to other meadow vole parents. When the foster pups became adults and had their own families, she measured the amount of parental care they gave to their second litters. The meadow vole pups that were raised by prairie voles gave more care to their own offspring than did meadow vole pups that had been fostered to other meadow vole parents. Cross-fostered females, for example, spent more time huddling over and nursing their young. The early experience of male voles influenced their parental behavior in much the same way as it affected the females' behavior. Although male meadow vole pups that were fostered to parents of their own species behaved as meadow vole males usually do in that they rarely entered the nest with the young, four of the eight meadow vole males raised by prairie vole parents nested with their mates and spent time in contact with the young. This cross-fostering experiment shows that the experience a vole has with its own mother will influence the way it treats its own offspring. However, not all behaviors were modified by early experience. Nonsocial behaviors such as food caching and tunnel building and overall activity level were unaffected by the species of the foster parents.

Thus, using the technique of cross-fostering, it is possible to determine whether a particular behavior is influenced by the parental environment.

ARTIFICIAL SELECTION

Artificial selection is another means of demonstrating that a behavior has a genetic basis. Artificial selection differs from natural selection in that an experimenter, not "nature," decides which individuals will breed and leave offspring. The rationale for artificial selection is that if the frequency of a trait in a population can be altered by choosing the appropriate breeders, then it must have a genetic basis since traits acquired by experience cannot be passed to offspring. Usually the first step is to test individuals of a genetically variable population for a particular behavior trait. Those individuals who show the desired attribute are mated with one another and those who lack the trait are prevented from breeding. If the character has a genetic basis, the alleles responsible for it will increase in frequency in the population because only those possessing them are producing offspring. As a result, the behavior becomes more common or exaggerated with each successive generation. When it proves possible to select for a trait, it is concluded that genetic inheritance is important to the expression of that trait. Furthermore, the environment is held constant, so it is reasonable to conclude that any change in the frequency of a behavior is due to a change in the frequency of alleles.

Humans often use artificial selection to create breeds of animals with traits that they consider useful. For example, even if you are not a dog lover, you have probably noticed that different breeds of dogs have different personalities. All dogs belong to the same species, but various behavior traits and hunting skills have been selected for in different breeds (strains). Terriers, for example, are fighters. Aggressiveness was selected for so they could be used to attack small game. The beagle, on the other hand, was developed as a scent hound. Since beagles usually work in packs to sniff out game, they must be more tolerant with companions. Therefore, their aggressiveness was reduced by selective breeding. Shetland sheep dogs were bred for their ability to learn to herd sheep. Spaniels, used for hunting birds, are "people dogs." They are more affectionate than aggressive. These behavioral differences among breeds of dogs were brought about by people who placed a premium on certain traits and arranged matings between individual dogs that showed the desired behavior. In other words, the frequency of particular behavior patterns present in all breeds has been modified through artificial selection; new behaviors were not created (Scott and Fuller, 1965).

The most famous example of artificial selection is the production of strains of bright and dull rats (Tolman, 1924; Tryon, 1940). The first step in obtaining these strains was to test a population of rats in a particular maze and measure the number of errors made, the time necessary to negotiate the maze, and the number of perfect runs for each individual as an indication of its ability to learn the maze. As you see in Figure 3.7, the initial population was quite variable in its maze running ability. Then the bright rats, those that learned the maze quickly and made few mistakes, were mated with other bright rats. Likewise, the dull rats, dunces that made many mistakes, were allowed to breed with other dullards. Most of the rats fell somewhere between these extremes and they were forced to remain chaste. The progeny of all crosses were tested in the same maze, and again mating was arranged so that brightest mated with brightest and dullest bred with dullest. The procedure was repeated generation after generation. The two lines of rats became increasingly divergent in their maze-running ability, and by the seventh generation of selection, there was little overlap in the scores of the two groups (Figure 3.7). Because the rearing conditions were always the same, it is reasonable to assume that the change

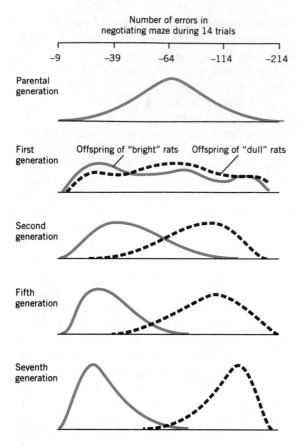

Figure 3.7 Artificial selection for good or poor maze-running ability in rats. The maze-running performance of rats in the original population was heterogeneous. Individuals that made few errors when they ran the maze were bred with other bright rats and those that made many errors during maze running were bred with other dull rats. After seven generations of selection, there was little overlap in the performances of the bright and dull strains. (Data from Tryon, 1940.)

in behavior in the two lines was due to a concentration of whatever alleles lead to enhanced maze-learning ability in one group and a loss of those alleles in the other group.

This experiment is cited not only because it is a classic and is useful for explaining the ra-

tionale of selection experiments, but also because it leads to another important point about artificial selection: Only a highly specific trait is usually selected. Although Tryon named the strains bright and dull because he thought he was selecting for general intelligence, he was actually selecting for traits that allowed animals to master only a specific type of maze quickly. Some years later, the descendants of the original strains were retested in the same maze and performed as expected; the offspring of bright rats did well and the progeny of dull rats did poorly. However, when they were tested in a slightly different maze, one that emphasized visual cues, the two populations did equally well (Searle, 1949).

HYBRIDIZATION

Hybridization is a third way to demonstrate a genetic influence on behavior. The usual procedure is to mate two individuals that display a particular type of behavior in distinct but different ways. Normally species are considered unable to interbreed successfully, but there are exceptions (for example, lions and tigers can produce tiglions—in the zoo but not in nature where they would be unlikely to encounter each other). In any case, once interbreeding has occurred, the behavior in the hybrid offspring is then observed. Depending on the number of genes influencing the behavior, the hybrids may perform the action in the same manner as one of the parents or as a combination of the parental types.

We will consider two examples of hybridization experiments. In the first, hygiene among bees, the hybrid offspring express the behaviors of only one of the parents. However, in the other example, the calling songs of crickets, the behavior of the hybrid offspring is a combination of the parental types. It does not matter which of these forms of inheritance is observed. Whenever the pattern of inheritance is always the same, a genetic basis for that behavior is implicated.

Hygienic Bees and Unhygienic Bees

As we know, personal hygiene is an important attribute, but for a social insect such as the honeybee (*Apis mellifera*), hive hygiene can be a matter of life and death, particularly if the hive has been infected with *Bacillus larvae*, a bacterium that causes American foulbrood. This disease kills larvae and pupae developing within the waxen cells of the comb. If the corpses are not removed, they accumulate, as do the bacteria within their bodies, and they serve as a source of infection that will destroy the entire colony. However, some strains of bees, the Brown and the Squires Resistant, are resistant to American foulbrood mainly because the workers open the cells containing dead individuals and remove them. On the other hand, two other strains, the Van Scoy and Squires Susceptible, are vulnerable to the disease because they allow the deceased brood to accumulate within the hive.

In an experiment crossing a hygienic strain with an unhygienic strain, none of the hybrids removes corpses from the hive. Thus, unhygienic behavior is said to be dominant over hygienic behavior. (In a heterozygous individual, the allele that is expressed is described as dominant. The allele that is not expressed is described as recessive.)

In another series of experiments (Rothenbuler, 1964a,b), the unhygienic hybrid offspring were mated with homozygous recessive individuals (a backcross). A backcross reveals the alleles present in an individual displaying the dominant form of the trait. Because the homozygous recessive parent can only contribute a recessive allele, the offspring will show the allele contributed by the other parent. Thus, if any offspring show the recessive form of the trait, in this case hygienic behavior, one parent must have been heterozygous. The backcross experiments produced four classes of offspring, each with approximately the same number of colonies:

1. In nine of the 29 colonies, w capped the cells containing dead bod they did not remove the corpses.

2. In six of the colonies, workers removed the dead bodies from cells that were already uncapped, but they would not uncap the cells themselves.

3. In eight colonies, workers were truly unhygienic; they neither uncapped cells nor removed corpses.

4. In the remaining six colonies, workers were completely hygienic and both uncapped cells and removed the dead bodies.

Apparently, then, hygienic behavior is controlled by two genes, one for uncapping (the U gene) and one for removing dead bodies (the R gene). The reasoning is, if the hygienic behavior were controlled by a single gene, a cross of a hybrid (and, therefore, unhygienic individual) with a homozygous recessive (and, therefore, hygienic individual) would result in only two classes of offspring, hygienic and unhygienic, with equal numbers of colonies produced. Four equally sized classes of offspring is what is predicted if two genes are involved in the expression of the trait (Figure 3.8).

The story is actually not so cut-and-dried. It turns out unhygienic workers sometimes perform hygienic behaviors, but not often. It seems likely, therefore, that the U and R genes are regulatory genes that control a group of genes whose activities result in uncapping or removal, respectively (Rothenbuler, 1964a,b).

Cricket Calling Songs

Hybridization experiments have also revealed an influence of genes on cricket calling songs. A male cricket's calling song, produced by rubbing the wings together, is used to attract a mate. A female can recognize a suitor of her own species because each species of cricket sings a different song. David Bentley and Ronald Hoy (1972) crossed two species of crickets with distinctive

a

	UR	Ur	ūR	ūr
UR	UURR	UURr	UūRR	UūRr
Ur	UURr	UUrr	UūRr	Uūrr
ūR	UūRR	UūRr	ūūRR	ūūRr
ūr	UūRr	Uūrr	ūūRr	ūūrr

b

Figure 3.8 Punnett Squares indicating the expected results from a cross between hygienic and unhygienic bees if (*a*) hygiene is controlled by a single gene and if (*b*) hygiene is controlled by two genes. In a Punnett Square, the possible allele combinations within gametes are indicated along the edges of the square. The possible allele combinations in the cells of the offspring are shown in the squares. The observed results of the cross had four classes of offspring indicating that the behavior is controlled by two genes, one for uncapping the cells (U) and one for removing the corpses (R).

songs, *Teleogryllus commodus* and *Teleogryllus oceanicus*, and analyzed 18 characteristics of the song of the hybrid offspring (Figure 3.9*a*).

It was discovered that when the parental species differed markedly in some characteristic of the song, that characteristic would be expressed in an intermediate form by the offspring (Figure 3.9*b*). This shuffling of components is believed to indicate that several genes are involved in producing each characteristic of the calling song.

Reciprocal crosses of the two parent species provided more interesting information. You recall that a reciprocal cross uses each species first as the mother and then as the father. The crosses here were a female *T. commodus* with a male *T. oceanicus* and a female *T. oceanicus* with a male *T. commodus*. In crickets, females have two X chromosomes and males have only one, therefore, a male's X chromosome always comes from his mother.* It turned out that three characteristics of the hybrid male's calling song (specifically, the duration of the interval between trills, the variability of the intertrill interval, and the variability of the phrase repetition rate) are more similar to the male calling song of his mother's species than to the song of his father's species. Only male crickets sing. As an example of one sex-linked characteristic of the song, the interpulse interval is shown as histograms in Figure 3.10. Notice that the intervals between pulses of the song of hybrids from both crosses resemble the intervals of the song of the mother's species. It was concluded that because certain features of the calling song are more similar to those of the maternal species' calling song, and because a male's X chromosome comes from his mother, the genes controlling these aspects of the calling song must be on the X

---

* During reproduction, a female cricket's pair of X chromosomes separate and each egg, then, contains an X chromosome. Because a male has only one X chromosome, half of his sperm will contain an X chromosome and the other half will not. If an egg is fertilized by an X-containing sperm, the resulting cell will be XX and will develop into a female. However, if the egg is fertilized by a sperm that lacks an X chromosome, the resulting cell will have a single X chromosome and develop into a male.

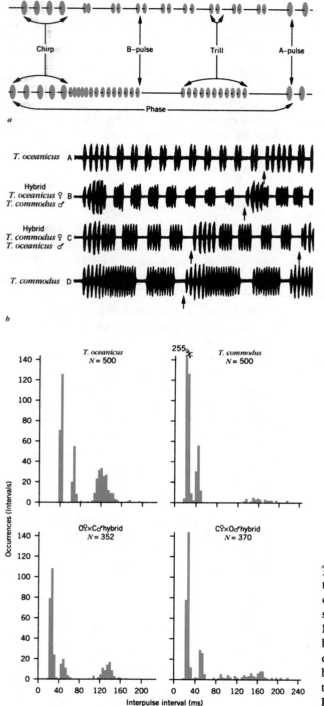

a

Chirp   B-pulse   Trill   A-pulse

Phase

T. oceanicus A

Hybrid
T. oceanicus ♀
T. commodus ♂ B

Hybrid
T. commodus ♀
T. oceanicus ♂ C

T. commodus D

b

Figure 3.9 Cricket songs *Teleogryllus* spp. and the terminology used to describe the structural components. (*a*) Diagram of the structural components of cricket songs and the terms used to describe various characteristics. (*b*) Patterns of sound pulses in calling songs of two species of crickets and the hybrids resulting from reciprocal cross of those species. The arrows indicate the beginning of repeating phrases. Notice that the parental calling songs differ in the number of pulses per trill, the number of trills per phrase, and the phrase repetition rate. The songs of the hybrids are intermediate in each of these characteristics, indicating that the behavior is controlled by a number of genes. (From Bentley & Hoy, 1972.)

Figure 3.10 Histograms of the interpulse intervals in the calling songs of two species of crickets *Teleogryllus* spp. and the hybrids resulting from reciprocal crosses of those species. Notice that in this song characteristic, the hybrid's song resembles that of the maternal species' song. The interpulse interval is controlled by a gene on the X chromosome. *N* indicates the number of individuals in the sample. (From Bentley & Hoy, 1972.)

chromosome. However, these genes do not work alone, because each of the sex-linked song characteristics differs slightly from the comparable characteristic in the song of the maternal species. Other genes, then, must influence those on the X chromosome.

Locating the Effects of Genes

The breeding experiments we have discussed so far were the primary tools used by classical behavior geneticists, but they are actually not very powerful tools. Inbreeding, selection, and hybridization may indicate that the behavior in question has a genetic basis and perhaps that it is controlled by few or many genes, but it tells us nothing about where or how the genes are exerting their effects. More modern techniques such as inducing single-gene mutations, mutation and mosaic analysis, and the use of recombinant DNA have filled in some of the gaps in our understanding of the relationship between genes and behavior.

SINGLE-GENE MUTATIONS

One way to find the links between behavior and genes is to induce a mutation in a single gene, screen the population for individuals who behave abnormally, and then search for the anatomical or physiological differences between the mutant and the normal organisms.

A mutation, a change in a gene's instructions for producing a protein, can be induced by agents that change the DNA bases. When organisms are exposed to mutagenic agents, some of them become behaviorally aberrant and can be separated from the population of normal individuals on the basis of their behavior. Appropriate genetic crosses can then determine whether the behavioral change is caused by an alteration in a single gene. Even a small change may result

in a difference in a specific aspect of an anatomical structure or a physiological process that mediates a behavior. Identifying the anatomical or physiological differences between mutant and normal individuals brings us closer to understanding how genes can influence behavior.

Studies on the avoidance reaction of the protozoan *Paramecium aurelia* and learning in the fruit fly, *Drosophila melanogaster*, have helped fill in a few of the "missing links" between genes and behavior.

Paramecium

Have you ever played on the bumper cars in an amusement park? What happens when you collide with another bumper car? First your car bounces backward a bit. Then you turn the wheel and try to move forward. If your maneuvering has not created a clear path, you have to repeat the same pattern of movement until the forward path of your car is unobstructed, and you can gleefully race on to another crash.

This pattern of movement is similar to that of a paramecium encountering a barrier or chemically unsuitable area. The avoidance reaction of this unicellular creature consists of backing up a body length or two, turning slightly, and swimming forward again. The actions are repeated until the protozoan can move forward (Figure 3.11).

Some of the cellular physiology underlying the forward swimming and avoidance response of *Paramecium aurelia* is known. The animal is propelled by cilia, short hairlike structures covering the body surface that beat like oars. The direction in which the cilia beat determines whether the paramecium moves ahead or backs up. When the cilia beat toward the posterior end, the animal moves forward; when they beat toward the anterior end, the animal swims backward.

It turns out that the concentration of free calcium ions within the cilia controls the direc-

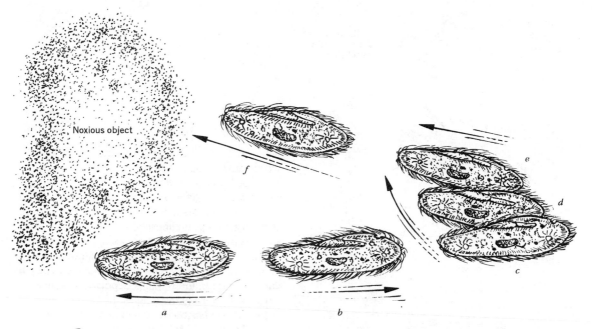

Noxious object

Figure 3.11 The avoidance response of *Paramecium*. When the animal encounters a barrier or a noxious area, it backs up for a short distance and turns a bit, about 30°, before swimming forward again. If its path is still not clear, the protozoan repeats the same pattern of movement.

tion of their beat (Eckert, 1972). When the concentration of calcium ions is low, the cilia beat toward the posterior end of the animal, but when the calcium ion concentration is elevated, the cilia beat toward the anterior end.

The concentration of calcium ions is, in turn, regulated by the responses of different regions of the cell membrane to external stimuli. When the concentration of calcium ions within the cell is low, the activity of the cilia beating toward the rear of the paramecium increases, thereby causing the organism to swim forward. When the animal bumps into a barrier or encounters a noxious area, the permeability of the membrane to calcium ions at its anterior end changes and more calcium ions enter the cell, causing the cilia at the anterior end to beat more vigorously, pushing the paramecium backward. The para-

mecium only backs up for a second or two because it begins to pump the calcium ions out again almost immediately. When the concentration of calcium ions returns to normal, the cilia once again beat toward the posterior end making the animal move forward (Figure 3.12).

Mutations in single genes in *Paramecium aurelia* have made it possible to correlate dramatic changes in swimming behavior with specific alterations in normal membrane characteristics. One mutant, *pawn* (named after the game piece in chess), cannot back up. If you imagine driving a car without a reverse gear, you will get an idea of how inconvenient this can be. Most of the time swimming forward is fine, but if a barrier or noxious area is encountered, the *pawn* mutant must plow right into it. The avoidance reaction is missing because this mutant has defec-

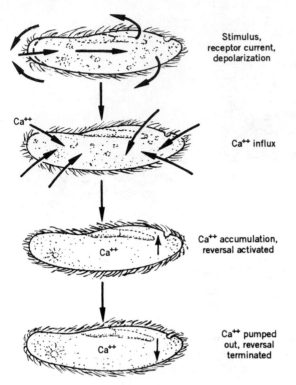

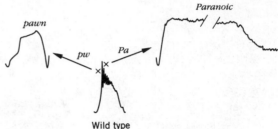

Figure 3.13 The changes in membrane potential accompanying the avoidance reactions of normal wild-type *Paramecium* and two mutants, *pawn* and *paranoiac*. The *pawn*'s membrane does depolarize when it encounters a noxious area, but the pattern of depolarization is different from that of a normal *Paramecium*, and it is not associated with an influx of calcium ions. The *pawn* mutant cannot back up because calcium ions do not enter the cell. The *paranoiac*'s membrane remains depolarized much longer than normal. As long as the membrane is depolarized, the *paranoiac* continues to back up. (From Kung et al., 1975.)

Figure 3.12 The physiology of *Paramecium*'s avoidance response. When a stimulus for the avoidance response is encountered, the cell membrane depolarizes and calcium ions enter the cell. The accumulation of calcium ions around the cilia causes them to beat in the reverse direction and the animal swims backward. Soon the resting potential of the membrane is restored and calcium ions are pumped out of the cell. As the concentration of calcium ions around the cilia is reduced, they begin to beat toward the posterior end so the animal swims forward again. (From Kung et al., 1975.)

tive "calcium ion gates" that do not allow calcium ions to enter the cell (Figure 3.13). Since there is no temporary increase in calcium ion concentration, the cilia never reverse their beating and the animal cannot back away from an obstacle. The loss of the avoidance reaction in the *pawn* mutant stems from an alteration in the gene involved in the production of the calcium gates. The protein produced by that gene may be part of the architecture of the gate or it might direct the assembly of the gate.

Another mutant, the *paranoiac*, behaves as its name suggests: It shows excessive avoidance behavior. This mutant may be swimming about and then it spontaneously switches into reverse and swims backward for awhile. It also swims backward much longer than normal in response to high sodium ion concentration.

Like the *pawn* mutation, the *paranoiac*'s behavioral change is related to alterations in membrane properties. This mutant begins to swim backward when the membrane depolarizes and calcium ions enter. It gets stuck in reverse because the calcium gates remain at least partially open, allowing calcium ion entry to persist at a low steady level. The prolonged increase in calcium ions around the cilia keeps the *Paramecium* swimming backward (Kung et al., 1975).

Drosophila

You have probably never met a fruit fly with remarkable intelligence, but some mutant strains are so poor at olfactory learning that they have earned the epithet *dunce*. Before considering the deficiency caused by the *dunce* mutation, olfactory learning in normal fruit flies and the way of demonstrating it should be described. Normal fruit flies can associate an odor with an unpleasant event, such as an electric shock, and learn to avoid that odor if it is encountered again. This should not be surprising because odors are important in the daily life of fruit flies for activities such as locating both food and appropriate mates. William Quinn and his colleagues (1974) demonstrated this avoidance conditioning by shocking a group of flies for 15 seconds in the presence of one odor, but not in the presence of a second odor. Then they presented the odors, one at a time, to the flies without shocking them to see how many would avoid each odor. Most normal flies avoid only the odor connected with electric shock, an association that lasts for 3 to 6 hours (Dudai, 1977).

Two mutants, *dunce*[1] and *dunce*[2], were isolated by exposing a population of *Drosophila melanogaster* to a chemical known to cause mutations and then screening the flies on the basis of their learning ability. The *dunce* mutations are alleles, or alternate forms, of a single gene on the X-chromosome.

In contrast to normal flies, *dunces* fail to learn to avoid odors associated with shock when they are taught with Quinn's experimental design (Dudai et al., 1976). Even larval *dunces* are deficient in olfactory learning (Aceves-Pina and Quinn, 1979).

Why don't *dunce* fruit flies learn as well as normal flies? The first guess, that the sensory system was defective so that the *dunce* flies could not detect either the odors or the shock, was incorrect. Experiments revealed that the mutants are able to detect both (Dudai et al., 1976). In spite of this, mutant flies are unable to remember the association between the shock and an odor.

Apparently the *dunce* mutants have a problem with the early stages of memory formation. If they are shocked in the presence of an odor without subsequent exposure to a second odor, they do associate the odor with the aversive stimulus and avoid it if tested immediately after training, but the association fades quickly. The association between the shock and an odor is short-lived and the experience with a control odor during the interval between the training and the test of learning seems to eliminate associations that may have formed. For these two reasons, it is concluded that an early stage of memory formation is defective in the *dunce* mutants. (Dudai, 1979).

What does the *dunce* gene do? It is either the structural gene for one of the soluble forms (form II) of cyclic AMP (adenosine monophosphate) phosphodiesterase or a regulatory gene that controls the activity of this enzyme. Cyclic AMP phosphodiesterase is the enzyme the degrades cyclic AMP, a mediator of many processes in different types of cells, to 5'-AMP. The conclusion that the *dunce* mutation affects cyclic AMP phosphodiesterase is based on several observations. First, the *dunce* mutation maps to the same position on the X chromosome as a gene causing a deficiency in form II of cyclic AMP phosphodiesterase and as a gene causing female sterility. Second, all three of these independently isolated mutants have reduced levels of the enzyme and, since the destruction of cyclic AMP is impaired, they also have elevated levels of cyclic AMP (Davis & Kiger, 1981). Furthermore, the flies possessing the female-sterile mutation perform no better than *dunces* on olfactory learning tasks (Byers, Davis and Kiger, 1981).

Let's put all this together. The *dunce* mutation causes a reduced level of cyclic AMP phosphodiesterase and, therefore, an increased level of cyclic AMP. In addition, the mutation impairs an early stage of memory formation in olfactory learning. This should lead you to suspect that the enzyme or cyclic AMP might play a role in this type of memory formation, an idea that has been tested by inhibiting cyclic AMP phosphodiesterase in normal flies and testing the olfactory learning abilities. When treated in this way, normal flies learn no better than *dunces* (Byers, Davis and Kiger, 1981).

Studies on the single-gene mutation creating *dunce* fruit flies have shown that alterations in the level of cyclic AMP phosphodiesterase accompany impairments in olfactory learning. Although it is not known how or why these two events are related, the relationship is not surprising given other studies showing effects of cyclic AMP on neuronal activity. Cyclic AMP is implicated as a factor that assists communication between certain neurons by enhancing both the calcium-mediated release of their neurotransmitters (O'Shea and Evans, 1979) and the response of neurons to certain neurotransmitters (Greengard, 1976). In addition, Marc Klein and Eric Kandel (1978) have shown that in the sea hare *Aplysia californica*, changes in the levels of cyclic AMP in certain nerve cells play a role in sensitization of the gill-withdrawal reflex. Sensitization, an increase in neuronal responsiveness to a stimulus due to previous exposure to another intense stimulus, might be considered a simple form of memory.

MUTATION AND MOSAIC ANALYSIS

In the study of *dunce* fruit flies, luck was on the side of the investigators. When a gene for the mutant behavior was mapped, it turned out to be in the same chromosome regions as a gene whose ultimate product was known. This helped to pinpoint cyclic AMP phosphodiesterase as a link between the mutant behavior and the gene. However, we are not always that fortunate, so other techniques must be used to determine how or where particular genes are acting.

Why is it so difficult to determine where a gene is acting? First, examining a population of mutants to look for the defective structure or physiological problem is like looking for the proverbial needle in a haystack. Where do you begin? It would be helpful to know which part of the haystack to search first. Furthermore, knowing where to look for a problem is crucial when the effect of the mutation cannot be seen directly, as would be the case if the gene's activity resulted in abnormal neural connections; an inability to absorb an important precursor; or an inability to produce a neurotransmitter, enzyme, or hormone. Knowing where to look for a defect is also helpful when there are several structures that logic would tell us could be affected by a mutation and cause a behavioral anomaly. For example, there is a mutant fruit fly called *wings up* that cannot fly because its wings are always held straight up. Think for a moment about where the fly's problem might be. Obviously the wings are the immediate problem, but what causes the wings to be held in an abnormal position? Is the defect in the wing structure itself, the wing muscles, the nerves that control the wing muscles, or in the messages sent from the brain? Not only is it difficult to determine which of several possible sites is the source of the problem, it is also hard to find the site of gene action when it is not in the structure that is outwardly affected. In the *wings up* fruit fly, for example, the wings are up because the indirect flight muscles that attach to them are either abnormal or missing (Hotta and Benzer, 1972). Another example, one in which a gene has its primary effect on a structure distant from the site of the problem, is seen in some humans. An inability of the cells of the gut to absorb enough vitamin A may cause the retina of the eye to degenerate, resulting in a sensory deficiency with obvious consequences for behavior. In this case, the gene's action is in the small intestines even though the defect is seen in the eye. In both of these examples, it would be easy to misidentify the site of gene action as the structure with the obvious defect.

It is possible to locate the primary site of gene action using mosaic analysis, which is a comparative study of mosaics, individuals composed of some sections of normal wild-type tissue and some pieces of mutant tissue. Although mosaic analysis does not always reveal what the gene is doing, in other words, the nature of the problem, it does locate where the gene is working. The procedure is to compare the distribution of normal and mutant tissue in all the mosaics displaying the mutant form of a behavior to identify the region of mutant tissue common to them all. Those regions are then implicated as the sites of action of that gene.

You have probably used the same reasoning used in mosaic analysis in solving other problems. Suppose, for example, you sit down to read in your favorite seat but, much to your frustration, the lamp next to the chair will not go on. How do you locate the source of the problem? First you might suspect that there was no power in the outlet you were using. So you plug the lamp into an outlet that you know is functional. The lamp still does not work. Next you check the bulb and find that it was not screwed in tightly, but screwing in the bulb does not solve the problem. In this analogy, the electrical connection between the bulb and the socket is considered defective or "mutant." However, correcting that problem or "mutation" did not make the light go on, so that defect is not the primary source of the problem. Next you replace the bulb with a new one. No luck. Finally, you try replacing the plug. The light goes on! Even if you could not see anything wrong with the plug, you know it was the cause of the problem because when it was replaced by a good plug, the light worked. What you did was to determine which of several components of the system was the source of the problem by replacing possible defective parts with good ones until you found the one that caused the problem.

When you want to determine which part of an animal showing mutant behavior is the source of the problem, you look at the distribution of normal and mutant tissue in mosaics and determine which structures must be mutant for the aberrant behavior to be seen. For example, suppose you compare all the mosaic fruit flies that show the mutant behavior and find that in some, but not all, the brain is composed of mutant tissue. In some mosaics, "nature" replaced a mutant part with a normal part. Since the animal still behaves abnormally, even when the brain is composed of normal tissue, clearly the brain cannot be the source of the problem. However, if the thoracic ganglion is composed of mutant tissue in every individual showing the aberrant form of the action, then the conclusion is that the thoracic ganglion is the structure altered by the gene. When the thoracic ganglion is defective, the behavior is abnormal. Conversely, when nature puts in a normal, rather than mutant, thoracic ganglion, the animal behaves normally. This is the equivalent to replacing the plug in the lamp analogy. Furthermore, as in the analogy, the site of the problem can be determined even if the specific anatomical or physiological defect in that structure cannot be detected without further testing. Mosaic analysis may not identify all the structures necessary for the expression of a particular behavior, but it does point out those that are always altered in individuals showing the mutant, rather than the wild-type, behavior. This technique has been useful in locating the sites of gene action in both fruit flies and mice.

Drosophila

When tracing the path between a gene and behavior by mosaic analysis, the first step must be to obtain mosaic animals. A common way to acquire a mosaic fruit fly (*Drosophila*) is by producing a gynandromorph, a fly that has some sections of male tissue and some pieces of female tissue. A *Drosophila* cell is female if it contains two X chromosomes. Normal male cells contain one X and one Y chromosome, but cells containing only a single X chromosome are also male. A gynandromorph results from an XX (female) zygote when one of the X chromosomes is lost from one of the two nuclei created by the first cell division. The chances of the X chromosome being lost are increased if it has an abnormal ring shape. When an X chromosome is lost, one of the resulting nuclei and all its descendants have two X chromosomes and express female characteristics, but the other daughter nucleus and all its descendants have only one X chromosome and express male characteristics. The fly that develops from this embryo is, therefore, half male and half female.

To produce a gynandromorph for behavioral analysis, a female containing a ring X chromosome that is also homozygous for a dominant allele producing the normal wild-type behavior

is mated with a male possessing a recessive mutation for that behavioral trait on his X chromosome. For example, a female with normal wings and flight might be mated with a male bearing the *wings up* mutation. Remember male fruit flies have only one X chromosome, so a male with a recessive mutation can be identified because he displays the mutant form of the behavior; in this example, he would be spotted by the position of his wings. It is necessary to have the female bear a dominant allele and the male a recessive allele so that female (normal) tissue can be distinguished from male (mutant) tissue in the mosaic. Notice in Figure 3.14 that when this male's X-containing sperm joins with an ovum containing a ring X, a female zygote results. If the ring X chromosome is lost during the first cell division, half the cells, those con-

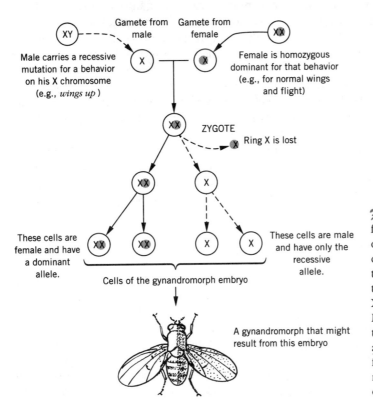

Figure 3.14 The development of a fruit fly gynandromorph due to the loss of a ring X chromosome during the first cell division of a female zygote. When the ring X chromosome is lost, one of the daughter nuclei then has only one X chromosome and is, therefore, male. Male cells are shaded. The other daughter nucleus has two X chromosomes and is female. The unshaded cells are female. These cells develop into a gynandromorph, an individual composed of half male cells and half female cells.

taining only the X chromosome that came from the father's gamete, are male, and half, those containing an X chromosome from the father in addition to the ring X chromosome from the mother, are female. The mutation can only be expressed in male cells because they lack an X chromosome with the dominant wild-type allele. None of the female cells of the gynandromorph expresses the mutation. In other words, the gynandromorph is a special type of mosaic, a fly with patches of both male (mutant) and female (normal) tissue.

Although all gynandromorphs are produced in the same way, by the loss of an X chromosome during one of the divisions of the embryo, they are not all identical. The male (mutant) and female (normal) tissue may be divided in a variety of ways because of the way insects develop. In Figure 3.15 you can see that daughter nuclei from each mitotic division tend to remain near one another. After several divisions, the nuclei migrate to the surface forming a single layer of cells around the yolk. Approximately half of this layer is female cells and the remainder is male. Because the nuclei stay near their neighbors even as they migrate, all the male cells are on one side of the embryo and the female cells on the other. However, the orientation of the dividing line between male and female cells varies because the first cell division can be in any direction on the surface of the egg. As a result, there are many ways that the embryo may be divided into male (mutant) and female (wild-type) parts.

Furthermore, as in all fruit fly embryos, the destiny of each cell in this single layer is determined by its position, so the orientation at the first cell division determines which structures will consist of male cells and which of female cells. A fate map indicating the structures that will eventually be formed is shown in Figure 3.16. Mentally draw lines separating the fate map in half. Notice that the orientation of this line will determine which structures will be com-

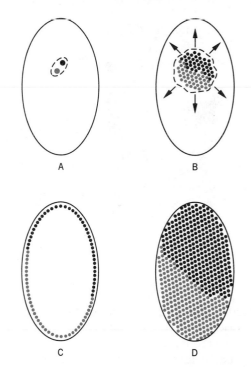

Figure 3.15 Early embryonic development of a fruit fly gynandromorph. (*a*) The products of the first nuclear division of the zygote. The male nucleus is indicated as a black circle and the female nucleus is indicated as a white circle. (*b*) A view of the surface of the early embryo. The nuclei continued to divide and migrated to the surface of the egg. During the divisions and the migration, nuclei tend to stay near their neighbors. As a result, all the female nuclei are on one half of the embryo and all the male nuclei are on the other half. (*c*) A cross section of a slightly older embryo. The nuclei are enclosed in cells at this point. The embryo is still divided so that one half has male cells and the other half has female cells. (*d*) A later stage of development. An irregular imaginary line divides the embryo into a male half and a female half.

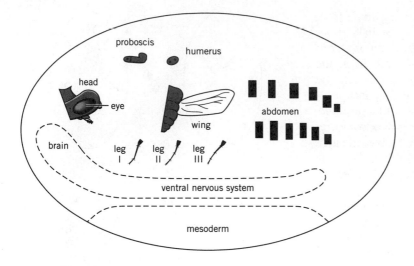

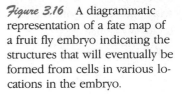

Figure 3.16 A diagrammatic representation of a fate map of a fruit fly embryo indicating the structures that will eventually be formed from cells in various locations in the embryo.

posed of male or female cells. In a real embryo, the orientation of the initial cell division varies, so there are innumerable ways that the adult fly may be divided into male and female chunks (Figure 3.17).

The sex of the cells making up each structure is important because it will determine whether the mutation under investigation can be expressed in those cells. If the cells of the anatomical structure in which the gene is normally expressed are female, the mutation will be hidden by the wild-type allele on the second X chromosome and the tissue will be normal. For example, the *wings up* mutation mentioned earlier will not be expressed in female cells so when the wing muscles consist of female cells, the gynandromorph has normal use of its wings. However, if the anatomical structure affected by the mutant gene falls in a male region of the adult, the mutation can be expressed and the tissue will be mutant. In the *wings up* example, the mutant allele can be expressed in the flight muscles of gynandromorphs that hold their wings up because those muscle cells are male. The next step is to select the gynandromorphs that show the abnormal behavior and compare

the distribution of male and female tissue in each animal. If the insect shows the mutant behavior, each anatomical structure altered by the gene must be composed of male cells. The structure that is male in every gynandromorph displaying the mutant behavior is considered to be the site of action for the gene in question.

In order to make this comparison, there must be a way to identify male or mutant cells. These cells can be distinguished if the male who fathered the gynandromorphs had another recessive allele on his X chromosome, one that produces a trait we can see in each male cell. Genetic variants of enzymes are particularly useful tissue markers. Although these variants catalyze the same reaction within the cell, they have different physical and chemical properties. Because each variant has different biochemical properties, it is sometimes possible to find a stain that reacts with only one form of the enzyme. Then the cells producing that genetic variant can be distinguished from cells producing other forms. Variants of enzymes such as acid phosphatase (Kankel and Hall, 1976) and succinate dehydrogenase (Lawrence, 1981) are examples of tissue markers useful for labeling in-

Initial nuclear division →

Resulting gynondromorph →

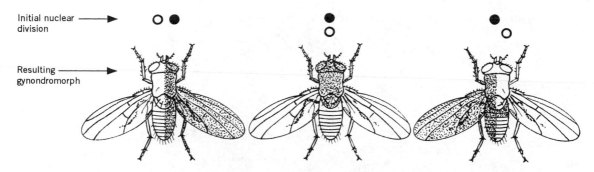

Figure 3.17 Diagrams showing the gynandromorphs that might develop if the first division of the zygote nucleus were oriented as shown above the individual.

ternal tissues. When histological markers such as these are used, the fly is sectioned and stained to highlight the form of the enzyme present in each cell. If the female cells have one form of the enzyme and the male cells another, the sex of the cells (and, therefore, the presence of the mutant tissue) can be determined at a glance. An example of a section of the brain of a gynandromorph that showed courtship behavior typical of males is shown in Figure 3.18. The stained cells are female and the unstained cells are male.

Before looking at examples of the use of mosaic analysis, it may be useful to summarize the steps in the procedure. A female whose ring-shaped X chromosome bears a dominant allele that produces a normal behavior is mated with a homozygous recessive male who behaves abnormally. If the ring X chromosome is lost during one of the first cell divisions of a female embryo, a gynandromorph is produced. Because the mutation can only be expressed in male cells, the gynandromorph is also a mosaic of mutant and normal tissue. The mutant cells are also labeled with a tissue marker so that they can be distinguished from normal ones. Then all the mosaic flies showing the mutant behavior are compared. The region that is male in every

gynandromorph that behaves abnormally is where the gene is acting.

Mosaic analysis has been used to locate the primary site of action of genes that influence a variety of behaviors. This information is often useful in directing future investigations to determine the exact nature of the anatomical or physiological cause for the mutation's effect on behavior. For example, mosaic analysis was used to pinpoint the brain as a site of action of the *per* gene that affects the circadian rhythm in *Drosophila* eclosion, that is, the emergence of adult flies from their pupal cases. In nature, this occurs around dawn each day. Any pupae that fail to develop enough to eclose during the dawn hours must wait until the following morning. Eclosion is described as rhythmic because flies in a population tend to emerge in a repeating pattern: many emerging at dawn and then very few until the following dawn. Interestingly, the pupae do not require environmental time cues to remain rhythmic. When pupae are maintained in a laboratory without normal time cues such as light or temperature cycles, they still emerge at approximately 24-hour intervals. Since the ability to measure the passage of time is not dependent on these obvious environmental cues, we attribute it to a biological

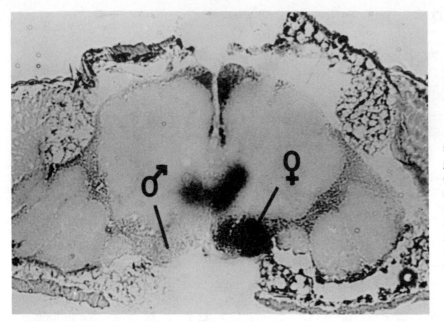

Figure 3.18 A section through the brain of a gynandromorph. Variants of an enzyme were used as tissue markers. The female cells were set up genetically to produce a form of the enzyme that leads to a staining reaction. The male cells are unstained because they carry only a mutant form of the enzyme gene.

clock (see Chapter 9). The biological clock is still running in adult fruit flies; their locomotor activity is cyclic with a 24-hour period. The *per* gene affects the biological clock governing both eclosion and activity.

The genetic roots of the biological clock of *Drosophila* have been studied by Ronald Konopka and Seymour Benzer (1971). They induced mutations and screened the population for individuals whose biological clocks had been altered. As you can see in Figure 3.19, one mutation resulted in flies whose biological clock runs fast, resulting in shorter period eclosion and activity rhythms. This mutation, *per*^{s}, results in rhythms with a 19-hour period. Another mutation, *per*^{l}, has the opposite effect—a rhythm with a long, 28-hour period. The third mutation, *per*^{o} creates arrhythmic flies; they emerge from their pupal cases and are active at random times throughout the day. These three mutations seem to be in the same functional gene, the *per* gene, which is located on the X chromosome. To determine where this gene has its primary effect,

mosaic flies were created. The head region in flies with aberrant rhythms was always composed of mutant cells, indicating that the *per* gene has its primary effect in the brain.

The knowledge that the brain was affected by the *per* gene told the investigators where to begin looking for more specific changes caused by the gene, thereby providing some hints as to how the brain might control circadian rhythms in fruit flies. One study suggested that the brain influenced rhythms through a hormonal, rather than a neural, output. Brains of *per*^{s} flies were transplanted into the abdomens of *per*^{o} flies. Four of the 55 survivors had short-period activity rhythms for at least three consecutive cycles. Presumably the transplanted brains degenerated in the 51 flies that were still arrhythmic after surgery. This suggests that the brain influences circadian rhythms by producing a humoral substance because the transplanted brains had no functional neural connections (Handler and Konopka, 1979). This suggestion is consistent with information gained from two additional

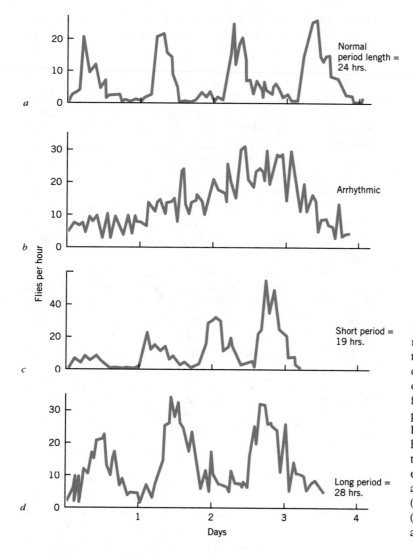

a Normal period length = 24 hrs.

b Arrhythmic

c Short period = 19 hrs.

d Long period = 28 hrs.

Flies per hour

Days

Figure 3.19 The eclosion rhythms of normal and three mutant strains of fruit flies studied in constant conditions. The top curve (*a*) shows the number of flies emerging each hour in a population of normal fruit flies. Peaks of eclosion occur every 24 hours. The lower curves depict the number of flies emerging each hour in a population of (*b*) arrhythmic flies, (*c*) short-period (19-hour), and (*d*) long-period (28-hour). (Data from Konopka and Benzer, 1971.)

studies. First, there are some neurosecretory cells in the brain that are often found in abnormal locations in *per^o* flies (Konopka and Wells, 1980). Furthermore, *per^o* flies produce only 25 percent of the normal amount of octopamine, a neurotransmitter (Livingstone, 1981). Whereas it is not yet clear how all these pieces of information fit together to explain how the brain influences circadian rhythms, such techniques show promise in helping to unravel the relationship between genes and behavior.

Mice

Although mosaic mice are used in the same way as mosaic fruit flies to determine where genes act to control behavior, they are generated through different procedures. A mosaic mouse, like a mosaic fruit fly, is composed of two genetically different types of cells, cells that were originally two different embryos. A mouse begins its life as a single cell, called a zygote, but that cell soon begins to divide rapidly and forms a cluster of cells called a morula. A morula can be removed from the mother's oviduct and separated into individual cells by treatment with the appropriate enzyme. To produce a mosaic mouse, a morula is removed from each of two females with different genetic makeups. There must be two differences in the genotypes of the donor females. First, they must differ in the alleles for the behavior under investigation. The genetic information in one of the morulae would cause a normal behavior pattern and the genes in the other would cause an aberrant action. In addition, the morulae would have to differ in some trait that could be used as a marker to identify which morula gave rise to each cell in the resulting mouse. The morula removed from each female is dissociated and the cells from both are mixed together. While this mixture is incubated for several hours, the cells reassociate to form a new morula composed of cells from both of the original morulae. Amazingly this new embryo continues its development. When it reaches a slightly more advanced stage of development, the blastocyst stage, the embryo is implanted in the uterus of another female where it will continue to develop into a genetically mixed-up mouse.

Mosaic mice differ from mosaic fruit flies in two ways. The first difference is in the percentage of normal and mutant cells in the mosaic. About half of the cells in a mosaic fruit fly are normal and the remainder are mutant. No such predictions can be made about the proportion of cells in a mosaic mouse that are derived from a particular morula. Although different morulae sometimes make equal contributions to a mosaic mouse, it is also possible that the resulting mouse is not a mosaic at all. Its cells may be all one genotype. Any mixture is possible. One rea-

son for the difficulty in predicting the percentage of cells derived from each morula is that some of the cells of the mosaic morula will develop into extraembryonic structures and others will give rise to the fetus. As a result, there is an even greater potential for variety in mosaic mice than there was in mosaic flies. A second difference in mosaics of these two species is in the way cells of different genetic composition are divided. Remember that mosaic fruit flies were made up of chunks of normal or mutant tissue. Usually a whole structure such as a limb or wing is composed of one genotype. In contrast, the cells from different morulae are more thoroughly intermingled in mosaic mice (Stent, 1981).

Mosaic analysis provides a way to locate the primary site of action of some of the genes that result in neurological defects in the cerebellum of mice. Nerve cells interact with one another more than cells of most other systems do, so it is particularly difficult to determine which cells in the nervous system are being affected by a particular gene. When a nerve cell degenerates or behaves in an abnormal manner, the problem could originate within that cell or within another cell that then fails to interact with it properly. Mosaic mice have been used to distinguish between these possibilities.

One recessive mutation that results in neurological difficulty for mice is called *pcd* because it causes *P*urkinje *c*ell *d*egeneration. Purkinje cells are large neurons that provide the only output from the cerebellum, the part of the brain that coordinates bodily movement. Animals that are homozygous for the *pcd* alleles seem normal at birth because they have Purkinje cells. However, these neurons gradually die, and as they degenerate the mice become uncoordinated. Richard Mullen (1977a) wondered whether the *pcd* mutation affects the Purkinje cells directly or whether the Purkinje cells are responding to an abnormality in some other cell. To answer this question, Mullen studied mosaic mice in which some cells were homozygous normal and other cells were homozygous for the *pcd* mutation. An enzyme tissue marker was used so that the normal cells could be distinguished from the *pcd* cells. Mullen reasoned that if the *pcd* mutation had its primary effect on the Purkinje cells, then any Purkinje cells bearing the mutation would degenerate and all those with the normal dominant allele would survive. However, if the *pcd* mutation directly affected some other cell type, making it incapable of interacting with Purkinje cells in the proper manner, then the population of surviving Purkinje cells would have mixed genotypes. When the brains of mosaic *pcd*/normal mice were sectioned and stained, Mullen found that all the surviving Purkinje cells were normal, but those with the *pcd* allele had degenerated. Therefore, the Purkinje cells were directly affected by the *pcd* gene. With the *pcd* gene, it seems that justice prevails. The fate of the Purkinje cells depends on their own genotype.

But the *reeler* mutation shows that life, even for a neuron, is not always fair; the fate of the Purkinje cells of *reeler* mice is determined by the genotype of other neurons. By the time *reeler* mice are two weeks old, they can be identified because they walk as if they are drunk, similar to the way normal mice move while they are recovering from anesthesia (Sidman, Green, and Appel, 1965). A *reeler* mouse has trouble walking because during its development, the Purkinje cells fail to migrate to the proper positions in the cerebellum. In contrast to the direct effect of the *pcd* mutation, mosaic analysis suggests that the *reeler* allele affects the Purkinje cells indirectly. Some of the Purkinje cells in the *reeler*/normal mosaic mice do migrate properly, but others do not. Of those that move to the correct locations, some have the *reeler* allele and others have the normal allele. In addition, many of the Purkinje cells that were in the proper position had the *reeler* genotype (Mullen, 1977b). The best explanation for this is that the *reeler* mutation has its effect on some other type of cell. This other cell may be necessary to stim-

ulate the migration of Purkinje cells (Caviness and Sidman, 1972), or it might be one that guides the Purkinje cell migration (Caviness and Rakic, 1978). If the latter suggestion is true, then whenever a Purkinje cell of either genotype encounters genetically normal "guiding" cells, it would be directed to the proper location, but whenever its directions come from guiding cells that bear the *reeler* mutation, it would be sent in the wrong direction and become twisted. Since the most obvious abnormality is seen in the Purkinje cells, it would have been difficult to pinpoint the guiding cells as the primary target of the *reeler* gene without mosaic analysis.

Recombinant DNA

The relationships between genes and behavior are now being considered by using techniques developed to produce recombinant DNA, that is, DNA from two different organisms. Specifically, it is possible to insert certain genes into organisms to determine their effect on behavior. In addition, recombinant research has revealed some of the ways that genes control behavior.

Of course, the first question to arise is, how do you get DNA from one organism into the DNA of another organism to produce recombinant DNA? This trick relies on the use of restriction enzymes, which are enzymes produced by bacteria that restrict the growth of viruses that normally infect them by chopping up the viral DNA. Restriction enzymes have been purified and can be used to fragment DNA from any source. Different restriction enzymes recognize different specific sequences of nucleotides and cleave the DNA at those spots. Some restriction enzymes do not clip the DNA straight across the two strands. Instead they cut one side of the DNA ladder a few rungs away from where the other side is snipped. Notice in Figure 3.20 that the restriction enzyme EcoR1 from *Escherichia coli*, a bacterium that dwells in your intestines,

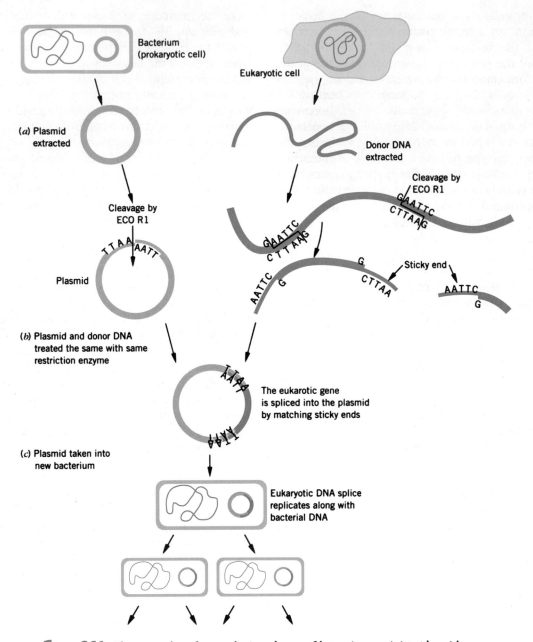

(a) Plasmid
extracted

Bacterium
(prokaryotic cell)

Eukaryotic cell

Donor DNA
extracted

Cleavage by
ECO R1

Cleavage by
ECO R1

GAATTC
CTTAAG

Plasmid

TTAA AATT

GAATTC
CTTAAG

G
CTTAA

Sticky end

AATTC
G

AATTC
G

(b) Plasmid and donor DNA
treated the same with same
restriction enzyme

The eukarotic gene
is spliced into the plasmid
by matching sticky ends

(c) Plasmid taken into
new bacterium

Eukaryotic DNA splice
replicates along with
bacterial DNA

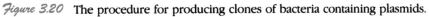

Figure 3.20 The procedure for producing clones of bacteria containing plasmids.

recognizes the sequence GAATTC and makes a cut between the guanine and adenine on each side of the DNA ladder so that each side of the DNA molecule is left with four unpaired bases. Because the pairing of bases is specific, adenine with thymine and cytosine with guanine, the single-stranded ends of the cleaved DNA molecule will recognize and pair with any piece of DNA that has a singled-stranded end with the appropriate complementary bases. Therefore, the cleaved ends are described as "sticky ends." A given restriction enzyme always cuts the DNA so that the sticky ends have particular base sequences. That is the key to the whole procedure—any two pieces of DNA cut with the same restriction enzyme can be rejoined at their sticky ends, even if the DNA comes from different sources. This, then, is how the DNA from two different organisms is prepared so the fragments will stick together.

So there are two kinds of DNA involved, one from the organism you want to study, the other from the organism that will "incubate" this DNA. First, then, you have to get the DNA of interest. There are two ways to do this. One way is to chop up the chromosomes by using a restriction enzyme. The other way is to synthesize pieces of DNA from mRNA molecules, forming what is called cDNA (complementary DNA). Since the sequence of bases on an mRNA molecule is translated into the sequence of amino acids comprising a protein, the cDNA that is produced from mRNA is a reconstruction of a gene. In either case, these pieces of DNA are then inserted into the DNA of another organism, most commonly into viral DNA or a bacterial plasmid. A plasmid is a small circular piece of DNA that is not part of the main circular chromosome of the bacterium. The same restriction enzyme is used to snip the DNA from the host organism as was used to cut out the segment with the gene of interest so that the pieces will have sticky ends with complementary base sequences.

Once the gene of interest is inserted into a host chromosome, the cell containing the recombinant DNA is allowed to multiply. Each time the cell, say an *E. coli* bacterium, replicates, it duplicates the plasmid as well as its chromosome. An incredible number of plasmids are manufactured because bacteria multiply so quickly.

INSERTING GENES—THE *DROSOPHILA PER* GENE

Now let us see how this technique has been used to study behavior. Recombinant DNA has made it possible to insert the per^+ allele into a per^o fly. A normal per^+ fly has 24 hour rhythms, but a per^o fly is arrhythmic. In these experiments, a fragment of DNA containing the *per* gene along with the dominant allele for rosy eye color was inserted into a plasmid. These recombinant plasmids were then injected into the pole cell region of early embryos of per^o flies that also had the recessive alleles for eye color. Some of the cells in the pole region develop into the cells that go on to form gametes when the embryo becomes an adult. The flies that developed from embryos treated this way were mated with arrhythmic (per^o) flies that were homozygous recessive for the rosy eye color gene. In the resulting offspring, rosy-eyed individuals could only have gotten their eye color from the DNA in the recombinant plasmid since both of their parents were homozygous recessive for that eye color gene. Therefore, the presence of rosy eyes indicates not only that the plasmid is present in those individuals, but also that the inserted genes could be expressed. The eye color serves as a marker that can be used to identify flies that have incorporated the plasmid. But our interest is really in the per^+ gene. Flies that incorporated the plasmid were not only rosy-eyed, they were also rhythmic. Their eclosion and activity rhythms resembled those of per^+ flies (Bargiello, Jackson, and Young, 1984). Using recombinant DNA is, as you can see, a rather precise way to demonstrate that a particular gene is necessary for the expression of a complex behavior pattern, in this case biological timing.

Is the *per* gene needed for the daily functioning of the clock or does it influence the clock during development, perhaps by creating a brain-damaged fly? This question has also been answered by using recombinant DNA techniques. Recall from the discussion at the beginning of this chapter that genes are not always active; they can be turned on or off. One way this may occur is through the activity of another gene, which is called a promoter gene. Using recombinant techniques, a segment of DNA was created that contained the *per*$^+$ gene and also a promoter gene to regulate its activity. When induced to do so by a temperature shock of 27°C, the promoter turned on the *per*$^+$ gene. This segment of DNA was injected into a *per*^{o} embryo, where, as in the previous example, it was incorporated into almost all the adult cells. When these adult flies were kept at 18°C, they were arrhythmic. However, when the temperature was raised to 27°C, the promoter turned on the *per*$^+$ gene, so the flies became rhythmic. The animals quickly became arrhythmic again when the temperature was lowered. This indicates that the product of the *per* gene is not very stable. Rhythmicity can be turned on or off by changing the temperature. The conclusion is that the product of the *per* gene must be present for the clock to function (Ewer, Rosbash, and Hall, 1988).

Another question that has been addressed by recombinant DNA techniques concerns the number of clocks present in a multicellular organism. As previously mentioned, mosaic analysis revealed that the *per* gene affects a certain cluster of cells in the brain. Certainly the brain is the site of one clock. Are there more? Research using recombinant techniques suggests that there are many clocks. A clock might be located wherever the *per* gene is active. Since there is no stain for the *per* gene product, the *per* gene was fused with a gene that would produce a product that could be stained, the B-galactosidase gene taken from the bacterium *E. coli*. The fused genes were inserted into a *per*^{o} embryo and became incorporated into adult cells. The activity of fused genes is linked, so the product of the bacterial gene, galactose, could be stained green to reveal the tissues that were expressing the *per* gene. In this way, it was shown that the *per* gene becomes active shortly before eclosion. In the adult, the *per* gene is expressed in many tissues, including the antennae, eyes, optic lobe, central brain, thoracic ganglia, Malphigian tubules, and ovarian follicle cells (Liu et al., 1988). Another study used antibodies specific for the *per* gene product to determine where the gene is expressed. Similar results were obtained (Siwicki et al., 1988). Because the *per* gene is expressed in so many tissues, it seems likely that there are several independent clocks.

MECHANISM OF CONTROL—*APLYSIA* EGG LAYING

The reproductive behavior of the sea hare *Aplysia* involves a number of fixed action patterns, such as the coordinated actions involved in the laying of eggs (Figure 3.21). When the snail is about to produce an egg string, it stops normal activity such as moving and eating. There are several physiological changes accompanying the cessation of these activities: The heart and respiratory rates increase, and the muscles of the reproductive ducts contract as the animal expels a string of eggs. The record-holding sea hare laid a string 17,520 cm long at a rate of 41,000 eggs per minute! When the egg string first appears, the snail grasps it in a fold of its upper lip and, waving its head in a stereotyped pattern, helps to pull the string out and winds it into an irregular mass. During this process, the egg string is coated with sticky mucus. Finally, the tangled egg string is attached to a rock or other firm support with a characteristic wave of the head (Kandel, 1976).

The neuroendocrine link between the genes and egg-laying behavior is known. Two small proteins, peptide A and peptide B, are produced by the atrial gland in the reproductive system. When these peptides are injected into an animal, they stimulate two clusters of neurons on top of

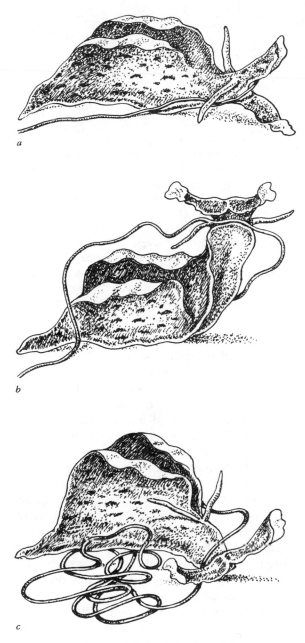

Figure 3.21 The egg-laying behavior of the sea hare *Aplysia*. (*a*) The snail grasps the end of the egg string in a fold of its upper lip. (*b*) Waving its head, it helps pull the egg string out. (*c*) The tangled mass of egg string is attached to a solid surface. (Scheller and Axel, 1984.)

the abdominal ganglion, called bag cells, to release other peptides (Figure 3.22). One of the bag cell products is egg-laying hormone (ELH), which acts as an excitatory neurotransmitter and increases the firing rate of neurons in the abdominal ganglion. In addition, ELH acts as a hormone that causes the smooth muscle of the reproductive ducts to contract and expel the egg string. It may also cause the changes in heart and respiratory rate. Simultaneously, the bag cells release alpha bag cell factor, an inhibitory transmitter that decreases the rate of specific neurons (L2, L3, L4, and L6), and beta bag cell factor, an excitatory transmitter that increases the firing rate of certain other neurons (L1 and R1). The stimulation or inhibition of particular neurons presumably coordinates the egg-laying behavior pattern (Scheller and Axel, 1984).

How do genes fit into this scenario? The ELH gene is translated into a large protein that is broken down into the shorter functional proteins, ELH, alpha bag factor, and beta bag factor. Thus a single gene is activated, and because its products act as neurotransmitters and hormones, that gene can orchestrate a complex behavioral sequence. Furthermore, if the gene is activated, all the components of the behavior pattern appear, as is typical of a fixed action pattern.

It was the sequence of bases on the ELH gene that revealed the mechanism of this gene's control of egg-laying behavior, but sequencing the ELH gene involved many steps. First the ELH gene had to be identified and cloned. To accomplish this an *Aplysia* DNA library was created. A DNA library is made up of clones of different recombinant plasmids containing fragments of all the DNA of an organism. Next the plasmids bearing the ELH gene had to be identified. This was done by making cDNA from mRNA in the bag cells and from mRNA in non-neural cells and then labeling each with a radioactive marker so that it could be identified. The cDNA from both sources was hybridized with the *Aplysia* DNA in each clone of recombinant plasmids. The

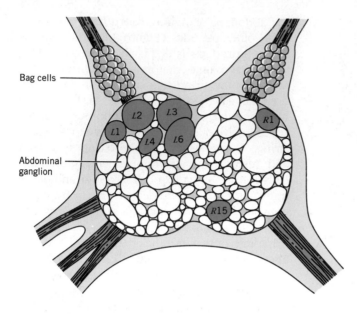

Bag cells

Abdominal ganglion

Figure 3.22 The abdominal ganglion and the bag cells of *Aplysia*. The ELH (egg-laying hormone) gene is expressed in the bag cells, and they produce several substances including ELH, alpha bag cell factor, and beta bag cell factor, that coordinate the actions involved in egg laying. Some of the large neurons in the abdominal ganglion that are affected by bag cell products are labeled. (Scheller and Axel, 1984.)

key to hybridization is the specificity of base pairing. When a DNA molecule is heated, the molecule unzips down the middle by breaking the bonds that hold the two strands together. This separates the base pairs and leaves two single strands of DNA. As the DNA cools, each strand usually finds another strand with the complementary sequence of bases, and these match up to reform double-stranded DNA. Because the bag cells produce ELH, they will have many mRNA molecules coding for ELH. Therefore, the cDNA from bag cells will have the gene for ELH. Non-neural cells do not produce ELH and consequently the cDNA made from their mRNAs will lack the ELH gene. As a result, the recombinant plasmids that hybridized with cDNA from bag cells, but not with cDNA from other cells, were identified as those containing the ELH gene. Next, the exact piece of DNA comprising the ELH gene was identified by hybridizing the clones known to contain the gene with mRNA from the bag cells. The predominant mRNA in the bag cells would code for ELH. The ELH gene could be spotted on an electron micrograph as

a region of hybridization. Once the gene was isolated, it was sequenced.

The sequence of bases in the ELH gene revealed that it codes for a large polyprotein precursor that is chopped into several functional proteins. The sequence of bases on the ELH gene codes for 271 amino acids, but ELH itself has only 36 amino acids. The sequence of bases that would be translated into ELH could be seen in the precursor, and it was bordered by signals that indicated that ELH would be clipped out of the larger protein (Figure 3.23). What came as a surprise was that the alpha bag factor and beta bag factor were also flanked by cleavage signals. In fact, the gene codes for 11 different proteins, each bordered by cleavage signals and four of these are known to be released by the bag cells. In other words, a single gene codes for a large protein that is broken down into at least three smaller active proteins with different roles in producing a behavior pattern. The role of the genes in generating this fixed action pattern is to produce peptides that act locally as neurotransmitters or distantly as hormones. These

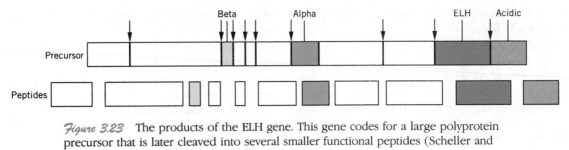

Figure 3.23 The products of the ELH gene. This gene codes for a large polyprotein precursor that is later cleaved into several smaller functional peptides (Scheller and Axel, 1984).

peptides stimulate or inhibit specific structures to coordinate a stereotyped behavior pattern (Scheller and Axel, 1984).

Summary

Genes control behavior by directing the synthesis of proteins. Many of those proteins influence anatomy or physiology, the foundation for behavior. Genes are DNA, a long molecule composed of two strands of nucleotides. Each nucleotide in DNA consists of a phosphate, a sugar (deoxyribose), and a nitrogenous base. The bases pair in a specific way: adenine with thymine and cytosine with guanine.

DNA is transcribed into RNA, which is similar to DNA except that it is single stranded, its sugar is ribose, and uracil substitutes for thymine. The RNA is formed along a single strand of DNA as adenine pairs with uracil and guanine pairs with cytosine. Thus the sequence of bases on the DNA orders the bases on RNA.

It is the sequence of bases that determines which protein will be formed. A sequence of three bases codes for an amino acid. Therefore, the instructions for making a protein, which is a long chain of amino acids, is the order of bases in the DNA.

If an animal has "a gene for some behavior," it means that the gene makes a difference in the performance of a behavior. However, even if an individual has a gene for a behavior, the action may not be observed. One reason is that not all genes are expressed all the time. They are turned on and off by several regulatory mechanisms. An important class of regulatory substances are the steroid hormones. Furthermore, the environment may influence the activity of genes. An example is the way that diet determines the caste of female ants. Environment and genes are continually interacting during the development of behavior.

The classical methods of experimentally demonstrating the genetic basis of behavior include inbreeding, artificial selection, and hybridization. Inbreeding is the mating of close relatives. When mating is arranged this way, the siblings of successive generations become increasingly similar genetically until 98 percent of their genes are identical. The genetic similarity of individuals in inbred strains is the key to their usefulness in behavioral genetics. When inbred strains that have been raised in similar environments are compared, any difference in behavior must be due to differences in their genes. Therefore strain comparisons are a means of demonstrating that genes can make a difference in the form of a behavior. The learning ability of inbred strains of mice shows considerable variability. If individuals of the same inbred strain are raised in different environments, any difference in behavior must be due to the environment. Thus inbred strains can also be used to demonstrate the importance of environment in the expression of a behavior.

Artificial selection is a different breeding regimen. Here individuals showing a desired behavior are bred with one another. If the frequency of the trait in the population increases when the appropriate breeders are mated, then the behavior must have a genetic basis. Artificial selection is the way breeds of dogs were created. The most famous laboratory study using artificial selection is the production of strains of maze-bright and maze-dull rats by mating the fast learners with one another and the slow learners with one another. A specific trait, not a general ability, is usually selected. However, when these strains of rats were tested in other mazes, the dull strain performed as well as the bright strain.

Hybridization is a third breeding system used to demonstrate that a given behavior has a genetic basis. Individuals that display the behavior in distinct but different ways are mated with one another and the behavior of the hybrid offspring is observed. Depending on the number of genes involved, the offspring's behavior may resemble that of one of the parents or it may contain components of both parental forms. The behavior of hybrid offspring of hygienic and unhygienic bees was the same as one of the parental forms. The behavior seems to be controlled by two genes: The U gene determines whether the cells containing dead larvae will be uncapped, and the R gene determines whether corpses will be removed from the cells. On the other hand, many genes are involved in determining the calling song of crickets.

Although these breeding programs demonstrate that genes influence the expression of many behaviors, they do not show where or how the genes act. Techniques that have been useful in pointing out what genes do to alter behavior include inducing mutations, mosaic analysis, and recombinant DNA.

A mutation is a change in the DNA. Mutations often result in the production of different proteins. Some mutations cause the individuals to show abnormal behavior patterns. When these mutants are separated from the general population, they can be studied to learn how anatomy or physiology has been altered to modify the behavior. The avoidance reaction of the protozoan *Paramecium* has been studied in this way. Unusual avoidance reactions are caused by altered membrane properties. The *pawn* mutant cannot back up because its membrane does not allow calcium ions to enter and cause the cilia to reverse their beating. The *paranoiac* mutant spends too much time backing up because its membrane does not repolarize and restore the normal resting potential. The poor memory of the *dunce Drosophila* mutant is thought to be related to its lack of an enzyme, cyclic AMP phosphodiesterase.

Mosaic analysis is another way to pinpoint the primary site of action of a gene. A mosaic is an individual composed of some mutant and some normal cells. When many mosaic animals are compared, the structure that is always composed of mutant tissue in the individuals that show the mutant form of a behavior is thought to be the primary site of action for that gene. The production of mosaics has been useful in studying certain behaviors in both fruit flies and mice. In fruit flies, for example, the *per^o* mutation that causes the bearer to be arrhythmic seems to affect the head region. The primary site of action of genes that cause neurological disorders in the cerebellum of mice has also been studied this way. The *pcd* mutation affects cerebellar Purkinje cells directly but the *reeler* mutation seems to have its primary effect on cells that direct the migration of Purkinje cells during development of the brain.

Producing and cloning recombinant DNA is another way that the relationship between genes and behavior has been clarified. For example, the *per* gene was shown to be an essential factor in the expression of circadian rhythmicity in fruit flies. When the *per$^+$* allele was isolated and inserted into arrhythmic flies, it restored the rhythms in eclosion and activity. The gene coding for egg-laying hormone (ELH) in *Aplysia* has also been cloned and identified. When it was sequenced, some interesting facts about genetic control of the fixed action pattern of egg laying were learned. The ELH gene codes for a polyprotein precursor that could be broken down into 11 separate proteins. Four of those are known to be released by bag cell neurons, cells that initiate egg laying. The roles of three of them, ELH, alpha bag cell factor, and beta bag cell factor, are known. They all act as neurotransmitters that either excite or inhibit specific cells in the abdominal ganglion. In addition, ELH serves as a hormone that causes the ducts of the reproductive system to contract and expel the eggs.

Natural Selection and Ecological Analysis of Behavior

*S*ex is the key to success.

Or, more precisely, sex is the key to success if sex results in reproduction and if success refers to evolutionary success. Evolutionary success, in the sense we will consider it, involves an individual's having a proportionally greater genetic representation in the next generation's gene pool (the collection of all alleles of all genes of all individuals of a population). Evolution is a change in the composition of the gene pool—that is, a change in the frequency of certain alleles relative to others. We can see that the best reproducers contribute more to the gene pool and are, therefore, winners in the game of evolution (Figure 4.1).

Evolution measures the value of every trait in terms of its ability to enable its bearer to produce offspring, offspring that will, in turn, produce more offspring. But that does not mean that the anatomy and physiology of reproductive systems or sexual behaviors are the only traits that matter. Almost every anatomical, physiological, or behavioral trait can have an effect on reproductive ability. For example, before an individual can leave descendants, it must survive to reproductive age. Therefore, traits that are important to finding food and appropriate habitat or evading predators will influence reproductive success. In some species, males must be capable of winning competition for a territory where mating can occur or where the young can be raised. Males may also have to compete for access to mates. However, even if an individual has traits that will enable it to build a better nest, find more food than others, or live to be 1000 years old, if it does not leave offspring, the alleles for those traits are not passed on; an individual may be the most magnificent specimen of its species, but without offspring it is generally an evolutionary failure.

Although the phrase "survival of the fittest" is often cited as a description of natural selection, it is often misunderstood. One reason is that the emphasis is too often on the "survival" part. Of course, survival is important, but evolutionarily, unless survival results in the opportunity to reproduce, it does not alter fitness. After all, what counts is ensuring that one's alleles remain in the gene pool. The individuals who leave more alleles than others do earn a larger chunk of the future gene pool.

Even the term "fitness" is often misinterpreted. Who is the fittest? Is it the biggest, strongest, most aggressive individual? Perhaps it is, but only if those traits enhance reproductive success. Although fitness is often thought of as physical fitness, in a discussion of evolution, fitness is the reproductive success of a gene or an individual relative to that of the other members of the species. The fittest individual, then, is the one that leaves the greatest number of surviving offspring or, put another way, individual fitness is a measure of the relative success an individual has in getting its alleles into the next generation.

This concept of fitness was expanded by William D. Hamiliton (1964). He said that because of common ancestry, close relatives share copies of alleles; therefore, by helping relatives to survive and reproduce, an individual increases its kind of alleles in the gene pool. In recognition of this, Hamilton broadened the term fitness to

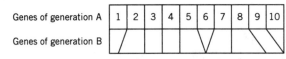

Genes of generation A	1	2	3	4	5	6	7	8	9	10
Genes of generation B										

Figure 4.1 A diagrammatic representation of hypothetical changes in allele frequencies. Ten individuals and their alleles are represented by each of the boxes in the top rectangle. Given the natural variation that exists in any population, some of the individuals will be better reproducers than others. Their alleles will occupy a larger chunk of the gene pool of the next generation. The alleles of less successful reproducers will dwindle or disappear. The alleles of the second generation are represented in the lower rectangle.

inclusive fitness—a measure of the proportion of an individual's kinds of alleles in the future gene pool. It does not stipulate by what means the alleles arrived there. It could be through an individual's own offspring or those of a relative. Thus, inclusive fitness counts all offspring, an individual's own and those of relatives, that exist as a result of that individual's effort. Of course close relatives have more alleles in common. Therefore, the offspring of relatives are devalued in their contribution to inclusive fitness according to the distance of relationship. At this point in our discussion, we will simply note the nature of inclusive fitness. It will be discussed more fully in Chapter 19.

Natural Selection

We can ask, then, what determines which individuals are more fit, and the broad answer is natural selection. Darwin's idea of natural selection can be summarized in the following words from *On the Origin of Species by Natural Selection or The Preservation of Favored Races in the Struggle for Life*, (1859, Chapter 1):

> As many more individuals of a species are born than can possibly survive and as, there is a frequently recurring struggle for existence, it follows that any being, if it vary however slightly in any manner profitable to itself, under the complex and sometimes varying conditions of life, will have a better chance of surviving, and thus be naturally selected. From the strong principle of inheritance, any selected variety will tend to propagate its new and modified form.

From this quotation, we can make several generalizations based on more recent observations:

1. Populations have a great reproductive potential. If every individual produced as many offspring as it possibly could, population growth would be explosive (exponential).

2. Populations tend to outgrow the resources available to them.

3. Individuals in the population tend to compete for the limited resources.

4. Variation exists among individuals of a population. Some of the variable traits have been inherited from parents and can be passed on to offspring. Among the inherited traits are a few that improve an individual's chances of surviving and reproducing.

5. Individuals possessing these beneficial inherited traits have a competitive edge so they are likely to survive and leave more offspring than other individuals. Since the offspring are likely to inherit the beneficial traits, the frequency of these traits increases in the population relative to the traits borne by less successful reproducers.

Natural selection, then, is differential reproduction. That is, nature "selects" those traits that enhance reproductive success. Such traits are selected because the individuals possessing them leave more offspring, including many that will have inherited the traits that contributed to their parents' reproductive success. For example, if males that are victorious in "head-to-head" conflicts with other males leave more offspring than do losers, as is the case among bighorn sheep (Figure 4.2; Geist, 1971), then alleles favoring the behavioral and anatomical traits leading to fighting success will increase in the gene pool while the alternative alleles of the losers decrease. If every individual made an equal contribution to the next generation, there would be no evolutionary change.

It should be made clear that although natural selection alters the frequencies of certain alleles in the gene pool, it does not operate directly on genotype (the genetic makeup) of individuals. It

Figure 4.2 Two bighorn rams clashing heads. The winner will have priority in mating and be more likely to leave offspring.

can only operate on phenotype, the expression of genotype. Thus, it has no direct effect on alleles for tallness, but it can operate on height—tall or short individuals being selected for, depending on the demands of the environment. It therefore cannot operate on recessive alleles in heterozygous individuals. Again, natural selection operates only on expressed characters.

We know that in some cases variation may be disjunct—variants falling into discrete classes, such as when a trait is due to the effect of a single gene. In other cases, variation is continuous. The continuum may exist as a gradient, changing gradually from one extreme to the other. It may also exist along a "normal," or bell-shaped, distribution, with most individuals falling about midway between the extremes, as in Figure 4.3. The reduced number of individuals at either extreme of a normal distribution is usually a result of stronger selection operating against those that are most different from the optimum. Under stable conditions, the optimum can be expected to remain unchanged as stabilizing selection operates (*a*). However, as we know, the optimum can shift as the environment changes. In such a case, those at one extremity

might come to be favored, and the curve representing the population might be shifted in that direction. The phenomenon is termed directional selection (*b* and *c*).

Although we now know more about the mechanism of heredity and the source of variation than did Darwin, his ideas still provide the framework for our understanding of natural selection. Any explanation of evolution by natural selection must include at least these two ideas: (1) Genetically determined variation exists within a population, and (2) some variants leave more descendants than others.

Genetic Variation

EXISTENCE OF VARIATION

All men are *not* created equal. Nor are all women. Nor dogs. Nor cats, slugs, or fruit flies. Not even the offspring of the same parents are identical, as is easily demonstrated by a glance at your little brother, or a look at your cat's kittens (Figure 4.4). In some cases, the differ-

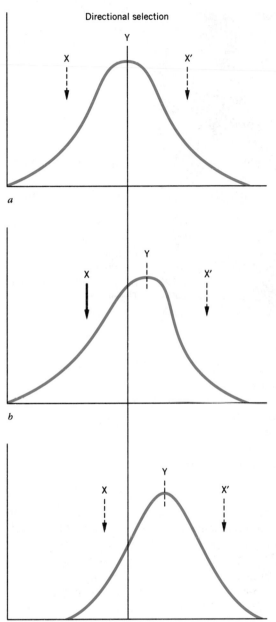

Directional selection

Y

X   X′

a

Y

X   X′

b

Y

X   X′

c Quantitative value for the character selected

Figure 4.3 Directional selection. As long as the selection pressures (X and X′) remain stable, the value for any trait in the population will tend to remain stable. When the pressures of X and X′ change, the optimum Y may shift, causing the population to fall in line with the new optimum.

ences are apparent, in other cases, less so. Nonetheless we are aware that variation generally exists within any population. In particular, we are interested in genetic variation, the ways the alleles differ from one individual to the next. So, the question may arise, what is the source of such variation?

SOURCES OF VARIATION

Where does genetic variability come from? There are two main sources: mutation and recombination. Mutation is a change in a gene or a chromosome that can be inherited. One type of mutation, a gene mutation, occurs when a mistake is made while the chromosome is being copied in preparation for cell division. The mistake may change the chromosome's instructions and result in an altered protein. Another type of mutation, a chromosomal mutation, occurs when there is a break in a chromosome or an incorrect number of chromosomes. Either type of mutation increases genetic variation.

Another source of genetic diversity is the recombination of alleles that occurs during meiosis, the type of cell division that results in the formation of gametes. During meiosis, the maternal and paternal chromosomes are shuffled. As a result, each sibling of the same parents will have a different mixture of chromosomes. Besides mixing up the assortment of entire chromosomes, meiosis scrambles alleles between homologous* chromosomes. Toward the beginning of meiosis (during the stage called prophase I) the homologous chromosomes lie close to one another and pieces of chromosomes containing alleles for the same genes are exchanged. As a result of this exchange, which is called crossing over, a chromosome inherited from one's mother may have some alleles replaced

---

* Homologous chromosomes are those that carry genes for the same traits. One was originally obtained from the mother and the other from the father.

Figure 4.4 Variation in offspring of the same two parents. During the meiotic cell division that forms gametes, the alleles of each parent are shuffled and recombined. Sexual reproduction would scramble any "perfect" combination of genes produced by natural selection.

by alleles from the chromosome inherited from one's father.

The resulting variability provides raw material on which natural selection can work. In more specific terms, an offspring may have a novel combination of maternal and paternal chromosomes and traits that allows it to survive and reproduce better than other members of the species.

FACTORS PRESERVING GENETIC VARIATION

Since natural selection favors those individuals with certain traits, why hasn't it eliminated from the population those individuals that bear other traits? Why do populations remain variable?

Frequency-Dependent Selection

One way that variation may be maintained is by frequency-dependent selection, in which an allele has a greater selective advantage when it is rare than when it is common. As a result, the frequency of the allele fluctuates—increasing until the allele is common and then decreasing. Thus, the fitness of an individual changes with the prevalence of its genotype. There are many examples of frequency-dependent selection (Ayala and Campbell, 1974). We will consider two types—frequency-dependent predation and frequency-dependent reproduction.

Frequency-dependent Predation   It is easy to see how frequency-dependent predation could maintain variation in a population of prey. Although predators usually have a varied diet, they often attack one prey type in excess of its abundance in nature. For example, when the members of a prey species differ in some characteristic, color for instance, a predator often concentrates on the most common form. Individuals with this characteristic are attacked until numbers decline in the population, carrying with them the alleles that coded for that trait. Then the predator often switches to the next most common form of prey, decreasing its numbers while the original variant restores its numbers.

There is evidence from many species that when there is variation in a characteristic of prey, predators often choose the most common type. For example, a common prey of the stickleback fish (*Gasterosteus aculeatus*) is the isopod *Ascellus*. The color of this isopod is variable, ranging from white to yellow to brown and black. Frequency-dependent predation was demonstrated in an artificial situation using only dark- and light-colored isopods in an aquarium. When equal numbers of dark and light isopods were presented to a stickleback, the fish showed no preference for either form. However, when isopod color morphs were offered in a ratio of 4:1, the stickleback chose an excess of the more common form. This was true whether the common color was dark or light (Maskell, Parkin, and Verspoor, 1977).

Frequency-dependent Reproduction In this type of mating, sometimes called the rare-male effect, a male with a rare phenotype enjoys more than the expected share of matings. This type of frequency-dependent mating would maintain variation because as alleles become uncommon, their bearer's mating success is enhanced, thereby increasing the frequency of his alleles. Bruce Grant and his colleagues (1974, 1980) demonstrated the rare-male advantage in the parasitic wasp (*Nasonia vitripennis*). When females were courted by males that differed in some way, either because they came from different inbred lines or because they were reared on different host species, whichever type of male was in the minority mated more frequently. For example, when the wild-type males outnumbered the mutant males 9:1, the mutant males mated significantly more often than would be expected. However, the tables were turned when the mutant males outnumbered the wild-type males 9:1; the wild-type males then mated more frequently than would be expected. When the courting males differed in the host species on which they were raised, the rare male again had a mating advantage.

Rare males have been shown to have a mating advantage in wild populations as well as in the laboratory. For example, the two-spot ladybird (*Adalia bipunctata*) has a variety of color forms ranging from black with two red spots to all red. The frequency of these color morphs varies among populations. John Muggleton (1979) determined the frequency of black (melanic) and red (nonmelanic) ladybirds in several populations. In addition, he recorded the number of each color form involved in mating. He concluded that regardless of color, the rare color morph mates more frequently.

Negative-Assortative Mating

Negative-assortative mating preserves genetic variation in a population. This, in essence, means "opposites attract." The term describes the situation in which individuals tend to choose mates with a different phenotype than their own. Obviously, if the phenotypic difference has a genetic basis, then genetic variability will be enhanced. Admittedly, such assortment is not common in nature. One example involves four color varieties of pigeons (*Columba livia*) in Britain. Here, matings are more common among different color birds (Murton, Westwood, Thearle, 1973). And in some strains of mice, individuals can determine whether others share the certain alleles by the smell of urine. They then choose mates that are different from themselves (Yamazaki et al., 1976, 1978). Patrick Bateson (1983) has suggested that some species are able to recognize kin, individuals sharing many common alleles, and then choose mates that differ from kin. Bateson's ideas are discussed further in Chapter 8.

Environmental Variation

Genetic diversity may also be maintained because some alleles are favored in one environment while others are more successful in other localities. Susan Riechert (1986a) has found, for instance, that there are genetically based differences in territorial behavior among populations of the funnel–web building spider *Agelenopsis*

Figure 4.5 A funnel-web spider at the entrance to the funnel that extends from its web. There are differences in the expression of territorial behavior in populations of this spider living under different ecological conditions.

aperta living in different habitats. These spiders compete for web sites and defend an area around the web as a territory (Figure 4.5). The territory must be large enough to provide sufficient food for survival and reproduction. This species occupies a wide variety of habitats in its range from northern Wyoming to southern Mexico. Spiders living in the relatively lush vegetation along the rivers and lakes in Arizona (riparian populations) allow neighbors to build webs closer to their own than will spiders in desert grassland populations of New Mexico (Figure 4.6). Furthermore, the intensity of territorial disputes between desert grassland spiders is greater than those between riparian spiders. Threat displays of grassland spiders are more likely to escalate into battles and the fighting more often results in physical injury (Riechert, 1979, 1981, 1982).

A series of interesting experiments has demonstrated the genetic basis of territorial behavior in these spiders (Riechert, 1986a). Spiders were collected from a desert grassland environment

Figure 4.6 Variation in ecological conditions in different regions of the range of the funnel-web spider. The desert grassland of New Mexico (top) has few suitable web sites and low prey abundance. In contrast, most of the woodlands near rivers in Arizona (bottom) have many suitable web sites and prey is plentiful.

in New Mexico and from a riparian environment in Arizona. Purebred lines were established by allowing individuals from a particular habitat to mate with one another. After the spiderlings emerged from the eggs, each was raised separately on a mixed diet of all they could eat. When they were mature, 16 females from each population line were placed into experimental enclosures where they could build webs. Just as is found in field populations, the average distance between laboratory-raised females from riparian populations was less than that between females from desert grassland populations. We see, then, that territory size is an inherited characteristic and not determined by recent feeding history or learned from previous territorial disputes.

Riechert (1986a) also examined the degree of genetic diversity between the two populations of spiders. To do this she used electrophoresis, a technique that reveals the number of alleles that exist for a given gene. This measure, in turn, can be used to estimate the degree of genetic variability among individuals in a population. In electrophoresis, molecules are placed in an electric field. Various components of a mixture separate because molecules of different sizes, shapes, and charges move at different rates in the electric field. (See Chapter 14 for a more thorough discussion of electrophoresis.) In this case, the molecules were proteins, enzymes that are known to exist in several different forms. The number of different forms (allozymes) of particular enzymes within a population can be used as a measure of genetic diversity because the specific structure of an enzyme is genetically determined. Riechert found that members of the two populations were genetically quite different from one another.

Those genetic differences and the differences in territorial behavior presumably arose as natural selection favored traits in the populations that suited the local environments. In relatively lush riparian areas found along rivers and lakes for instance, prey is more abundant and there are many more suitable web sites than in the desert grassland habitat. As a result, a spider living along a river can capture enough prey in a smaller area. In contrast, the desert grassland environment is severe. Prey is scarce and the scorching sun makes it too hot to forage during much of the day. Thus, larger territories are well suited to these stringent environmental conditions. It is interesting to note that desert grassland spiders defend large territories even when food is abundant. Territory size has been set genetically so that an individual's territory is large enough to provide sufficient energy during the times of low prey abundance, presumably because prey may be plentiful today but scant tomorrow (Riechert, 1986b). Why, we might wonder, are territorial disputes among grassland spiders so intense? The answer undoubtedly lies in the scarcity of web sites in the desert grassland habitat. Here, as in most places, the law of supply and demand increases the value of a web site. In this species, genetic diversity is increased as a response to the environment throughout its range.

Natural Selection and Adaptation

Variation implies inequality. As we have seen, in a variable population, certain individuals are likely to be more reproductively successful than others. Ecological conditions will determine which traits are favored during evolution. Therefore, as a result of natural selection, we would expect traits of individuals to become gradually fitted to the environment. The traits that allowed individuals to survive and reproduce better than their competitors are called adaptations.

In the previous section, we saw that the expression of territoriality in the funnel-web spider is well suited to the existing ecological conditions. We assumed that natural selection was responsible for the appropriateness of the behavior and, therefore, that territoriality is an adaptation. However, these were assumptions. For any characteristic to be an adaptation, individuals bearing the trait must leave more offspring than those lacking it. Although there are no data to prove that owners of large territories in desert grassland populations leave more offspring than individuals with small territories, there are data to support the suggestion that territory quality may influence reproductive success in another locality—a recent lava bed in central New Mexico. Here, spiders with quality web sites have 13 times the reproductive potential of their neighbors in poorer quality areas (Riechert and Tracy, 1975). Thus, it seems that our assumption of the adaptiveness of territoriality was correct.

When we ask questions about the adaptiveness of behavior, we are asking about its survival value. The aim in answering such a question is to understand why those animals that behave in a certain way survive and reproduce better than those who behave *in some other way*. In our consideration of territoriality in funnel-web spiders, the alternative forms of behavior were easy to identify. The fitness of spiders building webs in areas that are hot and have few prey could be compared to the fitness of others that build webs in areas with more favorable thermal conditions and features that attract prey. But the alternatives are not always this easy to identify because the losers of the competition may be long gone. Nonetheless, if we are to demonstrate adaptiveness, we must always identify the alternatives that natural selection had to choose from (Dawkins, 1989).

Questions about the evolution and adaptive significance of behavior have been central to ethology since its beginning. Recall that two of Tinbergen's four questions were, what is its function (survival value), and how did it evolve? These questions still form the roots and part of the supporting trunk of ethology. However, many who focus on questions of the function and evolution of behavior now prefer to think of their studies as branches of the ethological

tree, branches with labels such as "behavioral ecology" or "sociobiology." Many argue that any such branch ultimately deals with adaptation.

We must keep in mind that the notion that any specific trait is an adaptation is an assumption that must be tested. However, Stephen J. Gould and Richard Lewontin (1979) claim that the adaptiveness of a trait is too often accepted without testing. Indeed, they have ridiculed the adaptationist approach, which they claim breaks an individual into separate traits and assumes that each of those traits is adaptive. They named this assumption a Panglossian paradigm, after Dr. Pangloss in Voltaire's *Candide* who asserted, "Things cannot be other than they are. . . . Everything is made for the best purpose. Our noses were made to carry spectacles, so we have spectacles. Legs were clearly intended for breeches, and we wear them."

Although Gould and Lewontin suggest that Panglossian philosophy is inherent in the thinking of all adaptationists, Ernst Mayr (1983) argues that it is not. He asserts that adaptationists claim only that natural selection produces the best genotype possible given the many constraints placed on it, not that all traits are perfect. There are several reasons why a trait may not be adaptive. First, natural selection must act on the total individual, which is usually a mixed bag of traits—some good, some bad. Second, natural selection can only pick from the available alternatives. Those alternatives will depend on the constraints imposed by the individual's evolutionary history and its present conditions—ecological, anatomical, and physiological. The "perfect" mutation or allele combination may not have arisen. Finally, natural selection works in a given environment and conditions may vary from place to place or change over time.

Some evolutionary theorists restrict the term adaptation to those traits that have a genetic basis (Williams, 1966). We will follow their lead and exclude learned behaviors from this chapter. However, we will return to the role of learning in producing appropriate responses in Chapter 5.

The Maintenance of Nonadaptive Traits

We cannot expect that all observed characters, even when the data are in, will be found to be adaptive or at least neutral. There are undoubtedly nonadaptive characters in any population.

The question then arises, if certain traits are nonadaptive, how are they maintained in a population? One way nonadaptive traits can be maintained is through pleiotropy, multiple effects of a single gene. For example, a certain gene in mice that produces short ears also produces fewer ribs; and a single gene in *Drosophila* produces small wings, lowered fecundity, shorter life span, and other changes. If the *net* pleiotropic effect is beneficial, then nonadaptive traits can be maintained in the population. Nonadaptive characters may also continue in a population through linkage. In such a case, a nonadaptive gene may be located very close to a highly adaptive one on the chromosome. So, a nonadaptive allele may be carried along on the coattails of the adaptive one.

In other cases, nonadaptive traits are not being maintained. They are being selected out, and we simply observe them on their way to oblivion. Thus, when we look at any population, there are those on a winning track and those being nudged out.

We must keep in mind that because of the variation produced by mutation and meiotic recombination, winners and losers in the evolutionary game are constantly being generated. We should also keep in mind that variation produced by mutation is likely to be deleterious. After all, a random change in any finely tuned machine, whether it is a living body or an au-

tomobile engine, is unlikely to be an improvement. Furthermore, even if natural selection were somehow able to produce an individual with a perfect combination of alleles, that combination would be scrambled during the production of gametes and diluted in fertilization when the alleles are mixed with those of the other parent (Williams, 1975).

Nonadaptive traits may survive in spite of natural selection because of evolutionary time lag, the time between selection and its effects. Today's genetic combinations were selected to suit yesterday's conditions. But today will almost certainly differ from yesterday or tomorrow. Paradoxically, natural selection may actually be the reason that yesterday's adaptation is antiquated. Adaptations selected in one species will change the conditions for others (Van Valen, 1973). Maintaining the same advantage requires continuous change. The cat quick enough to snatch a bird successfully today might not be speedy enough to catch the next generation of birds because only the swiftest of yesterday's birds remained to leave offspring. Thus, natural selection always involves a lag as adaptation tracks conditions. It is like riding on the down escalator and trying to hug someone on the up escalator. To maintain the embrace, you would both have to change steps constantly.

The Search for Adaptiveness

Where do such problems leave the adaptationist approach? Right back where it started—with the idea that the hypothesized adaptiveness of any trait must be tested. If the hypothesis is incorrect, a new one can be generated and tested as science proceeds (Brown, 1982; Mayr, 1983).

The hypothesis that traits are adaptive has, in some instances, stimulated interesting research that might have been neglected if one readily assumed the nonadaptiveness of traits. Consider, for instance, Niko Tinbergen's observation of a seemingly nonadaptive behavior of black-headed gulls (*Larus ridibundus*). Tinbergen observed that the gull parent does not immediately remove the broken eggshells from the nest. He and his colleagues had already demonstrated that the presence of eggshells in the nest attracted predators such as herring gulls and carrion crows. But here was the black-headed gull sitting for hours among the conspicuous shell fragments. Tinbergen first thought that this delay must be dangerous and only explainable as a pleiotropic and nonadaptive effect associated with the removal behavior. Tinbergen warned, however, that such assumptions are, in essence, a refusal to investigate. He then observed the black-headed gull colonies more carefully, looking for any adaptiveness of the delay. He noted that chicks were commonly eaten by neighboring gulls and that cannibalistic neighbors took many more chicks when they were newly hatched and still wet than when they had dried

and become fluffy. Whereas a gull could swallow a wet chick within a few seconds, it took about 10 minutes to down a dry chick. One might imagine the difference as similar to trying to swallow a peeled grape as opposed to a cotton ball. Robber gulls were observed to snatch the wet young within a fraction of a second if the parent were distracted by a predator. In fact, one chick was gulped down while its parent was carrying off some eggshells. So, Tinbergen surmised that although removing the shells reduced predation by other species of birds, delaying removal until the chicks were dry decreased the likelihood of the chicks' being cannibalized by neighboring gulls while their parents were away on their chores (Tinbergen et al., 1962).

In this case, Tinbergen's new hypothesis was born of simple observation, but the search for adaptiveness of other behaviors has involved two intriguing avenues of research—the comparative approach and the experimental approach.

Figure 4.7 The nesting grounds of the herring gull. The nests of herring gulls are subject to predation because of their accessibility.

THE COMPARATIVE APPROACH

The comparative approach to the investigation of adapativeness involves comparing individuals of the same or related species that inhabit different kinds of environments. The rationale for this approach is that related species living in different ecological conditions experience different selection pressures. Differences in their behavior, therefore, may be due to natural selection molding their traits to suit their ecological conditions. For example, if one species suffers great risk of predation, and a closely related species has few predators, we might interpret behaviors shown only by the first species as antipredator adaptations. However, we must not leap to unwarranted conclusions. There may be other explanations for the behavioral difference that were not considered. Conclusions regarding the function of a behavior gain validity as more species with similar differences in selection pressure are compared and found to have similar behavioral differences.

The Comparative Approach in Action

Esther Cullen's (1957 a,b) comparisons of several gull species that descended from a common ground-dwelling ancestor but now live in very different ecological situations provide an excellent example of the comparative approach. She compared the behavior of typical ground-nesting gulls, the herring gull (*Larus argentatus*) and the black-headed gull (*L. ridibundus*), with that of a cliff-dwelling species, the kittiwake (*Rissa tridactyla*). Most gulls, including the herring gull, nest in low-lying areas or on offshore islands (Figure 4.7) and are subject to more predation than their cliff-dwelling cousins. Many aspects of herring gull behavior have been interpreted as adaptations to reduce predation. They hide their nests, which are composed of bits of twigs and grass piled together among clumps of shore grasses. The eggs and young are also cryptic—the young have certain behavioral characters that maximize the effect of their

Figure 4.8   Nesting area of the kittiwake. The simple shift in nesting sites from the ground to cliffside has rendered the kittiwake nests inaccessible to predators and has triggered a whole gamut of behavioral change.

camouflage. At their parents' warning call, they may quickly run under the nearest cover and thus may escape any predator that has been attracted to the nest; or they "freeze" and rely on their markings to render them inconspicuous.

The kittiwake nests on cliff ledges (Figure 4.8) and displays a wide range of behavioral traits not found in other gulls. Cullen concluded that many of these patterns have developed as adaptations to a life-style in which the bird truly is "on edge." For example, the kittiwakes fight by grabbing each other's bills and twisting, thus causing the opponent to fall off the ledge while ground-nesting gulls usually try to get above an opponent and peck down or pull at the body of the other (Figure 4.9). Probably because of their fighting patterns, ground nesters threaten by stretching into an "upright threat" posture. The kittiwake does not. Furthermore, female kittiwakes sit down to copulate, thus reducing the risk of falling off their narrow ledges during the process. Also, the kittiwake builds much tighter and deeper nests than do ground nesters because it would not be able to retrieve eggs that rolled out, as could its flatland fellows.

Later, when kittiwakes are nesting, conspicuous eggshells and droppings are left around the nest, a habit that in other gulls would attract predators. Young kittiwakes, of course, must remain in the nest until they can fly; therefore, all feeding is done at the nest. The young take the food directly from the parent, leaving no partly digested matter lying around the nest as a potential source of disease. Ground nesters, on the other hand, feed their free-roaming young anywhere in the nesting area, and they sometimes regurgitate food onto the ground near their young.

Also, the young of ground nesters practice flying by jumping high into the wind. It is a risky business living on a cliff ledge, so the young kittiwakes never jump very high and always face the cliff when they do jump.

The aim of Cullen's studies, and of all comparative studies, is to determine the function of a behavior by identifying the selective pressures that led to it. However, by itself the method merely identifies correlations between ecological variables and behavior. The investigator must then explain the correlation. Critics of the adaptationist approach have suggested that these explanations have no more scientific validity than Rudyard Kipling's *Just So Stories* that imaginatively explain how a camel got its hump or an elephant got its trunk (Gould and Lewontin, 1979).

The comparative approach becomes more convincing when it is used to predict behavior in a previously unstudied group (Hailman, 1967). For example, the nest site of the Galapagos swallow-tailed gull, *Larus (Creagus) furcatus*, is intermediate between the kittiwake and ground-nesting gulls. Its behavior was also intermediate between the two types, supporting the conclusion that the differences in behavior between the herring gull and kittiwake are due to selective pressures in their different environments.

If we accept that behavior in related species can diverge due to differences in selective pressure, then the converse should also be true—unrelated species experiencing the same selective pressure should display similar behavior. We might expect, for instance, that other species that experience high predation would show antipredator behaviors similar to those of the herring gull and black-headed gull. Mobbing, a communal attack on an intruder, is shown by both species of ground-dwelling gulls, which experience high predation, but not by the kittiwake, which suffers little predation. This comparison suggested that the function of mobbing is to drive away potential predators. The bank swallow (*Riparia riparia*) is unrelated to the gulls but experiences high predation, as do the herring gull and black-headed gull. Bank swallows also mob potential predators and successfully drive them away (Hoogland and Sherman, 1976).

Following the same reasoning, we might also expect that species that experience little predation might lack certain antipredator behaviors. The herring gull and black-headed gull remove broken eggshells from the nest after the chicks have hatched, but the kittiwake does not. It was hypothesized based on this comparison that the function of eggshell removal was to make the nest less noticeable to predators. Gannets (*Sulsa*

Figure 4.9 Aggression in herring gulls. The form of fighting is violent compared to that of kittiwakes.

bassana) are less closely related to kittiwakes than are the herring gull or black-headed gull. However, like the kittiwake they nest in places that are inaccessible to predators and also like the kittiwake, they leave broken eggshells in the nest (Nelson, 1967).

Limitations of the Comparative Approach

Although the comparative method can be helpful in the study of adaptation, it must be applied carefully (Clutton-Brock and Harvey, 1979, 1984; Gittleman, 1989). We should, for example, always consider, test, and rule out alternative hypotheses. Sometimes this can be done by listing the competing hypotheses and developing predictions for each. The confirmation of the predictions should lend more weight to some hypotheses than others.

We should also keep in mind that correlations between traits and ecological variables identified by the comparative approach do not demonstrate causality. Consider the difficulties in determining whether the diet was a cause or an

effect of sociality in John Crook's (1964) comparative study of over 90 species of weaver birds (Ploceinae). He noticed that species living in the forest generally eat insects and forage alone. In contrast, species inhabiting the savannah eat seeds and feed in flocks (Figure 4.10). Crook identified correlations between the degree of sociality and two factors—diet and predation. But which is cause and which effect? Seeds often have a patchy distribution and groups of foragers are more likely to find a patch that can feed them all. One might infer, therefore, that diet is the cause of flocking in weaver birds. However, living in groups is also a good defense against predators and seeds may be the only food source that could supply enough food for an entire flock. So, is sociality a cause or an effect of diet? The correlation does not answer the question.

In addition, we must be watchful for confounding variables and control for them. A confounding variable is a variable other than the factor of principal concern that may be a cause for the observed correlation. Common con-

Figure 4.10   Weaver birds. John Crook compared the diet and degree of sociality of weaver birds inhabiting different environments. He observed that solitary species, such as *ploceus nelicouri* (shown on the left; nest not shown in proportion) are usually insectivores that defend large territories in the forest. In contrast, social species, such as *ploceus phillipinus* (shown on right) eat seeds and live in the open savannah.

founding variables are age and body size. The failure to correct for these variables can lead to incorrect conclusions. For example, antler size is correlated with reproductive success among red deer (*Cervus elaphus*) stags. Antler size is also correlated with age and body size. When one controls for age by comparing reproductive success among stags of the same age, no association is found between antler length and reproductive success (Clutton-Brock and Harvey, 1979). So, antler size is not important to reproductive success, as it first appeared.

THE EXPERIMENTAL APPROACH

The comparative approach may suggest hypotheses to explain the function of particular traits that may then be tested using the experimental approach, in which conditions can be manipulated by the researcher.

Esther Cullen's (1957a,b) comparative study identified differences in the behavior of kittiwakes and ground-nesting gulls that were correlated with the degree of risk of predation. She found that kittiwakes, which have low predation rates, leave eggshell pieces in the nest, but ground-nesting gulls, which have high predation rates, generally remove broken eggshells. This correlation prompted Tinbergen to hypothesize that the survival value of eggshell removal was reduced predation on the young (Figure 4.11). However, several other hypotheses are possible: The sharp edges on shells might injure chicks or interfere with brooding; an empty shell might

Figure 4.11 A bittern removing pieces of eggshell from its nest. This behavior is typical of many bird species that rely heavily on nest concealment to reduce predation of the young. The white inner surface of the eggshell may make the nest more noticeable to a predator.

slip over an unhatched egg, encasing the chick in an impenetrable double layer of shell; or the egg remains might serve as a breeding ground for parasites, bacteria, or mold.

Tinbergen and his colleagues began experiments designed to test the hypothesis that eggshell removal reduced predation on chicks. The study began with observation. It was noted that the eggs, chicks, and nest were camouflaged and might be difficult for a predator to spot. However, the bright white inner surface of a piece of eggshell might catch a predator's eye and reveal the nest site. So they began by painting some black-headed gull eggs white to test the idea that white eggs might be more vulnerable to predators than the naturally camouflaged eggs. Of 68 naturally colored gull eggs, only 13

were taken by predators. However, 43 of the 69 white eggs were taken. The difference in predation rates lent credence to the idea that the white inner surface of eggshell pieces might endanger nearby eggs or chicks. Since all black-headed gulls remove eggshells, Tinbergen could not compare survival rates in natural nests with and without eggshell pieces. Instead, he created artificial variation to observe the effects of natural selection. He made his own gull nests with eggs and placed white pieces of shell at various distances from some of the nests. The broken eggshell bits did attract predators. Furthermore, the risk of predation decreased with increasing distance of eggshell pieces from the nest (Tinbergen et al., 1962).

Optimality

So far our discussion of adaptation has been one-sided. By focusing on the survival value of adaptation, we have considered only the benefits gained through certain behaviors. But most actions also have costs. For example, Tinbergen (1965) pointed out that in the black-headed gull, colonial living is advantageous as a means of predator defense. The drawback is that some birds lose their young to cannibalistic neighbors. Likewise, when a mother robin "decides" whether to search for one more worm to feed her young, the energy gained must be weighed against the risk of becoming cat food (Williams, 1966).

Optimality theory recognizes both costs and benefits and suggests that traits have evolved to be optimal—the best they can be under existing conditions. Natural selection might be seen to act as a person might when faced with an important decision in life. When choosing the best course of action, it is not uncommon for someone to create lists of the pros and cons of each alternative. The best choice is the alternative in which the advantages outweigh the disadvantages by the greatest amount. The lists are helpful because they identify factors that should be considered in the decision. But the decision may still be difficult because it may require integration of concerns along different dimensions. For instance, if you were deciding whether to get your own apartment or live with your parents, you might have to weigh factors representing two dimensions—money (the difference in rent) against freedom (gained by being on your own).

Optimality theory views natural selection as weighing the pros and cons, or to use the proper jargon, the costs and benefits, of each available alternative. First, natural selection translates all costs and benefits into common units—fitness. Then it chooses the behavioral alternative that maximizes the difference between costs and benefits (Figure 4.12). In economic terms, the alternative that maximizes the difference between costs and benefits is the one that yields the greatest profit. In terms of evolution, this is the choice that maximizes fitness and, therefore,

it is assumed that this is the alternative that would continue into the next generation.

Measuring costs and benefits in terms of fitness is difficult at best. It might be easy to count surviving offspring or those that are lost to predators. However, it is impossible to determine the number of offspring that never existed but could have *if* the parent had behaved in some other way. Could a mother robin have laid more eggs if she had been the earlier bird? Would she have saved any offspring from predation if she had stayed at the nest? The list of "what ifs" presents a problem for long-term optimality studies.

It has been argued, however, that animals might be expected to behave optimally, not just over the long haul, but also in daily activities. Consequently, optimality theory has used cost/benefit analyses to predict the form of many different short-term behaviors, such as degree of sociality, mating strategies, aggressiveness, and foraging. (Optimality theory has been extensively applied to foraging, and we will reconsider the concept in Chapter 12.) In each case, it is assumed that an animal's behavior maximizes the difference between costs and benefits. However, in studies of short-term optimality, profit is rarely considered in terms of fitness. Instead, costs and benefits are measured in terms such as time or energy, factors that are thought to be related to fitness but are more easily measured.

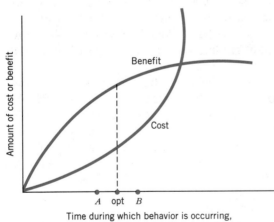

Figure 4.12 An optimality model for a hypothetical behavior. According to optimality theory, natural selection measures the costs and benefits of each behavioral alternative in units of fitness. It then chooses the alternative that maximizes the difference between cost and benefit and, thus, maximizes fitness.

OPTIMALITY AND TERRITORIALITY

As an example of how optimality models may be used to predict behavior, we will see how they have been applied to the study of territoriality. A territory, you may recall, is an area that is defended against intruders, generally in the protection of some resource. Optimality theory predicts that an individual should be territorial if the benefits from enhanced access to the resource are greater than the cost of defending the resource. The benefits of territoriality include increased mate attraction, decreased predation, protection of young or mates, reduced transmission of disease, and a guaranteed food source (Hinde, 1956). The costs of defending the territory include energy expenditure, risk of injury, and increased visibility to predators.

Optimality theory predicts that territoriality will evolve only when the benefits exceed the costs—that is, when the territory is economically defendable (Brown, 1964). If the resource is scarce, an individual may not gain enough to pay the defense bill. It may be economically wiser to look for greener pastures. Accordingly, the golden-winged sunbird (*Nectarinia reichenowi*) will abandon a territory when it no longer contains enough food to meet the energy costs of daily activities as well as defense (Gill and Wolf, 1975). On the other hand, if there is more than enough of the resource to go around, it would be energetically wasteful and economically unsound to defend it. Water striders (*Gerris remiges*) are among the species that will cease to defend territories if supplied with abundant food (Wilcox and Ruckdeschel, 1982). Likewise, we find that female marine iguanas do not bother defending territories with nest sites on most of the Galapagos Islands. They defend territories on only Hood Island, the only Galapagos island where nest sites are in short supply (Eibl-Eibesfeldt, 1966).

An important factor determining whether it is economically sensible to defend a resource is its distribution in space and time. Generally, animals are territorial if the resource occurs in patches or clumps. A pile of food, for instance, is easier to defend than food spread thinly over a large expanse, as long as the number of competitors anxious to contest ownership is not too great. Thus we find that male Everglades' pygmy sunfish defend territories near a clump of prey in an aquarium but are not territorial when prey are evenly dispersed (Rubenstein, 1981).

The degree of competition for the resource also influences its economic defendability. The more competitors, the greater is the cost of defense. Male fruit flies (*Drosophila melanogaster*) defend small areas of food that are suitable for oviposition, particularly if there are females in the vicinity. As would be predicted, however, the incidence of territoriality is reduced at higher densities of males (Hoffmann and Cacoyianni, 1990).

If animals are territorial when the benefits exceed costs, then territory holders should have greater fitness. Earlier in this chapter, it was mentioned that this is true for funnel-web spiders possessing higher-quality territories. Likewise, we find that larger male tree-dwelling iguanas (*Liolaemus tenuis*) defend larger trees, which are occupied by more females than are small trees. Thus, small males are usually monogamous and larger ones are polygamous. As you can see in Figure 4.13, the size of a male's harem, and therefore his fitness, depends on his size and the size of his tree (Manzure and Fuentes, 1979).

It is generally assumed that both the benefits and the costs increase with territory size. A larger territory might be expected to provide

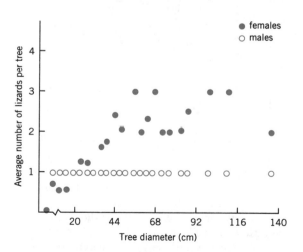

Figure 4.13 The relationship between territory (tree) size and harem size of male tree iguanas. A male's fitness generally increases with the size of its territory. (From Manzure & Fuentes, 1979.)

more food, for instance, but an animal would have to use more energy to patrol the increased border. Therefore, we might expect animals to adopt a territory size that would give them the greatest profit. However, there is no universally perfect territory size. The exact shape of the benefit and cost curves and their relative placement will vary among species and environments (Figure 4.14). The precise placements of these curves may alter or even reverse predictions of optimal territory size (Schoener, 1983).

Many studies that apply optimality theory to territoriality concentrate on feeding territories and measure costs and benefits as energy gains and expenditures. We might imagine at least two fundamental strategies. The first would be to spend the least amount of time possible to gain the bare minimum of necessary energy. This would be optimal, for example, if the risk of predation were high. In this case, it would be better to gather just enough food to get by and

return quickly to a safe location. Alternatively, some other activity, courting or nest building, for example, might be more important than eating. (Humans sometimes place little importance on eating. After all, fast food restaurants flourish because we often "eat just to live.") The second strategy would be to eat to maximize energy gains. For instance, it may be important to put on weight for migration or before hibernation.

We might predict, therefore, that in some species the size of an individual's territory would be adjusted to maximize energy gains. At least some individual rufous hummingbirds (*Selasphorus rufus*) appear to do this. During their southward migration, these birds pause for a few days in the mountain meadows of California to build the fat reserves needed to fuel the next leg of the journey. During this interval they feed on the nectar of the flowers of the Indian paintbrush. Each bird defends a group of flowers as a territory. The weight gain and territory size for a single individual are shown in Figure 4.15. As you can see, this bird adjusted the size of its territory so that it could gain weight as quickly as possible. This individual began with a small territory that contained few flowers, so its weight gain was minimal. It increased the territory size greatly on the third day. This territory had more flowers and the bird could obtain more energy, but it had to invest more energy in defense. Nonetheless, it gained somewhat more weight than possible on the smaller territory. Next, however, it reduced the size of its territory slightly on the fourth and fifth days, thereby cutting the energy costs of defense and maximizing its weight gain (Carpenter, Paton, and Hixon, 1983).

It is reasonable to assume that short-term optimality might influence long-term optimality. An animal that forages efficiently, for instance, might be expected to leave more offspring than one that does not. But maybe it wouldn't. Conceivably, it could become so intent on foraging to maximize energy gain that it would give up too quickly in an aggressive encounter to win a mate or it might fail to notice a stalking predator.

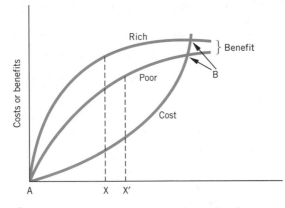

Figure 4.14 The hypothetical relationship between the costs and benefits of territoriality. Both costs and benefits increase with territory size. The shape of the benefits curve varies with the quality of the territory. It is profitable to defend the territory as long as the benefits exceed the costs, between points A and B. If the animal is to maximize its net gain, the optimal territory size is at point X or X', depending on the exact placement of the cost and benefit curves.

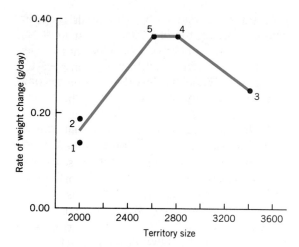

Figure 4.15 The relationship between territory size and weight gain for one rufous hummingbird. It is important for these birds to gain weight maximally during their stopovers along their migratory route. These data indicate the weight gained by a single territorial bird on five successive days. The numbers beside the data points indicate the day of the study. This individual adjusted its territory size so that weight gain was maximal. (From Carpenter, Paton and Hixon, 1983.)

Whereas short-term optimality studies focus on a single activity as if it were the only determinant of evolutionary success, long-term optimality considers the big picture. When we fail to get matches between predictions and observations in short-term optimality studies, it may be because natural selection must balance many different factors (Dawkins, 1986). Deviations from predictions are often surprisingly useful because they encourage us to probe for explanations. For example, it was the deviations from predictions based on Copernicus's model of the solar system that prompted the astronomer Percival Lowell to consider how gravity from another planet might affect the system, and he postulated the existence of a ninth planet 14 years before there was photographic proof of Pluto's existence. Likewise, altruistic behavior

seemed to be a deviation from the predictions of natural selection, and this deviation led to the development of the concept of inclusive fitness. Therefore, optimality remains a useful construct (Barash, 1982).

Evolutionarily Stable Strategies

We have been discussing optimality as if an individual's reproductive success depends only on its own behavior. In the examples mentioned this was generally true. Thus these are frequency-independent optimality models.

There are situations, however, in which the individual is not the sole master of its fate. Instead, the reproductive success gained by behaving a certain way depends on what the other members of the population are doing. In this situation, we would say that the optimal behavior is frequency-dependent because the fitness gains it yields depend on how many other members of the population are behaving the same way (Maynard Smith, 1976, 1982).

Rock–paper–scissors is a familiar game in which you win or lose depending on what your opponents do. Recall the rules: Rock breaks scissors; scissors cut paper; paper covers rock. Any one of the moves could win or lose depending on the actions of the other players. Play scissors and you win *if* your opponent plays paper. But play scissors too often and your opponent will catch on and defeat you by playing rock. The only way to avoid certain loss and have a chance of breaking even is to play all three strategies— rock, paper, and scissors—in random order with equal frequency (Maynard Smith, 1976).

Just as rock, paper, and scissors may be considered alternate strategies in a game, an animal's behavior may be described as a strategy. However, here strategy does not imply that the animal consciously plans the best course of action for maximizing reproductive success. "Strategy" is simply a description of what the animal does.

For some behaviors, alternate strategies may exist within a species. Consider the following examples: A male cricket may call to attract a mate and risk being parasitized, or he may intercept a female on her way to a caller. A female digger wasp may build her own nest or take over one that is already constructed. A male may fight over a female or display and not fight.

Which is the best strategy? In the previous examples and in many others, the best course of action depends on what others are doing. The optimal strategy when the payoffs depend on what others are doing is called an evolutionarily stable strategy (ESS). By definition, an ESS is a strategy that cannot be bettered and, therefore, cannot be replaced by any other strategy when most of the members of the population have adopted it. It cannot be bettered because when it is adopted by most of the members of the population, it results in maximum reproductive success for the individuals employing it. As a result, an ESS is unbeatable and uncheatable.

ESS theory is able to cope with situations in which the payoffs of an action depend on the actions of others, but it does not require that the individuals employ different strategies. An ESS may be pure, that is, consist of a single strategy, or it may be mixed, that is, consist of several strategies in a stable equilibrium. We have discussed an example of a pure strategy among black-headed gulls—the removal of eggshells from the nest. It is an ESS because there is no alternative behavior that will yield greater reproductive success.

A mixed strategy is actually a set of strategies that exists in an evolutionarily stable ratio. Consider, for instance, a hypothetical group of fish-catching birds. There are two strategies for getting dinner—catch your own fish or steal one from another bird. Natural selection is assumed to favor the strategy that maximizes benefit. Since the thief minimizes cost and gets full benefit, thievery is favored initially. However, as the

proportion of bandits in the population increases, so does the likelihood of encountering either another robber or a bird that had already had its fish stolen. Then honesty becomes the best policy. When hard-working birds become common, thievery once again becomes profitable (Dawkins, 1980).

As the relative frequencies of alternative strategies fluctuate, they reach some ratio at which both strategies result in equal reproductive success. That particular mix of strategies will be an ESS. Should either strategy become more common, its fitness will drop, bringing it back to the frequency at which the reproductive success equaled that of the other strategy.

Mixed ESSs can arise in two ways. First, different genotypes could be responsible for producing each strategy. In this case, each member of the population would always adopt the same strategy and the relative proportions of pure strategists would remain stable. Second, a mixed strategy could involve each individual varying its strategy and playing each one with a certain frequency. Using the rock–paper–scissors example again to illustrate this difference, we see that stability would result if one-third of the population played rock, one-third played scissors, and one-third played paper. Alternatively, each member of the population could play rock, paper, and scissors with equal frequency.

DIGGER WASP NESTING STRATEGY AS A MIXED ESS

The nesting behavior of female digger wasps (*Sphex ichneumoneus*) is a clear illustration of a mixed ESS (Brockmann, Grafen, and Dawkins, 1979). A female lays her eggs in underground nests that consist of a burrow with one or more side tunnels, each ending in a brood chamber (Figure 4.16). She lays a single egg in a brood chamber, after provisioning it with one to six katydids, a process that can take as long as 10 days.

To dig or not to dig, that is the question

Figure 4.16 A female digger wasp at the entrance to a burrow.

(Figure 4.17). On some occasions a female will dig her own nest and on others she will enter an existing burrow. These alternative strategies have been called "digging" and "entering." Digging has an obvious cost in time and energy since it takes a female an average of 100 minutes to dig a burrow. Furthermore, the investment is not risk free. There is no guarantee that she will not be joined by another female and if she is, she may lose her investment. In addition, there are temporary catastrophes, for instance, an invasion by ants or a centipede, that may force a female to abandon her burrow. Once the intruders have gone, the abandoned nest is quite suitable for nesting again. A female that finds an abandoned nest reaps the benefits without incurring the costs. So, it might seem that entering an existing burrow would be the favored strategy. Indeed it is—*if* the burrow is actually abandoned. Unfortunately, there is no way to determine whether the nest is abandoned or whether the resident is just out hunting. A female who is provisioning a nest is gone most of the time—it may be hours or days before she learns that another female is occupying the nest. Eventually

the two females will meet. When they do, they fight and the loser must leave behind all the katydids she placed in the nest. The winner takes

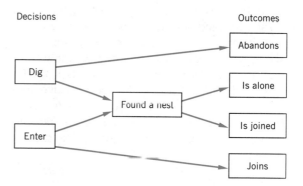

Figure 4.17 A female digger wasp's alternative nesting strategies and their outcomes. The two available strategies are to "dig" and to "enter." There are three possible outcomes to a decision to dig: she may remain alone and retain exclusive use of the burrow; she may be joined by another female; or she may have to abandon the nest due to a temporary catastrophe. If a female decides to enter an existing burrow, she may be alone or she may be joining another female. (From Brockmann et al., 1979.)

all. She lays her egg on the jointly provided supply of katydids.

Whether it is best to dig or to enter depends on what the other members of the population are doing. Entering saves time and energy. However, this strategy does best when it is rare. As entering becomes more common, there are fewer diggers. As a result, the chances of entering an occupied nest increase. Eventually, digging becomes a better strategy. The mixture of strategies should reach an equilibrium with an evolutionarily stable ratio of strategies. For the digger wasps studied in New Hampshire, the ratio was 41 enter decisions to 59 dig decisions (Figure 4.18).

If that mix of strategies is to be evolutionarily stable, the reproductive success for entering and digging must be equal. If they are not, the more successful strategy would become more prevalent in the population. To determine the reproductive success of these strategies, Jane Brockmann and her colleagues (1979) spent over 1500 hours observing 410 burrows. The reproductive success of dig and enter decisions was calculated. In a New Hampshire population, there was no significant difference in the reproductive success of the digging versus entering strategies. Whereas the reproductive success resulting from decisions to dig a nest was an average of 0.96 eggs laid per 100 hours, the reproductive success of decisions to enter an existing nest was 0.84 eggs per 100 hours. Therefore, the nesting strategies of female digger wasps are truly a mixed ESS.

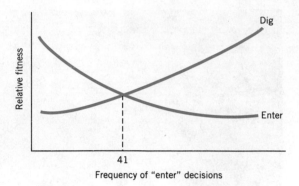

Figure 4.18 The digger wasp strategies to "dig" and to "enter" are a mixed ESS. The success of either strategy depends on the strategy adopted by other members of the population. For a New Hampshire population of digger wasps, the strategies have equal fitness when they exist in the ratio 41 "enter":59 "dig."

members of a population as opponents in a game. (The mathematical basis of game theory is similar to that underlying chess or checkers.)

Game theory models require that we take into account (1) all the existing strategies, (2) the frequency of each strategy in the population, and (3) the fitness payoff for each possible combination of strategies. This information is often presented in a matrix that indicates the payoff for each possible interaction, such as the one shown for the rock–paper–scissors game in Table 4.1. In this matrix, the payoff is +1 if you win, −1 if you lose, and 0 if you tie. Similar payoff matrices can be created to predict ESSs.

Game Theory

When the payoff depends on what others in the population are doing, we need more information than just the costs and benefits to predict the evolutionarily stable strategy. Game theory is a tool for analyzing such situations. It treats

GAME THEORY AND ANIMAL AGGRESSION

ESS theory and game theory have helped our understanding of the logic of animal combat, an obvious situation in which the best strategy depends on what competitors are doing. We begin by considering aggressive games in which there are only two strategies that might be employed. The "hawk" strategy is to fight to win. A hawk

Table 4.1

PAYOFF MATRIX FOR ROCK–PAPER–SCISSORS GAME

	ROCK	PAPER	SCISSORS
Rock	0	−1	+1
Paper	+1	0	−1
Scissors	−1	+1	0

will continue to fight until either it is seriously injured or its opponent retreats. The "dove" strategy (originally called "mouse") is to display but never engage in a serious battle. If a dove is attacked, it retreats before it can be injured seriously.

We can calculate the costs and benefits of each strategy, keeping in mind that the realized payoff will depend on the opponent's strategy. The expected payoffs are shown in Table 4.2. The possible benefit in all matches is the value of the resource to the victor. In a contest between two hawks, the cost is the injury that might result. The net benefit would be the value of the resource less the cost of injury. It is assumed that all hawks have an equal chance of winning. A hawk fighting another hawk will win half of its battles and be injured in the other half. Thus, when two hawks square off, the payoff is one-half of the net benefit. If a hawk meets a dove, the hawk always wins. Since a dove immediately

retreats, it gains nothing but it loses nothing. On the other hand, the hawk in a hawk–dove contest gets the full value of the resource at no cost. In a contest between two doves, the cost is the time spent displaying. Although the price of display is generally considerably less than the cost of a wound, both the winning and the losing doves must make this investment. It is assumed that all doves have an equal chance of winning. The benefit, therefore, is one-half the value of the resource. The payoff in a dove–dove contest is one-half the value of the resource minus the cost of displaying.

Dove–Dove Interactions—The "War of Attrition"

A dove, we have seen, displays, but it retreats if attacked. What happens, then, if the opponent is another dove? There is no attack. No one retreats. When two equally matched doves confront one another, the winner of this so-called "war of attrition" is the one that displays the longest. It seems obvious that this would result in a mixed ESS of varying display times because a pure ESS of a fixed display time could always be bettered by the strategy of a slightly longer display. The stable strategy is to be unpredictable, to display for a randomly chosen length of time. Since the cost of display increases with duration, shorter display times should be more common than longer ones (Maynard Smith, 1974).

Table 4.2

PAYOFF MATRIX FOR HAWK–DOVE GAME

PAYOFF RECEIVED BY	OPPONENT	
	HAWK	DOVE
Hawk	$\dfrac{\text{Value of resource} - \text{Cost of injury}}{2}$	Value of resource
Dove	0 (Gains nothing, loses nothing)	$\dfrac{\text{Value of resource}}{2} - \text{Cost of displaying}$

Conflicts Involving Hawks

When the cost of injury is greater than the value of the resource, as is usually the case, the ESS will always be a mixture of hawk and dove strategies. This is because the success of either the hawk or the dove strategy depends not only on its payoff, but also on the chances that an opponent will be playing the same or the alternate strategy. The greatest possible payoff, the full value of the resource without cost, goes to a hawk when it meets a dove. Therefore, as long as the population is mostly doves, a hawk fares quite well. As a result, the hawk strategy would spread through a population of doves and become more common, making encounters between two hawks more likely. However, since the cost of injury is greater than the value of the resource, hawks experience a decrease in fitness in hawk–hawk encounters. The fitness gained by a dove fighting another dove is meager compared to the fruits of victory enjoyed by a hawk conquering a dove, but at least there is no chance of losing fitness. The dove strategy could, therefore, invade and spread through a population of hawks. Thus, we see that neither the dove nor the hawk strategy is a pure ESS. Since the fitness gain of each strategy is frequency dependent, the ratio of strategies would fluctuate until a point was reached at which the fitness of hawk and dove was equal. This mixture would be evolutionarily stable. The precise mixture of hawk and dove strategies that would confer equal fitness to each would depend on the relative value of costs and benefits (Maynard Smith and Price, 1973).

Figure 4.19 Male rattlesnakes fighting. Game theory predicts that fights are not likely to escalate when the cost of injury is great. Each of these males could kill the other, but they do not bite. The males press against one another, belly to belly. Finally, the weaker individual yields and his head is pushed to the ground by the stronger animal.

Factors Influencing Strategy Success

Cost As the cost of fighting increases, the dove strategy should be favored. The potential cost of injury is greater in species that possess weapons, such as horns or sharp teeth. Therefore, the hawk–dove model predicts that contests between members of well-armed species should rarely escalate from display to battle (Figure 4.19). Conversely, the hawk strategy should be more common when the risk of injury is low. Accordingly, animals without weapons, such as toads (*Bufo bufo*), fight fiercely (Figure 4.20; Davies and Halliday, 1978). Although some individuals are injured, the risk of injury is much less than it would be if they had the prowess of lions.

Resource Value It seems reasonable that the intensity of fighting should increase with the value of the resource. After all, an animal that fights harder to win a richer prize will enjoy the same net benefit (value of resource minus cost of injury). This prediction seems intuitively obvious from human experience. You would probably put up more of a struggle if a mugger were stealing your life savings than you would to save a pocketful of change. We see this prediction fulfilled in many animal encounters as well. Offspring, a direct measure of fitness, are obviously of great value. It is not surprising, then, that a mother will fight fiercely to defend her young. It would also be expected that the most intense fights among males for access to females should take place during the interval when females are most likely to conceive. Indeed, red deer stags (*Cervus elaphus*) fight most fiercely and, as a result, are wounded most frequently during the period when most calves are conceived (Figure 4.21; Clutton-Brock, Guiness, and Albon, 1982). Territories also differ in value and it would be expected that animals would fight more intensely when the stakes are high. Likewise, among grassland populations of funnel-web spiders (*Agelenopsis aperta*), females expend more energy and risk greater injury in disputes over high-quality web sites (Riechert, 1986b).

As the value of the resource increases, we expect more hawklike behavior. Hawk can even become a pure ESS if the value of the resource

Figure 4.20 Two bullfrogs fighting. Game theory predicts that fights are more likely to escalate when the costs (risk of injury) are low.

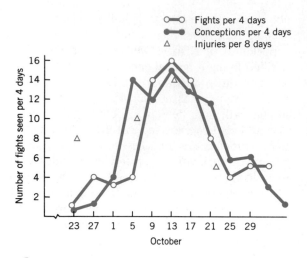

Figure 4.21 The number of fights among male red deer (open squares) compared to the number of conceptions during the rutting season. The number of fights and conceptions are indicated for 4 day intervals. The number of injuries per 8 day interval during the rut is indicated by stars. The number of fights and injuries peaks during the interval when conception is most likely. As predicted by game theory, male red deer fight harder when the value of the resource is greater. (From Clutton-Brock, Guiness and Albon, 1982.)

exceeds the cost of injury. Furthermore, when the value of victory is especially high, fatal fighting may evolve. To understand more completely why an animal might fight to the death over a resource, we must consider the lifetime consequences of a battle. What can be gained by retreating? One's future. In a sense, natural selection weighs the value of the resource against the value of the future. If the value of the future is close to or less than the value of the resource, the opponents have little to gain by giving up the fight. Thus, extreme hawklike strategies may evolve. When the value of the future is close to zero, the contestants have nothing to gain by retreating and will fight to the death. It should not be surprising, then, that fighting that leads to severe injury or death is usually seen when a

major chunk of the contestants' lifetime reproductive success is at stake (Enquist and Leimar, 1990).

Male elephant seals (*Mirounga angustirostris* and *M. leonina*), for instance, fight brutal and often bloody battles for the right to mate (Figure 4.22). Adult males bear many scars that serve as silent testimony to the intensity of combat. After a fight, when an adversary enters the sea, the water is often reddened with blood. The rivalry is intensified by the females. A female protests loudly when a male attempts to copulate with her (Cox and Le Boeuf, 1977). This attracts the attention of other males, who attempt to interfere. As a result, only the largest and strongest males are able to mate. (The benefit to the female is obvious.) All matings are performed by a few dominant males who defend harems of females. The duels between males are so strenuous that a male can usually be harem master for only a year or two before he dies. Thus, battles are titanic because a male's entire reproductive success is at stake (Le Boeuf, 1974; McCann 1981).

Asymmetries and Conditional Strategies

In the hawk–dove model we have been discussing, the strategies are played randomly with some frequency. It assumes that all hawks are equally matched competitors, as are all doves. But how often is this actually the case?

In real life, rivals are rarely true equals. Instead, they generally vary in some quality. Contests are, therefore, usually asymmetric.

Given that life is not fair and inequalities do exist, it seems reasonable that a competitor might be more successful if its strategy were adjusted to the situation—for instance, if a male played hawk to a puny opponent and dove to a mammoth rival. When a strategy is adjusted to the circumstances, it is called a conditional strategy. As might be expected, conditional strategies are highly successful and may invade and replace the mixed ESS of hawk–dove in which the strategies are played randomly.

Figure 4.22 Male elephant seals, scarred and bloody from battle. Fights between elephant seal bulls are brutal and often result in injury. Such titanic battles are predicted by game theory in situations where the value of the resource is very high. Only the most dominant bull will be harem master and leave offspring.

The inequalities, or asymmetries, among rivals can be grouped into three categories: (1) differences in the value of the resource to each contestant, (2) difference in the ability of each contestant to defend the resource, and (3) arbitrary differences between the opponents that affect the outcome even though they are unrelated to the resource value or ability to defend the resource (Maynard Smith and Parker, 1976). We will consider each type of asymmetry more closely to see how each might influence the outcome of a contest.

Asymmetry in the Resource Value It is not difficult to imagine situations in which the resource is more valuable to one of the contestants. Food, for instance, would be more precious to a starving animal than to a well-fed one. A female might vary in importance depending on whether she already had been inseminated. A territory might be more important to the resident because it knows the location of refuges or food sources there. As we have seen, game theory predicts that an animal should be willing to accept greater cost as the value of the resource increases, just as you might pay more for a more valuable item. The animal that considers the resource to be more valuable would be expected to fight harder in a hawklike battle or display longer in the war of attrition.

Food is a resource that often has different values to the contestants. A hungry animal may fight hard for its dinner, but after eating, it may be less willing to compete for a second course (Houston and McNamara, 1988). This prediction is borne out in northern harriers (*Circus cyaneus*), for instance. These small hawks catch small prey that requires some time to digest before they can consume another one. Once the owners of a territory have eaten, the value of prey temporarily decreases. During this interval, harriers are not as aggressive toward competitors that might steal food (Temeles, 1989)

Asymmetries in the value of food might also allow a ravenous, yet small, subordinate individual to win a contest over food from one that was larger but well fed (Popp, 1987). This has been shown to be true for bluethroats (*Luscinia s. svecica*) before migration. These small birds moult at their breeding site before beginning their long-distance flight. At this time the birds are generally quite lean and have little fat reserves. Body fat is important because during their flight food may be scarce and its availability is unpredictable. A fatter bird has a better chance of survival, and every bit of added fat is an extra safeguard. When food was provided at the moulting sites, the birds would fight frequently, with the winner chasing the loser from the food bowl. Individuals that had no body fat won more frequently than those that were close to their maximum recorded weight. Leaner birds were able to chase away larger birds that were already positioned at the feeding bowl. Presumably this is because the food was more important to the lean birds, so they were more highly motivated to win (Lindstrom et al., 1990).

A female may also be of different value to contestants, and her value may vary depending on whether she has been inseminated. The effort a male should put into a fight for a female depends on what he stands to gain in fitness. In bowl and doily spiders (*Frontinella pyramitela*), the winner in skirmishes over a female is the male who judges her to be worth the most. The judgment of a female's worth is based on the male's estimation of the number of eggs he can fertilize. The sperm of the first male to copulate with a female will fertilize her eggs and the percentage of eggs fertilized will increase, up to a point, with the duration of copulation. If copulation is interrupted, the next male can fertilize only the eggs remaining unfertilized. In this event, the first male loses only the unfertilized eggs. A virgin female generally has about 40 eggs. After 7 minutes of copulation, only about 4 eggs remain unfertilized, and after 21 minutes there may be only one unfertilized egg left. So, the value of the female decreases as copulation progresses. Only the copulating male would be able to assess her value, the number of unfertilized eggs, because an intruding male would not know the stage of copulation. Therefore, the intruder would assess the female to be worth 10 eggs—the average number of eggs fertilized in a male–female encounter (Figure 4.23).

The stage is set: The copulating male has an idea of how many eggs are left to be fertilized, some number between 40 and 0, but the in-

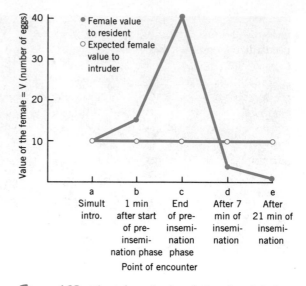

Figure 4.23   The value of a female bowl and doily spider to a resident and an intruding male. The value of the female is the male's assessment of the number of eggs she has left to be fertilized. This is an asymmetric contest since the value of the female to the resident and intruder is different. The value of the female to each male varies depending on the stage of copulation at which the intruder is introduced. (From Austad, 1983.)

truder assumes an average value of 10. If the contestants are deliberately chosen so that they are the same size, and therefore fighting ability, who wins? The data are consistent with the idea that each male first assesses the female's worth. Her value depends on the number of offspring he might expect to get from her. Then each male adjusts the intensity of his fight to reflect the female's value to him. Her worth to the intruder is constant because he does not know how long the other male has been copulating. To the intruder, the female is worth 10 eggs. For the other male the situation is different and constantly changing. He behaves as if her worth is judged by the length of time he has been copulating. The winner in a contest between two male bowl and doily spiders, therefore, depends on when the second male is introduced. The male who judges the female to be worth more, in terms of unfertilized eggs, is the winner. Steven Austad (1983) determined this by staging encounters. The first male would win until his assessment of the female's worth dropped below the intruder's estimation of her value, 10 unfertilized eggs. The results are shown in Figure 4.24. The first male won when the second male happened upon the scene immediately after the preinsemination period, when the first male had just determined that the female was sexually mature. However, if the intruder appeared after 7 minutes of copulation, the first male put up much less of a fight and generally lost. At this point he had already fertilized 90 percent of the eggs.

There are also examples in which a particular territory may have different values to the contestants. The worth of a territory may increase as the owner learns the location of good feeding

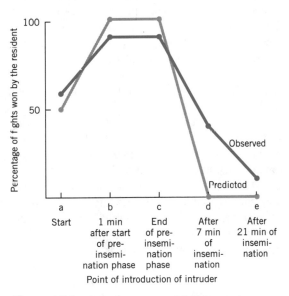

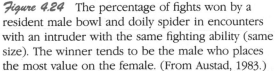

Figure 4.24 The percentage of fights won by a resident male bowl and doily spider in encounters with an intruder with the same fighting ability (same size). The winner tends to be the male who places the most value on the female. (From Austad, 1983.)

and nesting sites. We would predict the winner in a territorial dispute to be the individual that valued the property most. This expectation is supported in removal studies of male red-winged blackbirds (*Agelaius phoeniceus*). Territory owners were removed until replacement pairs moved in. The original owners were then released and they fought to reclaim their territory. We might assume that the newcomer would require a few days to become familiar with the territory. As he did, the property would become gradually more valuable to him. The original owner is already familiar with the benefits of the territory and places a high constant value on it. As expected, when the original territory holders were released after being held in captivity for up to 49 hours, the original owners nearly always won back their territories (Beletsky and Orians, 1987). However, when the original owners were retained for up to a week, the new residents usually defeated the former owners. The released owners were just as persistent in their attempts to recover their territories after they had been removed for 7 days as they were after 2 days. The difference in the outcome of the contests, therefore, seems to be due to a change in the behavior of the new residents. They were more willing to escalate contests as the territory became more valuable to them (Beletsky and Orians, 1989).

Asymmetry in Ability to Defend Resource In many instances, the contestants differ in their ability to defend the resource. One may be larger or heavier, have bigger weapons, or be a more skilled fighter. Characteristics such as these that bear on an opponent's ability to defend a resource describe its resource holding power (RHP). It seems intuitively obvious that contestants would increase their fitness by assessing their opponent's RHP and adjusting their fighting strategy to suit the occasion (Parker, 1974). It would be a bad idea, for instance, for Minnie Mouse to play hawk against Hulk Hogan. Size is one of the most important of the many qualities that could bear on an individual's ability to defend a resource. Larger animals are more likely to start a fight, persist longer, and eventually win the dispute than are smaller competitors. For example, female funnel-web spiders defend their territories from floaters who lack a territory and from neighboring territory holders. The contestants assess the weight of the opponent from movements on the web before any contact is made. The strategies played by each female depend on their relative weights. If the territory holder is larger, the intruder plays dove and immediately retreats. However, if the intruder is larger, she escalates the confrontation. The smaller web owner plays a strategy called retaliator—display unless attacked; if attacked, strike back (Riechert, 1984).

Weapons are also a component of RHP in those species that bear them. A male mountain sheep with small horns will defer to a competitor with larger horns (Geist, 1971). Likewise, male red deer judge the size of their competitor's antlers and retreat when then are outclassed (Clutton-Brock et al., 1982).

To minimize injury, animals should assess the RHP of their opponent before beginning a fight or very early in the fight (Parker, 1974). When the contestants know from the start the ability of their opponent to defend the resource, another fighting strategy, "assessor," becomes unbeatable (Maynard Smith, 1982). Assessor is a conditional strategy that follows the rule: "If one's RHP is greater than the opponent's, play hawk; if one's RHP smaller, play dove." However, assessor is only an ESS when the cost of assessing the opponent's RHP is less than the cost of losing a fight.

Arbitrary Asymmetry The differences between contestants that we have discussed so far have a fairly straightforward relationship to the fighting strategy of each contestant. We have seen that a competitor might follow a conditional strategy such as "play hawk if larger; dove, if smaller" or "persist longer when the resource is more valuable." The value of the resource and relative holding power are, therefore, called correlated asymmetries. The animal maximizes its fitness by responding to differences in resource value or RHP because the differences are likely to influence the outcome of the battle.

There are other differences between contestants that are not logically connected to the fighting strategy of the opponents, but which influence the outcome of the dispute. These arbitrary asymmetries (also called uncorrelated asymmetries) are, in essence, rules or conventions used to settle conflicts. Examples of arbitrary rules for settling differences among humans are flipping a coin or pulling straws. There is no reason that "heads" should win or that the "short straw" should lose. These are simply rules that are mutually agreed upon.

Prior ownership (or residency) is a common arbitrary asymmetry. Animals often appear to adhere to the principle that "possession is nine-tenths of the law." The bourgeois strategy sets rules for dealing with prior ownership: Play hawk if you had it first; otherwise, play dove. If the bourgeois strategy is added to hawk and dove strategies in a population, it does better than either. Bourgeois is, in fact, the only ESS in a population initially consisting of hawk, dove, and bourgeois strategies.

In the bourgeois strategy, the owner always wins, with barely a squabble, and the outcome of any dispute can be reversed by switching ownership. Studies of hamadryas baboons (*Papio hamadryas*) fit this bill. A male permitted to associate with a female for as little as 20 minutes was perceived as her "owner" by a second, newly introduced male. That the second male was deferring ownership was revealed when, several weeks later, he was permitted to associate with the female. When the tables were turned, the first male did not challenge the second male's ownership (Kummer, Gotz, and Angst, 1974). Another study demonstrated that prior ownership is more important in deciding the outcome of disputes than dominance. Generally, a dominant individual has an automatic right to the resource. However, when a subordinate hamadryas baboon has possession of a piece of food, a dominant individual will not attempt to take it. However, if food is thrown between two individuals, the dominant one does not permit the subordinate to have it (Sigg and Falett, 1985).

Male speckled wood butterflies (*Pararge aegeria*) also follow a bourgeois strategy when defending spots of sunlight that serve as mating territories. When an intruder approaches, a resident need only signal his ownership by an upward spiral flight. An intruder never challenges. When a resident is experimentally removed, he is almost instantly replaced by another male. Within minutes, the new male's ownership is established so firmly that he will win in an encounter with the former resident. An escalated contest occurs only when two males are experimentally tricked into joint "ownership" (Davies, 1978).

Although it may seem at first that the bourgeois strategy would always be used to settle territorial disputes, it is not. The territory holder may have won the property because of better fighting skill. In this case, the dispute is likely to be settled by asymmetries in RHP. In other cases, as we have seen, the territory may have different resource value to the competitors. The owner may know the location of abundant food, safe refuges, or nesting sites. This information would make the property more valuable to the resident, so the owner would be expected to fight harder than an intruder.

Summary

Evolution is a change in the composition of the gene pool. Alleles that enhance the reproductive success of individuals become more common in the gene pool because the individuals that leave the most offspring make a larger contribution to the gene pool. Individual fitness is a measure of the relative reproductive success of an individual. Inclusive fitness includes all offspring that exist because of the effort of a particular individual. Inclusive fitness, therefore, counts not only an individual's own offspring, but also relatives (devalued by the degree of relationship) that exist because of the individual's efforts.

Natural selection is differential reproduction. Individuals possessing genetic traits that enhance survival and reproduction leave more offspring than poorer competitors.

Some of the variation in populations is genetic. The sources of new variation are mutation and meiosis. Among the mechanisms that may preserve variation in a population are frequency-dependent selection, negative-assortative mating, and environmental variation. In frequency-dependent selection, the rare allele always has an advantage. Therefore, the allele increases in frequency until it is common, and then it decreases in frequency until it becomes rare enough to regain its advantage. These fluctuations maintain alternate alleles in the population. In frequency-dependent predation, for example, the most common form of prey is eaten in excess. This decreases the frequency of common alleles. In frequency-dependent reproduction, the rare male has a mating advantage. This increases the frequency of rare alleles. In negative-assortative mating, dissimilar individuals mate. If their differences are genetic, variation is preserved. Genetic diversity may also be maintained by variation in ecological conditions. If a species' range includes regions with different ecological conditions, some alleles may be favored in one environment while others are more successful in other regions.

Traits that allow individuals to survive and reproduce better than their competitors are called adaptations. Adaptive traits are appropriate for the prevailing environmental conditions.

Not all existing traits are adaptive. Some nonadaptive traits are maintained through pleiotropy, multiple effects of a single gene. The nonadaptive effects of the allele may be maintained in the population if the net effect of all its effects is beneficial. Other nonadaptive traits may be maintained because the genes responsible for them are located very close to a highly adaptive gene on the chromosome.

There are other reasons for the presence of nonadaptive traits. Natural selection may not have had time to eliminate them. Furthermore, natural selection is constantly being undermined by mutation and meiosis. Also, traits that are nonadaptive in existing conditions may have been adaptive in the previous environmental conditions.

Adaptation should always be treated as a hypothesis that must be tested. Methods of testing the hypothesis include comparisons and experimentation. In the comparative approach, traits of related species experiencing different ecological conditions are compared. Differences between such species are generally interpreted as the result of different selective pressures. Thus, the comparative approach identifies correlations between traits and ecological variables. The correlations must be explained and the hypotheses for the explanations tested by additional observations and/or experimentation.

Optimality theory considers both the benefits and the costs of actions. Natural selection is viewed as measuring costs and benefits in units of fitness and choosing the alternative that maximizes inclusive fitness. Fitness may be difficult for ecologists to measure. Therefore, in studies of optimality some factor assumed to be related to fitness, such as time or energy, is often measured.

An evolutionarily stable strategy (ESS) is one that cannot be bettered and, therefore, cannot be replaced by any other strategy once most of the members of the population have adopted it. A pure ESS is a single strategy adopted by all members of the population. A mixed ESS is a stable combination of several strategies.

It is important to think in terms of ESSs when the fitness payoffs for an action depend on what the other members of the population are doing. The mathematics underlying game theory is helpful in predicting behavior in situations in which the fitness payoff is frequency-dependent. Game theory models require that we take into account (1) all the existing strategies, (2) the frequency of each strategy in the population, and (3) the fitness payoff for each possible combination of strategies.

Game theory has been used extensively to analyze animal conflict. A simple model involves the hawk strategy (fight until seriously injured or until your opponent retreats) and the dove strategy (display but retreat if attacked). The benefit is the value of the resource. The cost is the risk of injury. A strategy's fitness payoff depends on the opponent's strategy. The precise mixture of strategies that would confer equal fitness to each depends on the relative values of costs and benefits.

The hawk–dove model assumes that the contestants are equally matched. In reality, this is rare. Contests are usually asymmetric. Conditional strategies, which are adjusted to the circumstances, are more successful in asymmetric contests. Asymmetries can exist in resource value, resource holding potential, or some arbitrary factor, such as prior ownership, that influences the outcome of the contest.

CHAPTER
5

Learning

*L*ive and learn. A truism, it would seem, but the truth is, some do and some don't. Some living things, tapeworms for instance, learn little in life. Others, for example, chimpanzees, learn a great deal throughout their lives, and if they are healthy, they are probably learning right up to the moment of their deaths. Without the ability to learn, some animals indeed would lose the struggle for survival in our changing world.

Learning and Adaptation

When we observe an animal in nature, we generally find that its behavior enhances its chances of surviving and reproducing. In the previous chapter, we focused on natural selection as a mechanism for producing behavior suited to environmental conditions. Some evolutionary theorists apply the concept of adaptation to only those circumstances in which there is evidence that natural selection shaped the behavior for the stated function (Williams, 1966). However, others define adaptation more broadly as all differences in traits that increase inclusive fitness. Adaptation may thus include not only traits with known genetic causes, but also the inherited potential for learning and even the learned behaviors themselves (Clutton-Brock and Harvey, 1979). Learning is a process in which the animal benefits from experience so that its behavior is better suited to environmental conditions (Rescorla, 1988). We will view learning here as adaptive and will consider some of the different ways in which it can shape behaviors to enhance an individual's ability to survive and reproduce.

Since learning ability is a product of natural selection, we should expect to see differences in the mechanisms and processes of learning among species (Kamil and Mauldin, 1988; Kamil and Yoerg, 1982). We would expect these differences because the environment and evolutionary history of a species will influence the degree to which a particular type of learning will alter fitness (the relative number of offspring left by an individual). In Chapter 2, we considered some biological constraints on learning and some ways that individuals of some species may be biologically prepared to learn certain things and not others.

Here we consider an example of species-specific differences in spatial learning and memory among Clark's nutcrackers (*Nucifraga columbiana*), pinyon jays (*Gymnorhinus cyanocephalus*), and scrub jays (*Aphelocoma coerulescens*), illustrating how ecology and evolution may influence a species' learning skills (Figure 5.1). These birds are among those that store seeds and later, when food is more difficult to find, recover them. So that their seeds will not be stolen by other animals, they hide them in small holes they dig in the ground and then cover over. The seeds are cached in this way in the autumn and used, as needed, through the winter and spring. Clark's nutcrackers may even use the cached seeds to feed to their fledglings in the summer (Vander Wall and Hutchins, 1983). This means that the birds must be able to return to cache sites months after the seeds were buried. Recovery of the seeds is quite an impressive feat for all three species, but particularly for Clark's nutcrackers, which may have 9000 caches covering many square kilometers (Balda, 1980; Vander Wall and Balda, 1977).

How do they find seeds buried so long ago? As surprising as it may seem to humans, many of whom have trouble remembering where they put their car keys, evidence is accumulating that the birds find the cached seeds simply because they remember exactly where they hid them (Balda, 1980; Vander Wall, 1982).

If natural selection shapes the learning ability of species, we might predict that those species that depend more heavily on cached food for survival would be better at recovering caches than other species that are less reliant on it. Clark's nutcrackers, pinyon jays, and scrub jays

a

b

c

Figure 5.1   Seed caching birds. Clark's nutcrackers (*a*), pinyon jays (*b*), and scrub jays (*c*) are birds that hide seeds in holes in the ground during the autumn and return to find the seeds during the winter and spring, when food is less plentiful.

are related species, members of the same family, that differ in their dependence on cached food. The nutcrackers live at high altitudes where there is little else but their stored seeds to eat during the winter and spring. Their winter diet consists almost entirely of cached pine seeds. Nutcrackers prepare for winter by storing as many as 33,000 seeds. Pinyon jays live at slightly lower altitudes where food is a little easier to find during the winter. Nonetheless, 70 to 90 percent of the pinyon jays' winter diet consists of some of the 20,000 seeds they cached in preparation for winter. In contrast, winter stresses are not this severe for a scrub jay. Scrub jays are smaller than the other two species and require less energy to maintain themselves. Furthermore, they live at much lower altitudes where food is somewhat easier to find in the winter. Scrub jays store only about 6000 seeds a year, and these account for less than 60 percent of the winter diet.

A study comparing cache recovery by Clark's nutcrackers, pinyon jays, and scrub jays found species differences in this ability that are correlated with their relative dependence on stored seeds (Balda and Kamil, 1989). In these experiments, the birds were permitted to store seeds in sand-filled holes in the floor of an indoor aviary. The floor had 90 holes, any of which could be filled with sand or blocked with a wooden plug. This arrangement allowed the experimenters to vary the position and number of the holes available for caching. Each bird's ability to recover caches was tested in two conditions. In one, only 15 holes were available for caching. In the other, all 90 holes were available. After a bird had placed eight caches, it was removed from the room. One week later, when it was returned to the aviary, all 90 holes were filled with sand. The bird then had to remember where it had buried the seeds. The accuracy of recovery was measured as the proportion of holes probed that contained seeds. All three species performed better than expected by chance alone. However, the pinyon jays and nutcrackers, the species that depend most heavily on finding their stored seeds to survive the winter, did significantly better than the scrub jays under both experimental conditions (Figure 5.2). The pinyon jays actually performed better than the nutcrackers. They were only slightly better when there were 15 holes available for caching, but much better when all 90 holes were available. It was suggested that this might be because pinyon jays tended to place their caches closer to-

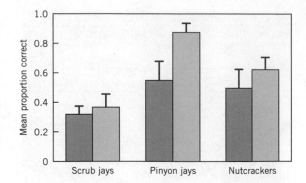

Figure 5.2 Histograms showing the accuracy with which scrub jays, pinyon jays, and Clark's nutcrackers find their caches. Each bird first hid seeds in sand-filled holes in an indoor aviary. The aviary floor had 90 holes, each of which could be filled with sand or plugged. In one experimental condition, indicated with the solid bars, there were 15 holes available for caching. In the other, indicated by striped bars, all 90 holes were available. After caching, the birds were removed from the room for a week. When they were returned, all 90 holes had been filled with sand. To recover the seeds the birds would probe the sand with their beaks. Accuracy was measured as the proportion of holes probed that contained seeds. Clark's nutcrackers and pinyon jays, the species most dependent on cached seeds for winter survival, were significantly better than scrub jays at recovering their caches. (From Balda and Kamil, 1989.)

gether than did the nutcrackers, resulting in their having to search a smaller area. The clumping of the caches was unexpected, but the researchers surmised that it might be an evolutionary result of differences in social structure. Pinyon jays are the only species of those tested that remain in flocks throughout the year, so caches are formed and recovered in the presence of others, which might prevent them from spreading their caches over a larger area. This idea could be tested by comparing related species that remain in flocks to see whether they, too, place their caches in clumps. In any case, the experiments suggested an influence of evolution and ecology on learning ability.

Definition of Learning

We can usually identify learning when we see it. However, learning can be difficult to define precisely. In the interest of simplicity, we begin with a broad definition and then refine it. First, then, let us say that learning is a process through which experience changes an individual's behavior. Clearly not all behavioral changes are due to learning. An athlete may run the last mile of a marathon at a slower speed that the first, but this is due to muscular fatigue, not learning. A person entering a dark movie theater from the sunlit street may at first trip over debris on the floor. However, within a few minutes the clutter is avoided. This change results from receptor adaptation. The receptors in the eyes had become less sensitive in the bright light, and it took a few minutes in dim light for their sensitivity to be restored so that the clutter could be seen. Maturation of the nervous system may also alter behavior. Our definition of learning must, therefore, specify that the behavioral changes it causes "cannot be understood in terms of maturational growth processes in the nervous system, fatigue or sensory adaptation" (Hinde, 1970). Instead, learned changes in behavior are due to experience (Thorpe, 1963). The experience may be with a certain stimulus situation (Hilgard, 1956) or practice (Kimble, 1961). Furthermore, most learning experiences must be repeated. Although some types of learning may occur with a single trial, most learning takes place gradually over several trials.

The behavioral change that results from learning may not be expressed immediately. For example, a person may memorize facts for a test and not demonstrate that learning has occurred until the day of the exam. The possibility of a time lag between the process of learning and the resulting change in performance requires that we modify our definition of learning once again. The change in behavior that results from learning is more accurately a change in the probability that a certain behavior will occur.

Categories of Learning

For convenience, learning is often grouped into different categories, but keep in mind that the relationship among them may be more complex than first appears. For example, they may overlap and the distinctions between them may not be clear-cut. As a result, those who study learning do not all agree on the nature of the categories or how many there should be. Nonetheless, categorizing types of learning has some usefulness, if only to assist communication. So, we will forge ahead, ignoring the arguments, and consider some of the distinguishing characteristics of some of the commonly recognized categories of learning. Our list of categories will not be complete. We will postpone discussions of examples of learning considered by some to represent distinct categories until later chapters. Imprinting, for instance, is discussed in the chapter on the development of behavior (Chapter 8).

HABITUATION

We usually imagine the change in behavior that results from learning as the addition of a new response to an individual's repertoire. However, in habituation, often considered the simplest form of learning, the animal learns *not* to show a characteristic response to a particular stimulus because, during repeated encounters, the stimulus was shown to be harmless. A bird must learn not to take flight at the sight of windblown leaves. Habituation is simply dropping responses instead of adding or changing them. It has been defined more precisely as a "relatively permanent waning of a response as a result of repeated stimulation which is not followed by any kind of reinforcement. It is specific to the stimulus" (Hinde, 1970). Persistent here means that the effects of habituation are rather long lasting, once they are acquired.

One finds habituation everywhere, occurring in simple unicellular protozoans as well as complex humans (Wyers, Pecker, and Herz, 1973). Consider just a few examples. The cnidarian *Hydra* living in a shallow dish of water will quickly retract its tentacles and shorten its body if the glass is tapped. But after a few such taps, its reactions slow and may stop (Rushforth, 1965). A snail crawling across a pane of glass will quickly retract if the glass is tapped. If the tapping continues, however, it will finally be ignored. Birds will at first be startled by a scarecrow but will come to ignore it or sit complacently on its arms, preening their feathers after a large meal of the farmer's grain.

Characteristics of Habituation

For an example that illustrates the essential characteristics of habituation we turn to the clamworm, *Nereis pelagica*. This marine polychaete lives in underwater burrows it constructs out of mud. *Nereis* partially emerges from the tube while feeding. However, certain sudden stimuli, such as a shadow that could herald the approach of a predator, cause it to withdraw quickly for protection. If the stimulus is repeated and there are no adverse consequences, the withdrawal response gradually wanes. It is obviously adaptive to withdraw when a predator is near, but a clamworm cannot capture its own prey while it is within its burrow. So, it is also adaptive to learn not to withdraw in the absence of danger.

The reduction of responsiveness of *Nereis* to recurring stimuli was easily demonstrated in the laboratory. A group of clamworms was maintained in a shallow pan of water and provided with glass tubes as burrows. A shadow was used to trigger the withdrawal response. The first time the shadow appeared, there was a high probability that each worm would withdraw. The sec-

ond time the shadow was presented, slightly fewer worms responded. The third presentation of the shadow elicited even fewer withdrawals. As shown in Figure 5.3, subsequent stimulus presentations resulted in a smaller and smaller percentage of the group responding. The animals began to learn that the stimulus was not associated with a threat (Clark, 1960).

If the clamworms were, indeed, learning that the stimulus was not harmful, then their loss of responsiveness was not because of sensory adaptation, the failure of the sensory receptors to register the presence of the stimulus. We know this because the time required for the clamworms to regain responsiveness is much longer than that needed for the receptors to recover. Sensory adaptation is a temporary condition that reduces the responsiveness of receptors during continuous stimulation. Receptor sensitivity is restored quickly after a few seconds' rest from stimulation. However, recovery from habituation, which is analogous to forgetting, generally takes much longer. Following 20 presentations of a moving shadow at 1-minute intervals, it took the clamworms 3 hours to completely recover the withdrawal response. Thus, the decline in

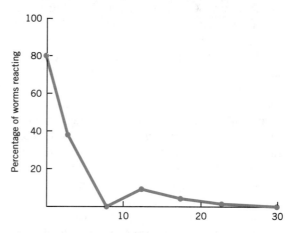

Figure 5.3 Habituation of the withdrawal response to a shadow in the marine worm, *Nereis*. In habituation, the simplest form of learning, the animal learns not to respond to frequently encountered stimuli that are not associated with reward or punishment. (Modified from Clark, 1960.)

responsiveness was not due to sensory adaptation.

Animals may also fail to respond to a repeated stimulus because of muscular fatigue. They may be simply too tired to do it again. Muscular fatigue as a reason for not responding can be ruled out by applying a different stimulus. Once again, the studies on habituation in *Nereis* are a good illustration of this. A group of worms that was habituated to a recurring shadow remained fully responsive to a mechanical shock and the reverse situation was also true. That is, when subjected to repeated mechanical shocks, the worms habituated and subsequent presentations of a moving shadow elicited the same degree of responsiveness as found in control animals that had been resting in the interim (Figure 5.4). Clearly, the worms were not simply too weary to withdraw since they remained responsive to a different stimulus (Clark, 1960).

The Adaptive Value of Habituation

The benefit of habituation is that it eliminates responses to frequently occurring stimuli that have no bearing on the animal's welfare without diminishing reactions to significant stimuli. Obviously it is important for *Nereis* to withdraw to the safety of its burrow when a shadow is that of a predator. However, if the shadow is encountered often without a predator's attack, it is probably caused by something harmless, perhaps a patch of algae repeatedly blocking the sun as it undulates with the waves. In this case, responding to the shadow each time it appears would leave the worm little time for other essential activities such as feeding or reproducing. These nonessential responses would not only compete with vital activities, they would also be a waste of energy. Habituation is one of the mechanisms that focuses attention and energy on the important aspects of the environment (Leibrecht and Askew, 1980).

As an example of how habituation is adaptive, consider the role it plays in shaping the natural pattern of escape responses of birds in the field.

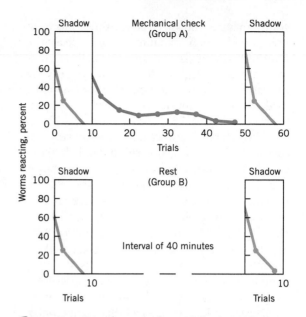

Figure 5.4 Stimulus specificity of the loss of responsiveness that occurs in habituation. Notice that both groups of worms habituated to the shadow during trials 1–10. A mechanical shock was then administered to group A at 1-minute intervals. The worms initially responded to the new stimulus, showing that the loss of the response to the shadow was not a result of fatigue. Group B rested for 40 minutes while group A received the mechanical shocks. Both groups were equally responsive to the shadow during the last 10 stimulus presentations. One can conclude that the animals learned not to respond to a specific stimulus, the shadow in this case. (Modified from Clark, 1960.)

The young of birds such as turkeys, chickens, and pheasants need not be taught escape behaviors, such as crouching, at the sight of objects moving overhead. Not unlike Chicken Little in the children's story, the chicks show protective responses to a great variety of objects, only a few of which are dangerous. On the other hand, the adults respond only to the image of a predator such as a hawk flying overhead. The narrowing of the responsiveness of adults is due to habituation. This occurs because innocuous stimuli are so much more common in an animal's environment than harmful stimuli. As the young mature, they are often exposed to harmless stimuli such as falling acorns, leaves, or nonpredatory birds, and they gradually learn not to cringe at the sight of them. However, because predators are rare, the young do not habituate to them and continue to respond to their appearance (Figure 5.5).

Although this explanation is intuitively logical, Schleidt (1961 a,b) verified that habituation could shape behavior in this way. He showed that models of various shapes—a gooselike silhouette, a circle, and a square—were all effective in eliciting alarm calls from young turkey chicks during the initial 2 days of testing. When any one of these models was presented frequently, the number of alarm calls it elicited decreased. The chicks remained responsive to stimuli they encountered only occasionally.

Aggressive behavior, as well as escape behavior, will habituate. This is adaptive because aggressive behavior uses a great deal of energy and exposes the animal to potential harm. For instance, when male stickleback fish (*Gasterosteus aculeatus*) first establish their territories, a great deal of aggressive display occurs at the boundaries. However, over time, the neighbors habituate to one another and the frequency of aggressive display decreases (Van den Assem and van der Molen, 1969). Similarly, the aggressive behavior of another territorial fish, the convict cichlid (*Cichlasoma nigrofasciatum*), has been shown to habituate (Peeke, Herz, and Gallagher, 1971). In laboratories, the aggressive response of a male Siamese fighting fish, *Betta splendens*, can be elicited by holding a mirror to the side of his tank. Seeing himself in the mirror, he responds with an aggressive display that makes him appear larger and more brilliant. If the mirror is left in place, however, and the image is not followed by an attack, he soon loses interest (Clayton and Hinde, 1968).

CLASSICAL CONDITIONING

Even if it were true that "you can't teach an old dog new tricks," it should be possible to teach it to perform an old trick under a new set of circumstances. We see such a relationship in classical conditioning.

Characteristics of Classical Conditioning

The principle of classical conditioning was first stated by Ivan Pavlov, a Russian physiologist whose main interest was actually the digestive activity of dogs. Pavlov's work began with the common observation that a dog salivates at the anticipation of food. Pavlov reasoned that it was unlikely that salivating before actually receiving the food was an inborn response. He thought, reasonably enough, that the animal had learned to associate the sight or smell of food with the food itself, and that the salivation was due to such learning. Thus, in anticipation of food, the dog would begin to salivate to a new stimulus that signaled the arrival of food—a connection had formed between an old response and a new stimulus.

Pavlov then began to experimentally investigate the association. In order to quantify the response, a small opening, or fistula, was made in the dog's cheek in such a manner that the saliva would drain into a funnel outside the dog's body where it was collected and could be measured. The hungry dog was then harnessed into position on a stand where it could be subjected to various stimuli. One stimulus tested was powdered food and, as expected, the dog salivated when the food was blown into its mouth. Then, immediately before the food was offered, Pavlov presented the dog with another stimulus, one that did not initially cause salivation, such as the sound of a bell. At intervals over several days, the dog was exposed to this pair of stimuli, first the sound and then immediately the food. Then Pavlov checked to see whether the dog salivated at the sound of the bell alone. It did. The results of this experiment are shown in Figure 5.6. Although the tone originally elicited no salivation, after 30 paired stimulations, the dog's saliva indeed would flow when the sound was presented without the food. As the number of trials increased, there was an

Figure 5.5 Narrowing of responsiveness by habituation as it might occur in nature. (*a*) The young of many species innately show escape responses to a wide variety of objects. (*b*) Most of these objects and organisms are harmless and because of the abundance of innocuous stimuli in the natural environment, the young animal gradually habituates to their presence. (*c*) Predators are rare and infrequently encountered, and their image still elicits an escape response.

increase in the amount of saliva secreted and a decrease in the length of time it took the dog to respond (Pavlov, 1927).

In more general terms, what happened? To begin, an animal has a particular inborn response to a certain stimulus. This stimulus is called the unconditioned stimulus (US) because the animal did not have to learn the response to it. A second stimulus, one that does not initially elicit the response, is repeatedly presented immediately before the US. After several pairings, the second stimulus is able to elicit the response. (The conditioned response actually differs slightly from the original response, but this point need not concern us.) This new stimulus is now called the conditioned stimulus (CS) because the animal's response has become conditional upon its presentation. The whole phenomenon is called a conditioned reflex. "Conditioned" is an unfortunate translation from the Russian. The term should be "conditional," implying that the behavior is conditional upon something happening.

The Conditioned Stimulus as a Signal

The order of the presentation of the US and CS is important. The CS must precede the US. The CS serves as a signal that the US will appear; a cue is of little value if it occurs after the fact. There is also an optimum interval between the CS and US if the CS is to serve effectively as a signal. Experiments have shown that an interval of 0.5 seconds between the stimuli is most effective (Hilgard and Atkinson, 1967).

Furthermore, useful signals are specific; they predict that a particular event or stimulus will follow. A signal is useless if it merely indicates that any one of a dozen events may follow. Therefore, it should not be surprising that for classical conditioning to occur, the CS must precede that US more often than it does other stimuli (Rescorla, 1967, 1988).

Extinction

If the CS (the tone in the experiment described) is presented frequently without being followed

Acquisition   Extinction

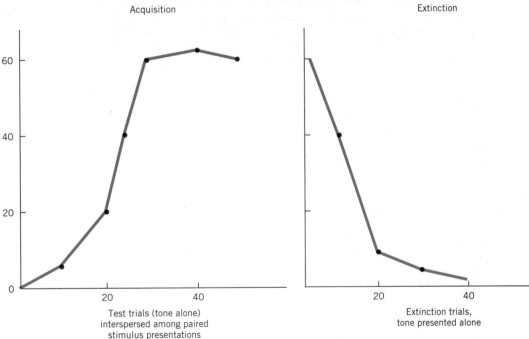

Test trials (tone alone)
interspersed among paired
stimulus presentations

Extinction trials,
tone presented alone

Figure 5.6 Classical conditioning of the salivary reflex in a dog to a tone. The dog is prepared so that the saliva it produces can be collected and measured. In this experiment, food serves as the unconditioned stimulus since it elicits the desired response (salivation) without prior training. During the acquisition trials, a tone is sounded immediately before the presentation of the food. Periodically, the animal's response to the tone alone is determined. Because of the repeated pairing of the tone and the food, the dog gradually begins to salivate in response to the tone alone. During extinction, the tone is sounded without food reinforcement. There is a progressive decrease in the volume of saliva produced. (Data from Pavlov, 1927.)

the previous chapter that game theory predicts
that damaging aggressive contests are most likely
to evolve when the value of the resource is great.
This prediction is borne out among blue gouramis. It is essential that males are successful in
territorial defense because females rarely mate
with a male without a territory. It is reasoned,
therefore, that any mechanism that might increase the likelihood of successful defense
would be favored by natural selection. A conditioned response to signals indicating the approach of a rival might prepare a male for battle
and give him a competitive edge. In nature, as
the rival approached, he would inadvertently
send visual, chemical, or mechanical signals that
territorial invasion was imminent.

Hollis has shown that male blue gouramis
that have been classically conditioned to a signal
that predicts an encounter with a rival are more
successful in aggressive contests. Pair members,
chosen to be of similar body size and level of
aggression, were housed on opposite sides of a
divided aquarium. One member of each pair
was trained so that a 10-second light (the CS)
preceded a 15-second viewing of a rival (the
US). The CS and US were also shown to the
other member of the pair, but their presentations were not paired. During the test trials, the
light signal was given before the barrier separating the members of the pair was removed,
allowing them to interact. The conditioned
males were superior in territorial defense. They
approached the territorial border with their fins
already erect in the frontal display. During the
territorial fights, they delivered significantly
more tailbeats and bites than did their competitors (Hollis, 1984). The conditioned males may
be the winners because the light (CS) increased
the level of androgens, male sex hormones
known to heighten aggressiveness in many species. A more aggressive male has a better chance
of winning the battle and defending his territory,
thereby increasing his chances of mating (Hollis,
1990).

OPERANT CONDITIONING

Characteristics of Operant Conditioning

When a behavior has favorable consequences, the probability that the act will be repeated is increased. This relationship may result because the animal learns to perform the behavior in order to be rewarded. This type of learning has been named operant conditioning to emphasize that the animal operates on the environment to produce consequences. Again, just as in classical conditioning, the timing of events is critical. In this case though, the behavior must be spontaneously emitted, not elicited by a stimulus as it is in classical conditioning, and the favorable result, or reinforcement, must follow it closely. In a sense, a cause-and-effect relationship develops between the performance of the act and the delivery of the reinforcer.

As long ago as 1855, before the term operant conditioning was coined, Herbert Spencer described an animal learning a response through operant conditioning. His account is interesting because it illustrates how operant conditioning may perfect motor skills in nature:

> Suppose, now, that in putting out its head to seize prey scarcely within reach, a creature has repeatedly failed. Suppose that along with the group of motor actions approximately adapted to seize prey at this distance . . . a slight forward movement of the body [is caused on some occasion]. Success will occur instead of failure . . . On recurrence of the circumstances, these muscular movements that were followed by success are likely to be repeated: what was at first an accidental combination of motions will now be a combination having considerable probability.

B. F. Skinner later devised an apparatus that is still used to study operant conditioning in the laboratory. Typically, a hungry animal is placed into a "Skinner box" and must learn to manipulate a mechanism that yields food. For example, a hungry rat placed in a Skinner box will move about randomly, investigating each nook and cranny. Eventually, it will put its weight on a lever provided in the box (Figure 5.7). When the lever is pressed, a bit of food drops into a tray. The rat will usually press the lever again within a few minutes. In other words, the rat first presses the lever as a random act, and when it is rewarded, the probability of the act being repeated increases. The apparatus has been modified for use with pigeons, which learn to peck a key for a food reward.

By definition, a stimulus that alters the probability of a behavior being repeated is called a reinforcer. In the experiments described so far, positive reinforcers were used. Positive reinforcers are those that increase the probability of a

Figure 5.7   A rat in a Skinner box. The hungry animal explores the box. Eventually it presses the bar. This automatically results in the delivery of a small food pellet to a tray. The rat quickly consumes the food. The food reward increases the probability that the rat will press the bar again.

behavior being repeated. Examples include food offered to a hungry rat or drink to a thirsty one. Negative reinforcers are those that increase the probability of a response when they are removed. If an unpleasant or painful stimulus stops when an animal performs a certain act, it is likely to repeat that action. For example, a rat will learn to push a panel to stop an electric shock (Mowrer, 1940) or push a bar to turn off a bright electric light for 60 seconds (Keller, 1941). We generally think of reinforcers as rewards. Skinner, however, preferred the term reinforcer because "reward" implies sensations that might be intuitively inferred but are not measurable. Reinforcement, then, is best defined operationally; it alters the probability of a response.

Shaping

Animals can be taught novel, and sometimes, complex acts in order to receive a reward. The process by which this occurs, called shaping, has been likened to the way in which a sculptor molds a lump of clay (Skinner, 1953). When creating a sculpture, the artist changes the formless mass into a final masterpiece through a series of minute changes in the former condition. In shaping, at first any gross approximation of the desired act is rewarded, but later reinforcements require closer and closer approximations to the desired goal.

For example, if you were to train a sea lion to jump from the water through a hoop, you would first reward it for approaching the hoop. When it learned to approach, then it would be rewarded only when it swam through the hoop. You would raise the hoop on successive trials until it was clear of the water and the sea lion would have to jump to receive the reward.

Animal trainers have used these techniques to create amazingly elaborate behaviors to titillate audiences. Years ago, there was a popular television show starring Priscilla the Fastidious Pig. Among Priscilla's tricks were turning on the radio and eating at the table. Unlike her slovenly cousins in the barn, Priscilla was exceedingly neat and could demonstrate this by picking clothes up from the floor and placing them in a hamper. When offered a variety of foods, she wisely chose the one produced by the sponsor of the show. Priscilla even participated in a quiz show by answering questions from the audience by turning on flashing lights that indicated "Yes" or "No." There was nothing exceptional about Priscilla's abilities; because pigs grow so quickly, a new Priscilla was trained and substituted every few months (Breland and Breland, 1966).

Human behaviors can also be shaped, sometimes unwittingly. In fact, much to their dismay, parents often unconsciously shape unpleasant behaviors in children. For example, many parents teach their children to shout and whine. A mother who is preoccupied with daily chores may ignore her three year old's quiet request. Typically, the child raises its voice slightly and repeats the request, until the mother responds. When she has adjusted to the new average intensity of her child's vocalizations, only louder requests are answered. Then the child learns that the mother responds even more quickly if whining is added. This little embellishment so irritates her that she responds immediately, just to make it stop. So now the child shouts and whines regularly.

Extinction

When reinforcement is withheld, the response rate will gradually decline, just as the strength of the conditioned reflex decreased when the CS was presented many times without the US. The process is called extinction.

Reinforcement Schedules

In real life, reward seldom follows every performance of an act. Instead the reward is usually intermittent. For example, a honeybee can only obtain nectar during the restricted interval each day that a particular flower is open. Therefore, once it has visited a flower, it must wait a day before it will be rewarded at that site again. The frequency with which rewards are offered is called the reinforcement schedule. Partial reinforcement schedules vary either the ratio of non-reinforced to reinforced response or the time period between successive reinforcements. Alternatively, reward may be doled out aperiodically (Ferster and Skinner, 1957).

Careful studies have revealed the effect of the schedule of reinforcement on the response strength. Each reinforcement schedule has predictable effects on the rate of response and on how long the animal will continue responding when it is no longer rewarded. We will highlight just a few examples. A continuous reinforcement schedule, in which each occurrence of the behavior is rewarded, is best during the initial training to establish and shape a response. A fixed ratio schedule, one in which the animal must respond a set number of times before reinforcement is given, usually results in very high response rates because the individual, in essence, determines how quickly it will be rewarded. The faster it responds, the sooner it completes the number of responses required to receive the reward. A fixed ratio reinforcement schedule is similar to piecework in factories, in which the employee gets paid when a certain number of items are completed. The very high production rate generated by piecework is a reason that employers like the system and unions generally oppose it. In a variable ratio schedule, the number of responses required for reinforcement varies randomly. This also generates very high response rates because the individual is rewarded for fast responding. In addition, the response tends to persist even if reward is withheld for awhile because the variability prevents any obvious response patterns from becoming established. An individual gambling with a slot machine experiences a variable ratio reinforcement schedule. The slot machine averages a certain level of payoff, but the reward is unsystematic and variable. This type of reinforcement schedule has been blamed for the persistence of gambling in those who have become "addicted" to games of chance.

LATENT LEARNING

We know that rewards increase the probability of an act being repeated. However, according to many learning theorists, there should be no change in the frequency of an unrewarded behavior. There are situations, though, in which animals learn without any obvious reward. If a rat that is neither hungry nor thirsty is placed in a maze that holds no reward, the rodent will still investigate the myriad paths, running down blind alleys and retracing its steps. Having been permitted to explore the maze, this rat will then learn to run the maze in fewer trials than inexperienced rats when food is finally offered as a reinforcement (Tolman and Holzik, 1930). Clearly the rat learned some of the characteristics of the maze during its unrewarded explorations, even though the knowledge was not put to immediate use; that is, it was latent.

Does latent learning deserve its own category? Thorpe (1963) defines it as "the association of indifferent stimuli or situations without patent reward." The essential difference between latent learning and trial-and-error learning (operant conditioning) is the absence of reward in latent learning. However, although this point was argued for many years (see Munn, 1950, for review), some learning theorists no longer consider reinforcement as essential for trial-and-error learning, thus eradicating the difference between these forms of learning. However, Hinde (1970) differentiated them by placing more emphasis on the unexpressed nature of latent learning. He suggested that both trial-and-error and latent learning involve memorization of characteristics of a situation but only the former translates this mastery into immediate performance.

The Adaptiveness of Latent Learning

However one chooses to categorize latent learning is academic; learning through exploration is possible and it is adaptive for animals in their natural setting. For example, a young coyote and its siblings venture out of their den to explore

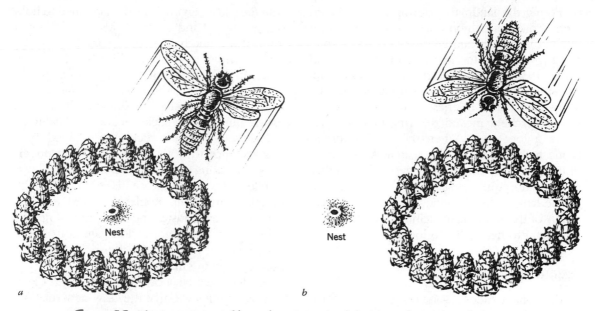

Nest

Nest

a

b

Figure 5.8   The importance of latent learning gained during exploration to the orientation of the female digger wasp. (*a*) While the wasp was within her nest, a ring of pine cones was placed around the nest hole. When she emerged, she made an aerial study of the nest area and flew off. During this orientation flight, she memorized the visual landmarks by which she would locate her nest upon returning. (*b*) While she was gone, the ring of pine cones was moved. The digger wasp returned to the center of the pine cone ring rather than to the nest hole. Obviously she had learned the landmarks for relocating her nest during her short flight immediately after leaving although that learning remained latent until her return. (Redrawn from Tinbergen, 1951.)

under the watchful eye of their parents. As they do so, they become familiar with the terrain. Where are the good places to find a meal? Where are hiding places that will ensure escape from a predator? As they wander around, looking and sniffing, they gain a detailed knowledge of their environment that increases the chance of survival later.

Although the value of latent learning seems intuitively obvious, laboratory studies have shown that familiarity with the terrain is an asset for survival (Metzgar, 1967). Pairs of white-footed mice (*Peromyscus leucopus*) were released into a room with a screech owl (*Otus*

asio). One of the duo had the opportunity to explore the room for a few days before the test. The other mouse had no experience in the room. On 13 of the 17 trials, the owl caught a mouse. Eleven of these were from the group that was unfamiliar with the room, clearly showing that knowledge of the environment increased the ability to evade the predator.

Exploration may also have functional significance in orientation. An animal that ventures out on long-distance foraging trips must be able to return home (Figure 5.8). For example, the digger wasp (*Philanthus triangulum*) is one of many organisms that learns the location of home by noting the landmarks during an orientation flight. This was demonstrated by placing a ring of 20 pinecones around the nest entrance while a wasp was within her nest (Tinbergen and Kruyt, 1938). When she left on a hunting trip, she spent the first 6 seconds flying around the immediate area of her nest. While she was gone, the researchers moved the ring of pinecones about a foot away. When the wasp returned home with her booty, she searched the center of the pinecone ring for her nest hole. On this, and all subsequent tests, the wasp was misled by the landmarks, showing that the brief exploration of the immediate vicinity of the nest was important to her ability to locate her nest.

INSIGHT LEARNING

In cartoons a sudden solution to a problem is suggested by a light bulb going on over someone's head. People often express the experience as, "Aha, I get it." You may have puzzled over a problem for several days, or weeks, and suddenly the answer flashes into your mind. Some learning theorists call this insight learning. It is characterized by its suddenness; it seems to occur too quickly to be the result of a trial-and-error process (Thorpe, 1963).

A famous example of insight learning comes from the behavior of Wolfgang Köhler's (1927) chimpanzees, particularly one called Sultan. In one experiment, Sultan first learned to use a stick as a tool to extend his reach and rake in a banana on the ground outside his cage. Having mastered this trick, he was given two sticks that when put together, end-to-end, were just long enough to reach the fruit. Sultan tried unsuccessfully to reach the reward with each of the sticks. He even managed to prod one stick with the tip of the other until it touched the banana, but since the sticks were not joined, he could not retrieve the fruit. For over an hour, Sultan persistently tried, and failed, to get the banana. Finally, he seemed to give up and began to play with the sticks. Later an intellectual flash apparently struck Sultan. As the chimp was playing

with the sticks (Figure 5.9), he happened to hold one in each hand so the ends were pointed toward one another. At this point he realized that the end of one stick could be fitted into the other, thus lengthening the tool. Immediately, he ran to the bars of his cage and began to rake in the bananas. As he was drawing the banana toward him, the two sticks separated. That Sultan quickly recovered the sticks and rejoined them was evidence to Köhler that the chimp understood that fitting two bamboo poles together was an effective way to increase his reach far enough to obtain the fruit. Köhler believed that the chimp displayed insight behavior in that he was able to apply the information gained from the experience of playing with the sticks to solve a problem, getting the bananas.

Köhler's explanation of chimps' problem-solving abilities was that they saw new relationships among events, relationships that were not specifically learned in the past, and they were able to consider the problem as a whole, not

Figure 5.9 Sultan playing with sticks. After playing with the sticks, Sultan gained insight into how to obtain the banana placed beyond his reach. The sticks could be fitted together end-to-end to increase his effective reach.

just the stimulus–response association between certain elements of the problem. It has been suggested, for instance, that chimps form a mental representation of the problem and then mentally apply trial-and-error patterns to it. An animal could be "thinking" through the possible responses and evaluating the possibility of success of each trial based on its past experience. The solution might seem sudden to observers because they do not have access to the animal's mental processes.

Other researchers explain sudden problem solving, such as that shown by Köhler's chimps, as the result of associations among previously learned components. It has been argued, for instance, that chimps that moved boxes and then climbed on them to reach a banana had previously acquired two separate behaviors—moving boxes toward targets and climbing on an object to reach another object. This idea was tested on pigeons. Pigeons do not usually do either of these actions, but they can learn to do them. When a similar insight learning situation was set up for pigeons, it was shown that *only* pigeons that had learned both of these two actions were able to solve the problem of reaching the banana. In one experiment, pigeons were reinforced for pushing a box toward a green spot on the floor, but not for pushing a box if there was no green spot on the floor. In a separate experiment the pigeons learned to sit on a box in order to peck at a banana that was suspended overhead. Pigeons trained to do both of these activities were then placed in a room lacking a green spot on the floor, but with both a box and a banana suspended from the ceiling. The behavior of these birds was remarkably similar to that of Köhler's chimps. Although they stretched and turned beneath the banana at first, they suddenly pushed the box under the banana, climbed on it, and pecked the banana. However, when pigeons that were trained to peck a banana but not to climb on a box in order to do it were placed in the same situation, they repeatedly stretched toward the banana but never reached it. So, learning to climb was an important component of the behavior. Other pigeons were trained to push a box in a certain direction for a reward, but they were never trained to climb and peck at a banana. When these birds were then placed in the same situation, they pushed the box around the room aimlessly, never climbing atop it to peck the banana. Therefore, learning to push a box in a certain direction for a reward was also an important component of the response. It was concluded, then, that seemingly insightful behavior might be built from specific stimulus–response relationships (Epstein et al., 1984). So we see that although no one disagrees about the importance of prior experience in insight learning, there is controversy about the role of prior experience.

Investigation of insight learning has proven very difficult. For one thing, it occurs very quickly and it is impossible to predict when it will happen. Thus, the experimenter can easily miss the moment of enlightenment. In addition, small changes in the arrangement of the problem situation can alter the outcome greatly. But an even greater problem is in the interpretation of the observation. All the observer sees is the solution to the problem. It is difficult to know *how* the animal achieved that solution. So, the interpretation of insightful behavior can lead to considerable controversy, as we will see later in the chapter.

LEARNING SETS

Sometimes an animal solves a problem in a flash as a result of prior experience with similar tasks. The animal seems to have learned the *principle* of that type of problem. This "learning how to learn" has been called a learning set. For example, humans establish learning sets in solving math problems when they have solved other problems of a similar type.

For an illustration of the formation of a learning set, we turn to the experiments of Harry Harlow (1949). Harlow repeatedly presented a monkey with two objects, each covering a small food well. However, the well under only one of these objects contained food. The positions of the objects were alternated randomly, but the monkey quickly learned which object covered the food. Then two new objects were introduced, and once again the animal had to learn which object hid the food. As new pairs of objects were introduced, the monkey was provided with new, but similar problems. With each new challenge, it took fewer presentations before the animal reliably chose the correct object. After several hundred object-discrimination problems of this sort, the monkey was able to choose the rewarded object by the second try about 97 percent of the time (Figure 5.10). The monkey had adopted the strategy of win–stay, lose–shift: If the object is rewarded, stick with it; but if it is not, choose the other one and stick with *it*. The monkey had formed a learning set.

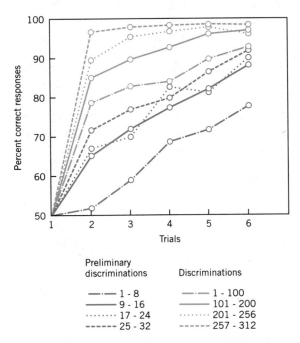

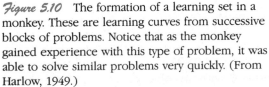

Figure 5.10 The formation of a learning set in a monkey. These are learning curves from successive blocks of problems. Notice that as the monkey gained experience with this type of problem, it was able to solve similar problems very quickly. (From Harlow, 1949.)

A variation on this learning set experiment is a repeated reversal problem. In this case, if the objects to be discriminated were a square and a circle, the square might be rewarded until the animal learned that association, and then the circle would yield the reward. Once the animal learned to choose the circle, the square would be designated "correct." The strategy for this type of problem would be win–shift, lose–stay.

The ability to form a learning set would be adaptive. In real life, an animal encounters a multitude of problems that it must learn to solve. Many of them are variations on a theme, just as the trials in the learning set experiments just described were. A learning set would reduce tremendously the amount of time wasted solving each similar problem separately.

SOCIAL LEARNING

Some organisms are able to learn from others. The possibility for learning in this way is much greater in social species because they spend more time close to others. Social learning by observation and imitation is assumed to be the basis of many human behaviors (Bandura, 1962; Meltzoff, 1988). You can probably think of many examples of this from your own family life. For instance, a child may mimic a parent's technique for caring for a younger sibling by handling a doll in the same way.

The adaptive value of social learning is fairly clear. It saves some of the time and energy that might be wasted as an individual learned about the business of survival by trial and error. Individuals of some species, Norway rats (*Rattus norvegicus*), for instance, learn what and where to eat from others of their kind (Galef, 1990). Others may learn to avoid dangerous situations by watching their fellows. For example, rhesus monkeys (*Macaca mulatta*) can learn to fear and avoid snakes by watching other monkeys show fear of snakes (Mineka and Cook, 1988). We see, then, that although each member of a population may have the capacity to learn appropriate responses for themselves, it is often more efficient and perhaps less dangerous to learn about the world from others (Galef, 1976). In other instances, interaction with adults is *critical* to learning appropriate behaviors by the young. An example would be juvenile Bewick's wrens (*Thryomanes bewickii*) that perfect their crude rendition of the song used for territorial defense by countersinging with a neighboring adult (Kroodsma, 1974).

There are many observations of birds and mammals in nature that seem to involve social learning. Some of the most commonly cited examples involve traditions, learned behaviors that appear in only one of several social groups of a single species in their natural habitat. For instance, a larcenous tradition was begun in England around 1921 when certain birds learned to break into milk bottles to steal the cream that, in those days, floated to the top. This nefarious technique spread throughout the continent as other birds acquired the habit (Fisher and Hinde, 1949; Figure 5.11). There are many examples of habits spreading through groups of

primates. Jane van Lawick-Goodall (1968) observed young chimpanzees in nature learning to use sticks and stems to gather termites from their holes by imitating their mothers or other adults. In addition, the chimps at the Delta Primate Center in Louisiana had poles to play with for several years before one chimp discovered that a pole could be angled against a wall and climbed to gain access to an object on a shelf. Others soon learned this new trick. The chimps could now avoid the electrified wires around trees and climb up the trees to get the leaves.

Although observations such as these are often interpreted as examples of social learning, we must not leap to unwarranted conclusions. Disagreements have developed over the mechanism of development of one of the most frequently cited examples of social learning, snow monkeys' habit of washing food. As the story goes, a young female snow monkey of Japan, named Imo, developed new techniques for the treatment of sweet potatoes and wheat, food provided by the researchers who study the social behavior of the snow monkeys. First Imo dis-

Figure 5.11 The tradition among birds of opening milk bottles to steal sips of cream spread rapidly from one area in England. This trick may have been spread by social learning.

Figure 5.12 The tradition of washing sweet potatoes in the sea was begun by a young Japanese snow monkey and it spread rapidly to other members of the troop.

covered that washing the sweet potato in the sea not only cleaned it, but also enhanced the flavor by lightly salting it (Figure 5.12). One of Imo's playmates observed her and followed suit. Then Imo's mother caught on. And so the tradition spread, usually from youngsters to mothers and siblings. When the youngsters became mothers, their offspring imitated the behavior as if food had always been cleaned this way. Several years later Imo started a new custom. The researchers spread wheat on the sand. The snow monkeys then had to painstakingly pick each grain from the sand. One day Imo tossed a handful of sand and wheat into the sea. The sand sank but the wheat floated so that it could be scooped up from the surface. This trick was also picked up by most monkeys in the troop during the next few years (Kawai, 1965; Kawamura, 1959).

Processes other than social learning have been suggested for the maintenance and spread of the food-washing habit in this troop of monkeys. The caretakers of the monkeys may have inadvertently rewarded food washing. The monkeys' only source of sweet potatoes is the caretaker. Since the food washing interests research-

ers and amuses tourists, the caretakers give more sweet potatoes to those members of the troop that were known to wash them than to those who did not. Thus, differential reinforcement may maintain the behavior. The habit may have spread because the monkeys near those who washed their sweet potatoes (who, by the way, were likely to be relatives) were also close to the caretaker and the source of reinforcement (Green, 1975). Although the habit clearly spread throughout the population, we do not know whether the snow monkeys learned to wash food by imitating others or because they were trained to do so by the caretaker.

Some traditions are due to social learning, but not because the individual observed another performing the activity and then imitated it. As we will see in Chapter 8, rats can learn dietary preferences from other rats. They learn what to eat, not by watching others, but by smelling their breath. In one experiment, a "demonstrator" rat ate food flavored with cocoa or cinnamon. The demonstrator was then anesthetized and placed 2 inches away from the wire cage of an awake "observer" rat. Although the demonstrator slept through the demonstration, the observer later showed a preference for the food the demonstrator had eaten (Galef, 1990).

There are surprisingly few examples of animals learning to do something just by seeing it done. In one such example, budgerigars learned a specific technique for removing the cover from a food dish. In this experiment, one budgie would watch a demonstrator remove the cover in one of three ways: using its feet, pecking with its bill, or pulling with its bill. When presented with a similar food dish, the observer used the same technique it had just witnessed (Dawson and Foss, 1965; Galef, Manzig, and Field, 1986).

PLAY

Play is fun. We know play when we do it and when we see animals doing it (Figure 5.13). We can usually identify animal play with little hesi-

Figure 5.13 Lion cubs playing. Play is a vehicle through which many things are learned.

tation. During a visit to the zoo, we even recognize play in species we have never seen before. This is true even though diverse forms of activity are classified as play. One type of play behavior is play fighting or play chasing. We have all been amused by the friendly tussles of kittens and puppies as they chase, wrestle, and pounce on one another. A second form of play is exercise. Foals kick up their heals and gallop. Young primates, including human children, may swing and roll and slide. Infant Mountain gorillas are particularly fond of sliding. They begin on their mother's body and then graduate to dirt banks and tree trunks. Polar bear cubs climb ashore only to leap back into the water. In the third form of play, objects are manipulated. When first presented with a novel object, a young animal typically explores it by touching, sniffing, or viewing it from different angles. After the initial sensory investigation, the object may become a toy (Fagan, 1981). Whichever way play is expressed, the attribute that pervades all play, and by which we most commonly identify it, is our subjective judgment of a lack of seriousness.

Although it is easy to spot play, it is difficult to define. One reason that play eludes a simple definition is that there is no specific behavior pattern or series of activities that exclusively characterize play. Play borrows pieces of other behavior patterns, usually incomplete sequences and often in an exaggerated form. The following are key features of play behavior that should be included in any definition:

1. **Intermingling of acts from functionally different types of behavior.** Play consists of elements drawn from other behavior patterns juxtaposed in new sequences. Polecats and badgers rapidly alternate prey-catching movements and aggressive behavior. During their play wrestling, the opponents may reverse roles often so that the submissive animal "tries out" dominance and vice versa (Eibl-Eibesfeldt, 1956). A playful mongoose mixes components from hunting and sexual behavior (Rensch and Ducker, 1959).

2. **Incomplete sequences.** In its normal context behaviors usually have several actions that follow one another in a predictable order. An adult cat would normally stalk a mouse and then strike, but a frolicking kitten may simply pounce on a leaf. An infant rhesus monkey may mount a companion without intromission. The play fights of polecats do not include the two extreme forms of attack ("sustained neck biting" and "sideways attack") or the two extreme fear responses ("defensive threat" or "screaming") (Poole, 1966). When a dog is aggressive, it bares its teeth and growls. Its hair stands on end, adding to its ferocious appearance. However, in a play fight, the snarl is not accompanied by the raising of hair.

But why do animals play? In other words, what function does this frolicking serve? The hypotheses for the long-term significance of play might be grouped into three categories (Bekoff and Byers, 1981):

1. Physiological: physical training for strength, endurance, and muscular coordination.

2. Social: practice of social skill such as grooming and sexual behavior; establishment and maintenance of social bonds.

3. Cognitive: learning of specific skills or improving overall perceptual abilities.

Since the focus of this chapter is learning, we will concentrate on what animals may learn through play. Animals learn a great deal while manipulating objects as toys. Remember that Köhler's chimp Sultan did not figure out how to retrieve a banana that was out of reach until he had played with the sticks. It was during play that the chimps at the Delta Primate Center learned to use a stick for pole vaulting, a skill that was later put to use for reaching leaves on tall trees and escaping from their confines. The information gained from manipulating "toys" may provide the experience that will be used for insight learning.

A common hypothesis for the adaptive function of play is that it affords animals the oppor-tunity to practice skills that will be essential to later survival (Caro, 1988). Hunting games, for instance, may help perfect the movements of prey catching such as stalking, throwing prey down, and shaking prey. Some examples of these actions are familiar. Kittens stalk leaves and pounce on balls of yarn. Puppies chase sticks and often "shake the life" from them as they would a prey. Even the neck bite used to kill prey may be practiced in play. For example, the neck bite of meerkats (*Suricata*, a South African viverrid related to the civet and mongoose) is the same whether the animal is playing, copulating, or killing its prey (Eibl-Eibesfeldt, 1975).

The play fighting of young animals may be practice for the battles of adults that establish dominance hierarchies and defend territories. In the fury of a play fight, there is no serious biting and no threat behavior. The cry of a polecat that was unintentionally bitten by its opponent will inhibit further aggression (Eibl-Eibesfeldt, 1975). Larger, older, and more dominant animals seem to handicap themselves in tussles with weaker playmates. Strength and skill are matched to that of the partner (Fagan, 1981). Some animals seem to practice territory defense as well. Irenaus Eibl-Eibesfeldt (1975) had tame polecats that would defend wastebaskets and hide under a blanket only to leap out at playmates that passed. Young deer and goats vie for possession of an area in a game reminiscent of king of the mountain (Darling, 1937).

Although it is frequently suggested that an adaptive function of play is to help develop adult predatory and fighting skills, the hypothesis is as yet unproven. Many observations of play, including the ones previously mentioned, suggest that animals play in ways and at times when we might expect them to if this were a function of play. In other words, it often looks as if play were designed for this function. The "argument by design" approach is probably the best supporting evidence we have. Other approaches to testing the idea, such as by experimentally altering early play experience or by looking for correlations between natural variation in play within populations and the expression of other types of behavior, have failed to provide clear answers. For instance, in one study the frequency of social play of each of four coyote (*Canis latrans*) pups was recorded when they were 20 to 34 days old. Differences among the pups in the amount of time they spent playing were noted. During the next 10 days, each pup was given 10-minute opportunities to kill a live mouse. The measure of predatory success was, therefore, the number of trials in which the mouse was killed. No correlation was seen between degree of playfulness and predatory success (Vincent and Bekoff, 1978). However, there are problems with this study, so the results are not conclusive (Martin and Caro, 1985). First, a sample size of four animals is too small to be sure that a significant correlation would not have been found if more pups were studied. Second, the natural differences in play may have been caused by additional, unconsidered factors. Third, the measure of predatory success, whether the mouse was killed, is so crude that it may have missed fine differences in predatory skill among the pups. Fourth, predatory success was determined within days of the play experience, perhaps too soon for effects to have become apparent. So, we see that the adaptive functions of play are still open to question.

Animal Cognition and Learning Studies

Beginning in the early 1980s, some scientists began to wonder whether animals have mental experiences—thoughts and feelings, for instance (Bateson and Klopfer, 1991; Griffin, 1981, 1982, 1984; Hoage and Goldman, 1986; Mellgren, 1983; Ristau, 1991). But how could we ever *know* whether other animals think or whether they are aware? Some scientists believe that

learning studies may shed some light on the issue of animal cognition, so we will reconsider some of the studies we have discussed.

Earlier in the chapter, the controversial nature of the interpretation of insight learning was mentioned. Why is this controversial? As we have seen, some workers believe that insight learning shows that the animal is *thinking*, and an animal that thinks about objects or events can be said to experience a simple level of consciousness (Griffin, 1991). An animal that thinks must also form mental representations of objects or events. Therefore, insight has been used as evidence of animal awareness or cognition. But not everyone agrees that animals, or even that some animals, might be aware. Some might be willing to accept the idea of awareness in a chimp, but not in a pigeon. However, in the experiments described earlier, if the chimp solved the problem by thinking, wasn't the pigeon also thinking?

Some researchers suggest that the ability to form a learning set might be evidence of awareness. According to Harlow (1949), learning set formation "transforms the organism from a creature that adapts to a changing environment by trial-and-error to one that adapts by seeming hypothesis and insight." Learning sets, he said, "are the mechanisms which, in part, transform the organism from a conditioned response robot to a reasonable rational creature."

Individuals of a few species have been able to learn certain concepts (abstract ideas). Surely this must show that animals form mental representations of objects and events. Pigeons (*Columba livia*) are able to form natural concepts such as "tree" or "water" or "human." They can recognize water, for instance, in various forms—a droplet, a river, a lake (Hernstein, Loveland, and Cable, 1976; Mallot and Siddall, 1972; Siegel and Honig, 1970).

Only one nonhuman animal has vocally demonstrated the ability to form concepts. This individual is Alex, an African Grey parrot (*Psittacus erithacus*) (Figure 5.14). We all know that parrots can be trained to talk, but most of us would

Figure 5.14 Alex, an African Grey Parrot who has learned several concepts. Alex knows the concept of same/different, an idea once thought to characterize only humans and their closest primate relatives.

guess that they are mimicking their trainers. This is certainly not true of Alex. He has learned to identify, request, refuse, or comment on more than 80 different objects. Furthermore, he has used language to show that he understands certain concepts. One such concept is quantitative. He can tell you how many items are in a group for collections of up to six items (Pepperberg, 1987a). Perhaps more surprising is his ability to understand the ideas of sameness and difference, an ability previously thought to characterize humans and maybe their close primate relatives (Premack, 1978). Alex demonstrated that he understood the concepts in experiments in which he was shown two objects at a time. The objects would differ in one of three qualities: color, shape, or material. He might be shown a yellow rawhide pentagon and a gray wooden pentagon or a green wooden triangle and a blue wooden triangle. Then Alex would be asked, "What's same?" or "What's different?" A correct answer to the first of these questions would be

to name the category of the similar characteristic. When he saw the first of the previous examples, Alex would have to answer "shape," not "pentagon." In the second example, a correct response to, "What's different?" would be "color," not "green." When shown objects he had seen before, Alex correctly identified the characteristic that was the same or different 76.6 percent of the time. He was also shown pairs of objects that he had never seen before and 85 percent of the time he correctly identified the characteristic the was the same or different (Pepperberg, 1987b). The experiments with Alex suggest that nonprimates can learn concepts and that we should continue our efforts to understand how other species might handle ideas.

And what about play? Are animals really having *fun*? Keep the question of animal cognition in mind as you become more familiar with other aspects of animal behavior. We will return to this issue again, especially in the discussion of animal communication.

Summary

Although learning does not always fit into distinct categories and the learning that occurs in nature may not fit into a pigeonhole, it is sometimes helpful to emphasize the unique characteristics of certain ways in which learning occurs by categorizing it. Some common classes are:

1. **Habituation.** The animal learns *not* to respond to a specific stimulus because it has been encountered frequently without important consequence. The loss of responsiveness can be distinguished from sensory adaptation and muscular fatigue. Habituation is adaptive because it conserves energy and leaves more time for other important activities.

2. **Classical conditioning.** The animal learns to give a response normally elicited by one stimulus (the unconditioned stimulus) to a new stimulus (the conditioned stimulus) because the two are repeatedly paired. The conditioned stimulus (CS) must precede the unconditioned stimulus (US). If the CS is presented many times without the US, the response to the new stimulus will be gradually lost. This is called extinction.

3. **Operant conditioning.** The frequency of some action is increased because it is reinforced. Novel behaviors can be introduced into the repertoire through shaping. During shaping, the reward is made contingent upon closer and closer approximation to the desired action. Sometimes not every response is reinforced. The frequency with which the reward is doled out is called the reinforcement schedule.

4. **Latent learning.** Latent learning occurs without any obvious reinforcement and is not obvious until sometime later in life. The information gained through exploration is an example.

5. **Insight learning.** This type of learning occurs rapidly and without any obvious trial-and-error responses. The animal seems to draw on information gained in previous similar situations to arrive at a solution to the problem.

6. **Learning sets.** During the formation of a learning set, the animal is learning how to learn. The subject learns to solve problems more quickly because it has an idea of the principle of the problem as a result of experience with other similar tasks.

7. **Social learning.** The animal learns from others. Social learning may occur by watching the behavior of another, but it may also occur by simpler means. Some customs spread rapidly throughout populations of animals by social learning, but some traditions may arise through individual learning.

8. **Play.** Play is expressed in a variety of ways: mock fighting and chasing, exercise, and manipulation of toys. Hypotheses for the function of play include that it improves physical condition, is important in developing social skills and bonds, and helps animals learn or perfect specific skills.

Many scientists have begun to wonder whether animals have mental experiences. Some studies of learning can be interpreted as supporting the idea that animals form mental representations of external objects or events and that they can form concepts.

CHAPTER
6

Physiological Analysis— Nerve Cells and Behavior

*J*ust as the house cat raises its paw to strike, the cockroach (*Periplaneta americana*) dashes across the floor and disappears into a tiny crevice. If we were to film this sequence and then replay it in slow motion, we would see that the cat's paw was still several centimeters away from the cockroach when the intended victim turned its body away from the cat and ran. How did the cockroach detect the predator in time to take evasive action? Kenneth Roeder, and later Jeffrey Camhi and colleagues, studied the escape response of cockroaches and discovered that these remarkable, albeit unlovable, house guests respond to gusts of air that are created by even the slightest of movements by their enemies (Camhi, 1984, 1988; Camhi, Tom, and Volman, 1978; Roeder, 1967). Cockroaches, it turns out, have numerous hairlike receptors that are sensitive to wind, and these receptors are located on two posterior appendages called cerci (singular, cercus; Figure 6.1). When these wind-sensitive receptors are stimulated, they alert the nervous system of the cockroach, and within a matter of milliseconds, the cockroach turns away from the direction of the wind and running behavior is initiated. Cockroaches, however, do not respond to all gusts of wind. Indeed, if they did, they would spend their entire lives on the move, because even as they walk, they themselves generate wind that could potentially stimulate their wind-sensitive receptors. How do cockroaches respond to biologically relevant stimuli, such as the small gusts of wind from a predator, and ignore nonthreatening stimuli, such as the self-generated wind caused by their own walking movements?

Before we can understand the selective behavioral responses of animals such as cockroaches to different stimuli in their environment, we must first understand how the nervous system is put together. What types of cells make up the nervous system and how do these cells transmit information? How does information received at specialized receptors (e.g., the cockroach's hairlike receptors on each cercus) travel through the nervous system, reach certain muscles, and produce a behavioral response such as turning away? What neural mechanisms control rhythmic behavior such as the running that begins soon after the cockroach's abrupt turn? Furthermore, how do animals determine the direction of a stimulus, so that, in the case of

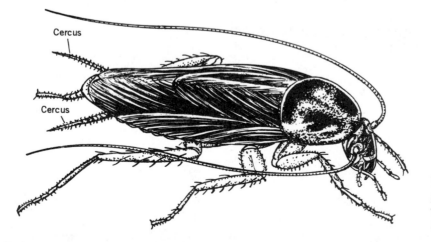

Cercus

Cercus

Figure 6.1 A cockroach, showing the two cerci, each with approximately 200 wind-sensitive hairs.

cockroaches, they run away from cats, rather than straight into them? These are some of the specific questions that we address in this chapter on the underlying neural basis of behavior. More generally, our examples should demonstrate that nervous systems have evolved so that animals can quickly (1) detect pertinent events in their environment, (2) choose appropriate responses to such events, and (3) coordinate the parts of their bodies necessary to execute the responses.

Concepts from Cellular Neurobiology

THE NEURON

The fundamental unit of the nervous system is the nerve cell, or neuron. Calling the neuron a fundamental unit is a bit misleading, though. When we say that the basic unit of a chimney is a brick, we mean that a chimney is constructed of many identical units called bricks. However, even though a nervous system is made up of neurons, those neurons are anything but identical (Figure 6.2). They range from a few microns (1 micron = 10^{-6} meters) in diameter to well over 100 microns in diameter. The largest neurons, in fact, are clearly visible to the naked eye.

Nerve cells can be classified on the basis of their function into three groups. Neurons that carry signals from a receptor organ at the periphery toward the central nervous system (in vertebrates, the brain and spinal cord, and in invertebrates, the brain and nerve cord) are called sensory or afferent neurons. Those that carry signals away from the central nervous system to muscles and glands are called efferent or motor neurons. Interneurons, found within the central nervous system, connect neurons to each other. Take a look at these three types of neurons in the cockroach (Figure 6.3). At the base of each of the wind-sensitive hairs on the cercus is a single sensory neuron that relays pertinent information from the external environment into the central nervous system. In the central nervous system, the sensory neuron makes contact with an interneuron; in this case, the interneuron is described as giant because of its exceptionally large diameter. This giant interneuron ascends the nerve cord to the head. Before reaching the head, however, the giant interneuron makes contact with an interneuron in the thoracic area which, in turn, makes contact with motor neurons that relay messages to the hind leg muscles. (An advantage of studying the neural basis of behavior in an invertebrate animal such as the cockroach is that it is sometimes possible to identify the individual neurons involved in a specific behavior, particularly a pattern of behavior associated with escape. Because escape requires fast action, the neurons involved in escape responses are often large in diameter to permit the rapid conduction of messages. The end result is that these large neurons are somewhat easier to identify than their smaller counterparts.)

Although there is no such thing as a typical neuron, it is possible to identify characteristics common to some neurons. We will use a motor neuron (Figure 6.4), in this case from a mammal, as our example. The nucleus of a motor neuron is contained in the cell body (soma), and small-diameter processes (neurites) typically extend from the cell body. In the traditional view, information enters a neuron via a collection of branching neurites and then travels down a single, long neurite to be passed on to other neurons. The neurites that receive the information are called dendrites; the single, long, cablelike neurite that transmits the information to other neurons is called an axon. In vertebrates, some axons have a fatty, insulated wrapping called myelin. In our example of the motor neuron, the axon ends on a muscle or a gland (an effector), which brings about the animal's behavioral response. Although the terms dendrite and axon are well established in the literature, it is now recognized that the flow of information through a neuron is often not so neatly divided into

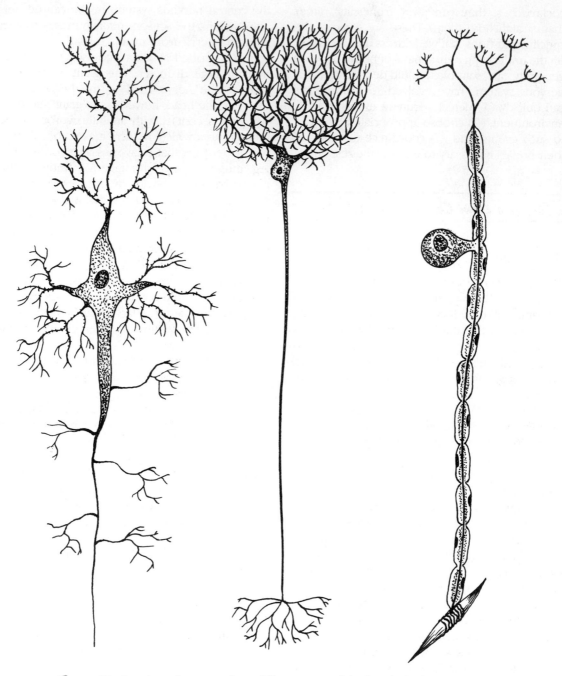

Figure 6.2   A variety of neurons from different parts of the human body.

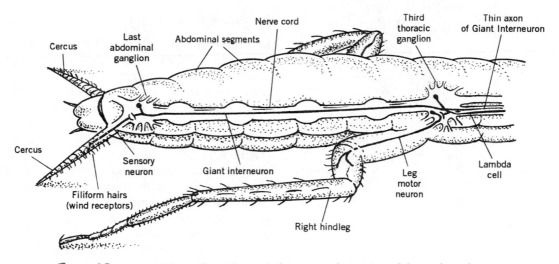

Figure 6.3 Some of the cells in the underlying neural circuitry of the cockroach escape response. Note that the sensory neuron, also called the wind-receptor neuron, leaving the cercus makes contact with a giant interneuron in the central nervous system which, in turn, makes contact with another interneuron that synapses with a motor neuron in the leg. (Modified from Camhi, 1980 with new information from Ritzmann and Pollack, 1986.)

separate receiving and transmitting processes. We will continue to use the terms, keeping in mind that in many cases, the specific direction of informational flow has not actually been demonstrated.

THE MESSAGE OF A NEURON

A neuron's message is an electrochemical signal caused by electrically charged atoms, called ions, moving across the membrane. Ions can cross the membrane of a nerve cell by means of either the sodium–potassium pump or ion channels. The pump uses cellular energy to move three sodium ions (Na^+) outward, while transporting two potassium ions (K^+) inward. An ion channel, on the other hand, is a small pore that extends through the membrane of a nerve cell. There are different types of channels, each type forming a specific passageway for only one or a few kinds of ions. Whereas some are passive ion channels that are always open, others are active ion channels (also called gated channels) that open in response to a specific triggering signal. Triggering signals may include the presence of chemicals (neurotransmitters) in the space between the membranes of neurons, changes in the charge difference across the membrane (the membrane potential), changes in the concentration of intracellular calcium ions (Ca^{++}), or any combination of these factors (see later discussion). Once an ion channel is open, the ion it admits may move across the membrane in response to either a concentration gradient (ions tend to move from an area where they are highly concentrated toward an area of lesser concentration) or an electrical gradient (because ions are charged atoms, they tend to move away from an area with a similar charge and toward an area of the opposite charge).

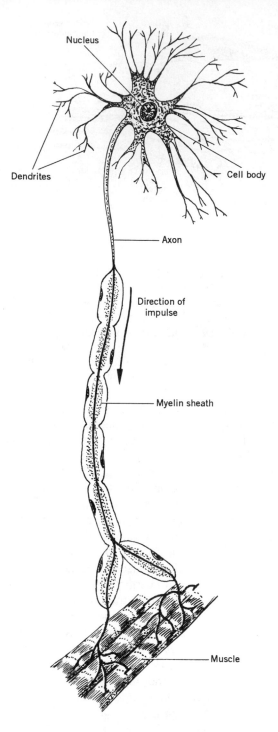

The Resting Potential

In a resting nerve cell, one that is not relaying a message, the area just inside the membrane is about 60 millivolts (mV) more negative than the fluid immediately outside the membrane. This charge difference across the membrane is called the resting potential of the neuron. A membrane in this resting state is described as polarized (Figure 6.5).

The resting potential results from the unequal distribution of certain ions across the membrane. The concentration of Na^+ is much greater outside the neuron than within. The concentration of K^+ shows just the opposite pattern, being greater inside the cell than outside. Certain large, negatively charged proteins are held within the neuron either because the membrane is impermeable to them or because they are bound to intracellular structures. These proteins are primarily responsible for the negative charge within the neuron.

Most of the active ion channels in the membrane of a resting neuron are closed, but passive channels are, of course, open. Because most of the passive channels are specific for K^+, the membrane is much more permeable to K^+ than it is to other ions. As previously mentioned, the concentration of K^+ is greater inside the cell than outside. This sets up a concentration gradient across the cell membrane, and as a result, some K^+ move toward the area of their lower concentration—that is, they pass through the channels in the membrane to the outside of the cell. With the movement of K^+ outward, the

Figure 6.4 A motor neuron. The soma or cell body maintains the cell. The dendrites are extensions specialized for receiving input from other cells. The axon is specialized to conduct the message away from the cell and to release a chemical that will communicate with another cell. In vertebrates, some axons are covered by a fatty myelin sheath.

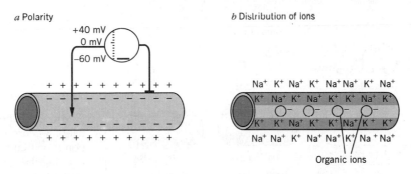

Figure 6.5 The resting potential. (*a*) In the resting state, the inside of an axon is more negative than the outside. (*b*) This charge difference is caused by the unequal distribution of ions inside and outside the cell. There are more sodium ions outside and more potassium ions inside. In addition, there are large negatively charged proteins (organic ions) held inside the cell, giving the interior an overall negative charge.

outside of the cell becomes more positively charged, and the inside more negatively charged. The initial concentration gradient has thus established an electrical gradient or separation of charge across the membrane. Any K^+ in the channels of the membrane will be repelled by the positive charge outside the cell and attracted to the negative charge inside the cell. At some point, the two forces—an electrical gradient drawing K^+ inward and a concentration gradient pushing K^+ outward—are equally balanced and there is no further net movement of K^+.

Why is Na^+ more concentrated outside the neuron? Although Na^+, like K^+, is attracted by the negative charge inside the neuron, the membrane is relatively impermeable to it, so only a few can leak through. Furthermore, the sodium–potassium pump actively removes Na^+ from within the cell, transporting it outward, against electrical and concentration gradients.

The Action Potential

The nerve impulse (action potential) is an electrical event that lasts for about 1 millisecond.

The action potential consists of a wave of depolarization followed by repolarization that spreads along the axon. The depolarization, or loss of the negative charge within, is caused by the inward movement of Na^+. However, the repolarization, or reestablishment of the negative charge within the neuron, is caused by K^+ leaving the cell (Figure 6.6).

Let's see how depolarization and repolarization occur. The membrane becomes slightly depolarized when some of the active Na^+ channels open and Na^+ enters the cell, drawn by both electrical and concentration gradients. The positive charge on Na^+ slightly offsets the negative charge inside the cell, and the membrane becomes slightly depolarized. If the depolarization is great enough, that is, if threshold is reached, voltage-sensitive sodium channels open, and sodium ions rush to the interior of the cell. At roughly the peak of the depolarization, about 0.5 milliseconds after the voltage-sensitive sodium gates open, they close and cannot reopen again for a few milliseconds. Almost simultaneously, voltage-sensitive potassium channels open, greatly increasing the membrane's permeability to K^+. Potassium ions then leave the cell,

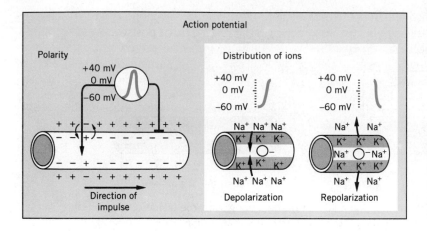

Figure 6.6 The action potential of a neuron. Depolarization is caused by sodium ions entering the cell and repolarization is caused by potassium ions exiting the cell.

driven by the temporary positive charge within and by the concentration gradient. The exodus of K^+ restores the negative charge to the inner boundary of the membrane. In fact, enough potassium ions may leave to temporarily make the cell's interior even more negatively charged than usual, a condition called hyperpolarization. Notice that although the original resting potential is eventually restored, the distribution of ions is different. This situation is corrected by the sodium–potassium pump.

This depolarization and repolarization of the neuronal membrane spreads rapidly along the axon, generated at each spot in the same manner that it was started. The local depolarization at one point of the membrane opens the voltage-sensitive sodium channels in the neighboring region of membrane, thereby triggering its depolarization. The net result is that a wave of excitation travels down the axon as each membrane section passes on its excitation, in the form of increased sodium permeability, to its neighbor.

The action potential that results from these quickly changing membrane permeabilities travels along an axon in much the same manner as a human wave travels around a sports stadium. A human wave is a kind of chain reaction that sports fans sometimes create to keep themselves

amused. It begins when people in one section of the stadium stand up, lift their hands above their heads, and then almost immediately sit back down. While these fans are in the process of sitting back down, the fans immediately to their left stand up, raise their arms, and then almost immediately sit back down. This behavior passes through every stadium section as each group becomes the stimulus for the group to its left, thus creating a traveling human wave.

Immediately after an action potential, the neuron cannot be stimulated to fire again for 0.5–2 milliseconds because the sodium channels cannot be reopened right away. At the start of this absolute refractory period, no amount of stimulation can generate an action potential. During the latter part of this interval, the relative refractory period, stimulation must be greater than the usual threshold value to generate an action potential. Although the refractory period is brief, it is biologically significant because it sets a maximum on the rate of firing.

The action potential is a neuron's long-distance signal. Once it is triggered, it travels along the axon with no loss in magnitude. Therefore, we describe the action potential as an all-or-none phenomenon.

If an action potential is always the same, how can differences in the intensity of stimuli be sensed? The intensity is encoded in the firing rate of the neuron and by the number of neurons responding. For example, if we place our hand on a hot stove rather than on a warm one, the firing rate of a neuron in our hand that responds to heat or pain may be increased. Also, touching a very hot stove will activate more neurons than will touching a warm stovetop because the thresholds of neurons that register heat vary—more neurons reach their thresholds and fire at higher temperatures.

SYNAPTIC TRANSMISSION

Now that we understand the basis of a neuron's message and how an action potential travels within a neuron, we will consider how that message gets from one neuron to another. The gap between neurons is called a synapse, and at a specific synapse, information is typically transferred in one direction, from the presynaptic neuron to the postsynaptic neuron. (The descriptor presynaptic or postsynaptic refers to the direction of information flow at a specific synapse—of course, a particular cell may be presynaptic with reference to one cell and postsynaptic with respect to another cell.) There are two major categories of synapses, electrical and chemical (Figure 6.7).

Electrical Synapses

In an electrical synapse, the gap between the neurons is small, only about 2 nanometers (1 nm = 10^{-9} m), and is bridged by tiny connecting tubes that allow ions to flow directly from one neuron to the other. When an action potential arrives at the axon terminal of the presynaptic neuron, Na^+ enters this terminal, causing a potential difference between the inside of this cell and the postsynaptic cell. As a result of the difference, positively charged ions, mostly K^+,

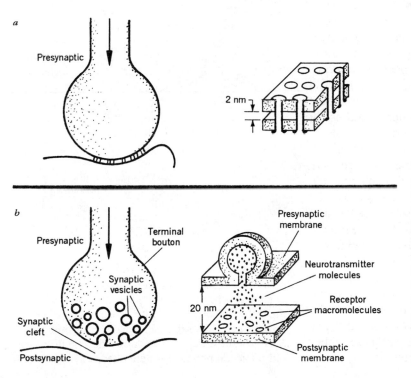

Figure 6.7 The gap between neurons is called a synapse. There are two types of synapses, electrical (*a*) and chemical (*b*). In an electrical synapse, the distance between neurons is very small, and ions can flow directly from one neuron to the next. In a chemical synapse, however, the distance between neurons is greater, and one neuron affects the activity of the other by releasing chemicals into the gap between them.

move from the presynaptic cell through the tiny tubular connections into the postsynaptic neuron. These newly arriving ions may sufficiently depolarize the postsynaptic cell to induce an action potential.

Electrical synapses are known for their speed of transmission. Whereas a signal may cross an electrical synapse in about 0.1 millisecond, durations on the order of 0.5 or 1 millisecond are typical at chemical synapses. Not surprisingly, then, electrical synapses are often part of the neural circuitry underlying patterns of behavior when shear speed is essential, such as the escape responses exhibited by animals when confronted by a predator.

Chemical Synapses

Chemical synapses are more common than electrical synapses and are characterized by a larger space between the membranes of the two neurons (typically 20–30 nm). Rather than information being transmitted from one neuron to the next via direct electrical connections, information is transmitted across a chemical synapse in the form of a chemical called a neurotransmitter. There are several steps in this process, and these steps account for the slower speed of transmission at a chemical synapse than at an electrical one. First, the action potential travels down the axon to small swellings (terminal boutons) at the end of the axon. At the terminal boutons, the action potential causes the neuron to release a neurotransmitter from small storage sacs called synaptic vesicles. It is thought that the release of neurotransmitter occurs because the action potential opens voltage-sensitive calcium ion channels. The calcium ions that flood to the inside initiate events that cause the synaptic vesicles to move toward the membrane of the terminal bouton. Once there, the vesicles fuse with the membrane and dump their contents into the gap between the cells (the synaptic cleft). The neurotransmitter then diffuses the short distance across the cleft and binds to receptors on either another neuron or a muscle cell.

The neurotransmitter that crosses the synaptic cleft will increase the permeability of the membrane of the postsynaptic cell to specific ions. This results in either excitation or inhibition of the postsynaptic neuron, depending on the particular ions involved. In the case of excitation, the neurotransmitter causes the opening of channels that allow both Na^+ and K^+ to pass through. Although some K^+ will move out, it is outnumbered by the Na^+ moving in. This causes a slight temporary depolarization of the membrane of the postsynaptic cell. This depolarization is called an excitatory postsynaptic potential or EPSP (Figure 6.8). If this depolarization reaches a certain point, the threshold, an action potential is generated in the postsynaptic cell.

On the other hand, a neurotransmitter may act in an inhibitory fashion, making it less likely that an action potential will be generated in the postsynaptic neuron. When the neurotransmitter binds to the receptors in an inhibitory synapse,

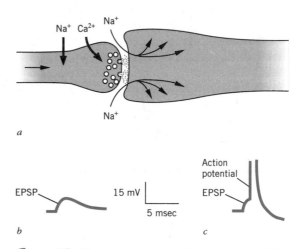

Figure 6.8 The events at an excitatory synapse. The neurotransmitter binds to a receptor and causes the opening of channels that allow sodium ions to enter the cell (*a*), thereby slightly depolarizing the postsynaptic cell. This slight depolarization is called an EPSP (*b*). If the depolarization reaches the threshold, an action potential is generated (*c*).

either K$^+$ channels or K$^+$/Cl$^-$ channels open. The exit of positively charged potassium ions or the influx of negatively charged chloride ions makes the inside of the postsynaptic membrane even more negative than usual. In other words, an inhibitory neurotransmitter momentarily hyperpolarizes the membrane. The hyperpolarization is called an inhibitory postsynaptic potential or IPSP (Figure 6.9).

As long as the neurotransmitter remains in the synapse, it will continue to excite or inhibit the postsynaptic cell. There are, however, a variety of ways in which the effect of a neurotransmitter can be halted (Cooper, Bloom, and Roth, 1978). In some cases, the neurotransmitter is broken down by an enzyme and its component parts are absorbed for resynthesis. In other instances, however, the molecules of neurotransmitter are released by the postsynaptic cell, absorbed intact by the presynaptic cell, and simply repackaged for subsequent release.

Integration

We have seen, then, that the signal that travels from a dendrite to the soma is actually a change in the charge across the cell's membrane. Most

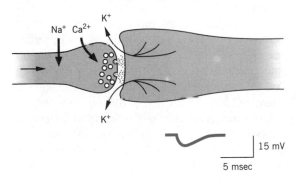

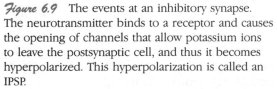

Figure 6.9 The events at an inhibitory synapse. The neurotransmitter binds to a receptor and causes the opening of channels that allow potassium ions to leave the postsynaptic cell, and thus it becomes hyperpolarized. This hyperpolarization is called an IPSP.

neurons receive input from many other neurons. In fact, a given neuron may communicate with hundreds or thousands of other cells. The slight depolarizations (EPSPs) and hyperpolarizations (IPSPs) that result from input from all the synapses are summed, both over time as one neuron sends a repeated signal and over space as many neurons send messages. In other words, the EPSPs and IPSPs combine with one another as they arrive at the soma. If these interacting changes in membrane potential combine to produce a large enough depolarization, voltage-sensitive Na$^+$ gates are opened and an action potential is triggered.

IONS, MEMBRANE PERMEABILITY, AND BEHAVIOR

Although ions and their movements through the channels of nerve cell membranes may seem, at best, to be only remotely related to an animal's behavior, we will see that this is not the case. Let us consider an example of how ions and changes in membrane permeability relate directly to what we observe an animal doing. The relationship between calcium ion permeability and the behavior of *Paramecia* was discussed in Chapter 3. Here we describe an example of how changes in membrane permeability caused by mutation in the Shaker gene of the fruit fly (*Drosophila melanogaster*) produce atypical behavior. Our specific example concerns the behavioral mutants that result from problems associated with the ion channels of nerve cells.

When fruit flies are anesthetized with ether, one occasionally sees a mutant fly that shakes its legs, wings, or abdomen. Among the mutations that result in shaking under ether anesthesia are *Shaker, Hyperkinetic,* and *Ether-a-go-go.* All this shaking is apparently a result of neurons with mutations that make them exceptionally excitable. The most is known about what makes the *Shaker* mutants so jittery, so we will concentrate on them.

It was first shown that the *Shaker* larvae were jittery because an excessive amount of neurotransmitter was released at the neuromuscular synapses (the chemical synapses between motor neurons and muscles), causing extreme muscle contractions. Furthermore, the muscle contractions were uncoordinated because the release of transmitter from different neurons was asynchronous (Jan, Jan, and Dennis, 1977). It was later shown, by recording from the neurons of adult *Shaker* flies, that the mutant neurons did not repolarize as quickly as normal neurons—in other words, their action potentials were very broad (Tanouye, Ferrus, and Fujita, 1981). Because the membrane does not repolarize quickly, an excess of calcium ions enters the neuron and causes the release of neurotransmitter to continue longer than is typical. Why doesn't the cell repolarize normally? Apparently, the *Shaker* gene codes for a protein that forms at least part of the potassium channels involved in repolarization; a mutation at the *Shaker* locus results in certain potassium channels not being formed and this disrupts the process of repolarization. That the behavioral defect, that is, shaking, results from the absence of K^+ channels was elegantly shown by experiments in which a functional *Shaker* gene was inserted into mutant flies. This experimental procedure resulted in a normal flow of potassium across the membranes of nerve cells and an end to the jittery behavior caused by the mutation (Zagotta et al., 1989).

NEUROMODULATORS

The changes in membrane potential that we have described occur in a matter of milliseconds. There are, however, voltage changes that occur over a slower time course—seconds, minutes, hours, and perhaps even days. The fast changes are brought about by "traditional" neurotransmitters, those chemicals, previously discussed, that open ion gates causing EPSPs or IPSPs. Slower changes are caused by neuromodulators, chemicals that alter neuronal activity in a different way—via biochemical means.

The effects of neuromodulators appear to be mediated by substances within the postsynaptic neuron called second messengers (Gillette et al., 1989). These second messengers (e.g., calcium and the cyclic nucleotides cAMP and cGMP), through one or more enzymatic steps, couple the membrane receptors of the postsynaptic cell with the movements of ions. Neuromodulators may, for example, upon reaching the receptor on the postsynaptic neuron, trigger the formation of the second messenger cyclic adenosine monophosphate (cAMP) within the neuron, which, in turn, activates an enzyme that changes the shape of proteins in certain ion channels. Once the ion channels have been altered in this manner, the permeability of the membrane to specific ions is also changed, thereby affecting the activity of the neuron. It is the relatively slow pace of the enzymatic activities that produces the typically prolonged effects of neuromodulators.

Functionally, neuromodulators appear to be intermediate to classic neurotransmitters and hormones (Lent and Watson, 1989). Whereas neurotransmitters are released at specific synaptic clefts and hormones are broadcast throughout the body via the bloodstream, neuromodulators are released in the general vicinity of their target tissue. It is, however, difficult to establish the precise point at which a neurotransmitter becomes a neuromodulator, and a neuromodulator a hormone. In fact, the same chemical may have different functions in different places. Some chemicals, dopamine for instance, act as neurotransmitters at some synapses and as neuromodulators at others. Likewise, a chemical may act locally in the nervous system as a neurotransmitter, while in other places in the body it is released into the bloodstream and acts at a distant site the way a hormone does (see Chapter 7 for a discussion of hormones). Despite some fuzziness in definition, there is no question that neuromodulators, through their actions on neurons, glands, and muscles, can produce profound effects on behavior. Consider, for example, the case of neuromodulation of feeding behavior in the leech.

The European leech (*Hirudo medicinalis*)

feeds on the blood of other animals, mammals in particular. Many aspects of leech feeding behavior are influenced by the neuromodulator serotonin (Lent, Dickinson, and Marshall, 1989). Before describing some of the neurophysiological aspects of feeding, let us first review the patterns of behavior involved (Figure 6.10). A hungry leech typically rests quietly at the water's edge, with both its anterior and posterior suckers attached to a surface. If aroused by shadows or ripples in the water, the leech releases its anterior sucker, and then the posterior sucker, and swims in an undulating fashion in the direction of the stimulus. Once the leech reaches the target, it explores the surface of the animal, moving in an inchworm-like manner. When a warm region is encountered, the leech bites with its three jaws lined with approximately 70 pairs of sharp teeth. If blood flows from the wound, ingestion begins. Rhythmic contractions of the muscular pharynx pump blood into the crop for the next half hour or so and then the leech, enormously distended, detaches from its host. (Leeches often ingest from seven to nine times their initial weight and then stop feeding when they sense the extreme distension of their body wall.) Following one of these massive blood meals, the leech crawls into a rock crevice to begin digestion, a process that may take up to one year to complete. Leeches in this satiated state behave very differently from their hungry counterparts, tending to crawl rather than to swim and when on a warm surface, to lift their heads rapidly and repeatedly rather than bite. Sometime in the weeks following a blood meal, the leech will reproduce.

Using a variety of techniques Charles Lent and colleagues (1989) demonstrated that serotonin modulates the physiology and patterns of behavior associated with feeding in leeches. Here we discuss a subset of their experiments—the experiments are meant not only to demonstrate how a particular neuromodulator influences behavior, but also to introduce some techniques that are commonly used when investigating nervous systems.

The first step in determining whether or not serotonin was involved in the feeding behavior of leeches entailed an examination of serotonin's effects on intact animals. Bathing leeches in serotonin (a method that does not cause injury to the leech and is therefore preferable to injecting serotonin) produced profound effects on feeding behavior. A serotonin bath reduced the time taken by leeches to initiate swimming toward potential prey, increased the frequency of biting by 40 percent and pharyngeal contractions by 25 percent, and increased the volume of blood ingested. Exposure to serotonin also increased both heart rate and mucus secretion, the first response being necessary to cope with the increased metabolic demands of swimming, and the second to aid in adhering to mammalian blood "donors." In addition, bathing in serotonin caused satiated leeches to bite (recall that satiated leeches typically do not bite, but instead lift their heads rapidly and repeatedly when on a warm surface). Clearly, then, serotonin influences the feeding behavior and physiology of *H. medicinalis*.

In another series of experiments, microelectrodes were inserted into identified neurons—neurons known to contain serotonin—in order to artificially cause these neurons to fire. A microelectrode consists of a fine wire inside a glass capillary tube that ends in a fine point so that it can pierce a neuron. The capillary tube is filled with a conducting solution that creates an electrical connection between the electrode and neuron. Microelectrodes can be used to record electrical activity from neurons or to stimulate neurons. Frequently researchers pair recording and stimulating electrodes—for example, action potentials can be stimulated and then recorded at a specific neuron (Figure 6.11). Alternatively, by stimulating one neuron while recording from a second, the functional relationship between the two neurons can be identified. In this particular experiment, electrical current was passed through a microelectrode, causing a specific serotonin neuron to become depolarized (that is, to fire), and then the researchers examined

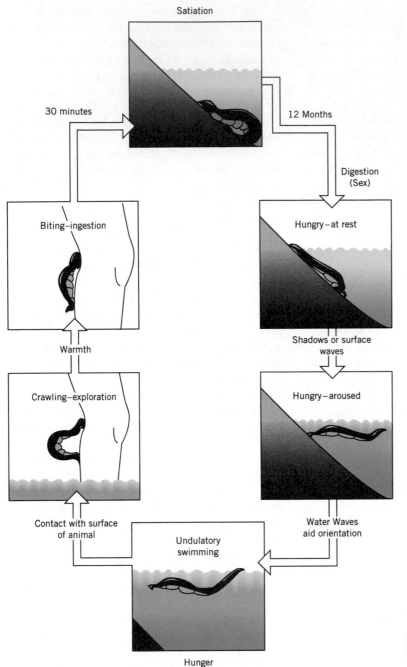

Figure 6.10 The feeding cycle of the leech. (Modified from Lent, Dickinson, and Marshall, 1989.)

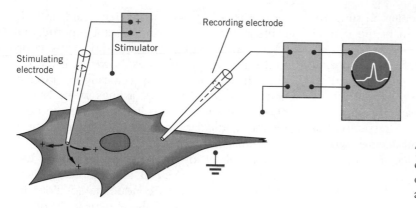

Stimulating electrode

Stimulator

Recording electrode

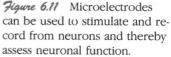

Figure 6.11 Microelectrodes can be used to stimulate and record from neurons and thereby assess neuronal function.

whether such firing evoked the physiological responses associated with feeding. The physiological responses were monitored by recording with a second microelectrode the membrane responses of the effector cells (salivary and muscle cells). Stimulation of the serotonin-containing neurons induced salivation, peristaltic (wavelike) movements of the pharynx, and twitches in jaw muscles. Thus, impulse activity of identified serotonin-containing neurons was sufficient for the major physiological components of leech feeding. It is important to note that because the membrane responses of the salivary and muscle cells occurred 5–10 seconds after the serotonin neurons were stimulated and then continued beyond the period of stimulation, Lent and coworkers (1989) inferred from this delay that serotonin was released some distance from the effector cells and not at discrete synapses with these cells. In other words, serotonin was acting as a neuromodulator and not as a neurotransmitter (remember that neurotransmitters are released at a specific synapse and produce their effects in a matter of milliseconds).

In a third group of experiments, the researchers showed that serotonin neurons were absolutely essential for feeding behavior and its physiological components. In this case, they examined the effects of selectively removing the serotonin neurons from the underlying neural circuits. Hungry leeches were treated with 5,7-DHT, a neurotoxin that specifically destroys serotonin within neurons, without destroying the neurons themselves (many neurotoxins actually destroy the nerve cells, a technique called lesion or ablation). When tested on a warm surface, these hungry toxin-treated leeches did not bite and instead lifted their heads and behaved as though they had just finished a blood meal. The leeches also failed to show any of the physiological responses associated with feeding. Removal of serotonin from leeches thus eliminated feeding behavior and its physiological components. When, however, serotonin was subsequently administered to these animals, biting, salivation, and pharyngeal peristalsis returned.

A final experiment revealed the interaction between feeding behavior and the neuromodulator serotonin. In this particular experiment, Lent and coworkers demonstrated that not only is leech feeding behavior driven by serotonin, but that the expression of feeding behavior (i.e., ingestion) decreases the amount of serotonin in the central nervous system. Hungry leeches were set up in pairs and then one member of each pair was fed to satiation. The amount of serotonin in the central nervous system of each member of the pair was then measured. Leeches that had been allowed to feed had less serotonin in their central nervous system than did hungry leeches. From these results we can see that serotonin drives feeding, and feeding, in turn, reduces the amount of serotonin in the central nervous system.

These experiments established the pivotal role of the neuromodulator serotonin in activating leech feeding. Not all of serotonin's effects on behavior, however, are activational. Serotonin inhibits leech sexual behavior, a response that makes sense given that reproductive activity would most likely reduce the feeding efficiency of these remarkable gluttons (Leake, 1986). The modulatory effects of serotonin on leech feeding are summarized in Figure 6.12.

One important technique that we did not consider in our discussion of the neuromodulation of leech feeding is the use of intracellular dyes. The technique of injecting dye into neurons revolutionized the study of nervous systems. Microelectrodes, as we have said, are valuable research tools in that they permit an understanding of the function of neurons. Prior to dye injection techniques, however, it was often difficult to identify anatomically the specific neuron from which recordings were made. Through the use of dye injection, we can now identify particular neurons and visualize their often amazing architecture and pathways.

Intracellular dyes cannot pass through the membrane of a nerve cell, are soluble in water, and are often used as conducting solutions in microelectrodes. Typically, a neuron is penetrated by a dye-filled microelectrode, and after the activity of the neuron is recorded, current is then passed through the electrode, expelling the dye into the neuron. (Alternatively, dye may be injected into a special region of the brain where it is taken up by axon terminals and transported back to the cell body of a neuron.) The dye moves throughout the neuron, unable to escape through the membrane. Dyes trapped inside either are visible or can be made visible by special treatment. Some commonly used dyes include lucifer yellow, cobalt, and horseradish peroxidase. Finally, the neuroanatomical techniques available today are not restricted to examining the static properties (e.g., structure) of neurons. Indeed, some dyes allow visualization of the physiological and chemical changes in neurons as well (Beltz, 1990). For example, a number of voltage-sensitive dyes change color when depolarization occurs. A neuron stained with such a dye will thus change color each time an action potential is fired.

We have completed our discussion of the basic concepts of cellular neurobiology. Armed with our understanding of how neurons transmit information, we now begin to trace the pathways by which information enters the nervous system, is processed, and ultimately results in behavioral output. Our immediate focus is on the reception by the nervous system of information from the environment—this reception takes place at the sense organs.

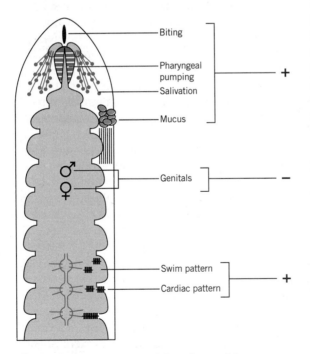

Figure 6.12 A summary of the effects of the neuromodulator serotonin on behavior and physiology of the leech. (Modified from Lent, Dickinson, and Marshall, 1989.)

Sense Organs

Animals constantly assess information about their environment. Much of this information comes from the diverse sources of environmental energy around them. Animals are exposed, for example, to thermal energy (heat and cold), mechanical energy such as sound, electromagnetic energy such as visible light, and forces associated with the earth's magnetic field. But before an animal can use this information, it must first detect the various forms of energy. The detection of environmental energy is the job of the animal's sense organs. We will describe the ears and eyes of animals in some detail, but first it would be helpful to consider the various classes of sensory receptors and some of the general principles of sensory reception.

CLASSES OF RECEPTORS

Sense organs are selective, containing specific receptor cells that monitor one particular form of environmental energy. In many cases, the receptor cells are sensory neurons, and thus information is conveyed along their axons from the periphery to the central nervous system. Sometimes, however, the receptor cells do not have axons, and in this case, they transfer their information to the central nervous system by making synaptic contact with another nerve cell whose axon transmits the information. Whatever their method of transmitting information, each sensory receptor is specialized to respond to a particular type of energy and therefore provides a different window on the world.

We can classify the receptors by the "view" they furnish, that is, based on the stimulus to which they respond. Thermoreceptors, for example, measure heat; they respond to changes in temperature. Mechanoreceptors respond to distortions in the receptor itself or in nearby cells; they are responsible for a variety of sensations—touch, hearing, equilibrium, and knowledge of body position. Photoreceptors detect light and are responsible for vision. Chemoreceptors respond to chemicals and are the basis of olfaction (sense of smell) and gustation (sense of taste). Electroreceptors, found in sharks, catfish, electric fish, and a few other species, are modified mechanoreceptors capable of detecting weak electric fields. Finally, some organisms apparently have magnetoreceptors that detect weak magnetic fields.

SOME GENERAL PRINCIPLES OF SENSORY RECEPTION

Soon we will see how diverse the sensory receptors of different animals can be, but for now let's focus on some features they have in common. Keep in mind that receptors simply detect stimuli; they do not interpret them. Perception, the interpretation of the input from receptors, occurs as the information from sensory receptors is received by the nervous system.

Receptors tell the animal about specific environmental stimuli by converting, or transducing, the particular form of stimulus energy they receive (e.g., mechanical energy) into electrochemical signals (nerve impulses). At the site of transduction, a change in membrane potential occurs, and because it occurs at a receptor, it is called a receptor potential. Receptor potentials, like action potentials, result from the redistribution of ions. For example, a stimulus to a mechanoreceptor may distort the membrane of the receptor, which, in turn, produces a change in the size of ion channels in the membrane, thereby causing a change in the distribution of ions. This redistribution of ions produces a change in membrane potential—this is the receptor potential. Although in most sense organs, receptor potentials are converted into action potentials that are then conducted along the axon for transmission to other neurons, in some sense organs, the vertebrate eye, for instance, the receptor potential itself is passively conducted along the axon.

As previously mentioned, both receptor potentials and action potentials result from the redistribution of ions. There is, however, an important distinction between the two types of potentials. Whereas the strength of an action potential is constant, not related to the strength of the stimulus, the magnitude of a receptor potential generally varies with the strength of the stimulus. A louder sound, for instance, may result in a larger receptor potential in the organ of hearing. This is important because as you can see in Figure 6.13, in some receptors, the strength and duration of the receptor potential determines frequency and duration of action potentials. The nervous system's interpretation of this information is the perception of stimulus intensity.

The response of some sensory receptors to a stimulus of a certain intensity also varies depending on how long the stimulus is present. When receptors are constantly stimulated, they may become less responsive. This waning effect is called sensory adaptation (sometimes called physiological adaptation). Sensory adaptation has at least two consequences. First, it means that receptors are, at some point, more responsive to changes in stimulation than they are to

constant stimulation. For example, when you first get into a jacuzzi, you may feel as if the water is too hot, but it quickly becomes comfortably warm. Second, prolonged stimuli may go unnoticed. Without sensory adaptation, for instance, you would be acutely aware of the touch of your clothing. Obviously, constant awareness of every stimulus would make it difficult to concentrate on a single stimulus of importance.

Now that we know some of the basic principles of sensory reception, let's explore auditory and visual receptors in more detail.

AUDITORY RECEPTORS

The Particle Movement and Pressure Components of Sound

Sound is caused by vibration. The vibration has two effects (Figure 6.14). One effect is to move the particles, or molecules, of the surrounding medium back and forth. When the vibrating object moves in one direction, it pushes the surrounding molecules in that direction, and when it moves back in the other direction, it draws the surrounding molecules back with it. This effect of vibration is called the particle movement, or displacement component, of sound.

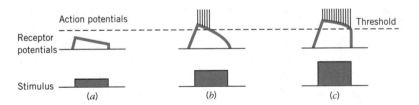

Figure 6.13 The response of a sensory receptor to stimuli. Sensory receptors transduce stimulus energy into the electrochemical energy of a receptor potential. In many receptors, the magnitude of the receptor potential varies with the strength of the stimulus. Note that the magnitude of the receptor potential increases with the strength of the stimulus from (a) to (c). The magnitude of the receptor potential and the length of time it remains above threshold determine the frequency and duration of action potentials.

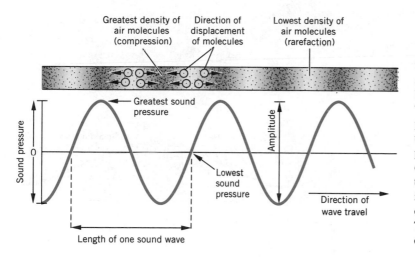

Greatest density of air molecules (compression)

Direction of displacement of molecules

Lowest density of air molecules (rarefaction)

Greatest sound pressure

Sound pressure

0

Amplitude

Lowest sound pressure

Direction of wave travel

Length of one sound wave

Figure 6.14 Characteristics of a sound wave. Sound is created by the vibration of an object. The vibration causes cyclic displacement of the molecules of the surrounding medium. Thus, particle displacement is one component of sound. The vibration also causes cyclic changes in the density of the molecules of the surrounding medium. This is the pressure component of sound.

The other effect is to generate waves of pressure fluctuations in the surrounding medium. As the vibrating object moves in one direction, it pushes the surrounding molecules closer together, thereby increasing the local pressure. When it moves in the opposite direction, the surrounding molecules return to their original position and the local pressure drops. In other words, sound can be considered to be a wave of alternating compression and rarefaction moving away from the vibrating object. This fluctuation in pressure is called the pressure component of sound. Audition, or hearing, is the detection of sound.

Some auditory receptors are designed to detect the movement of particles and others to detect the pressure changes caused by a vibrating object. Because all sound waves have both pressure and particle movement components, how might we determine which component the auditory receptors of a given species are specialized to detect? Towne and Kirchner (1989) asked this question while investigating hearing in honeybees (*Apis mellifera*). To answer it, they teased apart the two components and tested the responses of bees to each component. When you are given a hearing test in a doctor's office, you are asked to make some response, usually rais-

ing your hand, whenever you hear a sound transmitted through earphones. Instead of making bee-size earphones, these investigators constructed a tube that sound could be passed through. The tube contained sites where one of the components of sound was maximized and the other minimized. If the bees could hear the sound, they responded by raising their antennae and lowering their wings. Each bee heard a 265 Hz sound 10 times at a site where the particle-movement component was maximized and 10 times at a site where the pressure component was maximized. As you can see in Figure 6.15, the probability that the bees would respond to the sound was higher when they were tested at sites where the particle-movement component was maximized, indicating the honeybees are most likely responding to the particle-movement component of sound, rather than the pressure component. Now that we know something about the physical properties of sound, let us consider some examples of hearing receptors in invertebrates and vertebrates.

Selected Examples of Auditory Receptors

Invertebrates   Among the invertebrates, only arthropods have a sense of hearing (Camhi, 1984). Our particular examples of displacement-

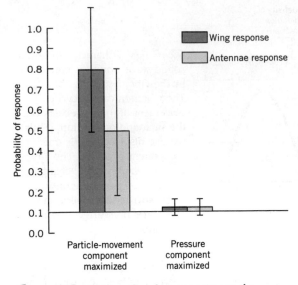

Figure 6.15 The results of an experiment determining whether the auditory receptors of honeybees respond to particle displacement or to pressure changes. (Drawn from the data of Towne and Kirchner, 1989.)

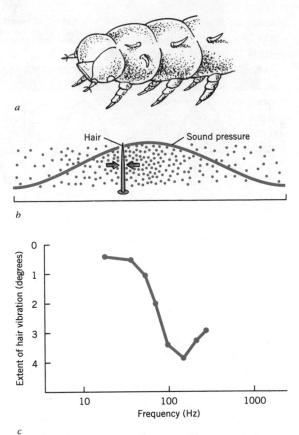

Figure 6.16 Auditory displacement receptors in a caterpillar. (*a*) The caterpillar *Barathra* has four auditory sensory hairs on each side of its body. (*b*) The movement of air particles associated with a sound wave moves the sensory hair. (*c*) The sensory hairs are maximally deflected by sounds of 150 Hz. This is the frequency of the buzz of the wings of the parasitic wasp, *Dolichovespula*, which preys on this caterpillar. (*a, c*; from Tautz, 1979; *b*; Camhi, 1984.)

and pressure-sensitive ears come from the insects.

All known displacement-sensitive receptor cells have at least one sensory hair that responds to the displacement of molecules in the surrounding medium. Such hairs are often tuned like piano strings to different biologically relevant frequencies by variations in thickness, stiffness, and length. For example, the displacement receptors of the caterpillar *Barathra* respond best to 150 Hz sound, the major frequency of the wingbeat of the parasitic wasp, *Dolichovespula*, which lays its eggs in the body of this caterpillar (Tautz, 1979). The displacement receptors are located in four sensory hairs projecting outward from each side of the caterpillar's body. The movement of particles associated with sounds of this frequency causes the hairs to be deflected back and forth, exciting the single sensory cell at the base of each hair (Figure

6.16). The wasp's buzzing wings can be detected within the distance of a meter. Low-intensity sounds of this frequency, which indicate that the wasp is still at a distance, cause the caterpillar to freeze in place, making it less noticeable to the wasp. However, as the wasp approaches, the buzz of its wings gets louder. High-intensity sounds of this frequency cause the caterpillar to drop from the branch on which it had been resting (Tautz and Markl, 1978).

Another type of auditory receptor found in insects is the tympanal organ. This organ responds not to the movement of particles, but to the pressure changes associated with sound. The tympanal structures consist of a membrane with sensory cells attached to the underside. When the membrane vibrates in response to sound, the sensory cells are stimulated and generate nerve impulses that are conducted to the central nervous system for interpretation.

Vertebrates The vertebrate ear is a pressure-sensitive receptor and is usually divided into three parts, the outer ear, the visible part with a connecting tube to the middle ear, where a vibrating membrane transfers impulses to one or a sequential arrangement of bones, and the inner ear that transfers the impulses to the brain. Among most fish, sound vibrations in water are conducted *directly* to the inner ear, but in the order Cypriniformes (minnows, catfish, and suckers), a set of tiny bones, the Weberian ossicles, connect the swim bladder to the inner ear. The air-filled bladder resonates with sound and moves the bones, and receptors in the inner ear are stimulated. A fish with such an apparatus can detect a much wider range of sound frequencies than those lacking it.

Land vertebrates have more complicated ears. The tympanic membrane, or eardrum, of amphibians and most reptiles is on the body surface. In birds and mammals, however, sound travels through an external ear before striking the eardrum. Most mammals have a pinna, the part of the ear that is visible from the outside, to catch the sound and funnel it to the auditory canal (Figure 6.17). Many mammals can move the pinna to gather the sound more effectively (Figure 6.18).

The auditory canal, then, channels sound to the tympanic membrane, which vibrates in response to the pressure changes of the sound wave. These vibrations cause movements in the bone or bones of the middle ear, called ossicles. There is a single ossicle (the stapes or columella) in amphibians, reptiles, and birds, but three (the malleus, incus, and stapes) in mammals. The ossicles increase the force of the sound wave so that the bone movements cause the oval window of the inner ear to vibrate and the pressure wave to be transmitted to the cochlear fluid. The result is that fluids inside the cochlea are shifted about, triggering impulses that are sent to the brain.

The mammalian cochlea is divided longitudinally by complex membranes, forming the vestibular and tympanic canals (see Figure 6.17). The actual organ of hearing, called the organ of Corti, is located on one of the membranes, called the basilar membrane. Sensory hairs (the auditory receptor cells) on the basilar membrane reach upward, touching an overhanging tectorial membrane. The vibrating fluid of the cochlea causes the basilar and tectorial membranes to draw together and pull apart, thus distorting the hair cells and causing a change in their membrane potential. The mechanical energy is transduced to electrochemical energy, and impulses are sent to the brain over the auditory nerve. The hair cells are so sensitive that a displacement as small as the diameter of an atom can trigger an impulse (Hudspeth, 1983). The basilar membrane is relatively narrow and stiff at the base near the oval window; at the other end, it is wider and more flexible. Frequency discrimination is possible because high-pitched tones stimulate the hairs at the narrow rigid base of the membrane; low-pitched tones are registered at the wider floppy apex (Khanna and Leonard, 1982; von Bekesy, 1956; Zwislocki, 1981).

Localization of Sound Source

It is often important for an organism to locate the source of sound. For example, the sound

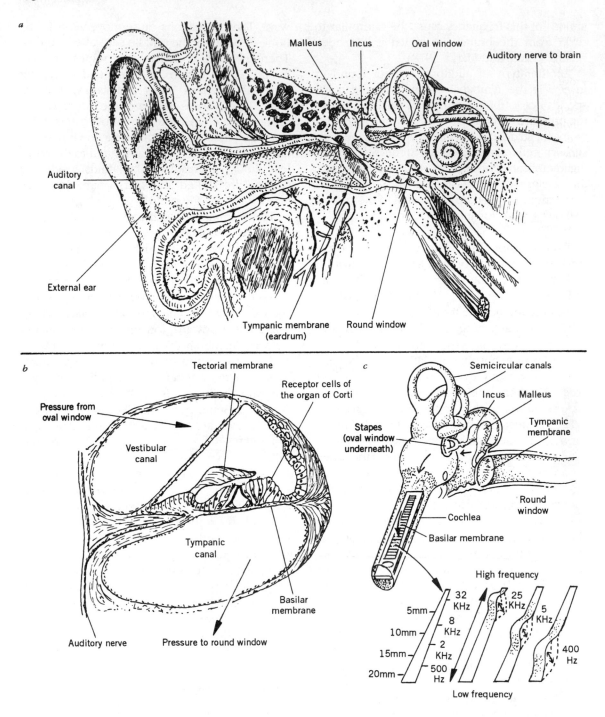

a

Malleus   Incus   Oval window

Auditory nerve to brain

Auditory canal

External ear

Tympanic membrane
(eardrum)   Round window

b

Tectorial membrane

Receptor cells of
the organ of Corti

Pressure from
oval window

Vestibular
canal

Tympanic
canal

Basilar
membrane

Auditory nerve   Pressure to round window

c

Semicircular canals

Incus   Malleus

Stapes
(oval window
underneath)

Tympanic
membrane

Round
window

Cochlea

Basilar membrane

High frequency

32
KHz

25
KHz

5mm

8
KHz

5
KHz

10mm

2
KHz

15mm

500
Hz   400
Hz

20mm

Low frequency

Figure 6.17 The human ear. The vibrating eardrum moves three bones that cause fluid to stimulate sensory areas in the cochlea (*a*), as described in the text. (*b*) Cross section of the cochlea. Sound waves move the basilar membrane and distort receptor cells that are anchored to the tectorial membrane. The distortion causes action potentials in the auditory nerve. (*c*) The cochlea, shown straightened. The areas of sensitivity to various frequencies are shown. Those near the oval window register the highest frequency sounds.

may be produced by a potential mate. A male mosquito finds a female by the sound of her beating wings. It would do a male little good to know that a female was present and be unable to locate her. Likewise, many predators determine their prey's position by localizing sounds generated by the prey. Locating the source of a sound has obvious importance to prey animals

as well—the cough of a leopard has fixed its position for many a wary baboon.

What properties of sound enable its source to be located? Actually, part of the answer is remarkably simple: Sounds can be localized by how loud they are, that is, by their intensity. A simple rule might be that sound seems louder when the receptor is closer to the sound. If only one ear is involved in locating the sound source, however, the rule may not hold (Camhi, 1984). Let's say that the left ear hears a soft sound—was the sound soft because it was produced by a weak source on the left side or by a strong source on the right side? In order to eliminate such confusion, both ears must be used in the sound localization process—this is called binaural comparison. Binaural comparison of sound intensity is used by some animals to locate the source of sounds.

Timing is also important in locating the

Figure 6.18 White-tailed deer turn their ears toward a sound in the forest.

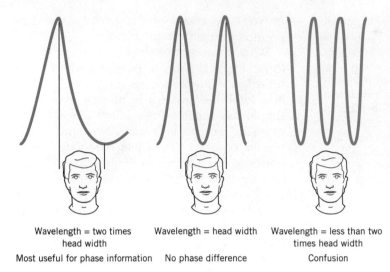

Wavelength = two times head width

Most useful for phase information

Wavelength = head width

No phase difference

Wavelength = less than two times head width

Confusion

Figure 6.19 Binaural comparison of phase. When the sound is prolonged, differences in the phase of the sound wave at each ear may indicate the direction of the source. The usefulness of this cue depends on the wavelength and the distance between the ears.

source of a sound. There are two differences in timing that could be of potential use, and both rely on binaural comparison. The first occurs at the onset of sound—sound begins and ends sooner in the ear that is closest to the source. The second difference in timing occurs during an ongoing sound. During a continuing sound, there are differences in the phase (the point in the wave of compression or rarefaction) of the sound wave reaching each ear. The extent of the phase difference will depend both on the wavelength of sound and on the distance between the ears (Figure 6.19). Now let's consider which of these three methods—binaural comparison of the intensity, onset, or phase of sound—might be used by barn owls to localize sound.

Silently and suddenly, a barn owl (*Tyto alba*) sweeps from the sky to strike its prey with astonishing accuracy (Figure 6.20). How does it find its prey? Although in nature the barn owl's keen night vision is important in locating prey, the sounds of a scurrying mouse are sufficient for the owl to strike with deadly precision. Laboratory tests have revealed that birds such as the barn owl are able to locate the source of sounds within 1 or 2 degrees in both the horizontal and

vertical planes (1 degree is approximately the width of your little finger held at arm's length). Due to its astounding ability to detect and locate the source of sound, this nocturnal predator can pinpoint its prey by the rustlings the prey makes and can precisely determine not only the prey's location along the ground, but also its own angle of elevation above the prey.

How do we know that the hunting owl uses the prey's sound? For one thing, we know that barn owls can catch a mouse in a completely darkened room (Payne, 1962). In experiments, a barn owl was able to capture a skittering leaf pulled along the floor by a string in a dark room (indicating that sight and smell are not involved), and if unable to see, it will leap into the middle of an expensive loudspeaker from which mouse sounds emanate.

The studies of barn owls' hearing offer an opportunity to learn how the sensory capacity of another species can be examined and how the cues relevant to that animal can be identified. Recall from our previous discussion that several sound cues are of potential use for localizing the source of the sound: binaural differences in loudness, onset, and phase. Just because a cue

Figure 6.20 A hunting barn owl. A barn owl can locate its prey using sound cues alone.

is available, however, does not mean that it is used. Therefore, when considering the barn owl's ability to localize sound, we must know which cues it is actually using.

It is possible to isolate the three types of cues for sound localization because they can be manipulated independently. It is easy to see how the intensity of sound can be eliminated as a potential cue: If the sounds arriving at the owl's ears are equally loud, the bird cannot use intensity differences to determine the source of the sound. Likewise, by playing the sounds through small speakers placed directly into the ears of the owl, the time of onset can be manipulated precisely. Finally, because the usefulness of binaural phase differences depends on the size of the head relative to the wavelength, it is possible to choose frequencies that eliminate phase differences as well.

By presenting sounds with various combinations of potential cues, it was shown that by itself, a binaural difference in the onset of sound is not enough information to allow an owl to accurately localize the source of a sound. Differences in the two ears in loudness and in the phase of a continuous sound, however, are important cues. Owls have difficulty orienting to sounds when they cannot use binaural differences in phase. When sounds of frequencies that provide ambiguous cues to an owl with respect to difference in phase in the two ears are played through a loudspeaker placed in the room, the owls have difficulty accurately localizing the sound (Knudsen, 1981).

Owls can and do use binaural differences in the phase of an ongoing sound to localize the source. This fact was demonstrated by playing sound through small speakers placed in the ear canals of owls (Moiseff and Konishi, 1981). Because an owl will turn its head to face the source of sound, it is possible to determine the direction from which the bird perceives the sound to be coming. It is also feasible to adjust ongoing delays in the phase of sound played in the owl's ears to create binaural differences as small as 1 microsecond. When sounds with ongoing differences in phase as small as 10 microseconds or as great as 80 microseconds were played in the ears of an owl, it quickly turned its head to the angle implied by the phase difference it perceived.

The owl's ability to locate a sound source is also aided by arrangement of the ear canals and facial feathers. Although the barn owl lacks pinna to catch sound and has only auditory canals to channel the sound toward the inner ear, these two canals are, oddly enough, situated asymmetrically such that the right one is higher than the left. This difference in placement helps the owl determine its own elevation above the sound source, information critical to an aerial predator. Also, the face of the barn owl is composed of rows of densely packed feathers called the facial ruff that act as a focusing apparatus for sound (Figure 6.21). There are troughs in the facial ruff that, like a hand cupped behind the ear, both amplify the sound and make the ear more sensitive to sound from certain directions.

The facial ruff assists the owl in localizing sound by creating differences in intensity of the sound in both ears. Loudness is a cue to localizing the sound in both the horizontal and vertical dimensions. Sound is generally louder in the ear closer to the source. Because of the structure of the facial ruff, the left ear collects low-frequency sounds primarily from the left side and the right ear collects low-frequency sounds from the right side. A comparison of the intensity of low-frequency sounds in each ear helps the owl determine from which side of the head the sound originates. However, the facial ruff channels high frequency sound to each ear differently depending on the elevation of the

Figure 6.21   The barn owl, a night hunter, has a facial disc of densely packed feathers that may gather sounds and aid in detecting their source.

sound source. As a result, the right ear is more sensitive to high-frequency sounds originating above the head and the left ear is more sensitive to high-frequency sounds from below the head. The owl compares the loudness of high-frequency sounds in each ear to determine its position above or below the sound source. As a sound source moves upward from below the bird to a position above the owl's head, the high-frequency sounds would first be loudest in the left ear and then gradually become louder in the right ear (Knudsen, 1981).

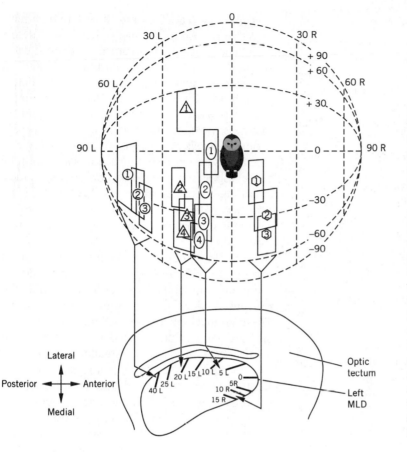

Figure 6.22 Auditory neurons in the midbrain area (MLD) of a barn owl. The top figure shows a hemisphere of space in front of the owl's head. Neurons in the MLD respond to sounds originating at different points. The numbered rectangles indicate 14 areas to which specific MLD neurons are tuned. The lower figure indicates the manner in which the auditory space of the owl is represented in the MLD. A horizontal section of the MLD is shown with bars indicating the position of specific neurons. The point in space to which that neuron responds is indicated. Notice that the neurons in the MLD are spatially organized. (From Knudsen and Konishi, 1978.)

Information on the timing and loudness of sounds in each ear is then sent to the central nervous system of the owl over the auditory nerve in a pattern of nerve impulses. In the major auditory area of the owl's midbrain, the mesencephalicus lateralis dorsalis (MLD), are certain neurons that respond selectively to specific degrees of binaural differences in sound (Figure 6.22). For example, there may be one MLD neuron that responds maximally to differences that correspond to a sound originating 30° to the right of the owl. The sound would arrive just a certain amount sooner and be a certain degree louder in the right ear than in the left. Those exact differences in timing and loudness stimulate that particular MLD cell. The degree of binaural difference varies with the location of sound, and the binaural difference that stimulates cells of the MLD varies from neuron to neuron (Knudsen, 1982).

PHOTORECEPTORS

Humans are highly visual animals, and hence we seem to have a special appreciation for sharp vision in other animals. Some of those other animals see well, indeed, by our standards, some see poorly, and some see a world that would be totally alien to us. Before we describe some of the visual systems of animals, let us first consider a few general principles of photoreception.

Some Principles of Photoreception

Visible light is a particular part of the spectrum of electromagnetic energy. Its wavelength ranges from about 300 to 10,000 millimicrons, but no animal can see more than a part of this range. The shortest wavelengths, such as X-rays and gamma rays, are not seen by any animal. Nor are the very long ones, such as radio waves.

Visual receptors (photoreceptors) contain pigments (colored substances) that absorb certain wavelengths of electromagnetic energy. When light is absorbed, the pigment molecule temporarily changes shape, resulting in the generation of a receptor potential. The number of visual pigments varies among species from one to several. Different visual pigments absorb different wavelengths. Only the wavelengths of light that are absorbed by the visual pigment can activate the receptor; therefore, an animal is literally blind to unabsorbed wavelengths. Night-flying insects, for instance, do not see yellow or red light. That is why lights of these colors, often sold as bug lights, do not attract insects the way white lights do. Because many nocturnal animals cannot see red light, but humans can, zoos can illuminate their habitats with white light during our night and red light during our day so that we can observe nocturnal animals during their active period.

Some Selected Examples of Photoreceptors

The environment is filled with peering eyes, each able to register those parts of the environment that are most important to the survival of the animal. Some register only light, helping the animal to find either well-lit places or the safety of darkness. Others can discriminate among different wavelengths (i.e., have color vision), while still others can detect the plane of polarization of sunlight and use this information for orientation. Here, then, we will consider some of these different photoreceptors in order to emphasize the range of visual abilities in the animal world.

Invertebrates   Some invertebrates have eyes that do not form images, but simply detect light. The unicellular organism, *Euglena*, has a simple light-sensitive eyespot that is shaded on one side so that the animal is able to detect roughly the direction of the light. Such information is useful for a photosynthetic organism that must position itself so as to gather the appropriate levels of sunlight. The planarian, a flatworm, has an improved directional system in its two eyespots that are shaded on opposite sides, an arrangement that more effectively detects the direction of light. In these animals, such information is important in helping them move away from light.

Some invertebrates have eyes that do form images. Among these species, the best vision is found among cephalopod molluscs such as the octopus (Figure 6.23*a*), squid and cuttlefish. These animals, of course, lack a protective shell and are vulnerable to predators, but they are also effective predators on smaller creatures that they locate by sight. There is a remarkable similarity between the optical system of the cephalopod eye (Figure 6.23*b*) and the vertebrate eye (see Figure 6.25), but the cephalopod photoreceptors actually resemble those of other invertebrates. Interestingly, the lens is not fixed in place as it is in all other invertebrates. Instead, cephalopods can change the position of the lens in relation to the retina (the layer containing photoreceptors). When the internal pressure of the eye increases, the lens moves forward, away from the retina, to permit viewing of distant objects. On the other hand, when the muscles around the eyes contract, the lens moves backward, toward the retina, for near vision. In addition, the cephalopod's cameralike eye is equipped with a wide-angle lens. In spite of this complex optical system, the cephalopod probably sees a grainy image because there are comparably few photoreceptors on the retina (Sinclair, 1985).

Most arthropods, particularly the insects and crustaceans, and a few species of polychaete annelids, see images with compound eyes, which are composed of numerous "eyes," or ommatidia (Figure 6.24). Each ommatidium is a tube with lenses at the outer end that focus light on photoreceptor cells deeper in the tube. The photoreceptor cells (called retinula cells) are sensory neurons that contain photopigments. The axons of each retinula cell pass through the base of the eye and make synaptic contact with neurons in the brain. When stimulated by light, the retinula cells produce a depolarizing receptor potential that may be converted into an action potential and passed along to the optic lobe of the brain. In some species with compound eyes, however, the receptor potential is not converted into an action potential; instead, the receptor potential itself is conducted along the axon.

The ommatidia cannot move to follow an image, so they stare blankly in their separate directions until something moves into their field

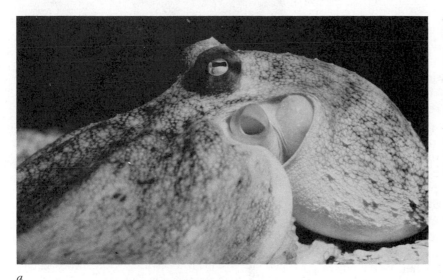

a

Figure 6.23 (*a*) Among invertebrates, cephalopods such as the octopus have the best vision. (*b*) The eye of the cuttlefish, a squidlike cephalopod. Light is focused by a lens onto a sensitive retina in a manner similar to the visual system of vertebrates.

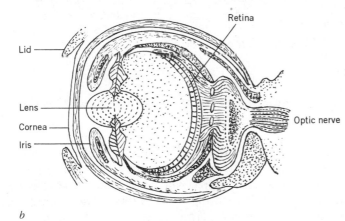

Lid

Lens

Cornea

Iris

Retina

Optic nerve

b

of vision. When an ommatidium is stimulated, it fires an impulse to the central nervous system. So, as an image moves across the compound eye's field of vision, it fires different receptors in turn. Thus, even very slight movements can be detected. The convex surface of such eyes may permit a visual field of 180 degrees.

The incredible sensitivity to slight movement that characterizes the compound eye is put to good use by the praying mantis (*Mantis religiosa*). This predator has two types of ommatidia in its compound eyes. The more central ones are in sleeves of pigment that shade adjacent ommatidia. As a result, these deadly predators are very sensitive to an insect slowly crawling along a leaf. Even the slowest movement will cause new ommatidia to fire, with no confusing effects caused by diffusing stimuli. However, since threshold intensity for each ommatidium must be reached, these receptors function best in good light. The more peripheral ommatidia, on the other hand, have transparent sleeves, so light passes from one ommatidium to another and can have an additive effect. Thus, these areas have high sensitivity but perhaps produce fuzzy outlines as light dances about the thresholds of various receptors simultaneously. Because light that stimulates one receptor is not blocked from reaching others, these receptors function especially well in dim light (Roeder, 1963).

The compound eye can form images. Light striking each ommatidium often varies in brightness, so the response of each differs. The ommatidia produce a mosaic of dots with varying brightness. The graininess of the final picture partially depends on the number of ommatidia in the eye. An underground worker fire ant has only nine ommatidia, so its view of the world is composed of only nine dots of varying brightness, probably a very grainy picture. In contrast, the eye of a dragon fly has 28,000 to 30,000 ommatidia, so the image may be much finer. Since the density of photoreceptors in the dragonfly eye is roughly the same as those in the human eye, it has been suggested that the image

Figure 6.24 The compound eye of most arthropods. (*a*) A close-up view of the compound eyes of a horse fly. Each facet is an ommatidium. (*b*) The ommatidia are arranged in a dome. Each ommatidium fires separately and with full force when stimulated. Note the extreme field of vision. (*c*) Light is focused onto the rhabdom, a structure formed by pigmented extensions of the photoreceptor cells (retinula cells) that encircle the rhabdom. (*b*) and (*c*) appear on the next page.

formed might be as refined as our own (Sinclair, 1985).

Vertebrates The eyes of vertebrates are spherical (Figure 6.25), filled with transparent fluid, and covered with a tough outer coat called the sclera. In most places the sclera is opaque (it does not permit light to enter), but at the

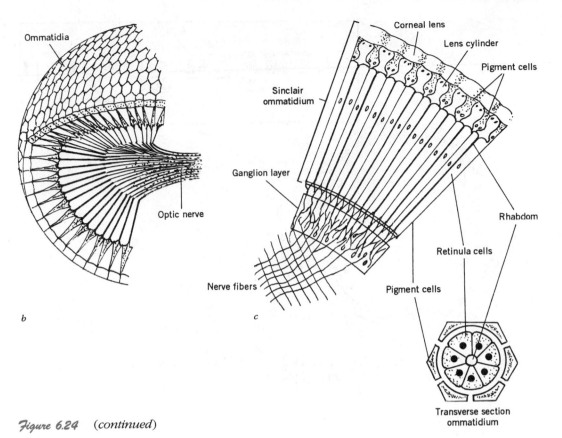

Figure 6.24 (continued)

front of the eye it is transparent and allows light to pass; and here it is called the cornea. Behind the cornea is the iris, a ring of muscle with a central opening called the pupil. By changing its size, the pupil regulates the amount of light entering the eye. Immediately behind the iris is the lens which focuses light from distant objects on the retina, the light-sensitive layer that lines the inner surface of the eye. The retina contains the photoreceptors. We will consider the structure of the retina in some detail because this information will establish a base from which we can discuss sensory filtering, the process by which sensory systems pass along only a fraction of the total information available to the animal.

Two types of photoreceptors are found in the mammalian retina: rods and cones. The rods are highly sensitive and function in dim light. Cones, on the other hand, function in daytime vision— they are responsible for color vision and for vision of the greatest sharpness or acuity. Cones are not stimulated in dim light. That is why humans lack acute vision and have a difficult time distinguishing color at night. Both types of receptor cells have molecules of photopigment in their extensively folded membranes.

The retina is divided into three main layers, the photoreceptive layer (the layer containing rods and cones), the bipolar cell layer, and the ganglion cell layer. Oddly enough, the photo-

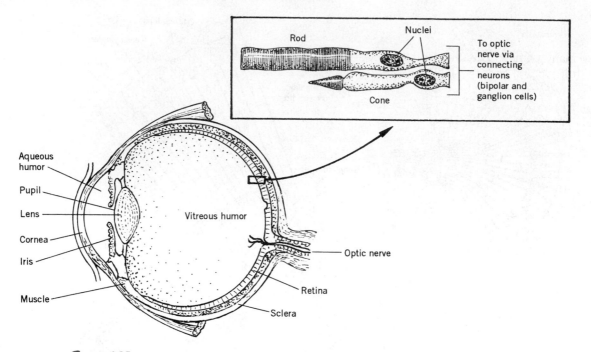

Figure 6.25 The human eye. Light is focused through a lens and falls on the retina composed of color-sensitive cones and light-sensitive rods.

receptive layer is at the back of the retina, and thus light must pass through the overlying layers (which are transparent) to reach the rods and cones. The rods and cones synapse with bipolar cells, which, in turn, synapse with ganglion cells, whose axons travel by way of the optic nerve to the brain. Two other types of cells—horizontal cells and amacrine cells—are also found in the retina. These cells transmit information parallel to the surface of the retina (as opposed to straight through the retina) and thus are involved in the interactions among adjacent photoreceptors.

The first step in transducing light energy into the energy of the nervous system involves the molecules of photopigment that are embedded in the membranes of the photoreceptor cells. When light hits the photopigment, the molecules

change configuration, and through a series of enzymatic reactions, they close the ion channels that are normally open in the receptor cell membrane. The decreased permeability of the receptor cell membrane to sodium ions causes the membrane to hyperpolarize (recall that the photoreceptors of invertebrates do just the opposite, they depolarize in response to light)—this change in the membrane potential is the receptor potential. Rather than being converted into an action potential, the receptor potential itself is conducted down the axon. The bipolar cells relay information to the ganglion cells, which, in turn, send nerve impulses along their axons (collectively called the optic nerve) to the brain. We will consider how visual information is processed, both in the retina and in the brain, in our discussion of sensory filtering.

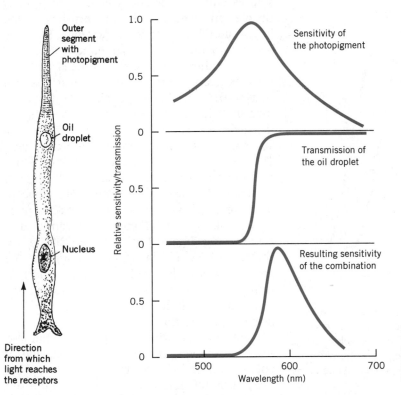

Color Vision

Many animals, including certain insects, fish, reptiles, birds, and mammals, are able to discriminate between different wavelengths; in other words, they have color vision. Color vision results when at least two receptor types respond differently to a given wavelength of light. The ratio of output from different types of receptors varies with the wavelength of light. The nervous system interprets this output as light of a particular color.

Color vision appears to have evolved several times. This notion is supported by the variation in the mechanisms by which receptors respond to certain wavelengths of light. In some animals the cones have different pigments; in others, color vision results from oil droplets in cones that selectively filter out or enhance certain

wavelengths of light before they strike the pigment.

In reptiles and birds, the light striking the pigment in a cone must first pass through oil droplets of different colors. Each color oil droplet filters out different wavelengths (Figure 6.26). Therefore, even when all cones have the same pigment, if different receptors have different oil droplets, the pigment will be struck by a different wavelength of light and the output of receptors will vary. As a result, the animal has some color vision. The pigeon, for instance, has the same pigment in all cones, but each cone has one of three colors of oil droplets. The spectrum of light reaching the pigment, therefore, varies with the color of the oil droplet in the cone and the relative responses of the cones differ (Martin and Muntz, 1978).

In certain primates, including humans, there

Figure 6.26 Example of the change in sensitivity to wavelengths due to an oil droplet in a cone. (From McFarland, 1981.)

are no oil droplets in the cones of the eye; instead, there are several pigments, each in one of three different types of cones (Wald, 1968). Each pigment absorbs a specific range of wavelengths. The wavelengths absorbed by each of the three cones of the human eye are shown in Figure 6.27. One pigment best absorbs wavelengths close to blue. The second maximally absorbs those around green, and the third absorbs red light the best. The wavelengths absorbed by the three types of cones overlap considerably. When light is absorbed, the pigment changes structure and a receptor potential is generated. When yellow light (550 nm) strikes the retina, the three kinds of cones respond differently. Blue cones do not absorb the yellow wavelength and do not respond at all. The red and green cones absorb this wavelength to different degrees and respond accordingly. The ra-

tio of the output from the three cone classes will vary for each wavelength of light. The ratio of impulses generated by these three cones, then, is interpreted as a particular color. The interpretation, in fact, begins before the impulses leave the retina. Recall that the cone cells synapse with a layer of bipolar cells, and these bipolar cells react differently to each type of cone cell. A given bipolar cell may be inhibited by the output of one cone type, but stimulated by the output from another type. For example, the interactions between the cones and the bipolar cells are such that the effects of blue and yellow light are opposite, as are the effects of red and green light (DeValois and Jacobs, 1968).

Detection of Polarized Light

When we look at the sky on a clear sunny day, it commonly appears uniformly blue. However, with careful scrutiny and a little training, most people can detect "Haidinger's brushes" in the bright blue sky. These look something like blue and yellow dumbbells oriented at right angles to one another. It is possible to see Haidinger's brushes because the skylight is polarized; that is, its waves move in a single plane. As they leave the sun, light waves vibrate in all directions, but that light is scattered and made to move in a single plane as it passes through the atmosphere. The blue section of Haidinger's brushes is oriented in the same direction as the plane of polarization. An easy way to sense the plane of polarization of skylight is to don a pair of polaroid sunglasses and tilt your head from side to side while gazing at different areas of the sky. You should see the sky appear to darken as the polaroid lenses block light vibrating in most directions. The sky appears to brighten when the polaroid filters in the lens are aligned with the plane of polarization of light.

Many animals, primarily arthropods, but also some vertebrates, can determine the plane of polarization of light. Such information can be used in orientation. Some animals use the sun

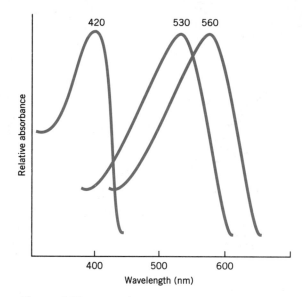

Figure 6.27 Spectral sensitivities of the three types of cones in the human retina. The pigment in each cone type varies and absorbs different wavelengths of light. The ratio of activity among the cone types and the integration of that output is responsible for our color vision.

as a compass for orientation. The plane of polarized light in the sky is related to the position of the sun, so animals that are sensitive to polarized light can view a small section of blue sky and determine the position of the sun, even when it is obscured by clouds (Waterman, 1961). (We will consider how some animals use polarized light for orientation in Chapter 11.) Because light reflected off water is strongly polarized, another possible use of sensitivity to polarized light is to detect bodies of water. Water bugs (*Notonecta*) dispersing from a pond use this cue to locate a new body of water. When the insect flies over an area from which horizontally polarized light is reflected upward, a landing response is triggered and the insect alights (Schwind, 1983).

In order to detect light that is polarized, a receptor must be more responsive to light vibrating in one direction than in others. Recall that in photoreceptors, light must be absorbed by a pigment molecule, which then changes its shape and initiates a receptor potential. Pigment molecules absorb light best when the direction of polarization is oriented in a certain manner relative to the molecule. The cell's response increases with the amount of light absorbed.

Insect eyes are amazingly sensitive to polarized light, but those of vertebrates are not. This difference in sensitivity is due to differences in how rigidly the pigment molecules are positioned in the membranes of the photoreceptors. In the vertebrate eye, pigment molecules are free to rotate in the membrane. Because they have no special orientation, pigment molecules in a vertebrate photoreceptor absorb light waves vibrating in any direction equally well. Therefore, there is no sensitivity to the plane of polarization. However, in the insect eye, the pigment molecules have a set orientation in the membranes of the receptor cells. As a result, insect eyes absorb light vibrating in one direction more than others (Figure 6.28). Insects apparently only detect polarized light in the ultraviolet wavelengths, and not all the ommatidia of

the compound eye can detect light polarized in a specific plane.

Because pigment molecules rotate freely in the membrane of the photoreceptors in a vertebrate's eye, it was surprising to learn that some vertebrates, including fish (Waterman, 1975), salamanders (Adler and Taylor, 1973; Taylor and Adler, 1973) and pigeons (Kreithen and Keeton, 1974) can detect changes in the plane of polarization of light. How is this possible? The mechanism is not fully understood in all species, but in salamanders at least, the structure responsible for sensitivity to polarized light is not the eye at all, but the pineal gland, a peculiar structure that lies near the top of the brain within the skull and contains photoreceptor cells.

Sensory Processing

In spite of the wide variety of stimuli bombarding an animal from its environment, it is able to detect only a limited range, and of those that it detects, it may ignore all but a few key stimuli. Recall, for example, that the "ears" (actually a few sensory hairs) of the caterpillar *Barathra* are particularly sensitive to sounds in the frequency produced by its archenemy, the parasitic wasp *Dolichovespula*. *Barathra*'s sensory hairs are more easily deflected by sounds around 150 Hz, the approximate frequency of the sound of the wasp's wingbeats, than by other frequencies—frequencies outside the 150 Hz range are most likely associated with events of less significance to the caterpillar than the arrival of the wasp. Recall, also, that the cockroach runs when exposed to a gust of wind coming from a predator, yet it ignores nonthreatening or irrelevant sources of wind, such as the wind that it creates itself while walking. How is such selectivity possible? In the laboratory, cockroaches run when exposed to wind with a peak velocity of only 12 millimeters per second, the approximate velocity of wind created by the lunge of a predator

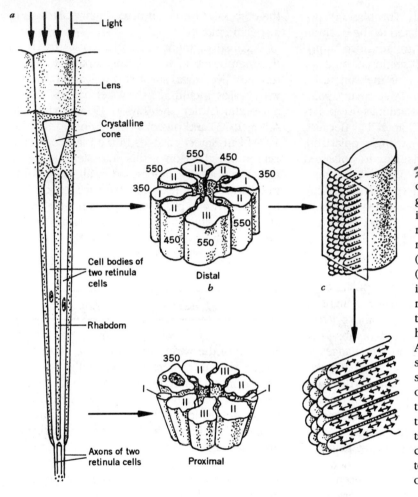

Figure 6.28 An ommatidium of a bee's eye. (*a*) In this longitudinal section, the rhabdom is visible in the center. The rhabdom is formed by pigment-containing projections (microvilli) of retinula cells. (*b*) Retinula cells are arranged in a circle. (*c*) The pigment molecules in the microvilli that form the rhabdom are held in a specific orientation. As a result, the eye is more sensitive to light vibrating in some planes than in others. In other words, the eye is sensitive to the plane of polarization of light. Interestingly, certain cells, particularly retinula cell 9, have greater sensitivity to polarized light than do others.

such as a toad. Cockroaches, however, do not respond to the 100 millimeters per second wind they create by normal walking. How is it that cockroaches manage not to respond to the relatively large wind created by walking and yet run when exposed to much smaller wind signals? It turns out that it is not the velocity of the wind that is the critical factor, but the acceleration (rate of change of wind speed), and a strike by a predator delivers wind with greater acceleration than the wind produced by the stepping legs of a cockroach. In fact, when cockroaches were tested with wind puffs all having the same peak velocity but differing in acceleration, they ran more frequently when exposed to winds of higher acceleration (Plummer and Camhi, 1981). Winds with low acceleration typically produced no response. Thus, cockroaches appear to pay particular attention to the acceleration of the wind stimulus, and this allows them to ignore irrelevant wind signals and to focus on important information in their environment.

We can see, then, that the job of an animal's sensory system is not to provide total information, but rather to be selective and provide only information that is vital to the animal's life-style or, more to the point, information that influences its reproductive success. How does an animal's nervous system enable it to focus on important events and to ignore irrelevant ones? We will consider the selective filtering action of nervous systems in the little skate and in the common toad.

STIMULUS FILTERING IN THE LITTLE SKATE

So far in our coverage of sensory systems, we have focused on two systems that are quite familiar to us—hearing and vision. Some animals, however, gather and process information from the environment using sensory systems that, from our perspective, can only be described as remarkable. The little skate (*Raja erinacea*) is one such animal, having receptors and brain structures that are specialized for detecting and analyzing weak electric fields. These fields are so weak that we, as humans, are totally unable to detect them. Little skates, however, derive a wealth of information from these fields; particulary critical is information relating to the whereabouts of their next meal. This specialized sensory ability of skates is called electroreception and it is the most recently discovered sensory modality (Bullock and Szabo, 1986).

The ability to detect weak electric fields is not unique to little skates, or even to other elasmobranch fish (sharks, skates, and rays). Indeed, electroreception seems to have evolved several times in a number of families of fish and amphibians (although its occurrence is still fairly rare), and it has been reported in at least one species of mammal, the duck-billed platypus (Scheich et al., 1986). What do these animals have in common? Electroreception, as you might guess from our list of animals, is associated with aquatic habitats, because water, unlike air, serves as a good conductor of electricity.

There are two general categories of electroreception, active and passive. Some animals,

such as the electric fish of Africa (Mormyriformes) and South America (Gymnotiformes), use specialized electric organs to generate an electric field around their body—this is active electroreception. Electric fish have two different types of receptor organs, called ampullary and tuberous receptors. Little skates, however, do not generate their own electric fields by means of special electric organs. Instead, skates respond to incidental electric fields in their environment such as those fields produced by prey or the earth's magnetic field—this is passive electroreception. Animals with this system possess only one type of receptor, the ampullary organ. In the case of elasmobranchs, ampullary organs are called ampullae of Lorenzini, after their seventeenth-century discoverer.

The ampullae of Lorenzini of the little skate are distributed in clusters over the head and expanded pectoral fins (Figure 6.29). Eighty percent of the receptors are located on the ventral body surface and are especially dense in the area of the snout, just in front of the mouth. Skates feed on bottom-dwelling invertebrates, and thus it makes sense that receptor organs are concentrated in areas where they permit accurate detection and localization of prey. Now let's consider the structure of ampullary organs in order to understand how they function.

Ampullary organs consist of a canal that opens by way of a pore on the skin surface. (Note that within a cluster of ampullary organs, the canals radiate in all directions; see Figure 6.29.) Each canal runs some distance beneath the skin and then terminates in a bulb-shaped structure, called the ampulla, which contains numerous pouches, each with hundreds of receptor cells. The canal is filled with a jellylike substance that is a good conductor of electricity; the walls of the canal are nonconductive. The jelly thus serves as a connection between the receptor cells at the base of the canal and the skin surface. The receptors can detect a difference in electrical potential between the water at the skin pore and the interior of the body, at the ampullae.

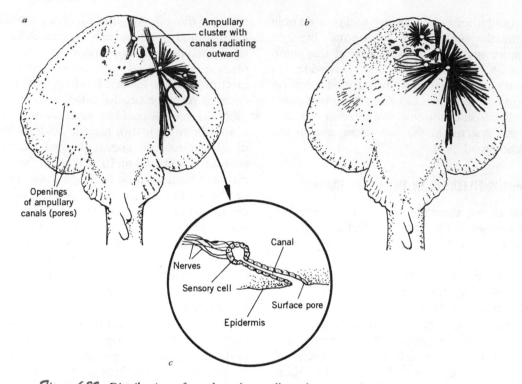

Figure 6.29 Distribution of canals and ampullary clusters in the thornback ray, a relative of the little skate. (*a*) Dorsal surface. (*b*) Ventral surface. (*c*) A single ampullary organ. Elasmobranch fish such as skates and rays use electroreception to locate prey—note the concentration of receptors on the ventral surface, especially around the mouth. (*a*, *b*; Redrawn from Montgomery, 1984.)

This property of ampullary organs permits the detection of electrical fields that are basically changes in electrical potential in space. Signals from the receptor cells are passed on to the anterior lateral line nerve (ALLN), which projects to the brain, specifically to an area of the medulla called the dorsal octavolateralis nucleus (DON). From here, messages are sent via ascending efferent neurons (AENS) to other areas of the brain.

One problem faced by little skates, as well as by other animals with similar electrosensory systems, is how to deal with self-generated "noise."

Although little skates do not generate electric fields around their bodies with specialized electric organs, their normal respiratory movements do, in fact, generate weak electric fields. These self-generated fields stimulate the skate's own electroreceptors, and in so doing, they could interfere with the detection of extrinsic electric signals, such as those produced by prey.

How is the nervous system of the little skate designed so that it filters out irrelevant information pertaining to the animal's own respiratory movements? Apparently, within the central nervous system of the skate, there is a neural mechanism that, in effect, cancels out signals concerning respiratory movements from opposite sides of the body (New and Bodznick, 1990). How does the cancellation occur? First, all receptors, regardless of their location on the body surface, respond in a similar fashion to the electric fields generated by respiratory movements; thus, the signals coming from receptors on each side of the body are equal. Second, as a result of the synaptic connections within the central nervous system, the signals from one side of the body are excitatory, while those from the other side of the body are inhibitory. As a result of this arrangement, the equal and opposite signals essentially cancel each other out. It is only when an electric field stimulates a localized area of electroreceptors on the skate's body, such as when a small crustacean edges along the ocean bottom, that the skate pays attention to the signal. The cancellation occurs in the DON because interference from respiratory movements is substantial in the ALLN fibers leading from the electroreceptors into the DON and is negligible in the fibers leaving the DON. By suppressing the electrosensory signals caused by respiratory movements, the neural mechanism in the DON allows the skate to focus on biologically relevant electrical stimuli, such as those associated with invertebrate prey.

SELECTIVE PROCESSING OF VISUAL INFORMATION IN THE COMMON TOAD

The nervous systems of amphibians, like that of the little skate, must filter out irrelevant information from the environment. In addition to this general filtering action, many nervous systems have cells that are specialized to detect specific, relevant features of certain stimuli. Frogs and toads, for example, recognize prey (e.g., earthworms, slugs, and beetles) and predators (e.g., large wading birds such as herons in the case of frogs, and hedgehogs in the case of toads) by sight, using, what may seem to us, like rather simple criteria to make this important distinction (see following discussion). Note that this distinction is made by specialized cells in the brain.

Over the past 20 or so years, Jorg-Peter Ewert and coworkers have examined how common toads (*Bufo bufo*) use visual information to recognize prey. Their approach combines behavioral observations with investigations of the underlying neural circuitry, and the results provide a remarkable example of how an animal's sensory system is comprised of specialized features that promote the detection of critical stimuli (Ewert, 1987; Ewert et al., 1990).

When a toad is sitting in a garden and a slug moves into its visual field, the toad turns toward the prey, fixes it in its sight, leans forward, flips out its sticky tongue, captures the prey, and draws it back into its mouth where it is swallowed with an eye-closing gulp. All this, of course, is followed by a ceremonial wipe of the mouth (Figure 6.30a). Part of this behavioral sequence—the initial orientation toward prey—was used by Ewert to determine exactly what stimulus parameters toads use to recognize prey. To get at this problem, he placed a hungry toad in a glass container and moved cardboard models that emphasized various stimulus characteristics horizontally around the toad (Figure 6.30b). If the toad interpreted the model as prey, it would orient toward the model, with the frequency of turns increasing as the resemblance between the model and prey increased. If, on the other hand, the toad interpreted the model as predator, it might freeze or crouch. As it turns out, toads distinguish between predator and prey simply on the relationship between two characteristics—the length of the object in relation to its direction of movement. Whereas a small stripe elongated in a horizontal direction (called the "worm configuration" by Ewert) is interpreted as prey, a stripe elongated in a vertical direction (the "antiworm configuration") is interpreted as predator, or at least as something inedible. In order to understand how such a

a

b

Figure 6.30 (*a*) The behavioral sequence of prey catching in the common toad. (*b*) Analysis of prey recognition in the toad. Here, the toad is confined to a glass container and a prey model (P) is moved horizontally around it. If the toad interprets the model as prey, it orients to the model. (*a*; Modified from Ewert, 1983; based on descriptions in Ewert, 1967; *b*; Modified from Ewert, 1969.)

distinction is made by toads, we will need to consider how visual information is processed, first in the retina, and then in the brain.

Processing of Visual Information in the Retina

Considerable processing of information occurs in the retina of the common toad. From our description of the visual system of vertebrates, you will recall that light energy hits the retina, which contains a layer of photoreceptors. Receptor potentials are then relayed to bipolar cells, which, in turn, send their messages to ganglion cells. The axons of the ganglion cells collectively form the optic nerve, which carries impulses to the brain (in the case of the toad, to the optic tectum and thalamus). How does the retina process visual information before passing it to the brain?

The ganglion cells within the toad's retina participate in one level of stimulus filtering. Such filtering can be examined by recording with microelectrodes the responses of individual ganglion cells to various stimuli moved in front of a stationary toad. Each ganglion cell receives input from several bipolar cells, and thus each ganglion cell receives information from a large number of receptors. The region of the retina from which receptors send messages to bipolar cells and then on to a specific ganglion cell is called the receptive field of that ganglion cell (Figure 6.31). Receptive fields in the retina of the toad are roughly circular in shape and have an outer inhibitory area that

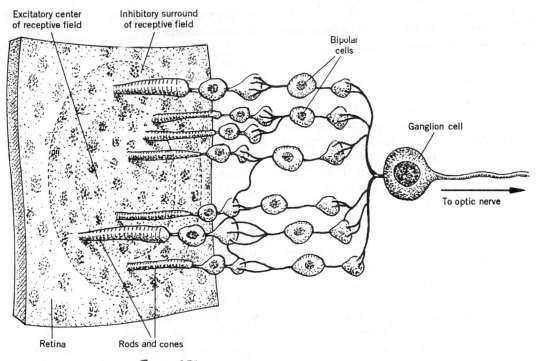

Excitatory center
of receptive field

Inhibitory surround
of receptive field

Bipolar
cells

Ganglion cell

To optic nerve

Retina

Rods and cones

Figure 6.31   The receptive field of a ganglion cell.

surrounds a central excitatory region. If a small object, let's say a slug, moves in front of a toad, the small image created on the retina will cover groups of receptors, some of which may comprise the excitatory region of a ganglion cell's receptive field. In this instance, then, the ganglion cell may fire and send impulses to the brain. If, however, the toad encounters a large stationary object, perhaps a garden wall, the image on the retina may cover both the inhibitory and excitatory regions of many ganglion cells and the ganglion cells probably will not fire. (When an inhibitory influence of a stimulus offsets an excitatory stimulus in a nearby region, the phenomenon is called lateral inhibition.) Because of their special properties, then, ganglion cells detect movement and pass this information

on to the brain where it will be used to determine whether such movement is associated with prey or predator.

Not all ganglion cells of the common toad transmit exactly the same type of information. Indeed, one of the discoveries from the recording experiments of Ewert and others is that in *B. bufo*, there are several different classes of ganglion cells. These classes of cells differ in the size of the centers of their receptive fields and in their responses to movement and contrast. Nevertheless, the important point here is that the ganglion cells filter the information provided to them by the bipolar cells, selecting only the most important aspects (e.g., information pertaining to movement, contrast, changes in illumination) and passing this information on to the brain for further processing. It is in the brain, and not in the retina, that the actual distinction between prey and predator is made.

Processing of Visual Information in the Brain

From the ganglion cells the toad's brain receives information on the basic parameters of the visual stimulus (e.g. dimensions, direction and rate of movement). In the brain, specifically in the optic tectum, there is a class of neurons (called Class T5[2] neurons) that responds only to certain combinations of features, those combinations that normally would identify a prey item. How were these feature-detecting neurons identified?

In a manner similar to that described for ganglion cells, microelectrodes can be used to monitor the responses of individual neurons in the brain of a stationary toad presented with various moving stimuli. Such recording experiments have shown that neurons in the optic tectum and thalamus can also be categorized into different classes, classes that are based, in part, on characteristics of their receptive fields (because neurons within the brain receive information from a cluster of ganglion cells, each neuron in the brain also has a receptive field—that is, an area of the retina monitored by its ganglion cells). Among the neurons that have been studied so far, the responses of the Class T5(2) neurons of the optic tectum show the best correlation with stimuli associated with prey recognition. Specifically, they show a greater response when the worm configuration is presented to the toad than when the antiworm configuration is presented. Thus, the T5(2) neurons respond specifically to combinations of visual characteristics that usually allow a toad to correctly distinguish between prey and predator or, more generally, between prey and inedible objects. We should point out, however, that the T5(2) neurons do not precisely identify worms, slugs, or beetles, but rather they respond to a specific set of characteristics that normally serves

to distinguish such delicacies from other objects in the environment.

Of course, the neural circuitry for prey-catching behavior in the toad does not end in the optic tectum. As is true for the neural circuitry underlying patterns of behavior in all animals, the sensory system must interface with the motor system to generate behavior. In toads, how is information about prey processed after the optic tectum? Where do neurons carrying information on the visual features of prey eventually interface with neurons involved in motor output? Although the answers to these questions are not entirely clear at present, it is known that the axons of some of the neurons in the optic tectum project to the medulla, an area of the brain near the spinal cord. By recording from neurons in the medulla, Ewert and colleagues have found some neurons that are responsive to visual stimuli associated with prey (Ewert et al., 1990; Schwippert, Beneke, and Ewert, 1990). Thus, this region of the brain may be an important premotor or motor processing station for information pertaining to prey. We see, then, that the neural basis for prey recognition in the common toad is actually a multilevel operation with processing of information occurring in the retina, optic tectum, and medulla. Now that we have some idea about how information picked up by sensory receptors is processed in the nervous system, let us turn our attention to a consideration of how such information results in motor output.

Motor Systems

Animals receive information about their environment via their sensory systems and then respond by means of their motor systems. Two principal components of motor systems are motor organs (typically muscles) and the neural circuits that control them. Like sensory receptors, muscles are often described as biological transducers. There are, however, some important distinctions between the two. Whereas sensory receptors transduce environmental energy such as light into the electrochemical signals of the nervous system, muscles convert the signals of the nervous system into the movements of the body. Furthermore, whereas sensory receptors are concerned with input, muscles are involved in output.

SOURCES OF NEURAL CONTROL IN MOTOR SYSTEMS

A particular movement, or behavior, is produced by muscles whose activity is controlled by motor neurons. Recall that motor neurons, in turn, usually receive their information from interneurons. Thus, each movement is ultimately controlled by the activity of a neural circuit. It turns out

that there are three major ways that neural circuits control and coordinate movement: (1) the sensory reflex, (2) the central pattern generator, and (3) motor command.

In sensory reflexes, sensory neurons initiate activity in motor neurons, sometimes through direct synaptic connections, but more typically, through connections with a small number of interneurons. A familiar example of a reflex is the knee-jerk reflex (also called the patellar reflex) in humans (Figure 6.32). This is the reflex that doctors sometimes check when testing for major damage to the spinal cord by tapping the patient's knee with a small rubber hammer. (The function of this reflex is actually to make small corrections in the position of a person's legs in order to maintain correct posture while standing.) The tap of the doctor's hammer stretches the patellar tendon. The patellar tendon pulls on and stretches the quadriceps femoris muscle,

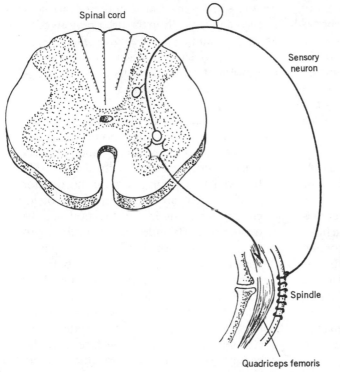

Figure 6.32 The patellar reflex. Excited stretch receptors in the quadriceps femoris muscle produce action potentials in sensory neurons whose processes extend to the spinal cord where they synapse with motor neurons. The activity of the sensory neurons excites the motor neurons innervating the quadriceps femoris muscle, causing the muscle to contract and the leg to extend.

the muscle responsible for extending the leg. Stretching the quadriceps femoris causes stretch receptors in the muscle to fire, and the excited stretch receptors produce action potentials in sensory neurons. At synapses in the spinal cord, the sensory neurons excite motor neurons. The motor neurons that these sensory neurons excite are the ones that cause contraction of the quadriceps femoris muscle. The result is a contraction of the leg muscle and an extension or kick of the leg. (Although not critical for the current discussion, the sensory neurons also synapse with interneurons in the spinal cord that inhibit motor neurons to the muscle responsible for the opposite action, that is, flexing the leg.) With so few neurons involved, the signal travels its entire route very quickly and with little interference from higher centers. This speed is one of the great advantages of the reflex.

In the case of central pattern generators, motor neurons are activated by rhythms generated within the central nervous system itself. More specifically, within the spinal cord of vertebrates or the nerve cord of invertebrates, there is a neuron or network of neurons that is capable of generating patterned activity in motor neurons, even when all sensory input has been removed from the system. The idea of central control of motor patterns is obviously quite different from that of peripheral control exemplified by the role of sensory neurons in reflexes.

Finally, when one or more interneurons descending from the brain initiate activity in motor neurons, the neural mechanism is called motor command. These interneurons, sometimes called command neurons, act in a switchlike manner, determining, on the basis of incoming sensory information, whether or not a particular pattern of behavior will be initiated. One job of command neurons appears to be the turning on and off of central pattern generators. Command neurons are involved in patterns of behavior such as the escape response of crayfish and the startle response of teleost fish.

We will describe each of the three sources of neural control in more detail in the context of the flying behavior of locusts. The precise role and relative importance of sensory feedback, central pattern generation, and motor command have been hotly debated for motor patterns in general, and for locust flight in particular (e.g., Pearson, 1987; Stevenson and Kutsch, 1987). A more detailed review of the neuromuscular basis for locust flight can be found in Young (1989).

LOCUST FLIGHT

Locusts are species of "short-horned" grasshoppers in the family Acrididae that exhibit legendary mass migrations. In fact, accounts of locust plagues date back to the Book of Exodus, written about 1500 B.C. (Williams, 1965). Swarms of locusts have also been recorded in recent times (Figure 6.33). Although representatives of the family are found throughout the world, the migratory species are found in the tropics or subtropics, and typically in the drier regions of these areas. As evidence of the amazing flight behavior of locusts, consider this account written by the entomologist C. B. Williams (1965, pp. 77–78) while working in East Africa:

A most spectacular flight occurred on 29th January, 1929, at Amani, in the Usambra Hills in north-eastern Tanganyika.... We received a telephone warning shortly after breakfast that an immense swarm of locusts was passing in our direction over an estate about six miles to the north.... An hour or so later the first outfliers began to appear—gigantic grasshoppers about six inches across the wings, and of a deep purple-brown. Minute by minute the numbers increased, like a brown mist over the tops of the trees. When they settled they changed the colour of the forest; by the weight of their numbers they broke branches of trees up to three inches in

Figure 6.33 A swarm of desert locusts in Ethiopia.

diameter; the noise of their slipping up and down on the corrugated iron roofs of the houses made conversation difficult.... The swarm was over a mile wide, over a hundred feet deep, and passed for nine hours at a speed of about six miles per hour.

The species of locust described in Williams's account was the desert locust (*Schistocerca gregaria*; Figure 6.34). Although this species has been used in some laboratory studies of flight behavior, the migratory locust (*Locusta migratoria*) is more commonly studied. Let us now take a closer look at the neural control of flight behavior in these remarkable insects.

The Motor Pattern

Locusts have two pairs of wings, the forewings, located on the second thoracic segment, and the hindwings located on the third thoracic seg-

ment. The wings of free-flying locusts move up and down, in a rhythmic manner, about 20 times per second. Because locusts maintained in the laboratory and tethered to a holder exhibit close to normal flight, it is possible to obtain a detailed analysis of the motor pattern (Figure 6.35*a*).

Figure 6.34 A desert locust.

During flight, the two pairs of wings do not move in a precisely synchronous manner, but rather the hindwings lead the forewings. The entire cycle of movement lasts about 50 milliseconds (Figure 6.35*b*).

Two sets of muscles act on each wing. One set, the elevators, raises the wing. The other set, the depressors, lowers it. Because these two sets of muscles have opposite effects on the wing, they are called antagonists (muscles with the same effects on a specific structure are called synergists).

How do signals from the locust's nervous system initiate activity in the flight muscles? A motor neuron and the muscle fibers (muscle cells) that it innervates comprise a motor unit; the junctions between motor neurons and muscle fibers are called neuromuscular synapses. When an action potential travels down the axon of a motor neuron, it moves across the neuromuscular synapse to the fibers in the motor unit. The resulting postsynaptic depolarization produces a muscle action potential that spreads across the surfaces of the muscle fibers and is then conducted to the insides of the fibers where it produces twitch contractions.

Patterns of activity in the flight muscles of a tethered locust can be recorded by inserting tiny wire electrodes into the muscles (again, see Figure 6.35*a*). The signals (a recording of muscle action potentials) from the elevator and depressor muscles of the flying insect are then displayed on an oscilloscope; this type of recording is called a myogram. Myograms have shown that the depressors are activated when the wings are up and the elevators are activated when the wings are down (Figure 6.35*c*). The relative timing of the activation of these muscles is critical, and we next examine how such timing may be controlled by the central nervous system.

The Role of the Central Pattern Generator

The flying behavior of locusts was one of the first motor systems for which a central pattern generator was shown to be an important source of neural control (Wilson, 1961). The role of the central pattern generator was highlighted by experiments in which the other two major sources of neural control, namely motor command and sensory feedback, were eliminated. We will consider the results of each of these experimental procedures in turn.

In the laboratory, a stream of air directed at the head of a tethered locust can initiate flight. Wind-sensitive hairs on the head are stimulated and the excitation is passed along brain interneurons that descend to ganglia in the thoracic area (a ganglion is an area of the central nervous system where large numbers of interneurons occur in closely packed aggregations). These interneurons that descend from the brain constitute the control system of motor command. Decapitation eliminates this system of neural control and results in an animal that can still fly when tethered. (Obviously, in the decapitated locust, flight cannot be initiated by blowing air on the head; flight can, however, be started by either gently pinching the abdomen of the animal or by removing the platform on which it is resting.) The important point here is that the basic system for controlling flight can operate independently of motor command (Wilson, 1961).

Locusts have mechanoreceptors on their wings that send information on wing position, via sensory neurons, to the central nervous system. This control system can be eliminated by sectioning the sensory nerves (e.g., Wilson, 1961) or in a technique developed more recently, by injecting the chemical phentolamine, which blocks the activation of mechanoreceptors without affecting the central nervous system (Ramirez and Pearson, 1990). The removal of sensory input by either technique is called deafferentation (recall that sensory neurons are also called afferent neurons). Although some changes in the flight pattern are observed after the surgery (e.g., the frequency of wing beats drops from 20 to about 10 per second), the basic pattern of wing movements is still present. The removal of sensory feedback from the wings can

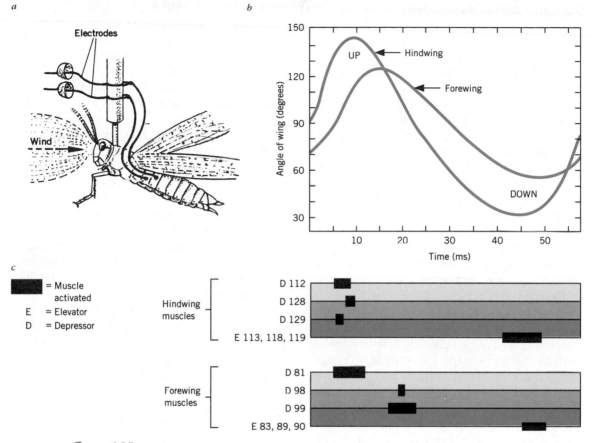

Figure 6.35 Analysis of the flight pattern of locusts. (*a*) By directing wind at the head, a tethered locust can be induced to fly. The activity of flight muscles of the second and third thoracic segments can then be measured by inserting small wire electrodes in them. (*b*) Pattern of wing movements during flight in a tethered locust. (*c*) Pattern of activity in individual flight muscles as recorded with microelectrodes. (*a*; Modified from Horsmann, Heinzel, and Wendler, 1983; *b*, *c*; Modified from Young, 1989; original data from Weis-Fogh, 1956 and Wilson and Weis-Fogh, 1962.)

be taken one step further than simply sectioning the sensory nerves. The wings and thoracic muscles can be completely removed. Even after such extensive dissection, a pattern of neural activity similar to that seen during flight occurs in the motor nerves that emerge from the thoracic ganglia. Taken together, these results suggest that the basic flight pattern of the locust is not dependent on sensory feedback but is generated within the central nervous system. In other

words, there is a neuronal circuit in the central nervous system that generates the alternating bursts of activity in motor neurons that innervate the elevator and depressor muscles.

At the level of the neuron, what is a central pattern generator? The central pattern generators underlying rhythmic patterns of behavior such as flying, walking, and breathing are called oscillators, and there are two basic types, cellular and network. A cellular oscillator is a neuron that generates rhythmic activity on its own. In contrast, a network oscillator is a collection of neurons (none of which alone qualifies as a cellular oscillator) whose interaction produces temporally patterned output. Sometimes central pattern generators combine the characteristics of cellular and network oscillators, and these are called hybrid or mixed oscillators. At present, the central pattern generator of locust flight appears to be a network oscillator.

How are neurons that are part of a network responsible for generating rhythmic output, such as the movements involved in locust flight, identified? The following three criteria must be met in order for a neuron to be considered part of a particular oscillator: (1) The neuron must have synaptic connections, either directly or indirectly, with the appropriate motor neurons—in the case of the locust, those motor neurons involved in flight; (2) the neuron must be active at the appropriate frequency—here, the flight frequency; and, in a relatively simple network such as that involved in locust flight, (3) when stimulated with current from a microelectrode, the neuron must be capable of resetting the flight rhythm. The motor neurons involved in flight do not meet these criteria. Some interneurons, however, have been discovered that do meet all three criteria, and these are considered to form part of the network of neurons comprising the central pattern generator (Figure 6.36; Robertson and Pearson, 1983, 1985). There remains, however, some debate as to whether separate oscillators exist to drive the movements of the forewings and hindwings, or whether the generator of the flight pattern, although distributed among several segmental ganglia, functions as a single unit (Robertson and Pearson, 1984; Stevenson and Kutsch, 1987; Wilson, 1961).

The Role of Sensory Feedback

Locusts have sensory receptors on their wings that monitor wing movement. At the base of each forewing and hindwing, for example, there is a stretch receptor. These receptors consist of a single sensory neuron that is attached to a strand of connective tissue running across the hinge of the wing. When the wings are raised, the receptors are stretched and stimulated. The axon of each sensory neuron sends branches to both the second and third thoracic ganglia and synapses with interneurons and motor neurons involved in flight.

Do stretch receptors meet the three criteria used to identify components of a rhythm-generating system? We have already mentioned that the sensory neuron of each stretch receptor synapses with the interneurons and motor neurons involved in flight—thus, stretch receptors meet the first criterion. With respect to the second criterion, stretch receptors are, in fact, active at the flight frequency. This can be shown by tethering a locust and preparing it in such a manner that both the wing position and the activity of a flight muscle are monitored. At the same time, the activity of a stretch receptor is examined by placing a fine wire around the sensory nerve that houses its axon. Such monitoring has revealed that stretch receptors produce rhythmic bursts of impulses that coincide with activation of the depressor muscles (Figure 6.37). In addition, the responses of stretch receptors are closely tied to wing movements. Specifically, the burst of impulses of the stretch receptor occurs near the peak of wing elevation. Thus, data pertaining to both wing position and wing muscle activity confirm that stretch receptors are active at the flight frequency (Mohl, 1985). The final criterion is also met—stretch receptors are capable of resetting the flight rhythm. This has been shown in experiments with deafferented locusts (animals in which all sensory nerves from the wings have been cut). Such animals, as previously mentioned, typically show a drop in their flight frequency from about 20 to 10 beats per second. Stimulation of the axon of a stretch receptor, however, can reset the normal flight rhythm (Pearson, Reye, and Robertson, 1983; Reye and Pearson, 1988).

Having met all three criteria, stretch receptors thus appear to play an integral role in the pattern

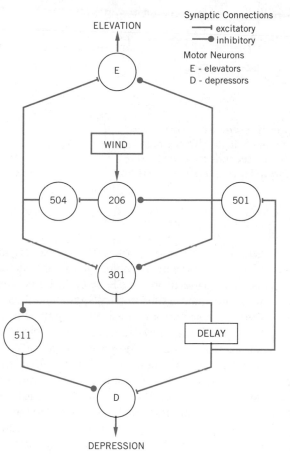

Figure 6.36 A schematic circuit showing the known elements of the central pattern generator for flight in the locust. The rhythmic activity produced by this network results from the interactions among component interneurons (identified interneurons have been assigned numbers). (From Robertson and Pearson, 1985.)

of motor output associated with flight in locusts. Importantly, because of their link with the outside world (e.g., sudden updrafts or downdrafts of air will cause changes in wing position that stimulate the receptors), stretch receptors impart some flexibility to the flight control system, allowing the animal to adjust its flight to prevailing environmental conditions on a cycle-by-cycle basis.

The Role of Motor Command

On the head of a locust, back between the compound eyes, can be found several fine hairs. As described earlier, these hairs are sensitive to wind, and when stimulated, they initiate and maintain flight. Within each hair can be found a sensory neuron whose axon projects to the brain where it synapses with interneurons whose axons extend down to the thoracic ganglia. These descending interneurons comprise the system of neural control called motor command. Responding not only to wind stimuli, but also to visual stimuli, these descending interneurons are described as multimodal because they are sensitive to stimuli from more than one sensory modality. It is thought that these descending interneurons are involved in modulating wingbeat during corrective steering maneuvers following air turbulence or errors in flight performance (Reichert, Rowell, and Griss, 1985; Rowell, 1989). (During steering, locusts actually appear

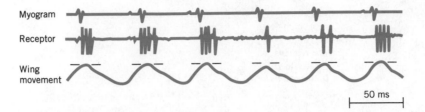

Figure 6.37 The activity of a forewing stretch receptor compared with forewing movements and the myogram of a hindwing depressor muscle. The forewing depressor muscle fires a few milliseconds after the hindwing depressor and is therefore in phase with the receptor burst. (From Mohl, 1985.)

to use information from several sources—their simple and compound eyes, antennae, cerci, and hairs on the head.)

This descending pathway plays an important role in the flight system of locusts, being responsible for turning on and off the central pattern generator and for modulating flight. The axons of the interneurons synapse with flight motor neurons and interneurons in the thoracic ganglia, and those interneurons that have been studied are rhythmically active during flight. In fact, because the locust's own wing movements produce air currents that stimulate its wind-sensitive hairs, the activity of the interneurons coincides with the flight frequency. Finally, activity of the interneurons can reset the flight rhythm. This has been demonstrated by either manipulating air currents directed at the head (Horsmann, Heinzel, and Wendler, 1983) or by electrically stimulating one of the interneurons (Bacon and Mohl, 1983). Thus, in their modulatory role, the descending interneurons, like the sensory neurons of the stretch receptors, ensure that the flight system is able to respond to environmental perturbations.

We see, then, that for the flight system of the locust, all three sources of neural control interact to influence the pattern of motor output. Although the basic rhythm of flight is generated by a central pattern generator in the thoracic

ganglia, information gathered by sense organs located on the wings enables the locust to adjust its flight pattern on a cycle-by-cycle basis. Sensory information thus imparts some flexibility to the system of motor control, allowing the animal to respond to air turbulence and other environmental uncertainties. The system of motor command plays a particularly important role in that the interneurons descending from the brain initiate (and modulate) activity in the central pattern generator. Interneurons descending from the brain appear to play a role in correctional steering, when the locust compensates for deviations in flight course. The combination of central and peripheral control observed in locust flight probably characterizes most patterns of rhythmic behavior in animals (Delcomyn, 1980).

Summary

The basic unit of the nervous system is the neuron, a cell that usually consists of three parts: a cell body, or soma, which maintains the cell; dendrites that receive information and conduct it toward the soma; and an axon that conducts the nerve impulse away from the soma. There are three functional classifications of neurons. A sensory neuron is specialized to detect stimuli or to receive information from sensory receptors and conduct it to the central nervous system. Interneurons, found within the central nervous system, link one neuron to another. A motor neuron carries the information from the central nervous system to a muscle or gland.

When a neuron is resting, that is, not conducting an impulse, it is more negative inside than outside. This resting potential is approximately -60 millivolts and results from an unequal distribution of ions. Many large, negatively charged proteins are held within the cell and give the interior the overall negative charge. The concentration of potassium ions (K^+) is roughly 30 times greater inside the neuron's membrane than outside. In contrast, sodium ions (Na^+) are more concentrated outside the membrane.

The message of a neuron is called an action potential or nerve impulse. The action potential is a wave of depolarization followed rapidly by repolarization that travels to the end of the axon with no loss in strength. Should a stimulus open ion channels that will allow sodium ions to cross the membrane, the ions will be drawn inside, attracted by the negative charge within and driven inward by their concentration gradient. The positively charged sodium ions depolarize the membrane. At the peak of depolarization, the sodium channels close and potassium channels open. Now potassium ions exit the cell, leaving that region of the membrane more negative on the inside once again (i.e., it is repolarized). The depolarization in one spot of the membrane triggers depolarization in the next adjacent region so that the impulse spreads along the axon.

Information travels from one neuron to the next by crossing the gap, or synapse, between them. There are two types of synapses, electrical and chemical. Electrical synapses involve such tight connections between the neurons that the impulse spreads directly from cell to cell. Electrical synapses are characterized by a rapid speed of transmission, and as a result, they are often part of the neural circuitry of escape responses. In contrast to the direct transfer of information that occurs at an electrical synapse, several steps are involved in transferring information across a chemical synapse. First, the action potential travels down the axon to small swellings called terminal boutons. Here, the action potential causes a neurotransmitter to be released from storage sacs (synaptic vesicles) into the gap between cells (synaptic cleft). The neurotransmitter diffuses across the gap and binds to a special receptor. When this occurs, ion channels open, changing the charge difference across the membrane. If the synapse is excitatory, the postsynaptic cell is slightly depolarized. However, if the synapse is inhibitory, the postsynaptic cell is slightly hyperpolarized, and no new action potential will be generated.

The membrane potential of the postsynaptic cell can also be changed by neuromodulators. These substances work more slowly than neurotransmitters and bring about changes via biochemical means. The effects of neuromodulators appear to be mediated by substances, called second messengers, within the postsynaptic neuron. These second messengers (e.g., cAMP and cGMP) activate enzymes that alter the ion channels of the nerve cell and thereby affect neural activity. Neuromodulators can have profound effects on behavior as demonstrated by the effects of the neuromodulator serotonin on feeding in the leech.

Animals receive information from the environment at their sense organs. Sense organs, such as eyes and ears, are selective, containing specific receptor cells that respond to a particular form of environmental energy. Sensory receptors transduce the energy of a stimulus into electrochemical energy of a receptor potential, a change in membrane potential caused by the redistribution of ions whose magnitude varies with the strength of the stimulus. When stimulation is prolonged, many receptors cease responding (sensory adaptation).

Despite the diverse array of stimuli bombarding an animal from its environment, it is able to detect only a limited range, and of those that it detects, it may ignore all but a few key stimuli. Sensory filtering may occur in the receptor itself (e.g., the sensory hairs of the caterpillar *Barathra* are particularly sensitive to sounds in the frequency produced by its archenemy, the parasitic wasp *Dolichovespula*) or at any step in the processing of information in the nervous system. Little skates have a neural mechanism in their central nervous system that filters out the electric fields generated by their own respiratory movements, allowing them to focus on fields generated by prey. In addition to filtering out irrelevant information from the environment, many animals have cells within their nervous system that are specialized to detect specific features of critical stimuli. In the brain of the common toad, there are neurons that respond only to certain combinations of features, those that normally identify prey.

Animals respond with their motor systems to information picked up by their sensory systems. Motor systems consist of motor organs (muscles) and the neural circuits that control them. There are three major sources of neural control for motor patterns: (1) the sensory reflex, (2) the central pattern generator, and (3) motor command. In sensory reflexes, sensory neurons initiate activity in motor neurons, sometimes through direct synaptic connections, but more typically, through connections with one or more interneurons. In the case of central pattern generators, motor neurons are activated by rhythms generated by a neuron or network of neurons within the central nervous system. Central pattern generators can activate motor neurons in the absence of all sensory feedback. Finally, motor command consists of interneurons descending from the brain that initiate activity in motor neurons. Such command neurons function to turn central pattern generators on and off. A close examination of the neural mechanisms underlying locust flight suggests that a combination of all three sources of control operates to produce rhythmic patterns of behavior.

CHAPTER
7

Physiological Analysis of Behavior — The Endocrine System

*L*ike peas in a pod, house mouse fetuses (*Mus musculus*) line the uterine horns of their mother (Figure 7.1). Each fetus has its own personal placenta (vascular connection to the mother) and floats within a fluid-filled compartment called the amniotic sac. Even before birth, the endocrine glands of these tiny individuals are producing hormones, chemical substances that may permanently alter not only their own, but also their neighbor's morphology, physiology, and behavior. In a fascinating series of experiments, Frederick vom Saal (1981) has shown that development of mouse fetuses can be modified by exposure to hormones secreted by contiguous littermates. As shown in Figure 7.1, there are three intrauterine positions that fetuses can occupy relative to siblings of the opposite sex. Females can be positioned between two male fetuses (2M females), next to one male (1M females), or not next to a male (0M females). But

what does intrauterine position have to do with adult patterns of behavior? vom Saal and Bronson (1980a) have shown that by day 17 of gestation, levels of testosterone (a steroid hormone secreted by the testes) are three times higher in the blood of male fetuses than in the blood of female fetuses. Even more intriguing is the finding that on this same day, 2M female fetuses (i.e., those females nestled between two male littermates) have significantly higher concentrations of testosterone in their blood and amniotic fluid than do female fetuses not next to males (i.e., 0M females). Apparently, hormones pass via either the amniotic fluid or uterine blood vessels to contiguous littermates. As a result of prenatal exposure to testosterone, adult 2M females display a host of traits that distinguish them from 0M females. Relative to 0M females, 2M females (1) are less attractive to males, (2) are more aggressive to female intruders, (3) mark a novel

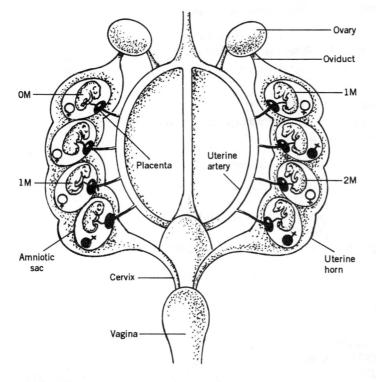

Figure 7.1 Mouse fetuses line the uterine horns of a pregnant female. Because the fetuses are in such close quarters, hormones from one fetus can influence behavioral development of contiguous fetuses. Female fetuses can occupy the following three positions relative to male fetuses: 2M = between 2 males; 1M = next to 1 male; 0M = not next to a male. 2M females differ substantially from 0M females in their adult behavior and physiology as a result of proximity to male fetuses in the intrauterine environment. (Modified from McLaren and Michie, 1960.)

environment at a higher rate, and (4) have longer and more irregular estrous cycles (vom Saal and Bronson, 1978, 1980a, 1980b). These differences in physiology and adult behavior exist despite the fact that after birth, testosterone levels do not differ between the two groups of female mice. Thus, behavioral differences in adulthood result from differential exposure to hormones in the intrauterine environment.

Position in the uterus also alters the adult behavior of male mice. Males that develop in utero between two male fetuses (2M males) tend to exhibit parental behavior when presented with a newborn pup, while males that develop between two female fetuses (0M males) usually kill the newborn (vom Saal, 1983). Differential exposure to testosterone is not responsible for these behavioral differences because concentrations of testosterone in the blood and amniotic fluid of 2M and 0M male fetuses do not differ. Instead, behavioral differences between 2M and 0M males in adulthood appear to result from differential exposure to estrogen, a steroid hormone secreted by the ovary and found in relatively high concentrations in female fetuses.

We can ask, then, just what are hormones and how do they produce such dramatic effects on behavior? Further, what role do hormones play in the development and display of behavioral differences between the sexes? Do hormonal effects on behavior vary as a function of species, season, experience, or social interactions? Can behavior, in turn, influence the presence of hormones? It is to these issues that we now turn.

The Endocrine System

DEFINITION OF ENDOCRINE GLAND AND HORMONE

We begin with a definition. Hormones are substances secreted in one part of the body that cause changes in other parts of the body. Hormones are secreted either by endocrine glands or by neurons. Unlike exocrine glands (e.g., sweat or salivary glands) that have specialized ducts for secretion of products, endocrine glands lack ducts and secrete their products directly into the bloodstream. Once in the blood, hormones travel along the vast network of vessels to virtually every part of the body. Hormones secreted by nerve cells are called neurohormones or neurosecretions. Rather than traveling in the circulatory system, neurohormones move along nerve axons. Most of the hormones of invertebrates are secreted by neurons rather than endocrine glands. Functioning as chemical messengers in an elaborate system of internal communication, hormones and neurohormones exert their effects at the cellular level, by altering metabolic activity or by inducing growth and differentiation. Changes at the cellular level can ultimately influence behavior.

HORMONAL VERSUS NEURAL COMMUNICATION

The endocrine system of an animal is closely associated with its nervous system. As mentioned, some hormones, in fact, are made by nerve cells. Additionally, neurons and hormones often work together to control a single process. For example, in some interactions, neurons respond to hormones, while in others, endocrine glands receive information and directions from the brain. Nervous and endocrine systems are so closely associated that they are often discussed as a single system, the neuroendocrine system. Despite this close association, neural and hormonal modes of information transfer serve different purposes within the body, and each system is essential in its own right.

In comparing communication via nervous or endocrine pathways, we should first briefly review how neurons transfer information. In the nervous system, information is transmitted along distinct pathways (chains of neurons) at speeds of up to 100 meters per second. After the impulse arrives at its destination within the body, neural information is transmitted via a series of electrical events that culminates in the release of neural transmitter near its target, such as an effector tissue—a muscle, for example. Neural transmitters are rapidly destroyed after they are secreted. As a result, information delivered by way of the nervous system usually produces a response that is rapid in onset, short in duration, and highly localized (Bentley, 1982). In contrast to neural communication, hormonal transfer of information tends to occur in a more leisurely, persistent manner. Typically, hormones are secreted slowly and remain in the bloodstream for some time. Rather than traveling to a precise location, these chemical messengers contact virtually all cells in the body, although only some cells are able to respond to the particular hormonal stimulus. In short, transfer of information by way of the endocrine system occurs more slowly than that of the nervous system, and it usually produces effects that are more general and long-lasting.

INVERTEBRATE ENDOCRINE SYSTEMS

In invertebrates, hormones regulate many physiological and behavioral aspects of reproduction as well as such processes as regeneration (flatworms and annelids), color change (crustaceans), and molting and metamorphosis (insects; Figure 7.2). There is substantial variation among invertebrate phyla in the structure and function of endocrine systems, and complete descriptions of each system are not possible here. We can, however, consider the neuroendocrine system of insects in some detail. We will begin with a general description of hormonal control of metamorphosis in insects and then describe the hormonal control of behavior in honeybees.

Hormonal Control of Growth and Metamorphosis in Insects

Because there are substantial interspecific differences in hormonal control of development among insects (a class with more than 750,000

Figure 7.2 A monarch butterfly emerging from a chrysalis (pupal case) during metamorphosis.

described species), we have to generalize a bit here. Growth in insects typically involves a series of steps in which the rigid nonexpansible exoskeleton is periodically discarded and replaced with a shiny, new, and most important, larger outfit. In addition, almost all insects undergo a process of metamorphosis, proceeding through a series of juvenile stages, each requiring the formation of a new exoskeleton and ending with a molt, before adulthood is reached. Some insects, such as grasshoppers and bugs, undergo a metamorphosis in which adult organs are grad-

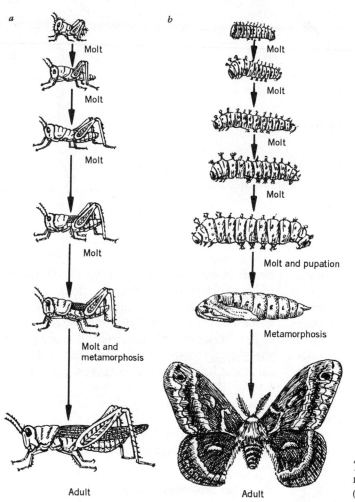

Figure 7.3 Metamorphosis in insects may be incomplete (*a*) or complete (*b*). (Modified from Gilbert, 1988.)

ually formed (a process called incomplete metamorphosis); other insects, such as flies, beetles, moths, and butterflies, undergo a sudden and dramatic transformation between larval and adult stages. During this latter process, termed complete metamorphosis, the old body of the larva is systematically destroyed during pupation and new adult organs develop from undifferentiated groups of cells. Figure 7.3 contrasts incomplete and complete metamorphosis. In most species of insects, hormones and neurosecretions that affect growth, development, and behavior are secreted at five locations: (1) neurosecretory cells located in the brain, (2) paired glands, corpora cardiaca, located just behind the brain, (3) paired glands, corpora allata, located alongside the esophagus, (4) single prothoracic gland located at the rear of the head (Figure 7.4), and (5) male and female gonads located toward the rear of the body.

Molting and metamorphosis are controlled by the interaction of two hormones (Figure 7.5), one favoring growth and differentiation of adult structures (MH = molting hormone; secreted

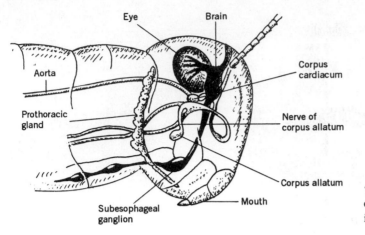

Figure 7.4 Neurosecretory cells and endocrine glands in the head region of an insect.

by the prothoracic gland) and the other favoring retention of juvenile characteristics (JH = juvenile hormone; secreted by the corpora allata, singular, corpus allatum). In the case of molting, some environmental factor (e.g., a temperature change) typically activates neurosecretory cells in the brain to secrete brain hormone (BH) into the corpora cardiaca (singular, corpus cardiacum) where it is stored. Once released from the corpora cardiaca, BH stimulates secretion of MH from the prothoracic gland and growth and molting ensue. In immature insects, the endocrine glands, called the corpora allata, secrete JH. At each larval molt, JH suppresses metamorphosis and ensures that the insect retains its immature characteristics. However, as the concentration of JH decreases over time, metamorphosis occurs and the insect is transformed from a larva into a pupa, the next developmental stage. Finally, once secretion of JH has ended altogether, the pupa molts and becomes an adult. Thus, metamorphosis to the adult form occurs when molting hormone acts in the absence of juvenile hormone. There are many behavioral changes associated with metamorphosis. In the tobacco hornworm (*Manduca sexta*), for example, a given individual changes from a crawling caterpillar to a flying adult moth in a

matter of weeks (see Chapter 8 for a discussion of how such dramatic changes occur).

Understanding the actions of juvenile hormone has proven beneficial to those interested in the production of insecticides that are highly specific and not harmful to the environment. Chemists have synthesized several potent analogs of JH that can be used to induce abnormal final molts or prolong or block development in many species of insects. Many flea sprays, for example, use an analog of JH to block development from immature to adult stages.

Hormonal Mediation of Behavioral Changes in Honeybees

Juvenile hormone (JH) also mediates morphological and behavioral changes in adult insects. In the eusocial insects in particular (i.e., all ants, all termites, and some bees and wasps), JH plays a role in the differentiation of distinct castes. Before examining hormonally mediated changes in the behavior of honeybee (*Apis mellifera*) workers, let us first briefly describe life history characteristics of eusocial insects. Eusociality will be covered in much greater detail in Chapter 19 in our discussion of altruism.

Eusocial insects must possess the following

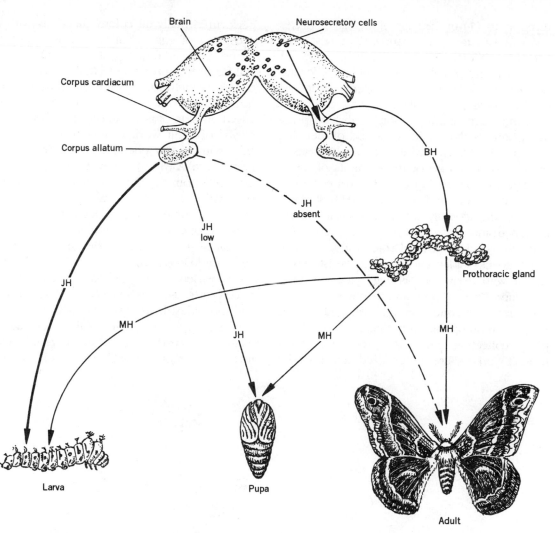

Figure 7.5 The neuroendocrine control of molting and metamorphosis in a moth. Neurosecretory cells in the brain secrete brain hormone (BH), which stimulates the prothoracic gland to secrete molting hormone (MH). In the immature insect, the corpora allata secrete juvenile hormone (JH), which suppresses metamorphosis at each molt. Metamorphosis to the adult form occurs when MH acts in the absence of JH. (Modified from Villee, Solomon, and Davis, 1985.)

three traits: (1) Individuals of the same species cooperate in caring for young; (2) a reproductive division of labor exists in which more or less sterile individuals work on behalf of fecund nestmates; and (3) there is an overlap of at least two generations in life stages that are capable of contributing to colony labor, so that offspring assist parents during some period of their life (Wilson, 1971). Honeybees, generally considered to occupy one of the summits of insect social evolution, live in colonies that consist of a single queen and up to 80,000 workers. Queens and workers differ dramatically in appearance and these differences are associated with divergent behavioral roles within the colony. Queens display a host of characteristics consistent with their role as egg-laying machines (e.g., larger overall body size than workers, large abdomens that contain greatly enlarged ovaries, reduced mouthparts and stingers, and lack of pollen-collecting and wax-producing structures). Workers are unfertilized females who

perform virtually all colony functions with the exception of egg laying (e.g., colony defense, brood care, and foraging) and have morphological features appropriate to their tasks.

Just as interesting as the differences among queens and workers in the honeybee is the age-based division of labor among workers. The behavior of adult worker bees changes dramatically with age, a phenomenon known as age polyethism. A typical honeybee worker that has emerged from the pupa as an adult winged insect will live for approximately 6 weeks and during this time moves through an orderly progression of tasks: Young adults (1–12 days old) specialize almost exclusively on cleaning nest cells; middle-aged bees (13–20 days old) perform a variety of tasks associated with brood and queen care, nest maintenance, and food storage; and the oldest bees (over 20 days of age) concentrate on foraging outside the nest. Age polyethism in worker honeybees is regulated by juvenile hormone (Figure 7.6); levels of JH in-

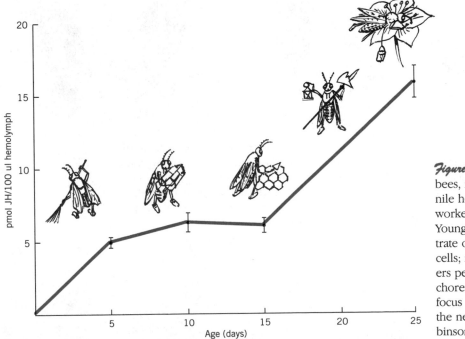

Figure 7.6 In honeybees, rising titers of juvenile hormone cause adult workers to switch tasks. Young workers concentrate on cleaning nest cells; middle-aged workers perform a variety of chores; and old workers focus on foraging outside the nest. (Data from Robinson et al., 1987.)

crease with worker age and mediate changes in work habits (Robinson, 1987; Robinson et al., 1987). However, rather than being a rigid system of control, endocrine mediation of the age-based division of labor in honeybees is a flexible process that permits ongoing adjustments in the proportions of workers involved in any one task. Rise in JH may be modulated by environmental conditions such as food availability or age structure of the colony, and workers with different genetic backgrounds respond somewhat differently to environmental stimuli (Robinson et al., 1989).

VERTEBRATE ENDOCRINE SYSTEMS

Our coverage of the hormones of vertebrates will concentrate largely on those associated with mammalian neuroendocrine systems. Although more than 40 hormones are secreted by the endocrine glands of mammals, we will focus our attention on hormones produced by the gonads (testes and ovaries), adrenal glands, pituitary gland, and pineal gland. Figure 7.7 shows locations of endocrine glands in the human, and Table 7.1 lists glands, the hormones secreted by each, and their function.

Testis

Gonads produce germ cells (eggs or sperm) and hormones. Thus, in the male, the testis is the site of formation and maturation of sperm and production of hormones called androgens (e.g., testosterone). Androgens are produced by the Leydig cells in the interstitial tissue of the testis (Figure 7.8) and are primarily responsible for development and maintenance of the male reproductive tract and secondary sex characteristics such as growth of body hair and enlargement of the larynx. Testosterone, as we will see later, plays an important role in regulating male sexual behavior and has also been implicated in certain forms of aggressive behavior.

At some level, the involvement of the testis

with male secondary sex characteristics, and sexual and aggressive behavior, has long been assumed. Thousands of years ago, the Chinese used castration as a means of punishment. In fact, throughout history, unceremonious removal of the testes has been performed on a variety of unfortunate people, including losers in battle and young singers in order to prevent deepening of the voice (Daly and Wilson, 1978). The testes have also been viewed as sources of youthful strength and vigor. The somewhat notorious experiment described next provides an example of the hopeful misconceptions surrounding the effects of male gonadal hormones on vitality.

In 1889, Charles Edward Brown-Sequard, a respected scientist with well over 500 publications, attempted to achieve self-rejuvenation at the age of 72 by 10 daily intramuscular injections of extract obtained from the testes of dogs (5 injections) and guinea pigs (5 injections). After several weeks of self-treatment, Brown-Sequard recorded in his notebook a number of promising signs, among them: (1) increased muscle strength, (2) improved bladder tonus (measured by how far he could urinate), and (3) greater resistance to mental fatigue (measured by enhanced ability to withstand long hours of work in the laboratory). Although quite certain of the revitalizing effects of testicular products, Brown-Sequard was anxious to have his findings verified in other subjects, and he urged scientists of his acquaintance to repeat the experiment. Despite his generous offer to provide the necessary material, volunteers were conspicuously absent. Brown-Sequard's public announcement of his findings in June 1889 was met with ridicule. The following passage describes the less than warm reception that his work on revitalization through injection of testicular products received:

Professor Brown-Sequard's audience appears to have received an impression of the intellectual capacity of the aged scien-

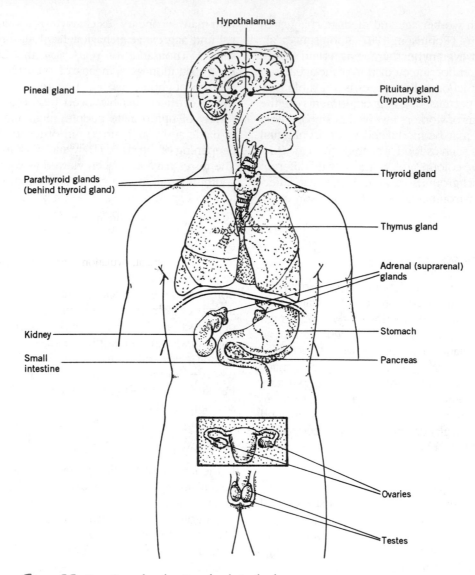

Figure 7.7   Location of endocrine glands in the human.

Table 7.1

SOURCES OF SOME OF THE HORMONES THAT AFFECT BEHAVIOR, HORMONES PRODUCED, AND SOME OF THEIR PHYSIOLOGICAL EFFECTS

SOURCE	HORMONE	PHYSIOLOGICAL EFFECT
Testis	Androgens	Development and maintenance of male reproductive tract and secondary sex characteristics
Ovary	Estrogens	Development and maintenance of female reproductive tract and secondary sex characteristics
	Progestogens	Gestation
Adrenal cortex	Steroids	Water and electrolyte balance, metabolism
Adrenal medulla	Epinephrine	Stress reaction
	Norepinephrine	
Posterior pituitary	Oxytocin	Milk ejection and parturition
	Vasopressin	Water balance
Anterior pituitary	Luteinizing hormone	Corpora lutea formation
	Follicle-stimulating hormone	Follicle development, ovulation
	Prolactin	Milk secretion
	Melanophore-stimulating hormone	Color change
Pineal	Melatonin	Annual reproductive cycle

tist very different from the one which he, in his elevated frame of mind, evidently expected to produce. This lecture must be regarded as further proof of the necessity of retiring professors who have attained their threescore years and ten. (transl. Steinach, 1940, p. 49)

Ovary

The ovaries secrete two groups of hormones, estrogens (e.g., estradiol) and progestins (e.g., progesterone). Within the ovary, estradiol is secreted by the Graafian follicle, a layer of epithelial cells that surrounds the ovum (Figure 7.9). Progesterone is secreted by a follicle that has ruptured, a structure called the corpus luteum. Maturation and eventual rupture of the follicle with release of the egg is termed ovulation. Estradiol is involved in the development and maintenance of the female reproductive tract and secondary sex characteristics such as breasts.

Progesterone, on the other hand, functions in preparing the uterus for implantation of fertilized ova and maintenance of pregnancy. In certain species of mammals, estradiol and progesterone have been implicated in sexual behavior; a rise in estrogen followed by a surge in progesterone appears to induce sexual receptivity in females of many species. Additionally, studies with the Norway rat (*Rattus norvegicus*) indicate that estrogen serves as a trigger to stimulate maternal behavior, and that both estrogen and progesterone are necessary to prime the female during pregnancy so that she responds to rising levels of estrogen (reviewed by Rosenblatt, Mayer, and Giordano, 1988).

Adrenal Gland

The adrenal glands sit on top of the kidneys. Each gland contains a central core (medulla) and an outer region (cortex). The medulla secretes epinephrine and norepinephrine, two hor-

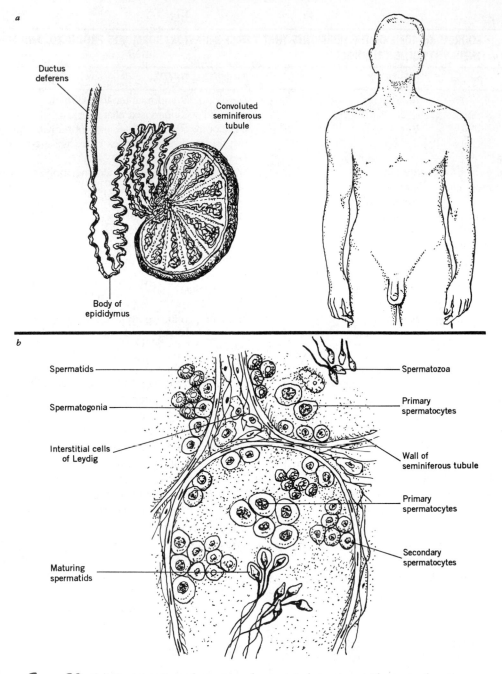

Figure 7.8 (*a*) Cross section of a testis and associated structures. The testes function to produce sperm and androgens. (*b*) Section through seminiferous tubules. Sperm form and mature within the seminiferous tubules and androgens are produced by the interstitial cells of Leydig. (Redrawn from Daly and Wilson, 1978.)

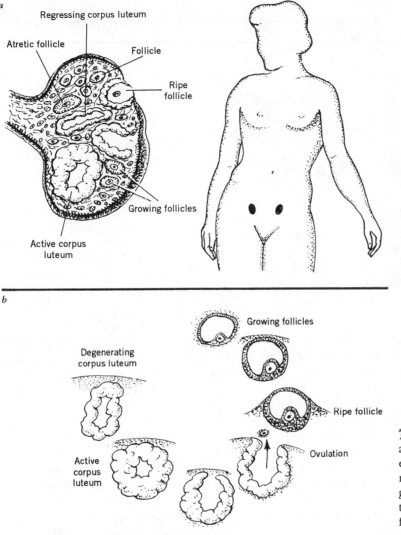

Figure 7.9 (*a*) Cross section of an ovary. The ovaries produce eggs and two groups of hormones, estrogens and progestogens. (*b*) Sequence of events in the ovarian cycle. (Redrawn from Daly and Wilson, 1978.)

mones that come into play during emergency situations. The cortex secretes at least 28 different hormones whose functions range from metabolism to resistance to infection.

Pituitary Gland

Like the adrenal glands, the pituitary gland consists of two sections that are quite different in both structure and function. The posterior pituitary is called the neurohypophysis because of its close connection to the nervous system (Figure 7.10). As evidence of the close relationship between the hypothalamus (a part of the brain located just above the pituitary gland) and the posterior pituitary, the posterior pituitary ceases to function if the nervous connection to the hypothalamus is severed. Two neurohormones secreted by the posterior pituitary, vasopressin and oxytocin, are actually synthesized in the hypo-

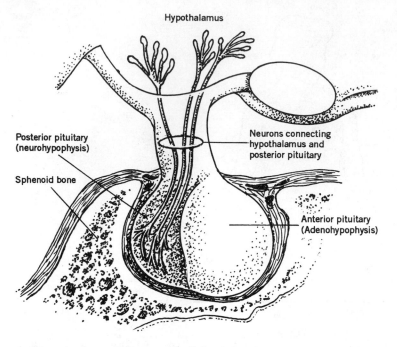

Hypothalamus

Posterior pituitary
(neurohypophysis)

Sphenoid bone

Neurons connecting
hypothalamus and
posterior pituitary

Anterior pituitary
(Adenohypophysis)

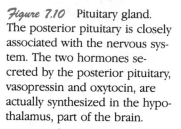

Figure 7.10 Pituitary gland. The posterior pituitary is closely associated with the nervous system. The two hormones secreted by the posterior pituitary, vasopressin and oxytocin, are actually synthesized in the hypothalamus, part of the brain.

thalamus. After synthesis in cell bodies of neurons within the hypothalamus, vasopressin and oxytocin travel down the axons into the posterior pituitary where they are released into the bloodstream. Whereas vasopressin promotes water absorption by the kidneys, oxytocin is involved in uterine contractions at birth and secretion of milk from mammary glands. There is some evidence that oxytocin may be involved in the induction of maternal behavior in rodents.

The anterior pituitary or adenohypophysis secretes an array of hormones. Some hormones produced by the anterior pituitary produce direct effects on target tissues (e.g., somatotropin stimulates growth of bone and muscle), while others stimulate other endocrine glands to secrete their hormones (e.g., adrenocorticotropin stimulates the adrenal glands to secrete their hormones, and follicle-stimulating hormone, FSH, stimulates the ovary to secrete estrogen). Three important hormones released by the anterior pituitary are luteinizing hormone, prolactin, and melanophore-stimulating hormone. Whereas luteinizing hormone (LH) regulates reproductive cycles in females and androgen secretion in males, prolactin is involved in parental care in birds and milk secretion in mammals (Figure 7.11). In some mammals, prolactin may also be correlated with paternal behavior (e.g., Gubernick and Nelson, 1989).

Melanophore-stimulating hormone (MSH) is secreted from a section of the anterior pituitary that borders the posterior pituitary. MSH affects the dispersion of pigment granules in color cells and is responsible for producing color change in vertebrates. Pigment may be present within scales, hair, feathers, or the skin itself, in cells called chromatophores. Chromatophores in the skin commonly contain a black or brown pigment called melanin and thus are referred to as melanophores. Many vertebrates can alter their coloration in response to environmental and behavioral needs. Such changes may take place relatively slowly (e.g., some mammals change coat color over a period of weeks as the seasons change) or relatively rapidly (e.g., certain reptiles turn pale at night and dark during the day), but in each case these changes are influenced

Figure 7.11 A young manatee obtains milk from its mother.

by MSH. Secretion of MSH produces color change by causing pigment granules to disperse. Seasonal change in coat color in some mammals may be inhibited by melatonin, a hormone secreted by the pineal gland. With the arrival of spring, MSH increases, and a brown coat replaces the white coat of winter (Figure 7.12*a*), as pigment granules disperse within the color cells. However, as the fall approaches, MSH secretion is inhibited by melatonin from the pineal gland. The pigment granules of new hairs remain clumped, and the new coat is essentially unpigmented or white (Figure 7.12*b*).

As in the posterior pituitary, functioning of the anterior pituitary is closely tied to the hypothalamus. This time, however, the interaction is based largely on a circulatory rather than nervous connection (Figure 7.13). The hypothalamus contains specialized neurosecretory cells that receive neural input and respond by discharging chemicals into a network of capillaries (the portal system) leading to the anterior pituitary. These chemicals from the hypothalamus are termed releasing factors, and each stimulates the

production and secretion of certain hormones from the anterior pituitary, which may, in turn, cause the release of hormones from other glands. For example, the hypothalamus produces a releasing factor that stimulates the anterior pituitary to secrete FSH, which, in turn, stimulates the ovary to secrete estrogen. A surplus of estrogen in the circulatory system then causes the anterior pituitary to inhibit the production of FSH. This system of hormonal regulation is termed a negative feedback mechanism.

In recent years, chemical compounds known as endorphins have been found in the brain, gastrointestinal tract, and pituitary gland of vertebrates. Thought to function as neurotransmitters and hormones, endorphins are chemically similar to morphine and appear to exert a pain-relieving or analgesic effect in the body. Acupuncture, a medical procedure that has long held an aura of mystery in the Western world, may relieve pain by stimulating certain peripheral nerve fibers to release endorphins. Once the body's naturally occurring opiates are released, an analgesic effect is produced. In addi-

a

b

Figure 7.12 Seasonal changes in the coat color of the snowshoe hare. (a) The white coat of winter changing to the brown coat of spring and summer. (b) The coat in midwinter.

tion to their role in relief of pain, endorphins have also been linked to behavioral activation, memory and learning, sexual behavior, and mental illnesses such as depression and schizophrenia (see Koob, LeMoal, and Bloom, 1984, for review).

Pineal Gland

The pineal gland is present in most vertebrates (it is reportedly absent in some animals including hagfish, crocodiles, and whales; Bentley, 1982). The location and primary function of the pineal vary according to taxon. In fish, amphibians, and reptiles, it is positioned at the top of the brain, just below the skull. In most species within these groups (not including the snakes), the pineal contains photoreceptor cells and functions as a "third eye."

In contrast to its sensory function in fish, amphibians, and reptiles, the pineal gland has largely an endocrine function in birds and mammals. In mammals, the pineal lies within the brain and lacks functioning photosensory cells; here such cells have been modified to form secretory cells. Among mammals, melatonin is perhaps the most important product secreted by the pineal gland. This hormone is produced during periods of darkness and has an inhibitory effect on the gonads. Melatonin is thought to mediate annual patterns of breeding in mammals by translating seasonal changes in photoperiod (the number of hours of daylight during a 24 hour period) into physiological effects on the reproductive organs (Reiter, 1980). The role of the pineal gland in setting the annual rhythm of reproduction is covered in more detail in our discussion of biological clocks in Chapter 9.

TYPES OF HORMONES AND THEIR MODES OF ACTION

Animals produce two types of hormones, peptides and steroids. Although differing in struc-

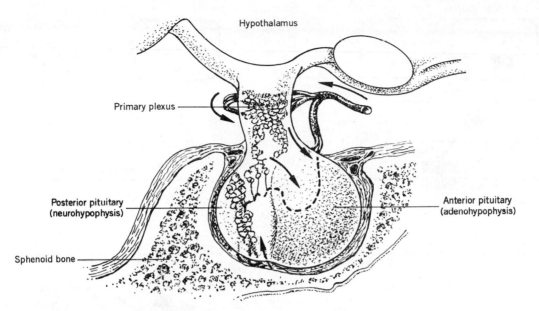

Figure 7.13 Pituitary gland. Releasing factors from the hypothalamus reach the anterior pituitary via a network of capillaries called the primary plexus. These chemicals from the hypothalamus stimulate the anterior pituitary to produce and secrete certain hormones.

ture and mode of action, both peptide and steroid hormones interact with the machinery of gene expression and protein synthesis to eventually influence behavior.

Peptide Hormones

Peptide hormones are protein molecules ranging in size from 3 to 300 amino acids (Roberts, 1984). These hormones are water soluble and usually affect cells at specific receptor sites at the cell surface (Figure 7.14). Through a complex sequence of molecular interactions, including use of a secondary messenger (cAMP), peptide hormones create short-term changes in cell membrane properties and long-term changes in protein synthesis. Examples of peptide hormones are luteinizing hormone (LH) and folli-

cle-stimulating hormone (FSH) produced by the anterior pituitary.

Steroid Hormones

Steroids are a group of closely related hormones secreted primarily by the gonads and adrenal glands in vertebrates. The four major classes of steroids include progestins, androgens, estrogens, and corticosteroids. The first three classes are secreted by the gonads and are often referred to as the sex steroids. All steroid hormones are chemically derived from cholesterol and hence are highly fat soluble. As a result of their solubility in lipids, steroid hormones move easily through the lipid boundaries of cells and into the cell interior or cytoplasm (Figure 7.15). Once inside a cell, steroids combine with recep-

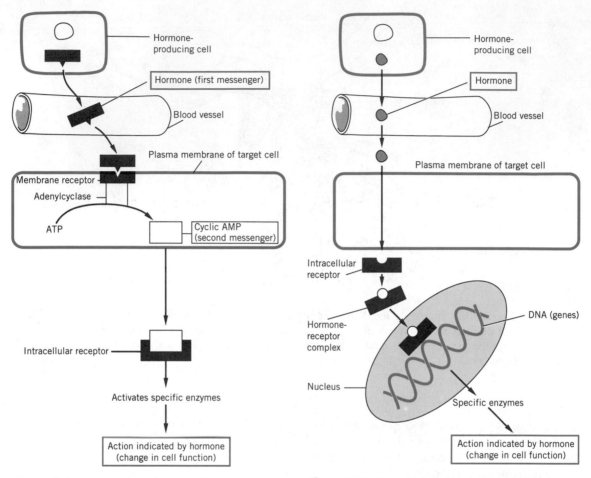

Figure 7.14   Proposed mechanism of action of peptide hormones.

Figure 7.15   Proposed mechanism of action of steroid hormones.

tor molecules, move to the nucleus of the cell where they attach to DNA, and affect subsequent protein synthesis and gene expression.

Because we will be discussing the role of steroid hormones in reproductive behavior, two points about the sex steroids should be emphasized here. First, although different hormones produce different effects, the sex steroids are chemically very similar (Figure 7.16). Some hormones lie along the pathway of synthesis of others. Testosterone, for example, is an inter-

mediate step in the synthesis of estradiol. As a result of the common structure of these two steroid hormones, some behavior patterns may be activated by injections of either testosterone or estradiol. In other words, there is a certain degree of substitutability associated with steroid hormones. A second point to emphasize is that hormonal output is not rigidly determined by sex. Females generally have small amounts of "male hormones" such as testosterone, and males typically have low levels of "female hor-

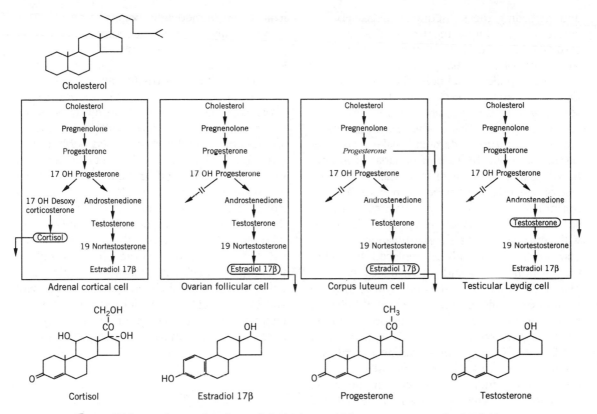

Figure 7.16 Biochemical pathways by which steroid hormones are synthesized. Note that many of the steroid hormones are chemically very similar. (From Tepperman, 1980; Bentley, 1982; modified from Daly and Wilson, 1978.)

mones." However, we often think of the testes only as sources of androgens and overlook their ability to secrete estrogens. (Obviously, then, we should use the terms "male" and "female" hormones with caution!)

Methods of Studying Hormone-Behavior Relationships

Several techniques are available for the study of hormonal influences on behavior. Here we ex-

amine two experimental approaches to questions in behavioral endocrinology. The first approach involves the removal of a gland followed by hormone replacement therapy. In the second approach, researchers look for changes in behavior that parallel fluctuations in hormone levels.

REMOVAL OF GLAND AND HORMONE REPLACEMENT

Fairly conclusive evidence of the function of a hormone can be gained by removing the source of the hormone, that is, the endocrine gland,

and recording the subsequent effects. For example, if removal of the testes of a male rat results in decreased copulatory behavior, then we might conclude that gonadal hormones are involved in male sexual behavior in this species. One could argue, however, that this decline in sexual activity is not related to the removal of the testes per se, but to the general stress of surgery. How can researchers make a stronger statement about the role of gonadal hormones in the sexual behavior of male rats? A stronger test of the relationship between male gonadal hormones and copulatory behavior would involve following surgical removal of the testes with replacement of testicular hormones. Testicular hormones could be replaced by either transplanting new glands or by injecting or implanting synthetic testosterone. If male gonadal hormones do indeed influence sexual behavior in the rat, then copulatory behavior should decline after castration and return to presurgery levels once the gland or hormone is replaced. Let us now consider an example of the use of castration and androgen replacement therapy in the study of hormonal control of sexual and aggressive behavior in lizards.

David Crews examined the neuroendocrine control of reproduction in several species of reptiles (Crews, 1979, 1987). Among his favorite subjects is *Anolis carolinensis*, the green anole. This small iguanid lizard inhabits the southeastern United States and displays a social system in which males fiercely defend their territories against male intruders. The territory of a single male often encompasses the home ranges of two or three females. As you might expect from these living arrangements, male anoles have an interesting repertoire of aggressive and sexual behaviors (Figure 7.17).

Both agonistic and sexual displays of male *A. carolinensis* share a species-typical bobbing movement that is made all the more dramatic by extension of the red throat fan or dewlap. When confronted by a male intruder, a resident male anole immediately begins to display. This show usually involves the homeowner's compressing his body and adjusting his posture in such a way as to present the intruder with a lateral view of his impressive physique. As if this were not enough, the resident then lowers his hyoid apparatus (a structure in the back of the throat responsible for movements of the tongue) and exhibits a highly stereotyped bobbing pattern. The display ends at this point if the intruder rapidly nods his head, thereby acknowledging his subservient position. However, if the intruder fails to display the submissive posture, the display of the resident male escalates to ever increasing frequencies. In the heat of confrontation, the two combatants acquire a crest along the back and neck and a black spot behind each eye. A wrestling match ensues as the resident and intruder circle, with locked jaws, in an attempt to dislodge the other from the prized perch. For male anoles, courtship behavior is

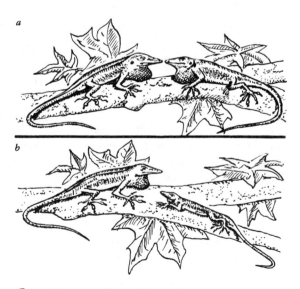

Figure 7.17 Displays of the male green anole. (*a*) Aggressive posturing between two males. (*b*) Courtship display directed by a male to a female (smaller individual). (Redrawn from Crews, 1979.)

very similar to aggressive behavior. Typically however, the body is not laterally compressed, and the bobbing dewlap display is less stereotyped in courtship; in effect, each male has his own version of how best to attract females.

Once Crews had documented the display repertoire of the feisty *A. carolinensis*, he set out to examine hormonal control of male aggressive and sexual behavior through castration and androgen replacement therapy. In the case of sexual behavior, removal of the testes led to a sharp decline. Administration of testosterone implants reinstated sexual behavior to precastration levels (Figure 7.18; Crews, 1974; Crews et al., 1978). Thus, Crews and coworkers concluded that testosterone regulates courtship and copulation in the male green anole. The relationship between testosterone and aggressive behavior was not so simple. If a male was castrated and returned to his home cage, he continued to be aggressive toward intruders. However, if the male was castrated and placed in a new cage, his aggressive behavior declined in a manner similar to that noted for sexual behavior. Thus, unlike sexual behavior, aggressive behavior appears to be only partially dependent on gonadal hormones and

subject to influence by experiential factors such as residence status.

CORRELATIONAL STUDIES

Hormonal influences on behavior can also be approached by correlational studies. In using this approach, researchers look for changes in behavior that parallel fluctuations in hormone levels. Consider, for example, a correlational study that revealed the relationship between level of testosterone and aggressive behavior in a songbird.

John Wingfield has examined the behavioral endocrinology of birds under natural conditions (for a review, see Wingfield and Moore, 1987). In one study of song sparrows (*Melospiza melodia*; Figure 7.19), Wingfield (1984a) captured males in mist nets or traps baited with seed, collected a small blood sample from the wing vein, and marked each individual with a unique combination of leg bands. Birds were released at the site of capture and seemed relatively unperturbed by the sampling procedure; in fact, some individuals sang within 15–30 minutes of release. A given male was sampled from 5 to 10

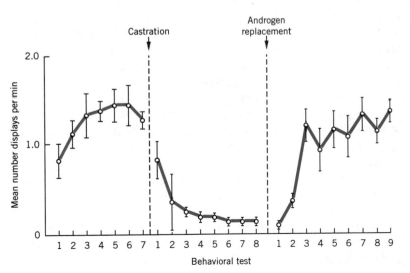

Figure 7.18 Effect of castration and testosterone replacement therapy on courtship behavior of the male green anole. (Modified from Crews, 1979.)

Figure 7.19   Male song sparrow.

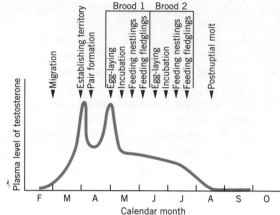

Figure 7.20   Changes in circulating levels of testosterone in free-living male song sparrows as a function of stage of breeding cycle. (From Wingfield, 1984a.)

times during a single breeding season, and each sample was analyzed for testosterone. Wingfield also observed their behavior during this period. As shown in Figure 7.20, he found a close correlation between peak levels of male territorial and aggressive behavior and maximum levels of testosterone. Male song sparrows defend their territories most intensely during initial establishment of the territory and during the period when their mate is laying the first clutch. At the time of egg laying, females are sexually receptive and males aggressively guard them from other would-be suitors. Testosterone reaches peak levels during the initial period of territory establishment and during the laying of eggs for the first brood. Interestingly, testosterone does not peak during the period when the female is sexually receptive and laying the second clutch. Behavioral observations corroborate testosterone profiles. The enthusiasm with which a male guards his mate during the second laying period is less than that exhibited during the first period.

This pattern may be related to the fact that while the female is laying the second clutch, males are often responsible for feeding fledglings from their first clutch. Wingfield speculates that high levels of testosterone and the resulting heightened levels of male aggression would be incompatible with parental behavior. Field studies of the song sparrow demonstrate that circulating levels of testosterone wax and wane in parallel with changing patterns of male territorial aggression. This correlational evidence strongly suggests that testosterone regulates aggressive behavior in this species.

Although simple in concept, correlational studies have become quite sophisticated as a result of major advances in techniques for measuring hormones. For example, cannulation techniques now allow administration of minute amounts of hormone to specific brain regions. In some species, direct introduction of gonadal hormones into appropriate areas of the brain induces sexual behavior in gonadectomized (i.e., castrated or ovariectomized) adults. Other advances utilize techniques whereby hormones are labeled with radioactivity and their paths traced through the body. The discovery of anti-hormones, drugs that can temporarily and reversibly suppress the actions of specific hormones, has also aided investigation of hormonal influences on behavior. Another example of a technique used in studies of behavioral endocrinology is cross-transfusion (Terkel, 1972). This technique allows blood to be cross-transfused between two freely moving animals (Figure 7.21). In a now famous study, Terkel and Rosenblatt (1972) cross-transfused blood between a mother rat and a virgin female. They found that by linking blood supplies, maternal behavior could be induced in the virgin rat. These data were instrumental in demonstrating the hormonal basis of maternal behavior. Finally, in recent years, it has become possible to monitor hormone levels through analysis of urine and feces rather than blood. These less invasive procedures are often used in field studies and when repeated sampling is necessary.

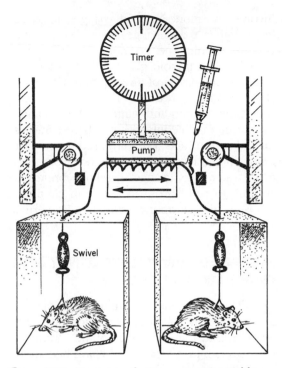

Figure 7.21 Cross-transfusion apparatus used by Terkel and Rosenblatt (1972) to demonstrate hormonal basis of maternal behavior in rodents. (From Terkel, 1972.)

How Hormones Influence Behavior

There are several pathways through which hormones can influence behavior. Generally speaking, hormones modify behavior by affecting one or more of the following: (1) sensory or perceptual mechanisms, (2) development or activity of the central nervous system, and (3) effector mechanisms important in the execution of behavior.

EFFECTS ON SENSATION AND PERCEPTION

Hormones influence the ability to detect certain stimuli as well as preferences for particular stimuli. For example, Diamond, Diamond, and Mast (1972) found that the ability of human females to detect a test light varied across the menstrual cycle. Visual sensitivity was greatest around the

time of ovulation, remained high up until menses, and then declined abruptly at time of menstruation. Following menstruation, visual sensitivity again began to increase. Significant changes in visual sensitivity were not found in men or in women using oral contraceptives (oral contraceptives typically maintain progesterone at consistently high levels and estrogen at consistently low levels). These results suggest that cyclic hormonal changes associated with the menstrual cycle are responsible for changes in visual function. Another example of hormones influencing the ability to detect certain stimuli involves the finding that olfactory sensitivity in female rats varies as a function of stage of estrous cycle (Pietras and Moulton, 1974). A final case involves hormone-induced changes in preference, rather than changes in ability to detect certain stimuli. The hormone thyroxin is secreted by the thyroid gland and plays a critical role in the behavior of three-spined sticklebacks (*Gasterosteus aculeatus*). These fish normally spend the autumn and winter in the sea and in the spring migrate into rivers to breed. Successful migration and breeding require, among other things, a switch in preference from salt to fresh water. This change in salinity preference appears to result from increases in circulating levels of the hormone thyroxin (Baggerman, 1962).

EFFECTS ON DEVELOPMENT AND ACTIVITY OF THE CENTRAL NERVOUS SYSTEM

Circulating hormones can influence behavior by altering the morphology, physiological activity, or neurotransmitter function of the central nervous system. In fact, hormones have been found to affect a variety of characteristics of different regions of the brain, including such things as (1) volume of brain tissue, (2) number of cells in brain tissue, (3) size of cell bodies of neurons, (4) extent of dendritic branching of neurons, and (5) percent of neurons sensitive to particular hormones. One example in which hormones influence the morphology of the brain involves the development of singing behavior in birds. In the zebra finch (*Taeniopygia guttata*), sex differences in the brain nuclei that control song are established around the time of hatching. Early exposure to hormones regulates the size of song nuclei (Gurney and Konishi, 1980) and the number of neurons contained within these areas (DeVoogd, 1990; Nordeen, Nordeen, and Arnold, 1986). Thus in the zebra finch, sex differences in adult singing behavior (males sing and females do not) are linked to morphological differences that are established in the brains of males and females as a result of the hormonal milieu soon after hatching. (See Chapter 8 for a detailed discussion of bird song.) Finally, as previously mentioned, some hormones influence behavior by altering activity or neurotransmitter function of the central nervous system. Certain hormones secreted by the adrenal cortex, for example, are known to affect behavior by influencing brain excitability and neurotransmitter metabolism (e.g., Joels and de Kloet, 1989; Rees and Gray, 1984).

EFFECTS ON EFFECTOR MECHANISMS

Hormones can influence behavior by affecting motor neurons and muscles. Consider, for example, two cases of sexually dimorphic patterns of behavior—copulatory movements in rats and calling behavior in frogs—that illustrate hormonal influences on effectors.

The levator ani/bulbocavernosus muscles control copulatory reflexes in male rats (Hart, 1980). Although these muscles are present in both sexes at birth, they are completely absent in adult females. Breedlove and Arnold (1983) have shown that the levator ani/bulbocavernosus muscles shrink and fold inward in females unless supplied with androgen. Lack of androgen in females during the perinatal period (time surrounding birth) also results in the death of the motor neurons supplying these muscles. Thus, sex differences in the copulatory movements of adult rats result, in part, from early hormonal influences on growth and maintenance of the specific muscles and motor neurons involved in mating.

A second example of sex differences in behavior resulting from hormonal influences on effector mechanisms involves the calling behavior of the clawed frog, *Xenopus laevis* (Kelley, 1988; Kelley and Gorlick, 1990). Clawed frogs occur in sub-Saharan Africa where they inhabit shallow, and often murky, bodies of water, ranging from ponds and lakes to sewers. Males of this species attract and excite females by emitting metallic-sounding calls in which they alter-

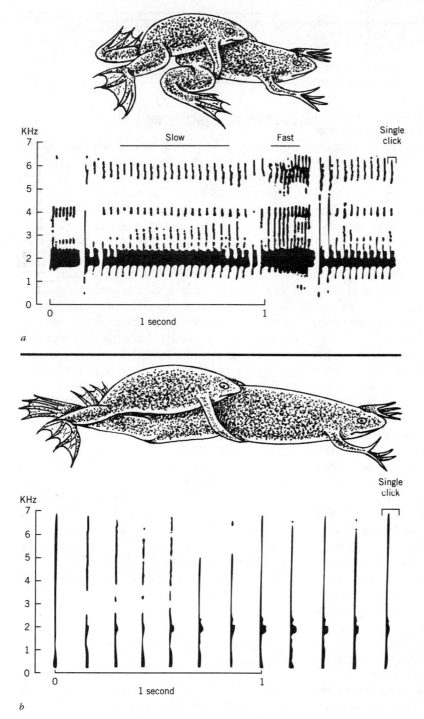

Figure 7.22 (*a*) Male clawed frog clasping a sexually receptive female that responded to his mate call (shown below). The male's mate call consists of slow and fast trills. (*b*) Male clasping a sexually unreceptive female that responds to his clasp by emitting the ticking call (shown below). Differences in the calling behavior of male and female clawed frogs result from the effects of hormones on the muscles of the larynx. (From Kelley and Gorlick, 1990.)

nate fast and slow trills (Wetzel and Kelley, 1983). These mate calls allow females to find males in their typically soupy locations. Sexually receptive females do not vocalize, and they permit males to clasp them around the waist, in a position termed amplexus, for several hours while their eggs are released and fertilized. Although receptive females are silent, females that are not sexually ready produce a ticking call that consists of slow, monotonous clicks. Females that are declining or terminating clasping attempts by males also typically produce the ticking call (Weintraub, Bockman, and Kelley, 1985). The calls of male and female clawed frogs are summarized in Figure 7.22.

How does it come about that male clawed frogs produce mating calls that consist of rapid trills, while females are only capable of producing the slow ticking call? Darcy Kelley and co-workers demonstrated that characteristics of the muscles and neuromuscular junctions of the larynx are responsible for sex differences in the rate at which calls are produced (in males the muscles of the larynx contract and relax 71 times per second as compared to 6 times per second in females). Adult males have 8 times as many muscle fibers in their larynx as do females. Additionally, male muscle cells are of the fast-twitch, fatigue-resistant type, whereas most muscle cells in the larynx of females are slow-twitch and fatigue-prone. How do these changes in laryngeal muscles come about? Sassoon and Kelley (1986) demonstrated that at the time of metamorphosis (change from tadpole to adult frog), the number of muscle fibers in the larynx of males and females is identical to the number in the larynx of adult females. Apparently, as males mature and their levels of androgens rise, new fibers are added. In addition to increasing the number of muscle fibers, androgens also influence the type of fiber, promoting expression of the fast-twitch cells. In short, sex differences in the calling behavior of *X. laevis* can be traced, in part, to hormone-induced changes in the muscles of the larynx.

Organizational and Activational Effects of Hormones—Sexual Behavior: A Case in Point

DEFINING THE DICHOTOMY

The modes by which steroid hormones influence behavior may be classified as organizational and activational (Phoenix et al., 1959). In organizational effects, steroids act to organize neural pathways responsible for certain patterns of behavior. Organizational effects occur early in life, usually just before or after birth, and tend to be permanent. This permanence implies structural changes in the brain or other long-term cellular changes, such as in the responsiveness of neurons to steroid hormones (Arnold and Breedlove, 1985). Alternatively, steroid hormones may influence behavior by activating neural systems responsible for mediating specific patterns of behavior. In contrast to organizational effects, activational effects usually occur in adulthood and tend to be transient, lasting only as long as the hormone is present at relatively high levels. In keeping with their impermanence, activational effects are thought to involve subtle changes in previously established connections (such as slight changes in neurotransmitter production or release along established pathways) rather than gross reorganization of neural pathways. At this point, we will consider the organizational and activational effects of steroid hormones as they relate to the development and display of sexual behavior in the Norway rat.

AN EXAMPLE—SEXUAL BEHAVIOR OF THE NORWAY RAT

Historically, the sexual behavior of rodents has been the most popular research topic in behavioral endocrinology and the trend continues today (Svare, 1988a). Because of the large amount of information available on the Norway rat, we will consider the effects of sex steroids on the behavior of this species. First we will consider sex differences in copulatory behavior, then the organizational effects produced by testosterone secretion in young animals, and finally the activational effects produced by secretion of sex steroids in adults.

Not surprisingly, adult male and female rats differ in their sexual behavior (Figure 7.23). Whereas mounting, intromission, and ejaculation typify mating in males, behavioral patterns associated with solicitation and acceptance characterize the sexual behavior of females. The lordosis posture, for example, is a copulatory position that female rats assume when grasped on the flanks by an interested male. The intensity of the lordosis response varies across the ovulatory cycle, being most pronounced when mature eggs are ready to be fertilized. The sexual behavior of female rats also includes a variety of solicitation behaviors, such as ear wiggling and a hopping-and-darting gait, that typically precede display of the lordosis posture (Beach, 1976). Although mounting is almost always associated with males and lordosis with females, these behavior patterns occasionally occur in the other sex. Every once in a while females will mount other females, and similarly, males will occasionally accept mounts from their cagemates. However, by and large, males display mounting and females assume the lordosis posture. These differences in patterns of adult copulatory behavior are due to differences in the brains of male and female rats, differences that are induced by the irreversible actions of androgens in late fetal and early neonatal life. Let's now consider the organizational effects of gonadal steroids on sexual behavior.

First, it is clear that testosterone in the bloodstream of neonatal rats produces organizational effects. During perinatal life, male and female rats possess the potential to develop neural control mechanisms for both masculine and feminine sexual behavior. Certain neurons within the brains of both males and females have the capacity to bind sex hormones. During a brief

Figure 7.23 Male and female Norway rats differ in their sexual behavior. Whereas mounting is characteristic of males, the acceptance posture, termed lordosis, is characteristic of females. These sex differences in adult behavior are established through the action of steroid hormones around the time of birth.

period starting about 2 weeks after conception and extending until approximately 4–5 days after birth, however, testosterone secreted by the testes of developing males is bound to receptors in the target neurons. Once there, testosterone initiates the production of enzymes that will switch development onto the "male track." The neonatal testosterone causes males to (1) develop the capacity to express masculine sexual behavior and (2) lose the capacity to express feminine copulatory behavior.

Experiments involving castration and hormone replacement techniques have demonstrated the organizational effects of early secretion of testosterone. Removal of the testes in a rat soon after birth results in an adult with a reduced capacity to display masculine patterns of sexual behavior and an enhanced capacity to display feminine patterns. These males are capable of high levels of female solicitation and lordosis as adults. However, if removal of the testes is followed by an experimental injection of testosterone before 5 days of age, and the proper male hormones are administered in adulthood, the rat will display normal male sexual behavior. Normal female fetuses produce low levels of testosterone, so the male developmental pattern is not initiated. A single injection of testosterone into a female rat soon after birth, however, produces irreversible effects on her adult sexual behavior. Relative to normal females, the testosterone-treated female shows fewer feminine and more masculine patterns of copulatory behavior. Thus, the sexually undeveloped rat brain is "female," and development of a "male" brain requires the presence of testosterone around the time of birth. The effects of perinatal testosterone secretion on adult sexual behavior are organizational in that they occur early in life and involve permanent structural changes in the brain. Figure 7.24 summarizes sexual differentiation in the brain and behavior of the neonatal rat.

Before moving to the activational effects of sex steroids, we should mention that masculinization of the brain may be somewhat more complex than just described. In laboratory rats, testosterone appears to be only an intermediate chemical in the process, and it is estradiol, a hormone usually associated with females, that actually directs development along the masculine track. Testosterone enters neurons in specific regions of the rat brain and is converted intracellularly to estradiol which in turn causes masculinization. Look again at Figure 7.16 to see that steroid hormones are chemically very similar and that testosterone lies along the pathway of synthesis of estradiol.

The main question that arises from estradiol's role in the masculinization process is, why doesn't estradiol in young female rats have the same effect? To begin with, the levels of estradiol in young females are very low. In addition, during this critical period of brain development, an estrogen-binding protein called alpha-feto-protein is produced in the livers of rat fetuses. This protein, found in the cerebrospinal fluid of newborn males and females, persists in ever decreasing amounts during the first three weeks of life. During this time, alpha-feto-protein prevents estradiol from reaching target neurons in the brain. In female rats, then, alpha-feto-protein binds any circulating estradiol and thereby prevents it from initiating the male pattern (McEwen, 1976). Alpha-feto-protein does not, however, bind testosterone. Thus, in male rats, testosterone produced by the testes can reach the brain, be converted to estradiol, and result in sexual differentiation. The extent to which this information can be generalized to other mammalian species remains to be established.

In adulthood, steroid hormones produce activational effects on sexual behavior in male and female rats. Female rats with high blood levels of estrogen and progesterone display feminine sexual behaviors in the presence of a sexually active male, but these patterns rarely occur when levels of these ovarian hormones are low. In fact,

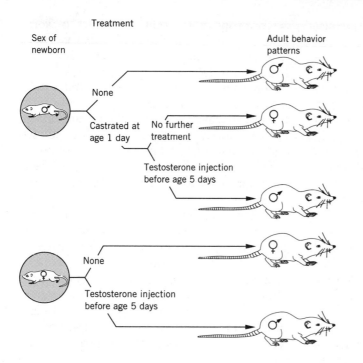

Treatment

Sex of
newborn

Adult behavior
patterns

None

Castrated at
age 1 day

No further
treatment

Testosterone injection
before age 5 days

None

Testosterone injection
before age 5 days

Figure 7.24 Pattern of sexual differentiation in the brain and behavior of the Norway rat.

an adult female whose ovaries have been removed will not copulate unless she receives injections of estrogen and progesterone. Similarly, removal of the testes in an adult male eventually eliminates copulatory behavior, unless he is given injections of testosterone. In these cases, the effects of steroid hormones on sexual behavior are termed activational because estrogen and progesterone in females and testosterone in males presumably exert their effects by activating existing neural pathways. High levels of the gonadal steroids activate specific patterns of sexual behavior. Thus, in contrast to permanent changes in sexual behavior caused by administration of testosterone during the neonatal period, only a transient activational effect on copulatory behavior is produced by sex steroids in adulthood.

One final point will help to distinguish organizational and activational effects of steroid hormones on sexual behavior. Males and females that have had their reproductive organs removed in adulthood generally cannot be induced to behave like members of the opposite sex. For example, a female rat that has had her ovaries removed in adulthood cannot, through injections of testosterone, be induced to show mounting behavior. By adulthood, the nervous systems of adult males and females have already differentiated (i.e., the organizational effects of early steroid secretion have long since occurred), and the mature brains are not capable of responding to hormonal signals of the opposite sex.

QUESTIONING THE DICHOTOMY

In 1985 Arthur Arnold and S. Marc Breedlove questioned the traditional division of steroid influences on behavior into organizational and activational effects. They reviewed experimental findings from the previous decade and concluded that the organizational–activational distinction was too restrictive. How would one classify, for example, effects produced by steroid hormones that were both organizational and activational in nature, such as the production of permanent effects in adulthood? Additionally, while acknowledging the wealth of behavioral evidence supporting the organizational–activational dichotomy, their attempts to uncover biochemical, anatomical, or physiological evidence of two fundamentally different ways in which steroid hormones act on the nervous system were unsuccessful. In their opinion (Arnold and Breedlove, 1985), failure to find specific cellular processes uniquely associated with each type of effect further blurs the organizational–activational distinction. Although it is important to keep such concerns in mind when discussing steroid influences, we believe that the traditional distinction of organizational and activational effects is still useful in categorizing hormonal effects on behavior.

THE DIVERSITY OF MECHANISMS UNDERLYING SEXUAL BEHAVIOR

We have just described how mating behavior of the adult Norway rat is associated with sex hormones. Now we might ask, do gonadal hormones activate mating behavior in all vertebrates? Or do the hormonal mechanisms underlying mating behavior differ across species?

David Crews (1984, 1987) compared patterns of reproduction in a wide variety of vertebrates and found numerous exceptions to the "rule" of hormone-dependence of mating behavior. In his survey, Crews considered relationships among the following three components of the reproductive process: (1) production of gametes, (2) secretion of sex steroids by the gonads, and (3) timing of mating behavior. Amid

the diversity of reproductive tactics of vertebrates, the following three general patterns of reproduction emerged: (1) associated, (2) dissociated, and (3) constant (Figure 7.25).

Some animals, such as the Norway rat, exhibit a close temporal association between gonadal activity and mating—specifically, gonadal growth and an increase in circulating levels of sex steroids activate mating behavior. This pattern of gonadal activity in relation to mating has been termed an associated reproductive pattern (Figure 7.25a) and has been found in numerous laboratory and domesticated species. Other species, however, exhibit a dramatically different pattern of reproduction in which mating behavior is completely uncoupled from gamete maturation and secretion of sex steroids. In species exhibiting the dissociated reproductive pattern (Figure 7.25b), gonadal activity occurs only after all breeding activity for the current season has ceased, and gametes are thus produced and stored for the next breeding season. Gonadal hormones may not play any role in the activation of sexual behavior in species that display the dissociated pattern. Typically, these species inhabit harsh environments in which there is a predictable, but narrow, window of opportunity to breed, and a specific physical or behavioral cue triggers mating behavior. Consider, for example, the red-sided garter snake (*Thamnophis sirtalis parietalis*), a species that ranges farther north than any other reptile in the Western Hemisphere. The courtship behavior of adult male garter snakes is activated by an increase in ambient temperature following winter dormancy, rather than by a surge in testicular hormones (reviewed by Crews, 1983).

Garter snakes in western Canada emerge in early spring from subterranean limestone caverns, where they have spent the winter hibernating in aggregations of up to 10,000 individuals. Males emerge first, en masse, and congregate at the den opening. Females emerge singly or in small groups over the next 1–3 weeks and

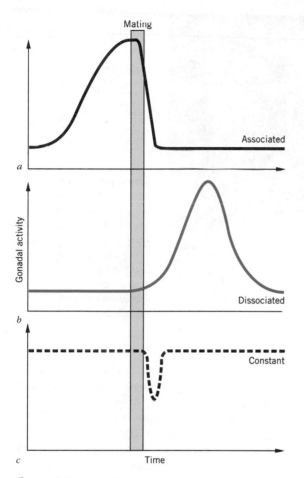

Figure 7.25 Vertebrates exhibit three general patterns of reproduction. In species exhibiting the associated reproductive pattern (*a*), mating occurs at the time of maximum gonadal activity as measured by maturation of gametes and peak levels of sex steroids. In species exhibiting the dissociated pattern (*b*), mating occurs at a time of minimal gonadal activity. In species exhibiting the constant reproductive pattern (*c*), gonadal activity is maintained at or near maximum levels at all times. (From Crews, 1987.)

mate with males hanging out at the entrance. This difference in the timing of emergence of males and females results in males greatly out-

numbering females at the den opening, sometimes 500 to one (Figure 7.26)! In view of these odds, it is not unusual for a writhing mass of snakes called a "mating ball" to form, in which over 100 males attempt to mate with a single female. Against all odds females usually mate with only one male and immediately disperse to summer feeding grounds where they give birth to live young in August. Males, on the other hand, remain at the den opening and move to feeding grounds only after all the females have emerged. Testicular activity is minimal in male garter snakes during the period of emerging and mating. In fact, it is 5–10 weeks later, after the males have left the den site and will no longer court females, that the testes grow and androgen levels increase. Sperm produced during this time is stored for use the next spring.

Numerous experiments utilizing castration and replacement techniques and destruction of temperature-sensing areas of the brain have revealed that rather than relying on surges of sex steroids, the neural mechanisms that activate sexual behavior in male garter snakes are triggered by a shift in temperature. This is not to say that courtship behavior is completely independent of hormonal control, or that sex steroids do not play an organizational role in the development of sexual behavior, but rather that mating is temporally dissociated from maximum gonadal activity. Like the male, the female red-sided garter snake mates when her gonads are small, gametes immature, and circulating levels of sex steroids low. In the case of the female, however, changes in sexual attractivity and receptivity are mediated by physiological changes that occur as a consequence of mating. Thus, although both male and female red-sided garter snakes display a dissociated pattern of reproduction, they differ in type of stimulus that triggers breeding behavior. Whereas change in ambient temperature triggers courtship behavior in males, stimuli associated with mating appear to activate physiological changes in females.

Figure 7.26 Male red-sided garter snakes wait at den openings for emerging females. Dense mating aggregations form as females emerge singly or in small groups. Activation of sexual behavior in this species is independent of sex steroids.

The third type of reproductive tactic, described by Crews (1987) as the constant reproductive pattern (Figure 7.25c), is characteristic of species that inhabit harsh environments, such as certain deserts, where suitable breeding conditions occur suddenly and unpredictably. In the case of desert-dwelling animals, reproduction is often initiated by rainfall. While waiting for suitable circumstances in which to breed, these species maintain large gonads, mature gametes, and high circulating levels of sex steroids for prolonged periods of time. The zebra finch, *Taeniopygia guttata castanotis*, lives in deserts of Australia where rainfall occurs rarely and unpredictably. Through droughts that can last for years, males and females maintain reproductive systems in a constant state of readiness. No matter how long the wait, each sex is poised, prepared for the opportunity to breed (Serventy, 1971). Courtship among males and females in desert populations begins shortly after the rain starts to fall, copulation occurs within hours, and

nest building can begin as early as the next day. In order to maintain this accelerated pace, both males and females carry material to the nest. It is interesting that in more humid areas of their range where the reproductive needs are not so immediate, males and females exhibit the division of labor characteristic of finches—that is, the male alone carries grass to the nest, and the female waits at the nest for each delivery and arranges the new material as it arrives (Immelmann, 1963).

We should mention that in some vertebrates, males and females of the same species exhibit fundamentally different reproductive strategies (Crews, 1984, 1987). In painted turtles (*Chrysemys picta*), for example, males show a dissociation between gonadal activity and mating behavior, whereas females exhibit an associated pattern (Licht, 1984). Musk shrews (*Suncus murinus*; Figure 7.27), insectivores from tropical regions of Asia and the Far East, show just the opposite pattern; gonadal activity and sexual be-

Figure 7.27 Musk shrews exhibit mixed reproductive patterns. Whereas males exhibit an associated reproductive pattern (i.e., mating occurs at peak gonadal activity), females display a dissociated reproductive pattern (gonadal activity is minimal at time of mating).

havior are associated in males and dissociated in females (Dryden and Anderson, 1977; Rissman, 1990). Such mixed reproductive strategies have been reported in mammals, reptiles, amphibians, and fish.

Here, then, we have seen that there is a great diversity in reproductive patterns among vertebrates, and that reproductive patterns may vary with levels of gonadal hormones in complex ways. Next we will see that species and sex are only two of the factors that influence hormone–behavior relationships.

Factors That May Influence Effects of Hormones on Behavior

Hormonal influences on behavior are not always as straightforward and invariable as we might assume. Indeed, the effects of hormones on behavior can vary across different species, strains,

and individuals, as well as with season and experience. Here, then, we will consider a few of these factors that enter into the hormone–behavior relationship.

SPECIES EFFECTS

Not surprisingly, the precise effects of hormones on behavior may vary according to species. Such differences make it difficult to extrapolate from the hormonal mechanisms of one species to those of another. Here we consider differences among rodent species in the effects of adrenal hormones on early development.

Early studies with Norway rats and house mice focused on the effects of administering high doses of adrenocortical hormones, called glucocorticoids, to pups during the first week or two of life. This research demonstrated that manipulation of levels of glucocorticoids shortly after birth resulted in suppressed brain and body growth and impaired behavioral development in both rats and mice (e.g., Howard, 1974; Schapiro, Salas, and Vukovich, 1970). Despite the consistent findings for rats and mice, a more recent study by Prohazka, Novak, and Meyer (1986) with microtine rodents revealed dramatic species differences in response to early treatment with a cortical hormone.

Prohazka and colleagues examined the effects of a single injection of hydrocortisone on day 2 of life in meadow voles (*Microtus pennsylvanicus*) and pine voles (*M. pinetorum*), two closely related species of North American rodents. The investigators compared, among other things, neuromuscular development in experimental pups (hydrocortisone-treated) and control pups (saline-treated) of the two species. As in earlier studies with rats, neuromuscular development was assessed by measuring age-dependent changes in swimming behavior. Approximately every 5 days, beginning on day 5 and running

through day 35, each pup was dropped 12 centimeters into an aquarium filled with warm water. The ability of the pups to stay afloat for 1 minute was recorded on videotape; the less distinguished swimmers (that is, pups whose noses went under water for 10 seconds) were immediately removed from the tank. Maturation of swimming behavior in meadow and pine voles was rated on a scale, modified from an earlier system used with rats (Schapiro, Salas, and Vukovich, 1970). On the scale, complete inhibition of front paw movement and ability to maintain the nose above water level received the highest marks. Thus, neuromuscular maturation in both species of voles involved a progressive elevation of head/nose position, along with a gradual inhibition of front paw movement. Figure 7.28 illustrates the 5-point rating scale used in scoring the young swimmers.

The results suggested species differences in

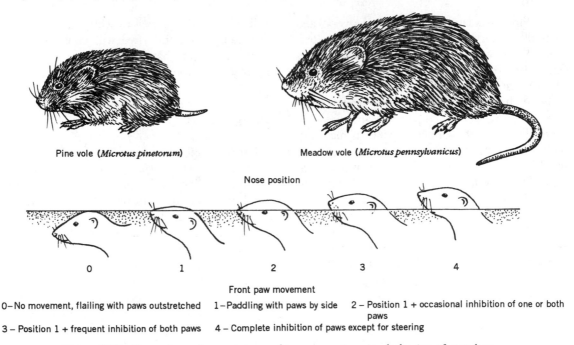

Pine vole (*Microtus pinetorum*)   Meadow vole (*Microtus pennsylvanicus*)

Nose position

Front paw movement

0–No movement, flailing with paws outstretched   1–Paddling with paws by side   2 – Position 1 + occasional inhibition of one or both paws

3 – Position 1 + frequent inhibition of both paws   4 – Complete inhibition of paws except for steering

Figure 7.28 Five-point rating system used to score swimming behavior of meadow and pine voles. Swimming ability serves as an index of neuromuscular development. (Rating system redrawn from Prohazka, Novak, and Meyer, 1986.)

response to early hormone administration (Figure 7.29). Hormone-treated pine voles exhibited delayed swimming development relative to controls, a pattern consistent with the earlier reports on rats and mice. However, when compared with saline controls, hydrocortisone-treated meadow voles showed accelerated attainment of adult swimming patterns. What might account for such dramatic species differences? Perhaps, as suggested by the authors, differences in the responses of meadow and pine voles are related to the stage of development of the two species at the time of hormone administration on day 2. Although no information exists for rate of prenatal development, meadow voles develop more rapidly than pine voles after birth (McGuire and Novak, 1984). Different rates of postnatal development might be accompanied by similar variation during the prenatal period, and thus meadow and pine voles could have been at different ontogenetic stages at the time of hormone treatment on day 2. Although the basis for species differences in response to hydrocortisone treatment remains to be determined, this experiment illustrates the important point that even closely related species may respond differently to the same hormonal treatment.

Our second example of species differences in hormone–behavior relationships involves differential effects on survival of testosterone administration in song sparrows and cowbirds (*Molothrus ater*). This time, however, differences in social and territorial systems, rather than differences in rate of development, appear to form the basis for interspecific variation in response to hormone treatment. Dufty (1989) reported that adult male cowbirds, subcutaneously implanted with testosterone for extended periods of time, showed decreased survival relative to control males that were either implanted with empty capsules or not implanted at all. Whereas only 1 of 16 (6.3%) testosterone-implanted males returned to the breeding grounds the following year, 33 of 81 (40.7%) males with-

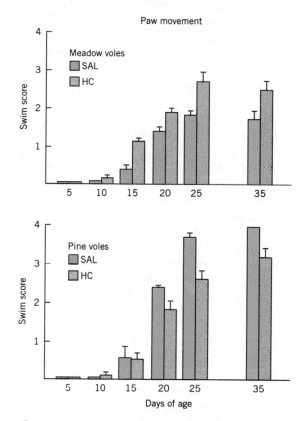

Figure 7.29 Changes in swimming behavior (paw movement) in saline-treated (SAL) and hydrocortisone-treated (HC) meadow and pine voles. Response to early hydrocortisone treatment differed between the two species; whereas hormone-treated meadow voles showed accelerated attainment of adult swimming patterns, hormone-treated pine voles exhibited delayed swimming development. (From Prohazka, Novak, and Meyer, 1986.)

out implants and 3 of 6 (50.0%) males with empty implants returned. Testosterone-implanted males also suffered more injuries than did controls. In contrast to Dufty's findings for cowbirds, Wingfield (1984b) reported no difference in survival of testosterone-implanted and control song sparrow males. What factors might explain the difference in response between cow-birds and song sparrows?

Song sparrows and cowbirds differ dramati-cally in their system of mating and territoriality, and according to Dufty (1989), these differences might help explain differential effects of testos-terone on survival. Song sparrow males are ter-ritorial and territory boundaries may provide a "psychological" barrier to intruding males, something similar to the home court advantage in basketball. In addition, testosterone-im-planted males of this species do not differ from control males in frequency of intrusion into neighboring territories (Wingfield, 1984a). In short, the system of territoriality in song spar-rows may shield males from high risks associ-ated with aggressive interactions with other males. Cowbird males, on the other hand, do not defend territories but establish dominance hierarchies instead. The increased aggression in-duced by testosterone in the absence of the buff-ering effects of territory boundaries could result in escalation of aggressive interactions and in-creased risk of injury. In addition, given the ex-istence of dominance hierarchies in male cow-birds, a male whose behavior is inconsistent with his position within the hierarchy may be a favorite target for aggression from other males. Species differences in territoriality may thus ex-plain the differential effects of testosterone on survival in song sparrows and cowbirds.

In these two examples, a given hormone pro-duced different effects on behavior of different species. In other examples of interspecific vari-ation, different hormones may affect the same behavior. For example, in fish, the relative im-portance of steroid hormones in influencing fe-male sexual behavior depends on the mode of reproduction employed by the species in ques-tion (reviewed by Stacey, 1987). In species such as the guppy (*Poecilia reticulata*) that utilize internal fertilization, estrogen plays an important role in regulating female sexual behavior. How-ever in externally fertilizing species such as the goldfish (*Carassius auratus*), estrogen does not appear to play a critical role. Instead, prostaglan-dins, local hormones that are derived from fatty acids, stimulate sexual behavior of goldfish fe-males.

STRAIN EFFECTS

Even within a species the same hormones can produce different effects on behavior. For ex-ample, identical hormone treatments may pro-duce different results in different strains of the same species. One example of strain differences in hormone–behavior relationships concerns the effects of progesterone on aggressive behav-ior of pregnant house mice of two inbred strains, C57BL/6J and DBA/2J.

Bruce Svare (1988b) conducted three exper-iments that (1) explored the nature of preg-nancy-induced aggressiveness in C57BL/6J and DBA/2J strains and (2) determined the extent to which any differences in aggressive behavior could be attributed to strain-dependent differ-ences in either circulating progesterone or sen-sitivity to progesterone. In the first experiment, 30 pregnant females of each strain were tested for aggression on days 16, 17, and 18 of an approximately 21-day gestation period. During aggression tests, an adult male (of a third strain called Rockland-Swiss or R-S) was placed into the home cage of the female for 3 minutes. The number of attacks on R-S males was recorded. Ninety-three percent of DBA/2J females exhib-ited aggression during at least one of the three tests as compared to only 56 percent of C57BL/6J females. Were strain differences in pregnancy-induced aggressive behavior related to proges-terone levels?

Experiment 2 examined changes in aggres-sive behavior and levels of progesterone throughout the gestation period in females of both strains. Fifteen females of each strain were tested for aggressive behavior on days 6, 9, 12, 15, and 18 of gestation. Within each strain, sep-

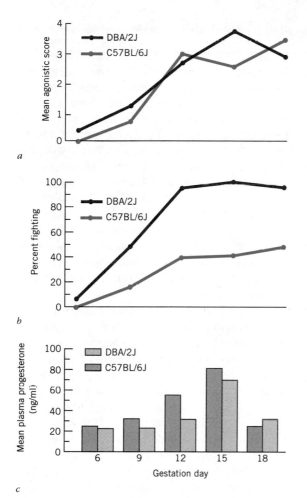

Figure 7.30 Changes across gestation in (*a*) scores for pregnancy-induced aggression, (*b*) percentage of females fighting, and (*c*) serum progesterone in DBA/2J and C57BL/6J strains of house mice. Strain differences in pregnancy-induced aggression do not result from differences in circulating levels of progesterone. (From Svare, 1988b.)

arate groups of females (N = 10 per group) were sampled for progesterone in blood serum on each of the 5 points in gestation. Figure 7.30 depicts pregnancy-induced aggressive behavior and progesterone levels of the two strains throughout gestation. Note that aggression and

serum progesterone increase in both strains as pregnancy advances and that there is a sharp decline in progesterone, but not aggression, near parturition. The fact that aggression continues at a high level during late pregnancy while progesterone concentrations are at a low point suggests that the hormonal requirements for the initiation and maintenance of aggressive behavior may be different (Svare, 1988b). Most important, however, Figure 7.30 shows that strain differences in behavior are not related to levels of serum progesterone.

Given that different levels of circulating progesterone could not explain strain differences in aggressive behavior, experiment 3 examined whether differential neural sensitivity to progesterone might explain the strain differences. In this experiment, virgin C57BL/6J and DBA/2J females (N = 60 per strain) were randomly divided into three groups (N = 20 per group) within each strain. A capsule containing either 25 milligrams of progesterone, 12.5 milligrams of progesterone, or sesame oil (the controls) was subcutaneously implanted under the nape of the neck of each female. Following surgery, the animals were tested for aggression every 3 days for a maximum of five tests or until fighting occurred. Figure 7.31 shows the cumulative percentage of virgin females that became aggressive after having been implanted with progesterone or oil. Although females of both strains became aggressive to male intruders after treatment with progesterone, DBA/2J mice appeared to be more responsive to the hormone than C57BL/6J females—a higher percentage of progesterone-implanted DBA/2J females exhibited fighting behavior.

Other work supports the idea that these two strains of house mice differ in their sensitivity to progesterone (Broida and Svare, 1982, 1983). Nest-building behavior during gestation is a progesterone-dependent response, and pregnant DBA/2J females exhibit much higher levels of this behavior than do pregnant C57BL/6J fe-

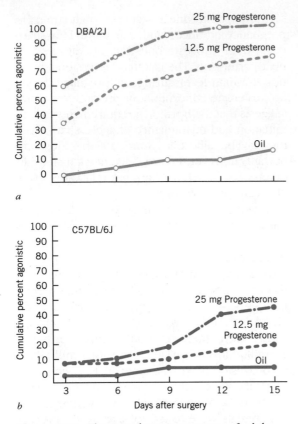

Figure 7.31 The cumulative percentage of adult virgin female house mice of strains DBA/2J (*a*) and C57BL/6J (*b*) that exhibited fighting behavior after being given implants of 25 milligrams of progesterone or 12.5 milligrams of progesterone or oil (controls). Although females of both strains become aggressive to intruders following administration of progesterone, DBA/2J mice were more responsive to the hormone treatment than were C57BL/6J females. Thus, differential sensitivity to progesterone appears to underlie strain differences in pregnancy-induced aggression. (From Svare, 1988b.)

males. Additionally, when progesterone implants are administered, virgin DBA/2J females are much more responsive to the treatment and build larger, more complex nests than do similarly treated C57BL/6J virgins.

INDIVIDUAL DIFFERENCES

We began this chapter with an example of how individual differences in behavior can arise. Remember that female mouse fetuses located in *utero* between two male littermates (2M females) are exposed to higher concentrations of testosterone than are female fetuses not next to males (0M females). You may recall that this difference in exposure to testosterone in the intrauterine environment produces individual differences in the behavior of adult females. For example, 2M females are more aggressive to female intruders, mark novel environments at higher rates, and are less attractive to males than are 0M females. Now we will consider another example of individual differences in hormone–behavior interactions.

Frank Beach (1976) outlined the following three components of female sexual behavior in mammals: (1) attractivity—a female's stimulus value in evoking sexual responses from males, (2) proceptivity—reactions of a female toward a male that constitute her taking the initiative in establishing or maintaining sexual interactions, and (3) receptivity—responses of the female that are necessary for a male's successful copulation. Beach also noted that the components of female sexual behavior are dependent upon more than just the female's hormonal condition, and that individual differences constitute important nonhormonal influences. For example, when presented with several equally receptive females, male dogs (*Canis familiaris*) often display strong individual preferences for particular females. This is perhaps best illustrated by LeBoeuf's (1967) studies of beagles. LeBoeuf monitored the visiting behavior of male beagles when they were presented with several estrous females that were tethered in an open field. Clear-cut individual preferences emerged in patterns of male visitation. Some males spent approximately 80–90 percent of their test time with certain females and 20 percent or less with the remaining females. Other males showed equally strong preferences but spent most of their time near females that were apparently unattractive to the first males. Beauty in beagles, it seems, truly is in the eye of the beholder. A second group of experiments demonstrated that male dogs are not alone in having strong individual preferences. Many female beagles in estrous exhibit selective receptivity by readily permitting some males to mate while avoiding or attacking other suitors (Beach and LeBoeuf, 1967). Figure 7.32 shows selective receptivity in three female beagles and the differences in individual preferences among them. See Chapter 14 for further discussion of mate choice.

SOCIAL/EXPERIENTIAL EFFECTS

The interaction between behavior and steroid hormones is a dynamic one. Whereas steroid hormones can activate specific forms of behavior, social or behavioral stimuli can, in turn,

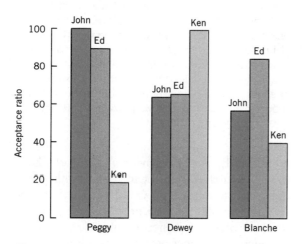

Figure 7.32 Percentage of sexual attempts by three male beagles that were accepted by three female beagles. Females exhibit selective receptivity by readily permitting some males to mount while aggressively discouraging other males with nips and growls. Note also that preferences for particular males vary among the individual females. (Data from Beach and LeBoeuf, 1967.)

induce rapid changes in levels of those steroids. For example, sexual stimuli have been shown to trigger rapid increases in plasma androgen levels in many male vertebrates. The marine toad (*Bufo marinus*), a native of Central and South America, is an explosive breeder and the first amphibian species for which activation of hormonal state by sociosexual cues was demonstrated. Orchinik, Licht, and Crews (1988) studied two populations of marine toads in Hawaii, where the species breeds year-round with bursts of mating activity following heavy rainfall. During these breeding explosions, males typically compete to clasp the limited number of females, and mating involves prolonged amplexus. Orchinik, Licht, and Crews (1988) found that when males were allowed to clasp stimulus females for 0, 1, 2, or 3 hours, concentrations of androgens (testosterone and a form of testosterone called 5-alpha-dihydrotestosterone or 5-DHT for short) were positively correlated with hours spent in amplexus (Figure 7.33). In addition, in field-sampled males, androgen concentrations were higher in amplexing males than in unpaired "bachelor" males. The apparent rise in androgens during amplexus suggests that mating behavior induced the hormonal response rather than vice versa.

Social factors also interact with hormone levels and mating in squirrel monkeys (*Siamiri sciureus*). During the nonbreeding season, the social organization of this species consists of a cohesive nucleus of females and a group of peripheral males. As the breeding season approaches, both males and females undergo "fatting," a phenomenon that is characterized by increases in body weight of up to 35 percent! At the same time, the sex-segregated pattern of living begins to change, and males and females engage in more social activities. Several studies of squirrel monkeys have revealed interesting interactions between social and hormonal factors and seasonal patterns of reproduction. For example, in group-housed animals, manipulation of the female's physiological state during

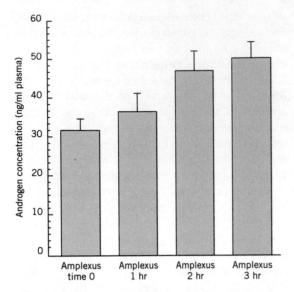

Figure 7.33 Changes in plasma androgen in male marine toads as a function of time spent in amplexus. The rise in androgens during amplexus suggests that mating behavior induced the hormonal response rather than vice versa. (From Orchinik, Licht, and Crews, 1988.)

the nonbreeding season through the administration of hormones is sufficient to initiate behavioral and physiological changes in both males and females that normally occur only during the breeding season (Jarosz, Kuehl, and Dukelow, 1977). Another experiment demonstrated that changes in social environment can produce different effects on the behavior and physiology of males and females. As mentioned, during the breeding season squirrel monkeys typically live in social groups that contain individuals of both sexes and all ages. Not surprisingly then, in the laboratory, this species reproduces more successfully when housed in large mixed-sex groups rather than in male–female pairs. Mendoza and Mason (1989) attempted to induce breeding activity in pair-housed monkeys by forming new heterosexual pairs just prior to the start of the breeding season. The change in social partners induced breeding readiness, as

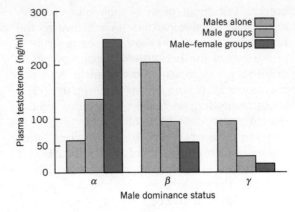

Figure 7.34 Mean levels of plasma testosterone in male squirrel monkeys during each of three phases of social group formation. Males are grouped according to their rank in the dominance hierarchy, α being highest and γ being lowest. Although relative dominance status was an important factor in determining a male's physiological response to changes in the social environment, testosterone levels during the phase of individual housing could not be used to predict the subsequent dominance status of the males. Thus, social interactions among the males in each group influenced individual levels of testosterone rather than levels of testosterone dictating social behavior. The influence of dominance status on levels of testosterone was especially evident when five females were added to each group: Testosterone increased in dominant males but decreased in subordinate males. (From Mendoza et al., 1979.)

measured by social behavior and gonadal hormones, in males, but not in females. Although the social stimulation provided by a single novel female was sufficient to produce increases in levels of plasma testosterone in males, levels of estrogen and progesterone were not altered in females by presentation of a new male. However, the male response to a new female was transitory, lasting only a few days. This transient response on the part of males is in sharp contrast to the changes in sexual behavior induced by group formation that we will now consider.

The type of social group appears to be an

Table 7.2
_____

INFLUENCE OF SEASON ON SEXUAL RECEPTIVITY OF FEMALE GREEN ANOLES (*A. CAROLINENSIS*)

SEASON	TREATMENT	N	PERCENTAGE RECEPTIVE
Summer	Estradiol	6	17
	Estradiol, progesterone	8	75
Winter	Estradiol	26	0
	Estradiol, progesterone	28	0

Source: From Wu, Whittier, and Crews, 1985.

important influence on hormone–behavior interactions in squirrel monkeys. Mendoza et al. (1979) examined the physiological response of adult male squirrel monkeys that were first housed individually, then in groups of three males, and finally in heterosexual groups formed by the addition of five females to each group of males. When males were moved from individual housing into all-male groups, a linear dominance hierarchy was established within each group; the alpha male (α) was the most dominant individual, the beta male (β) was next in line; and the gamma male (γ) was the most subordinate group member. The relative dominance status that emerged following formation of male groups was critical in determining the physiological response of each male to group formation (Figure 7.34). The alpha male in each group had the highest level of plasma testosterone and the gamma male the lowest. However, testosterone levels prior to group formation (that is, when the males were individually housed) could not be used to predict the subsequent dominance status of the males. In short, the social relationships among males appeared to influence each individual's level of testosterone, rather than the individual's physiological state determining the nature of social relationships. The impact of dominance status on testosterone levels was further accentuated when females were added to each group. Whereas dominant males showed dramatic increases in testosterone, beta and gamma males exhibited declines (Figure 7.34). The physiological re-

sponses of males to group formation lasted several weeks.

SEASONAL EFFECTS

In considering seasonal influences on hormone–behavior relationships, we return to our friend, the green anole. Females of this species undergo a seasonal cycle in which their ovaries become active in spring, produce eggs for approximately 12 weeks, and regress in late summer. Female sexual behavior is linked to this ovarian cycle and is induced by a rise in estrogen followed by a surge in progesterone. There is, however, a seasonal change in the sensitivity of the female brain to the stimulatory effects of estrogen and progesterone on sexual receptivity. Whereas ovariectomized females given estrogen and progesterone treatments in the summer are receptive to the sexual advances of courting males, similarly treated females in the winter show no signs of estrus (Table 7.2; Wu, Whittier, and Crews, 1985). In short, the same hormonal treatment given at different times of the year produces different results. These data raise the possibility that in green anole females, an internal system exists that regulates sensitivity to steroid hormones during different periods of the breeding cycle. Such an endogenous system would allow the green anole female to be sexually receptive during certain times of the year and unreceptive at others, irrespective of external influences.

Summary

Animals have two systems of internal communication, the nervous system and the endocrine system. These two systems are so closely associated that they are often referred to as a single system—the neuroendocrine system. Despite this close association, neural and hormonal modes of information transfer differ in many respects. Typically, transfer of information by way of the endocrine system occurs more slowly than that of the nervous system, and the effects that are produced are more general and long lasting. Whereas neural information is transmitted via a series of electrical events, communication by way of the endocrine system occurs through hormones, chemical substances that are secreted by either endocrine glands or neurons. Both hormones and neurohormones produce changes at the cellular level that ultimately influence behavior. Two types of hormones exist. Peptide hormones are water soluble protein molecules that exert their effects at the cell surface. In contrast, steroid hormones are derived from cholesterol, and because they are fat soluble, they move through the cell boundaries and produce changes in the interior of the cell.

The neurohormones of invertebrates influence such diverse processes as reproductive behavior, regeneration, color change, molting, and metamorphosis. Hormones also influence a wide variety of physiological and behavioral processes in vertebrates; those that affect behavior are produced primarily by the gonads (ovaries and testes), adrenal glands, pituitary gland, and pineal gland. The mechanisms by which hormones influence behavior include alterations in (1) sensation or perception, (2) development and activity of the central nervous system, or (3) effector mechanisms. Hormonal effects on behavior can be studied by the technique of removal of gland and hormone replacement or by correlational studies. In the latter, researchers look for changes in behavior that parallel fluctuations in hormone levels.

Traditionally, the effects of steroid hormones on behavior have been divided into organizational and activational effects. Organizational effects occur early in life and tend to be permanent. In contrast, activational effects occur in adulthood and tend to be transient, lasting only as long as the hormone is present in relatively high concentrations. In this case, steroids serve only to activate existing neural pathways responsible for a specific behavior rather than to organize neural pathways. The traditional distinction between organizational and activational effects has been questioned in recent years due to the lack of biochemical, anatomical, and physiological evidence for two fundamentally different ways in which steroids produce their effects.

Hormone–behavior relationships are influenced by factors such as season and experience and vary across species, strains, and individuals. Social factors, in particular, demonstrate the reciprocal nature of the hormone–behavior relationship. In some instances, hormones initiate changes in behavior, while in others, behavior causes changes in levels of circulating hormones.

CHAPTER
8

The Development of Behavior

A mallard duckling (*Anas platyrhynchos*) hatches. After spending approximately 4 weeks in the egg, the youngster spends only a single day free from the confines of its eggshell, yet still beneath its mother, before confronting the world beyond the nest. Should a predator, sometime during this first day, wander into the area around the nest, the duckling, in concert with its eight or so siblings, responds rapidly to the alarm call of its mother by "freezing"—crouching low and ceasing all movement and vocalization (Miller, 1980). If we assume a happy ending to this encounter (that is, the nest and its contents go unnoticed), the very next morning the duckling responds promptly to yet another one of its mother's calls, this time the assembly call, by following her from the nest to a nearby pond or lake (Miller and Gottlieb, 1978). Here, the duckling will string along behind its mother and siblings for some time to come. As the weeks pass and the young bird continues to associate with family members, it learns the characteristics of an appropriate mate (Schutz, 1965). This information, although obtained early in life, will not come in handy until the first breeding season. Indeed, the duckling learns, soon after hatching, most of the things it needs to survive and reproduce. We see, then, that the changes that occur during development may contribute to fitness immediately (as in the duckling's freeze response to its mother's alarm call) as well as in adulthood (as in mate preference).

Several questions arise from this brief description of early behavioral development in mallard ducks. How does the genetic makeup of the duckling interact with its internal and external environment to produce such behavior? How are the nervous and endocrine systems involved in behavioral development? What role do visual, auditory, or social stimuli play in the development of freezing, following, and sexual behaviors? Are experiences prior to hatching important to the development of posthatching behavior? What happens when a behavior, such as

the following response, ceases to be a part of the individual's behavioral repertoire? If we look beyond the single duckling in our example and consider the species as a whole, does behavioral development always proceed in a predictable and reliable fashion? As we see, there are a number of important questions associated with the development of behavior in animals.

Causes of Behavioral Change during Development

Patterns of behavior come and go throughout development. A behavior may appear in an animal's repertoire, only to disappear or change shortly thereafter. Here we consider four of the major causes of behavioral change during development. Keep in mind, however, that these causes are not mutually exclusive.

DEVELOPMENT OF THE NERVOUS SYSTEM

Behavior sometimes changes in response to the development of the nervous system. Often, correlations between the state of the nervous system and the behavior of a developing animal are most obvious during embryonic life. Consider, for example, the neural and behavioral development of embryonic Atlantic salmon (*Salmo salar*; Abu-Gideiri, 1966; Huntingford, 1986). Stages in the development of this fish are depicted in Figure 8.1.

The first movements of the embryo are seen in the feeble twitches of the heart. Movements in the dorsal musculature soon follow. It is interesting that the earliest twitches of the heart and dorsal musculature occur before the nervous system has formed. These movements are thus myogenic, or muscular in origin. The impulse begins in the muscle itself. Approximately halfway through embryonic life, the major motor

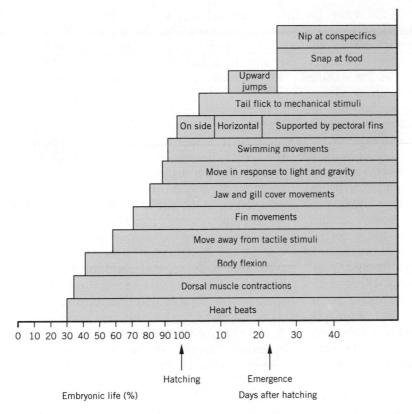

Figure 8.1 Behavioral development in the Atlantic salmon. Patterns of behavior emerge in parallel with the development of neural structures necessary for their performance. (Modified from Huntingford, 1986; drawn from the data of Abu-Gideiri, 1966; Dill, 1977.)

systems appear in the spinal cord. A short time later, the motor neurons make contact with anterior muscles, giving the embryo the ability to flex its body. Soon, with the development of neural connections at different points and on both sides of the body, the embryo displays the first undulating movements associated with swimming. Development of the sensory system of the trunk and its connection to the skin occurs a short time later; after this, the embryo can move in response to tactile stimulation. Finally, the neural circuits underlying both fin and jaw movements are complete, allowing independent and coordinated movement of these structures. Neural and behavioral development continues (in fact, the young salmon in our story has not even reached hatching yet). But we can see from these examples that changes in the nervous sys-

tem underlie the appearance of new patterns of behavior.

An associated question is, what happens to neural circuits during development when an old behavior disappears from an animal's repertoire? In some cases, the neural circuits are dismantled or permanently altered once the behavior ceases to occur (see the example of the tobacco hornworm described later). This alteration, however, does not occur in the patterns of behavior associated with hatching in chickens (*Gallus domesticus*).

Under normal circumstances, hatching behavior occurs only once in the life of a chicken, typically during a 45- to 90-minute period at the end of incubation. During hatching, the chick escapes from the confines of its shell through a highly stereotyped series of movements involving rotation of the upper body and thrusts of the head and legs. Since patterns of behavior associated with hatching disappear from the repertoire of chickens, Anne Bekoff and Julie Kauer (1984) became interested in the fate of the neural circuitry underlying, in particular, the leg movements of hatching. Their method was an interesting one. Rather than asking the age-old question, "Which came first, the chicken or the egg?," they asked, "What happens when you put a chicken back into an egg?" Bekoff and Kauer placed posthatching chicks up to 61 days of age (chicks at this age have molted their fuzzy down, are fully feathered, and basically resemble small chickens) in artificial glass eggs and recorded their behavior and muscle movements. Each chick was gently folded into the hatching position and placed into a ventilated glass egg of the appropriate size. Within 2 minutes of being placed in the artificial eggs, chicks of all ages began to produce a behavior that qualitatively and quantitatively resembled that of normal hatching. Rather than being dismantled or permanently altered after hatching, the neural circuitry for the leg movements of hatching clearly remains functional in older chickens. The question of why the circuitry remains functional is intriguing.

CHANGES IN HORMONAL STATE

Unlike the situation described for hatching behavior in chickens, the neural connections associated with some patterns of behavior are dismantled or permanently modified once the particular behavior is eliminated from an animal's repertoire. Furthermore, such modifications to the nervous system are often regulated hormonally. The tobacco hornworm, *Manduca sexta*, is an insect that undergoes complete metamorphosis. During this dramatic transformation, the nervous system of the hornworm must sequentially control three very different stages: the larva or caterpillar, the pupa, and the adult moth (Weeks, Jacobs, and Miles, 1989).

Metamorphosis, as you might well imagine, entails a remarkable change, not only in the animal's morphology, but also in its behavioral repertoire. Although some patterns of behavior are exhibited in all three stages (e.g., behavior associated with shedding of the cuticle), many behaviors are restricted to a single stage (e.g., crawling in the larva and flight in the adult). In the case of stage-specific patterns of behavior, the neural circuitry controlling these behaviors is assembled and dismantled during development. In the tobacco hornworm, much of the remodeling of the nervous system is controlled hormonally. Figuring prominently in this remodeling process are the ecdysteroids, hormones that are released by the prothoracic glands to drive each molt (see Chapter 7 for a review of hormonal control of insect metamorphosis). Response of the nervous system to ecdysteroid hormones includes production and programmed death of neurons and alterations in the structure of neurons, specifically the growth and regression of dendrites (Weeks et al., 1989; Weeks and Levine, 1990). A number of questions are associated with such changes. How do these hormonally induced changes in the nervous system relate to changes in behavior? Is the production of neurons involved in the appearance of new behaviors? Is neuronal death responsible for the disappearance of outmoded patterns of behavior? Finally, how does the growth or regression of dendrites relate to behavioral change? We can approach such questions by examining how the hormonally mediated modification of the nervous system is responsible for the loss of specific behaviors in the tobacco hornworm.

Caterpillars of the tobacco hornworm possess abdominal prolegs (Figure 8.2*a*) that act in simple withdrawal reflexes and more complex behaviors such as crawling, shedding of the cuticle, and helping the animal grasp the substrate. Although these behaviors are important to the caterpillar, they are not to the pupa; and the proleg behaviors gradually disappear during the larval–pupal transformation. The question is, then, what causes their disappearance?

b Larval muscles of prolegs

c Changes in structure of motor neurons

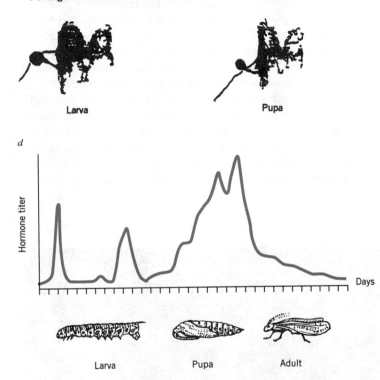

Figure 8.2 During metamorphosis in the tobacco hornworm, modifications of the nervous system that produce or eliminate patterns of behavior are controlled hormonally. (*a*) The abdominal prolegs of the caterpillar are involved in withdrawal reflexes, crawling, and grasping of the substrate. These behaviors are important to the larva but disappear from the animal's repertoire once it reaches pupal stages. (*b*) Cutaway view showing the retractor muscles of the prolegs. (*c*) The dendrites of the motor neurons that innervate the retractor muscles regress during the larval–pupal transformation. (*d*) The peak in ecdysteroid hormones just before transformation to the pupa triggers dendritic regression. (Modified from Weeks et al., 1989; data from Weeks and Truman, 1984; Bollenbacher et al., 1981.)

Most proleg movements are accomplished by retractor muscles (Figure 8.2*b*) that are innervated by motor neurons with densely branching arbors or dendrites. During the larval–pupal transformation, substantial regression of the dendrites of the motor neurons occurs (Figure 8.2*c*). Many of the motor neurons die, and the proleg muscles degenerate and become nonfunctional. We now know that the demise of the proleg neuromuscular system, and hence proleg behaviors, is caused by a peak in ecdysteroid hormones just before the transition to the pupal stage (Figure 8.2*d*). At this time, high levels of ecdysteroids trigger regression of the dendrites of the motor neurons innervating the proleg muscles. As a result of dendritic regression, the motor neurons are removed from behavioral circuits and proleg behaviors are lost in the pupa. Metamorphosis in the tobacco hornworm thus provides an example of how hormonally induced changes in the nervous system produce behavioral change in developing animals.

CHANGES IN NON-NEURAL MORPHOLOGY

Sometimes behavioral change is driven by morphological changes that are not neural. In this case, the development of a particular behavior may parallel the development of specific structures necessary for its performance. Consider the changes in feeding behavior that occur in the paddlefish *Polyodon spathula*. This fish, perhaps best known for its large paddle-shaped snout that, in some individuals, can be almost half the length of the body, occurs in the Mississippi and Ohio river drainages of North America. Although larval paddlefish feed by chasing and selectively plucking individual zooplankton from the water, adult paddlefish are indiscriminate filter feeders, dropping their lower jaw and consuming all material strained from the water as they plow through their environment (Figure 8.3).

Changes in the feeding behavior of paddlefish parallel the development of gill rakers (Rosen

and Hales, 1981). These bony structures, comblike in appearance, project from the gill arches into the oral cavity and strain food particles from the water. Keeping in mind that some paddlefish grow to over 2 meters in length, gill rakers first appear as small protuberances along the midsection of gill arches when the fish are approximately 100 millimeters long (about 4 inches). At this point in development, the fish still seek out and capture individual prey. By the time individuals are about 125 millimeters in length, the "buds" of gill rakers line all the gill arches and the projections themselves have increased in length. The young paddlers, however, still are not capable of efficient filter feeding. Paddlefish have well-developed gill rakers once they are approximately 300 millimeters in length, and it is at this stage that feeding behavior takes on the fully adult pattern of indiscriminate filter feeding. Here, then, we have an example of how changes in behavior are coordinated with the development of specific anatomical structures.

Figure 8.3   A paddlefish exhibiting the adult feeding pattern, filter feeding. Rather than indiscriminate filter feeding, larval paddlefish selectively pick zooplankton out of the water column. The eventual development of the adult mode of feeding parallels development of gill rakers in the young fish.

EXPERIENCE

Experience can be a major cause of behavioral change during development. Indeed, much of the rest of this chapter is devoted to the important role that experience, particularly that occurring early in life, has in shaping behavior. First, however, let us consider the influence of social experience on the development of food preferences in Norway rats (*Rattus norvegicus*).

Free-living Norway rats are extremely social creatures. Typically, individuals live in colonies and each colony occupies a separate burrow system. When foraging, colony members leave the burrow system, feed, and then return to the safety of their burrow. The menu of food items available to hungry rats is often quite varied, not to mention dangerous, considering our efforts to poison the rascals. So how do young rats know what foods to eat and what foods to avoid?

For more than 20 years Bennett Galef and coworkers have used the development of diet choice in rats as a means to explore the role of social experience in behavioral development. Their findings tell us much about the ways in which individuals can use the behavior of others to guide their own development. In particular, the menu selections of rats can be profoundly influenced by their social interactions with conspecifics that have recently eaten. Galef and Wigmore (1983) housed two rats together for a few days and then removed one of the rats, called the "demonstrator" rat, to a separate room and provided it with either a cinnamon-flavored diet or a cocoa-flavored diet. The rat remaining in the original cage was called the "observer" rat (this name is somewhat misleading because the observer rat cannot actually see the demonstrator rat eating, but we will go along with the authors' terminology here). Immediately after the demonstrator rat finished eating, it was returned to the cage containing the observer rat and the two were allowed to interact for 15 minutes. After this, the demonstrator was removed from the cage, and the observer was offered an opportunity to eat from either of two food cups, one containing a preweighed amount of the cinnamon-flavored diet and the other a preweighed amount of the cocoa-flavored diet.

As can be seen in Figure 8.4, the experience of interacting with an individual that has recently eaten can have dramatic effects on the food preferences of rats. Here, those observers whose cagemates ate the cocoa-flavored diet ate a much greater percentage of the cocoa diet than did those rats whose cagemates ate the cinnamon diet. This effect was apparent at least up until 60 hours after the social interaction occurred. It appears, then, that an observer rat can get information from its cagemate that allows it to identify the food eaten earlier by its cagemate. Furthermore, this information, exchanged during social interactions, biases the observer's subse-

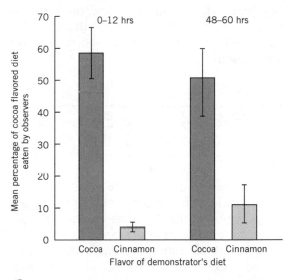

Figure 8.4 Social experience with a recently fed conspecific influences feeding behavior of Norway rats. The amount of cocoa-flavored diet eaten by observer rats was much greater if their demonstrator cagemate had previously eaten the cocoa-flavored, rather than the cinnamon-flavored, diet. The experience of smelling the breath of the demonstrator rat appears responsible for the observer's subsequent diet selection. (Modified from Galef and Wigmore, 1983.)

quent selection of diets, in a manner much like that of a person's dining out and ordering a meal recommended by a friend. Although in the human analogy, information concerning the meal is probably transmitted via verbal communication, in the case of the Norway rat, it is the experience of smelling the breath of their recently fed cagemate that causes the subsequent food preferences. Under natural conditions, eating what others are eating should, on average, be a safe bet. A young rat that smells the breath of an adult returning to the home burrow can, in all likelihood, safely use this information to select its next meal. On the basis of the older rat's safe return, it has apparently not consumed a lethal quantity of poisonous food.

We see, then, that factors such as experience or changes in the animal's nervous system or hormonal state can cause behavioral change during development. Our next topic, the development of singing behavior in birds, illustrates the combined influences of neural, hormonal, and social control.

The Role of Genes and Environment in the Development of Bird Song

---

All of us have probably, at one time or another, been struck by the beautiful and often complex songs produced by the birds around us. Consider, for example, the rich and melodious soliloquies of the canary (*Serinus canarius*), or the odd mix of rattles, whistles, and clicks woven throughout the song of the starling (*Sturnus vulgaris*), a talented bird, indeed, but one that goes largely unappreciated. The question arises, then, how does bird song develop?

The development of singing behavior in birds illustrates the complex interaction between genetic and environmental factors in the development of behavior. Bird song development has been examined from the standpoint of evolution

and ecology as well as from the more mechanistic approach of assessing neural, hormonal, and social influences on behavior. Together, these approaches have revealed the continuous interplay between the developing animal and its internal and external environment. As we will see, the factors that influence development of song range from interactions between cells to those between individuals. We begin by describing the development of singing behavior in the zebra finch (*Taeniopygia guttata*), an extremely social species native to Australia.

GENETIC, HORMONAL, AND NEURAL CONTROL OF SONG IN THE ZEBRA FINCH

As with most songbirds, among zebra finches, only the male sings and the song is thought to stimulate reproductive behavior in females (Slater, Eales, and Clayton, 1988). Although males of many species of birds also use song to advertise the boundaries of their territories, zebra finches breed colonially and males seem to tolerate other birds even short distances from their nests. Sexual differences in the singing behavior of zebra finches is a reflection of sex differences in neuroanatomy, and researchers have now identified the relatively large portion of the brain that controls song (Figure 8.5*a*). The song pathway consists of nine areas of the brain, most are located in the forebrain and one innervates the vocal organ called the syrinx. Male zebra finches have markedly larger song control areas than do females (Nottebohm and Arnold, 1976; Figure 8.5*b*). Sex differences in neuroanatomy are also evident at the cellular level; male song regions have more and larger neurons than do the same regions in females (Gurney, 1981).

How, then, do such differences arise? Not surprisingly, the chromosomal difference between the sexes forms the basis for the dramatic differences in male and female brain structure. In birds, males have two X chromosomes and females have one X and one Y chromosome. Absence of the Y chromosome results in produc-

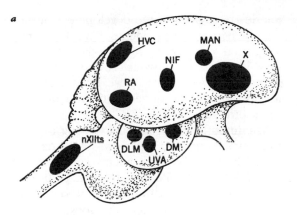

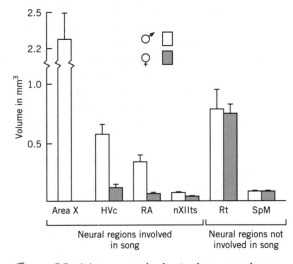

tion of estrogen by the embryonic male gonads, and this hormone appears to be the organizing substance that stimulates growth and differentiation of the regions of the brain involved in song (Gurney and Konishi, 1980). A remarkable group of experiments illustrates the powerful effects of estrogen on the neural network controlling singing behavior. If pellets containing estrogen are implanted in newly hatched male zebra finches, they have no effect (i.e., in adulthood the brains and song of estrogen-treated males are similar to those of untreated males). In contrast, treatment of nestling females with estrogen results in enlarged song areas in the brain; although these areas are larger than those of untreated females, they are not as large as corresponding areas in the male brain (Figure 8.6). Estrogen, acting either directly or indirectly, appears to masculinize the brain.

Other experiments have revealed that the effects of hormones on the brain and singing be-

Figure 8.5 (*a*) Areas in the brain that control song. (*b*) Sexual dimorphism in the brains of zebra finches. Note that four of the brain regions involved in song (Area X, HVc, RA, and nXII) are substantially larger in males than in females. No such difference is found in two regions (Rt and SpM) that are not involved in song. Sex differences in singing behavior (males sing and females do not) are thus reflected in brain structure. (*a*; After Kroodsma and Konishi, 1991; *b*; Nottebohn and Arnold, 1976.)

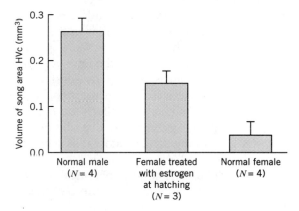

Figure 8.6 Treatment with estrogen at hatching results in enlarged song areas in the brain of female zebra finches. Here, the volume of the song area HVc is substantially larger in estrogen-treated females than in normal females, but it is still somewhat smaller than that of normal males. Estrogen is thus involved in the masculinization of the brain of songbirds. (Modified from Gurney and Konishi, 1980.)

havior of zebra finches do not end around the time of hatching. Although the typical male zebra finch has a masculinized brain, elevated levels of testosterone in the bloodstream are necessary to stimulate singing in adulthood. Similarly, those females that receive estrogen implants soon after hatching, and thus develop enlarged song areas, do not sing in adulthood unless testosterone is administered (Gurney and Konishi, 1980). Females that do not receive estrogen implants at hatching, but receive testosterone in adulthood, do not sing. Apparently, early exposure to estrogen establishes a sensitivity to testosterone in the brains of experimental females (and normal males) and exposure to testosterone in adulthood stimulates song. Whereas estrogen early in life "organizes" the development of a male song system, a high circulating level of testosterone in adulthood appears necessary to "activate" singing behavior (see Chapter 7 for a further discussion of the organizational and activational effects of steroid hormones).

The brains of birds are not always sensitive to the organizing effects of estrogen. For example, administering estrogen to adult females has no effect on the size of song areas in the brain (Gurney and Konishi, 1980). Thus, there appears to be a critical period, around the time of hatching, when the brain is particularly sensitive to hormonal influences. In this window of time, estrogen exerts its powerful effects on the developing nervous system. After that, the neural pathway controlling song cannot be switched to the male track.

We will discuss the concept of critical periods in behavioral development in more detail later in the chapter.

ROLE OF LEARNING IN DEVELOPMENT OF SONG IN THE WHITE-CROWNED SPARROW

We have seen that genetic information, the hormonal milieu, and development of the nervous system all combine to produce sexual dimorphism in singing behavior. We can now ask what role learning plays in the development of bird song. Whereas zebra finches have frequently been used to study the physiological bases for song development, white-crowned sparrows (*Zonotrichia leucophrys*) have been a favorite subject for those curious about the role of learning in bird song.

The importance of learning to song in adult male white-crowned sparrows was demonstrated by isolation experiments conducted by Peter Marler and colleagues (Marler, 1970; Marler and Tamura, 1964). Under natural conditions, young male sparrows hear the songs of their father and other adult males around them. During the first few months of life, the young male produces only subsong, a highly variable, rambling series of sounds with none of the syllables typical of the full song of adult males. As time passes, however, the song produced by the young male becomes more and more refined and soon contains elements recognizable as white-crowned sparrow syllables; this vocalization is called plastic song. Finally, the song of the male crystallizes, and he produces the full song characteristic of adult male white-crowned sparrows.

Young male sparrows reared in isolation develop abnormal song. However, if isolated males hear tapes of white-crown song during the period from 10 to 50 days of age, they will develop their normal species song. Young birds that hear white-crown song either before 10 days of age or after 50 days do not copy the song. Thus, under the conditions of social isolation and "tape tutoring," the critical period for song learning occurs between 10 and 50 days posthatching. There is some flexibility in the timing and duration of this critical period (see later discussion). The important point, though, is that in order for song to develop normally, male sparrows must be exposed early in life to the song of adult males of their species.

In addition to hearing adult conspecific song, young male white-crowned sparrows must also hear themselves sing if they are to produce normal song. A series of classic experiments revealed the critical role of auditory feedback in song development (Konishi, 1965). Birds that are exposed to the songs of adult males early in life and then deafened prior to the onset of subsong produce a rambling and variable song. Apparently, in order to develop full song, a bird must be able to hear his own voice and compare his vocal output to songs that he memorized some months previously. Further studies revealed that isolated males with intact hearing produce songs that are more normal in structure than those produced by birds deafened prior to subsong (Figure 8.7). Thus, if a bird can hear himself sing, he can produce a song with some

of the normal species-specific qualities. Finally, although auditory feedback is important for song development, it is not as essential for the maintenance of song. Birds deafened after their songs have crystallized continue to produce relatively normal song for some time thereafter.

The results of these and other laboratory studies suggested that song development in sparrows consists of two phases, a sensory phase during which songs are learned and stored, and a sensorimotor stage when singing actually begins. In the sensory phase, sounds that are heard during the first few weeks of life are stored in memory for weeks or months without rehearsal. During this time, young males produce only subsong, which does not involve retrieval or rehearsal of previously learned material. In fact, young males probably use subsong in learning how to sing.

In most sparrows, the sensorimotor phase begins at approximately 5 months of age. At this time, birds retrieve learned song from memory and rehearse the song, constantly matching their sounds to the sounds they memorized. When males reach about 7–9 months of age, song patterns crystallize, and they begin to produce full song. The final adult song will remain virtually unchanged for the rest of the bird's life. The phases of song development are outlined in Table 8.1.

Earlier, we saw how sex differences in singing behavior are reflected in the neuroanatomy of songbirds. In most species it is only the male that sings and song control areas in the brain are typically much larger in males than in females. One might wonder, then, do these song control areas also change during the phases of song development in males? Interestingly, the anatomy of the neural system controlling song indeed undergoes dramatic change over the course of song development (see review by Nordeen and Nordeen, 1990). In fact, neurons in two song control areas (HVc and Area X) reproduce rapidly in male zebra finches during the period when songs are being learned, and their

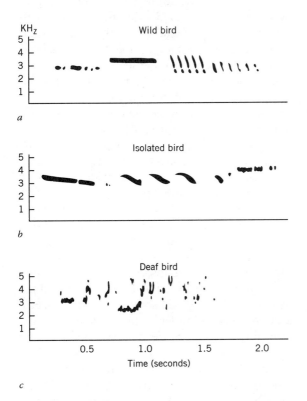

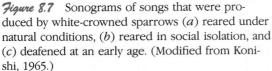

Figure 8.7 Sonograms of songs that were produced by white-crowned sparrows (*a*) reared under natural conditions, (*b*) reared in social isolation, and (*c*) deafened at an early age. (Modified from Konishi, 1965.)

Table 8.1

PHASES OF SONG DEVELOPMENT IN SPARROWS

1. Sensory phase
   a. Acquisition—song learned
   b. Storage—song retained in memory
2. Sensorimotor phase
   a. Retrieval and production—motor rehearsal of learned song
   b. Motor stabilization—song crystallized into final adult song

Source: Marler, 1987.

proliferation may actually help define the critical period for vocal learning. Specifically, the increase in the number of neurons in HVc and Area X occurs during the peak period of the sensory phase when young males listen to and memorize songs that they hear around them. The number of neurons in these two areas does not change during the sensorimotor phase when males rehearse material acquired earlier and eventually crystallize their song.

Now we will move from the level of the neuron to the level of behavior in our consideration of critical periods in song learning.

CRITICAL PERIODS IN SONG LEARNING

We know from experiments on male white-crowned sparrows raised in isolation and tutored with tape recordings of white-crown song that the critical period for song learning extends from 10 to 50 days of age. How does this compare with what is known about the time course for learning in other songbirds or in white-crowned sparrows reared under different conditions?

Length and timing of the critical period for song learning vary greatly across species. Canaries, for example, have been called "lifelong learners" as a result of their ability to continually revise their song throughout adulthood. White-

crowned sparrows and zebra finches, however, have a restricted period of learning that occurs during the first few weeks or months of life; these species are referred to as "age-limited learners."

The critical periods for song learning not only vary from one species to another, but within a species as well, from one individual to the next. Individual differences in the timing of song learning are commonly reported in both field and laboratory studies. In the laboratory, the length of the critical period can be manipulated by depriving male birds of optimal social and auditory stimuli, such as a singing adult male of their own species. Young male zebra finches copy sounds from adult males ("tutors") with whom they are housed when they are between 35 and 65 days of age. Males that are not permitted to hear any song during this period, however, retain the capacity to learn songs well into adulthood (Eales, 1985). Additionally, males may reproduce song that they hear before 35 days of age if the tutor they are exposed to after this time provides inadequate stimulation (e.g., is of a different species or is not visible to the young male; Slater, Eales, and Clayton, 1988). In some species, length of critical period depends on the method of sound presentation. Several studies have shown that live, socially interactive tutors provide stronger stimuli for learning in laboratory birds than do tape recordings of song. For example, in white-crowned sparrows, males that were exposed to live tutors of their own species learned songs beyond the 10–50 day critical period found with taped song (Baptista and Petrinovich, 1984). Stimulation from a tape recording can thus reveal a more narrow critical period than stimulation from a live tutor. Aspects of the young birds' physical environment can also influence the timing of critical periods. In the marsh wren, *Cistothorus palustris*, the time course for vocal learning is influenced by photoperiod (Kroodsma and Pickert, 1980). All these examples show that both among and within species, there is variation in the timing and duration of the critical period for song learning.

OWN-SPECIES BIAS IN SONG LEARNING

When given a choice, most young male birds learn their song from members of their own species. For example, male white-crowned sparrows copied only the songs of their own species when tutored with tapes of both song sparrow and white-crowned sparrow songs (Marler, 1970). This preference is called own-species bias, and it also influences the speed and accuracy with which young males learn. Typically, birds learn the song of their own species more rapidly and accurately than they learn the song of a different species, although this tendency varies across species and with method of song presentation. When a young male is exposed to tape recordings of conspecifics, live tutors of another species can sometimes override the bias for learning conspecific song. When young white-crowned sparrows are presented with visible singing song sparrow tutors and can hear only the songs of white-crowned sparrows in the background, they learn the song of the song sparrow (Baptista and Petrinovich, 1986). That young male sparrows learn the songs of the adult that they are allowed to interact with, even if this adult belongs to another species, emphasizes the importance of social interactions in the song learning process.

CHOICE OF TUTOR

In nature, male songbirds learn their song from older conspecifics. One wonders, then, do sons keep it in the family and learn their song from their father, or do they learn from neighboring males? The issue of identity of a song tutor is related to the timing of critical periods. If the critical period for song learning occurs soon after hatching, then fathers are likely song tutors. Alternatively, if the critical period occurs (or ex-

tends for) several months after hatching and includes the time when young males are establishing territories of their own, then neighboring males may serve as tutors. As an example, we can consider how tutors are chosen in zebra finches.

In the wild, zebra finches breed colonially and because breeding is triggered by rainfall, pairs within a colony often breed synchronously (Slater, Eales, and Clayton, 1988). The young fledge at approximately 3 weeks of age, gain independence from their parents at about 5 weeks, and then join flocks of other juveniles and nonbreeding adults. Thus, young males experience a complex social environment in which they are exposed to many conspecifics both before and after fledging. We can ask, then, from which older male or males within this vast social network do young zebra finches learn their songs? Jorg Bohner (1983) examined this question by giving young male zebra finches the choice of using either their father or another male as a song model. Two pairs of zebra finches were placed in the same cage but were separated into different compartments by means of a wire lattice. The adult males of these pairs had very different songs, although both songs were well within the bounds of typical zebra finch song. Young males hatched in these cages were raised by their parents and then at fledging were housed in a separate cage adjacent to that of their parents and neighbors. From this vantage point, the young males could see and hear their father and the neighboring male. When the young males were 100 days of age, their songs were recorded and compared with the song of their father and that of their neighbor. The results are summarized in Table 8.2. Of the 11 males raised by Bohner, 9 produced songs in which the majority of elements came from the father's song, 1 male copied all his song elements from the neighbor, and one male produced a song that contained mostly new elements. Note that none of the males produced a

Table 8.2

COMPOSITION OF THE SONGS OF MALE ZEBRA FINCHES GIVEN THE CHOICE OF THEIR FATHER OR MALE NEIGHBOR AS SONG TUTOR

MALE	PERCENTAGE OF ELEMENTS COPIED FROM THE FATHER'S SONG	PERCENTAGE OF ELEMENTS COPIED FROM THE NEIGHBOR'S SONG	PERCENTAGE OF NEW ELEMENTS
1	100	—	—
2	100	—	—
3	67	—	33
4	100	—	—
5	75	—	25
6	83	—	17
7	90	—	10
8	—	100	—
9	100	—	—
10	17	—	83
11	64	—	36

Source: Data from Bohner, 1983.

"mixed song," that is, a song containing elements from both the father's and neighbor's songs. Thus, when given the choice of using their father or a neighboring male as a song tutor, male zebra finches selected only one of the two available males, and the majority chose their father. In a later study, Bohner (1990) demonstrated that young zebra finch males exposed to their father for only 35 days posthatching developed copies of their father's song that were as complete as those of males that had contact with their fathers for 100 days. These results suggested that zebra finch males could learn all details of their species-specific song within the first 35 days of life when still dependent on their parents. Fathers were thus a likely candidate for the role of tutor in the zebra finch.

Further evidence of the role of fathers in song learning came from a study conducted by N. S. Clayton (1987). She raised young males with their parents, and once the males had fledged, she housed each male with two adult zebra finch tutors. One tutor sang a song like that of the father and the other sang a song that was quite different from the father's song. The young males chose as their model for song learning the song most similar to that of their father, and again, they selected only one tutor to learn from.

Unfortunately, all the evidence is not so straightforward. Although the studies by Bohner (1983, 1990) and Clayton (1987) pointed to the father as the likely tutor in zebra finches, very different results were obtained in a fourth study on choice of tutor in this species. Heather Williams (1990) examined choices of song models by young male zebra finches raised in an environment that more closely simulated the colonial social conditions of zebra finches in the wild. Young males were raised in an aviary that contained 10 breeding pairs of zebra finches and two nonbreeding males. Under these conditions, sons did not preferentially copy the songs of their fathers and in fact, most young males copied syllables from at least two adult males in the aviary population (Figure 8.8). The difference in the findings of this study and those of the previous investigations highlights the sensitivity of song development to experimental design in general, and to social environment in particular. Although sometimes it is hard to learn the les-

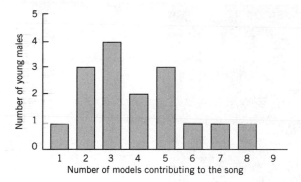

Figure 8.8 Number of song models from which male zebra finches raised in an aviary with several breeding pairs copied syllables for their songs. Note that in this complex social environment, males typically copied syllables from more than one model. (Modified from Williams, 1990.)

sons suggested by such discrepancies, it is, at the very least, clear that song development is probably more complex than was first suggested.

NONAUDITORY EXPERIENCE AND SONG DEVELOPMENT IN THE COWBIRD

Young males of other species may learn to sing not only by listening to the songs of other males, but also through direct social interaction. Brown-headed cowbirds (*Molothrus ater*) are brood parasites. Females of this species lay their eggs in the nests of other bird species, thereby relinquishing all parental duties to foster parents. If young male cowbirds relied on learning characteristics of their songs by listening to male birds in the vicinity of the nest during their first few weeks of life, they would most likely sing the song of their foster parents. But field and laboratory observations indicate that this is not the case at all; when male cowbirds become sexually mature, they sing cowbird song. Unlike the white-crowned sparrows previously described, male cowbirds that have never heard other cowbirds sing produce the correct cow-

bird song. Does this mean that experience is unimportant in the development of cowbird song? No. Learning, as it turns out, does play an integral role in shaping cowbird song. In a fascinating series of experiments, Meredith West and Andrew King have shown that the songs of male cowbirds are shaped, in part, by male–male dominance interactions and the response of females to their song. Let us take a look at some of their experiments.

Interest in the role of social interactions in development of song in cowbirds grew from the startling finding that although the songs of male cowbirds reared in isolation from male conspecifics were nearly identical to those of normally reared males, they were substantially more appealing to female cowbirds (Figure 8.9), as measured by their ability to release the female copulatory posture. In fact, the songs of males raised in either auditory and visual isolation from male conspecifics or visual isolation from male conspecifics were twice as effective in releasing the copulatory posture in female cowbirds as those of males raised either in groups containing males and females or in visual contact with such groups (Figure 8.10; West, King, and Eastzer, 1981). Why are the songs of isolates more potent than those of normal males? The answer came from observations of what happened when group-reared males and isolation-reared males were individually introduced into an aviary with an established colony of cowbirds. Group-reared males kept a low profile in their new environment; they did not sing in the presence of resident males and were not attacked. In contrast, the isolates repeatedly sang their high-potency song in the presence of resident males and females, and the resident males responded by attacking and sometimes killing the energetic songsters. Apparently, males living in mixed-sex groups, such as the flocks that occur in nature, learn to suppress their song and thereby avoid attack from dominant males.

What role might female cowbirds play in shaping the songs of males? In order to examine

Figure 8.9 Female cow-
birds prefer the songs of
males reared in isolation
to those of males reared
socially.

this question, King and West (1983) used two
different subspecies of cowbirds, one found in
the area of North Carolina (*M. ater ater*) and the
other from around Texas (*M. ater obscurus*). We
will refer to these subspecies of cowbirds as
subspecies A and subspecies O, respectively.
Male cowbirds of subspecies A were hand reared
in acoustic isolation until they were 50 days old.
At this time they were housed for an entire year
with (1) individuals of another species (canaries
or starlings), (2) adult female cowbirds of subspecies A, or (3) adult female cowbirds of subspecies O. At the end of this period, the researchers examined the songs of the males. The
three groups of males of subspecies A had developed remarkably different songs (Figure
8.11). Males that were housed with canaries or
starlings had diverse repertoires, singing a mix
of song from the two cowbird subspecies, along
with imitations of the songs of other species in
their cage. Whereas males housed with cowbird
females of subspecies A sang all A songs, those
housed with cowbird females of subspecies O

sang predominantly O songs. Male cowbirds
thus develop song repertoires that are biased
toward the preferences of their female companions.

It follows that because female cowbirds do
not sing, their influence on male song development cannot be through song. Instead, they
may encourage the male's singing by a simple
display called the wing stroke, in which one or
both wings are moved rapidly away from the
body (West and King, 1988). These wing strokes,
however, are not doled out in an undiscriminating fashion: Approximately 94 percent of the
songs of males produce absolutely no visible
change in the behavior of female cowbirds, and
on average, a single wing stroke occurs for every
100 songs. It seems, then, that male cowbirds
adjust their songs to the whims (or is it wings?)
of their audience, molding their song structure
so as to avoid conflict with dominant males and
to stimulate available females. Just how they
achieve this delicate balance in natural flocks is
an interesting question.

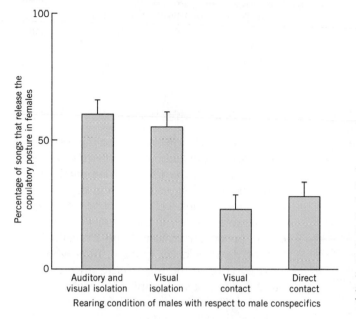

Figure 8.10 The effects of interactions with dominant males on development of song in cowbirds. Males that do not have visual or direct contact with dominant males produce more potent songs than do males that have visual or direct contact. In cowbirds then, young males learn to modify their song to avoid attack; this is an example of nonauditory stimuli influencing song learning. (Drawn from the data of West, King, and Eastzer, 1981.)

FUNCTIONAL SIGNIFICANCE OF SONG LEARNING

Much of our discussion of the development of singing behavior in birds has been about the role of learning. Before leaving the topic of bird song, however, we might ask, why must birds learn to sing? Wouldn't it be simpler, and indeed safer evolutionarily speaking, for male zebra finches, white-crowned sparrows, and cowbirds to produce a song that could not be changed by experience? Apparently not, but the precise reasons are still being hotly debated.

Three hypotheses have been proposed for the function of song learning. First, song learning may have evolved as a means of transferring complex patterns of behavior from one generation to the next (Andrew, 1962). This hypothesis is based on the commonly held, but unverified, belief that more information can be stored in the memory than in the genes. Learning, it is believed, allows for all the exquisite details associated with species-specific song.

A second hypothesis states that learning al-

lows individuals to conform more precisely to the particular social setting in which they find themselves. As a specific example of the advantage of learning to match the songs of neighbors, Petrinovich (1988) suggested that conforming songs might elicit fewer aggressive attacks than would completely novel songs.

A third hypothesis is that song learning may function as a population recognition mechanism (Marler and Tamura, 1962) and therefore must be strongly influenced by learning. This hypothesis was developed to explain the existence of regional song dialects, acoustic themes that are shared among males in the same geographic area and believed to be important in mate choice. According to this view, birds of both sexes could learn the song of their father and then in adulthood, females would choose as mates males that sing that song (Baker and Cunningham, 1985). In this way, learned regional dialects could function in the maintenance of local populations that are well adapted to their particular breeding habitat. This notion has been

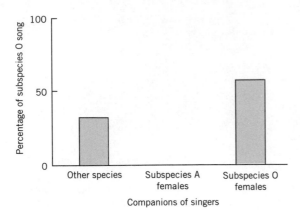

Figure 8.11 Female cowbirds influence the development of song in males. When males of subspecies A are housed with either subspecies A or subspecies O females, they mold their song to the preferences of their female companion. Note that subspecies A males housed with females of their own subspecies do not sing any subspecies O song, but those housed with subspecies O females sing subspecies O song predominantly. (Drawn from the data of King and West, 1983.)

debated, however. For example, white-crowned sparrows in the field do not show a simple relationship between type of song and mate choice (Petrinovich, 1988). Obviously, a greater emphasis on the study of song development in free-living birds is essential to our understanding of the functional significance of song learning.

The Concept of Critical Periods

Earlier in this chapter we described a critical period around the time of hatching when the hormone estrogen exerts a powerful effect on the developing nervous system of zebra finches. We also described the critical period associated with song learning in white-crowned sparrows, a time when young males are particularly sensitive to the songs of older conspecifics around them. We should add now that critical periods are important in many other kinds of learning. Because such windows of opportunity for learning are so obviously important in the development of behavior, let's take a closer look at the phenomenon.

HOW CRITICAL PERIOD WAS DEFINED

Konrad Lorenz (1935) borrowed the term "critical period" from embryology where it was used to describe times in early development that were characterized by rapid changes in organization. During these brief, well-defined periods, experimental interruption of the normal sequence of events produced profound and irreversible effects on the developing embryo. Thus, when first used by Lorenz, the term "critical period" implied a phase of susceptibility to environmental stimuli that was brief, well defined, and within which exposure to certain stimuli produced irreversible effects on subsequent behavior.

Recently, terms such as "sensitive period," "sensitive phase," "susceptible period," and "optimal phase" have been used in place of critical period, but we will continue to use the traditional term here. We should be aware, however, that the newer terms indeed reflect certain modifications in the definition of this period in light of more recent research. In fact, many of Lorenz's basic precepts have now been modified; some of these more contemporary ideas should be familiar from our discussion of critical periods in bird song. Specifically, we now know that critical periods (1) are fairly extended, (2) are not sharply defined but gradual in their onset and termination, (3) differ in duration between species, individuals, and functional systems, and (4) depend on the nature and intensity of environmental stimuli both before and during the critical period (Immelmann and Suomi, 1981). Additionally, most phenomena based on critical periods are not irreversible. Instead, patterns of behavior developed during critical periods can usually be altered or suppressed under certain conditions, especially those associated with high levels of stress. Deprivation (e.g., rearing an animal in isolation or in darkness) can reverse or destroy a pattern of behavior established during a critical period. It is important, however, not to overemphasize the reversibility of patterns of behavior established during critical periods. Conditions such as rearing in complete isolation or total darkness are unlikely to be encountered by most animals outside the laboratory environment. Furthermore, even in the laboratory, behaviors established in critical periods are usually more resistant to change than those learned at other times (Immelmann and Suomi, 1981).

TIMING OF CRITICAL PERIODS

In most animals, as with zebra finches and white-crowned sparrows, critical periods occur early in life. Why is this so? We usually assume that this is the time when animals have the greatest opportunity to gain knowledge from parents and close relatives, knowledge that is particularly important in species recognition. Later, they might not interact with them so intimately, and in some cases, they will, in fact, be exposed to intense stimuli from other species (Immelmann and Suomi, 1981). For example, in some species of birds, young remain in the nest for only a few weeks after hatching and then leave to join mixed-species flocks. It would not be surprising if young of these species learn to recognize conspecifics during a brief critical period before leaving the nest. Otherwise, choosing an appropriate mate could later be a confusing exercise indeed, since birds that waited too long to learn the defining qualities of their species might very well learn the plumage and song characteristics of another species.

Some animals have little or no contact with their parents or other close relatives after birth or hatching. One might wonder, then, would critical periods occur early in development in these species? Consider the case of Pacific salmon in the genus *Oncorhynchus*. Adult salmon spawn in freshwater, usually streams, and depending on the species and population, they may or may not die after spawning. When the juveniles emerge from their gravel nests, they may reside in their home streams for awhile, but in many species the young fish migrate thousands of kilometers to oceanic feeding grounds. After a time at sea, virtually all surviving fish return to their home stream to spawn. It is a remarkable feat of navigation and when they reach the freshwater inlets, they unerringly swim up the appropriate tributaries, making all the correct decisions at every fork until they reach the very stream where they were born, and they seem do it by smell. Apparently, before their migration to the sea and during a critical period, juvenile salmon learn the odor of their home stream. Upon returning to the vicinity of their home stream, they are stimulated to swim upstream by the familiar odor (Scholz et al., 1976; for a more detailed discussion of salmon homing see Chapter 11). Why is it important that salmon learn the precise location of their natal stream? The answer is that each population is finely adapted to its home water, so much so that salmon experimentally introduced into other streams show a higher mortality rate than locally adapted individuals (Quinn and Dittman, 1990). The period of early learning thus ensures that it is the odor of the fish's own birthplace that is remembered. We see, then, that during critical periods, animals may learn the appropriate cues, not only of conspecifics, but of the physical environment itself.

Onset of Sensitivity

It is clear that animals have a heightened sensitivity to certain environmental stimuli such as the song of a neighboring male or the smell of a home stream. One might ask, then, what causes this increased sensitivity to certain cues? It seems that onset of sensitivity may be due both to endogenous (internal) and exogenous (external) factors (Bateson, 1979). Increases in sensitivity generally begin as soon as the relevant motor and sensory capacities of the young animal are developed. Changes in internal state, such as fluctuations in hormone levels, may also influence sensitivity. Then, endogenous factors interact with environmental variables to produce the start of the critical period. For example, although the visual component of filial imprinting in birds (the response of newly hatched young to follow their mother; see later discussion) begins once hatchlings are able to perceive and process optical stimuli, experience with light appears to interact with the internal conditions to initiate this particular critical period (e.g., Bateson, 1976). Here, then, we have endogenous factors (ability of the nervous system to deal with optical stimuli) interacting with environmental factors (exposure to light) to produce the start of the critical period.

Decline in Sensitivity

Two hypotheses, the internal clock hypothesis and the competitive exclusion hypothesis, have been proposed to account for the termination of critical periods. The internal clock hypothesis assumes that the decline in sensitivity is under endogenous control. According to this idea, the termination of critical periods is determined by the ticking of some internal clock and has little, if anything, to do with external factors such as experience. In other words, some physiological factor, intrinsic to the animal, ends the period of receptivity to external stimulation.

In the competitive exclusion hypothesis, filial imprinting is viewed as a self-terminating process in which external factors play a critical role. According to this idea, a given external experience and its impact on neural growth prevent subsequent experiences from modifying the behavior (Bateson, 1979, 1987). Suppose that a behavior, such as the response of ducklings to follow their mother, is based on the growth of neural connections into an area of finite size. When the growth of these connections has pro-

ceeded to some point, the first experience (e.g., a view of the mother) is better able to direct the behavior than is a later form of input (e.g., a view of other ducks on the pond). The competitive exclusion hypothesis states that the first experience effectively excludes subsequent experiences from affecting the behavior, and in so doing, it terminates the critical period.

IMPORTANCE OF CRITICAL PERIODS IN BEHAVIORAL DEVELOPMENT AND SOME EXAMPLES

Now that we have discussed the definition, adaptiveness, and control of critical periods, we should tell you that some people believe they do not exist. In fact, over the past several years, there has been a vigorous argument over the importance of early experience in shaping subsequent behavior (reviewed by Klopfer, 1988). Although strong proponents of critical periods believe that such periods characterize all developmental processes and that exposure to environmental stimuli during the sensitive phase produces irreversible effects on behavior (e.g., Scott, 1962), critics downplay the influence of the early environment on subsequent behavior, claiming that virtually any effect of early experience can later be modified (e.g., Clarke and Clarke, 1976). In this latter view, events that occur early in an organism's life are not of pivotal importance; instead, it is the accumulation of experiences, particularly those that happen time and again, that influences behavioral development. As is often the case, the truth seems to encompass both explanations. In some instances, early experience produces virtually unalterable effects on behavior, while in others, environmental effects on behavior can be modified at a later point in time. Although critical periods do not characterize all developmental processes, they do play an important role in many behavioral phenomena. Now let's consider selected examples of behavioral development

Figure 8.12 Young Canada geese following their mother.

that depend, to varying degrees, on specific experiences during a window of time.

Filial Imprinting in Birds

Anyone who has ever watched chicks in a farmyard or ducklings and goslings on a pond knows that the young generally follow their mother wherever she goes (Figure 8.12). How does such following behavior develop? Konrad Lorenz (1935), working with newly hatched goslings, was the first to systematically study this behavior. In one experiment, Lorenz divided a clutch of eggs laid by a greylag goose (*Anser anser*) into two groups. One group was hatched by their mother and, as expected, these goslings trailed behind her. The second group was hatched in an incubator. The first moving object these goslings encountered was Lorenz, and they responded to him as they normally would to their mother. Lorenz marked the goslings so that he could determine in which group they belonged and placed them all under a box. When the box was lifted, liberating all the goslings simultaneously, they streamed toward their respective "parents," normally reared goslings toward their

mother and incubator-reared youngsters toward Lorenz. The goslings had developed a preference for characteristics associated with their "mother" and expressed this preference through their following behavior. The attachment was unfailing and from that point on Lorenz had goslings following in his footsteps (Figure 8.13).

Because social attachment evidenced by following seemed to be immediate and irreversible, Lorenz named the process *Prägung*, which means stamping. The English translation is imprinting. Used in this context, the term suggested that during the first encounter with a moving object, its image was somehow permanently stamped on the nervous system of the young animal. Today the process by which young birds develop a preference for following their mother is called filial imprinting. It is the archetypal example of the role of early experience in behavioral development. The biological function of filial imprinting is probably to allow young birds to recognize close relatives and thereby distinguish their parents from other adults that might attack them (Bateson, 1990). We will focus our discussion on the develop-

Figure 8.13 Goslings following their "mother," Konrad Lorenz. Lorenz was one of the first scientists to study imprinting experimentally.

ment of the following response in mallard ducklings.

Mallard ducklings, like most young within the orders Anseriformes (ducks, geese, and swans) and Galliformes (chickens, turkeys, and quail) are precocial, quite capable of moving about and feeding on their own just a short time after hatching. Filial imprinting is usually studied in species with precocial young. The following response is nonexistent—or much less evident—in species, such as the songbirds discussed earlier, whose young are altricial, virtually helpless and therefore dependent upon their parents for the first few weeks after hatching. We begin our coverage of the following response in mallard ducklings with a brief description of reproduction and early development in this species.

Upon finding a suitable nesting site, typically a shallow crevice in the ground, the mallard hen begins to lay her eggs, at the rate of one egg per day (Miller and Gottlieb, 1978). The average clutch size is 8–10 and after the last egg is laid, the hen begins incubation, a process that lasts approximately 26 days. Encouraged by the warmth of the mother's body, the embryo inside

each egg begins to develop. Two to three days before hatching, each embryo moves its head into the air space within its egg and begins to vocalize; these vocalizations are called contentment calls. About 24 hours later, the embryos pip the outer shell and then take another day to break through the rest of the shell and hatch. Most of the ducklings in a clutch will hatch within an interval of 10 hours. The hen broods her young for a day and then leaves the nest and emits calls to encourage the ducklings to follow her. Although she calls during incubation and brooding, the frequency increases dramatically at the time of the nest exodus. Prompted by their mother's assembly calls, the young leave the nest and follow her to a nearby pond or lake where they will paddle behind her. What characteristics of the mother form the basis for the ducklings' attachment? Do the ducklings imprint on the mother's call, her physical appearance, or some combination of the two? What role might siblings play in the development of the following response? Finally, is there a critical period during which exposure to certain cues must occur for the normal development of filial behavior?

You may recall that we asked some of these questions in our introduction to this chapter. Many of the answers come from the laboratory of Gilbert Gottlieb. Over the past 30 or so years, Gottlieb and coworkers have examined the development of the following response in Peking ducks, a domestic form of the mallard (despite their domestication, Peking ducks are quite similar to their wild counterparts in their behavior). In one experiment with ducklings that had never had contact with their mother, Gottlieb (1965) examined (1) the relative importance of hen auditory and visual cues in the development of the following response and (2) whether the parental call of a duckling's own species would be more effective than parental calls of other species in inducing and maintaining attachment behavior. As part of the study, 224 eggs were hatched in an incubator; thus, the ducklings never came in contact with their mothers. The ducklings were divided into four groups and tested for their following response. In all four groups, the ducklings were tested with a stuffed replica of a Peking hen as it moved about a circular runway. Individuals in one of the groups, however, were tested with a silent hen, while individuals in the other three groups were tested with hens that emitted assembly calls through a speaker concealed on their undersides. Of the ducklings that heard hen assembly calls, one group was exposed to mallard calls (Peking calls), one group to wood duck calls, and one group to domestic chicken calls. Each duckling was given a 20-minute test to determine if it would follow the stuffed model around the circular arena. The results are presented in Figure 8.14. As you can see, the maternal call of the mallard was far more effective than the maternal calls of the wood duck or chicken in inducing following by the ducklings, and all conditions with calls were much more effective than no call at all.

In a second experiment, incubator-reared ducklings were given a choice of following either a stuffed Peking hen emitting mallard calls

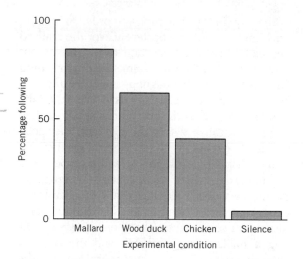

Figure 8.14 A stuffed Peking hen emitting mallard calls was more effective in eliciting the following response in incubator-reared Peking ducklings than a hen emitting either wood duck or chicken calls (Peking ducks are a domestic form of the mallard duck). All hens with calls were more effective than a silent hen. Thus, auditory cues from the mother are important in controlling the early behavior of ducklings, and ducklings respond selectively to the call of their own species without previous exposure to it. (Drawn from the data of Gottlieb, 1965.)

or a stuffed Peking hen emitting chicken calls. When placed in this simultaneous choice situation approximately 1 day after hatching, the majority of ducklings (76%) followed the model emitting the mallard call. Taken together, these experiments demonstrate that auditory stimuli from the mother play an important role in controlling the behavior of newly hatched ducklings and ducklings respond selectively to the maternal assembly call of their own species without any previous exposure to it (remember that all ducklings were reared in incubators and therefore had no contact with hens).

These findings do not, however, suggest that experience is unimportant in the development of a preference for the assembly call. Experience, as it turns out, is critical, but it occurs prenatally (before hatching). In order for Peking ducklings to exhibit a preference for the mallard hen call, they must hear their own contentment calls or those of their siblings before hatching (Gottlieb, 1978). If ducklings are reared in isolation (and therefore are not exposed to the calls of their siblings) and made mute just before they begin to vocalize within the egg (and therefore are not exposed to their own calls), they no longer display their highly selective response to the maternal call of their species. When these ducklings without normal embryonic auditory experience are placed in a test apparatus equidistant between two speakers, one speaker emitting a mallard maternal call and the other a chicken maternal call, they choose the chicken call almost as often as they choose the mallard call 48 hours after hatching. In contrast, ducklings with normal embryonic auditory experience always choose the mallard call over the chicken call (Table 8.3). Thus, auditory experience prior to hatching is important to the development of the following response because it fine tunes early call preferences in young mallards.

Is there a critical period during which exposure to contentment calls must occur for ducklings to exhibit a preference for the call of the mallard hen? If ducklings heard contentment calls after hatching, would they display the normal preference for mallard calls? Gottlieb (1985) set out to answer these questions. He began by raising ducklings in isolation and making them mute before they began to vocalize within the egg (again, these manipulations ensure that the ducklings are not exposed to their own calls or the calls of their siblings). One group was exposed to contentment calls during the embryonic period (approximately 24 hours prior to hatching) and another group was exposed to contentment calls during the postnatal period (approximately 24 hours after hatching). Ducklings in each group were then tested for their preference for the mallard hen call; this time, however, they were given the choice of approaching a speaker that emitted normal calls of a mallard hen or a speaker that emitted artificially slowed calls of a mallard hen. As shown in Table 8.4, ducklings must be exposed to contentment calls prior to hatching if they are to show a preference for the normal mallard hen call; exposure after hatching is ineffective in pro-

Table 8.3

THE EFFECTS OF EMBRYONIC AUDITORY EXPERIENCE ON CALL PREFERENCES OF PEKING DUCKLINGS 48 HOURS AFTER HATCHING

GROUP	N	MALLARD CALL	CHICKEN CALL	BOTH
Vocal-communal	24	24	0	0
Devocal-isolated				
First experiment	22	12	9	1
Replication	21	14	6	1
Total	43	26	15	2

Ducklings raised in the vocal-communal group could hear themselves and the calls of siblings prior to hatching; devocal isolated ducklings had no such auditory experience. N = number of ducklings that responded to calls.
Source: Data from Gottlieb, 1978.

Table 8.4

EFFECTS ON PREFERENCES OF DEVOCAL-ISOLATED DUCKLINGS OF EMBRYONIC VERSUS POSTNATAL EXPOSURE TO CONTENTMENT CALLS

TIME OF EXPOSURE TO CONTENTMENT CALLS	N	PREFERENCE		
		NORMAL MALLARD	SLOWED MALLARD	BOTH
Embryonic	32	21	8	3
Postnatal	37	14	20	3

N = number of ducklings that responded to calls.
Source: Data from Gottlieb, 1985.

ducing the preference for the normal call of their species. Here, then, we have an example of an embryonic critical period.

Another question is, what role, if any, might visual stimuli play in development of the following response? Several experiments have shown that two conditions must be simultaneously met for visual imprinting on a mallard hen to occur in the ducklings. The ducklings must be reared with other ducklings and allowed to actively follow a mallard hen, or a model, if they are to prefer the appearance of a hen of their own species to that of a hen of another species at later testing (e.g., Dyer, Lickliter, and Gottlieb, 1989; Lickliter and Gottlieb, 1988). Ducklings that are reared in isolation and given passive exposure to a stuffed mallard hen (i.e., housed with a stationary model) do not develop a preference for the mallard hen. These results suggest that under natural conditions, ducklings learn the visual characteristics of their mother after leaving the nest and following her around. Thus, whereas auditory stimulation from the mother appears largely responsible for prompting the ducklings to leave the nest and for influencing their earliest following behavior, the hen's appearance becomes important after the nest exodus. The importance to filial imprinting in mallards of auditory cues, visual cues, and active following was well summarized by Konrad Lorenz (1952, pp. 42–43). In an experiment with

Peking ducklings, Lorenz found himself in a rather embarrassing position for one destined to become a Nobel Laureate. In his words:

The freshly hatched ducklings have an inborn reaction to the call-note, but not to the optical picture of the mother. Anything that emits the right quack note will be considered as mother, whether it is a fat white Peking duck or a still fatter man. However, the substituted object must not exceed a certain height. At the beginning of these experiments, I had sat myself down in the grass amongst the ducklings and, in order to make them follow me, had dragged myself, sitting, away from them. So it came about, on a certain Whit-Sunday, that, in company with my ducklings, I was wandering about, squatting and quacking, in a May-green meadow at the upper part of our garden. I was congratulating myself on the obedience and exactitude with which my ducklings came waddling after me, when I suddenly looked up and saw the garden fence framed by a row of dead-white faces: a group of tourists was standing at the fence and staring horrified in my direction. Forgivable! For all they could see was a big man with a beard dragging himself, crouching, round the meadow, in figures of eight, glancing constantly over his shoulder and quacking—but the ducklings, the all-revealing and all-explaining ducklings were hidden in the tall spring grass from the view of the astonished crowd!

To summarize, the experiments of Gilbert Gottlieb and coworkers have demonstrated that filial imprinting in Peking ducklings results from a complex interaction of auditory, visual, and social stimuli provided by the hen and siblings. Several important generalizations about early behavioral development arise from their work. First, we can no longer think of experience as occurring only after birth or hatching; embryonic experience can also influence behavior. In the case of Peking ducklings, listening to the contentment calls of siblings before hatching is critical to their preferring the maternal call of their own species after hatching. The experiments with ducklings also demonstrate that a variety of stimuli may be involved in the development of a single pattern of behavior and that the relative importance of different stimuli may change as the young animal matures. Although the following behavior of ducklings soon after hatching is influenced largely by auditory cues from their mother, only a few days later, visual stimuli (in combination with auditory stimuli) become important in the following response. The relative priorities of different cues in the development of the following response matches the timing for development of the auditory and visual systems; in ducklings, as in all birds, the auditory system develops in advance of the visual system (Gottlieb, 1968). These results emphasize the close interaction between physical maturation and experience in early behavioral development. Finally, the study of filial imprinting in Peking ducklings illustrates that we must be open-minded when trying to sort out just which experiences influence a given behavior. Who would have thought that listening to sib-

lings prior to hatching or interacting with siblings after hatching would be critical in development of the ducklings' attachment to their mother? We now know that such nonobvious experiential factors are indeed essential to the development of following behavior in this species.

Sexual Imprinting in Birds

We have seen that early experience influences a duckling's attachment to its mother, an attachment that the young bird demonstrates by trailing behind her in the days following the exodus from the nest. Early experience also has important consequences for the development of mate preferences in birds. In many species, experience with parents and siblings early in life influences sexual preferences in adulthood. The learning process in this case is called sexual imprinting. Typically, sexual preferences develop after filial preferences, although the critical periods may overlap to some degree (Bateson, 1979). Whereas filial imprinting is indicated by the following response of young birds, sexual imprinting is typically shown in the preferences of sexually mature birds for individuals of the opposite sex. It is important to note that unlike filial imprinting, sexual imprinting occurs in both altricial and precocial birds.

One dramatic demonstration of the importance of early experience to subsequent mate preference came from cross-fostering experiments with finches. Klaus Immelmann (1969) placed eggs of zebra finches (*Taeniopygia guttata*) in clutches belonging to Bengalese finches (*Lonchura striata*). The Bengalese foster parents raised the entire brood until the young could feed themselves. From then on, young zebra finch males were reared in isolation until sexually mature. When they were later given a choice between a zebra finch female and a Bengalese finch female, zebra finch males courted Bengalese females almost exclusively. In a second study, cross-fostered zebra finch males were again separated from their foster Bengalese parents, but this time they were provided with a conspecific female and nesting supplies. Most of these males eventually mated with conspecific females and successfully produced young. When they were tested several months or years later, however, the males still displayed a preference for Bengalese females. Thus, brief contact with foster parents early in life exerted a more powerful, longer-lasting influence on mate preference than did long-term social contact in adulthood (Immelmann, 1972).

A second example of the impact of early experience on later mate preference also involves the manipulation of foster rearing. In this case, however, we see that birds can develop strong attachments for individuals that are quite different from themselves, providing they have had early exposure to such individuals. We return again to the writings of Konrad Lorenz, who discovered the strength of such attachments when he became the object of a male bird's fancy—a bird, as it turns out, with a particular penchant for the ritual of courtship feeding. Here Lorenz (1952, pp. 135–136) describes his ordeal with a male jackdaw (a bird similar to a crow) that had become imprinted on him:

The male jackdaw became most importunate in that he continually wanted to feed me with what he considered the choicest delicacies. Remarkably enough, he recognized the human mouth in an anatomically correct way as the orifice of ingestion and he was overjoyed if I opened my lips to him, uttering at the same time an adequate begging note. This must be considered as an act of self-sacrifice on my part, since even I cannot pretend to like the taste of finely minced worm, generously mixed with jackdaw saliva. You will understand that I found it difficult to co-operate with the bird in this manner every few minutes! But if I did not, I had to guard my ears

against him, otherwise, before I knew what was happening, the passage of one of these organs would be filled right up to the drum with warm worm pulp, for jackdaws, when feeding their female or their young, push the food mass, with the aid of their tongue, deep down into the partner's pharynx. However, this bird only made use of my ears when I refused him my mouth, on which the first attempt was always made.

At first glance, it might seem curious that an individual has to learn to identify an appropriate mate. Wouldn't it be a safer evolutionary strategy for zebra finches and jackdaws to have a mating preference that was unmodifiable by early social experience? Apparently not. For now, though, we can only theorize about the importance of early learning in choosing mates. One idea, put forth by Patrick Bateson (1983), provides an interesting explanation for the functional significance of sexual imprinting. Bateson suggested that animals learn to identify and selectively respond to kin. Armed with information on what their relatives look like, individuals then choose mates similar, but not identical, to their family members. Given that both extreme inbreeding and outbreeding may have costs (see Chapter 14), sexual imprinting provides information that allows animals to strike a balance between the two. Bateson used evidence from studies of quail to support his argument.

Japanese quail (*Coturnix coturnix japonica*) prefer to mate with individuals that are similar to, yet slightly different from, members of their family (Bateson, 1982). In one study, chicks were reared with siblings for the first 30 days after hatching and then socially isolated until they became sexually mature. At 60 days of age, males and females were tested for mate preference in an apparatus that permitted viewing of several other Japanese quail (Figure 8.15a). The birds that were viewed were of the opposite sex of

a

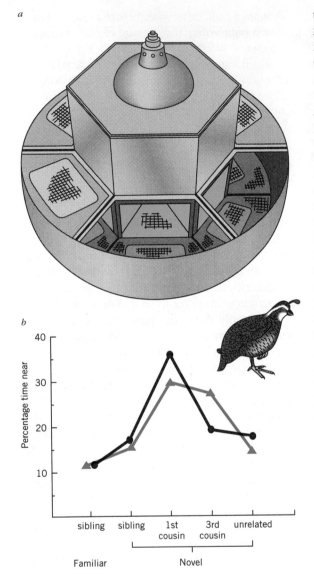

b

the test animals and belonged to one of the following five groups: (1) familiar sibling, (2) novel (unfamiliar) sibling, (3) novel first cousin, (4) novel third cousin, or (5) novel unrelated individual. Similarity in plumage between Japanese quail is considered to be proportional to genetic relatedness, and thus test animals could presumably judge genetic distance on the basis of plumage characteristics. As we see in Figure 8.15*b*, both males and females preferred to spend time near first cousins. The sexual preferences displayed by quail reflect a choice slightly displaced from the familiar characteristics of relatives: Siblings are too familiar, unrelated animals too different, but first cousins are the perfect mix of familiarity and novelty. Bateson suggested that such fine-tuned preferences resulted from a combination of two learning processes, imprinting and habituation. Whereas imprinting would tend to restrict preferences to the familiar, habituation would tend to reduce responsiveness to familiar animals. When one form of learning is superimposed upon the other, the result is a preference slightly displaced from the familiar (see Chapter 5 for a further discussion of these two types of learning).

The results from studies of Japanese quail certainly seem to support Bateson's contention that through sexual imprinting, some young animals learn the characteristics of their close relatives, and then in adulthood, they choose a mate similar, but not identical, to their family members. A note of caution, however, is in order. As Bateson (1983) himself points out, we must be careful in generalizing results from laboratory studies to animals in their natural environment. Clearly, laboratory conditions of rearing and testing animals are vastly different from the natural conditions under which animals live and choose mates. We should also be aware that our choice of measures, such as the number of approaches an animal makes toward another, may not reflect mate preference or actual mating in nature. In the best of all possible worlds, we

Figure 8.15 Japanese quail prefer to spend time near individuals that are similar to, yet slightly different from, members of their family. (*a*) Apparatus used by Patrick Bateson to test preferences of adult quail. (*b*) Both male (shown as triangles) and female quail (shown as circles) prefer first cousins, possibly striking an optimal balance between inbreeding and outbreeding. (*a*; From Bateson, 1982; *b*; Modified from Bateson, 1982.)

would compare results from the controlled environment of the laboratory with field observations on the impact of early experience on subsequent mate choice. In the field, however, it is often difficult, if not impossible, to know the precise genetic relationship of animals and to track animals from birth to adulthood, chronicling their early experiences and subsequent mating behavior. Although many factors preclude, or at least make very difficult, the study of sexual imprinting and later mate choice in animals in their natural environment, we should make every attempt to design laboratory experiments and to evaluate our results from such experiments in the context of what is known about a species' environment and social system.

Imprinting-like Processes in Mammals

Most of the research on imprinting has been done with birds, but the phenomenon is not restricted to birds. Just as ducklings will string along behind their mother, a litter of young shrews of the genus *Crocidura* will line up behind their mother, each youngster grasping the fur of the shrew in front, as the mother leads the caravan from place to place. As is typical of imprinting in most mammals, olfactory cues, rather than visual or auditory information, provide the relevant stimulus. From 8 to 14 days of age, young shrews imprint on the odor of their mother while nursing. If a shrew is nursed by a foster mother during this time period, it imprints on the odor of the substitute mother. When returned to its biological mother on day 15, the young shrew will not tag along behind her or any of its siblings that remained with her. The youngster will, however, follow a cloth impregnated with the odor of its foster mother (Zippelius, 1972).

It is likely that odor preferences change with age, and mammals may not always find the odor of their mother and siblings quite so attractive. Christopher Janus (1988) examined the development of olfactory preferences in spiny mice, *Acomys cahirinus* (Figure 8.16*a*). The genus *Acomys* is unique among murid rodents (the family Muridae includes rodents such as New and Old World rats and mice, hamsters, gerbils, and lemmings) in that newborn are precocial—they are born relatively well developed. In contrast to altricial rodents that are born naked, with an incompletely developed olfactory system and closed eyes and ears, newborn spiny mice can hear and smell at birth. Shortly after that, their eyes open and they are capable of coordinated movement. Hair appears shortly thereafter. As a result of their rapid rate of physical development and tendency to be active at night when visual cues are limited, early development of attachment to specific olfactory cues might be of great importance to spiny mice. Janus tested spiny mice from 1 to 35 days of age in a multichoice preference apparatus: Mice had the option of spending time near dishes that were empty or near dishes that contained shavings that were either clean, soiled by family members, or soiled by unrelated conspecifics. An interesting pattern in the development of olfactory preferences emerged (Figure 8.16*b*). Youngsters from 1 to 10 days of age showed a strong preference for familiar home cage odor. This preference declined with age and by day 25, pups no longer found the familiar odor of parents and siblings attractive. At about this time, pups began to show an increasing preference for odor from unfamiliar conspecifics.

Janus speculated that decreased attachment to familiar odor and increased attachment to unfamiliar odor facilitate dispersal from the home area and emigration to other social groups. In addition, increased preference for odors different from those of parents and siblings might help spiny mice avoid breeding with close relatives. We have already mentioned the relationship between early exposure to family cues and subsequent mate preference in the section on sexual imprinting in birds. Now we see that experience with parents and siblings

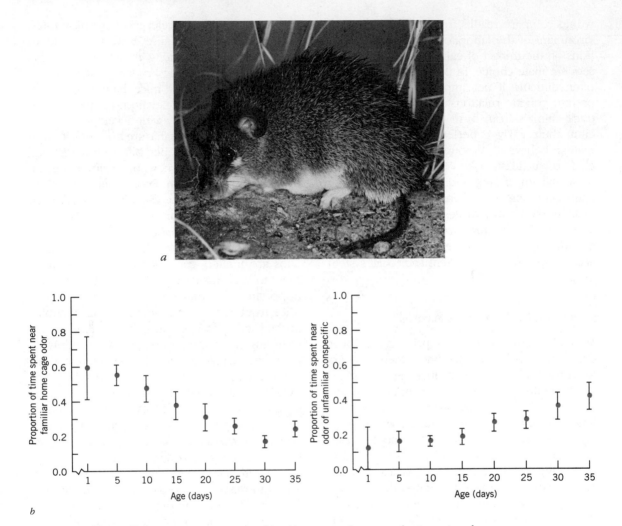

Figure 8.16 (*a*) A spiny mouse. (*b*) Olfactory preferences of spiny mice change over time. Preference for odors from family members declines with age, but preference for odors from unfamiliar conspecifics increases. (Modified from Janus, 1988.)

can also influence adult sexual preferences in mammals.

Finally, another imprinting-like process that occurs in mammals involves maternal attachment. In some species of ungulates, a lasting mother–young bond is established rapidly after birth and results in the mother's directing her care exclusively toward her own offspring. The rejection of the young of other mothers is a nononsense affair and can often be quite violent. Consider the development of maternal attachment in the domestic goat (*Capra hircus*), a species known for its intolerance of alien young. Young goats (kids) are precocial, capable of wandering away from their mother soon after birth. Thus, it is important that a bond between mother

Figure 8.17 A mother goat rejecting an unfamiliar kid. Mother goats become attached to their own young shortly after birth and butt away alien young.

and offspring be established early. Furthermore, kids will initially approach any mother to nurse, even though there is no reciprocity among mothers for care of their offspring. It also pays then for a mother to be able to tell her own young from others (Gubernick, 1980). Although the formation of mother–young relationships in goats depends on mutual bonding between the mother and her offspring, we will focus our discussion on the development of the mother's attachment to her kids.

Domestic goat mothers, also called nannies, become strongly attached to their offspring and butt away young that are not theirs (Figure 8.17). Peter Klopfer (1971) investigated formation of the maternal bond in this species. In one experiment, he examined the influence of early contact with offspring on subsequent acceptance or rejection of own and alien young after a period of separation. There were two treatment groups, each containing 15 mother goats. In the first group, kids were removed from mothers immediately after birth and were wrapped in toweling to keep them warm during the separation. Mothers in the second group were permitted 5

minutes of contact with their firstborn kids before the young, and all subsequent littermates, were removed. Kids were presented to females after a 1-, 2-, or 3-hour separation and were introduced one at a time for a 10-minute test period. Females were scored according to whether they accepted or rejected the kid. Acceptance was characterized by the mother's licking the youngster and allowing it to nurse; rejection typically involved withdrawing from the kid and butting it when it persisted in trying to nurse. There were no differences in the effects of separations of 1, 2, or 3 hours, so these subgroups were combined and the analysis limited to whether or not mothers had 5 minutes of contact with their young after birth. Fourteen of the 15 mothers who were allowed contact with their firstborn for 5 minutes after birth accepted and nursed all their own offspring, while vigorously rejecting all alien young. Of the 15 mothers that had been immediately separated from their kids at birth, only 2 allowed their kids to nurse and all rejected alien young. It seemed, then, that a short period of contact with young just after birth was essential for development of maternal attachment in goats. We see, in this example, an exception to the rule that critical periods occur early in life. Here, it is the adult mother goat that "imprints" on her young.

Klopfer's early research indicated that maternal attachment in goats is specific, rapidly formed, and fairly stable. Maternal attachment to young seemed to occur during a critical period just after birth in which only 5 minutes of contact with at least one kid was necessary for development of the maternal bond. More recent findings, however, suggest that a longer period of initial mother–young contact is necessary for maternal imprinting in goats. In these studies, mothers given 5 minutes of contact with their own kid immediately after birth often failed to later discriminate between their own and alien young after a 1-hour separation; 62 percent of the goats accepted both their own and alien young; and only 14 percent accepted their own and rejected alien kids (Gubernick, 1980; Gubernick, Jones, and Klopfer, 1979). What could account for the mothers' acceptance of alien kids in these studies? The important factor turned out to be whether or not the alien kids had been kept with their mothers during the experiment. Alien kids that had never had contact with their own mother were accepted; alien young that had been kept with their mother for at least 8 hours were routinely rejected by other mothers. These results suggested that mothers in contact with their kids label them in some way, perhaps by permeating them with their own scent, and that labeled kids are subsequently rejected by other mothers. Additional studies revealed that mother goats do indeed label their youngsters, directly through licking and indirectly through their milk (Gubernick, 1981). The alien kids used in Klopfer's early study had all been kept with their own mothers and thus maternal labeling could explain why they were subsequently rejected by mothers given only 5 minutes of contact with their newborn. The more recent findings suggest that if Klopfer had used kids that had not had contact with their mother after birth (and thus had not been labeled), the mothers would probably have accepted the alien young.

In summary, then, mother goats must associate with their kids for a certain interval of time after birth in order for reliable recognition and discrimination to subsequently occur. During this critical period the mother appears to label her kids and concurrently to learn the label. Although labels may initially be olfactory, we can expect that, in time, they may change to include other features of the young such as vocalizations and appearance. We might also expect different labels to be used for recognition at close quarters as opposed to at a distance. Finally, in line with current views on critical periods in behavioral development, the sensitive period for maternal attachment in goats does not appear to be as short and well defined as would have been previously believed. Indeed, mothers will accept alien kids that have not had contact with their own mother for at least 1 hour after birth.

Development of Social Behavior in Ants

Up until now we have been considering the effects of experience during specific windows of time on the development of social behavior in birds and mammals. We can now ask, what evidence have we of critical periods in behavioral development in other classes of animals? Our next example comes from the insects, a class that contains roughly three-quarters of the approximately 1 million known species of animals on earth. The insect that will concern us here is the ant.

The life of an ant can often be quite complicated. In many species, individuals live in complex societies in which labor is divided among colony members, and their roles can change dramatically and repeatedly over the course of a few weeks. Given their intricate social system and frequent changes in job status, it is not surprising that ants have become popular subjects for researchers interested in behavioral development. Of particular interest is the neotropical ant *Ectatomma tuberculatum*, a species in which the individual's job changes with age.

Annette Champalbert and Jean-Paul Lachaud (1990) examined the role of early social experience in the development of behavior of *E. tuberculatum* workers. The two researchers were interested in what effects a 10-day period of social isolation would have on behavioral development in workers of different ages. They began their task on a coffee plantation in southern Mexico where they collected four ant colonies from the bases of different coffee trees, each colony containing from 200 to 300 individuals. Each colony was placed into an artificial nest made of plaster that had several interconnected chambers and a foraging area. A glass pane permitted the ants to be observed. A few hours after their emergence from cocoons, 15 workers from each of the four colonies were individually labeled with a small numbered tag glued to their thorax and then reintroduced into their respective colonies. These were the control workers. Other ants, the experimental workers, were labeled in the same manner and then isolated in a glass tube equipped with food and water. In one colony the isolation period began at emergence, while in the second, third, and fourth colonies, the period of isolation began when workers were 2, 4, or 8 days old. Like the controls, there were 15 workers in each of the four isolation groups. Champalbert and Lachaud recorded the behavior of control and experimental ants for 45 days after emergence.

The question was, what effects, if any, would social isolation have on the behavioral development of workers in the experimental groups? Also, would the effects of isolation depend on when they were isolated? During the first week after emergence, the control workers spent most of their time feeding or being groomed by other colony members. By the second week, however, work began, and the ants started to specialize in nursing activities, first caring for larvae, then cocoons, and finally eggs. Sometime during the third week, the workers changed their job status

again and began to explore the nest and engage in domestic tasks. Finally, about a month after emergence, the control workers focused on activities related to their new careers as either guards at the nest or foragers outside the nest. Thus, over the course of only a few weeks, *E. tuberculatum* workers typically underwent several changes in behavioral specialization, and in a very specific sequence.

So, how did the behavior of ants in the experimental groups compare with that of controls? What were the effects of isolation immediately at emergence, or 2, 4, or 8 days later? Interestingly enough, the workers isolated at 2 days after emergence showed the most abnormal behavioral development. They differed from controls in the order of appearance of their activities and in the level of performance of their tasks, being particularly lax in the department of brood care (Figure 8.18). In contrast, behavioral development of workers isolated either at emergence or 4 or 8 days later was more similar to that of controls. In the case of workers isolated at emergence, behavioral development was simply delayed. The reintroduction of these workers into society seemed to correspond to a second emergence; although 11 days old at the time of reintroduction, these workers behaved like newly emerged ants. Soon after reintroduction into the colony, however, the various specializations in activities appeared progressively in the same sequence as that of the control workers. Only minor abnormalities were noted in the workers isolated at 4 days, and the behavioral development of those isolated at 8 days was almost identical to that of controls. In summary, the extent of behavioral abnormalities observed in *E. tuberculatum* workers as a result of a 10-day period of social isolation depended on the age of workers at the time of isolation, with the most serious abnormalities occurring in ants isolated 2 days after emergence.

Champalbert and Lachaud (1990) concluded from their observations of *E. tuberculatum* worker ants that there is a critical period during

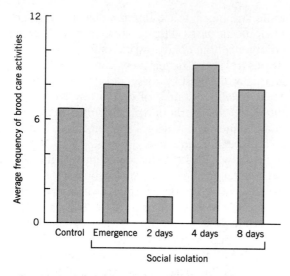

Figure 8.18 A critical period for the development of brood care behavior in worker ants. Here, a 10-day period of social isolation beginning 2 days after emergence produces more severe disruptions in the performance of brood care activities by *Ectatomma tuberculatum* workers than do periods of isolation beginning either at emergence or 4 or 8 days later. (Drawn from the data of Champalbert and Lachaud, 1990.)

the first 4 days after emergence in which exposure to social stimuli affects the establishment of behavior, especially that related to brood care. One might ask, then, why would workers isolated at emergence (and therefore removed from social stimuli during the 4-day critical period) display fairly normal behavioral development? The authors suggest that when workers are isolated immediately after emergence, their development is simply put on hold. The reintroduction of these workers into the colony after their period of social isolation mimics an emergence. During the 4 days following reintroduction, these workers receive the social stimuli important in the development of brood care activities. Put another way, it is not the physiologically defined age of 0–4 days that is important for the development of brood care behavior,

but rather the first 4 days of social contact with colony members. These 4 days may follow either natural emergence or the artificial second emergence produced by reintroduction of workers into the colony after isolation. Here, again, we see an example of flexibility associated with critical periods in behavioral development; in ants isolated at emergence, lack of exposure to appropriate social stimuli delays closure of the critical period until a few days after reintroduction to the colony. In view of the poor performance of workers isolated at 2 days, it also appears that the entire sensitive period must occur in an uninterrupted fashion in order to ensure normal behavioral development. Because these workers received some portion of the necessary social stimulation during their first 2 days following emergence, their developmental system could not be put on hold until reintroduction into the colony.

Developmental Homeostasis

Throughout this chapter we have described the effects of experience on behavioral development in a variety of animals. Just as smelling the breath of a recently fed cagemate influences subsequent selection of food in Norway rats, listening to the songs of adult male conspecifics affects vocal development in zebra finches and white-crowned sparrows. Whereas stimuli associated with parents and siblings influence filial behavior in shrews and mallard ducklings, early contact with colony members influences brood care behavior in worker ants. Experience certainly does seem to play an important role in behavioral development. Having said this, we might also note that the manipulations used to demonstrate the sensitivity of the developing animal to external influence are often quite severe, typically involving rearing in isolation or rearing by another species, two conditions that are not likely to be encountered by most animals under natural conditions. Indeed, despite the fact that in their natural environments, individuals within a species often develop under a diverse array of physical and social conditions, the vast majority of adults display species-typical, normal behavior. Under field conditions, after all, most white-crowned sparrows sing white-crown song, most mallard ducklings follow their mother, and most worker ants perform their brood care responsibilities in an admirable fashion. Why do such complex patterns of behavior develop so reliably when there seems to be so much room for error?

The developmental processes appear capable of buffering themselves against potentially harmful influences to produce functional adults. This buffering capacity is called developmental homeostasis. A buffer, of course, is something that dampens drastic changes. We can look around us and see that despite diverse experiences, behavioral development in most individuals proceeds in a very predictable and reliable manner. In effect, the developmental process demonstrates a certain stability and resilience in the face of a host of constantly changing variables including, as we have seen, many of those we introduce experimentally. You may have already been struck by the virtually normal development of worker ants that were isolated from colony members for a 10-day period either at emergence or 4 or 8 days later. Despite being removed from the hustle and bustle of normal colony life and placed in separate small glass tubes, individuals in these groups, for the most part, exhibited normal behavioral development. As we saw, only ants isolated 2 days after emergence displayed highly abnormal behavior, suggesting that the first 4 days of social contact with colony members are critical to normal development. Isn't it remarkable that in ants, highly social animals that undergo complex changes in behavioral specializations, only 4 days of contact with colony members are sufficient to produce normal behavioral development? As we see in our next example concerning social develop-

ment in rhesus monkeys, often only a bare minimum of experience is necessary to promote virtually normal behavioral development, even under the most harsh of circumstances.

SOCIAL DEVELOPMENT IN RHESUS MONKEYS

The ability of animals to exhibit normal development when provided with very little critical experience can be seen in the results of rearing condition research with rhesus monkeys (*Macaca mulatta*). In this research, experimenters devised highly abnormal environments in which to rear infant monkeys and then examined the effects of such environments on the subsequent social development of their subjects. In most cases (see following discussion), these highly contrived laboratory environments produced severe effects on social behavior. The environments were virtually devoid of the features necessary for the development of normal social behavior. The ability of rhesus monkeys to compensate for unfavorable rearing conditions was then shown by experiments that revealed just how little critical experience is needed to produce normal behavior.

Psychologists Margaret and Harry Harlow pioneered rearing condition research with rhesus monkeys by examining the effects on social development of rearing infant monkeys in either total or partial isolation, or with a surrogate mother. In the case of rearing in total isolation, rhesus monkey infants were raised alone in laboratory cages. Deprived of physical, visual, and auditory contact with other monkeys for the early part of their lives (typically for either 3, 6, or 12 months), these youngsters saw only the hand of the experimenter at feeding. Rhesus monkeys reared under these conditions exhibited stereotyped rocking movements and often clutched and pinched or bit themselves when huddled in a corner of the cage (Harlow, Harlow, and Suomi, 1971). When they were a few years old (still immature) and were placed with infants, the experimental monkeys would some-

Figure 8.19   Rhesus monkey female that was reared in isolation during infancy ignoring her infant.

Figure 8.20   Rhesus monkey infants reared in partial isolation can see and hear other monkeys but receive no physical contact from conspecifics. These so-called partial isolates show all the severe social deficits associated with total social isolation, thus emphasizing the importance of physical contact to normal social development in this species.

times freeze in apparent terror and yet at other times explode in bouts of extreme aggression. Once mature, they exhibited inappropriate sexual and parental behavior (Harlow, Harlow, and Suomi, 1971). Rhesus monkey females reared without mothers or peers were usually indifferent to their infants (Figure 8.19) or downright abusive, although they sometimes improved when caring for their second offspring.

In the partial isolation condition, infants were raised alone in wire cages where they could see and hear other monkeys but had no physical contact with them (Figure 8.20). Despite the provision of some social stimulation, partial isolates developed a behavioral syndrome similar to that of total isolates (Harlow, Harlow, and Suomi, 1971). These findings underscore the importance of physical contact to the development of normal social behavior in rhesus monkeys.

In the third rearing condition, infant rhesus monkeys were reared with inanimate surrogate mothers. Surrogate mothers typically consisted of a wire cylinder, often covered in terry cloth, surmounted by a wooden head (Figure 8.21). These surrogates were designed by Harry Harlow to provide bodily contact comfort for rhesus

infants and gained him the prestigious title of "father of the cloth mother" (Harlow, Harlow, and Suomi, 1971). Although infants became quite attached to their surrogate mother and exhibited less self-directed activities (e.g., clutching and biting themselves) than did total isolates, they still failed to show appropriate social behavior.

We must ask, then, what is so special about being raised by a rhesus monkey mother? In order to tease apart the important factors in the bonding of infant to mother, surrogate mothers were modified in various ways to provide milk, warmth, and rocking movements. In one set of studies, infants preferred a nonlactating cloth

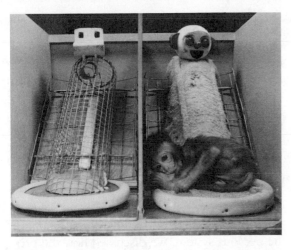

Figure 8.21 Rhesus infant with inanimate surrogate mother.

mother to a lactating wire surrogate, suggesting that contact comfort is more important in formation of the infant to mother bond than is the satisfaction of nutritional needs. This finding caused Harlow, Harlow, and Suomi (1971) to proclaim, "There is more than merely milk to human kindness." The bottom line remained, however, that rearing by an inanimate surrogate mother, no matter what her qualities, produced severely disturbed monkeys.

These experiments demonstrated that monkeys reared in highly abnormal physical and social environments display highly abnormal patterns of social behavior. Rearing by inanimate surrogate mothers or in total or partial isolation provided none of the critical social experience necessary for normal social development. The important question then became, what level of social contact would be necessary to produce normal social development? What would be the bare minimum of social experience that would lead to normal monkeys—monkeys that played with other monkeys rather than huddling and rocking in a corner, that displayed normal social and sexual interactions rather than extreme

aggression or lack of interest, and that cared for their own infants rather than ignoring or abusing them? (It is important to keep in mind here that even the social settings in the laboratory that produce "normal" monkeys—i.e., settings in which infants are reared with their mother and also exposed to peers—are probably less complex than what most infants would experience in the field. Normal behavior is thus in reference to laboratory, and not field, standards.) Amazingly enough, infants that were housed alone with a surrogate mother and provided with only 20 minutes of daily contact with peers exhibited the full range of social behavior with no developmental delay (Harlow and Harlow, 1962). Thus, although rhesus infants in nature are almost constantly exposed to conspecifics, even a few minutes of daily contact with peers in the laboratory are sufficient to produce normal social behavior. Obviously, a little social experience goes a long way. The ability of rhesus monkey infants to exhibit normal social development when provided with only a few minutes of daily contact with peers again illustrates the resilience of the developmental process.

Another interesting finding that arose from this research was that rearing with peers produces more normal social development than does rearing with mothers alone (Harlow and Harlow, 1962). This finding must have come as a shock to those in the scientific community who, along with the famous psychoanalyst Sigmund Freud, believed the relationship between an infant and its mother to be the most important component of the early environment. It is not that mothers are unimportant to infant monkeys, but they are not, in the case of rhesus monkeys, as essential as peers to normal social development. Indeed, experience that would seem to be critical to the development of a particular behavior sometimes turns out to be less important or not important at all. We emphasize this point in the next section as we discuss the development of swimming behavior in an amphibian.

NEUROBEHAVIORAL DEVELOPMENT IN AMPHIBIANS

Behavioral Development in the Absence of Embryonic Experience

In a manner not unlike that of the embryonic salmon described at the beginning of this chapter, embryos of the South African clawed frog, *Xenopus laevis*, show certain behaviors prior to hatching. The onset of movement in the embryos begins about 25 hours after fertilization, and alternating flexure of the body and swimming behavior begin 10 and 20 hours later, respectively. A tadpole hatches approximately 50 hours after fertilization. Are embryonic behaviors necessary for the performance of normal swimming behavior by tadpoles? Lanny Haverkamp and Ronald Oppenheim (1986) sought to answer this question by immobilizing *Xenopus* embryos prior to the onset of movement and then observing the behavior of tadpoles once the period of immobility had ended. Immobilization was accomplished by immersing embryos in drug solutions (lidocaine or chloretone) or by injections of alpha-bungarotoxin prior to the onset of embryonic movement; all three treatments suppressed activity in the nervous system and thus all behavior in the embryo.

How did the complete absence of embryonic neural activity and behavior affect patterns of swimming after hatching? Amazingly enough, once the suppressing drugs were removed, the swimming behavior of experimental tadpoles was virtually equivalent to that of normally reared tadpoles. In addition, the anatomical and neurophysiological substrates of swimming in the experimental tadpoles were indistinguishable from those of normal tadpoles (Haverkamp, 1986). Thus, although embryonic experience may sometimes be critical to the development of species-typical postnatal behavior (e.g., mallard ducklings must hear the contentment calls of their siblings prior to hatching in order to follow the maternal assembly call after hatching), we cannot always assume that this is the case. In *X. laevis*, embryonic experience, in particular the patterns of behavior that lead up to swimming, are not necessary for the normal performance of swimming behavior by tadpoles. So, if at first glance we assumed that embryonic swimming movements were necessary precursors to the development of swimming behavior in tadpoles, we would certainly have been wrong.

Resilience in the Developing Nervous System

We began this chapter with examples of how behavioral change accompanies development of the nervous system. We discussed how the production and death of neurons, as well as structural changes in neurons (e.g., dendritic growth and regression), relate to the appearance and disappearance of behaviors from the repertoire of the tobacco hornworm. Are there examples of developmental homeostasis at the level of the nervous system that might help to explain the tendency for behavior to develop normally? We will consider the same basic paradigm as that just described for *Xenopus* embryos, this time, however, focusing on the effects on nervous system structure and function of eliminating all neural activity.

The ability of neurons to develop normally in the absence of neural activity was elegantly demonstrated by the transplantation experiments of William Harris (1980). Harris took advantage of the fact that tetrodotoxin (TTX), a powerful neurotoxin that blocks all propagation of neural impulses, is a normal constituent of the body fluids of the California newt, *Taricha torosa*. The newt's tissues, as you would expect, are insensitive to the effects of TTX. This is not the case, however, for the neural tissues of another amphibian, the Mexican axolotl, *Ambystoma mexicanum*. Upon exposure to TTX, all neural activity ceases in the axolotl. Harris transplanted an embryonic eye of the Mexican axolotl into the head region of newt embryos. The transplanted eye was a third eye to the newt, rather

Figure 8.22 Newt with third eye from an axolotl. The transplantation experiments of William Harris show that neural connections from the transplanted eye reach their appropriate targets in the newt brain despite the complete inhibition of neural activity caused by tetrodotoxin in the tissues of the newt. (From Harris, 1980.)

than a replacement eye. In the newt, all neural activity associated with the axolotl eye was inhibited. Once the newts had matured (Figure 8.22), Harris examined the neural projections associated with the transplanted eye. Despite the complete inhibition of neural activity, projections from the transplanted eye found their way to their proper targets in the newt brain and made functionally effective connections there.

The normal development of optic nerve fibers is all the more surprising when we consider that the neural tissue was transplanted from a different genus and species, and that the transplanted tissue was in direct competition with the normal host eye at the target sites in the brain. The ability of the neural connections of the axolotl eye to develop normally in the face of all of these environmental perturbations illustrates developmental homeostasis. This is not to say that development of the nervous system can always proceed in a normal fashion in the absence of neural activity. Indeed there are some developmental processes in which neural activity plays an important role. We simply use this example as a means of demonstrating the resilience associated with the developmental process, a resilience reflected in the structure and function of the nervous system as well as in the behavior of developing animals.

Summary

Patterns of behavior appear, disappear, and alter in form as animals develop. There are several causes of behavioral change, and one of the most significant is the development of the nervous system. During embryonic life, patterns of behavior often emerge in parallel with development of the sensory and neural structures necessary for their performance. Sometimes neuronal growth, death, or alterations in structure can be linked to developmental changes in a particular behavior. Changes in hormonal state can also trigger behavioral change, as when the tobacco hornworm loses behaviors associated with the prolegs as a result of a peak in the level of ecdysteroid hormones. This peak in ecdysteroids causes regression in the underlying neural circuitry and hence loss of proleg behaviors. Behavioral change may also come about through the development of specific morphological structures. In developing paddlefish, the change in feeding behavior from chasing and picking select items out of the water column to indiscriminate filter feeding parallels the development of gill rakers. Finally, behavioral change may occur through experience, as when a young rat alters its selection of food on the basis of smelling the breath of a conspecific that has recently eaten. Neural, hormonal, morphological, and experiential causes of behavioral change often interact during the continuous interplay between the developing animal and its internal and external environment.

Genetic factors also enter into the complex interaction between the developing animal and its environment. The interaction between genetic and environmental factors is perhaps best illustrated by the development of sexually dimorphic patterns of behavior, such as singing in birds. Only male birds sing, a fact reflected in their neuroanatomy. The developmental basis for sex differences in the brains and vocal behavior of songbirds is the chromosomal difference between the sexes. This genetic difference dictates patterns of secretion of steroid hormones that, in males, lead to growth and differentiation of the regions of the brain involved in song. In addition to neural and hormonal influences, experience plays an important role in the development of singing behavior. If they are to produce normal song in adulthood, young males must listen to the songs of adult conspecifics and must be able to hear themselves sing.

In male songbirds, the pulse of steroid hormone that affects the developing nervous system and the experience of hearing conspecific song are not always capable of influencing the development of singing behavior. Indeed, in order to be effective, the change in hormone level and the experience of listening to conspecifics must occur during somewhat restricted periods of time when the young bird is particularly sensitive to such influences. These periods of enhanced sensitivity to environmental stimuli are called critical periods, and they characterize a diverse array of behavioral phenomena. In some cases, critical periods occur prior to birth or hatching, and experience during the prenatal period influences development of species-typical postnatal behavior. An embryonic critical period has been demonstrated for the development of filial imprinting, the response of ducklings to follow their mother. While still in the egg, mallard ducklings must hear the contentment calls of their siblings in order to respond to the assembly call of their mother after hatching. Although usually occurring early in life, critical periods occasionally occur in adulthood, as when mother goats learn the characteristics of their offspring just after birth, a learn-

ing process that results in the acceptance of their own young and the rejection of alien kids. There are tremendous differences among species in the timing of critical periods for the development of a specific behavior, and even within a species, there is flexibility in the time course of critical periods. Social interactions, in particular, can influence the timing and duration of critical periods.

We have said that experience, especially that occurring during a critical period in the young animal's life, can profoundly influence behavioral development. Under normal circumstances, however, animals encounter a wide variety of physical and social conditions, and most develop normally. The ability of development to proceed in a normal fashion in the face of environmental perturbation is called developmental homeostasis. The buffering capacity of the developmental process can sometimes be seen in the surprisingly normal development of animals reared under the most deprived laboratory conditions. Even more striking, however, is how little critical experience is sometimes sufficient to induce normal behavioral development. In their natural environment, rhesus monkeys are usually exposed to numerous conspecifics, and yet in the laboratory, only 20 minutes of daily contact with peers is sufficient to promote normal social development, at least by laboratory standards. The resilience associated with the developmental process can also be seen at the level of the nervous system. Here, in the face of apparently severe environmental change, such as that imposed by the complete inhibition of neural activity and by transplantation of neural tissue from one species to another, neural development proceeds in a predictable and reliable manner. The ability of the developing animal's behavior and nervous system to withstand the vagaries of environmental change results, in the vast majority of cases, in the production of fully functional adults.

Behavior of the Individual Surviving in Its Environment

CHAPTER
9

Biological Clocks

*I*magine for a moment that you have a pet hamster, a friendly fellow who quietly shares your bedroom. However, one Thursday night you are up late studying for a big exam and you find it hard to concentrate because a hamster is most active during these late night hours and his running wheel is squeaking. So, you place the hamster and his cage in your closet and continue studying. The next morning, you take the test and then leave for the weekend. On Monday night you are back and notice a squeaking in your closet. You have forgotten about the hamster. He has had plenty of food and water, but he has been in the dark for 3 days. As you retrieve him, you notice that he begins to run on his wheel at about the same time he normally did.

How did he know what time it was? We do not know. We glibly attribute the ability to measure time without any obvious environmental cues to an internal, living clock. When any hamster is sequestered in the constant darkness and temperature of the laboratory so that each turn of its running wheel can be recorded automatically for months or even years, a record similar to the one shown in Figure 9.1 usually results. Notice that the hamster woke up almost exactly 12 minutes later each day during the entire study. Its bouts of activity alternate with rest with such regularity that it is often described as an activity rhythm. The ability to measure time is common not just among hamsters, but also in most animals. In fact, biological clocks have been found in every eukaryotic organism tested (Hastings, Rusak, and Boulos, 1991).

The rhythmical nature of life may come as a surprise to some, but should not in the light of evolutionary principles. Life evolved under cyclical conditions, and the differences in phases of the cycles are often so pronounced as to place a high adaptive value on life being able to accommodate itself as specifically as possible to each phase.

Every living thing is subjected to the regularly varying environmental conditions on earth orchestrated by the relative movements of the heavenly bodies: the earth, the moon and the sun. As the earth spins on its axis relative to the sun, life is exposed to rhythmic variations in light intensity, temperature, relative humidity, barometric pressure, geomagnetism, cosmic radiation, and the electrostatic field. The earth also rotates relative to the moon once every lunar day (24.8 hours). The moon's gravitational pull draws the water on the earth's surface toward it causing it to "pile up," thus resulting in high tide. These tidal cycles result in dramatic changes in the environment of intertidal organisms—flooding followed by desiccation associated with changes in salinity and temperature. The relative positions of the earth, moon, and sun will result in the fortnightly alternation between spring and neap tides, as will be explained shortly. The moon revolves about the earth once every lunar month (29.5 days), generating changes in the intensity of nocturnal illumination and causing fluctuations in the earth's magnetic field. Finally, the earth circles the sun, resulting in the progression of the seasons with its sometimes dramatic alterations in photoperiod and temperature.

Although the environmental modifications may be extreme, they are predictable. Often it is advantageous to gear an activity to occur at a specific time relative to some rhythmic aspect of the environment. So, biological clocks are generally thought to have evolved as adaptations to these environmental cycles (Daan and Aschoff, 1982; Enright, 1970; Hoffmann, 1976).

Rhythmic Behavior

Indeed, rhythmicity in behavior and physiology is so common that it must be considered by anyone studying animal behavior. An animal is not perpetually the same. Instead its behavior

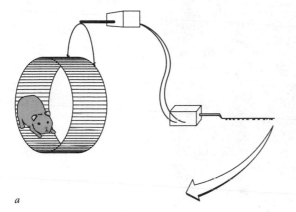

Figure 9.1 (*a*) A hamster in a running wheel equipped to record each turn of the wheel shows periodic bouts of activity that alternate with rest. (*b*) In constant dim light, the cycle length of this activity rhythm is slightly longer than 24 hours. For each rotation of the running wheel, a vertical line is automatically made on a chart at the time of day when the activity occurred. In this record the bouts of activity were so intense that the vertical lines appear to have fused, forming dark horizontal bands. Notice that although the animal had no light or temperature cycle as a clue to the time of day, it would awaken about 12 minutes later each day.

a

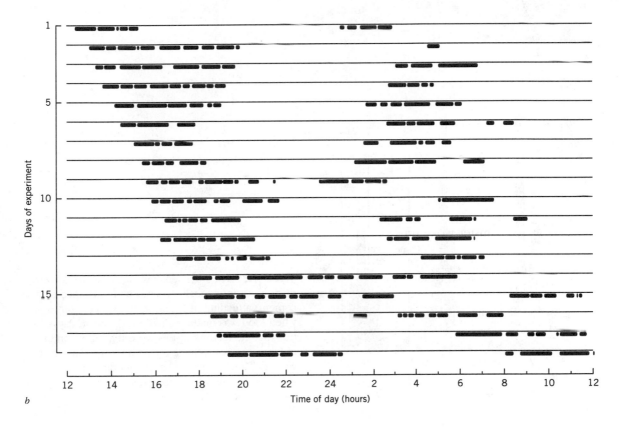

b

a

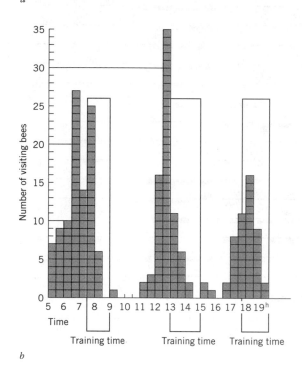

b

Figure 9.2 The time sense in bees. (*a*) Bees were marked for individual recognition and trained to come to a feeding dish only at the specific times when food was made available. (*b*) After 6 days of training, the feeding dishes were left empty and the number of bees arriving throughout the day was recorded. The bees arrived at the feeding station only at the times when food had been previously present. (Modified from Beling, 1929.)

may fluctuate so that it is appropriate to the time of day, or the state of the tides, or the phase of the moon, or the season of the year. A description of a behavior at one point in a cycle may be totally inaccurate at another time. To familiarize you with the variety of behaviors that are rhythmic, a potpourri of biological rhythms is presented.

DAILY RHYTHMS

Most animals restrict their activity to a specific portion of the day. The hamster, for example, is busiest at night, as are many other creatures such as cockroaches, bats, mice, and rats. Other species, songbirds and humans for instance, are active during the day.

Among the best "timers" are bees. Their time sense was experimentally demonstrated during the early part of this century by marking them for the purpose of individual recognition and offering them sugar water at a feeding station during a restricted time each day, between 10 A.M. and noon. After 6 to 8 days of this training, most of the bees frequented the feeding station only during certain hours. The real test, however, was on subsequent days when no food was present at the feeding station. As seen in Figure 9.2, the greatest number of bees returned to the empty feeding station only at the time when food had been previously available (Beling, 1929). In subsequent tests, it was found that the bee time sense is astonishingly accurate. Bees can be trained to go to nine different feeding stations at nine different times of the day. They are able to distinguish points in time separated by as little as 20 minutes (Koltermann, 1971). The adaptiveness of such abilities for bees is clear. Flowers have a rhythm in nectar secretion, producing more at some times of day than at others. The biological clock allows bees to time their visits to flowers so that they arrive when the flower is secreting nectar. This means that the bees can gather the maximum amount of food with a minimum effort.

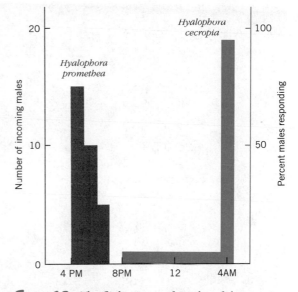

Figure 9.3 The flight times of moths of the species *Hyalophora cecropia* and *Hyalophora promethea*. The difference in the time of activity of these species helps prevent the males from mating with a female of the wrong species. (Modified from Wilson and Bossert, 1963.)

Clocks are important not only in feeding, but in mating as well. For example, the fruit fly, *Dacus tryoni,* mates only during the evening twilight. Such a rhythm ensures reproductive synchrony between members of the same species, thereby increasing the chances of finding an appropriate mate. Other species of genus *Dacus* mate at different times of the day (Tychsen and Fletcher, 1971). Thus, timing may also avoid matings with other related species that may be mating at that season, and time, energy, and gametes are not wasted on doomed reproductive efforts. Even in moth species in which the female entices a male by emitting a potent sex attractant, the biological clock plays an important role (Figure 9.3). Female moths of the genus *Hyalophora* produce a sex attractant pheromone that is attractive to both *H. cecropia* and *H. promethea* males. Costly reproductive mistakes are avoided because these species are ac-

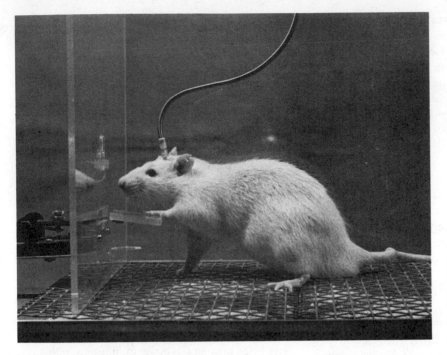

Figure 9.4 A rat with an electrode implanted in the "pleasure center" of the brain. When the rat presses the lever, the pleasure center is stimulated and a record is made of the action.

tive at different times of the day (Wilson and Bossert, 1963).

Certain patterns of learned behavior may be under temporal control. When an electrode is placed into certain brain regions (areas of the hypothalamus or certain midbrain nuclei) of a rat, the animal will quickly learn to press a lever in order to get electrical stimulation (Figure 9.4). An observer is tempted to conclude that the effect of the stimulation is the rat's equivalent of some form of ecstasy since animals with electrodes in these brain areas would rather stimulate themselves than eat, drink, or even copulate. Some will press the lever as often as 5000 times an hour. When the electrodes are first implanted, a rat will typically begin a 2-day marathon of bar pressing at a high rate, resting infrequently for intervals of only a few minutes. After this time, quiet periods, during which the rat may still respond but with lower frequency, alternate with periods of rapid pressing and soon a 24-hour cycle appears (Figure 9.5; Terman and Terman,

1970). The animals have apparently learned to press the bar to receive a reward, but the reward has more value at some times of the day than at others. Likewise, the operant behavior of a chimpanzee responding for a food reinforcement has been shown to be rhythmic (Ternes, Farner, and Deavors, 1967). It has been suggested that since identical stimuli may vary in their effectiveness as reinforcers with the hour of the day, the ease of learning should also fluctuate daily.

LUNAR DAY RHYTHMS

Although the 24-hour daily rhythm in light and dark is probably the most familiar environmental cycle, there are many others of importance. For example, the interaction of the gravitational fields of the moon and the sun create other pronounced environmental changes—those associated with the tides.

As the moon passes over the surfaces of the earth, its gravitational field draws up a bulge in

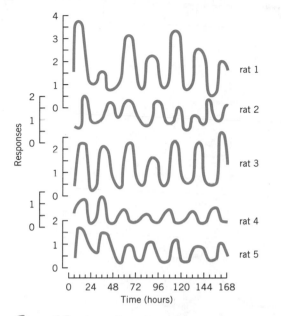

Figure 9.5 The cyclic self-stimulation of the pleasure center by rats. Notice periods of rapid stimulation alternate with rest in cycles of approximately 24 hours even when the rat has no obvious time cues. (From Terman and Terman, 1970.)

the ocean waters. One bulge occurs beneath the moon and another on the opposite side of the earth. These bulges sweep across the seas as the earth rotates beneath the moon, thus causing high tides when they reach the shoreline. Since there are two "heaps" of water, there are usually two high tides each lunar day, one every 12.4 hours. The tides may cause some rather dramatic changes in the environment, particularly for organisms living on the seashore.

The activity of the fiddler crab, *Uca pugnax*, a resident of the intertidal zone, is synchronized with the tidal changes. Fiddler crabs can be seen scurrying along the marsh during low tide in search of food and mates. Before the sea floods the area, the crabs return to their burrows to wait out the inundation. When the fiddler crab is removed from the beach and sequestered in the laboratory, away from tidal changes, its behavior remains rhythmic (Figure 9.6). Periods of activity alternate with quiescence every 12.4 hours, the usual interval between high tides (Figure 9.7; Palmer, 1990).

Figure 9.6 Fiddler crabs maintained in constant conditions in the laboratory. Each crab is placed in an actograph that will allow the crab's activity to be monitored. An actograph consists of a box into which the crab is placed. The box is centered over a fulcrum so that the box teeters as the animal scurries back and forth. Each time the box tips in one direction, it depresses a microswitch and causes a mark to be made by a recording device.

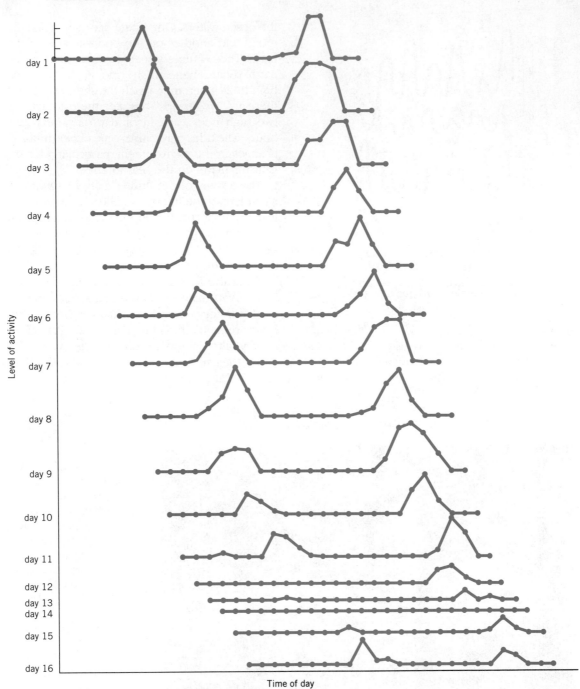

Figure 9.7 An activity rhythm in a fiddler crab (*Uca pugnax*). Although the crab was maintained in constant darkness and temperature (20°C), the animal was active at approximately the times of low tide at its home beach. (From Palmer, 1990.)

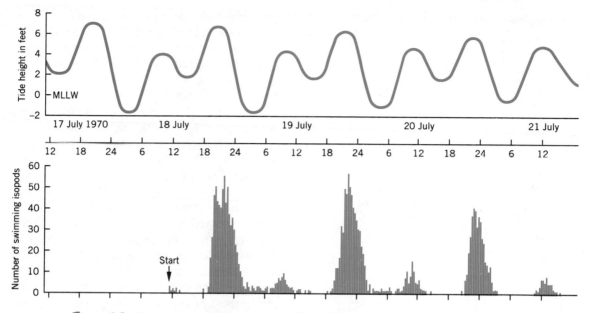

Figure 9.8   The activity pattern of isopods (*Excirolana chiltoni*) maintained in the laboratory. The pattern of activity mimics that of the height of the tide in the area of collection. (Modified from Klapow, 1972.)

Living higher in the intertidal zone than the fiddler crab is a small isopod (a crustacean), *Excirolana chiltoni*. The open beach habitat of this organism is periodically exposed to the swirling surf. *Excirolana* lives buried in the sand during low tide but as the pounding waves reach its abode, it emerges and begins swimming and feeding in the breaking waves. However, it is not simply that inundation presents an ultimatum to sink or swim. When the animals are brought into the laboratory and are maintained in a sandy-bottomed beaker of seawater, they begin to tread water only at the times of high tide on their local beaches (Figure 9.8). One isopod displayed a tidal rhythm for 65 days in constant conditions (Enright, 1972).

SEMILUNAR RHYTHMS

The height of the tides is also influenced by the gravitational field of the sun. In fact, the highest

tides are caused by the gravitational fields of the moon and the sun operating together. At new and full moons, the earth, the moon, and the sun are in line, causing the gravitational fields of the sun and the moon to augment each other (Figure 9.9). This is why the earth experiences the highest high tides and lowest low tides at new and full moons. These periods of greatest tidal exchange are referred to as the spring tides. At the quarters of the moon, the gravitational fields of the moon and the sun are at right angles to one another. Since their pulls are now antagonistic, the tidal exchange is smaller than at other times of the month. These periods of lowest high tides and highest low tides are called the neap tides. Some organisms possess a biological clock that allows them to predict the times of spring, or neap tides, and gear their activities to these regular changes.

The grunion, *Leuresthes tenuis*, is a small silvery fish living in the waters off the coast of

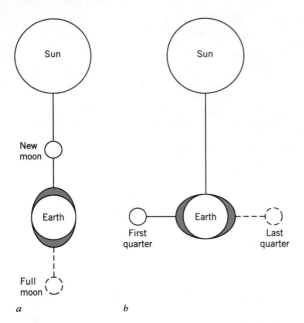

Figure 9.9 The effect of the relative positions of the earth, moon, and sun on the amplitude of tidal exchange. (*a*) At the times of new and full moons, the gravitational fields of the moon and the sun assist one another, causing the spring tides. (*b*) During the first and last quarters of the moon, the gravitational fields of the moon and the sun are perpendicular to one another. This results in the smallest tidal exchange, the neap tides.

California that precisely times its reproductive activities to occur at fortnightly intervals synchronized with the spring tides. This fish is the only one in the sea to spawn on land. Beginning in late February and continuing until early September, for 3 or 4 nights just after the greatest spring tides, the adult grunion gather in the waters just offshore to wait until the tide has reached its peak and is just beginning to recede. Then they ride the waves ashore for a night of mating. At the height of the mating frenzy, the beach is carpeted with their quivering bodies. A female digs, tail first, into the sand so that only her head, from the gills up, is exposed. Several males may wrap around her and release sperm-containing milt that seeps through the sand and fertilizes the eggs discharged by the female (Figure 9.10). After she has laid all her eggs, the fish catch a wave and disappear into the swirling sea, leaving the eggs buried in the sand to develop into the next generation.

The timing of this event is exquisite. Since the spring tides have the highest high tides of the fortnightly cycle, by waiting until the nights just after the spring tide to mate, the grunion guarantee that, unless there is a storm, their eggs will have about 10 days to develop undisturbed in the sand before the pounding surf reaches the eggs buried on the beach and triggers the hatching. If the fish mated before the peak of the spring tides, the next night's high tide would wash the eggs away. The same fate would meet the eggs if the grunion did not wait until the turn of the tide on their mating night. Because the nighttime tide always crests higher than the daytime tide, not only are the eggs safer from the surging waves, but the fish themselves are less vulnerable to predators while they mate (Ricciutti, 1978).

Another example of a fortnightly rhythm is seen in the tiny chironomid midge, *Clunio marinus*. In *Clunio* it is the end of development, the emergence of adults from their pupal cases, that is programmed to coincide with tidal changes. These insects live at the lowest extreme of the intertidal zone so that they are exposed to the air for only a few hours once every 2 weeks, during each spring low tide. When the tide recedes, the males are first to break free of their puparia, the cases in which they developed. Each one locates a female and assists her emergence. They have little time to waste for their habitat will soon be submerged again, so copulation follows quickly. Then the winged males carry their mate to the place where she will lay her eggs. All these activities must be precisely timed so that they occur during the short, 2-hour period during which the habitat is exposed.

Dietrich Neumann (1976) has found that if a population is brought into the laboratory and

Figure 9.10 Mating grunion. The female buries her posterior end in the sand and releases her eggs while a male wraps himself around her and releases his sperm.

maintained in a light–dark cycle in which 12 hours of light alternate with 12 hours of darkness, emergence from the puparium is random. If, however, one simulates the light of full moon by leaving a dim (0.4 lux) light on for 4 consecutive nights, the emergence of adults from the puparia becomes synchronized. Now, just as in nature, emergence occurs at approximately fortnightly intervals for about 2 months (Figure 9.11).

MONTHLY RHYTHMS

The interval from full moon to full moon, a synodic lunar month (29.5 days), corresponds to the length of time it takes the moon to revolve once around the earth. Some organisms possess a clock that allows them to program their activities to occur at specific times during this cycle.

An example of a clearly adaptive monthly rhythm is the timing of the reproductive activities of the marine polychaete *Eunice viridis*, the Samoan Palolo worm. Unlike the organisms previously mentioned in which rhythms in reproduction are precisely synchronized with propitious environmental conditions, the Palolo worm restricts its procreation to a specific time so that the population will release gametes simultaneously, increasing the probability of fertilization in the water.

The Palolo worm lives in crevices of the coral reefs off Samoa and Fiji. In preparation for reproduction, it elongates by budding new segments that become packed with gametes. This section of the worm, the epitoke, may be a foot long. Around sunrise on a day at, or near, the last quarter of the moon during October and November, the epitokes synchronously tear free from the rest of the body and rise to the surface, churning the water. Then they explode in unison, liberating gametes into the sea where fertilization takes place. The timing of this self-destructive orgy is so predictable that the epicurean natives of Samoa and Fiji calculate the day of the event in advance and arrive at the coral reefs at dawn and await the frenetic swarming so that they can scoop up the epitokes, which

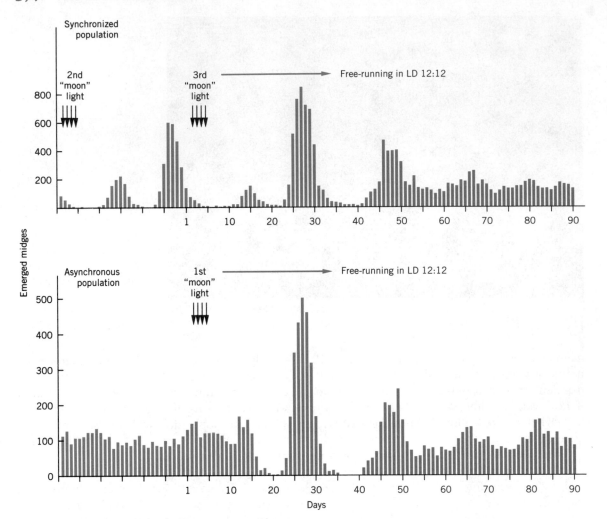

Figure 9.11 A fortnightly rhythm in emergence of a chironomid midge (*Clunio marinus*). Emergence and mating are programmed by the biological clock to occur during the short interval during the neap tides when their habitat is exposed. The rhythm continues in the laboratory if the insects are exposed to cycles of 12 hours of light alternating with 12 hours of darkness and to a simulated moon for 4 consecutive nights. (From Neumann, 1976.)

they eat either raw or roasted (Figure 9.12; Burrows, 1945).

The ant lion (*Myrmeleon obscurus*) shows another sort of monthly rhythm. A lazy hunter, it builds a steep-sided conical pit in the sand and then lies in ambush at the base, with all but its immense pincers covered with sand, waiting for some small arthropod, such as an ant, to slide into the pit toward its outstretched jaws. The ant lion then sucks out the prey's body fluids (Figure 9.13). Perhaps the most interesting observation on the ant lion's behavior is that it is

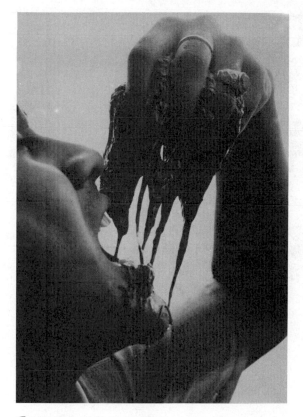

Figure 9.12 A feast of Palolo worms. The parts of the Palolo worms (*Eunice viridis*) specialized for reproduction (epitokes) break free from the rest of the worm and swarm synchronously to the surface of the sea releasing their gametes. The monthly rhythm in swarming increases the probability of fertilization and allows the local human population to predict the event in advance so that they can scoop up handfuls of epitokes to eat.

different at full moon, similar to the pattern of blood sucking by vampires in fictional (one hopes) literature. It constructs larger pits at the time of full moon than at new moon. Careful daily measurements of the size of pits of ant lions cloistered in constant conditions in the laboratory have revealed that this is a clock-controlled rhythm and not a simple response to some aspect of the environment such as the

amount of moonlight (Youthed and Moran, 1969).

There are numerous other monthly rhythms. The level of activity of the isopod *Excirolana* (Enright, 1972), rats with thalamic lesions (Richter, 1975), and honeybees (Oehmke, 1973) varies rhythmically over each month. The color sensitivity of the guppy (Lang, 1964, 1967) and of man (Dresler, 1940) also fluctuates during each month. In addition, while planarians will always move away from the light, the angle they assume as they move away from two perpendicular light beams will vary regularly during the month (Brown and Park, 1975; Goodenough, 1978, 1980). Furthermore, there is a monthly fluctuation in the day-to-day variation of the initial bearings of homing pigeons (Larkin and Keeton, 1978).

ANNUAL RHYTHMS

The seasonal changes in the environment can be quite dramatic, especially in the temperate zone. As the days shorten and the temperature drops, plants and animals prepare themselves for severe and frigid weather. An example of such a behavior that is controlled by an annual biological clock is the hibernation of the golden-mantled ground squirrel, *Citellus lateralis*. When this squirrel is maintained in the laboratory at a constant temperature and with an unvarying day length (12 hours of light alternating with 12 hours of darkness), it is deprived of the most obvious cues for the onset of winter or arrival of spring. Still, it will enter a period of hibernation at approximately yearly intervals. This was true even when laboratory-born animals were subjected to constant cold (3°C) and darkness; such animals still showed an annual rhythm of alternating activity and hibernation. Some of them even continued the pattern for 3 years (Figure 9.14).

Associated with hibernation but separable from it is a seasonal cycle in feeding. If a ground squirrel is subjected to a relentless "summer"

a

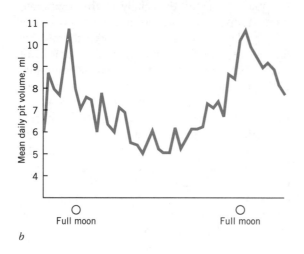

b

Figure 9.13 A monthly rhythm in the pit size of the predatory ant lion (*Myrmeleon obscurus*). (*a*) The ant lion waits at the bottom of a self-constructed pit with only its pincers exposed. When a small arthropod, such as this ant, slips into the pit, the ant lion sucks the prey's body juices. (*b*) Monthly rhythms in the pit size of 50 ant lions maintained under normal daylight conditions in the laboratory. Each of the predators was fed one ant a day. Larger pits were constructed at full moon than at new moon. (From Youthed and Moran, 1969.)

by maintaining the laboratory temperature at an unvarying and torrid 35°C (95°F), it cannot enter hibernation. However, it will still reduce its food and water consumption during the assumed winter and begin to eat again and gain weight in the spring (Pengelley and Asmundson, 1971).

Numerous other clock-controlled annual rhythms have been described, including the migratory restlessness and molt of the willow warbler, *Phyloscopus trochilus* (Gwinner, 1971), growth of antlers in the sika deer *Cervus nippon* (Goss, 1969), testes size of the European starling,

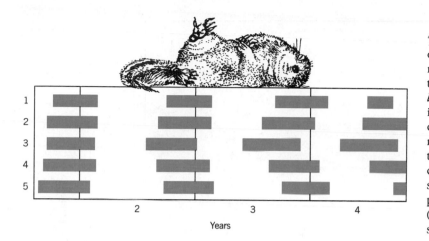

Years

Figure 9.14 The 4-year records of the annual hibernation rhythms of five golden-mantled ground squirrels (*Citellus lateralis*). These animals were isolated from birth and kept continuously in darkness at a near-freezing (3°C) temperature. Although they had no cues for the changing of seasons, they hibernated at approximately yearly intervals. (From Pengelley and Asmundson, 1970, 1971.)

Sturnus vulgaris (Schwab, 1971), the molt and reproduction in the cave crayfish, *Orconectes pellucidus* (Jegla and Poulson, 1969), and the growth and development of the marine cnidarian related to *Hydra, Campanularia flexosa* (Brock, 1974).

One must be cautious in describing a behavior or physiological process that fluctuates annually as one that is controlled by an annual clock. Many of the seasonal changes in behavior are controlled by the changing photoperiod, the shortening of days during the winter months and the increasing daylight periods of the spring and summer. A daily clock appears to be involved in measuring the interval of darkness to regulate photoperiodically controlled behaviors and processes. Unlike a response governed by photoperiod, a rhythm that is controlled by an annual clock will continue to be rhythmic even in the absence of changing day length.

Rebecca Holberton and Kenneth Able (pers. comm.) are among the few scientists who have studied annual rhythms in constant lighting conditions rather than in a constant photoperiod. They have found that the reproductive and migratory rhythms of dark-eyed juncos (*Junco hyemalis*) continue in constant dim light and constant temperature for at least 3 years.

Advantages of a Clock

We have seen that there are many behavioral rhythms that match the prominent geophysical cycles—a day, a lunar day, a lunar month, and a year. The geophysical cycles generate rhythmic changes in environmental conditions. One might wonder, then, why biological clocks exist at all. If the clocks cause changes that are correlated with environmental cues, why not just respond to the cues themselves?

ANTICIPATION OF ENVIRONMENTAL CHANGE

One reason for timing an event with a biological clock rather than responding directly to environmental fluctuations is that it allows an animal to anticipate the change and to allow adequate time for the preparation of the behavior. For example, in nature, adult fruit flies, *Drosophila*, emerge from their pupal cases during a short interval around dawn. At this time the atmosphere is cool and moist, allowing the flies an opportunity to expand their wings with a minimal loss of water through the still permeable cuticle. This procedure takes several hours to

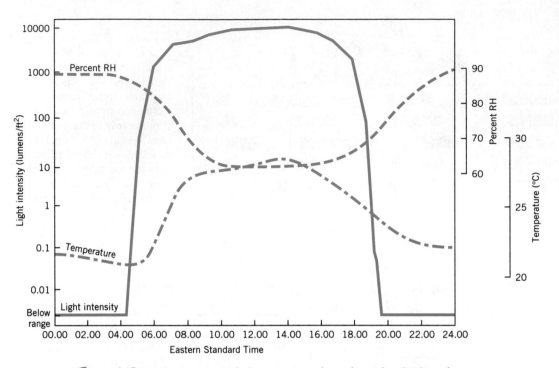

Figure 9.15 Daily changes in light intensity, relative humidity (RH), and temperature. If the fruit fly (*Drosophila*) were to wait for a reliable cue of dawn, such as sunrise, it would not have time to prepare for and complete its emergence from the pupal case before the relative humidity dropped. In arid daytime air, the insects wings might dry before they were properly expanded. (Modified from Tychsen and Fletcher, 1971.)

complete. As you can see in Figure 9.15, the relative humidity drops rapidly after the sun rises. If the flies waited until there was a change in light intensity, temperature, or relative humidity before beginning the preparations for emergence, they would emerge later in the day, a time when the water loss to the arid air could prevent the wings from expanding properly.

SYNCHRONIZATION OF BEHAVIOR WITH AN EVENT THAT CANNOT BE SENSED DIRECTLY

Another advantage to the clock control of an event is that it allows a behavior to be synchronized with a factor in the environment that the animal cannot sense directly. An example is the timing of bee flights to patches of flowers open only during restricted times of the day (Figure 9.16). The flowers visited for nectar gathering may be far away from the hive, so the bee could not use vision or olfaction to determine whether the flowers were open.

CONTINUOUS MEASUREMENT OF TIME

Sometimes an animal may consult its clock to determine what time it is. As we have seen, this information is necessary to anticipate environmental change or to synchronize behavior with other events. However, there are other times when a clock is consulted in order to measure an interval of time. This, then, is a third benefit of a biological clock.

The ability to measure the passage of time continuously is crucial to an animal's time-compensated orientation. For example, a worker honeybee (*Apis* spp.) indicates the direction to a nectar source to recruit bees through a dance that informs them of the proper flight bearing relative to the sun. Since the sun is a moving reference point, the honeybee must know not only the time of day when it discovered the nectar, but also how much time has passed since then. Her biological clock provides this information. The use of the sun as a compass will be explored in more detail in Chapter 11.

The Clock versus the Hands of the Clock

When we study biological rhythms we actually look at the rhythmic processes and make inferences about the clock itself. However, it is important to keep in mind that the biological clock is separate from the processes it drives. Perhaps an analogy to a more familiar timepiece will emphasize this important point. The clock mechanism of your alarm clock is distinct from the hands of the clock, although it is responsible for their movements. If you are particularly vindictive about being awakened from a dream by the clamor of your alarm and tear the hands from the face of your clock, the internal gears will continue to run undaunted. And so it is with internal clocks. You can alter the rhythmic process without affecting the mechanism.

J. Woodland Hastings (1970) performed the biological equivalent of tearing the hands from the clock. The experiment was performed on *Gonyaulax*, a dinoflagellate that displays a rhythm in bioluminescence (giving off light). The intensity of emitted light varies regularly during each day. This glow was completely stopped by the addition of puromycin, an inhib-

Figure 9.16 Honeybees can use their biological clock to time their visits to distant patches of flowers so that they arrive when the flowers are open and nectar is available.

itor of protein synthesis. However, when the inhibitor was washed out of the cells, the rhythm reappeared in phase with untreated controls (Figure 9.17), demonstrating that although the process of luminescence had been halted, the clock was running accurately the entire time. Therefore, like the hands on a clock, the rhythmic process, bioluminescence in this case, is separate from the clock mechanism. Processes are made to be rhythmic because they are coupled to and driven by a biological clock.

Clock Properties

Although we do not yet know the molecular mechanism for the clock, it is possible to list some of its characteristics. Any hypothesis for

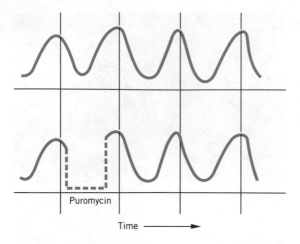

Figure 9.17 Diagrammatic representation of the effect of a pulse of puromycin (an inhibitor of protein synthesis) on the bioluminescence rhythm in *Gonyaulax*. Although luminescence was prevented during the treatment, the clock was still running so that after the inhibitor was washed out, the rhythm resumed in phase with untreated controls. This demonstrates that the biological clock is separate from the rhythmic process. (From Brown, Hastings, and Palmer, 1970.)

how rhythms are timed must be consistent with these known properties.

PERSISTENCE IN CONSTANT CONDITIONS

A defining property of clock-controlled rhythms is that cycles continue in the absence of environmental cues such as light–dark and temperature cycles. This means that the external day–night cycles in light or temperature are not causing the rhythms. Instead we attribute the ability to keep time without external cues to an internal, biological clock.

However, in the constancy of the laboratory, the period (the interval between two identical points in the cycle) of the rhythm is rarely exactly what it was in nature; it becomes slightly longer or shorter. This change in the period length is described with the prefix circa. So, a

daily rhythm, one that is 24 hours in nature, becomes circadian (circa—about; diem—a day) in constant conditions (Figure 9.18). A lunar day (tidal) rhythm becomes circalunidian; a monthly rhythm, circamonthly; and an annual rhythm, circannual.

In other words, a laboratory hamster kept in constant conditions may begin running a little later every night. If it starts running 10 minutes later in each cycle, after 2 weeks its activity will

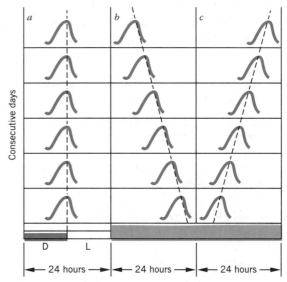

Figure 9.18 Diagram illustrating a biological rhythm in the entrained and free-running state. (*a*) The clock is entrained to the light–dark cycle indicated by the bars at the bottom of the column. Entrainment is the establishment of a stable phase relationship between the rhythm and the light–dark cycle, thus ensuring that the activities programmed by the clock occur at the appropriate times. When an organism is placed in constant conditions, the period length of its rhythms is rarely exactly 24 hours. Depending on the organism, the light intensity, and the temperature, the period length may be slightly longer than 24 hours (*b*) or slightly shorter than 24 hours (*c*). The adjective circadian is used to described this change in period. (From Brown, Hastings, and Palmer, 1970.)

be about 2½ hours out of phase with the actual daily cycle.

When an animal is in constant conditions, the period length of its rhythms generally deviates from that observed in nature. An assumption is made that the period length of the rhythm is a reflection of the rate at which the clock is running. Sometimes this point is emphasized by describing the circadian period length as free-running, implying that it is no longer manipulated by environmental cycles.

STABILITY OF PERIOD LENGTH

Many of the properties of the biological clock are those that would be demanded of any timepiece. One such property is precision. If any clock is to be useful, it must be precise and the biological clock is no exception.

If an animal is cloistered in unvarying conditions and the free-running period length of its activity rhythm is determined on successive days of several months, the measurements are usually found to be extremely consistent. For some animals the precision is astounding. For example, the biological clock of the flying squirrel, *Glaucomys volans*, measures a day to within minutes without external time cues. In fact, the precision of the clock in most animals is greater than one might expect. The daily variability in the free-running period length is frequently no more than 15 minutes and is almost always less than 1 hour (Saunders, 1977).

ENTRAINMENT BY ENVIRONMENTAL CYCLES

Period Control: Daily Adjustment of the Free-running Period to Natural Day–Night Cycle

Isn't it annoying when your watch runs slowly? If you did not reset it every day, your schedule would soon become a shambles. And so it is with biological clocks.

As we have just seen, the clocks themselves do not run on a precise 24-hour cycle. In nature,

however, the period length of biological rhythms is strictly 24 hours. This is possible because the clock is entrained by (locked onto) the natural light–dark cycle. The clock is reset during each day by changes in light intensity as the sun rises and sets. For example, if the free-running period length of a mouse housed in constant darkness were 24 hours and 15 minutes, its clock would have to be reset by a quarter of an hour every day, or it would awaken progressively later each evening. These adjustments must be made so that the onset of activity, or any event programmed by the clock, will occur at the appropriate time each day.

Phase Control: Adjustment to a New Light–Dark Cycle

We have seen that one function of the biological clock is to time certain activities so that they occur at the best point of some predictable cycle in the environment. To be useful, then, they must be set to local time. But some species travel great distances during their lives. Clearly there must be a way to set the clock during long-distance treks or those activities would occur at inappropriate times. In other words, biological clocks, like any clock, cannot be useful unless there is some way to set them (adjust the phase). If you were to fly from Cape Cod, Massachusetts, across three time zones to Big Sur, California, the first thing that you would want to do is to set your watch to the local time. It is obvious that if your biological clock is to gear your activities to the appropriate time of day in the new locale, it too must be reset.

When you first step off the plane after any flight across time zones, your biological clock is still set to the local time of home. The clock will gradually adjust to the day–night cycle in the new locale (Figure 9.19). However, this shift cannot occur immediately; it may take several days. The length of time required for the biological clock to be reset to the new local time increases with the number of time zones traversed. To

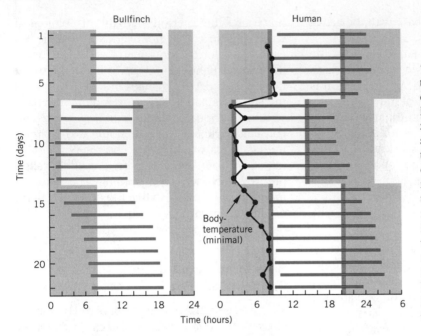

Bullfinch

Human

Time (days)

Time (hours)

Body-temperature (minimal)

Figure 9.19 The resetting of the biological clock by light–dark cycles. Immediately following a trip across time zones, the biological clock is still set to home time rather than that of the vacation locale. It may take several days for the clock to be reset so that it has the proper phase relationship with the new light–dark cycle. In addition, various rhythms (clocks) may rephase at different rates. During the interval of readjustment, the individual suffers from jet lag and may not feel well. (From Aschoff, 1967.)

make matters worse, not all your body functions readjust at the same rate so that the normal phase relationship among physiological processes is upset. Therefore for a few days after longitudinal travel, your body time is out of phase with local time and your rhythms may be peaking at inappropriate times relative to one another. During this time you often suffer psychological and physiological disturbances. The syndrome of effects, which frequently includes a decrease in mental alertness and gastric distress, is referred to as jet lag.

For circadian rhythms the most powerful phase-setting agent is a light–dark cycle, although temperature cycles are also effective in plants and poikilothermic animals (those whose body temperature tends to be near that of their surroundings and are commonly called cold blooded). With a few exceptions, temperature cycles are generally not very effective in setting the clocks of birds or mammals (Hastings, Rusak,

and Boulos, 1991). When a rhythm becomes locked onto (synchronized with) an environmental cycle such as light–dark changes, we say that it is entrained to the cycle.

In the laboratory the biological clock can be reset at will by manipulating the light–dark cycle. If a hamster is kept in a light–dark cycle with 12 hours of light alternating with 12 hours of darkness such that the light is turned off at 6 P.M. real time, its activity begins shortly after 6 each evening. The light–dark cycle might then be changed so that darkness begins at midnight. Over the next few days the clock would be gradually reset so that at the end of about 5 days, the activity would begin shortly after midnight real time. If, after several weeks of this lighting regime, the hamster is returned to constant conditions, its activity rhythm would have a period that approximates 24 hours and, more important, the rhythm would initially be in phase with the second light–dark cycle.

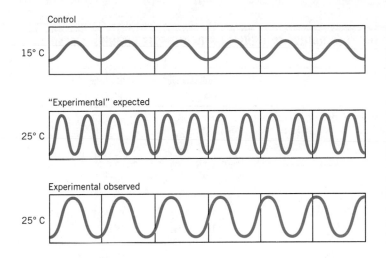

Control

15° C

"Experimental" expected

25° C

Experimental observed

25° C

Figure 9.20 A diagram of the expected and observed change in period length of a typical biological rhythm when an organism is subjected to a 10°C increase in temperature. If the biological clock responded as do most chemical reactions, its rate would be at least doubled. This would be reflected in a halving of the period length. However, a 10°C change in temperature usually has a modest effect on the rate at which the clock runs, so that the period is only slightly shorter than the control's. (From Brown, Hastings, and Palmer, 1970.)

TEMPERATURE COMPENSATION

The biological clock remains accurate in spite of large changes in environmental temperature. This is somewhat surprising if one assumes, as most workers in the field do, that the timing mechanism is rooted in the cell's biochemistry. As a general rule, chemical reactions double or triple in rate for each 10°C change in temperature. However, the effect of an equal temperature rise on the rate at which the clock runs is usually minor, rarely as large as 20 percent (Figure 9.20). This effect may be described by a temperature coefficient or Q_{10} value (calculated as the period at T°/the period at T° + 10°). The examples shown in Table 9.1 are characteristic of the effect of temperature on the clock of poikilothermic animals and plants. You can see that the Q_{10} values usually fall between 0.8 and 1.04, unlike the temperature coefficients for most chemical reactions, which are typically in the range of 2.0 to 4.0. This insensitivity to the effects of temperature suggests that the clock somehow compensates for the effects of temperature. It should be apparent, however, that if the clock were as sensitive to temperature changes as most other chemical reactions are, it

would function as a thermometer, indicating the ambient temperature by its rate of running, rather than as a timepiece.

Attempts to Discover the Clock's Mechanism

CHEMICAL MANIPULATION

In searching for the mechanism of biological clocks, one approach is to attempt to interfere chemically with processes involved in timing. A chemical, preferably one with a known effect on some cellular activity, is continuously administered to an organism living in constant conditions, and if there is a change in the period length of the rhythm, one conclusion is that the substance has changed the rate at which the clock is running. Sometimes, however, the substance to be tested is too toxic for sustained administration or the determination of the effect of the chemical at different points in the circadian cycle may be required. In those cases, the chemical is added for a short period of time and

Table 9.1

SELECTED EXAMPLES SHOWING THE TEMPERATURE COMPENSATION OF THE FREE-RUNNING CIRCADIAN PERIOD.

ORGANISM	RHYTHMIC PHENOMENON	TEMPERATURE RANGE, °C	PERIOD RANGE, IN HOURS	Q_{10}
Euglena gracilis (alga)	Phototaxis to test light in constant darkness	16.7–33	26.2–23.2	1.01–1.1
Oedogonium cardiacum (alga)	Sporulation	17.5–27.5	20–25	0.8
Gonyaulax polyedra (dinoflagellate)	Bioluminescence	18–25	22.9–24.7	0.9
	Cell division	18.5–25	23.9–25.4	0.85
Neurospora crassa (fungus)	Zonation of growth in dim red light	24–31	22–21.7	1.03
Phaseolus multiflorus (bean)	Leaf movement in constant darkness	15–25	28.3–28.0	1.01
Leucophaea maderae (cockroach)	Locomotor activity	20–30	25.1–24.3	1.04
Schistocerca gregaria (locust)	Deposition of daily growth layers in cuticle, constant darkness			1.04
Drosophila pseudoobscura (fruit fly)	Pupal eclosion in constant darkness	16–26	24.5–24.0	1.02
Lacerta sicula (lizard)	Locomotor activity	16–35	25.2–24.2	1.02

Source: Saunders, 1977.

then removed. If this pulse causes a phase shift in the rhythm relative to untreated controls, it is concluded that the drug reached the clock and affected its timing. When a chemical is found to influence biological timing in one of these ways, the process affected by the chemical is judged to be somehow involved in the functioning of the clock.

Unfortunately, interpretation of the results of chemical manipulation experiments is not always as easy as described. In some experiments, a chemical found to have an effect on the clock has so many effects on a living system that no conclusions can be drawn regarding specific cellular processes involved in timing. Furthermore, even if a chemical has only one primary effect, it may have side effects on other cellular reactions.

Even when it is reasonably certain that a chemical is influencing the clock by altering a specific cellular function, it is still difficult to know whether that process is directly or only indirectly involved in the timing mechanism. As an example of this problem, consider the interpretation of the results of experiments using cycloheximide (an inhibitor of protein synthesis within the cytoplasm). Because cycloheximide has been shown to affect biological timing in some organisms, it has been suggested that protein synthesis may be an integral part of the timing mechanism. On the other hand, the inhibitor could be affecting the clock indirectly. Perhaps interrupting protein synthesis eliminates a particular protein that is essential for transport of a critical substance across the membrane. In that case, protein synthesis would be indirectly involved in the functioning of the clock.

Partially as a result of these difficulties, we are still far from knowing the molecular basis for circadian rhythmicity. Even so, chemical manipulation experiments have begun to offer hints of some processes that may be directly or indirectly involved in the timing mechanism (reviewed in Feldman, 1989; Hastings, Rusak, and Boulos, 1991).

Since so many cellular processes are influenced by information in the genes, workers naturally investigated the possibility that genetic material was an essential component of the clock mechanism. In order for the information in a gene to affect a cellular process, the DNA composing that gene must first be transcribed into messenger RNA (mRNA). The mRNA then migrates from the nucleus to the cytoplasm and is translated into a protein on the ribosomes (refer to Chapter 3 for more details of gene expression).

The steps in this process that are important in biological pacemaking have been identified by chemical manipulation experiments. Transcription, forming mRNA, does not appear to be involved in the clock's mechanism. In general,

chemicals that affect transcription have no consistent or reproducible effect on the period or phase of any of the rhythms that have been monitored. Likewise, inhibitors of protein synthesis occurring within subcellular structures generally have no effect on biological timing. However, the synthesis of proteins within the cytoplasm appears to be important to the functioning of the clock. This generalization is based on the observation that inhibitors specifically blocking cytoplasmic protein synthesis affect the period of phase of several circadian systems (Sargent et al., 1976).

Membranes are another common denominator of cellular organization. Chemical manipulation experiments have provided some primarily circumstantial evidence that membrane structure and/or membrane transport are involved in generating circadian oscillations. Ethanol interacts with lipids in membranes by changing their fluidity, and it has been shown to have an effect on the biological clock driving some rhythms. Valinomycin is an antibiotic that facilitates the movement of potassium ions across membranes. Pulses of valinomycin or potassium ions have caused phase changes in some rhythmic systems. Likewise, lithium, an ion that can substitute for potassium ions in some reactions, has occasionally been found to disturb timing (Sweeney, 1976).

Calcium ions and calmodulin, chemicals important in certain aspects of cell regulation, might also be involved in the clock's mechanism. An increase in the concentration of calcium ions within the cell is the first step in bringing about the effects of many animal chemical messengers, including neurotransmitters, growth factors, and hormones. The calcium ions may affect the activities of certain enzymes directly or by first binding to calmodulin. A chemical that alters the movement of calcium into the cell also causes phase shifts. In the bread mold *Neurospora*, the direction of the phase shift, that is whether the rhythm is advanced or delayed, is determined by whether calcium is entering or leaving the cell. In addition, substances that affect the activity of calmodulin also cause phase shifts in certain organisms, including *Neurospora* (Goto, Laval-Martin, and Edmunds, 1985; Nakashima, 1986).

GENETIC STUDIES

There are two classes of mutants that may offer information on clocks—those with metabolic deficiencies and those with altered clock properties. Those with metabolic deficiencies sometimes provide the opportunity to manipulate cell structure or function in ways that turn out to affect the clock. For instance, some studies have used a mutant strain of *Neurospora* that is unable to synthesize its own fatty acids and, therefore, must obtain them from its surroundings. These studies have led to speculations that some fatty acid–containing component of the cell, presumably the membranes, is involved in the circadian timing mechanism (Brody and Martins, 1979).

Mutants with known metabolic deficiencies can be used in yet another way to explore the biological clockworks. Here, one looks for altered clock properties associated with a known metabolic problem. *Neurospora* again provides an example. One mutation has been found that affects both the function or structure of the mitochondrial membrane and the period length of a rhythm. Thus, this work suggests that the mitochondrial membrane might be important to the clock mechanism (Dieckmann and Brody, 1980).

The most exciting genetic studies of the biological clock today are those that focus on mutants with altered clock properties. Clock mutants have been isolated in the fruit fly *Drosophila* (Konopka and Benzer, 1971; Rosbash and Hall, 1989), *Neurospora* (Dunlap, 1990; Feldman and Dunlap, 1983), *Chlamydomonas*, a unicellular photosynthetic organism (Bruce, 1972, 1974; Mergenhagen, 1984), and the hamster (*Mesocricetus auratus*; Ralph and Menaker, 1988). Mutants are generally obtained by expos-

ing the organism to an agent that causes random changes in the genetic material and then looking for individuals with altered biological clocks.

Ronald Konopka and Seymour Benzer (1971) isolated an arrhythmic mutant *Drosophila* as well as mutants with periods of 19 hours and 28 hours. These mutations are all in the same gene, the *per* gene. Since the *per* gene was first described, other *per* mutations and other clock genes have been identified (Dunlap, 1990). (Details of some studies on genetic analysis of the *per* gene were discussed in Chapter 3.) The protein produced by the *per* gene seems to affect circadian rhythms by acting in the brain, where the levels of mRNA specific for its production are known to be rhythmic (Hardin, Hall, and Rosbash, 1990). In fact, the level or activity of *per* product affects the levels of *per* mRNA. This shows that the *per* protein is important in determining the period length of rhythms in these flies and is, therefore, either part of the clock or very close to it.

We now know that the *per* protein is a proteoglycan (a type of protein with highly branched side chains of sugars), but we are not sure how it acts to make the fruit fly rhythmic (Jackson et al., 1986; Rosbash and Hall, 1985). We do know, however, that a protein similar to the *per* product is necessary for neural development in *Drosophila* (Crews, Thomas, and Goodman, 1988). Some scientists have shown that it may alter the ease with which cells can communicate over specialized tight junctions, called gap junctions (Bargiello et al., 1987). This idea is consistent with the suggestion that the *per* protein links (couples) several shorter period clocks so that a longer, circadian period is produced. Indeed, analysis of arrhythmic *per* mutants has revealed activity rhythms with period lengths of about 10 to 15 hours. One interpretation of this finding is that these short period clocks are normally coupled to produce a circadian period, but the coupling is defective in the *per* mutants (Dowse, Hall, and Ringo, 1987).

There are at least seven different genes that affect the biological clock in the bread mold *Neurospora*, and one of these, *frq*, has at least eight different mutations (Dunlap, 1990). The *frq* gene was first isolated by Jerry Feldman and coworkers (Feldman and Hoyle, 1973; Feldman and Wasser, 1971). Most *frq* mutations have altered period lengths, but one has lost its ability to compensate for changes in environmental temperature. The amount of *frq* gene protein within a cell can be manipulated by creating special cells (heterokaryons) that contain more than one nucleus. When this was done, it was found that the rate at which the *Neurospora* clock ticks depends on the amount of *frq* gene protein produced by that cell. So this protein seems to be important in the organization of the clock (Gardner and Feldman, 1980).

Circadian Organization

Moving up from the molecular level, we can begin to ask questions about where in a cell the biological clock might be found. The existence of clock-controlled rhythms in unicellular organisms and plants tells us that a complex nervous system or endocrine system is not an essential component of the biological clock. A single cell may contain the necessary equipment for the biological timing. As incredible as it seems, an intact cell is not even a prerequisite for rhythmicity. Consider, for example, the large unicellular alga called *Acetabularia*. The clock continues to tick in this cell even if the nucleus is removed or if it is cut into pieces (Karakashian and Schweiger, 1976; Mergenhagen and Schweiger, 1973, 1975; Sweeney and Haxo, 1961; Woolum, 1991).

MULTIPLE CLOCKS

Since the clock can exist in single cells, does this mean that every cell in a multicellular or-

ganism has its personal timepiece? It is clear that although not all the cells in an animal have their own clocks, some cells do. One way to demonstrate that a group of cells has its own clock is to remove the tissue, grow it in culture, and see whether the rhythm persists. For example, when the adrenal glands of a hamster are grown in tissue culture, they continue to secrete their hormone, corticosterone, rhythmically for 10 days (Shiotsuka, Jovonovich, and Jovonovich, 1974). Likewise, pineal glands of chickens that are organ cultured in continuous darkness display a circadian rhythm in the production of the hormone melatonin (Figure 9.21; Takahashi, Hamm, and Menaker, 1980) and in N-acetyl transferase, the enzyme that controls the melatonin rhythm (Deguchi, 1979; Kasal, Menaker, and Perez-Polo, 1979). In fact, the chick pineal cells can be separated from one another and not only will the rhythm in melatonin continue, but it can be entrained by a light–dark cycle (Robertson and Takahashi, 1988a,b). If different glands or organs have clocks that continue running even when removed from the body, it follows that a multicellular organism must have several clocks.

Another way to demonstrate that an organism may have more than one clock would be to somehow get different clocks in one individual to run independently, each at its own rate. Occasionally this happens when an animal is kept in constant conditions. When humans lived in underground shelters without any time cues, about 15 percent of them had a sleep–wakefulness rhythm with a different period length than their body temperature rhythm (Aschoff, 1965). Notice in Figure 9.22 that one person's body temperature rhythm had a period length of 24.7 hours while the sleep–wakefulness rhythm had a period of 32.6 hours. Similar desynchronization of rhythms has been shown in a squirrel monkey (*Saimiri sciureus*) maintained in constant conditions (Moore-Ede and Sulzman, 1977). Following the reasoning that the period length of a rhythm in constant conditions is a reflection of the rate at which the clock is run-

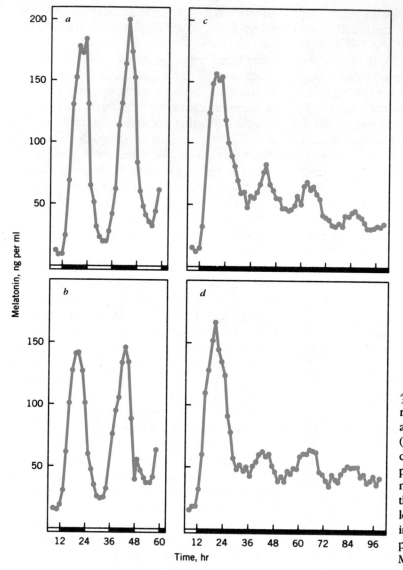

Figure 9.21 Rhythmic melatonin release in a light–dark cycle (*a* and *b*) and in constant conditions (*c* and *d*) from chicken pineals cultured *in vitro*. The rhythmic production of melatonin from pineals in tissue culture indicates that the pineal gland is a biological clock capable of generating rhythms without external input. (From Takahashi, Hamm and Menaker, 1980.)

ning, this may be taken as evidence that these processes are controlled by different clocks.

A HIERARCHY OF CLOCKS

How then are an individual's many clocks synchronized so that all the rhythmic processes occur at the appropriate time relative to one another and the environment's cycles? It appears that there are one or more "master" clocks in the brain that are entrained by the light–dark cycle and regulate other clocks through the nervous and/or endocrine system. Therefore, we can consider three questions: (1) What photoreceptors are responsible for entrainment? (2) Where is the master clock? and (3) How does

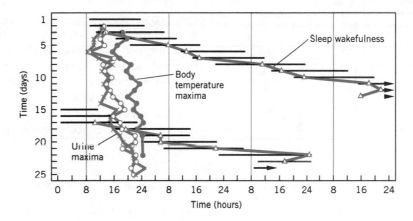

Figure 9.22   Desynchronization of rhythms of a man living in constant conditions. The maxima of body temperature (closed circles), urine volume (open circles), and potassium excretion (crosses) occur at approximately 24.7 hour intervals, while the sleep–wakefulness rhythm (lines) and calcium excretion (open triangles) had a period of 32.6 hours. The fact that these rhythms have different period lengths is evidence that the processes are controlled by different clocks. (Modified from Aschoff, 1965).

the master clock regulate the other clocks in the body? We will approach these questions by considering the "clockshops" of silkmoths, cockroaches, birds, and rodents.

Circadian Organization in Silkmoths

In silkmoths, the photoreceptor for entrainment and the clock controlling eclosion (emergence of the adult from the pupal case) are located in the brain, which controls eclosion by producing a hormone. A series of elegant experiments revealed the details of the neuroendocrine control of the circadian eclosion rhythm (Truman and Riddiford, 1970). The moths (*Hyalophora cecropia*) emerged at the usual time of day, just after sunrise, even if the nerves connecting the eyes and the brain were cut and certain parts of the nervous system (such as the subesophageal ganglion, corpora cardiaca, the corpora allata) or the developing compound eyes were removed. Apparently this treatment did not stop the clock or blind it to the light–dark cycle. However, when the brains were removed from

pupae, the adults emerged at random times throughout the day. When a brain was implanted in the abdomen of a pupa whose brain had been removed, the adult emerged at the customary time of day. So, the brain is needed if the clock is to tick; but how does it "know" when the sun rises or sets?

The search for the photoreceptor for entrainment involved removing the brains from 20 pupae and implanting a brain into the head end of 10 of these individuals and into the abdomen of the remaining 10. The pupae were then inserted through holes in an opaque board so that the two ends of each pupa could be presented with light–dark cycles that were 12 hours out of phase with one another. In other words, both ends of the pupae experienced 12 hours of light alternating with 12 hours of darkness, but the anterior end experienced light (day) while the posterior end was exposed to dark (night) and vice versa. The time of eclosion was determined by the light–dark cycle that the brain "saw," whether it had been implanted into the head or abdomen (Figure 9.23). So we see that the brain is the

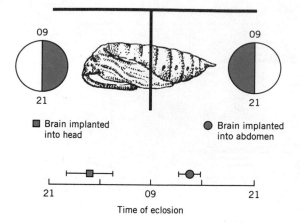

■ Brain implanted into head

● Brain implanted into abdomen

Time of eclosion

Figure 9.23 Eclosion of two groups of loose brain cecropia moths. The anterior end of each pupa was exposed to light from 9 A.M. to 9 P.M., while the posterior end was exposed to light from 9 P.M. to 9 A.M. The pupae in one group had their brains removed and reimplanted in the anterior end. The pupae in the other group had their brains removed and reimplanted in their abdomens. The moths emerged from the pupal case at the correct time of the light cycle to which the brain was exposed. (Modified from Truman, 1971.)

photoreceptor for the clock, and since there were no longer any nerves connected to the brains in these pupae, the brain must be hormonally linked to the eclosion process.

Although the previous experiments demonstrated that the brain is essential for rhythmicity, they did not conclusively show that it was the clock. An alternative explanation for these results is that the brain is responding to a signal from the clock by releasing a hormone necessary to initiate eclosion. Evidence that the brain is the clock would be that the implantation of a brain transferred some clock property, such as a characteristic period or phase. This was done by exchanging the brains of two species of moths, *H. cecropia*, which ecloses just after dawn, and *Antheraea pernyi*, which emerges just before sunset. The moths emerged at the time of day appropriate for the species whose brain

they possessed (Figure 9.24). Since it is the clock that determines the appropriate time for eclosion, the transplantation of this phase information indicates that the brain contains the clock. Later experiments point to the cerebral lobes of the brain as the more specific site of the clock and the medial neurosecretory cell cluster as the source of the hormone initiating eclosion (Truman, 1972).

Circadian Organization in Cockroaches

The circadian system in cockroaches is somewhat different from that in moths. Unlike moths, in which the brain is the photoreceptor for entrainment, in cockroaches, such as *Leucophaea maderae* and *Periplaneta americana*, light information reaches the clock through the compound eyes (Roberts, 1965). The location of the biological clock has been pinpointed to a region of the optic lobes of the brain. It is perhaps more precise to say that the cockroach has two clocks, since there are two optic lobes and each can time independently (Page, 1985; Roberts, 1974). The clock regulates activity in the cockroach via nerves, not hormones as in the moths. We know this because cutting the nerves leading from the optic lobes causes cockroach activity to become arrhythmic. However, the neural connections will reform within about 40 days if they have been cut or if optic lobes have been transplanted into a cockroach, and activity once again becomes rhythmic (Page, 1983).

Circadian Organization in Birds

Birds have photoreceptors in their brains that are capable of providing information to the clock about the lighting conditions. The circadian system in birds seems to contain three interacting clocks: the pineal gland, the suprachiasmatic nuclei of the hypothalamus, and the eyes. The importance of each of these clocks varies among species. This system of interacting clocks controls other clocks within the body through

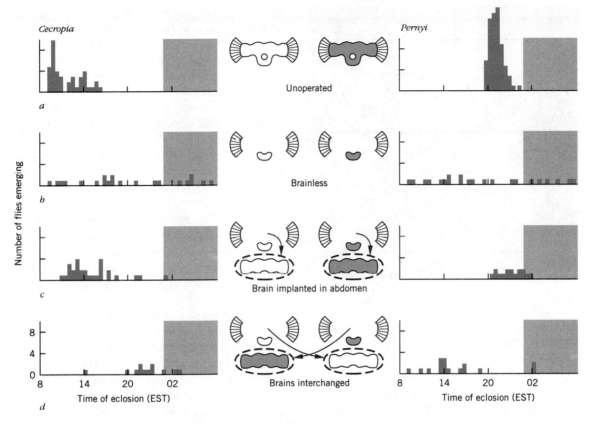

Figure 9.24 The eclosion patterns of *Hyalophora cecropia* and *H. pernyi* pupae showing the effects of various experimental treatments. The shaded area indicates the time of darkness. (*a*) The eclosion of normal intact pupae of these moth species. *Cecropia* moths usually emerge just after dawn and *Pernyi* moths emerge shortly before sunset. (*b*) Brainless pupae eclose randomly throughout the day. (*c*) Eclosion is still rhythmic in pupae in which the brains have been removed and reimplanted in the abdomen. (*d*) When the brains of pupae of these species are interchanged, they eclose at the time characteristic of the species whose brain they possess. These experiments demonstrate that the clock controlling eclosion is in the brain and that it is hormonally coupled to the eclosion process. (From Truman, 1971.)

the pineal's rhythmic output of the hormone melatonin.

Although the eyes can provide lighting information to a bird's clock, there are also photoreceptors in the brain of a bird that can cause entrainment (McMillan, Keatts, and Menaker,

1975). Light, it seems, can penetrate through skin, skull, and brain tissue to reach photoreceptors within the brain itself (Figure 9.25). In one experiment, a blinded house sparrow (*Passer domesticus*) was exposed to cycles of 12 hours of dim light (0.2 lux) alternating with 12

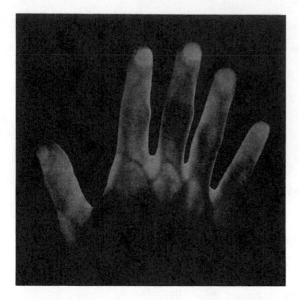

Figure 9.25 Flesh is amazingly transparent to light. The brain of some species contains photoreceptors that help coordinate the biological rhythms of some species of birds with the day–night cycle.

hours of darkness. The light was so faint that most birds did not entrain to this cycle. However, when the feathers were plucked from a bird's head so that the amount of light reaching the brain was increased by several orders of magnitude, the bald bird entrained to the light–dark cycle. As the feathers grew back, entrainment failed. Replucking restored entrainment until India ink was injected beneath the skin of the head. The ink allowed only one-tenth of the light to reach the brain and entrainment was lost. Sensitivity to the light cycle was regained when some of the scalp and the underlying India ink were scraped off. From the results of this experiment (shown in Figure 9.26), one can conclude that direct illumination of the brain can cause entrainment (Menaker, 1968).

Where in the brain are the photoreceptors? We know that the pineal gland of at least some birds is sensitive to light because the rhythm in the hormone melatonin from pineals removed

from chickens and kept alive in tissue culture will still entrain to light–dark cycles (Takahashi, Hamm, and Menaker, 1980). But there is more to the story than this. The perch-hopping and body temperature rhythms of house sparrows (*P. domesticus*) whose pineals and eyes have been removed still entrain to light–dark cycles (Menaker, 1968). So, there must be photoreceptors outside the pineal. Studies have tried to find these photoreceptors, but most have met with frustration because even the most focused light beam spreads outward within the brain tissue. Nonetheless, there do seem to be receptors within the hypothalamus and nearby regions (Silver et al., 1988). We still do not know whether the eyes, pineal, and other brain photoreceptors are doing the same job or are playing different roles in conveying light information to the clock.

We now know that there are several interacting clocks in the circadian system of birds: the pineal, the suprachiasmatic nuclei of the hypothalamus (SCN), and the eyes. The search for the site of the circadian pacemaker began with the discovery that the pineal gland is an important master clock in some species. When the pineal is surgically removed from the brain of a house sparrow or a white-crowned sparrow (*Zonotrichia leucophrys*), the bird's perch-hopping activity and body temperature are no longer rhythmic in constant darkness (Gaston and Menaker, 1968). This demonstrates only that the pineal is necessary for the expression of rhythmicity. But a later experiment indicated that the pineal is a circadian pacemaker. Two groups of donor house sparrows were entrained to light–dark cycles that were 10 hours out of phase. The pineal glands of these sparrows were transplanted into the anterior chamber of the eye of arrhythmic pinealectomized sparrows that had been maintained in continuous darkness. Within a few days following a successful transplant, the recipients, still housed in constant darkness, became rhythmic. The phase of the

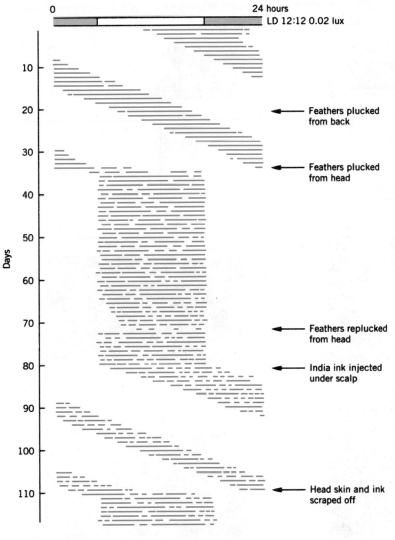

Passer domesticus

0   24 hours
LD 12:12 0.02 lux

Days

← Feathers plucked from back

← Feathers plucked from head

← Feathers replucked from head

← India ink injected under scalp

← Head skin and ink scraped off

Figure 9.26 Activity rhythm of a blind house sparrow kept in a 24-hour light–dark cycle with 12 hours of dim (0.2 lux) light. The time of darkness is indicated by the shaded bars at the top of the figure. Activity is represented by the blue horizontal bars. When the feathers were plucked from the bird's back there was no effect on the rhythm. The bird was not entrained to the light–dark cycle. Then the feathers were plucked from the bird's head, thereby increasing the light reaching the brain. This resulted in entrainment until the feathers began to grow back. When the feathers were plucked once again, entrainment was re-established. Next, India ink was injected under the skin of the scalp to decrease the light intensity reaching the brain. Entrainment was lost until the ink and the skin of the scalp were scraped off (Modified from Menaker, 1968.)

newly instilled rhythm was that of the donor bird (Figure 9.27). This strongly suggests that the pineal gland of house sparrows contains a clock essential for the persistence of rhythmicity in the absence of external time cues (Zimmerman and Menaker, 1979).

The importance of the pineal gland in the circadian system depends on the species of bird.

We have seen its importance as a major clock in house sparrows, and the same is true in European starlings, *Sturnus vulgaris* (Gwinner, 1978). However, it is not as important in the circadian system of Japanese quail (*Coturnix coturnix japonica*; Simpson and Follett, 1981; Underwood and Siopes, 1984).

In some species of birds, the SCN has been

Donors

Recipients

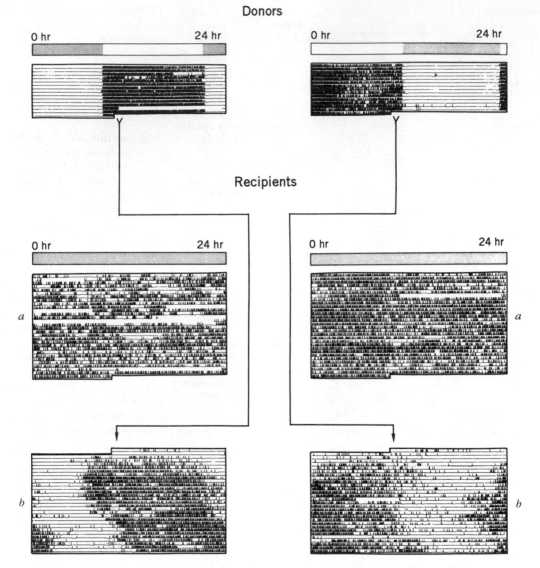

Figure 9.27 The activity records from house sparrows used in a pineal transplantation experiment. There were two groups of donors, each kept on a different light–dark cycle. The shaded portion of the bar above each record indicates the time of darkness. The recipients were arrhythmic pinealectomized sparrows maintained in constant darkness. The presurgery activity records of the recipients are shown in (*a*). The records shown in (*b*) are from birds that have a pineal gland transplanted into the anterior chamber of one of their eyes. The recipients are now rhythmic in constant darkness. The initial phase of the activity record is that of the donor's rhythm. (Modified from Zimmerman and Menaker, 1979.)

shown to be an important biological clock. The perch-hopping activity of the house sparrow, the Java sparrow (*Padda oryzivora*), and Japanese quail is severely disrupted following the destruction of the SCN ((Ebihara and Kawamura, 1981; Simpson and Follett, 1981; Takahashi and Menaker, 1979). The results shown in Figure 9.28 are typical of those of eight house sparrows in which at least 80 percent of the SCN was destroyed. Although the animals were arrhythmic in constant conditions when they lacked their suprachiasmatic nuclei, they did entrain to light–dark cycles. Lesions that spared the SCN, that only partially destroyed them, or that were unilateral did not abolish the free-running activity rhythm.

In some species of birds, the eyes may also be involved in the circadian system. Quail, for instance, remain rhythmic after pinealectomy, unless they are also blinded (Underwood and Siopes, 1984). Blinding also affects rhythmicity in pigeons (*Columba livia*; Ebihara, Uchiyama, and Oshima, 1984), but not in sparrows or chickens (Hastings, Rusak, and Boulos, 1991). In Japanese quail, the eye has been shown to be a clock in its own right (Underwood, Barrett, and Siopes, 1990a).

How do the clocks in a bird's circadian system interact? The SCN is thought to communicate with the pineal gland over a neural pathway. The pineal may "talk back" to the SCN through its rhythmic production of melatonin. The SCN is known to bind melatonin (Vaněček, Paulik, and Illnerová, 1987), which is a good indication that it responds to the hormone. It may be that the eyes also play a role in the circadian system by rhythmically secreting melatonin, much in the same way as the pineal gland. The eyes of chickens, quail, and sparrows have rhythms in melatonin production (Binkley, 1988). The eyes of quail produce about a third of the melatonin in the blood, and the pineal produces the rest. In spite of this, the link between a quail's eyes and the rest of its circadian system is a neural pathway, not their rhythmic melatonin secretion. Continuous administration of melatonin does not affect the free-running period of activity (Simpson and Follett, 1981). Furthermore, cutting the optic nerve does not disrupt the rhythmic production of melatonin, but it does have a dramatic effect on the activity rhythm. In fact, the effect of cutting the optic nerve on the activity rhythm is the same as blinding the bird (Underwood, Barrett, and Siopes, 1990b).

How does this multiclock system exert its control over the rest of the circadian system? The experiment in which the pineal placed into the anterior chamber of the eye restored rhythmicity to a pinealectomized host demonstrated that the pineal must exert its influence via hormones because the nerve connections of the gland were obviously eliminated.

Melatonin, a hormone produced by the pineal in greater amounts during the night than during the day, appears to be the hormone through which the pineal exerts its effect. If a capsule that continually releases melatonin is implanted in an intact house sparrow, the bird either becomes continuously active or the period length of its rhythm is shortened. In other words, when melatonin release is continuous rather than periodic (as it normally is), circadian rhythmicity is disturbed (Turek, McMillan, and Menaker, 1976).

Furthermore, periodic release of melatonin can entrain the activity rhythm of a bird. This has been demonstrated by pinealectomizing a group of European starlings, *Sturnus vulgaris*, and giving half the birds a daily injection of melatonin and the other half a daily injection of sesame oil (to ensure that any effect of the clock was caused by the hormone and not the injection). Almost all, 21 out of 22, of the starlings receiving melatonin injections entrained, while only 1 of the 10 control birds did (Gwinner and Benzinger, 1978). The daily injections seem to have mimicked the effect of the rhythmic release of melatonin by the pineal gland.

a

b

0   24   48

Figure 9.28 The activity records of house sparrows before and after lesions of the suprachiasmatic nuclei. (*a*) The effect of a lesion on a bird in constant darkness. The surgery was performed at the time indicated by the arrow on the left. The distinct free-running activity rhythm that was present before the operation was absent in the lesioned bird. (*b*) Entrainment of a sparrow with a lesion in the suprachiasmatic nuclei. The light–dark cycle is indicated in the arrows at the top of the record. The arrow points up to indicate lights on and down to indicate lights off. In spite of the lesion, the sparrow's activity was confined almost entirely to the light. At the time indicated by the arrow on the left the bird was transferred to constant darkness. The rhythm in activity begins to disappear, but for several days it remained concentrated at the time that had been light in the previous light–dark cycle. This suggests that the bird was entrained to the light cycle. (From Takahashi and Menaker, 1979.)

Circadian Organization in Rodents

The circadian organization in rodents differs from that in the birds we have discussed in two ways: Mammals seem to lack extraretinal photoreceptors for entrainment, and the master clock seems to exert its influence through nerves rather than hormones.

The eyes are the photoreceptors for light entrainment in rodents. The information about the lighting conditions reaches the clock through the retinohypothalamic tract (a bundle of nerve fibers connecting the photosensitive layer of the eye, the retina, with the hypothalamus). The importance of the retinohypothalamic tract was elucidated in a roundabout way. This tract cannot be severed without destroying the SCN, an act that results in arrhythmicity. However, entrainment to a light–dark cycle continues following the lesions of the other tracts known to leave the eye (the primary optic tract and the accessory optic tract), so it is believed that the light information must reach the clock via the retinohypothalamic tract (Moore, 1979).

There is evidence that the SCN is a master clock in rodents that holds several "slave" clocks in the proper phase relationship. Destruction of this area causes disturbances in the rhythms of the rat (Moore and Eichler, 1972; Raisman and Brown-Grant, 1977) and the hamster (Stetson and Watson-Whitmyre, 1976). Comparable destruction elsewhere in the brain has not so far destroyed rhythmicity. Since cuts in the nervous connections after, but not before, the SCN affect rhythms, it is presumed that this center governs the circadian system via a neural output. The assumption that the SCN is an independent clock in rats gained support from other types of experiments as well. In one experiment, the SCN was isolated from neural input and its activity remained rhythmic. Neural activity in the SCN and in one of several other brain locations was recorded simultaneously to be certain that the observations were specific to the SCN. In an intact animal, neural activity in all regions of the brain was always rhythmic. Then, a knife cut was made around the SCN, creating an isolated island of hypothalamic tissue that included the SCN. Following this treatment, the neural activity in the hypothalamic island remained rhythmic in constant darkness while activity in other brain regions was continuous (Figure 9.29). This strongly suggests that the SCN is a self-sustaining oscillator that instills rhythmicity in other brain regions through neural connections (Inouye and Kawamura, 1979).

If an SCN is transplanted into the brains of rats or hamsters that have been made arrhythmic by destroying their own suprachiasmatic nuclei, their activity becomes rhythmic once again. The transplanted tissue is from fetal or newborn animals so that new neural connections may be made in the host's brain. The individuals whose activity rhythms were restored after they received a new SCN had formed some neural connections so that the grafted SCN received input from the eyes and could communicate with other regions of the brain (Lehman et al., 1987;

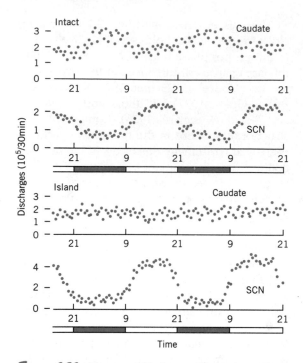

Figure 9.29 Nerve activity in specific brain regions (the caudate and SCN) of a rat before and after isolating an "island" of brain tissue containing the suprachiasmatic nuclei. (*a*) The activity in a normal intact rat is rhythmic in both regions of the brain. After a region of the hypothalamus containing the suprachiasmatic nuclei was isolated as an "island," the nerve activity was rhythmic within the SCN but not outside the "island" in the caudate region. This supports the idea that the suprachiasmatic nuclei are self-sustaining oscillators. (Courtesy of S.T. Inouye.)

Sawaki, Nihonmatsu, and Kawamura, 1984). Furthermore, the period length of the restored activity rhythm matches that of the transplanted SCN, not that of the recipient (Ralph, Davis, and Menaker, 1988; Sollars and Kimble, 1988). This result is what one would expect if the SCN were, indeed, the clock providing timing information and not just a component needed to make the host clock function.

A master clock should regulate both the period and the phase of circadian rhythms. In experiments in which the SCN is destroyed, it is impossible to know what effect this has on the period or phase of any rhythm since the animal is arrythmic. Instead, the SCN can be electrically stimulated to see whether this changes the period or phase of rhythms. In one experiment, rats and hamsters were first blinded, the surgical way to keep the animals in constant darkness. Then, brief pulses of current were applied through electrodes implanted in or near the SCN. This stimulation caused changes in the phase of the rhythms monitored, activity in hamsters and feeding in rats. Depending on the circadian time of the stimulation, the pulses of current caused either phase advances or delays in the rhythms. In some animals, the period length was also changed. When brain regions other than the SCN were stimulated, there were no effects on the animals' rhythms (Rusak and Groos, 1982).

Our understanding of circadian organization is complicated by the fact that there is a suprachiasmatic nucleus in each half of the brain. There is evidence supporting the hypothesis that each of these centers is an independent oscillator (or contains a population of oscillators). In a normal hamster the two nuclei are mutually coupled and produce an activity rhythm with a single well-defined band of activity during each cycle. When a nocturnal animal such as the hamster is maintained in constant light for an extended period, its activity pattern occasionally splits into two components that have different free-running periods until they become stabilized approximately 180 degrees out of phase with one another. Some workers have interpreted the splitting phenomenon as evidence for two circadian oscillators controlling activity that have become uncoupled as a result of the continuous illumination. There is some experimental support for this interpretation. When one of the suprachiasmatic nuclei in hamsters whose activity pattern had become split was destroyed, the split in the activity was eliminated and replaced by a new single activity rhythm, presumably the one controlled by the remaining supra chiasmatic nucleus (Pickard and Turek, 1982).

Summary

Life evolved in a cyclic environment. The environmental changes in some factors are extreme. However, they are also predictable because they are associated with the relative movements of the earth, sun, and moon. Often it is advantageous to gear an activity to occur at a specific time relative to some rhythmic aspect of the environment. Thus clocks evolved as adaptations to the environmental cycles.

There are many examples of rhythmic processes that match the basic geophysical periods: a day (24 hours), the tides (12.4 hours), a lunar day (24.8 hours), a fortnight (14 days), a lunar month (29.5 days), and a year (365 days). Many of these processes remain rhythmic when the individual is isolated from the obvious environmental cycles that might be thought to provide time cues. For instance, many daily rhythms persist when the individual is kept in the laboratory without light–dark or temperature cycles. Therefore, we say that the rhythms are caused by an internal biological clock.

There are several reasons why it may be advantageous to have a biological clock to measure time rather than responding directly to environmental changes: (1) anticipation of the environmental changes with enough time to prepare for the behavior, (2) synchronization of the behavior with some event that cannot be sensed directly, and (3) continuous measurement of time so that time-compensated orientation is possible.

The biological clock is separate from the rhythms it drives. Processes become rhythmic when they are coupled to the biological clock.

In the constancy of the laboratory, the period length of biological rhythms deviates slightly from the one displayed in nature. In this case, they are described with the prefix circa, meaning "about." Thus, they are called circadian, circalunidian, or circannual. The period length in constant conditions is described as the free-running period and is assumed to reflect the rate at which the clock is running. The free-running period is generally kept constant, which indicates that the biological clock is very accurate.

Although the period length of a biological rhythm is "circa" in the constancy of the laboratory, in nature it matches that of the geophysical cycle exactly because the clock is entrained to (locked onto) an environmental cycle. Entrainment adjusts both the period length and the phase of the rhythm. Daily rhythms can be entrained to light–dark cycles, and in some species, to temperature cycles.

Environmental temperature has only a slight effect on the rate at which the clock runs. This property is described as temperature compensation.

One approach to discover the clock's mechanism has been through chemical manipulation. A drug or inhibitor is administered to an organism that shows a rhythm in constant conditions to see whether it affects the clock. Indications that the clock is affected are a change in the free-running period caused by constant administration of the chemical or a change in the phase of the rhythm caused by a pulse of the chemical. Chemical manipulation experiments have suggested that protein synthesis, membranes, calcium ions, or calmodulin may be important in the mechanism of the clock.

Some genetic studies may attempt to correlate metabolic deficiencies in certain mutants with alterations in their biological clock. Such studies have suggested that membrane and energy metabolism may be involved in the mechanism of the clock.

Other genetic studies use mutants with altered clock properties. The two most thoroughly studied mutations are in the period (*per*) gene of the fruit fly *Drosophila* and the frequency (*frq*) gene of the bread mold *Neurospora*.

Biological clocks can exist in single cells and there may be many clocks in a single individual. It seems that there is a hierarchy of clocks with one or more master clocks regulating the activities of other slave clocks.

We are beginning to understand the circadian system in several species. A silkmoth has photoreceptors for entrainment and a clock located in its brain, which then regulates eclosion by producing a hormone. In cockroaches, light information reaches the clock through the compound eyes. There are two biological clocks, one in each optic lobe, that regulate activity through nerves.

Birds have photoreceptors in their brains capable of providing information to the clock about lighting conditions. The circadian system in birds seems to contain three interacting clocks: the pineal gland, the suprachiasmatic nuclei of the hypothalamus, and the eyes. The importance of each of these clocks varies among species. This system of interacting clocks controls other clocks within the body through the pineal's rhythmic output of the hormone melatonin.

The photoreceptors for entrainment in rodents are the eyes. The information reaches the suprachiasmatic nucleus via nerves (the hypothalamic tract). The master clock control for certain rhythms is in the SCN, which regulates activity through nerves.

CHAPTER
10

Orientation in Space

*M*any of us have been moved on a crisp autumn day while standing enveloped in the reds, yellows, and browns of the season and watching formations of ducks or geese fly against a steely sky. An observer might have noticed that if it is early in the day, the flocks may be heading almost due south. If it is nearing dusk or if fields of grain are nearby, they may be temporarily diverted to resting or feeding areas. But when they resume their flight, they will head southward again.

The following spring we may stand beside a swift-moving river in the Pacific Northwest and watch salmon below a dam or a fish ladder. As they lie in deeper pools resting before the next powerful surge that will carry them one step nearer the spawning ground, they all face one way—upstream.

Both the birds and the fish are responding to a complex and changing environment by positioning themselves correctly with respect to it and by moving from one particular part of it to another. Although the feats of migration are astounding, they are no more crucial to survival than are mundane daily activities such as seeking a suitable habitat, looking for food and returning home again, searching for a mate, or identifying offspring. These actions also depend on the proper orientation to key aspects of the environment. Indeed, an animal's life depends on oriented movements both within and between habitats. In this chapter we will explore some of the mechanisms by which animals orient in space.

Types of Orientation

SIMPLE ORIENTATION RESPONSES

If you look into your aquarium and see a fish belly up, you know something is wrong. This is not the normal posture for fish. Furthermore, if you found one of them on your desk, you would know for certain that something had gone awry, even if the fish were somehow propped into its proper swimming position. Not only does each species have a customary stance, but each is usually found in a particular type of habitat. The term orientation is an umbrella that covers all the reactions that guide an animal into its correct posture, into the proper habitat, or during long-distance travel such as migration and homing (Schöne, 1984). We will consider some of the simpler responses first.

Kinesis

Perhaps the simplest response by which animals may end up in a suitable location is a kinesis, a response that is not directed toward or away from the stimulus, but nonetheless results in the individual settling in a suitable location. A kinesis, then, is an orientation mechanism in which an animal's rate of movement or the amount of turning in its path varies with the strength of the stimulus at each step of the movement (Benhamou, 1989; Benhamou and Bovet, 1989). To illustrate, consider the humidity kinesis of wood lice. These small isopods are commonly found under rocks or flower pots or in leaf litter (Figure 10.1). It is adaptive that they cluster in moist areas such as these because they will die if they are exposed to dry air for very long. However, they are not found in damp areas just because those that stray into dry areas die. This can be demonstrated experimentally by distributing wood lice over the floor of a chamber that has both humid and dry areas; the wood lice will soon be found congregated in the humid part of the floor. The explanation is rather simple: They are more active in the drier areas and they scramble about until, quite by accident, they encounter a more humid part where activity slows down until many of them become motionless (Edney, 1954; Gunn and Kennedy, 1936). The effect of humidity on activity has also been demonstrated by determining the percentage of

Figure 10.1 Wood lice such as these (*Oniscus asellus*) tend to accumulate under rocks. The wood lice are guided to favorable locations by a humidity kinesis. They are more active in dry areas than in moist ones.

wood lice (*Porcellio scaber*) that were motionless under various constant humidity conditions: More than 70 percent of the animals were motionless at 90 percent humidity, but an average of only 20 percent were motionless at 30 percent humidity (Gunn, 1937). Thus, the wood lice aggregate in the moist area because they move more slowly there, just as cars accumulate when drivers slow down to rubberneck at the scene of an accident. Furthermore, just as most drivers do not seek out traffic jams, wood lice do not specifically orient their movements toward dampness. Since a kinesis is a nondirectional response, the receptors need only be able to register variations in the intensity of the stimulus.

Kineses may also be used by certain vertebrates when locating an appropriate habitat. For a larva of the brook lamprey, *Lampetra planeri*, a suitable position is buried head downward in the muddy bottom of a stream or lake. A kinetic response to light helps them settle in this environment. These larvae have a light receptor near the tip of the tail. When the larva burrows deeply

enough into the mud to cover the light receptor, it becomes motionless. However, if the light receptor is exposed, the larva becomes agitated and squirms about with its head pointed downward until it is once again covered. In one experiment the larvae were placed in an aquarium with a bottom that prevented burrowing. The aquarium was illuminated on one end and graded into a darkened portion. The lamprey became very active at the lighted end. The activity increased with the intensity of light to which the animals were exposed, but the activity was independent of the direction of the light. Since the larvae were not able to burrow as they randomly moved about, they eventually ended up in the darkened end where activity was reduced (Jones, 1955).

Taxis

Of much more interest biologically are movements that are oriented with respect to the stimulus. The simplest directed orientation mechanism is a taxis, a mechanism that generally moves the animal toward or away from the stimulus (Song and Poff, 1989). A taxis, then, is based on a determination of the direction of the stimulus gradient (Benhamou and Bovet, 1992).

A taxis can be described in different ways. One denotes the stimulus to which the animal is responding. Examples include a phototaxis to light, a geotaxis to gravity, a chemotaxis to a chemical, a rheotaxis to a current, or a phonotaxis to a sound. Another describes whether an animal moves toward or away from the stimulus. Thus, an animal that moves toward light and away from gravity is positively phototactic and negatively geotactic.

Different types of animals may use different methods to determine the direction of stimulation, depending on their perceptual abilities and the nature of the stimulus. An animal with a single receptor may determine the direction of stimulation by moving the receptor around and sampling in various positions. The closer to the source, the more intense is the stimulus. For example, a house fly maggot (*Musca domestica*) moves into a dark place when it is time to pupate by moving its "head" from side to side and sampling the light intensity (Figure 10.2). If the light is brighter on one side, it moves to the other (Mast, 1938).

Some animals with two receptors make a simultaneous comparison of intensity of each side and then orient so that the receptors are stimulated equally. If such an animal displays a positive phototaxis and is blinded in one eye, it will continuously circle to the side of the good eye.

Some species have receptors that can determine both the direction to and the intensity of the stimulus. In this case, if an animal showing a positive phototaxis were blinded in one eye, it would still be able to move directly toward the light source.

More Complicated Light Reactions

Many aquatic animals orient to light in a manner that helps keep them swimming horizontally.

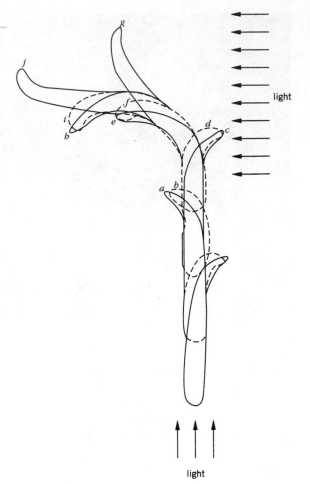

Figure 10.2 A taxis guides a fly maggot to a dark site for pupation. The maggot turns its body to measure the light intensity in various locations with its single photoreceptor. It moves in the direction of lowest light intensity. (From Fraenkel and Gunn, 1961.)

Some of them, fish for instance, may show a dorsal light reaction, one in which their dorsal (back) side is kept toward the light. Others, such as the brine shrimp *Artemia*, show a ventral light reaction in which the ventral (belly) side remains toward the light.

We all know that animals do not always move directly toward or away from a stimulus. Animals showing the light compass reaction assume an angle with a light source. An animal that makes short excursions from the nest may assume a fixed angle with the sun on its outward journey. This orientation angle would be reversed during the return trip. An ant (*Lasius niger*), for instance, may use the sun as a reference point during its foraging foray. The ant's bearing may be at some angle relative to the sun during its outward trip, say 45 degrees to the right. Assuming the trip were short, a path with the sun 135 degrees to the left would bring the ant back to its nest (Santschi, 1911). However, if it is important that the same compass bearing be maintained for a long time, the animal must change its orientation to the light as the sun moves across the sky. Using its biological clock, the animal measures the passage of time and appropriately adjusts the angle of its movement relative to the sun. We will consider the use of the sun as a compass again later in the chapter.

ORIENTATION AND NAVIGATION

Animals often travel long distances to and from home, but they do not all accomplish this feat in the same way. Strategies for finding the way during long-distance travel are often grouped into three levels of ability (Able, 1980; Griffin, 1955; Terrill, 1991). One level is piloting, the ability to find a goal by referring to familiar landmarks. The animal may search either randomly or systematically for the relevant landmarks. Although we usually think of landmarks as visual, the guidepost may be in any sensory modality. As we will see in the next chapter, olfactory cues guide salmon during their upstream migration.

A second level, called compass, or directional, orientation is the ability to head in a geographical direction without the use of landmarks. As we will see in the next chapter, the sun, the stars, and even the earth's magnetic field may be used as compasses by many different animal species.

Most migrating birds use compass orientation

(Emlen, 1975). They simply fly in a particular compass direction and stop when they have flown a preprogrammed distance.

Similarly, the white butterfly (*Peiris rapae*), known as the cabbage butterfly in the United States, heads in one compass direction while migrating during some seasons. In other seasons, it flies in roughly the opposite compass direction and ends up near its starting point (Baker, 1978).

One way to demonstrate that an animal is using compass orientation is to move it to a distant location and determine whether it continues in the same direction or compensates for the movement. If it does not compensate for the relocation, compass orientation is indicated (Figure 10.3). When immature birds of certain migratory species, such as European starlings (Perdeck, 1967), were displaced experimentally, they flew in the same direction as the parent group that had not been moved, and importantly, they flew for the same distance. In other words, they paralleled the parent group. In some cases, this meant that they ended up in ecologically unsatisfactory places (Figure 10.4). In another experiment, storks were taken from nests in the Baltic region, reared in West Germany, and released after the local storks had left (Schüz, 1949). The experimental group had a strong tendency to head south–southeast, the direction of their par-

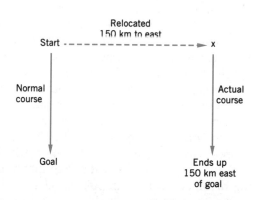

Figure 10.3 Experimental relocation of an animal that is using compass orientation causes it to miss the goal by the amount of its displacement.

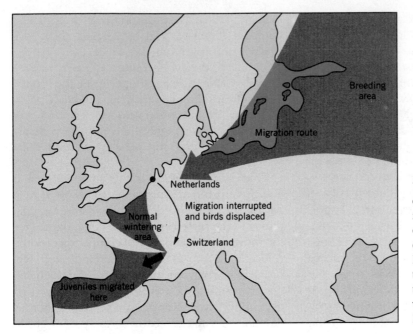

Figure 10.4 Immature starlings captured in the Netherlands and released in Switzerland did not compensate for the relocation during their autumn migration. Instead they traveled southwest, their normal migratory direction, and ended up in incorrect wintering areas. (Modified from Perdeck, 1958.)

ent population, even though the local West German storks were heading southwest.

There is growing evidence that the migratory direction of birds is genetically controlled (reviewed in Berthold, 1991). Many bird species, in addition to the storks we just discussed, find their way without help from others of their species. Observations such as these suggest that migratory direction may be inherited. Individual birds held in the laboratory flutter in the direction in which they would be flying if they were free. When their cousins in nature have completed their migratory journey, the captive birds also cease their directional activity. Furthermore, many species, particularly those that fly from Central Europe to Africa, change compass bearing during their flight. Garden warblers (*Sylvia borin*) and blackcaps (*Sylvia atricapilla*) held in the laboratory change the direction in which they flutter in their cages at the time that free-flying members of their population change direction (Gwinner and Wiltschko, 1978; Helbig, Berthold, and Wiltschko, 1989). Crossbreeding

studies have also shown the inheritance of migratory direction. Andreas Helbig (1991) crossbred members of two populations of blackcaps that had very different migratory directions. The orientation of the offspring was intermediate between those of the parents.

An animal may also use compasses in another type of navigation, one that may be used to find its way home after carrying out daily activities, such as foraging. In this type of navigation, called dead reckoning, the animal appears to know its position relative to home through the sequence of distance and direction traveled during each leg of the outward journey (Figure 10.5). The estimates of distance and direction are often adjusted for any displacement due to current or wind. Then, knowing its location relative to home, the animal can head directly there, using its compass(es). A compass may also be used to determine the direction traveled on each leg of the outward journey, or direction may be estimated from the twists and turns taken, sounds, smells, or even the earth's magnetic field. Once close to home, where landmarks are familiar, piloting may be used to pinpoint the exact location of home.

There is evidence that many types of animals use dead reckoning to find their way around. Consider, for example, the desert ant (*Cataglyphis bicolor*). During its foraging forays, this insect wanders far from its nest over almost featureless terrain. After prey is located, sometimes 100 meters away from the nest, roughly the distance of a football field, the ant turns and heads directly toward home. It appears that the ant knows its position relative to its nest by taking into account each turn and the distance traveled on each leg of its outward trip. To obtain this information, an ant must make the trip itself. If an ant that is leaving its nest is moved to a new spot by an experimenter, it does not seem to know where it is and searches randomly for its nest (Wehner and Flatt, 1972). Furthermore, when an ant is captured as it is leaving a feeding station headed for home and is relocated to a distant site, the ant's path is in a direction that would have led it home, if it had not been experimentally moved (Wehner and Srinivasan, 1981).

A third level of orientation, sometimes called true navigation,* is the ability to maintain or establish reference to a goal, regardless of its location, without the use of landmarks. Generally, this implies that the animal cannot directly sense its goal and that if it is displaced en route to the goal, it compensates for displacement by changing direction, thereby heading once again toward the goal (Figure 10.6).

We see, then, that true navigation requires that the animal have both a compass and a map that indicates position relative to a goal. Imagine

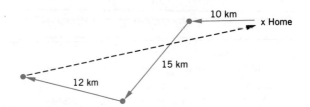

Figure 10.5 Navigation by dead reckoning. This involves determining one's position using the direction and distance of each successive leg of the outward trip. A compass can then be used to steer a course directly toward home.

yourself abandoned in an unfamiliar place with only a compass to guide your homeward journey. Before you could head home, you would also need a map, so that you could know where you were relative to home. Only then could you use your compass and orient yourself correctly.

Only a few species, most notably the homing pigeon (*Columba livia*), have been shown to have true navigational ability. Two other groups of birds, including oceanic seabirds and swallows, are also known to home with great accuracy (Able, 1980; Emlen, 1975).

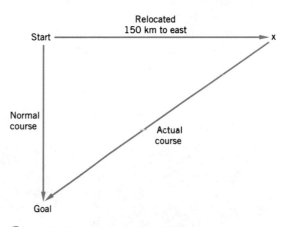

Figure 10.6 An animal that finds its way using true navigation can compensate for experimental relocation and travel toward the goal. The term true navigation also implies that the animal cannot directly sense its goal and that it is not using familiar landmarks to direct its journey.

---

* True navigation is an unfortunate term since it carries with it the implication that other ways finding one's way from place to place are not real methods of navigating. This is certainly not true. Nonetheless, we will use the term simply to distinguish this method of maintaining a course from the others.

Migration

We will consider migration to be the movement of animals away from an area and the subsequent return to that area. Animals usually migrate between breeding areas and overwintering, or feeding, areas. We will see, however, that migration may involve movement between other sites as well.

FEATS OF MIGRATION

Some migratory feats seem incredible. If we did not have indisputable evidence, many of these instances would be denied on the basis of common sense. The migratory habit has evolved in so many different species that it would be impossible to describe them all here. Instead, we will highlight some of the more impressive ventures.

Arthropods

Among the arthropods that migrate are lobsters. Spiny lobsters (*Panulirus argus*), for instance, march single file from the shallow water around the Bahamas, where they summer, to the deeper water in the fall (Figure 10.7; Herrnkind and Kanciruk, 1978).

One of the most impressive insect migrators is the monarch butterfly (*Danaus plexippus*). Each fall, spectacular numbers of monarchs leave southern Canada and the northeastern United States and fly up to 3000 kilometers to high-altitude overwintering sites in fir forests of the Sierra Madre mountains of central Mexico (Figure 10.8). Some of these monarchs migrate back to their breeding sites in the spring—a round trip of 6500 kilometers (Malcolm, 1987).

Fish

Many species of fish are also outstanding migrators. European eels (*Anguilla*) typically take two-and-a-half years to migrate to Europe from the Sargasso Sea, in the middle of the Atlantic Ocean, where they hatch. It is thought that they return to the Sargasso Sea to spawn and die. Salmon also leave the stream in which they hatched, and depending on the species, they may spend years at sea before returning to the very stream in which they hatched.

Reptiles

Among the many migrating reptiles is the green turtle (*Chelonia mydas*; Figure 10.9). Female green turtles feed in the warm marine pastures off the coast of Brazil and then swim roughly

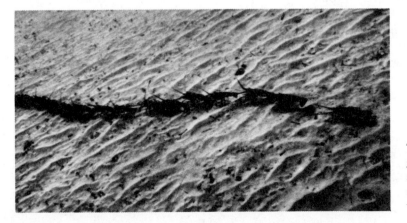

Figure 10.7 Spiny lobsters migrating single file between their summer homes in shallow water and their winter homes in deeper water.

a

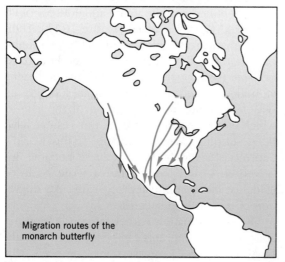

Migration routes of the
monarch butterfly

b

1800 kilometers to the sandy shores of Ascension Island, where they lay about 100 eggs, each the size of a golf ball. Ascension Island is a tiny target, only 8 kilometers long, in the middle of the Atlantic. During World War II, Ascension Island was used by airplane pilots to stop to refuel. Recognizing that a small navigational error could easily cause them to miss their critical pit stop, these pilots would comment, "If you miss Ascension, your wife gets a pension." The turtles, however, have no problem finding the island (Lohmann, 1992).

Birds

Birds are perhaps the best known group of migrators, but, that said, we will still stand amazed at their accomplishments. Consider the Arctic terns (*Sterna paradisaea*), which nest within 10 degrees of the North Pole in the barren Canadian Arctic. Where do they go to spend their winter? Some tropical retreat? No, indeed. Incredibly enough, they travel along the west coast of Africa to winter in the Antarctic. In the spring, they leave these chilly barrens to fly halfway around the world to the same nest sites they left in the Arctic. The distance is equal to the circumnavigation of the globe, and it is done by these fragile birds once each year.

Other long-distance migrators include the greater shearwater (*Puffinus gravis*) and bristle-thighed curlew (*Numenius tahitiensis*). The greater shearwater ranges over the vast expanses of the entire Atlantic Ocean and returns to the tiny isles of Tristan de Cuhna, more than 2400 kilometers from the nearest land. The bristle-thighed curlew nests in coastal Alaska and winters on Pacific islands 9700 kilometers away.

Figure 10.8 (*a*) Monarch butterflies roosting on trees in their overwintering sites in the mountains west of Mexico City. (*b*) The monarchs travel in dense clouds from northeastern North America to overwinter in Mexico or from central California to the coast.

Figure 10.9 The green turtle. One population migrates 1800 kilometers from its feeding ground off the coast of Brazil to Ascension Island.

The speed of some migrators is astounding. For example, a lesser yellowlegs was banded one August day in Massachusetts and was shot in Martinique 6 days later. This means that the bird traveled at least 444 kilometers per day.

Mammals

When we think of migrators, birds usually come to mind, but many mammals are also remarkably successful migrators (Figure 10.10). It is interesting that the journeys of migratory land mammals, including wildebeest, caribou, elk, and bison, are generally shorter than those of aquatic mammals. The mammal record for migratory distance is held by the gray whale, *Eschrichtius robustus*, which travels 10,000 kilometers from its feeding areas above the arctic circle to its breeding grounds near Mexico's Baja California.

METHODS OF TRACKING MIGRANTS

One of the first questions that arises regarding migratory animals is, where do they go? The best answers come from banding or tagging experiments. The animals are labeled in some fashion and the sites where they are recovered are noted. This technique has been used on almost every type of animal, including insects, fish, reptiles, birds, and mammals. An animal is caught and tagged or banded so that it can be individually identified and the name and address of an interested agency is given (Figure 10.11). Then the animal is released. When it is found, caught, or shot, the finder is requested to send the band or tag to the proper address. Information begins to build up, indicating the animal's range, its seasonal habitat, and something of the rate at which it can travel. Some of the results have been startling. Many members of the world's enormous army of birders have become involved in bird banding, and a large body of evidence has accumulated regarding the migration of certain bird species.

Other information comes from watching the animals. Some migrants can be followed visually. Daytime (diurnal) migrants may be seen with the naked eye or with the use of telescopic aids. At night, migrating birds have been counted by

Figure 10.10 Migrating caribou marching through Alaska in July. During the spring, many caribou breed in the tundra. Beginning in July, they migrate south where food will be available through the winter.

focusing a telescope on the moon and recording the birds that crossed the brightly lit background.

Some animals can be tracked using small transistorized transmitters that send radio signals to receiving equipment. In this way the exact movements of individual animals can be followed. It can be used on a wide variety of animals—from sea turtles (Figure 10.12) to polar bears. In some studies, satellites have been used to pick up signals from specially equipped animals and track their movements (Fancy et al., 1988; Nowak and Berthold, 1991).

One of the most valuable sources of information is radar. The "angels" that plague members of the military or flight controllers are often groups of migrating birds, and although no species identification can be made, migratory routes can be described as well as the heading, speed, altitude, and approximate number of the migrators.

ORIGINS OF MIGRATION

It might be well to ask how migration started. A common answer is that migration was initiated

a

b

Figure 10.11 We have learned a great deal about the migratory feat of animals from banding studies. Animals, such as this butterfly (*a*) or this bird (*b*), are banded or tagged, and the locations where they are later recovered are noted.

Figure 10.12 The movements of animals, such as this loggerhead sea turtle, can be tracked by equipping them with radio transmitters.

as a means of escaping undesirable conditions. For example, some say that some bird species may have evolved in northern regions when the climate there was tolerable all year. However, when ice-age glaciers began advancing and changing the climate drastically, the birds began their southward migration. They may then return to the northern breeding areas each year when the climate is more moderate (Sauer, 1963).

It is interesting that the same constant compass headings that guide birds from their North American breeding grounds to their wintering grounds in Central or South America today would have been even more effective 9,000 to 18,000 years ago, during the time when the glaciers were drastically changing the climate. So, it seems that these compass bearings may have evolved during the periods of full glaciation (Williams and Webb, pers. comm.).

Another hypothesis is that migratory pathways arose in response to continental drift. A species' breeding area originally might have been close to the areas where other activities took place. However, as the land masses drifted apart, the individuals and their descendants had to make longer and longer trips (Wolfson, 1948). However, this hypothesis is not easily applied to the origin of bird migration. The major movements of continents, those of a scale great enough to account for most avian migratory paths, took place millions of years before modern species of birds and their migration routes evolved (Amadon, 1948). The suggestion is more plausible as an explanation for the origin of the migratory circuit of other species, such as the green turtle. It has even been suggested that the migration of monarch butterflies originated with changes in the North American land mass due to continental drift. In this case, however, it is not thought to be the gradual separation of breeding and feeding areas, but the changing distribution of the exclusive food source of monarch larvae, milkweed, that is the causative factor of their migration (Urquhart, 1987).

It has also been suggested that migration in

some species is an adaptation that permits the exploitation of temporary or moving resources. As we have seen, monarch larvae feed only on milkweed. In regions of the eastern United States, however, the milkweed plants grow only during the spring and summer months (Urquhart, 1987). Certain species of insectivorous bats may migrate in response to the size of the insect supply. Mexican free-tailed bats (*Tadarida brasiliensis*), for instance, leave the southwestern United States as the harsh winter climate causes the insect supply to dwindle. They migrate to regions of Mexico where insects are available throughout the winter (Fenton, 1983). Wildebeest on the Serengeti also migrate following temporary resources. During the wet season, they graze in the open grasslands. The short-grasses found there are more easily digested and have higher concentrations of calcium and protein than taller grasses found elsewhere. However, the wildebeest must have drinking water and the water holes of the open grasslands evaporate during the dry season. At this time, then, the wildebeest migrate to the wooded grasslands where water is available (Kreulen, 1975; Maddock, 1979). It has also been suggested that certain birds evolved in the tropics and gradually extended their ranges into the north. However, they may have had to return to southern areas in the winter, when the northern climate became too harsh (Lincoln, 1950). For reviews of the hypotheses of how migration may have developed, see Dorst (1962), Cox (1985), or Terrill (1991).

WHY MIGRATE?

Why should an animal bother traveling hundreds or thousands of kilometers to one location only to return to its starting point half a year later? There are probably many answers to this question. No single explanation could apply to the diverse species of migrators. However, the simplicity of one explanation hides its profundity: Animals migrate because they leave more

offspring if they do. Not all species will have more reproductive success by migrating, and not all species migrate.

Although the actual costs and benefits of migration vary among species, the benefits must result in the production of more offspring than would be possible if the migrator stayed put. To explore this question further, we will consider some of the possible costs and benefits that might accompany migration.

Costs of Migration

Migration takes a tremendous toll. Only half of the songbirds that leave the coast of Massachusetts each year ever return and less than half of the waterfowl in North America that migrate south each fall return to their breeding grounds (Fisher, 1979).

One reason for these enormous losses is that traveling such long distances requires a great deal of energy. For instance, a bird uses about 6 to 8 times more energy when flying than when resting. Imagine yourself running 4-minute miles continuously for 80 hours. This would require the same amount of energy as a blackpoll's nonstop transoceanic flight of 105 to 155 hours (Williams and Williams, 1978). It is a good thing the blackpoll is not walking, though. It turns out that the energy costs of traveling a certain distance vary with the mode of locomotion. If each obtained the same amount of energy from each gram of fat, using 1 gram, a mammal could walk 15 kilometers, a bird could fly 54 kilometers, and a fish could swim 154 kilometers (Aidley, 1981). However, even if it were easier for a bird to travel great distances than it would be for you, migration is still metabolically demanding. It is not unusual for ships at sea to have migrants land on deck, exhausted and sometimes dying.

Many weary migrants fall to predators. For example, songbirds are "fast food" for the Eleanora's falcon. The songbirds, worn out by their flight across the Mediterranean, land in the nesting area of this falcon, which times its breeding so that its young will hatch when the songbirds arrive (Walter, 1979).

The cost of migration is high in many cases because large areas of inhospitable terrain must be passed. Not only do terrestrial birds pay a large cost for crossing the sea, but when a land area is unfamiliar, an animal may have problems finding food or water. And then there is the matter of obstacles. Birds often crash into tall structures such as lighthouses, skyscrapers, and TV towers. In a single night, seven towers in Illinois felled 3200 birds (Fisher, 1979).

Migration generally occurs during the spring and fall, times of notoriously unstable weather that can drastically raise the cost of migration. Severe rain- and snowstorms kill millions of migrating monarch butterflies. It is not uncommon to see thousands of dead or dying monarchs on the shores of Lake Ontario or Lake Erie following a severe storm (Urquhart, 1987). A flock of Lapland longspurs migrating through Minnesota encountered a sudden night snowstorm. The next morning, 750,000 of these small birds were found dead within 1 square mile (Emlen, 1972). In Manitoba, Canada, a spring storm killed so many warblers that dead birds were found about 8 inches apart (Fisher, 1979). Butterflies are even more sensitive to freezing temperatures. Although the monarch's Mexican wintering grounds usually remain above freezing, one frigid night killed over 2 million monarchs (Calvert and Brower, 1986).

In addition to such dangers, territorial animals, such as birds, must relinquish the rights to a hard-won territory each year and compete vigorously to become reestablished the following year.

So what rewards could possibly override such disadvantages?

Benefits of Migration

Energy Profit   We can intuitively understand the advantages of animals moving from approaching arctic winters to the sunny tropics. We can even see why animals might move shorter distances, for example from Nebraska to Oklahoma, for the winter. Only a cursory familiarity with the elements of nature could also convince us of the advantages of simply moving from a mountain top to a valley every winter. In each case the animals are trading a less hospitable habitat for a more hospitable one.

The severe weather during the northern winters has several consequences that may favor migration. Survival at low temperatures requires more energy. Marine mammals, such as whales, leave the frigid Arctic and Antarctic waters to winter in warmer tropical waters. Although food is not a great deal more abundant here, less energy is spent to maintain body temperature. The energy saved by reducing heat loss at least partially compensates for the energy spent on migration (Kinne, 1975). A slight variation of this argument has been offered to explain why monarch butterflies migrate to areas in Mexico where the temperature is usually just slightly above freezing, between 4° and 11°C, even though there are warmer locations. These overwintering sites do not reduce the monarchs' heat loss, but they do reduce metabolic rate. Warmer temperatures would unnecessarily elevate their metabolic rate (Calvert and Brower, 1986).

As winter approaches, increasing the animals' energy need, the food supply drops and forces any resident species into more severe competition for such commodities. So, in spite of the energy required for migration, such movement may result in an overall energetic savings. For example, a study of the dickcissel (*Spiza americana*) revealed that despite the energy costs of migration, this bird enjoys an energetic advantage from both its southward autumn migration and its northward spring migration. Studies of the junco (*Junco hyemalis*), white-throated sparrow (*Zonotrichia albicolis*), and the tree-sparrow (*Spizella arborea*) show that by avoiding the temperature stresses of northern winters, these species compensate for at least some of the energy spent on migration (reviewed in Dingle, 1980).

The question arises, then, if there is so much food in the warmer winter habitats, why do species migrate from such areas? Why do they return to their summer homes at all?

Reproductive Benefits   One answer might be that there are certain important advantages to rearing broods in the summer habitats. For example, days in the far north are long, and the bird's working day can be extended—they can bring more food to their offspring in a given period of time and perhaps rear the brood faster. Another result of long days is that more food is available for offspring and more young can be raised (Figure 10.13). It is known that, generally, the farther north from the tropics a species breeds, the larger is its brood (Welty, 1975).

In some species, migrations might enhance reproductive success by bringing members of the opposite sex together, increasing the chance of mating. For instance, this is a partial explanation for the return of salmon from the sea to their freshwater streams where they hatched. Other species migrate to areas that provide the necessary conditions for breeding or that offer some protection from predators. Gray and humpback whales, for example, breed in coastal bays and lagoons that provide the warmer temperatures needed for calving and help protect the calves from predation. The need for protected rookery sites may prompt seal, sea lion, and walrus migrations, as they come ashore on their traditional beaches after months at sea.

Reduction in Competition   Another advantage in returning to the temperate zone is that of escaping the high level of competition that exists in a warmer, more densely populated area. The annual flush of life in the temperate zones provides a predictable supply of food that can be exploited readily by certain species (Lack, 1968).

Figure 10.13   The arctic tern, a champion migrator, travels 36,000 kilometers each year in migratory flights. It breeds in northern regions where the days are long, allowing a parent to gather more food for its young.

Reduction in Predation   A third advantage to returning to temperate zones to breed lies in escaping predation. If predators are unable to follow herds of migratory ungulates, for instance, there would be fewer deaths due to predation. Thus, escape from predation has been suggested as the reason that the number of migratory ungulates is so much greater than the number of nonmigratory ungulates (Fryxell, Greever, and Sinclair, 1988).

In the far north, breeding periods are very short because of the weather cycles. This can be an advantage to nesting birds, which are in danger of falling prey to predators. The short season results in a great number of birds' nesting simultaneously. Thus, the likelihood of any single individual being taken by a predator is reduced. Also, since there is not an extended period of food availability for predators, their numbers are kept low. By leaving certain geographical areas each year, migratory species deprive many parasites and microorganisms of permanent hosts to which they can closely adapt. Long harsh winters in the frozen north act to reduce further the numbers of these threats.

Summary

Orientation to key aspects of the environment is critical to an animal's survival. The term orientation includes all the reactions that guide an animal into its correct posture, into its proper environment, or during migration and homing.

In some species, simple orientation responses maintain the individual's correct posture or guide it to the suitable surroundings. A kinesis, such as the humidity kinesis of wood lice, is an undirected response that brings the individual into a suitable environment. The response is proportional to the intensity of the stimulus. A taxis is a response in which the animal moves directly toward or away from the stimulus. The animal moves in the direction of most favorable stimulus intensity.

A dorsal light reaction keeps an animal's dorsal (back) side toward the light. A ventral light reaction orients an animal so that its ventral (belly) side is toward the light. In a light compass reaction, the animal assumes a certain angle with a light source.

Navigation describes an animal's ability to orient over long distances. One level of orientation, called piloting, is the ability to locate a goal by referring to landmarks. A second level of orientation is compass orientation, in which an animal orients in a particular compass direction without referring to landmarks. This is the type of navigation used by most bird migrants. A third level of orientation describes an animal's ability to locate the goal without the use of landmarks, even if it is released in an unfamiliar location. True navigation requires a map to determine location and a compass to guide the journey.

Migratory habits have evolved in a wide variety of species including arthropods, fish, reptiles, birds, and mammals. The migratory paths can be tracked by tagging or banding migrants, or by following them either visually or with the assistance of radar or radio transmitters.

Although we have no way of knowing how migration originated, there are many hypotheses that attempt to explain it. Animals may have moved south to escape the advancing ice-age glaciers. It has been suggested that animals may have migrated because of continental drift. Although this idea is unlikely to explain bird migration, it may account for the migration of green turtles. Birds may have moved south to avoid severe weather, or they may have moved north from temperate regions to avoid competition.

Whatever the origins of migration, it can only be maintained if the individuals that migrate leave more offspring than those that do not. Among the costs of migration are energy expenditure, predation, crossing unfamiliar and inhospitable terrain, and risk of exposure to severe weather. The advantages may include a favorable net energy balance. In spite of the energy cost of migration, an animal may gain both by escaping the metabolically draining harsh temperatures and by avoiding the increase in competition that accompanies a reduction in food supplies. They may also gain by a reduction in predation or parasitism.

CHAPTER
11

Mechanisms of Orientation

*T*he feats of migration are indeed astounding—an arctic tern circumnavigating the globe, a monarch butterfly fluttering thousands of miles to winter in Mexico, a salmon returning to its natal stream after perhaps years in the open sea. How do they do it? This falls into the category of a "good question," and there is no simple answer. Different species may use different mechanisms and any given species usually has several navigational mechanisms. Indeed, common themes in orientation systems are the use of multiple cues, a hierarchy of systems, and transfer of information among various systems (Able, 1991). When one mechanism becomes temporarily inoperative, a backup is used. Furthermore, a navigational system may involve more than one sensory system. These interactions can be quite complex, but we will simplify matters by considering each sensory mechanism separately.

Visual Cues

Visual mechanisms of orientation include the use of landmarks; celestial clues such as the sun, stars, or moon; and polarized light.

LANDMARKS

Landmark recognition is perhaps the most obvious way that vision may be used for orientation or navigation. Humans use landmarks frequently when giving directions: "Turn left before the bank" or "Make a right just after the gas station." Because the use of landmarks is so familiar to us, it is probably not too surprising to learn that many animals also use them to find their way.

Demonstrating Landmark Use

There are various ways to show that landmarks play a role in orientation. One way is to move the landmark and see whether this alters the orientation of the animal. Niko Tinbergen demonstrated that the digger wasp *Philanthus triangulum* relies on landmarks to relocate its nest site after a foraging flight. While a female wasp was inside the nest, a ring of 20 pine cones was placed around the opening. When she left the nest, she flew around the area, apparently noting local landmarks, and then flew off in search of prey. During her absence, the ring of pine cones was moved a short distance (1 foot) away. On each of 13 observed trips, the returning wasp searched the middle of the pine cone ring for the nest opening. However, she did not find the nest opening until the pine cones were returned to their original position (Tinbergen and Kruyt, 1938).

Another way of determining how important landmarks are to orientation is to prevent the animal from using them by moving the animal to an unfamiliar area. If landmark recognition is important to navigation, then it is expected that an animal transported into a previously unexplored area would have more difficulty finding home than it would within familiar terrain. It has been shown, for example, that the percentage of bank swallows (*Riparia riparia*) that return home is smaller when the birds are released 80 kilometers (50 miles) from home than when they are released closer to home (Sargent, 1962).

Animals can also be prevented from using landmarks by clouding their vision. Consider, for example, the ingenious way that Klaus Schmidt-Koenig and Hans Schlichte demonstrated that homing pigeons do not require landmarks to return to the vicinity of their home loft—they created frosted contact lenses for the pigeons (Figure 11.1). Through these lenses, pigeons could only vaguely see nearby objects and distant ones not at all. Nonetheless, the flight paths of these pigeons were oriented toward home just as accurately as those of control pigeons (Schmidt-Koenig and Schlichte, 1972). Thus, the pigeons cannot be depending on familiar landmarks to guide their journey home. Note that

Figure 11.1 Homing pigeons wearing frosted contact lenses are unable to use landmarks for navigation. However, these pigeons head home just as accurately as pigeons with normal vision. Therefore, although pigeons may use landmarks if they are available, they do not require them to home.

this does not mean that pigeons do not use landmarks when they are available, just that they can determine the homeward direction without them. Also, although pigeons with frosted lenses get to the general area of their home loft, they often cannot find the loft itself. Landmarks, then, may be important in pinpointing the exact loft location, but are not necessary for determining the direction of home.

Models of Landmark Use

Knowing that an animal uses landmarks to find its way does not tell us *how* those landmarks are used. Do other animals use landmarks as humans do, as part of a mental map of the area? Perhaps some species do, but others might use landmarks in different ways. A simple model of landmark use is that the animal stores the image of a group of landmarks in its memory, almost like a photograph of the scene. Then it moves about the environment until its view of nearby objects matches the remembered "snapshot."

Rudiger Wehner (1981) suggested that a whole series of memory snapshots might be filed in the order in which they are encountered. He added that invertebrates might be able to use landmarks by comparing the successive images of surrounding objects with a series of memory snapshots of the landmarks along a familiar route. For this technique to work, however, the animal must always travel along a familiar path, as do desert ants (*Cataglyphis bicolor*) (Figure 11.2). They follow one path during the outbound journey and another for the return, using a different set of memory images on the two legs of their trip (Wehner and Flatt, 1972). The use of two sets of memory images would be predicted by the model since the ant could never match the memorized images of landmarks viewed from a particular spot along its outward path with the image of the same landmarks viewed while returning. Furthermore, if an ant is experimentally displaced to a new spot within a familiar area, it does not head directly home. Instead it wanders about until it accidentally finds its usual route and then follows it home (Weh-

Figure 11.2 The desert ant uses a remembered sequence of landmark images to find its way home in a familiar area.

ner, 1983; Wehner and Raber, 1979). Hover flies also seem to use landmarks by following a series of memorized images (Collett and Land, 1975) .

Alternatively, animals might use landmarks by learning their relative positions, thereby forming a "cognitive map" of the area for reference. In this case, the animal could return directly home from a site *over a route it has never traveled* because it has a mental map of the area based on remembered relationships among familiar landmarks. This is a technique often used by humans. Even if you always walked to a friend's house along the same path, if you were dropped off in an unfamiliar spot but could see a familiar landmark, you would remember the position of your destination relative to that landmark and take a new trail to the house. Many other vertebrates are thought to use landmarks in this manner as well. As discussed in Chapter 5, rats that are permitted to explore an area without a food reward remember important cues and use them in subsequent rewarded tests (Tolman, 1948). Even certain fish may have mental maps of their habitat. For instance, the goby *Bathy-*

gobius soporator is a fish that lives in the intertidal zone and often becomes stranded in a pool during low tide. Although a stranded fish cannot see other tide pools, when it jumps between tide pools, its movement appears to be oriented toward an adjacent tide pool, as if it has a mental image of the relative positions of tide pools in its habitat. This image of the local topography could be learned as the fish swims over the area during high tide (Aronson, 1971).

More recently, James Gould (1986) questioned whether certain invertebrates, specifically the honeybee (*Apis mellifera ligustica*), might also form cognitive maps. He trained bees to follow a specific route to a feeding station. Foragers regularly visiting the feeding station were captured at the hive entrance and transported in the dark to a new location the same distance from the hive as the feeding station, but in a different direction. When released, the bees headed in the direction of the feeding station, exactly as would be predicted if they had mental maps of the landmarks in the area (Figure 11.3). Furthermore, the flight times of the bees indicated that they flew directly to the feeding station.

However, we should not be too quick to conclude that bees form mental maps. Similar experiments led Fred Dyer (1991) to conclude that the bees use landmarks associated with routes previously traveled. Furthermore, when Randolf Menzel (1989, Menzel et al., 1990) performed similar experiments with honeybees, he interpreted his results to indicate that bees use the sun for orientation. So, next we will consider the sun compass orientation of bees and a host of other animals.

THE SUN

Some animals are able to use the sun as a reference point when traveling to and from home by assuming some angle between their path and the sun. In a now classic mirror experiment, it was found that the ant *Lasius niger* orients its foraging forays in this manner (Santschi, 1911).

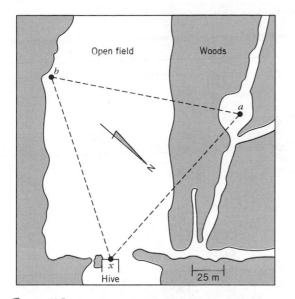

Figure 11.3 Bees may form a cognitive map of an area in which the relationship among the landmarks is known. To explore this possibility, bees were first trained to feeding site *a*. Later, they were captured as they left the hive and transported in the dark to site *b*. When released, the foragers flew directly to site *a*, even though it was hidden from view. This has been interpreted as indicating that bees have a mental map of the landmarks in a familiar area. (From Gould, 1986.)

The sun was screened from the ant's view and a mirror was used to reflect the sun from a different direction. The ant then shifted its path so that the path was at the same angle with the reflected sun as it had been with the real sun.

Many animals use the sun as a celestial compass, not just as a reference point. In other words, these animals can determine compass direction from the position of the sun. Because of the earth's rotation, the sun appears to move through the sky at an average rate of 15 degrees per hour. In the temperate latitudes of the Northern Hemisphere, for example, the sun rises in the the east and moves across the sky to set on the western horizon. The specific course that the sun appears to take varies with the latitude of the observer and the season of the year, but

it is predictable (Figure 11.4). Therefore, if the path and the time of day are known, the sun can be used as a compass.

Knowledge of one compass bearing is all that is necessary for orientation in any direction. Consider this simplified example. Suppose you decided to camp in the woods a short distance north of your home. As you headed for your campsite at 9 A.M., the sun would be in the east, so you would keep the sun on your right to travel north. However, during your homeward trek the next morning, you would keep the sun on your left to travel south.

The use of the sun for orientation is complicated by its apparent motion through the sky. The sun appears to move at an average rate of 15 degrees an hour. Therefore, an animal heading straight for its goal and navigating by keeping a constant angle between its path and the sun would, after 1 hour, be following a path that would be off by 15 degrees. Some species take only short trips, so errors due to the sun's apparent motion are inconsequential. These species do not adjust their course with the sun's. But if the sun is to be used as an orientation clue for a prolonged period, the animal must compensate for the sun's movement. To do this it must be able to measure the passage of time and correctly adjust its angle with the position of the sun. At 9 A.M. an animal wishing to travel south might keep the sun at an angle of 45 degrees to its left. By 3 P.M., however, the sun will have moved approximately 90 degrees at an average rate of 15 degrees an hour. To maintain the same southward bearing, the animal must now assume a 45-degree angle with the sun on its right. Time is measured by using a biological clock (see Chapter 9).

Evidence for Time-compensated Sun Compass

The first work on sun compass orientation was done on birds and bees in the laboratories of Gustav Kramer (1950) and Karl von Frisch (1950) respectively. Although these two investigative groups worked at the same time, neither

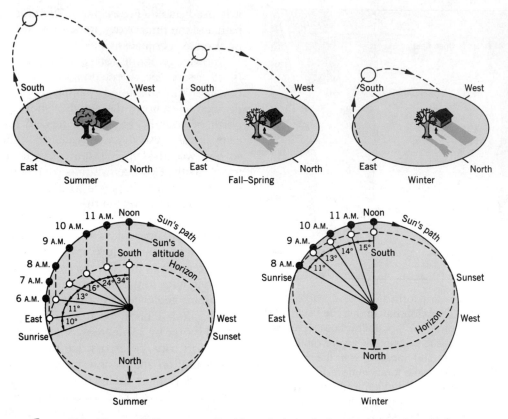

Figure 11.4 The sun follows a predictable path through the sky that varies with latitude and season. If the sun's course and the time of day are known, the sun's bearing (azimuth) provides a compass bearing. The sun appears to move across the sky at an average rate of 15 degrees an hour. Therefore, if the sun is to be used as a compass for a long time interval, the animal must compensate for its movement.

knew of the other's work. In spite of this, they often used similar experimental designs to reveal the details of sun compass orientation. We will take a closer look at the experiments of Gustav Kramer here, but if you want to compare these studies to those of von Frisch, consult von Frisch's fascinating book, *The Dance Language and Orientation of Bees* (1967).

Gustav Kramer (1949) began his studies by trapping migrant birds and caring for them in cages. He then noticed that they became restless at their normal migration season. Furthermore, most of their activity took place on the side of

the cage corresponding to the direction in which the birds would be flying if they were free to migrate. This activity has been aptly named migratory restlessness. In noting these tendencies, Kramer set the stage for a series of experiments that would yield valuable evidence in the quest for the navigational mechanisms of birds.

The indication that birds migrating during the day use the sun as a navigational clue was that the orientation (directionality) of migratory restlessness was lost when the sun was obscured from view. Kramer (1951) set up outdoor experiments with caged starlings, *Sturnus vulgaris*

Figure 11.5 Starlings are daytime migrators and were the subject of Gustav Kramer's pioneering work on bird navigation.

(Figure 11.5), which are daytime migrators, and found that they oriented in the normal migratory direction unless the sky was overcast, in which case they lost their directional ability and moved about randomly. When the sun reappeared, they oriented correctly again, suggesting that they were using the sun as a compass.

Then Kramer devised experiments in which the sun was blocked from view and a mirror was used to change the *apparent* position of the sun. The birds reoriented according to the direction of the new "sun."

Because migration occurs during limited periods in the fall and the spring, experiments using migratory restlessness to study orientation mechanisms are limited to two brief intervals a year. To eliminate this problem, Kramer devised an orientation cage in which there were 11 iden-

tical food boxes encircling a central bird cage (Figure 11.6). Kramer and his students trained birds to expect food in a box that lay in a certain compass direction. This ring of food boxes could be rotated so that a bird trained to get food in a given compass direction would not always be going to the same food box. This eliminated the possibility that the bird might learn to recognize the food dish by some characteristic, such as a dent. As long as the birds could see the sun, they would approach the proper food box. However, on overcast days the birds were often disoriented, as would be expected if they were using a sun compass.

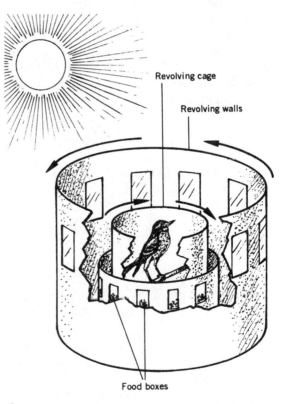

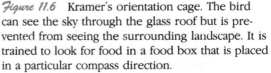

Figure 11.6 Kramer's orientation cage. The bird can see the sky through the glass roof but is prevented from seeing the surrounding landscape. It is trained to look for food in a food box that is placed in a particular compass direction.

The results of experiments with birds in Kramer's orientation cages not only confirm those on migratory restlessness (Kramer, 1951), but also indicate that the birds compensate for the movement of the sun. Actually, the idea of time-compensated sun compass orientation began when Kramer noticed that the birds in his orientation cages were able to orient in the proper direction even as the sun moved across the sky. When the real sun was replaced with a stationary light source, the birds continually adjusted their orientation with the stationary sun as though it were moving. The orientation with the artificial sun changed at a rate of about 15 degrees an hour, just as it would to maintain a constant compass bearing using the real sun.

The birds are able to compensate for the sun's apparent movement; therefore, they must possess some sort of independent timing mechanism. As we saw in Chapter 9, the biological clock that allows birds to compensate for the movement of the sun can be reset by artificially altering the light–dark regime. Initially the birds are placed in an artificial light–dark cycle that corresponds to the natural lighting conditions outside, the lights are on from 6 A.M. to 6 P.M. The light period is then shifted so that it begins earlier or later than the actual time of dawn. For example, if the animal is exposed to a light–dark cycle that is shifted so that the lights come on at noon instead of 6 A.M., the animal's biological clock is gradually reset. In this case, the animal's body time would be set 6 hours later than real time. Therefore, if the biological clock is used to compensate for the movement of the sun, orientation should be off by the amount that the sun had moved during that interval. In this example, orientation should be shifted 90 degrees (6 × 15°) clockwise (e.g. west instead of south) (Figure 11.7).

One of Kramer's students, Klaus Hoffmann (1954), was the first to use the clock-shift experiment to demonstrate the involvement of the biological clock in sun compass orientation. Af-

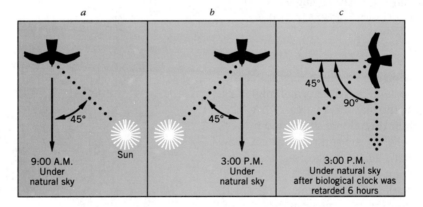

Figure 11.7 A clock-shift experiment demonstrates time-compensated sun compass orientation. (*a*) The flight path of a bird flying south at 9 A.M. might be at an angle 45 degrees to the right of the sun. (*b*) By 3 P.M., the sun would have moved roughly 90 degrees, so to continue flying in the same direction, the bird's flight path might be at an angle 45 degrees to the left of the sun. (*c*) If the bird's biological clock were delayed by 6 hours, and the bird's orientation tested at 3 P.M. (when the bird's body time was 9 A.M.), it would orient to the west. The flight path of the bird would be determined by the bird's biological clock. The flight path would, therefore, be appropriate for 9 A.M. and orientation would be shifted by 90 degrees clockwise. (From Palmer, 1966.)

ter resetting the internal clock of starlings by keeping them in an artificial light–dark cycle for several days, the birds' orientation was shifted by the predicted amount.

Development of the Sun Compass

In some species, at least, the relationship among time of day, the position of the sun, and compass direction is learned. Young homing pigeons reared in a light–dark cycle that is 6 hours delayed from the natural one and allowed to see the sun during the afternoon (morning according to their biological clocks) do not show the expected 90-degree clockwise deviation in their orientation. Apparently, they calibrate their sun compass according to their biological clock. As long as they are tested with their biological clock set 6 hours behind real time, they orient correctly. However, when they are exposed to a natural photoperiod for 5 to 6 days and then tested, their orientation is shifted by 90 degrees counter clockwise. The natural photoperiod is 6 hours ahead of the setting on their internal clock. The birds had learned the relationship among body time, the position of the sun, and geographic direction (W. Wiltschko, Wiltschko, and Keeton, 1976).

How does a bird learn that the sun rises in the east? What reference system is used to link the position of the sun with geographic direction so that it can be used as a compass? We do not know the answer yet, but there are suggestions. One idea is that the pattern of polarization of light in the sky is used to calibrate the sun compass. As will be described in more detail later in this chapter, this pattern of polarization depends on the position of the sun and, therefore, seems to move through the sky as the sun does. The centers of rotation of the polarization pattern are at the north and south poles and, therefore, these centers could serve as reliable indicators of geographic direction (Brines, 1980; Phillips and Waldvogel, 1988). Another suggestion is that the earth's magnetic field is the ref-

erence system by which the sun compass is set. As we shall see later in this chapter, there are several potential ways that geographic direction could be obtained from the earth's magnetic field. Wolfgang Wiltschko and coworkers have provided some support for the idea that the earth's magnetic field is involved in the learning process that sets the sun compass. They raised homing pigeons in a magnetic field that was rotated either 60 or 120 degrees clockwise. The young birds could see the sun only in an abnormal relationship to the magnetic field. When the pigeons were released for their first flight in sunshine, their orientation was shifted slightly clockwise (W. Wiltschko et al., 1983).

Ecological Considerations

Although sun compass orientation has now been demonstrated in a host of both invertebrate and vertebrate species, we should not assume that it is the primary means of orientation in all species. Rather, sun compass orientation must be demonstrated for each species independently because even closely related species may use quite different orientation mechanisms. The specific mechanism is generally influenced by the kind of environment in which the species lives.

We see the influence of the environment on the evolutionary choices of orientation mechanisms in two species of shore-dwelling isopods, *Tylos latreillii* and *Tylos punctatus* (see Herrnkind, 1972). Both species make daily migrations perpendicular to the shoreline between their burrows and feeding areas, but their movements are in opposite directions. *T. latreillii* burrows are in moist areas and they move into the drier sandy areas to feed. When displaced into a dry area, *T. latreillii* will escape toward the water, their movements guided by a celestial compass (Pardi, 1954). In contrast, *T. punctatus* show no sun compass orientation. When they are displaced, they too move toward home. However, their escape movements are directed uphill; they move up a slope as slight as 3 degrees (Hamner,

Smyth, and Molford, 1968). *T. punctatus* lives lower on the beach where the action of the waves and tides ensures a consistent slope. The higher stretch of beach where *T. latreillii* is active is so uneven that the general slope of the land is masked, making the sun a more reliable orientation cue.

The Sun and Nocturnal Orientation

We have seen how the sun is used for orientation during the daytime, but it is interesting to note that some nocturnal migrants select their flight direction by the point of sunset. They then associate this point with other cues that will be available to guide their flight throughout the night, such as the wind direction, stars, or the earth's magnetic field (reviewed in Moore, 1987a).

Although Kramer (1950) noticed that the nocturnal restlessness of his birds was oriented more precisely if they were put into their cages before sunset, the first experiments designed to determine the role of the setting sun were performed by Frank Moore (1978, 1980) on the savannah sparrow (*Passerculus sandwichensis*). He found that the nocturnal restlessness of these birds was well oriented if they could see the sunset, even if they were prevented from viewing the stars. However, orientation deteriorated when they could see the stars, but not the sunset. The activity of several other nocturnal migrants is also more oriented when they are permitted to see the sunset (white-throated sparrows, *Zonotrichia albicollis* [Bingman and Able, 1979; Lucia and Osborne, 1983]; tree sparrows, *Spizella arborea* [Cherry, 1984]; and the European robin, *Erithacus rubecula* [Katz, 1985]).

STAR COMPASS

Many species of bird migrants travel at night. Even if they set their bearings by the position of the setting sun, how do they steer their course throughout the night? One important cue is the stars. This was first demonstrated by Franz and Eleonore Sauer (Sauer, 1957, 1961; Sauer and Sauer, 1960). They exposed several species of sylviid warblers in cages to the autumn night sky of Germany. The Sauers' birds had been hand reared indoors and had never seen the sky, but after one glimpse at the stars, they oriented southward. When the test was repeated in the spring, they oriented northward. Here was evidence that the night sky could provide navigational clues.

Franz Sauer then began a series of experiments aimed at discovering just which objects in the nighttime sky the birds use as clues. He moved his caged birds inside the Bremer planetarium so that the nighttime sky could be controlled. He first lined up his planetarium sky with the sky outside and found that the birds oriented themselves in the proper migratory direction for that time of year. Then the lights were turned out and the star pattern of the sky was rotated. The birds continued to orient according to the new direction of the planetarium sky. When the dome was diffusely lit, the birds were disoriented and moved about randomly. In some experiments, even though the moon and planets were not projected, the birds oriented correctly; apparently taking their bearings from the stars. Since the Sauers first demonstrated the existence of star orientation, the list of species known to navigate in this way has grown considerably.

We know the most about the mechanism of star compass orientation in the Indigo Bunting (*Passerina cyanea*). Our knowledge has been gained primarily through Stephen Emlen's systematic planetarium studies. These indicate that the Indigo Bunting relies on the region of the sky within 35 degrees of Polaris (Figure 11.8). Since Polaris is the pole star, it shows little apparent movement and, therefore, provides the most stationary reference point in the northern sky. The other constellations rotate around this

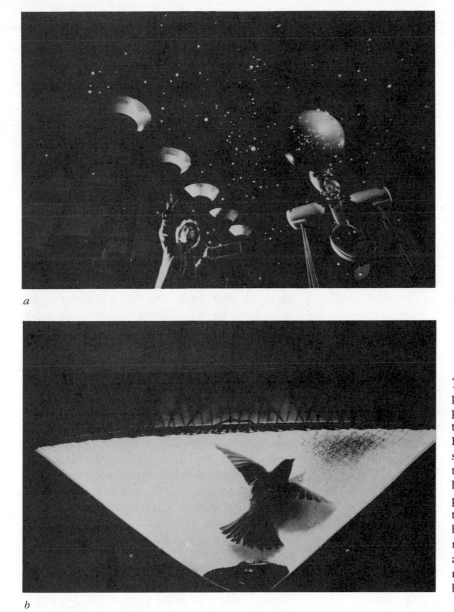

a

b

Figure 11.8 (*a*) Star compass orientation was explored by exposing nocturnal migrants, Indigo Buntings, to a planetarium sky. During the normal time of migration, caged birds will flutter in their proper migratory direction if the stars are visible. (*b*) In some studies, a record of activity was created on the sides of a funnel-shaped cage by the bird's inky feet.

point (Figure 11.9). The stars nearer Polaris move through smaller arcs than do those farther away, nearer the celestial equator. The birds learn that the center of rotation of the stars is in the north, information that is used to guide their migration either northward or southward. The major constellations in this region are the Big Dipper, the Little Dipper, Draco, Cepheus, and Casseopeia. Experiments have shown that it is not necessary that all these constellations be vis-

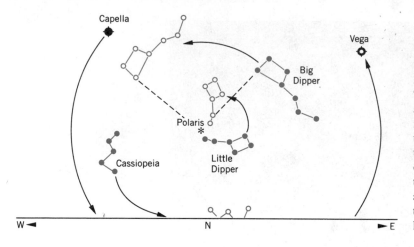

Figure 11.9 The stars rotate around Polaris, the North Star. The center of rotation of the stars tells birds which way is north. The changing star positions in the northern sky during the spring are shown here. The closed circles indicate star positions during the early evening and the open circles indicate the positions of the same stars 6 hours later.

ible at once. If one constellation is blocked by cloud cover, the bird simply relies on an alternative constellation (Emlen 1967a,b).

Young birds learn that the center of rotation of stars is north. The axis of rotation then gives directional meaning to the configuration of constellations. Once their star compass has been set in this way, the birds need not see the constellations rotate. Simply viewing certain constellations is sufficient for orientation. This was first demonstrated by exposing groups of young Indigo Buntings to normal star patterns in a planetarium sky. One group saw a normal pattern of rotation of stars, one that rotated around Polaris. The other group viewed the normal pattern of stars, but instead of rotating around Polaris, they rotated around Betelgeuse, a bright star closer to the equator. When the birds came into migratory condition, their orientation was tested under a stationary sky. Although each group was headed in a different geographic direction, both groups were well oriented in the appropriate migratory direction relative to the center of rotation they had experienced, either Betelgeuse or Polaris (Figure 11.10). In other words, in the autumn, when the birds would be heading south for the winter, those that had experienced Betelgeuse as the center of rotation interpreted the position of that star as north and headed away from it (Emlen, 1969, 1970, 1972).

The development of the star compass has been studied in only a few species other than the Indigo Bunting. The garden warbler (*Sylvia borin*; W. Wiltschko, 1982; W. Wiltschko et al., 1987c) and the pied flycatcher (*Ficedula hypoleuca*; Bingman, 1984) also learn that the center of celestial rotation indicates north.

MOON COMPASS

Since the moon is much more prominent in the night sky than are the stars, it might seem that it is the most obvious object by which to orient during the night. However, there are problems associated with the use of the moon for orientation. First, it is visible on only half of the nights each month. Second, it is a moving beacon and, as such, requires constant compensation. Third, it moves more slowly than does the sun; a lunar day is 24 hours and 50 minutes per cycle. Therefore, any animal that also used the sun would have to have two biological clocks.

Only a few species are known to have a moon compass. The best known are beach hoppers, such as *Talitrus saltator* (Papi and Pardi, 1963), *Orchestoidea corniculata*, and *Talorchestia martensii* (Enright, 1972). These small crustacean amphipods live near the shore (Figure 11.11). They bury themselves in the damp sand to escape the drying sunlight of midday. When the

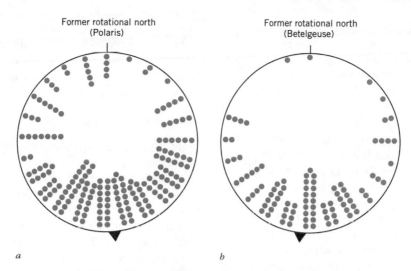

Former rotational north
(Polaris)

Former rotational north
(Betelgeuse)

a   *b*

Figure 11.10   The orientation of Indigo Buntings to a stationary planetarium sky after exposure to different celestial rotations. During their first summer, Indigo Buntings learn that the center of celestial rotation is north. This was demonstrated by exposing a group of young birds to a planetarium sky that (*a*) rotated around Polaris (the North Star) or (*b*) around Betelgeuse. During their first autumn, when they would be migrating south, they were exposed to a stationary planetarium sky. Each dot is the mean direction of activity for a single test. The arrow on the periphery of the circle is the overall mean direction of activity. Each group oriented away from the star that had been the center of rotation. (Modified from data from Emlen, 1970; Able and Bingman, 1987.)

Figure 11.11   The beach hopper, *Talorchestoidea*, displays time-compensated moon compass orientation.

humidity is higher in the late afternoon, they emerge and hop about on the beach, often traveling inland the length of a football field. Then, before the morning sun can dry them, they must find their way back to the moist sands near the shoreline. This journey is oriented to the sun by day and the moon by night.

The moon compass has been demonstrated by collecting the beach hoppers when they were inland and placing them in a glass vessel that was screened so that the sky was visible but the landscape was not. The dryness of the vessel stimulated escape movements, which were directed toward the sea on moonlit nights. However, on moonless nights or when the moon was blocked from view, the beach hoppers dispersed randomly. When mirrors were used to alter the apparent position of the moon, the beach hoppers reoriented appropriately. They even compensated for the movement of the moon across the night sky at the proper rate to remain headed toward the sea.

There is also some evidence of moon compass orientation in a terrestrial isopod (*T. latreillii*; Pardi, 1954), the wolf spider (*Arctosa variana*; Tongiorgi, 1970), the yellow underwing moth (*Noctua pronuba*; Sotthibandhu and Baker, 1979), the southern cricket frog (*Acris gryllus*; Ferguson, Landreth, and Turnipseed, 1965), and sockeye salmon (*Oncorhynchus nerka*; Brannon et al., 1981). Evidence that birds can use the moon as a compass is limited and circumstantial. Some species, such as savannah sparrows (*P. sandwichensis*), do not seem to use the moon to maintain a preferred direction (Moore, 1987b).

POLARIZED LIGHT AND ORIENTATION

One of the puzzling facets of sun compass orientation is that many animals continue to orient correctly even when their view of the sky is blocked. How is this possible? For at least some of these animals, there is another celestial orientation cue available in patches of blue sky. This cue is polarized light. Before considering orientation to polarized light, it might be useful to become familiar with the nature of polarized light and how the pattern of skylight polarization depends on the position of the sun.

The Nature of Polarized Light

Light consists of many electromagnetic waves, all vibrating perpendicularly to the direction of propagation (Figure 11.12). As a crude analogy, think of a rope held loosely between two people as a light beam. The rope itself would define the direction of propagation of the light beam. If one person repeatedly flicked his or her wrist, the rope would begin to wave or oscillate. These oscillations would always be perpendicular to the length of the rope, but they could be vertical, horizontal, or at any angle in between. The same is true of light waves. Most light consists of a great many waves vibrating in all possible planes perpendicular to the direction in which the wave is traveling. Such light is described as being unpolarized. In fully polarized light, however, all waves vibrate in only one plane. Our rope light beam, for instance, would become vertically polarized if one person's wrist was flicked only up and down.

As sunlight passes through the atmosphere, it becomes polarized by air molecules and particles in the air, but the degree and direction of polarization in a given region of the sky depend on the position of the sun. In other words, there is a pattern of polarized light in the sky that is

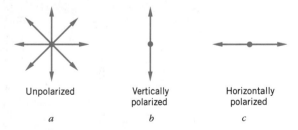

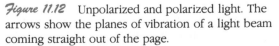

Figure 11.12 Unpolarized and polarized light. The arrows show the planes of vibration of a light beam coming straight out of the page.

a

b

Figure 11.13 The pattern of skylight polarization at (*a*) 9 A.M., (*b*) at noon and (*c*) at 3 P.M. The darker region of the sky is the band of maximum polarization. The diagrams below show the pattern of polarization (*d*) with the sun on the horizon (*e*), at 45 degree elevation, and (*f*) at zenith. The arrows indicate the direction of the plane of polarization. The small circle denotes the position of the sun. The pattern of polarization depends on the position of the sun. The blue sky provides an orientation cue for animals that can perceive the plane of polarization.

c

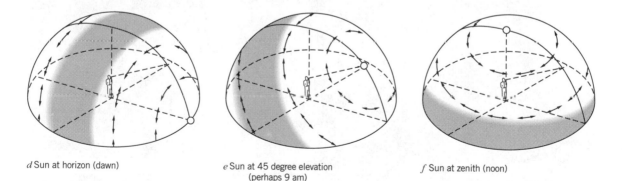

d Sun at horizon (dawn)

e Sun at 45 degree elevation (perhaps 9 am)

f Sun at zenith (noon)

directly related to the sun's position (Figure 11.13). One aspect of this pattern is the degree of polarization. To picture the pattern of polarization, think of the sky as a celestial sphere with the sun at one pole and an "antisun" at the other. The light at the poles is unpolarized, but it becomes gradually more strongly polarized with distance from the poles. Thus, between the sun and the antisun, there is a band where the light in the sky is more highly polarized than in other regions. This region is described as the band of maximum polarization. But there is more to the pattern than this—the direction of the plane of polarization (called the e-vector) also varies according to the position of the sun. The plane of polarization of sunlight is always perpendicular to the direction in which the light beam is traveling. If you were to draw imaginary lines of latitude on the celestial sphere so that they formed concentric circles around the sun and antisun, these lines would indicate the plane of polarization at any point in the sky. Since the entire pattern of polarization of light in the sky is determined by the sun's position, the pattern moves westward as the sun moves through the sky (Waterman, 1989).

Uses of Polarized Light in Orientation

Polarized light reflected from shiny surfaces, such as water or a moist substrate, is used by certain aquatic insects to detect suitable habi-tat. Indeed, polarized light may actually attract them (Schwind, 1991). For the backswimmer, *Notonecta glauca* (Figure 11.14), the horizontally polarized light reflected from the surface of a pond is not only a beacon that helps the insect locate a new body of water during dispersal, it also triggers a plunge reaction that brings the insect closer to a new home (Schwind, 1983).

There are two possible ways in which the plane of polarization of the light in the sky is used as an orientation cue. First, polarized light is used as an axis for orientation. In other words, an animal might move at some angle with respect to the plane of polarization. Many animals use polarized light in this way. Salamanders living near a shoreline, for instance, can use the plane of polarization to direct their movements toward land or water (Adler, 1976). When a predator approaches from the shoreline, the grass shrimp (*Palaemonetes vulgaris*) darts quickly away from the shore toward deeper water, an

Figure 11.14 Many aquatic insects, such as this backswimmer, use polarized light reflected from water or a moist surface to locate an appropriate habitat. A backswimmer spends almost its entire life underwater. These insects are commonly seen in ponds, suspended beneath the water surface, as this one is. Adults can fly, however, and may disperse to a new pond before laying the second batch of eggs of the season.

escape response that is at least partially guided by the celestial pattern of polarization (Goddard and Forward, 1991). Polarized light may also be used by young sockeye salmon to maintain their proper migratory direction while traveling through rivers or lakes (Dill, 1971). Indeed, many species of fish are able to detect polarized light, although we do not yet know the ways they use this ability (Hawryshyn, 1992).

A second way the pattern of polarization of sunlight might be used is to determine the sun's position when it is blocked from view. This method is possible because the pattern of polarization depends on the position of the sun. Honeybees (*A. mellifera*) and desert ants do this in an amazingly simple manner. The receptors for polarized light are located in a small region near the top of the rim of the compound eye. The response of a given receptor depends on its alignment with the plane of polarization. The receptors are arranged so that if the insect is headed directly toward or away from the sun, the combined responses of these receptors are greater than when the insect is at any other angle relative to the sun. Therefore, all the insect must do is move around until the summed output of its receptors is maximum to be headed toward or away from the sun. The next step, determining which end of its body is facing the sun, is usually fairly easy. Having located the sun, the insect can then orient accordingly (Wehner, 1987).

The polarization of light in the sky could also provide an orientation cue at dawn and dusk, when the sun is below the horizon. It is known that many birds that migrate at night set their bearings at sunset. Apparently, the pattern of skylight polarization at sunset (Able, 1982) and at sunrise (Moore, 1986) is a cue assisting the orientation of birds migrating at these times because experiments have shown that the birds' directional tendencies are altered when the plane of polarized light to which they are exposed is experimentally shifted by rotating polarizing filters.

Magnetic Cues

CUES FROM THE EARTH'S MAGNETIC FIELD

Our earth acts like a giant magnet. To picture the geomagnetic field around the earth, imagine an immense bar magnet through the earth's core from north to south. However this bar magnet is tilted slightly from the geographic north–south axis, and the magnetic poles are shifted slightly from the geographic, or rotational, poles (Figure 11.15). The difference between the magnetic pole and the geographic pole is called the declination, or variation, of the earth's magnetic field. Because the declination is small in most places, usually less than 20 degrees, magnetic north is usually a good indicator of geographic north. The declination is, of course, greatest near the poles.

Several aspects of the earth's magnetic field vary in a predictable manner and could, therefore, provide directional clues. One aspect is polarity. The magnetic north pole is dubbed the positive pole and magnetic south, the negative pole. The second aspect is the course of the lines of force. These leave the magnetic south pole vertically, curve around the surface of the earth, then become level with the surface at the magnetic equator, and reenter the magnetic north pole going straight down. As a result, the lines of force have both horizontal and vertical components. The horizontal component runs between the magnetic north and south. The vertical component is the angle that the line of force makes with gravity. The vertical component, often referred to as the inclination, or dip, in the field is steepest near the poles and near zero at the equator. The third aspect that varies predictably is the intensity (or strength) of the geomagnetic field. It is greatest at the poles and least at the equator.

Thus we see that the polarity, inclination, and intensity of the earth's magnetic field vary systematically with latitude, providing three poten-

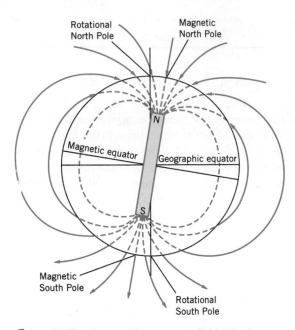

Figure 11.15 The earth's magnetic field. The lines of force leave magnetic south vertically, curve around the earth's surface and enter magnetic north heading straight down. The geomagnetic field provides several possible cues for navigation: polarity, the north–south axis of the lines of force, and the inclination of the lines of force. The magnetic compass of most animals appears to be an inclination compass. They determine the north–south axis from the orientation of the lines of force but assign direction to this by the inclination of the force lines. In the Northern Hemisphere, north is the direction in which the force lines dip toward the earth.

tial orientation cues. Which of these are used? Our own experience with compasses immediately brings polarity to mind. When the needle on your compass points north, it is responding to the polarity of the earth's field. Although a few animals, including salamanders (Phillips, 1986), salmon fry (Quinn, 1980), and bobolinks (Beason and Nichols, 1984), seem to respond to polarity, they are in the minority.

More commonly, animals appear to use the magnetic field inclination. They distinguish between "poleward," where the lines of force are steepest, and "equatorward," where the lines of force are parallel to the earth's surface. Although the horizontal component of the earth's field tells the animal the north–south axis, the vertical component (or inclination) is the cue that reveals whether it is going toward the pole or toward the equator.

The classic experiments on birds that demonstrate the use of an inclination compass are typical of those that have been done on other species. It has been shown, for example, that the migratory restlessness of European robins (*Erithacus rubecula*) remains oriented in the proper direction even when the birds have no visual cues. When the magnetic world that the birds experienced was reversed by switching the north and south poles of an experimental field, there was no effect on their orientation. However, the birds reoriented if the inclination in the experimental field was altered. It is interesting that these birds were not able to orient according to horizontal field lines, such as those that might occur around the equator. A bird could determine the north–south axis in a horizontal field, but without the inclination, it would not know which direction to travel to head toward the equator. Just how a migrant deals with this problem while in the equatorial regions is not known (W. Wiltschko and R. Wiltschko, 1972).

The results of an experiment on free-flying homing pigeons are also consistent with the idea that a bird's magnetic compass is based on the inclination of the magnetic lines of force. Small Helmholtz coil hats were fitted onto the heads of homing pigeons (Figure 11.16a). A Helmholtz coil is a device that generates a magnetic field when an electric current runs through it. The magnetic field experienced by the birds can be altered by reversing the direction of current flow through the coil. On cloudy days, when the pigeons relied on magnetic cues rather than their sun compass, they oriented as if they considered

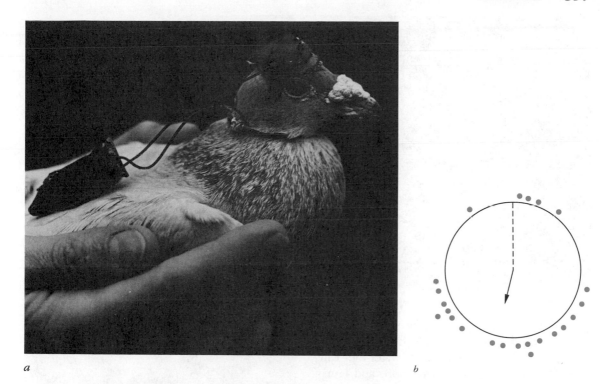

a

b

Figure 11.16 (*a*) A pigeon with a Helmholtz coil, a device that generates a magnetic field, on its head. (*b*) The magnetic field experienced by the pigeon can be altered by changing the direction in which the electric current runs through the coil. On overcast days, when the birds could not use the sun as a compass, the magnetic field influenced their orientation. They oriented as if they interpreted north as the direction in which the magnetic lines of force dip into the earth. Each dot indicates the direction in which a bird vanished from sight after being released. The arrow in the center indicates the mean vanishing bearing. (Modified from data of Walcott & Green, 1974.)

north to be the direction in which the magnetic lines of force dip into the earth. Those birds that experienced the greatest dip in the magnetic field in the north, as it is in the normal geomagnetic field, headed home. In contrast, the birds that experienced the greatest dip in the magnetic field in the south were misdirected by the reversed magnetic information and headed directly away from home (Figure 11.16b; Visalberghi and Alleva, 1979; Walcott and Green, 1974).

There are also some indications that animals respond to the small differences in the intensity of the geomagnetic field (Keeton, Larkin, and Windsor, 1974; Kowalski, Wiltschko, and Fuller, 1988; Southern, 1978). If changes in magnetic intensity can be sensed, the gradual increase in strength between the equator and the poles could also serve as a crude compass.

WIDESPREAD MAGNETIC EFFECTS ON ORIENTATION

Influences of magnetic fields on orientation are known in a wide variety of organisms. (See the Appendix for a discussion of the sensory receptor for magnetoreception.) Consider, for exam-

Figure 11.17   *Tritonia*, a marine mollusk, can detect and orient to weak magnetic fields.

ple, the mud-dwelling bacteria (*Aquaspirillum*) that follows the magnetic lines of force downward to the muddy bottom of the water they call home (Blakemore, 1975; Blakemore and Frankel, 1981; Frankel, 1981; Frankel and Blakemore, 1980).

Many invertebrates have also been shown to be sensitive to magnetic fields (reviewed by Brown, 1971). Among these is the marine snail, *Tritonia* (Figure 11.17). In the laboratory, these mollusks orient eastward in a normal geomagnetic field but move randomly when the field is canceled (Lohmann and Willows, 1987).

Among the invertebrates, however, the best known effects of magnetism are on honeybees. Magnetism influences the directional information conveyed in the dances that honeybee scouts perform to send other foragers off in the direction to the food source (Lindauer and Martin, 1968, 1972; Martin and Lindauer, 1977), and foraging bees pay attention to magnetic stimuli at the feeding site (Walker and Bitterman, 1985).

It has also been suggested that fish and mammals may navigate through the open waters using magnetic cues. Sockeye salmon, for instance, appear to have a magnetic compass (Quinn, 1980, 1982; Quinn and Brannon, 1982). As a

Figure 11.18   A hatchling loggerhead sea turtle. These turtles may use the earth's magnetic field to guide their travels through the open ocean.

loggerhead sea turtle (*Caretta caretta*; Figure 11.18) makes its way across the featureless Atlantic Ocean from the coast of Florida (perhaps to the Sargasso Sea and back), it may be guided by the earth's magnetic field, in addition to the pattern of waves (Lohmann, 1992). Laboratory experiments have revealed that loggerhead hatchlings use the earth's magnetic field for orientation. The hatchlings swim toward *magnetic* northeast in the normal geomagnetic field and continue to do so when the field is experimentally reversed (Figure 11.19; Lohmann, 1991).

The ocean floor is striped with bands of solidified lava that bear the record of the earth's magnetic reversals. Running generally parallel to the mid-ocean ridge, these bands form mag-

netic highways that could guide long-distance treks by marine organisms. It has been suggested, for instance, that the magnetic landscape may guide whale voyages. Studies on whale stranding in four diverse countries (Canada, New Zealand, England, and the United States) reveal a good correlation between the areas where apparently healthy whales make a navigational error serious enough to strand themselves alive and sites of magnetic anomalies (Kirschvink, Dizon, and Westphal, 1986; Klinowska, 1985, 1986). Of course, correlations are

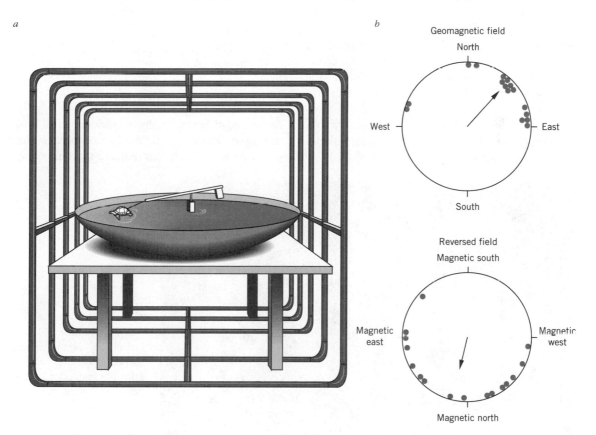

Figure 11.19 A demonstration of the ability of loggerhead sea turtle hatchlings to orient to magnetic fields. (*a*) A sea turtle is harnessed in a small tank so that its swimming direction can be determined. A coil that can alter the magnetic field experienced by the turtles surrounds the tank. (*b*) In the earth's magnetic field the turtles orient toward magnetic northeast. When the field is reversed, the hatchlings still orient to magnetic northeast, even though this is in the opposite geographic direction. (From Lohman, 1991.)

not firm evidence that whales use magnetic orientational cues, but the studies do lend credence to the suggestion that areas with confusing magnetic information may misguide whales. Certain terrestrial mammals, European woodmice (*Apodemus sylvaticus*) for example, also have a magnetic sense of direction (Mather and Baker, 1981).

Perhaps the most controversial claim of magnetic orientation is Robin Baker's (1980, 1981, 1985) report of the ability in humans. Baker took blindfolded people on a tortuous bus ride and then asked them to indicate the direction toward their starting location by written or verbal reports or by pointing. Altered magnetic fields created by bar magnets strapped to their heads or by Helmholtz coil helmets influenced the direction in which the subjects indicated the starting location (Figure 11.20). However, other researchers who attempted to repeat Baker's experiments reported that they did not find a similar ability of humans to orient, whether or not they were wearing magnets ((Fildes et al., 1984;

Gould, 1985; Gould and Able, 1981; Westby and Partridge, 1986; Zusne and Allen, 1981). This added fuel to the fires of skepticism. Baker's statistics and methodology were criticized by some (Dayton, 1985). Others accepted Baker's positive results but suggested alternative ways to explain the data (Adler and Pelkie, 1985). Thus challenged, Baker reanalyzed the data on human navigation and magnetoreception from all studies that had been performed. Based on his analysis of the combined data, Baker (1987) reaffirmed his conclusion that humans can orient without visual cues and that this ability is at least partially based on magnetic cues. The arguments and experimentation are likely to continue, but they are part of the excitement of science.

MAGNETISM AND BIRD ORIENTATION

As we have seen, there is evidence that birds use the earth's magnetic field as a compass. They seem to use the horizontal component of the

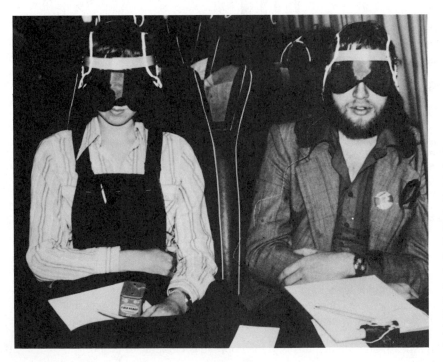

Figure 11.20 Humans wearing helmets with Helmholtz coils in experiments testing the influence of changed magnetic fields on human orientation. Robin Baker's claim that humans do not require visual cues for orientation and that the geomagnetic field may be the basis for this navigation has proven to be quite controversial.

field to identify the north–south axis and the vertical component (that is, the angle between the lines of force and gravity) to determine whether they are headed toward the pole or the equator. However, in spite of intensive research on geomagnetic sensitivity, the issue is still quite controversial. So, it may prove informative to consider the history of the idea.

The History of the Idea of Geomagnetic Sensitivity in Birds

When it was first suggested that birds might use the earth's magnetic field as an orientation cue (Yeagley, 1947, 1951), most people scoffed at the idea. However, their curiosity stimulated, many researchers explored the possibility. After one initial report that magnets attached to the wings of birds impaired their performance (Yeagley, 1947), there was a series of reports that failed to find evidence that birds could use magnetic cues for navigation (Gordon, 1948; Matthews, 1951; van Riper and Kalmbach, 1952; Yeagley, 1951).

The idea of magnetic orientation fell into disrepute until it was demonstrated that caged European robins could orient in the proper migratory direction without visual cues (Merkel and Fromme, 1958) and that this orientation was lost in a steel vault where the geomagnetic field was greatly reduced (Fromme, 1961). When the cages were enclosed in Helmholtz coils and the magnetic field shifted, the orientation of the birds was predictably altered (Merkel, 1971; Merkel and Wiltschko, 1965; W. Wiltschko, 1968). However, even this was greeted with skepticism because the effect of magnetism on orientation was apparent only after the data had been analyzed statistically.

Around this time William T. Keeton (1971) demonstrated that magnets could interfere with pigeon homing. He found that although pigeons released with bar magnets glued to their backs were frequently disoriented under cloudy skies, birds with brass bars of equal weight were not (Figure 11.21). Under sunny skies, magnets had no effect on the orientation of experienced birds, but they had a slight and erratic effect on first-flight pigeons. Thus, it seems that pigeons use the geomagnetic field as a backup compass when the sun is not available. Keeton noticed that besides the availability of the sun as a cue, factors such as the degree of homing experience of the birds, familiarity with the release site, and the distance to the release site influenced the degree to which the magnets disoriented the pigeons. These observations planted the seed of an idea that has now come to full flower: Animals have many orientation cues available for use under different circumstances. These cues are arranged hierarchically and interact in ways we are just beginning to understand (Able, 1991).

There is now a considerable body of research on the effects of magnetism on bird orientation. Much of it has yielded positive results, but there have also been some disturbing inconsistencies. Perhaps the most unsettling of these is that the results of Keeton's later studies do not confirm those of his initial experiments. When Bruce Moore (1988) analyzed the unpublished data collected by Keeton between 1971 and 1979, he could not find an effect of magnetism on the accuracy or speed of homing. It is not clear why Keeton's later data are inconsistent with his earlier studies. Moore suggests two possible explanations: (1) Some unknown factor masked the magnetic effect in the later studies; or (2) pigeons do not sense magnetic fields and, therefore, Keeton's initial results were due to some other factor. Either of these explanations could be correct, but there is also a third possibility. We now realize that pigeons use a variety of navigational cues that interact in some rather complicated ways. It will be some time before we completely understand the interactions among and hierarchy of navigational cues. Moore lumped birds with varying degrees of experience and familiarity with the release site together in his analysis. Did this treatment wash out the effects of magnets?

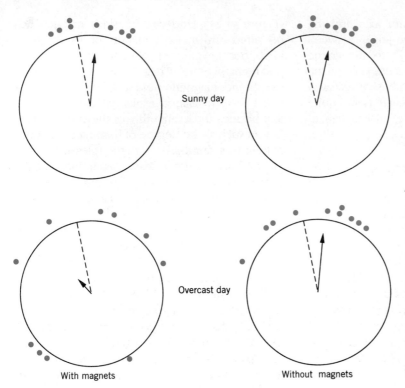

Sunny day

Overcast day

With magnets   Without magnets

Figure 11.21   The influence of magnetism on pigeon homing. Pigeons were released with either a magnet or a brass bar of the same weight on their back. On sunny days, the pigeons used the sun as a compass and homed accurately with or without a magnet. However, on cloudy days, the magnets disoriented the birds. Each dot represents the birds vanishing direction. (Modified from Keeton, 1971.)

Development of the Magnetic Compass of Birds

The magnetic compass of birds develops when they are very young. In fact, on their first flight, pigeons seem to rely on it exclusively. As the birds gain flying experience, however, they learn to use other cues. They begin to use the sun compass when they are about 3 months old, provided they have been able to see the sun (R. Wiltschko, 1983). Once established, however, the sun compass becomes the primary orientation mechanism (R. Wiltschko and W. Wiltschko, 1981), and the magnetic compass serves mainly as a backup system for cloudy days when the sun cannot be seen (Keeton, 1971). Even on sunny days, however, the magnetic compass is not turned off completely. Instead there is some interaction between the two systems (Visalberghi and Alleva, 1979; Walcott, 1977).

Wolfgang and Roswitha Wiltschko (1988a,b)

have noted that either the magnetic compass or the celestial rotation of the stars is sufficient as a directional cue in some birds. Using only magnetic information, handraised garden warblers (*Sylvia borin*; W. Wiltschko and Gwinner, 1974), pied flycatchers (*Ficedula hypoleuca*; Beck and W. Wiltschko, 1982), and savannah sparrows (*P. sandwichensis*; Bingman, 1981) were able to orient in their migratory direction. However, it has also been demonstrated that two of these species, the pied flycatcher (Bingman, 1984) and the garden warbler (W. Wiltschko, 1982), are able to orient properly without meaningful magnetic information if exposed to celestial rotation of the stars during early development. So a bird has two cues—magnetism and celestial rotation—to tell it which way is north.

Most of the time, these two cues are in agreement. However, recall that there are some places, particularly near the poles, where geographic and magnetic north do not coincide (that is, where the declination is large). When geographic north (which might be determined by celestial rotation) and magnetic north are not close together, the bird's two primary compasses are pointing in different directions. How does it know which way is really north? Apparently it resets its magnetic compass using visual information from either the day or night sky (Able and Able, 1990).

It is interesting to note that the sensitivity of the magnetic compass of birds corresponds to the strength of the earth's magnetic field. A bird generally does not respond to magnetic fields that are much stronger or weaker than the earth's magnetic field that is typical of the area in which it has been living. In fact, the range of intensities to which a bird may respond is usually narrower than those that it might experience during migration. However, it seems that the range of sensitivity may be adjusted by exposure to a field of a new strength for a period of time. Thus, responsiveness may be fine-tuned during migration (W. Wiltschko, 1978).

Is There a Magnetic Map?

In the previous chapter, we mentioned that true navigation requires not only a compass, but also a map. The map is necessary to know one's location relative to home and a compass is needed to guide that homeward journey. So far, we have discussed reasons for thinking that the earth's magnetic field serves as a compass for birds. But a bird such as the homing pigeon also needs a map. Although the nature of the pigeon's map is unknown, it has been popular to speculate on the possibility that the earth's magnetic field might be its basis (Gould, 1980, 1982; Lednor, 1982; Moore, 1980; Walcott, 1980, 1982). Indeed, there are characteristics of the geomagnetic field that could serve as a map.

Some of the magnetic effects on pigeon homing seem to be more than interference with the magnetic compass and, therefore, may support the idea of a magnetic map. One example is the disorientation of pigeons released in magnetic anomalies, places where the earth's magnetic field is extremely irregular. We might expect that a pigeon relying on the predictable changes in the geomagnetic field would become confused in areas where the field is abnormal. We do find that some magnetic anomalies disorient pigeons even under sunny skies, when presumably they would be using the sun as a compass (Frei, 1982; Frei and Wagner, 1976; Kiepenheuer, 1982; Wagner, 1976; Walcott, 1978). A perfect compass (the sun) cannot help if the map is messed up. This suggests that the geomagnetic field may be more than just a compass. As you can see in Figure 11.22, some birds released at magnetic anomalies appear to follow the magnetic topography, usually preferring the magnetic valleys where the lower field strength is closer to home values.

Another observation consistent with idea of a geomagnetic map is the shift in the initial bearings of pigeons that occurs when the geomagnetic field increases during magnetic storms (Keeton, Larkin, and Windsor, 1974; Frei, 1982). If the regular increase in the earth's magnetic field from the equator to the poles somehow told a bird its latitude, we would expect a magnetic storm would cause a bird to read its magnetic map incorrectly and use its compass to head off in the wrong direction.

As you can see, so far the evidence for a magnetic map is scanty and suggestive. Indeed, when all the evidence is considered, some people still doubt that the homing pigeon, at least, has a magnetic map (Walcott, 1991). However, research will continue, and it is the unanswered questions that make the study of animal behavior so fascinating.

Chemical Cues

In this section we will focus on the use of olfactory cues for orientation during homing. We will

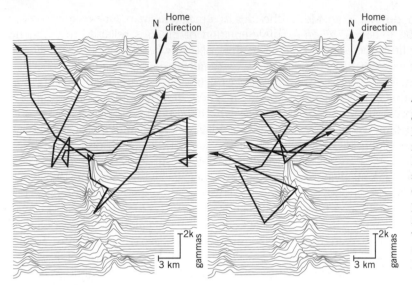

Figure 11.22 The flight paths of pigeons in magnetic anomalies. In some places the geomagnetic field is highly irregular. Pigeons released in these areas may be completely disoriented, even on sunny days. The paths of these pigeons seem to follow the magnetic valleys, where the field strength is closer to the value at the home loft. (From Gould, 1980.)

see that salmon are guided to their natal stream by chemical landmarks, and we will examine the more recent suggestion that pigeons may also use olfactory cues when homing.

OLFACTION AND SALMON HOMING

One of the most remarkable stories in the annals of animal behavior concerns the travels of the salmon. Salmon hatch in the cold, clear, fresh water of rivers or lakes and then descend from their natal streams and swim to sea, fanning out in all directions. Once they reach the ocean, depending on the species, they may spend 1 to 5 years until they reach their breeding condition. Now large, glistening, beautifully colored creatures, they head from their feeding grounds back through the trackless sea to the very river from which they came. When they reach the river, they swim upstream, always turning up the correct tributary until they reach the very one where they spent their youth.

Although navigation in the open seas appears to depend on the integration of several sensory cues, including magnetism, sun compass, polar-ized light, and perhaps even odors, navigation up the rivers is primarily based on olfactory cues (reviewed in Brannon, 1982). During their upstream migration, the salmon follow a chemical trail back to their natal tributary. When they come to a fork in the river, they may swim back and forth across the two confluent streams. If they mistakenly swim up the wrong branch and lose the scent of the home stream, they drift downstream until the scent is encountered again. Then, they usually take the correct route.

Sensory deprivation experiments have demonstrated the importance of olfaction in salmon homing. Blinding the fish had no effect, but plugging their nasal cavities impaired their ability to home correctly. Coho salmon (*Oncorhynchus kisutch*) were trapped shortly after they had made their choice of forks in a Y-shaped stream. The nasal cavities of half of those caught in each branch were plugged. The other half were untreated. All the fish were then released downstream from the fork and allowed to repeat their upstream migration. Whereas 89 percent of the control fish returned to the branch where they were originally captured, only 60 percent of the fish with nose plugs made the correct choice (Wisby and Hasler, 1954). In another study, a fish with its nose plugged swam with others of its kind to the opening of its home pond. However, unable to smell the special characteristics of its home waters, it did not enter the pond (Cooper et al., 1976).

Although it seems certain that salmon use olfactory cues when migrating upstream, the source of the specific stream odor is still debated. Two hypotheses have been suggested. One is that fish are guided to their natal stream by its characteristic fragrance from the rocks, soil, and plants. The fish imprint on (learn) the unique combination of stream odors when they are young and follow the scent to their natal stream (Hasler, Scholz, and Horral, 1978; Hasler and Scholz, 1983; Hasler and Wisby, 1951). This hypothesis was later extended by the suggestion that salmon actually learn the sequence of stream odors encountered during their outward migration (Brannon and Quinn, 1990; Harden-Jones, 1968). An alternative hypothesis is that the odors are pheromones (communicatory chemicals, see Chapter 17), mucus, or fecal material from other fish of the same species. In this case, the fish would not learn the characteristic odor of the home stream, but rather that of their colleagues (Nordeng, 1971, 1977).

In a series of experiments designed to test the imprinting hypothesis, Arthur Hasler and co-workers demonstrated that salmon are able to imprint on a chemical not usually found in natural waters and that they use this cue to locate their home stream. Groups of young salmon (smolts) were exposed to minute quantities of either morpholine or PEA (phenethyl alcohol). A third group, the controls, were not exposed to either chemical. The fish were marked and released in a lake that was equidistant from the two streams that would later contain the test chemicals. Eighteen months later the mature fish were ready to begin their upstream migration. Morpholine was then added to one of the rivers

and PEA to the other. All 19 of the streams along 200 kilometers of shoreline were monitored for marked fish. Only a small percentage of any of the marked fish returned, but those that did returned to the correct stream. Over 90 percent of those that returned to a morpholine- or PEA-scented stream had been previously exposed to that chemical. In contrast, the number of control fish visiting one of the scented streams was never more than 31 percent and was usually much lower (Scholz et al., 1976). In other studies, individual fish were followed during migration (Johnson and Hasler, 1980; Madison et al., 1973; Scholz et al., 1975). As you can see by the fish tracks shown in Figure 11.23, fish that had been imprinted on morpholine stopped and milled about in an area that had been treated with morpholine, but continued migrating through the same area when morpholine was absent. Fish that had not been previously exposed to morpholine continued swimming through the area even when morpholine was present. The same was true when PEA was used as the chemical label.

Other studies, however, make the pheromone hypothesis quite plausible. Coho salmon can discriminate their population (Quinn and Tolson, 1986) and even their siblings (Quinn and Busak, 1985; Quinn and Hara, 1986) using odor cues, and salmon are attracted to water bearing the scent of their own species (Hoglund and Astrand, 1973) or their own population (Selset and Doving, 1980). Water containing odors from spawning conspecifics is particularly attractive (Newcombe and Hartman, 1973). Furthermore, breeding Atlantic adult salmon (*Salmo salar*) suddenly appeared in a previously barren stream shortly after fry were added (Solomon, 1973). These studies suggest that odors from conspecifics may be one part of the symphony of odors characterizing the home stream, but they do not demonstrate that odors from other fish are important or necessary in guiding the upstream migration.

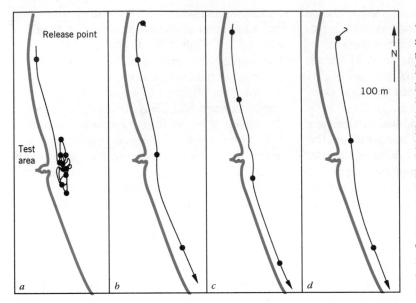

Figure 11.23 The paths of salmon under different conditions. The track of a salmon that had been imprinted on morpholine when it encounters the mouth of a river containing morpholine (*a*) and when morpholine was absent (*b*). The path in (*c*) is that of a salmon that was not imprinted on morpholine when encountering the mouth of a morpholine-containing river, (*d*) shows the movements of a morpholine-imprinted salmon when encountering the mouth of a river that contained another chemical. (From Hasler, Scholz and Horral, 1978.)

When both cues are present, salmon tend to return to the site they experienced as juveniles instead of an area containing the odor of conspecifics. In an experiment, groups of juvenile coho salmon were released as smolts. As adults, they returned to their release site. In order to reach this site, they had to swim within about 100 meters of a hatchery containing their siblings as well as other conspecifics. Although the hatchery's effluent contained odors from its captive fish, the migrating salmon ignored these conspecific odors and swam past the hatchery to their release site. Likewise, Atlantic char could not be lured into returning to a non-natal stream by the presence of adults (Black and Dempsey, 1986). Thus, site-specific odors seem to be more important to homing salmon than odors from conspecifics.

OLFACTION AND PIGEON HOMING

Although no one denies that olfactory cues are of paramount importance during the upstream migration of salmon, suggestions that olfactory cues form the map used by homing pigeon have been quite controversial (Papi, 1986; Schmidt-Koenig, 1987).

Models of Olfactory Navigation

Two models for olfactory navigation have been suggested. According to Floriani Papi's "mosaic" model, pigeons form a mosaic map of environmental odors within a radius of 70–100 kilometers of their home loft. Some of this map would take shape as the young birds experienced odors at specific locations during exercise and training flights. More distant features of the map would be filled in as wind carried faraway odors to the loft. One odor might be brought by wind from the north and another by wind from the east. The bird would associate each odor with the direction of the wind carrying it. When the wind shifted direction, the odors that arrived first would be closer than those that took longer to arrive (Papi et al., 1972). For instance, a hypothetical pigeon might learn that the sea is to the west, an evergreen forest is south, a large city is north, and a garbage dump is east. If the bird in this example smelled pine needles at its release site, it would assume that it was in the forest south of its loft and would use one of its compasses, perhaps the sun or the earth's magnetic field, to fly north.

Hans G. Walraff (1980, 1981) has suggested a "gradient" model of olfactory navigation that assumes there are stable gradients in the intensity of one or more environmental odors. Then, wherever it was, the bird would determine the strength of the odor and compare it to the remembered intensity at the home loft. Unlike the mosaic model, which requires only that the bird make qualitative discriminations among odors, the gradient model demands that the bird make both qualitative and quantitative discriminations. Reconsider the previous example. The smell of the ocean might form an east–west gradient and the fragrance of the evergreen forest might generate a north–south gradient. If the bird in the previous example smelled the air at a release site and determined that the scent of the sea was stronger but the smell of the forest was weaker than at the home loft, it would determine that its current position was northwest of home.

Tests of the Models

These models of olfactory navigation have stimulated intensive research, but it is still not clear how important odors are in the navigation of homing pigeons. Let us see how various researchers have approached the question.

Depriving Birds of Their Sense of Smell
One approach to testing olfactory hypotheses is to deprive the pigeon of its sense of smell and observe the effect on its orientation and homing success. These smell-blind (anosmic) pigeons are usually less accurate in their initial orientation and fewer return home. Olfactory deprivation disrupts orientation only when the bird is released at an unfamiliar release site. Once a

bird has made as few as one or two flights in a location, anosmia has no effect. Regardless of its effect on orientation, olfactory deprivation always delays the bird's departure from the release site.

Interestingly, the extent of the effect of olfactory deprivation varies. The homing of Italian birds is greatly impaired by olfactory deprivation (Benvenuti et al., 1973; Ioalè, 1983; Papi et al., 1971, 1972). Some German birds are affected (Snyder and Cheney, 1975), but most in Frankfort are not (R. Wiltschko and W. Wiltschko, 1987; W. Wiltschko, R. Wiltschko, and Jahnel, 1987a). American birds with an impaired sense of smell are less successful in reaching the home loft, but they usually head in the proper direction when leaving the release site (Keeton, Kreithen, and Hermayer, 1977; Papi et al., 1978).

Why are there differences in the effect of olfactory deprivation in different countries? Do Italian pigeons "smell" better than those in Frankfort or America? Do Italian pigeons form more vivid olfactory maps because they can rely on the presence of an odor, such as the scent of the sea? Or do the variations in the conditions at the lofts where they are raised and live make the difference? The Wiltschkos addressed this issue by raising birds of the same stock in the same area, but in different types of lofts (R. Wiltschko and W. Wiltschko, 1989; W. Wiltschko, R. Wiltschko, and Walcott, 1987b). In Italy, pigeons are usually kept in a rooftop loft where they are exposed to the winds. When the Wiltschkos maintained their birds "Italian" style, olfactory deprivation shifted the initial bearing of the pigeons so that they were less homeward oriented. However, when the birds were kept in ground level garden lofts that were sheltered from the wind, following the usual Frankfurt procedure, olfactory deprivation did not affect the orientation of pigeons (Figure 11.24).

It is tempting to conclude from the studies comparing the effects of olfactory deprivation on pigeons in different countries that olfactory

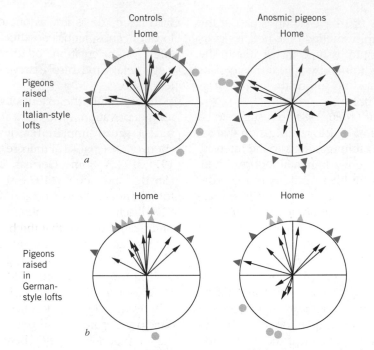

Figure 11.24 Orientation of control and anosmic (smell-blind) pigeons that had been maintained in wind-exposed Italian-style lofts (top) and wind-sheltered German-style lofts (bottom). Each arrow shows the mean flight vector during a single release. The arrow points in the mean direction of the flight path and indicates by its length the amount of variability in that path. The symbols at the periphery denote the direction and type of orientation during a release. Light blue triangles = homeward oriented; dark blue triangles = biased orientation. (Circles show responses that were not statistically significant.) Notice that the birds raised in Italian-style lofts were disoriented when deprived of olfaction, but those from German-style lofts were not. Notice also that the anosmic and control birds raised in Italian-style lofts made the same orientation mistakes, that is, showed the same release site biases. (From R. Wiltschko and W. Wiltschko, 1989.)

cues are more important to pigeons from Italian-style lofts than to those from Frankfort lofts. Indeed, this is a common conclusion (Waldvogel, 1989). However, nagging doubts remain. When this experiment was repeated in Italy following the same procedures used by the Wiltschkos, the results were different. Regardless of the conditions under which they were raised, none of the pigeons that had its sense of smell experimentally impaired oriented toward home

(Benvenutti et al., 1990b). Furthermore, the Wiltschkos (1989) question this interpretation because of a paradox in their results. Regardless of how they were raised, all their control birds, those with a normal sense of smell, made the same navigational mistakes. If olfaction is essential to pigeon navigation and if "Italian" pigeons rely on olfaction more than others, why didn't their sense of smell improve their homeward orientation? Furthermore, if "Italian" pigeons use olfactory cues more than "Frankfort" pigeons, we would expect the "Frankfort" pigeons to orient like the anosmic "Italian" pigeons. Paradoxically, they orient like the "Italian" controls. Besides its effect on the pigeon's sense of smell, perhaps olfactory deprivation affects another behavior, one not primarily controlled by olfaction, and this other behavior alters homing performance.

It may be that the difference in the effect of olfactory deprivation between American and Italian birds is due to regional differences in the navigational cues available during development. When eggs of American pigeons were hatched and raised in Italy, their homing was just as severely affected by anosmia as was the homing of Italian birds raised in the same loft. Thus, there do not seem to be genetic differences in the orientation mechanisms between Italian and American pigeons. The difference in their behavior, then, is likely to be due to differences in the cues available during development (Benvenutti et al., 1990*a*).

Distorting the Olfactory Map   Another way of testing olfactory hypotheses is to manipulate olfactory information to distort the bird's olfactory map. This has been done by deflecting the natural winds to make it appear that odors come from another direction. The deflector lofts used in these experiments typically have wooden baffles that shift windflow in a predictable manner (Figure 11.25). For instance, wind from the south might be deflected so that it seemed to come from the east. A pigeon in this loft would form an olfactory map that was shifted counterclockwise by 45 degrees. When it was released south of its loft, we would expect it to interpret the local odors as those east of its loft and fly west to get home.

Unlike the olfactory deprivation studies previously discussed, deflector loft experiments have shown consistent shifts in the orientation of homing pigeons in every country in which

tests have been conducted (Baldaccini et al., 1975; Kiepenheuer, 1978; Waldvogel and Phillips, 1982; Waldvogel et al., 1978). However, there are reasons to believe that the shift in orientation observed in pigeons from deflector lofts might be due to something other than a distorted olfactory map. We would expect pigeons that were temporarily prevented from smelling at the time of their release to be unable to read their olfactory map and to orient randomly. But this is not the case—the orientation of anosmic pigeons from deflector lofts is still shifted (Kiepenheuer, 1979). In light of this, it has been suggested that the baffles in these lofts also deflect sunlight and that the consistent shift in pigeon orientation is caused by an alteration in the sun compass (Phillips and Waldvogel, 1982; Waldvogel and Phillips, 1982; Waldvogel, Phillips, and Brown, 1988).

Manipulating Olfactory Information
Although the interpretation of olfactory deprivation and deflector loft experiments is quite controversial, the experiments in which olfactory information predictably alters the orientation of pigeons remain as unshaken support for an olfactory hypothesis. In one set of experiments, pigeons were exposed to air from one of four future release sites. After 2 hours of exposure, their sense of smell was blocked with a local anesthetic applied to the olfactory mucosae, and they were transported to one of the release sites. Only the pigeons that had been exposed to the air from their release site oriented toward home. Pigeons that had smelled the air from one of the other release sites disappeared randomly. These results make sense if we imagine that the pigeon assumes that the last odor it sniffed before its sense of smell was blocked came from its release site. Because the odor of that release site had been previously incorporated in the bird's olfactory map, it knew the correct direction to the home loft (Kiepenheuer, 1986).

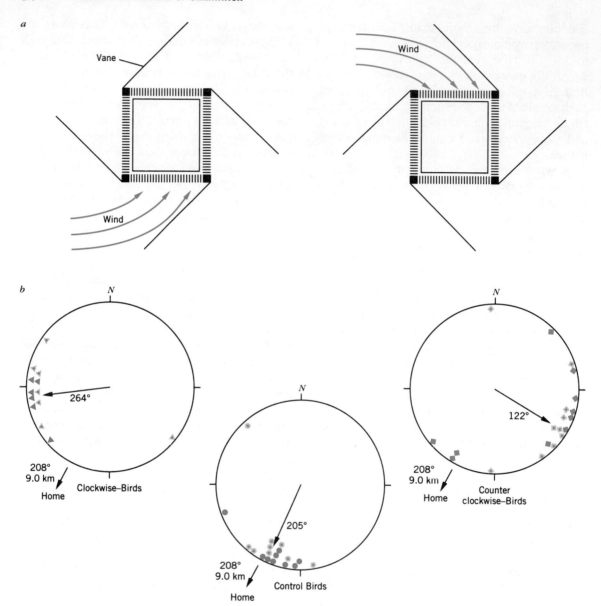

Figure 11.25 Deflector lofts shift the orientation of pigeons. (*a*) Deflector lofts have baffles that shift the apparent direction of the wind by 90 degrees. Pigeons living in deflector lofts should form shifted olfactory maps. (*b*) The vanishing directions of these pigeons is shifted by about 90 degrees. The dots at the periphery of the circle denote the direction in which the pigeon flew out of sight. The arrow within the circle indicates the mean bearing of all birds. The light and dark data points refer to the responses of bird released on different dates. Although a shift in orientation is reported in all deflector loft experiments, this shift may be due to deflection of light rather than a shift in the olfactory map. (Data from Baldaccini et al., 1975.)

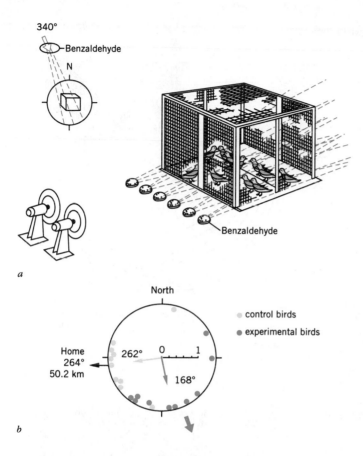

a

b

Figure 11.26 The results of an experiment manipulating a pigeon's olfactory information. (*a*) The experimental pigeons were kept in a loft that was exposed to natural odors as well as a breeze carrying the odor of benzaldehyde from a source northwest of the loft. Control birds were exposed to only natural odors. While they were transported to the release site, all birds were exposed to the odor of benzaldehyde. (*b*) The orientation of the experimental birds, but not the control birds, was altered by exposure to benzaldehyde. The initial orientation of control birds was homeward. However, the initial orientation of experimental birds was toward the southeast, as would be expected if they interpreted the odor of benzaldehyde as an indication that the release site was northwest of the loft. The experimental birds oriented as if they formed an olfactory map containing an area with the odor of benzaldehyde. (Data from Ioalè, Nozzolini, and Papi, 1990.)

In a similar experiment, the orientation of pigeons was influenced by their experience with an unnatural odor, benzaldehyde (Figure 11.26). Pigeons were kept in lofts where they were fully exposed to the wind. The experimental birds were exposed to an air current coming from a specific direction and carrying the odor of benzaldehyde in addition to the natural breezes. We would expect these pigeons to incorporate this information into their olfactory maps. The control birds were exposed to only the natural winds, so they would not have an area with the odor of benzaldehyde in their olfactory map. All the birds were exposed to benzaldehyde while they were transported to the release site and at that site. The experimental birds took off in a direction opposite to that from which they had experienced benzaldehyde at the loft. In other words, they oriented as if they used an olfactory map containing an area scented with benzaldehyde. If the release site did not smell of benzaldehyde, the experimental birds were homeward oriented. The control birds were not confused by the smell of benzaldehyde at the release site and flew home. Since benzaldehyde was not part of their olfactory map, they did not associate it with a particular direction. They used other cues to guide them home (Ioale, Nozzolini, and Papi, 1990).

Sound Cues and Echolocation

The world is a noisy place and most animals take advantage of some sounds for orientation.

Many, in fact, intentionally produce sounds that help others pinpoint the sender's position. Others unintentionally make noises as they move around and thereby reveal their location. For instance, the sounds of a scampering mouse guide the deadly strike of a barn owl.

Sounds from nonbiological sources may also convey directional information. Waves crashing on the shore or wind whistling through a mountain range might help tell an animal where it is.

Most animals must be fairly close to the source to be able to use these sound cues. However, such sources also generate low-frequency sound, infrasound, that could theoretically carry information on distant features of the landscape over thousands of kilometers to the animals that can hear these deep tones. If a migrant bird can hear infrasound, it could put this ability to good use. While flying high above the Mississippi Valley, it could conceivably hear a thunderstorm in the Rockies or the waves pounding the Atlantic coast. We know that pigeons are among those birds that can hear infrasound (Kreithen and Quine, 1979; Schermuly and Klinke, 1990), but we do not know whether they use this information for orientation.

Perhaps the most amazing use of sound for orientation is echolocation—a process by which the animal makes sounds and analyzes the returning echoes to create an acoustic picture of its surroundings (Figure 11.27). Because it is among the most fascinating uses of sound for orientation, we will focus on it in more detail.

ECHOLOCATION IN BATS

When making a list of things that go bump in the night, do not include bats. Many of us have marveled at bats darting between trees at daredevil speeds, catching tiny insects on the wing while cloaked in the darkness of night (Figure 11.28). Although not all species of bats use echolocation, most of those in the United States, Canada, and Europe do.

As a bat cruises through the air in search of a meal, it emits a series of high-frequency (ul-

Figure 11.27 An echolocating bat produces loud pulses of sound (indicated here as semicircles radiating outward from the bat's mouth) and then analyzes the returning echoes (indicated here as semicircles radiating from the insect toward the bat) to get an acoustic picture of its surroundings.

Figure 11.28 Bats dominate the night sky. In spite of the darkness of night, insectivorous bats perform intricate maneuvers to catch small insects in flight. They detect their prey and avoid colliding with obstacles by echolocation.

trasonic) pulses of sound. We cannot hear sounds at such a high pitch, so the hunting bat seems silent to us. Bat cries generally fall somewhere in the frequency range of 10 to 200 kilohertz. These very short wavelengths are ideally suited for detecting small- or medium-size insects because the wavelengths are not much longer than the prey. However, the use of high-frequency sounds is an evolutionary trade-off. They allow the bat to locate small prey but limit the range over which the prey can be detected because high-frequency sounds do not carry as well as those with low frequencies.

Although we cannot hear them, the cries of most species of bats are intensely loud. The softest (least intense) sound that most humans can hear is about 0.0002 dyne (a unit of energy) per centimeter[3]. Among the quietest of bats is the whispering bat, whose cry is about 1 dyne per centimeter[3]. This is roughly as loud as someone whispering at a distance of 10 centimeters. In contrast, the cries of other bats may be as loud as a nearby jet engine, 200 dynes per centimeter[3] (Fenton, 1983). So our quietest night might be a cacophony for other species.

A bat alters the rate of its cries during a hunt (Griffin, Webster, and Michael, 1960). When it is searching for prey, it produces about 20 sound pulses per second. This rate of pulsing allows the bat to get a general picture of its surroundings. But once an object is detected, the rate of pulsing is increased to roughly 50 to 80 cries per second and each pulse is shortened. By increasing the pulse rate, the bat gains more information characterizing its prey and can track it more accurately (Suga, 1990). Finally, just before the prey is captured, the pulse rate increases to 100 or even 200 pulses per second, depending on the species (Figure 11.29).

There is great variety in the structure of cries

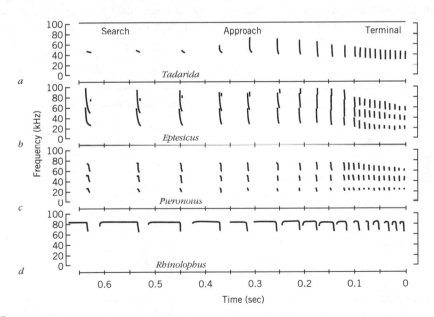

Figure 11.29 The increasing rate of bat echolocation calls at different stages of the hunt. During the search phase, the bat scans the area, producing about 20 sound pulses per second. When prey is detected, the pulse rate more than doubles so the bat can get a better picture and track the prey more accurately. During the final stages, the pulse rate increases dramatically to about 100 to 200 cries per second. (From Simmons, Fenton, and O'Farrell, 1979.)

from different species of bats, but the pulses fall into three general categories (Figure 11.30). One category is the frequency modulated (FM) calls. These sweep downward through a broad range of frequencies, usually covering at least one octave. The more frequencies in the outgoing pulse, the greater is the number of altered frequencies in the echo that can be analyzed to provide more information about the target. In other words, the amount of information about the target conveyed by the echo increases with the bandwidth of the call. The FM calls are, therefore, well suited to determine the size, shape, and surface texture of the target. Constant frequency (CF) calls, which contain a single frequency, comprise the second category of signals. Although this type of call does not reveal as many details about the target, it allows the bat to detect an object and determine whether it is moving toward or away from the bat and at what rate. CF calls allow a bat to detect an insect at greater distances than do FM calls. In addition, they allow bats to identify objects as insects by

their fluttering wings. The third category includes calls that combine a constant frequency component with a frequency modulated component. Typically these have a constant tone followed by a downward sweep (Neuwiller, 1984).

These types of calls are analyzed differently. The interval between the emission of an FM pulse and the return of the echo tells the bat the distance to the target. If the interval gets shorter, the bat knows it is getting closer (Simmons, 1973, 1979). In contrast, when a bat produces a CF pulse, it determines whether it is closing in on its prey by measuring the Doppler shift. The Doppler shift is the apparent change in frequency of a sound source that is moving relative to the listener (or vice versa). You have undoubtedly experienced the Doppler shift listening to the whistle of a passing train or the whine of a race car at the track. Although the actual pitch of the sound remains constant, it appears to become higher as the source moves toward you and to drop as the source moves away from you. This is because the pitch of the

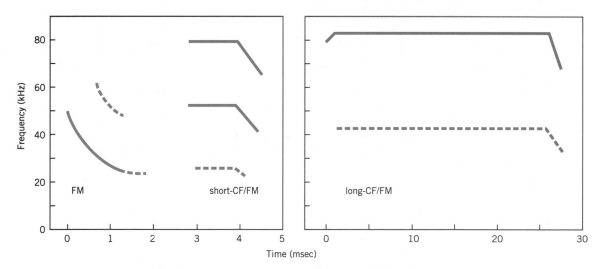

Figure 11.30 Three categories of bat echolation sounds. The sounds emitted by bats differ according to the species and the situation. (From Simmons, Howell, and Suga, 1975.)

sound you hear depends on the number of sound waves per second that strike the eardrum. You can imagine the sound waves bumping into one another and becoming compressed as the source moves toward you. For this reason you experience a higher frequency (pitch) sound as the source moves closer. As the source moves away, the sound waves are stretched out and the apparent pitch is lower. Bats using CF calls usually adjust the pitch of their cry so that the Doppler-shifted echo that returns is in the range of frequencies to which their ears are most sensitive (Schuller, Beuter, and Rubsamen, 1975; Simmons, 1974).

Most species of bats specialize in producing a single type of pulse, one suited to the style of foraging typical of the species, but some can adjust their echolocation strategy to the situation. The Mexican free-tail bat (*Tadarida brasiliensis*) forages in open spaces where insects can be detected against an uncluttered background and chased with little danger of bumping into obstacles. When searching for prey, this bat uses a CF pulse that is well suited for simply detecting the prey. The pulse becomes increasingly frequency modulated once the prey is detected. The FM signal provides information characterizing the insect that helps the bat decide whether to attack, and if the answer is yes, it guides the pursuit. Species that forage within vegetation must distinguish the echo of an insect fluttering by from the clutter of echoes from twigs and leaves. These bats typically use a signal composed of a long CF pulse with one or two FM components. With a pulse of this structure, the bat can identify its prey as a fluttering target amidst a background of stationary objects and track its movements. Other bats are gleaners that pick up prey from a surface. They must distinguish a slowly moving or stationary object against a highly reflective background. The fishing bat, *Noctilio leporinus*, detects ripples in the water caused by small surface-feeding fish. Its echolocation signals switch from CF or CF/FM pulses while searching for prey to an FM cry

once a fish has been detected. This echolocation strategy helps it notice a disturbance against a uniform background and then determine whether that disturbance is caused by a fish (Neuwiller, 1984, 1990; Simmons, Fenton, and O'Farrell, 1979; Simmons, Howell, and Suga, 1975).

We are most familiar with echolocation in bats, but there are several other groups of animals that share this ability (reviewed in Vaughan, 1986): the toothed whales and dolphins, several species of shrews, and a few species of birds (Oilbirds and Cave swiftlets).

Electric Cues and Electrolocation

Electric cues have a variety of potential uses for those organisms that can sense them. As we will see in the next chapter, certain predators use the electric cues given off by living organisms to detect their prey. In addition, electric fields generated by nonliving sources, such as the motion of great ocean currents, waves and tides, or rivers, could provide cues for navigation. Although there is currently no evidence that migrating fish such as salmon, shad, herring, or tuna are electroreceptive, there is some evidence that electric features of the ocean floor may help guide the movements of bottom feeding species, such as the dogfish shark (Waterman, 1989).

Although most living organisms generate weak electric fields in water, only a few species have electric organs that generate pulses that create electric fields that can be used in communication (see Chapter 17) and orientation. The electric organs of weak electric fish (mormyrids), located near their tail, for instance, generate a continuous stream of brief electric pulses. The result is an electric field around the fish in which the head acts as the positive pole and the tail as the negative pole. Any distortions in the field are detected by numerous electro-

Figure 11.31 Weakly electric fish, such as this African knife fish (*Gymnarchis niloticus*), generate an electric field that is used for communication and orientation.

receptors in the lateral lines along the sides of the fish. A weakly electric fish keeps its body rigid, a posture that simplifies the analysis of the electric signals (Figure 11.31).

The electric sense of these fish tells them what's up or, more accurately, what's down—the bottom of their watery home (Feng, 1977; Meyer, Heiligenberg, and Bullock, 1976). This information helps them keep the correct postural relationship with their environment.

These fish also examine their surroundings using their electric sense. Since they live in turbid waters where vision is limited, electrolocation is quite useful. Objects whose electrical conductivity differs from that of water disturb this electric field. An object with greater conductivity than that of water, another animal for instance, directs current toward itself. Objects that are less conductive than water, such as a rock jutting into its path, deflect the current away (Figure 11.32). The distortion varies depending on the location of the object relative to the fish. In this way, the fish can tell where the object is located (Bastian, 1981). Some species, such as *Gnathonemus*, get a sense of depth perception by alternately firing the electric organs on the right and left sides of their bodies (Chichibu, 1981).

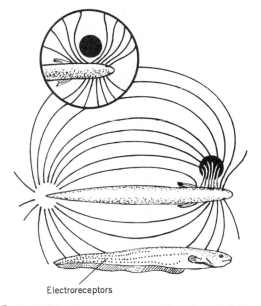

Electroreceptors

Figure 11.32 Electroreception. The electric field generated by this fish is distorted by objects. A conductor draws the lines of force together and an insulator spreads them out. The fish senses the changes in the electric field to "picture" its environment.

The electric sense is quite precise. Laboratory studies have shown, for instance, that *Gymnarchis* can distinguish a glass rod that is 4 millimeters in diameter from one that is 6 millimeters in diameter using its electric sense (Lissmann, 1963).

There are some features of the electric sense that are disadvantageous when compared to vision. First, the electric sense allows the fish to detect only nearby objects because the intensity of the field drops off quickly with increasing distance from the animal. Second, there is no lens to focus an electric image, so the pictures are blurred. However, certain electric fish sharpen the image by curving their body and tail around a novel object while swimming back and forth, encircling the object (Heiligenberg, 1977).

Summary

Animals have access to and use many different cues for orientation and navigation. The sensory modality of the primary cue for orientation varies among species and many species have a hierarchy of cues. Although the interactions among cues can be complex, we have considered each sensory basis separately.

Visual cues include landmarks; the sun, stars, or moon; and the pattern of skylight polarization. Methods of demonstrating that landmarks are used in navigation include moving the landmark to see whether the animal reorients or becomes disoriented, moving the animal to an unfamiliar area, or impairing the animal's vision so that landmarks cannot be used. Some species use landmarks by matching the objects viewed with the remembered image of the array of landmarks. When landmarks are used in this way, the animal must always follow the familiar path. Other species are able to form a mental map of an area by remembering the relationships among landmarks. These animals can follow a route that they have not previously traveled.

The sun may be used as a point of reference by assuming some angle relative to it during the journey and then reversing the angle to get home. Alternatively, since the sun follows a predicable path through the sky, if the time of day is known, the sun's position provides a compass bearing. If the sun is used as an orientation cue over a long interval of time, the animal must compensate for the sun's movement. Animals must learn to use the sun as a compass. The point of sunset is also an orientation cue that some nocturnal migrants use to select their flight direction, which is then maintained throughout the night by using other cues.

The stars provide an orientation cue for some nocturnal avian migrants. Birds such as the Indigo Bunting learn that the center of celestial rotation is north. This gives directional meaning to the constellations in the circumpolar area. Since the spatial relationship among these constellations is constant, if one is blocked by cloud cover, the birds can use the others to determine the direction of north.

Few animals use the moon as a compass. Beachhoppers, such as *Talitrus* and *Orchestoidea* orient by the moon and compensate for its movements during their nighttime journeys.

Sunlight becomes polarized as it passes through the atmosphere. The pattern of polarization of light in the sky varies with the position of the sun. Polarized light may provide an axis for orientation or it may allow animals to locate the sun from a patch of blue sky even when their view of the sun is blocked.

The earth's magnetic field provides several cues that could be used for orientation. Most animals appear to have an inclination compass. They distinguish between equatorward (where the magnetic lines of force are horizontal) and poleward (where the lines of force dip in toward the earth's surface). The ability to use magnetic cues is widespread. The magnetic compass is one of the first orientation cues used by homing pigeons and many species of bird migrants. The sun compass later becomes the primary orientation cue of homing pigeons. Some researchers have suggested that, in addition to serving as a compass, the geomagnetic field may serve as the homing pigeon's map, revealing its location relative to home.

Chemical cues can also be used for orientation. Salmon are guided to their natal stream by chemical cues. One hypothesis is that young salmon learn (imprint on) the characteristic odors of the stream in which they hatched and then follow the odor trail back to their natal stream. The alternative hypothesis is that fish are innately attracted to pheromones, communicatory chemicals, produced by conspecific fish. When both cues are present, salmon tend to return to the site they experienced as juveniles instead of an area containing conspecifics.

It has also been suggested that homing pigeons rely on olfactory navigation. The results of olfactory deprivation studies are inconsistent. The results of deflector lofts are consistent but may not be due to a shifted olfactory map. However, the results of experiments in which olfactory information is manipulated are consistent with an olfactory basis for pigeon navigation. The role of olfaction in pigeon homing remains controversial.

There are many ways that sound can be used for orientation. Most species orient to sound by listening and then locating the source. Some species, the best known of which are bats, echolocate. They produce bursts of sound and then analyze the echoes to develop a picture of the environment. The echolocation cries of bats are very high pitched and those of most species are intensely loud. The rate of pulsing increases during a hunt, helping a bat characterize its prey and track it more accurately. There are three categories of bat cries: frequency modulated (FM), constant frequency (CF), and those that are constant frequency with a frequency sweep at one end. The echoes from FM pulses provide the most information about the target. A bat analyzes an FM call by measuring the interval between the pulse and the return of the echo to determine the distance to the target. CF pulses allow a bat to detect prey from a greater distance than do FM calls and to determine whether the target is approaching or moving away. A CF pulse is analyzed by the Doppler shift. Most bat species produce only one type of call and its structure is suited to their style of foraging.

Some aquatic species can detect electric fields. These could potentially be of use in navigation. A few species have electric organs that can generate electric fields that can be used in communication and navigation. The weak electric fish generate a stream of electric pulses and sense objects by the disturbance created in this symmetrical field.

CHAPTER 12

Foraging

---

*T*he silence of a marsh on the eastern sea-coast is broken when a hawk flushes a flock of ducks as it suddenly strikes and carries off a field mouse to feed to its young. The mouse's cousins scampered to safety but will soon return to gather seeds. The ducks, too, will come back and continue feeding on aquatic plants. Meanwhile, a sparrow quickly makes a meal of an insect feeding on the tall grass. The "leftovers" may go to others—crayfish perhaps.

Indeed, life on this earth is tied together in an intricate web in which energy is exchanged between those that eat and those that are eaten. Although the names of the players may differ, the deadly pageant that, paradoxically, sustains life is played out on all the earth's various stages—marsh and river, tundra and forest, desert and steamy jungle alike.

Obtaining Food

One of the hallmarks of life is the transference of energy. Once life has captured energy in molecular bonds—whether as a tiny flowering plant on an alpine hillside or as some little known life form on the ocean floor—that energy is destined to be transferred from one kind of organism to another as one is eaten by the next.

Essentially, what we will consider here is the transference of the energy in molecular bonds from any organism to some animal. Therefore, we will consider the foraging, or food-getting, behavior of both herbivorous (plant-eating) and carnivorous (flesh-eating) animals, keeping in mind that most species are omnivorous and eat a variety of food types.

FILTER FEEDING

Many aquatic animals feed by straining food from the surrounding water. As examples, bivalve mollusks, such as clams, are able to filter food particles from their water environment by pumping water across feathery gill structures and mucous membranes. Although filter feeding is more common in invertebrates than in vertebrates, there are some interesting vertebrate examples. One is the flamingo. Its beak is modified for filtering crustaceans and small mollusks from the mud of shallow ponds and lakes. Another significant example is one of the largest known life forms: the baleen whales. Epidermal plates in the roof of the mouth have become specialized into filter structures that strain small crustaceans and other zooplankton from the water.

PLANT EATING

Plants have a variety of parts—leaves, stems, fruits, and flowers to name a few—each of which may be used by different species of animals. Giraffe may browse on leaves and wildebeast graze on grasses. The larvae of many insects are among those who are leaf-miners and chew tunnels through the nutritious photosynthetic tissue of the leaf that is found between the upper and lower surfaces. Other species, bumblebees for instance, collect nectar from flowers for a living. Others, including the fruit bat, eat fruit and yet others gather seeds.

A variety of animals cultivate some or all of the food they need. For example, leaf cutter ants (*Acromyrmex* spp.) cut fresh leaves and carry the pieces back to the nest under the ground (Figure 12.1). There they alter the plant material, encouraging the growth of a special fungus on the leaves. This fungus, whose existence is unknown outside the ant nests, is thought to be the primary food source for these ants (Weber, 1972). The ants actively prepare their fungus gardens. Pieces of suitable leaves are first licked on both sides, a process that removes the waxy layer covering the leaf and reduces the population of microorganisms that might compete with the desired fungus. The leaf fragments are then

Figure 12.1 Leaf cutter ants transporting leaves to fertilize their fungus gardens. These ants cultivate a special fungus that serves as their primary source for food.

chewed to a pulp, placed in the fungus garden, and inoculated with hyphae of fungus. When prepared in this way, the leaves are a richer source of nourishment for the fungus. The ants are, indeed, good farmers. Fungus gardens in abandoned nests degenerate quickly as they become overrun with other microbes (Quinlan and Cherrett, 1977).

Cultivation of plants also occurs in mammals. Prairie dogs pull out the grasses that they do not like to eat, resulting in the proliferation of the grasses they do like around their burrows. Gorillas are known to rip down large plants for no apparent reason, the result being a surge in the growth of young, fast-growing vegetation, encouraged by the new light.

HUNTING

Modes of Hunting

We find many styles of hunting, among the many species that eat other animals. *Octopus cyanea*, for instance, does not look for prey and then chase it. This octopus has membranes between its tentacles, which make it resemble an umbrella. When hunting, it stretches its webbed tentacles over coral rocks or algae where prey are likely to be found and then feels beneath the web to see if prey have been captured. If it is unsuccessful, it simply tries again elsewhere (Curio, 1976).

Other animals, such as birds that pick up insects from among the leaves of trees, spend much of their foraging time searching for prey. However, once an insect is spotted, little time is needed to chase, capture, and eat it. Most insects just sit there or move slowly about, so the capture is easy and most are swallowed in a single gulp.

Pursuit   Other predators pursue their prey. Many of these rely on speed, endurance, power, or brute strength. A cheetah's strategy, for example, is usually to stalk its prey so that it can draw within a few hundred yards. It then chases its quarry, sometimes running at 70 miles per hour, but such speed cannot be maintained for more than a brief burst. Average running speeds are generally less than 40 miles per hour (Frame and Frame, 1980).

Traps   Some predators are able to trap their prey, that is, to manipulate objects or alter their environment, in such a way as to capture, or at least restrain, prey. It is interesting that such predators are often arthropods. In the summertime you may notice the traps of orb-web weaving spiders in the doors and windows of old houses and barns, or in grasses or trees, or any area where their sticky filaments can intercept flying insects. The resident spider usually positions itself so it will not be obvious to the prey. Then, when the web vibrates at a given frequency, the spider rushes out. The vibration is the signal that prey is in the web and must be secured with additional silk before it can escape.

Humpback whales (*Megaptera novae-angliae*) build bubble "nets" to trap, or more accurately to corral, their prey. Beginning about 50 feet deep, the whale begins blowing bubbles by forcing bursts of air from its blowhole while swimming in an upward spiral. The big bubbles, along with a mist of small ones, form a cylindrical net that concentrates krill and small fish (Figure 12.2). The whale then swims upward through the center of the bubble net with an open mouth and devours the prey in a single gulp (Earle, 1979).

Some predators have specialized structures that are used as lures to attract prey. The alligator snapping turtle holds its mouth wide open while it lies on the bottom of muddy streams and lakes. The tongue has a wormlike outgrowth that is wiggled, attracting fish that the turtle then snaps up. It is interesting that lures can be used where vision is normally not possible. The deep sea angler fish lives at extreme ocean depths where virtually total darkness prevails. Nonetheless, the female has a long, fleshy appendage attached to the top of the head that is luminous and acts as a lure, bringing curious fish within the reach of its jaws (Wickler, 1968).

Spiders of the genus *Argiope* lure certain flying insects to their traps. They decorate their webs with bars and crosses that reflect ultraviolet light (Figure 12.3). These web adornments actually seem to attract prey. Studies have shown that webs with ultraviolet reflecting decorations catch more insects in an hour than do undecorated ones (Craig and Bernard, 1990).

Lures may also be specific actions that induce prey to draw closer to their predators. A certain

Figure 12.2   The humpback whale traps krill and other small prey in cylindrical bubble "nets" that it constructs by forcing air out of its blowhole. In this photo the bubble net is seen as a ring of bubbles on the surface of the water. Using a bubble net, a whale can corral the fast-swimming krill that would otherwise scatter and consume them in a single gulp.

The Intricate Web of Life — Survival and Reproduction Depend in Large Part on an Animal's Ability to Obtain Food.

Green plans harness solar energy and manufacture the food on which virtually all animals ultimately depend. Animals may avail themselves of almost any part of the plant.

As this rhinoceros stands grazing among water hyacinths, a cattle egret rides on the rhino's back and feeds on insects in the surrounding vegetation.

A mother orangutan shares fruit with her infant. In this way, the infant may learn to select appropriate foods.

A hummingbird generally takes nectar while hovering in front of a flower. Its grooved tongue, which can be extended far beyond the end of its beak, is tipped with brushy borders that collect nectar by capillary action. Within the beak, the nectar from the bristles is sucked through the groove and swallowed.

An aphid uses its piercing mouthparts to tap the plant juices and excretes a sugary liquid called honeydew. Some ant species, such as these red ants (*Myrmica puginoides*), ingest the honeydew. The ants trigger the excretion of honeydew by tapping an aphid on its abdomen. The ants vigorously protect the aphids from their predators.

The Deadly Pageant of Life.
To prey, perchance to eat — predators use a variety of methods to secure food.

This bullfrog sits in ambush in a seasonal pond in Botswana, ready to snatch prey that wander by.

The fish swimming near the surface has accidentally brushed against the deadly stinging tentacles of a Portuguese man-of-war (*Physalia*) and has become paralyzed by the toxin in the stinging cells on the tentacles. *Physalia's* tentacles may hang down several meters below its gas-filled float. The fish will then be drawn up to the mouths of the feeding polyps.

Alaskan brown bears congregate in good fishing areas.

The fringed-lipped bat is alerted to the presence of a frog by the frog's calls and then zeroes in on it using echolocation.

In the complete darkness of the abyssal ocean depths, the viperfish lures its prey within striking distance with the light emitted from photophores. Photophores are special organs that contain luminescent bacteria.

Jumping spiders have excellent eyesight and stalk their prey until they are close enough to jump with unerring precision. The legs extend, not just by muscle power, but also because blood is pumped into them, causing rapid extension.

The red fox is another carnivore that actively hunts its prey. It is seen here jumping after a mouse, but it also eats insects, birds, and birds' eggs.

A chameleon uses its protrusible tongue to catch its prey while anchored to a branch with its prehensile tail and plier-like feet. The tongue can snatch an insect at a distance of over half the chameleon's body length. The tongue's tip bears an indented sticky pad that attaches to the insect, in this case a grasshopper, and allows it to be drawn back into the chameleon's mouth.

The preying mantis is a carnivorous insect that eats a variety of other insects and sometimes even others of its own species. Slight movement of the prey will cause the mantis to strike with lightning quickness. Here, a preying mantis eats a monarch butterfly.

Finding Food in Other Ways.

These gooseneck barnacles are filter feeders. Their feathery appendages, called cirri, are unrolled and extended out from the exoskeleton. The cirri may be spread out like a fan or waved to capture small organisms in the vicinity.

A giant octopus (*Octopus dofleini*) is scavenging a dogfish shark that had been killed and thrown back by fishermen. Although it accepts a windfall meal, the octopus more commonly hunts for its prey, usually shellfish such as crabs.

The housefly has sensory hairs on its feet that detect suitable food substances. The fly thereby senses the nature of whatever it is walking on. When a sugary substance is detected, proboscis extension is triggered. The proboscis also has sensory hairs that respond to food. When these are stimulated, the fly feeds.

Figure 12.3 Ultra-violet decorations on the webs built by *Argiope* spiders lure certain flying insects into the trap.

Figure 12.4 The predatory jumping spider *Portia* performs a vibratory display that mimics the courtship display of another species of jumping spider, *Euryattus*, on a female's leaf nest. This action lures her out of her nest where she can be attacked.

jumping spider of Queensland, Australia, *Portia fimbriata*, mimics the vibratory courtship display of another type of jumping spider (*Euryattus* sp.) on which it feeds. Unlike most other jumping spiders, the *Euryattus* spiders nest within rolled-up leaves that dangle from rocks or vegetation by heavy guylines. A courting *Euryattus* male stands on the suspension nest of a female and performs a vibratory display called shuddering. This causes the female to come out of her nest. *Portia* predators shudder in a manner reminiscent of that of a courting *Euryattus* male (Figure 12.4), causing the female to emerge from her leaf nest. If she retreats before capture, the hunter waits and repeats the action. In experimental tests in which *Portia* females encountered suspension nests of female *Euryattus*, the predator successfully attacked and killed the prey 40 percent of the time (Jackson and Wilcox 1990).

Ambush   Some animals literally "lie in the weeds." That is, certain predators employ a hide-

and-wait strategy in acquiring prey. This strategy, ambush, involves hiding motionless until the prey draws near enough to attack. The larvae of the water scavenger (*Hydrochara obtusata*), for instance, prey on small aquatic organisms such as tadpoles. However, these beetle larvae are very poor swimmers. So, instead of chasing prey through the water, a beetle larva sits on vegetation and ambushes prey that swim by, snatching them right out of the water (Formanowicz, Bobka, and Brodie, 1982). Certain vertebrates also employ ambush strategies. The African lion may wait in specific places, such as water holes, where its odds of finding prey are increased as thirsty herds make their way to drink (Schaller, 1972).

Antidetection Adaptations

In some cases the ability of the predator to draw near is enhanced by camouflage, rendering the predator more difficult to detect. Camouflage can involve body colors, markings, or even the specialization of body parts. For instance, the spotted coats of leopards and jaguars provide excellent camouflage among bushes and forest canopy. Leopards become difficult to see on low tree branches because their coat pattern blends in with the variegated light created by sunlight as it penetrates the branches (Figure 12.5). African lions, on the other hand, lack intricate patterns and discrete colors. However, their tawny coat color provides excellent camouflage in the grassy savannahs they normally inhabit.

Examples of specialized body parts providing camouflage can be found throughout the animal kingdom. The sargassum fish, for example, has fins and body filaments that imitate the leaves and bladders of the weed beds it inhabits. Its fins have become so specialized in imitating the sargassum weed that the fish can barely swim (Figure 12.6). However, it is so perfectly concealed that it has become an ultimate ambush predator of tiny fish. Many predatory insects are also camouflauged. An insectivorous mantid (*Hymenopus coronatus*) has body parts that look like the petals of the orchid on which it rests. Not only does it resemble a plant, but

when it walks, it moves in uneven motions so that it looks like leaves being blown by the wind (Wickler, 1968).

Certain predatory species are disguised as other, beneficial species. Such aggressive mimicry allows these predators to draw within striking distance of their unsuspecting prey. For example, by mimicking beneficial cleaner fish, certain blennid fish are able to get within easy striking distance of prey. The cleaning wrasse (*Labroides dimidiatus*) removes external parasites, diseased tissue fungi, and bacteria from other fish (Figure 12.7a). Indeed, parasite-laden fish line up at coral reef "cleaning stations" much like cars at a car wash to avail themselves of the services of the cleaning wrasse. Some cleaning stations may have hundreds of patrons each day. The cleaning wrasses generally advertise their services with distinctive swimming motions performed above the cleaning station. Customers show their willingness to be cleaned by postures that are characteristic of the species. The blennid fish (*Aspidontus taeniatus*), a phony

Figure 12.5 Leopard in tree. The leopard's markings help hide it in the spots of light created as sunlight penetrates the leaves of bushes or trees.

Figure 12.6 The markings and special appendages of the sargassum fish make it nearly invisible in the sargassum weed in which it waits to ambush its prey.

cleaner, looks and behaves like the cleaning wrasse (Figure 12.7b), but when it is invited to approach, it takes a bite out of the customer (Wickler, 1968).

Adaptations for Detection of Prey

Most prey species have little interest in the welfare of foragers, and we find them at least not cooperating with, if not outright avoiding, the forager. Therefore, most foragers must search for food items. It should not be surprising, then, to find that they have sensory specializations that increase the chances of finding an edible item. As we look around the animal kingdom, we find many sensory specializations for prey detection that may seem unusual to us because the sense is poorly developed or lacking among humans.

Although you may be able to sense the end of the ski lodge in which the fire is glowing, you would find it difficult to find a hot meal by its warmth alone. In contrast, female mosquitoes in search of a blood meal in preparation for reproduction are activated to begin a hunt by increased levels of carbon dioxide given off by the victim, but they actually locate their host by its body heat (Herter, 1962). Perhaps this explains some experiences you may have had with these charming creatures. For instance, you may have noticed that mosquitoes are more attracted to you than to a friend sitting nearby, a situation that breeds resentment toward your friend. This is likely not your friend's fault, though. It is just that you are warmer. Mosquitoes find a warmer person more easily. In addition, it explains why mosquitoes are more bothersome on a cool evening than on a hot afternoon. They can find you more easily when the temperature difference between your body and the air is greater.

The snakes in two large families, the Crotalidae, or pit vipers (e.g., rattlesnake, water moccasin, and copperhead), and the Boidae (e.g., boa constrictor, python, and anaconda) also use their prey's body heat to help guide their hunt.

a

b

Figure 12.7 (*a*) A cleaning wrasse removing parasites from another fish. The service of cleaner fish is beneficial because it removes external parasites, diseased tissue, fungi, and bacteria. The cleaner wrasse has distinctive markings and behavior patterns that advertise its services. (*b*) *Aspidontus* is disguised as a cleaning wrasse and mimics its behavior. As a result, fish in need of cleaning service allow the phony cleaner to approach. *Aspidontus* then takes a bite out of the would-be customer.

They have special receptors that are so sensitive to infrared radiation (heat) that these snakes can locate their warm-blooded prey even in the darkness of night (Figure 12.8). A rattlesnake, for instance, can detect a mouse whose body temperature is at least 10°C above air temperature within a range of 40 centimeters. Furthermore, the detection is swift, within half a second,

a

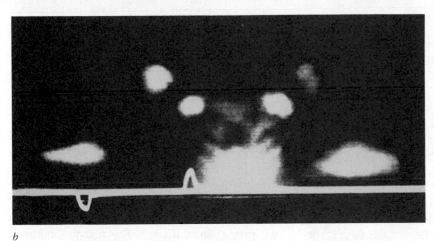

b

Figure 12.8 (*a*) The heat sensors of the black-tailed rattlesnake, located in pit organs between each eye and nostril, (the dark circular structures in this photo) allow it to detect its warm-blood prey in complete darkness. Snakes of the families Crotalidae and Boidae have infrared heat sensors that help them find their prey. (*b*) A rattlesnake's infrared picture of a mouse.

as it must be if the snake is to strike successfully at a mouse scampering by, hidden in the night's blackness.

These snakes can locate the source of infrared radiation with amazing accuracy. Laboratory studies have shown that, even with both eyes covered, a rattlesnake can strike a warm object, such as a soldering iron, within 5 degrees of dead center of the source (Figure 12.9). This accuracy, which may be deadly for a mouse, can

be traced to the structure of the pit organs that house the infrared sensory endings. Because the opening of the pit organ of a rattlesnake is less than half the area of the sensory epithelium, a small warm object within half a meter of the pit opening will stimulate no more than a quarter of the epithelium. The location of the stimulated portion of the membrane will be interpreted by the brain to reveal the location of the source. Furthermore, the great density of heat-sensitive

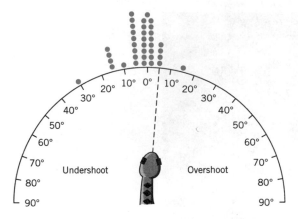

Figure 12.9 A rattelsnake's heat sensors can direct an amazingly accurate strike. Each circle indicates the angular error of a strike at a warm soldering iron by a blindfolded rattlesnake. The average error of the strikes was less than five degrees. (From Newman and Hartline, 1983.)

endings in the membrane permits accurate determination of the position of the stimulated patch. By turning its head from side to side, the snake can locate the boundaries of the heat source, thereby determining the size of the warm object. In addition to allowing the snake to select prey of an appropriate size, a mouse and not a moose, for example, knowledge of the boundaries of the potential meal allows the snake to strike with great accuracy (Gamow and Harris, 1973; Newman and Hartline, 1982).

The sand scorpion (*Paruoctonus mesaensis*), a nocturnal predator of the Mojave Desert, detects and locates prey by the vibrations they generate in the sand. The scorpions can detect disturbances up to 30 centimeters away and can locate the distance and direction to the source of vibration virtually perfectly at distances of 10 centimeters or less (Brownell, 1984).

Sharks have a well-deserved reputation as consummate predators (Figure 12.10). Many senses are involved in the shark's astounding ability to detect and track down prey. It can hear prey from a distance of over nine football fields

and will turn and swim toward it. When the shark is within a few hundred meters of its prey, its nose can help direct its search. As the shark nears its prey, its lateral line organs detect small disturbances in water caused by the swimming motion of prey, so the shark knows the location of the prey even if it cannot see it. Within close range, the shark can see its prey. Perhaps the most amazing of all the shark's senses is its ability to detect small electric fields.

Nearly all living organisms generate electric fields when they are in sea water that could reveal their presence, type, and perhaps even their physiological condition to an interested predator. The slightest movement, even the beating of the heart, could betray the quarry to an electrosensitive predator, such as a shark. Even if the prey could remain perfectly motionless and will its heart to stop beating, the voltage difference between its body fluids and seawater or between different parts of its body would send an electric message revealing its location to its predator. Wounded prey are an easy mark because even a minor scratch can double this voltage difference and send an even stronger signal (Bullock, 1973).

Several species of predators other than sharks are known to use electric cues to locate their prey. Other electroreceptive fish and some amphibians locate live prey this way. It is interesting that the platypus (*Ornithorhynchus anatinus*), an egg-laying Australian mammal, has electroreceptors on its ducklike bill that help in locating prey (Figure 12.11). The platypus hunts at night in murky streams. As it probes with its bill in the mud or amidst stones, it stirs up small animals such as freshwater shrimp. When disturbed in this way, a shrimp moves away with a series of rapid tail flicks that cause electric fields and tip off the platypus to its presence (Griffiths, 1988; Scheich et al., 1986).

Why would someone have ever suspected that an electric sense, so foreign to human abilities, exists, and once suspicion was aroused, how might an animal's ability to detect a weak

Figure 12.10 A shark devouring its prey. The shark has a multitude of sensory specializations that allow it to detect and locate prey.

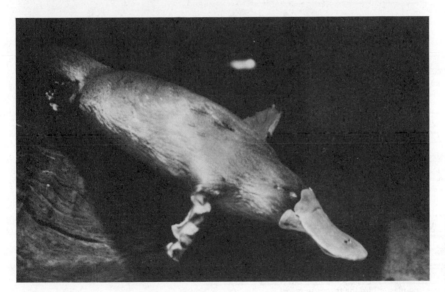

Figure 12.11 The duck-billed platypus is among the predators that can use electric fields generated by its prey to locate them.

electric field be demonstrated? Let's consider the progression of the study in sharks. In 1935, Sven Dijkgraaf noticed a shark (*Scyliorhinus canicula*) respond oddly to a rusty steel wire—the otherwise fearless warrior swam directly away. Even when blindfolded, the shark showed oriented escape reactions (Dijkgraaf and Kalmijn, 1962). Suspecting that the galvanic currents generated by the wire in water might be causing this reaction, Adrianus Kalmijn (1966) sent a weak current into the water through electrodes. Sure enough, the sharks showed the same oriented escape behavior.

Later Kalmijn wondered about the biological significance of the shark's sensitivity to weak electric fields and demonstrated that one use of this sense is to locate prey by the small electric fields caused by action potentials within the victim's body (Figure 12.12). He buried a flatfish, a plaice (*Pleuronectes platessa*) under the sand or encased in an agar chamber; the shark was still able to make a directed attack. Thus it was shown that visual cues are not essential. What about olfaction? To test the roles of olfactory cues, pieces of whiting were placed in a nylon bag within an agar chamber that had water flowing through it. The shark looked for its victim at the end of the outlet tube where the odor molecules were detected, instead of striking at the agar chamber as it had previously done. However, when a thin sheet of polyethylene film, which blocks electric current, was placed

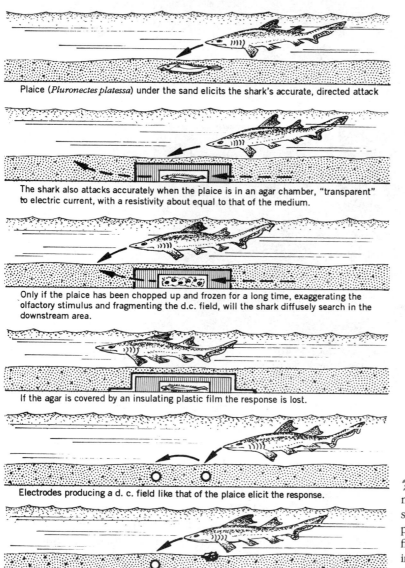

Plaice (*Pluronectes platessa*) under the sand elicits the shark's accurate, directed attack

The shark also attacks accurately when the plaice is in an agar chamber, "transparent" to electric current, with a resistivity about equal to that of the medium.

Only if the plaice has been chopped up and frozen for a long time, exaggerating the olfactory stimulus and fragmenting the d.c. field, will the shark diffusely search in the downstream area.

If the agar is covered by an insulating plastic film the response is lost.

Electrodes producing a d. c. field like that of the plaice elicit the response.

Figure 12.12 The experiments demonstrating that sharks can locate their prey using the electric field generated by any living organism in sea water. (From Kalmijn, 1971.)

around a plaice within the same agar chamber, the shark could no longer find its prey. The shark searched diligently because the odor of fish was spread around the bottom of the tank, but the search was in vain. Finally, when electrodes producing a current of about the strength of that produced by a living plaice (4μA) were buried in the sand and the shark was aroused by the odor of fish, it made a direct attack on the electrodes. The shark attacked the electrodes even when a piece of fish was sitting openly on the bottom of the tank (Kalmijn, 1971).

Hunting Cryptic Prey

Prey animals have developed several means of reducing the probability that they will be detected by their predators, as we will see in the next chapter. One such technique is crypsis (or camouflage)—blending in with the background.

Clearly any method that a predator can use to increase its ability to detect its cryptic prey will be beneficial. One method might be the formation of a "search image," in which the predator learns characteristic features of prey and searches for these when hunting. This concept is a familiar one. If we are looking for our car keys, we do not look for something large, round, and blue; instead, we keep our eyes peeled for something small and shiny. Our search image, then, is centered on the characteristics of the keys. Similarly, an animal's search image is thought to be a mechanism, such as filtering out unimportant stimuli or focusing attention on the important stimuli, that increases a predator's ability to find cryptic prey.

Luuk Tinbergen (1960) employed the concept of search image formation to explain the change in diet of the birds he observed feeding their young in the Dutch pinewoods. He noted that the abundance of particular species of insect varied throughout the season. A species might be scarce at first, become plentiful later, and then disappear. However, the birds did not con-

sume all perfectly palatable prey types in direct proportion to their abundance. There was usually a delay between the appearance of a given insect species in the environment and the time when the birds began to sample them. But once they did, there was often a sharp increase in the rate of predation on the new prey. Tinbergen suggested that the sudden change in diet might have occurred because the predators learned to see their cryptic prey through the use of search images as they learned the prey's traits.

We do, indeed, find fairly sudden changes in the diet of several kinds of predators and that acceptable prey species are not always consumed in proportion to their abundance. But are these changes in menu due to the formation of a search image? To find out, one would need to devise an experiment in which several types of prey were equally acceptable and all equally easy to catch and to handle.

There have been several approaches to testing the hypothesis that a search image is responsible for the change in a predator's diet, but we will discuss only two. One uses operant conditioning. Birds respond to images projected on a screen by pecking at keys in a Skinner box (see Chapter 5). The birds do not eat the prey they detect. Instead they receive a standard food reward for a correct response. Thus, palatability and ease of capture and handling are constant. The other approach involves "free-feeding" under controlled conditions. The type of food item is kept the same, but its crypticity is altered by varying its color or that of the background so that they match or contrast.

Alexandra Pietrewicz and Alan Kamil (1981) used the operant conditioning approach to simulate predator–prey interactions between blue jays (*Cyanocitta cristata*) and one of their normal prey types, underwing moths (*Catocala* spp.). Hungry blue jays were given a slide show in a Skinner box theater (Figure 12.13). Some slides contained a moth. Some did not. If a moth was present and the bird saw it, it could peck at the slide and receive a mealworm reward. After

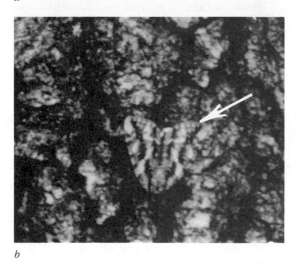

Figure 12.13 (*a*) A blue jay "foraging" for cryptic prey in a Skinner box. The blue jay was shown slides, some of which contained a cryptic moth. (*b*) Examples of cryptic moths shown to the blue jays on slides. *Catocala relicta* is on the right and *C. retecta* on the left. If the bird was shown a slide that contained a moth and pecked the appropriate key after spotting it, the bird received a meal worm reward.

a short interval, the next trial began. When no moth was present, the blue jay could abandon its search of that area and look elsewhere by pecking at a key that would then advance the projector to the next slide. There are two types of mistakes the bird could make. One is a false alarm—incorrectly responding as if a moth were present. The other is a miss, in which the bird failed to see the cryptic moth and pecked at the advance key. Either of these mistakes resulted in a delay before the hungry predator could search for prey in the next slide.

Pietrewicz and Kamil (1979, 1981) used this operant conditioning procedure to determine whether or not the birds form search images. The search image hypothesis would predict that recent experience with a prey type would allow the birds to learn the key characteristics of that prey type and prime them to look for others of that type. It was, therefore, reasoned that encounters with one species of cryptic moth would help jays find that type more accurately and quickly, but experience with another species would interfere. The jays were shown a series of slides, half of which contained a moth. They were tested under three conditions: (1) The moth-containing slides were of *Catocala retecta* only, (2) the moth-containing slides showed only *Catocala relicta*; and (3) the two moth species were shown in random order. The results, shown in Figure 12.14, are exactly what would be predicted by the search image hypothesis. The jays' ability to detect one prey type improved with consecutive encounters. However, although both prey types were cryptic, prey detection did not get better when the two species were encountered in random order. Thus, we

see that experience with one type of cryptic prey improved the predator's ability to find that type of prey, but not other kinds of cryptic prey.

Marion Dawkins's (1971) investigations on search image used domestic chickens freely feeding on colored rice grains. In some trials, these rice grains were made conspicuous by placing them on a contrasting background. In others, they were made cryptic by presenting them on a matching background. Each chick was observed feeding under both conditions. The chicks quickly spotted the conspicuous grains and consumed them at a high rate. Although the chicks took several minutes to detect the cryptic grains, by the end of a test period, they were eating them at the same rate as they did the conspicuous grains (Figure 12.15). Thus, the results of this experiment are also consistent with the idea of search image formation.

Although it is clear that predators get better at finding cryptic prey with experience, not everyone agrees that this is accomplished by forming a search image. It has been suggested, for instance, that changes in search rate may account for the increased ability to find cryptic prey (Gendron and Staddon, 1983, 1984). The general idea of the search rate hypothesis is that after a predator first finds a cryptic prey item, it may scan the area more slowly and look more carefully for others. If more time is spent viewing an area, the predator is more likely to notice camouflaged prey than if it only glances at the spot. However, if there is no prey hidden in the area, the time spent looking for it is wasted. Therefore, the predator faces a speed/accuracy trade-off. As a result, the predator might optimize its foraging by adjusting its search rate to the degree of crypticity and abundance of the prey. If conspicuous food items are abundant, then it is best to inspect the area quickly. But if there are many cryptic prey to be found, then a slower search rate may actually increase the rate at which food items are discovered because fewer will be missed.

It seems that birds do modify their search

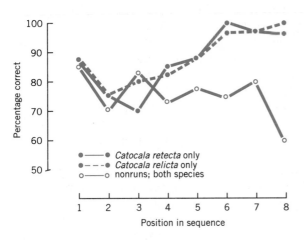

Figure 12.14   The percentage of correct responses by blue jays when shown moth slides in sequences of the same species (runs) or a random sequence of both species. After experience with one cryptic species of moth, blue jays became better able to detect that species. However, if the species were shown in random order, the jays' performances did not improve. This is consistent with the search image hypothesis. (Data from Pietrewicz and Kamil, 1981.)

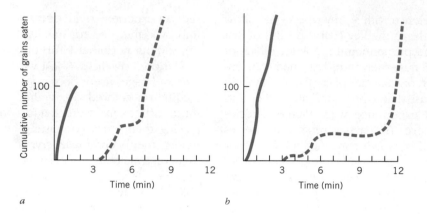

Figure 12.15 The cumulative number of rice grains consumed by chicks foraging for cryptic grains (dashed line) or conspicuous grains (solid line). In (*a*) the chicks were looking for orange rice grains on an orange (cryptic) or green (conspicuous) background. In (*b*) the chicks were looking for green rice grains on a green (cryptic) or orange (conspicuous) background. Notice that chicks always spotted the conspicuous grains and fed on them quickly. Although they took longer to notice the cryptic grains, the birds eventually began to eat them at the same rate as conspicuous grains. (Data from Dawkins, 1971.)

rate when the degree of crypticity of their prey changes. Observers of birds foraging in nature often notice that birds appear to hunt more slowly when their prey is cryptic (Goss-Custard, 1977; Smith, 1974). However, it is difficult to measure crypticity and search time simply by watching birds hunt. Quantification is easier under controlled conditions, and in studies where this has been possible, the results confirm that search rate varies with the degree to which the prey is camouflaged. For instance, measurements were made of the search rate of bobwhite quail (*Colinus virginianus*) hunting artificial prey in an aviary. The degree of camouflage of the prey, wormlike pellets made from dough, was manipulated by changing their color. The birds reduced their hunting speed as their prey became more cryptic (Gendron, 1986).

When a predator becomes more efficient at discovering cryptic prey, we might wonder whether it is forming a search image or adjusting its search rate. The increased efficiency would

be predicted by either hypothesis. Similarities in predictions of the two hypotheses have triggered an argument that the results of the studies on search image previously described, as well as others, are equally consistent with the search rate hypothesis (Guilford and Dawkins, 1987). If the simpler search rate hypothesis can explain the change in accuracy of prey detection, should we toss aside the idea of search image as an unnecessary complication? This question led to a reexamination of the search image hypothesis.

New evidence for search image formation is provided by the work of Patricia Blough (1989, 1991), who used operant conditioning techniques and a rather abstract foraging situation. In her experiments, the "prey" were letters of the alphabet displayed on a screen. The predators were pigeons (*Columba livia*) trained to look for a particular target letter of the alphabet in a display of several letters. When they pecked at the correct letter, they received a food reward. The degree of crypticity could be enhanced by increasing the number of nontarget letters in the display. Blough reasoned that if a predator knows that cryptic prey are available, its attention can be focused on finding them, and it will find them more efficiently. This priming of attention might be analogous to a search image. She tested this idea by providing a cue, for instance vertical screen borders, that would prime the pigeon to expect a target letter in the display that followed. The results show that pigeons that were cued to expect a target did find it more quickly (Blough, 1989).

The best way to determine whether predators use search images is to observe their behavior in situations in which the search image and search rate hypothesis predict different responses. One such situation is when there are several types of cryptic prey hidden in the hunting area. The search rate hypothesis predicts that a predator in this situation will find all types of cryptic prey with equal ease. If recent experience with one type of cryptic prey causes a predator to slow its search for others, it will be more likely to find any cryptic prey of any type. However, a search image is specific for one type of prey and would, therefore, be less likely to help a predator detect other prey types. The search image hypothesis assumes that, because of experience with one type of cryptic prey, a predator forms a search image based on the key characteristics of that particular prey. Therefore, a search image for one type of cryptic prey might even interfere with the detection of other types.

In some experiments, Blough simulated a situation in which there were two forms of cryptic prey by making the target either of two letters. She also provided two types of cues—one signaled that either of the target letters would be present, the other signaled a specific letter. She found that a pigeon primed to expect a particular letter found it more quickly than if the cue simply informed the bird that one of the two letters would appear in the display. Furthermore, when the pigeon was misinformed about the identity of the expected target, it took longer to find it, even if it was conspicuous, than to find a cryptic target that was correctly cued (Blough, 1991). This specificity is just what would be expected if a search image were formed.

Another prediction that would distinguish between the search rate and search image hypotheses involves changes in search rate as a result of experience with cryptic and conspicuous prey. Search rate is most easily measured as response time, since the animal is presumed to respond as soon as prey is detected. The search rate hypothesis is that predators learn to adjust their search rate to the degree of crypticity of common prey. Therefore, it would predict that response time would increase as a result of experience with cryptic prey. It should take longer to detect prey and to recognize that there is no prey in that patch of the environment. The search image hypothesis would predict no change or a slight improvement in response time.

These predictions were tested by measuring the response times of pigeons detecting the presence of wheat or beans on backgrounds that altered the degree of crypticity. On a sandy background, beans were more cryptic than wheat. The search rate hypothesis, therefore, would predict increases in response time during and following successive presentations of cryptic beans, but decreases during and following successive presentations of the more conspicuous wheat. This was not observed. Instead, as the search image hypothesis predicts, the birds detected the beans more quickly as they gained experience in detecting beans through successive encounters (Bond and Riley, 1991).

So we see that both ideas seem to have merit. There is evidence that predators form search images, at least in some circumstances, and they may also adjust their search rate. Perhaps this should not be surprising. A forager's life depends on finding food, so we might expect it to use any technique that will enhance its success.

Optimal Foraging

A foraging animal may have a variety of potential food items available to it. Some are easier to find; some are easier to handle and digest; some are easier to capture; and some have a higher nutritive value than others. So, how does an animal decide which to eat?

As we discussed in Chapter 4, optimality theory assumes natural selection will favor the optimal behavioral alternative. In other words, when an animal is faced with several possible courses of actions, the one whose benefits outweigh its costs by the greatest margin will be favored by natural selection.

Foraging has been a favorite subject for testing optimality theory for several reasons. One is that the costs and benefits are easily translated into energy units. This is important because optimal foraging theory attempts to predict the foraging patterns of animals on the basis of net energy gain per unit of time. Some of the factors considered include the caloric value of the food, the amount of energy required to digest the food, the amount of time it takes to find the food, and the amount of time it takes to handle the food. Another reason that foraging is often used to test optimality theory is that it is relatively easy to think of foraging as a series of decisions and then to focus on one decision at a time. Examples of some of these decisions are what to eat, where to look for food, how long to search one area before moving on, and what sort of path to take through the area. Finally, foraging behavior is suited to tests of optimality theory because once an aspect of foraging has been chosen for study, it is relatively easy to identify the behavioral alternatives the individual has available.

According to theory, individuals that forage optimally are likely to be fitter, that is, to leave more offspring, than those that do not. It seems intuitively logical that success in getting food is going to be translated into other kinds of successes because food provides the energy for other activities. In the few species in which finding food efficiently is a matter of life and death, the connection is obvious. Certain shrews, for instance, digest their food within 3 hours and cannot live for more than 3 to 5 hours without eating (Matthews, 1971), and in the winter, a titmouse must catch an insect every 3 seconds in order to survive (Gibb, 1960). But what about the majority of species? As long as an individual's stomach is full, does it really matter whether it foraged in the most efficient manner? To assume that foraging success need only be adequate would be to ignore that foraging competes with other activities, even in species that have less stringent energy demands than do shrews or titmice. An individual's "spare" time can be spent on other activities related to survival and reproduction.

We should keep in mind that the link between fitness and optimal foraging is difficult to verify because fitness is measured as lifetime reproductive success and foraging success is usually measured in much more immediate terms. Nonetheless, certain circumstantial evidence indicates a relationship between foraging success and fitness. In some species, dogwhelks (*Nucella lapella*) for instance, growth can serve as an indirect measure of fitness. It has been shown that individual differences in foraging activities of dogwhelks, both in the laboratory and the field, result in differences in growth (Burrows and Hughes, 1990). Thus, the premise that foraging and fitness are related seems sound. Although optimal foraging theory seems reasonable, we should nonetheless examine it with a critical eye.

FOOD SELECTION

The first step in acquiring a meal is generally deciding what to eat. As will we see, this may involve several types of decisions.

Distinguishing between Profitable and Unprofitable Prey

An optimal forager must be able to distinguish between food that is worth eating and food that requires more energy to process than it provides. So, we might begin our consideration of optimal foraging by looking for examples of animals that are able to distinguish between profitable and nonprofitable food items. Some evidence comes from crows feeding on shellfish along Canada's western seacoast (Figure 12.16).

Northwestern crows (*Corvus caurinus*) search along the water line during low tide for whelks, which are large mollusks. When a suitable whelk is found, the crow carries it in its beak toward the shore, flies almost vertically upward, and then drops the whelk. If the shellfish breaks open, the crow eats it. If not, the bird retrieves the whelk, flies upward, and drops it again, repeating the procedure until it does break.

Reto Zach wondered whether the crows distinguish profitable from nonprofitable whelks. He reasoned that foraging crows must make an energetic profit or they would not do it. In other words, they must gain more calories from the whelk than they use obtaining it. He thought about factors that might influence the profit margin. He asked first whether the crows were selective in their choice of whelks. He was curious whether there were differences in the ease of handling whelks of different sizes. One might expect that flying upward would be the most energetically costly part of eating whelk. Then he asked whether crows adjust their cost according to the expected caloric value of the meal. How many times would a crow continue to drop a whelk that was difficult to crack open?

Zach (1978) found that the crows preferentially prey on larger whelks. He demonstrated this in several ways. First, he collected broken pieces of whelk shells, and by comparing the size of the base of the shells to those of living whelks, he estimated the size of the whelks that had been eaten. The whelks selected by the crows appeared to be among the largest and heaviest on the beach. Larger whelks provide more energy, so this observation was consistent with the hypothesis that the crows could distin-

Figure 12.16 Northwestern crows choose profitable over nonprofitable prey. Although this crow is eating a mussel, they often feed on whelks, which they break open by dropping onto rocks from a height of about 5 meters.

guish profitable prey. However, Zach was cautious in his interpretation. Perhaps, he reasoned, the pieces of smaller shells were more easily washed out to sea by wave action, leaving the larger pieces over represented in his sample. So, Zach offered each of three pairs of crows equal numbers of small, medium, and large whelks and, at hourly intervals, recorded the number of each size taken. Crows selected large whelks (Table 12.1). Even after the crows had eaten most of the large whelks, they continued to ignore smaller ones. Was this because large whelks are more palatable? No. By removing whelks from their shells and presenting equal numbers of each size class to crows, Zach demonstrated that all size classes were equally palatable.

Why, then, do crows accept only the largest and heaviest whelks on the beach? One reason seemed simple enough—larger whelks have a higher caloric value. However, Zach wondered whether whelk size affects the ease of breaking the shell as well. To answer this question, he collected whelks of different sizes and dropped them onto rocks from different heights. He found that large whelks were more likely to break than the medium and small whelks; the large ones required fewer drops from any given height.

We see in Figure 12.17 the total height to which a crow would have to fly, on the average, to break whelks of the three different sizes at different heights. Notice that the probability of a whelk breaking increases with the height of the drop. The lower the dropping height, the more drops are required to break it. The total vertical height needed to break the shell can be determined by multiplying the number of drops by the height of the drop. There is a certain minimum altitude, slightly more than 5 meters, needed to break a shell. Flying above this height would be a waste of energy for the bird. Zach (1979) found that the crows dropped whelks from a mean height of 5.23 meters (± 0.07m) and that the crows dropped the large whelks from a height that minimized the amount of energy needed to get the food out of the shell, thereby maximizing the net energy gain.

Given this information, Zach was able to calculate a crow's expected energy profit from whelks of different sizes. He found that the crows are likely to gain 2.04 kilocalories from a large whelk, but use 0.55 kilocalories to obtain it. Thus, the net energy gain for a large whelk is 1.49 kilocalories per whelk. Medium and small whelks, on the other hand, require more energy to handle because they are harder to break. They also contain fewer calories. Zach calculated that

Table 12.1

NUMBERS OF SMALL, MEDIUM, AND LARGE WHELKS LAID OUT AND CUMULATIVE NUMBERS OF WHELKS TAKEN OVER THE SUBSEQUENT 5 HOURS[a]

	TYPE OF WHELK		
	SMALL	MEDIUM	LARGE
Laid out	75	75	75
Taken after 1 hr	0	2	28
2 hr	0	3	56
3 hr	0	4	65
4 hr	0	4	68
5 hr	0	6	71

[a] Results from three pairs of crows were homogeneous (replicated goodness of fit test) and therefore combined. Right from the start whelks were taken nonrandomly ($p < .005$, single classification goodness of fit test).
Source: Zach, 1978.

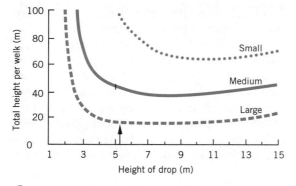

Figure 12.17 The total height of the drop required to break the shells of different-sized whelks. North-western crows choose large whelks and drop them from a height of about 5 meters (indicated by the arrow). This minimizes the energy used in obtaining food and yields an energy profit. The crows would use more energy than they obtain by feeding on small or medium whelks. (From Zach, 1979.)

the net energy gain for a medium whelk would be − 0.30 kilocalories. We see, then, that the bird would use more energy attempting to eat a medium whelk than it would get in return. Obviously, small whelks, which are even harder to break and contain relatively few calories, would be even more costly to eat. So crows are able to distinguish between profitable and nonprofitable food items.

Optimal Diet

The previous example shows us that animals can distinguish prey that will give them a net increase in energy from prey that will actually cost them energy. This is not surprising since individuals of any species that do not make an energy profit while foraging would soon starve to death.

However, when faced with a variety of types of food, all of which would yield a net energy gain, should the individual eat some of each? Optimal foraging theory answers no. There exists an optimal diet, one that maximizes energy gain (Schoener, 1987).

To determine the optimal diet, we must first calculate the profitability of each food item. Profitability, in this case, is calculated by dividing the net energy gain by the handling time.

profitability = net energy gain/handling time

The net energy gain represents the number of calories the animal can get from the food minus the number of calories required to process that food, due to handling or digestion, for instance. Handling time includes both the amount of time the animal needs to pursue the food item and the amount of time needed to prepare it so that it can be swallowed.

Obviously, the food item that has the greatest profitability should be eaten whenever it is encountered. But if a forager finds the second most profitable food item, should it continue searching for the "best" item or accept this one? It turns out that this decision depends on the abundance of the most profitable prey, not on the abundance of the second-best food item. If the best type of food is rare, it will take a long time to find. In this case, the animal will increase its overall energy intake by accepting any food that provides an overall energy gain. But as the abundance of the best food increases, the animal loses valuable time attending to other, less profitable items, when it could be dealing with only the best ones. Consider the following analogy. Assume for the moment that lobster is your favorite food. Although hot dogs may stop your hunger pangs, you are not fond of them. Obviously, if you can get enough lobster to satisfy your needs, you should not bother with hot dogs, no matter how plentiful they are.

We see, then, that optimal diet theory leads to at least two testable predictions (Pyke, Pulliam, and Charnov, 1977):

1. The acceptability of a food item depends on the abundance of more profitable food items and is independent of its own abundance.

2. As high-ranking items become more common, less profitable items will be eliminated from the diet.

Observations that are qualitatively consistent with these predictions have been obtained in a field study of redshanks (*Tringa totanus*), shorebirds that feed on worms (Figure 12.18). Although large worms provide more energy than do small ones, they are difficult to find in some locations. The rates at which redshanks preyed on large and small worms at various study sites with different abundances of prey sizes were compared. As the abundance of the larger, more profitable prey increased, the birds became

more selective. Whenever large worms were common, they were eaten instead of the small worms, even if the small ones were most plentiful (Goss-Custard, 1977).

In field studies such as this one, the rate at which the subject encounters different prey types is largely out of the control of the experimenter. To eliminate this difficulty, John Krebs and coworkers (1977) designed an apparatus that would allow them to control the abundance of two different food items presented to great tits (*Parus major*). The birds were placed in cages in which they obtained food from a conveyor belt that ran past the cage. This allowed Krebs to control both the type of prey encountered and the abundance of the different prey. The prey items were small and large pieces of mealworms. The energy value of large mealworms was greater than that for small ones, but the handling time was equal for the two prey types. Krebs found that when there were few mealworms of any size, large or small, available, the birds ate both kinds of prey. They showed no preferences. However, when the large mealworms were made more plentiful, the birds became selective and tended to ignore the small ones. As long as the abundance of the large mealworms was kept at this level, the birds selected them, even if the small mealworms became more abundant than the large ones.

Nutrient Constraints

So far we have assumed that the only thing a predator has to consider is maximizing its energy intake. But if obtaining calories were the only nutritional concern, we could eat nothing but our favorite food—say, hot fudge sundaes or Spam. There would be no need for mothers to insist that we eat our vegetables.

The commonly touted nutritional guideline for humans, "Eat a variety of foods," may also apply to other animals. It has been shown, for instance, that nestlings of the European bee-

Figure 12.18 A redshank, a shorebird that feeds on worms. Large worms are more profitable than small ones. When redshanks forage in areas where large worms are abundant, they are more selective and eat more large worms than they do in areas where large worms are rare.

eater (*Merops apiaster*) convert food to body weight more efficiently if they are fed a mixture of bees and dragonflies than if they receive only bees or only dragonflies (Krebs and Avery, 1984).

Animals may have specific nutritional requirements, and they may have to balance their diet so that the essential dietary component is supplied along with sufficient energy. The moose (*Alces alces*) living on Isle Royale in Lake Superior provide an example (Figure 12.19). They must obtain enough energy for the growth and maintenance of their huge bodies, but they also have a minimum daily requirement for sodium. The leaves of deciduous trees on the shore contain more calories than aquatic plants, so to maximize energy intake, the moose should eat only land plants. However, land plants have a low sodium content. In contrast, low-calorie aquatic plants have a higher concentration of sodium. A moose balances these needs by eating a mixture of plants such that energy intake is as great as it can be while sufficient sodium is still obtained (Belovsky, 1978).

SEARCHING FOR FOOD

So far we have used optimality theory to help us understand how an animal chooses a given type of food. Optimality theory can also help explain how animals actively search for food.

Choosing a Site

Once a forager has decided what to eat, it must then decide where to look for it. Since food is typically found in patches rather than evenly spread across the environment, an animal searching for food is likely to encounter areas where food is easier to obtain than in other locations.

Assessing Site Quality   Optimal foraging theory would predict that animals will choose

Figure 12.19 A foraging moose. Moose must obtain enough energy for growth and maintenance, and the leaves of deciduous trees contain more energy than do aquatic plants. However, moose must also obtain a minimal amount of sodium. So, although aquatic plants provide less energy, moose must eat some low-calorie plants. This is an example of how nutrient requirements can constrain optimal foraging.

the best area (patch) for foraging. Operant conditioning techniques have been useful in testing this prediction. In one experiment, patch choice of caged rats was explored by using bar pressing for a food reward to simulate foraging. The abundance of food could be varied by altering the size of the food pellet reward, and foraging costs could be changed by increasing or decreasing the number of bar presses required to obtain a reward. When permitted to choose between levers that differed in food abundance or the ease of obtaining food, rats fed more frequently and ate larger meals at more profitable patches (levers). This was true whether the difference in profitability was in the size of the pellet or in the work required to obtain it (Johnson and Collier, 1989). So, as predicted, the animals chose the best patch in which to forage.

In order to determine which of several possible foraging sites is the most profitable, a forager must sample them periodically. It has been demonstrated that great tits sample feeding patches and use this information to choose the most profitable patch. In this experiment, the food patches were grids containing different numbers of meal worms. The birds quickly learned to feed from the one with the highest density of meal worms, but they continued to sample less profitable patches. The value of sampling became apparent when the density of meal worms in the most profitable patch was reduced suddenly. The tits then switched to the grid with the second highest prey density (Smith and Sweatman, 1974).

The astute reader has probably noticed that sampling is not without cost. The time spent sampling different patches before concentrating on the most profitable one lowers the average rate of food intake. Compromises must be made. Optimality theory has permitted predictions about the conditions under which a forager should explore patches other than the one currently being exploited. Consider, for example, an individual with a choice of two foraging sites. One patch reliably provides a moderate amount of food. The other fluctuates in richness reaching extremes far above and below the stable one. It is clearly beneficial to exploit the fluctuating site during times of peak abundance, but it is detrimental to waste time there when food is less plentiful than at the stable site. To determine the level of food in the fluctuating site, it must be sampled—but how often?

One model predicts that the fluctuating site should be sampled more often when the value of the stable one falls below the average value of the fluctuating one (Stephens, 1987). Chipmunks are well suited to test this prediction because much of their natural food supply, seeds from deciduous trees, is found in patches of fluctuating abundance (Figure 12.20). The supply of seeds below a particular tree may be here today and gone tomorrow, varying with the schedule of ripening, amount of wind, and activities of other animals. As a result, chipmunks must decide how often to check the seed supply at other trees to determine whether it might be beneficial to switch foraging locations. The amount of time individual chipmunks spent foraging and exploring alternative sites was monitored to determine what factors influence how frequently alternative locations were sampled. In one experiment, the chipmunks (*Tamias striatus*) foraged from artificial patches—trays of sunflower seeds. The value of a given patch was manipulated by varying the number of seeds on the tray. It was found that the chipmunks spent more time sampling the food density at other locations as the quality of the patch being exploited decreased (Kramer and Weary, 1991).

Sensitivity to Variability in Food Abundance Some species live in an environment in which the food supply is unpredictable. As a result, an individual may have to choose between a site that reliably supplies a moderate amount of food and one that fluctuates between a rich and poor food supply. The individual that chooses a variable site could get lucky and find plenty of food quite easily, but there is always

Figure 12.20 A chipmunk foraging. Chipmunks generally forage on seeds beneath deciduous trees. Thus, their food source exists in patches of fluctuating abundance. As the quality of their current feeding location declines, they spend more time sampling the food abundance at other locations.

the risk that food will be scarce. In the jargon of foraging theory, the term risk refers to variability in food abundance. Some species seem to consider the degree of variability in food supply when deciding where to forage. These species are described as being risk-sensitive. Our discussion so far has assumed that animals are indifferent to risk, that they behave as if all they care about is maximizing energy gain. It has been presumed that animals will always choose the patch that provides the highest average yield despite differences in the fluctuation in food supply. This view is too simple, however. Studies have shown that many animals do discriminate between reliable and risky sites, even when the average amount of food is the same at both locations. Some animals are gamblers and choose the variable site. They are called risk-prone. Others, those who are risk-averse, tend to choose reliable sites where they are more or less guaranteed of finding a mediocre amount of food (Stephens and Krebs, 1986).

Why might it be advantageous to consider the variability of food supply when deciding where

to look for food? One hypothesis for the function of risk-sensitivity is that it reduces the chances of starvation (Stephens and Charnov, 1982). The reasoning is as follows: An animal that fails to find a certain minimal amount of energy each day will die of starvation. If enough food can be found at the site that provides a stable food supply, then there is no benefit to gambling on finding sufficient food at the variable location. However, if the stable site does not provide enough food to prevent starvation, then the only chance for survival is to forage in the location where the food supply is variable and hope for the best.

The hypothesis that risk-sensitivity minimizes the risk of starvation predicts that individuals will consider the variability in food supply in addition to the average amount of food available at two locations. A preference for stable sites should change to one for risky patches as the individual get closer to starvation. This prediction has been met in studies of shrews (Barnard and Brown, 1989) as well as several species of birds, including juncos (Caraco, 1981), white-crowned sparrows (Caraco, 1983), humming-birds (Stephens and Paton, 1986), and warblers (Moore and Simm, 1986).

Risk-sensitivity in bumblebees (*Bombus* spp.) has been the basis of several experiments and has led to disagreement over the function of risk-sensitivity (Figure 12.21). In one experiment, Leslie Real (1981) permitted bumblebees to choose between two types of artificial flowers, one that provides a constant supply of nectar and one with a variable supply, but both yielding the same average amount of nectar. Eighty five percent of the bees' visits to flowers were to those that provided a constant source of nectar. Thus, they were risk-averse. When the average

Figure 12.21 A bumblebee foraging. Bumblebees are risk-sensitive foragers, that is, they behave as if they consider whether the foraging location provides a stable food supply or one that varies in richness. When the colony has sufficient energy reserves, the bees chose to forage at flowers that provide a stable, but mediocre, amount of nectar. However, when colony energy reserves are low, they prefer to gather nectar from flowers that fluctuate in nectar abundance.

volume of nectar supplied by variable flowers was greater than in the constant sources, the bees would visit variable sites more frequently (Real, Ott, and Silverfine, 1982).

A reanalysis of the data from these experiments led to the suggestion of an alternative hypothesis for risk sensitivity in bumblebees. It has been proposed that when the mean reward of reliable and variable sources is similar, bumblebees prefer reliable sources of nectar so that their net rate of energy intake is maximized (Harder and Real, 1987). It turns out that when the time needed to handle flowers and to fly between them is taken into account, the expected rate of net energy intake is lower at variable sites.

Ralph Carter and Lawrence Dill (1990) distinguished between these hypotheses by keeping the expected net energy gain the same in variable and constant flowers and observing the bees' preference as the colony's energy reserves were manipulated. A bumblebee colony's energy source is nectar stored in open-topped honey pots, so the level of nectar in the honey pots could be altered to change the colony's energy reserves. If risk-sensitivity serves to make it more likely that bees will obtain some minimal amount of energy, then their behavior should change with changes in the amount of nectar in the honey pot. However, if risk sensitivity in bumblebees is a result of lower rewards at variable flowers, then there should be no change when the energy reserves are altered. It was found that when reward rate was kept constant, energy reserves did affect the bees' flower choice. Risky variable flowers were preferred when the honey pots were experimentally drained of nectar, but constant flowers were preferred when the honey pots were experimentally supplemented with nectar. Thus, the hypothesis that risk-sensitivity functions to increase the likelihood of obtaining some minimal amount of energy is supported.

In the case of bumblebees, however, it seems

that this energy requirement is necessary to minimize the probability of reproductive failure rather than to prevent starvation. An individual forager could easily obtain enough energy to sustain itself with the first few flower visits, but if the colony does not have enough energy, it may be unable to repel parasites and will have a longer period of brood development* (Carter and Dill, 1990). Thus, the function of risk sensitivity may depend on how the forager uses the energy it obtains. Different models of risk-sensitive foraging can be developed for cases in which the energy is used for reproduction and those in which the energy is used to minimize the risk of starvation (McNamara, Mermad, and Houston, 1991).

Deciding when to Leave a Patch—Marginal Value Theorem

As an animal forages within a particular patch, food may become more difficult to obtain. This may be because the supply becomes depleted as it is eaten, because the best sources were utilized first leaving only poorer quality sources, or because prey begin to take evasive action. Regardless of the cause, at some point it becomes advantageous for the animal to move to a new patch where food will be easier to find.

According to optimality theory, there are several factors for an animal to consider when deciding whether to leave a patch. One is the abundance of food at the current site relative to other locations. Clearly, it would be wise to move on if one could do better somewhere else. This could be determined by comparing the food abundance at other feeding locations to the expected food abundance at the current location. To maximize energy gain, the individual should look for a new patch when its rate of intake at

the current feeding location drops below the average value for all other areas. If the food availability at the current location is below the average value of other locations, the animal has a better than even chance of moving to a better patch.

In deciding whether to leave its present site, the animal must also consider the difficulty in getting from one patch to another. The more difficult it is to get to a new location, measured in time or effort, the longer the individual should continue feeding at its current spot. After all, a new location could be below average. (The grass may only *look* greener on the other side of the fence.) In that event, time would have been better spent by searching for food within the old patch.

The marginal value theorem, which describes behavior that maximizes long-term rate of return for resources that are patchily distributed, has often been applied to foraging situations (Charnov, 1976). Put simply, this model predicts that as travel time between patches increases or as the average quality of the overall habitat decreases, the animal should spend relatively more time in any given patch. If the animal leaves a patch too soon, its overall rate of energy intake will decrease because it will be wasting time traveling, and if the animal remains in the patch too long, it will waste time searching for food in a depleted area.

We might ask, then, what rule of thumb a forager uses when deciding whether to leave a patch. The decision rules most frequently suggested are (1) remain in the patch for a certain length of time; (2) remain in the patch until a certain number of food items have been obtained; and (3) remain in the patch until the rate of food capture declines to some threshold level. As might be expected, different species appear to follow different rules.

In laboratory experiments, the blue-gill sunfish (*Lepomis macrochirus*) seems to use the rate of prey capture as a measure of the quality of

---

*Worker bees are sterile. Their reproductive gains are in the sisters they help to raise. See chapter 19 for a discussion of how this might have evolved.

the patch (DeVries, Stein, and Chesson, 1989). However, these fish remain in a patch longer than predicted by the marginal value theorem (Crowley, DeVries, and Sih, 1990). Although there are several other possible explanations why blue-gills linger at a depleted site, one is that the costs of traveling to a new area make it worthwhile to linger at the current location.

Travel time does appear to be a factor in determining how long the individuals of some species remain in a foraging patch. One such species is the eastern chipmunk. Chipmunks gather food to bring back to their nests and, therefore, provide an example of central place foraging. In an experiment similar to one previously described, the food patches were trays of sunflower seeds. By varying the distance of the tray from the burrows, travel time of the chipmunks was altered. It was found that the amount of time spent at the trays (patch time) increased as the distance between the seed tray and the burrow increased (Giraldeau and Kramer, 1982).

CONSTRAINTS ON OPTIMAL FORAGING—PREDATION AND COMPETITION

In our discussion so far, we have considered examples of individuals foraging in a manner that allows them to make the largest possible energy profit. However, this is not always true. Other factors, such as the presence of predators or competitors, may require the forager to compromise.

The ultimate challenge is to eat without being eaten. Therefore, a forager may have to sacrifice some energy gain for safety. This trade-off may be made in a variety of ways. One way is to minimize the time spent foraging. Since it is difficult to simultaneously watch for predators and search for food, risk generally increases with foraging time. In some circumstances, such as when it is not necessary to gain weight in preparation for hibernation or reproduction or to bring food to offspring, additional calories are

not worth the risk. Such is the case for a non-breeding golden-winged sunbird (*Necterinia reichenowi*). After gathering just enough food to sustain itself, it retreats to the safety of a sheltered location (Pyke, 1979).

Another way to make the trade-off between predation risk and energy gain is to feed in safer locations, even if food is not plentiful. Pika, for instance, are easy pickings for several predators (Figure 12.22). They generally do not risk venturing too far from the safety of their homes amid rockslides and boulder piles in the tundra. There are several lines of evidence suggesting that foraging by collared pikas (*Ochotona collaris*) is influenced by predation. First, they generally find what they can in the overgrazed areas near their homes rather than gambling their lives on the chance to browse in the rich meadowlands farther away. Second, juveniles, which are easiest for predators to catch, remain closer to their home than do adults. On the other hand, lactating females, with an increased need for food, forage farther from home than do pregnant females. The meadows farther from home are richer, but the risk of predation is also greater. Third, the length of time that pika refrain from foraging after a predator, such as an ermine, has been in the vicinity depends on how close the ermine had come to capturing the pika. Finally, when the habitat was experimentally modified by creating long narrow rows of rocks that could serve as refuge for pika facing the threat of predation, individuals would venture farther from home (Holmes, 1991).

We might expect that the balance between safety and energy gain might be influenced by increases in either the value of the food or degree of danger. Indeed, several studies have confirmed these expectations. One such study examined the importance of these factors to foraging young coho salmon, *Oncorhynchus kisutch*. During the first two years of their lives, coho salmon live in streams where they feed on small invertebrates that drift by. They wait for suitable prey in a central location where they

Figure 12.22 Pika (*Ochotona princeps*) gathering roots and plants near their burrows. Pika generally forage in meadows close to the safety of their homes even though the food supply is meager compared to that in meadows farther away. It is thought that optimal foraging is constrained by the presence of predators.

blend with the gravel background and are difficult for predators to spot. When a prey item appears, the fish swims to gulp it. While moving, the fish is much more visible to predators, so risk increases with the distance traveled.

The value of food was manipulated by varying the size of the prey, the degree of hunger, and the amount of competition for the food item. Juvenile coho salmon will take greater risks when the importance of the food is greater. For example, they will travel longer distances to obtain larger prey (Dunbrack and Dill, 1983). Furthermore, a hungrier fish will travel farther for food than a satiated fish. The fish also weighs the importance of obtaining the food item against the chance of losing it to a competitor. In a field situation, the prey drifting along will be gulped down by the first fish to reach it. So, a fish that does not risk swimming toward its prey may lose it to a neighboring territory holder. In one laboratory study, mirrors reflected the fish's own image, thereby increasing the apparent density of competitors. Again, the

fish increase the distance they travel from their territory to obtain prey.

However, if the fish sees a predator and, therefore, perceives that the potential danger is high, its attack distance decreases. This was demonstrated by using a pulley system to threaten the fish periodically with a model of a natural predator, a photograph of a rainbow trout. The more frequently the model was presented, the shorter the attack distance (Figure 12.23; Dill and Fraser, 1984).

It is often the case that values change with maturity so it should not be surprising that the relative weights given to foraging success and predation risk may change with age. This has proven to be true among certain colonial web-building spiders (*Metepeira incrassata*). Position within the colony influences both foraging success and the risk of predation or parasitism. Individuals on the edges of the colony are more successful at capturing prey, but they are also at greater risk. Prey capture rates are 24 to 42 percent higher for individuals located on the pe-

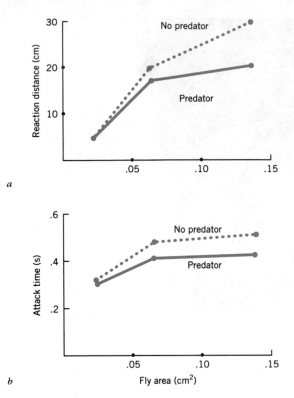

a

b

Fly area (cm²)

Figure 12.23 The risk of predation constrains optimal foraging in young coho salmon. When the young salmon swim away from their covering vegetation to attack a prey, they are more visible to predators. Larger flies, which are more valuable prey, cover a larger area. Notice that the fish will risk swimming farther from its protective resting place for larger flies. However, it will not swim as far in the presence of a predator, regardless of the size of the fly. (Data from Dill and Fraser, 1984.)

riphery of the colony. Unfortunately, spiders in these locations are also subject to 3 times as much predation as individuals in core locations. Thus, there is an inevitable trade-off between foraging success and predation risk. The outcome of this trade-off will influence the number of offspring an individual is likely to leave. Reproductive success is greater in the central lo-cations where the risk of predation and egg parasitism is lower. Predators always prefer large reproductive females over smaller spiders, and since predators approach the web from the edges, the large females living there are at particularly high risk. Furthermore, if the eggs are left unguarded at the periphery, there is an increased chance of cocoon parasitism. As a result, most individuals hatch and begin life in the central regions of the colony. Here, prey is scarce so larger, and older spiders have an advantage in contests over prey. Therefore, younger, smaller spiders may take a chance on living on the edges of the colony where they can obtain more food and grow faster, thereby increasing the odds of reaching sexual maturity. But as the spiderlings mature, values change and safety becomes more important than foraging success. So, the larger spiders that have reached sexual maturity prefer the core positions (Raynor and Uetz, 1990).

Summary

There are many ways an animal can obtain food. Many aquatic animals filter food from the surrounding water. Certain species work as farmers and cultivate their specialized or preferable food. The leaf cutter ant, for instance, maintains fungus gardens and ensures a supply of a special fungus that is its primary food source.

Animals that hunt for prey employ special techniques that enhance their prey capture success. Some build traps. Others lure their prey with specialized structures or behaviors that mimic the courtship displays of the prey species. Still other species employ a hide-and-wait strategy and ambush their prey.

Certain hunters also have adaptations that reduce the likelihood that they will be spotted by their prey, thereby allowing them to come within striking distance of their prey. Camouflage, for instance, makes the predators more difficult to see because they blend in with the environment. Certain other predators are disguised as species that are beneficial to the prey.

There are many sensory specializations that enhance a predator's ability to detect its prey. Some animals, such as mosquitoes and certain snakes (the Crotalidae and the Boidae), detect the body heat that emanates from their warm-blooded victims. The sand scorpion, on the other hand, is exceedingly sensitive to vibrations in the sand created by its prey. From a human perspective, perhaps the most unusual sensory specialization is the ability of certain animals, including sharks, to detect the electric fields that living organisms generate when they are in sea water. The electric sense was shown to exist in sharks by eliminating other sensory cues that could be used for prey localization and determining whether the shark could still locate its prey. Although visual and olfactory cues may be used in nature, they are not necessary for prey detection. However, a shark will attack electrodes that produce a weak electric field, even if acceptable food is visible.

We often find that predators become more adept at finding cryptic prey following experience with them. Two hypotheses have been presented to explain this observation. The search image hypothesis is that the predator learns key features of the cryptic prey and then focuses on them while hunting. The search rate hypothesis is that after a predator finds a few cryptic prey, it slows the rate at which it searches for prey and, as a result, it misses fewer cryptic food items. Studies have shown that predators use both techniques to locate cryptic prey.

According to optimality theory, natural selection favors the behavioral alternative whose benefits outweigh its costs by the greatest amount. The optimal foraging pattern, then, is the one that maximizes net energy gain.

One of the first decisions a forager must make is what to eat. Studies have shown that animals

such as northwestern crows choose food items that provide a net energy gain (large whelks) and not those that require more energy to obtain than they provide (medium or small whelks).

When faced with a variety of foods, all of which would yield an energy gain, the forager should choose the optimal diet, the one that maximizes energy gain. The profitability of a food item is defined as the energy it provides divided by the time it takes to find, capture, prepare, and digest that item. According to theory, the acceptability of any particular food item depends on the abundance of the most profitable food item, rather than the abundance of that particular item. When the most profitable items are abundant, less profitable items will be eliminated from the diet. The results of field studies on redshanks feeding on worms and laboratory studies of great tits feeding on mealworms are consistent with the predictions of optimal diet theory.

When we find animals choosing less profitable food items, it may be due to specific nutritional requirements. The individual may have to eat some less energetically profitable items in order to obtain an essential dietary component.

Food is often easier to find in some places than in others. Optimal foraging theory predicts that animals will choose to forage where food is most easily obtained. In order to determine which area is most profitable, many locations must be sampled. But sampling different patches before concentrating on the best one reduces the average rate of food intake. Optimal foraging theory predicts that the time spent sampling should increase as the quality of the patch being exploited decreases.

As an animal forages within a patch, the food may become more difficult to obtain. The marginal value theorem predicts that (1) an individual should leave its current feeding location when the abundance of food there drops below the expected average for the environment, and (2) an individual should remain at its current feeding location longer the more difficult it is to reach a new location.

Individuals of some species may have to choose between a feeding location that supplies a moderate, but constant, supply of food and one that fluctuates between a rich and a poor food supply. Species that seem to prefer or avoid variable feeding sites are called risk-sensitive.

Risk-sensitivity may function to lessen the chances of failing to obtain enough energy to stay alive or for reproduction.

When we observe animals that are not foraging optimally, it may be because they are sacrificing energy gain to avoid predation or because they are competing with other individuals for food.

CHAPTER 13

Antipredator Behavior

*A*nimals are never safe from predators, and monarch butterflies (*Danaus plexippus*) are no exception as they fall prey to birds and small mammals. In light of the constant threat of predation, these butterflies have developed an impressive array of devices to outsmart their enemies, and their protective strategies appear to work, at least some of the time.

Like many animals, monarch butterflies may use a combination of color pattern and behavior to avoid being eaten. Their boldly patterned orange, black, and white wings warn potential predators that they taste bad, their unpalatability being due to their assimilation of noxious chemicals from food plants. In particular, monarch larvae feed on milkweed plants (Asclepiadaceae) and incorporate toxins, called cardiac glycosides, into their own tissues (Brower et al., 1968). Predatory birds that eat one of these insects, even as adults, experience severe vomiting and tend to avoid butterflies of similar appearance in the future. However, one might wonder what good it is to be filled with toxins if the individual must be eaten before the poisons will work. The advantage is that many predators release unharmed prey that are brightly colored and bad tasting. Poisons, stolen from plants, may thus deter some predators. However, not all milkweed plants contain cardiac glycosides. Butterflies reared on plants that do not contain the poison are quite palatable, although they still may be avoided by predators who have had experience with noxious members of the species. Even butterflies of other species try to cash in on the defense system of *D. plexippus*. By resembling the sometimes unpalatable monarch, individuals of other species deceive predators into avoiding them as well. Such deception is a common component of antipredator strategies.

No defense system works all the time. Even with the monarch butterfly, the effectiveness of protective strategies often varies with species of predator, season, and context of the predator–prey encounter. Each fall, monarch butterflies, some from as far away as the northeastern corner of the United States, migrate to the mountains of central Mexico where they spend the winter in densely packed aggregations of tens of millions of individuals. The months spent in Mexico, however, are far from a winter vacation (Figure 13.1). Birds of two species, the black-backed oriole (*Icterus galbula abeillei*) and the black-headed grosbeak (*Pheucticus melanocephalus*), have penetrated the monarch's chemical defense system; these two species eat an estimated 4,550 to 34,300 butterflies per day in some overwintering colonies (Brower and Calvert, 1985). The oriole selectively strips out relatively palatable portions of the butterflies' bodies (e.g., the thoracic muscle and abdominal contents), and the grosbeak appears insensitive to the cardiac glycosides (Fink and Brower, 1981). All, however, is not lost for the monarch. Facing each winter with the prospects of an avian feeding frenzy, the butterflies reinforce their antipredator system by converging in great numbers (Calvert, Hedrick, and Brower, 1979). By forming dense aggregations, the predation risk to any one individual is dramatically diluted. Additionally, because predation is most intense at the periphery of the colony, central positions are highly sought after and are quickly assumed by the first individuals to arrive. In the life of a monarch butterfly, it just does not pay to be fashionably late in arriving at the overwintering site.

Since predation is a pervasive theme in the pageant of life, we can ask how other animals cope with its constant threat. Which devices aid in escaping detection by a predator (primary defense mechanisms) and which come into play once a prey animal has been detected and capture seems all too imminent (secondary defense

Figure 13-1 Avian predators penetrate the chemical defense system of monarch butterflies at their overwintering sites in Mexico.

mechanisms)? Does membership in a group always confer antipredator privileges?

Crypsis

Animals that are camouflaged to blend with their environment are said to be cryptic; "I am not here" is the message of cryptic species. Crypticity may range from simple markings that break the body contour to devices that render the animal almost invisible. There are many well-known examples of crypticity, including reed-nesting marsh birds that sit immobile with bills pointed upward, moths that seem to disappear on the bark of a tree, and frogs that are almost invisible against the leaves of the forest floor (Figure 13.2). By blending with their background, these animals may escape the notice of visually hunting predators.

AVOIDING DETECTION THROUGH COLOR AND MARKING

Breaking the Contour

Many animals avoid being seen by simply matching their background color, but sometimes such coloration is not enough because visually hunting predators may recognize prey by their body contour. Some animals break their outline by developing bizarre projections that conceal their body's contour. Other species have bold markings that break their body outline. Termed disruptive coloration, this latter antipredator device is perhaps best illustrated in the black-and-white vertical stripes of the zebra (Figure 13.3).

Countershading

Countershading is another option for avoiding the unwanted attentions of predators. Because light normally comes from above, the ventral

Figure 13.2 Frogs of Malaysia, among dead leaves.

Figure 13.3 The principle of disruptive coloration. To see how it works, place the book in an upright position and then back away. See which disappears first, the zebra or the pseudozebra. The bold stripes of the zebra should obscure the outline of the body best. (After Cott, 1940.)

surface of the body is typically in shadow and predators may cue in on darkened bellies. Many animals appear to obscure the ventral shadow by being paler ventrally and darker dorsally (Figure 13.4). For animals that rest upside down, we would expect just the opposite color pattern, that is, the dorsal surface would be pale and the ventral surface dark.

Direct evidence for countershading as a concealing adaptation is embarrassingly meager. Until such evidence is obtained, we cannot discount the possibility that the combination of dark backs and light bellies is totally unrelated to crypsis and is involved instead in biological functions such as thermoregulation or protection against ultraviolet radiation (Kiltie, 1988).

Transparency

Some animals are cryptic by simply being transparent. Although no animal is completely transparent, organisms such as cnidarians (e.g., hydroids and jellyfish), ctenophores (e.g., comb jellies), and the pelagic (open ocean) larval stages of many fish achieve near transparency by such means as high water content of tissues, small size, and reduced number of light-absorbing molecules or pigments (McFall-Ngai, 1990).

Frequently neglected in discussions of cryptic mechanisms, transparency is probably the dominant form of crypsis in aquatic environments. It is more common in aquatic than in terrestrial habitats for two reasons. The first is based on

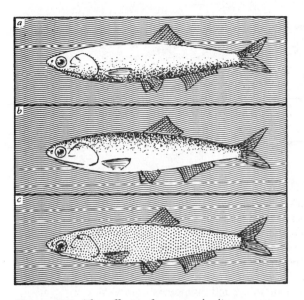

Figure 13.4 The effects of countershading on conspicuousness. Light normally comes from above, and under these circumstances fish that are uniformly colored have a conspicuous ventral outline (*a*). A countershaded fish is darker dorsally then ventrally (*b*; shown here, illuminated from all sides), and thus its body outline is obscured when light comes only from above (*c*). (After Cott, 1940)

differences in the refractive indices (the angle at which light bends when passing from one medium into another) of water and air. Because animal bodies are largely water, when light travels from the surrounding water into the tissues of an aquatic animal, the angle of light is virtually unchanged; and in the absence of light-scattering or light-absorbing elements, the animal appears transparent (light, then, is basically passing from water into water). In contrast, in a terrestrial environment, light must pass from air into the water-filled tissues of an animal. The difference in the refractive indices of air and the terrestrial animal's tissues creates an obvious body outline, greatly diminishing transparency. The second reason transparency is rarely used as a camouflaging mechanism by terrestrial animals has to

do with the deleterious effects of ultraviolet radiation on land. In aquatic habitats, much of the ultraviolet radiation is filtered out within a few meters of the water's surface and thus animals living beyond this distance are not subject to the same radiation damage as terrestrial organisms.

Color Change

Usually, cryptic animals are camouflaged in some habitats but not in others, and thus their occurrence is often restricted to those particular areas where they are best concealed. One way some species get around this restriction is by changing color as they change backgrounds. Although the chameleon is perhaps the most familiar example, the cuttlefish (*Sepia officinalis*) is the true master of color change. (The cuttlefish is a cephalopod mollusc related to such creatures as squid, octopus, and nautilus. Chances are that you have actually seen part of the white skeletal support of this animal hanging in a bird cage. Called cuttlebone, the material is not bone at all, but it serves to help parakeets and other birds keep their beaks sharp.) Among naturalists, the cuttlefish is best known for the speed with which it changes color to match its background, each pattern rendering it virtually invisible to both its predators and its prey. William Holmes (1940), working in a laboratory in England, documented the magnificent changes in color and pattern (Figure 13.5). According to Holmes, the swimming cuttlefish adopts disruptive coloration in the form of a brown-and-white zebra pattern. When resting on the bottom, however, *Sepia* adjusts its color to that of the substrate at hand. Within a matter of seconds the dorsal color can change from dark brown to sandy brown to almost white as the cuttlefish settles on one background after another. Not surprisingly, the cuttlefish also demonstrates countershading. Typically darker dorsally than ventrally, the cuttlefish thus obscures the ventral shadow created by light from above. When turned on its back in the water, however, its

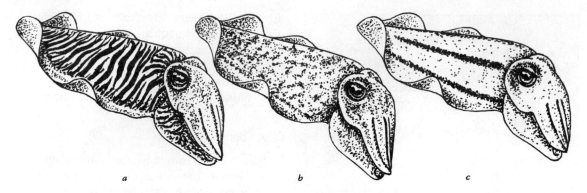

a b c

Figure 13.5 Many cryptic animals are restricted to portions of their habitat in which they are well concealed. The cuttlefish gets around this restriction by changing its color to match the particular background on which it rests. When free-swimming, the cuttlefish has disruptive marks reminiscent of the zebra pattern (*a*), but within a matter of seconds of settling on a particular substrate, its dorsal pallor can change to either dark brown, white, or a light mottle pattern characteristic of sand (*b*). Pattern (*c*) occurs just before the cuttlefish ejects a cloud of ink and disappears from view completely. (After Holmes, 1940.)

ventral surface darkens and its dorsal surface pales.

Color Polymorphism

Many insects possess the ability to change color. Typically, the transition from one color to the next does not occur instantaneously as the individual moves from one background to another (as with cuttlefish), but rather at specific times in their natural history, such as molting. Joy Grayson and Malcolm Edmunds (1989) examined the causes of color and color change in caterpillars of the poplar hawkmoth (*Laothoe populi*), a species that normally proceeds through four larval instars (stages before pupation and metamorphosis to the adult moth). Final instar caterpillars of this species can be yellow-green, dull green, or white. The dull green color is genetically determined, and yellow-green and white are environmentally induced polymorphisms (i.e., different forms or "morphs" whose color depends on the immediate surroundings). Apparently, the main factor

that determines whether a caterpillar becomes white or yellow-green is the surface upon which it rests and feeds during its first two or three instars. It is neither the color (wavelength of light) nor the nutritive qualities of the leaf that are critical. Instead, it is the intensity of the light reflected from the leaf and perceived by the young caterpillar: If a larva sees white, then it becomes white; but if it should see green, grey, or black, then it becomes yellow-green.

Environmentally induced polymorphisms in body color also occur in grasshoppers that inhabit the savanna of Africa (Hocking, 1964). Fires, either natural or set by humans, sweep through these areas annually or sometimes every second or third year. Following a fire, some grasshoppers change color in a matter of days so as to blend with the blackened ground. Other grasshoppers cope with periodic burning of the vegetation by using two color forms, one green and one black, each of which seeks its appropriate burnt or unburnt background.

A similar situation has been reported for fox squirrels (*Sciurus niger*) in the eastern United

Figure 13.6 Some color morphs of the fox squirrel. Note the variation in percentages of dorsal black and pattern around the head and ear region. (After Kiltie, 1989.)

States. Fox squirrels have been described as the most variable in color of all mammals in North America (e.g., Cahalane, 1961). Color varies both among and within populations. Dorsal coloration may range from gray or tan to black, and coloration on the head and ear region is often distinctive (Figure 13.6). Even within a single litter, both melanistic (black) and nonmelanistic young can be found. Intrigued by the variation in coat color of fox squirrels, Richard Kiltie (1989) examined close to 2000 museum specimens of this species. He determined the percentage of dorsal black for each skin and compiled information on the occurrence of wildfires in the eastern United States. Taken together, his data on coat color and fires show that the incidence of melanistic individuals is correlated with the frequencies of wildfires over the total range of the squirrels. Both wildfires and melanistic squirrels are more common in the southeastern United States (Figure 13.7).

In fox squirrels, the melanistic polymorphism in coat color may thus be maintained by the periodic blackening of the ground and lower portions of tree trunks by wildfires. One would imagine that dark squirrels are less conspicuous to hawks than are light or variably colored individuals against a blackened background. The advantage, however, does not remain with the black squirrels for long. As rainfall and new plant growth convert a charred area into a less uniformly black substrate, fox squirrels with variable amounts and patterns of black dorsal coloration would be more cryptic than uniformly black individuals against the patches of light and dark background. Finally, when the period of regrowth of the pine and oak forest is almost complete, the advantage may shift to squirrels that are uniformly light in coloration. Thus, variable coat color in fox squirrels may result from the alternating cryptic superiority of light and dark individuals in an environment that periodically burns and regenerates. Development of black coloration in response to fire is termed fire melanism.

AVOIDING DETECTION THROUGH BEHAVIOR

The vast majority of animals lack the ability to rapidly adjust color and pattern to their immediate surroundings. Since individuals that cannot change color will be conspicuous if they rest in the wrong place, selection of the appropriate

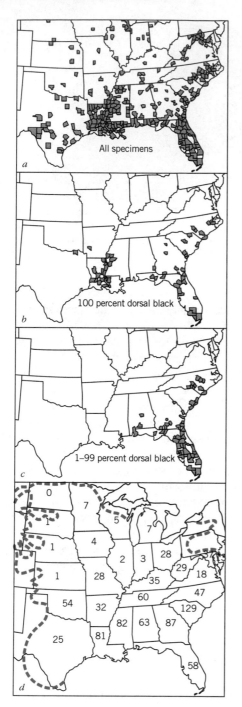

Figure 13.7   In the fox squirrel, the incidence of melanism is correlated with frequency of wildfires. (*a*) Most of the counties in the eastern United States from which Kiltie (1989) examined museum specimens of fox squirrels. (*b*) Counties from which specimens with 100 percent dorsal black were noted. (*c*) Counties from which specimens with intermediate levels of dorsal black (1–99%) were recorded. (*d*) Average number of wildfires per state in protected forestlands during the years 1978–1982; values have been normalized to take into account the area of land under wildfire surveillance (dashed lines depict limits of fox squirrel range). Note that melanistic fox squirrels occur primarily in the southern portions of the species range (*b*) and that individuals with intermediate levels of dark coloration are limited to the eastern Gulf and Atlantic coastal plains (*c*). These areas in the southeastern United States are also the areas in which wildfires are most common (*d*). (From Kiltie, 1989.)

background, and correct orientation on that background, are critical.

The Peppered Moth: A Case in Point

In Europe, the peppered moth (*Biston betularia*), comes in both typical and melanic forms (Figure 13.8). The typical form (known as *typica*) is whitish gray with a sprinkling of black dots and is cryptic when resting on lichen-covered trees, and the most extreme melanic form, known as *carbonaria*, is almost completely black and is well concealed on the bark of dark trees. Before 1850, less than 1 percent of the peppered moths around Manchester, England, were melanic, but by 1895 melanic moths made up 98 percent of the Manchester population (Howlett and Majerus, 1987).

The incredible success of the black form and the relatively poor showing by the light form have been correlated with the extreme industrialization in England. In the mid-1880s England, converting to an industrial economy, started burning soft coal in its factories. The

Figure 13.8 Two forms of the peppered moth. Above is the typical form and below is the melanic form. Typical moths survive better in unpolluted woods with lichen-covered trees, but melanic moths fare better in polluted areas with soot-covered trees.

lichens that covered the trunks of trees were destroyed, and the greenery of the English countryside quietly submitted to its new layer of soot. The tables had turned. Now, the melanic moths were hidden against the soot-covered vegetation and the typical form was conspicuous to predators.

Bird predation was a major selective force in the population shift from the light to the dark form. This was demonstrated by a series of experiments conducted by H. B. D. Kettlewell (1955, 1956). At one heavily polluted site and one less polluted site, he marked, released onto tree trunks, and then recaptured light and dark individuals, each time recording the number of survivors of each morph. Kettlewell also observed the behavior of insectivorous birds feeding on the trees where moths had been released. The results, summarized in Table 13.1, demonstrate that the dark forms survived better at the heavily polluted site and the light forms fared better at the less polluted location. Apparently, when placed on the bark of soot-covered trees, the dark form is less conspicuous to birds than is the light form and suffers less predation. The reverse is true in relatively pollution-free areas where the light form is almost invisible against the bark of lichen-covered trees.

Over the years, as England's countryside has returned to a less polluted state, many researchers have attempted to predict changes in the frequencies of dark and light forms by using computer models. In most instances, the observed frequencies of moths did not agree with those predicted, and dark forms were more common in unpolluted areas than the models projected. As it turns out, *Biston betularia* does not usually rest in exposed positions on tree trunks, but rather in more hidden locations on the underside of branches (Howlett and Majerus, 1987). Given its dark appearance, the melanic form apparently gains a greater advantage from resting in the shadows of branch joints than does the typical form with its salt-and-pepper look. Thus, in *B. betularia* survival depends not only on body color but on behavior, that is, choice of location on a tree.

Selection of the Appropriate Background

Many animals appear to select "correct" backgrounds, and once there, they exhibit behavior that maximizes their crypsis. The California yellow-legged frog (*Rana muscosa*) inhabits swift flowing streams in the woodlands of southern California. The light gray granite boulders that

Table 13-1

INDUSTRIAL MELANISM IN PEPPERED MOTHS

LOCATION	TYPICALS		MELANICS	
	NUMBER RELEASED	NUMBER (PERCENTAGE) RECAPTURED	NUMBER RELEASED	NUMBER (PERCENTAGE) RECAPTURED
Dorset (unpolluted)	496	62(12.5)	473	30(6.3)
Birmingham (polluted)	137	18(13.1)	447	123(27.5)

Source: Data from Kettlewell, 1955, 1956.

line the streams seem conspicuous resting spots for the yellow-brown frog. Below the water, however, these same boulders are covered by a yellow-brown layer of algae. At a moment's notice, *R. muscosa* leaps into the water and lies motionless against a background to which it is perfectly matched (Norris and Lowe, 1964).

Although the mechanisms for color matching among most cryptic animals have gone largely unexplored, it appears from experiments with insects that choice of a background in some species results from a behavioral matching response (i.e., animals compare certain parts of their bodies with their backgrounds; Sargent, 1968). Remember that cryptic animals not only select appropriately colored backgrounds, but they also assume positions that maximize the effectiveness of their crypsis. For example,

Figure 13.9 Many moths that exhibit barklike crypsis orient in such a manner that their markings match up with the lines and ridges of the bark. This moth has oriented so as to be virtually indistinguishable from the bark upon which it rests.

moths that exhibit barklike crypsis orient themselves on tree trunks in such a way that their markings are aligned with the direction of the lines and ridges of the bark (Figure 13.9). Although some moth species appear to use tactile cues associated with the immediate substrate to determine the proper resting position, other species adopt resting attitudes on the basis of cues, such as gravity, that are independent of the bark (Sargent, 1969). Now that we have some ideas about how background selection occurs in at least a few species, we turn our attention to a test of why it occurs.

Is the combination of cryptic coloration and background selection adaptive? If it is, then prey should experience less predation when sitting on substrates they tend to select as resting spots than when sitting on other surfaces. Blair Feltmate and D. Dudley Williams (1989) tested this idea using rainbow trout (*Salmo gairdneri*) as predators and stonefly nymphs (*Paragnetina media*) as prey. Background color preference of stoneflies, stream insects that are dark brown to black in color, was first tested by placing each of 24 nymphs into its own aquarium along with one dark brown and one light gray commercial tile on the aquarium bottom. Nymphs were left to settle for 24 hours and then at 1400 hours (2 P.M.) the researchers recorded whether nymphs rested on the dark brown or light gray substrate. The experiment was then repeated with recordings of nymphal position at either 0200, 0600, 0800, and 2100 hours to test whether selection of substrate varied as a function of time of day (lights in the laboratory were on timers and were off from 1900 to 0700 hours). Thus, independent replicates of the experiment were run at five different times of the day, three in the dark (2100, 0200, and 0600 hours) and two in the light (0800 and 1400 hours). The results, depicted in Figure 13.10, demonstrate that stoneflies selected the dark brown substrate rather than the light gray one at 0800, 1400, and 2100 hours; no selection was observed at 0200 or 0600 hours. Although stonefly nymphs selected the dark over the light substrate, this selection

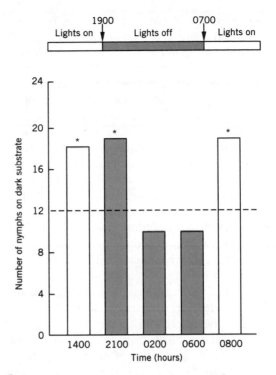

Figure 13.10 Substrate selection in stonefly nymphs at various times of the day (white bars represent data when lights are on in the laboratory and dark bars when lights are off) when given the option of resting on dark brown or light gray tiles. During the lights on period (0800 and 1400 hours) and shortly after the lights go off (2100 hours), a larger number of nymphs were observed on dark tiles, a background upon which they were cryptic, than on light tiles. Substrate selection was not apparent during the two remaining dark observations (0200 and 0600 hours). The dashed line represents expected results if no selection occurred; the asterisks indicate selection for dark brown tiles. (From Feltmate and Williams, 1989.)

ceased approximately 2 hours after the lights in the laboratory went off and resumed within 1 hour of the lights' being turned on.

In the next experiment, Feltmate and Williams examined whether stoneflies resting on the light substrate were more vulnerable to predation by rainbow trout. As before, each stonefly was introduced into its own aquarium. This time, however, the tank contained either light or dark tiles (not both as in the first experiment). A trout was released into each tank after the nymphs had 2 hours to adjust to their new surroundings. Twenty-four hours after releasing the nymphs, the authors recorded the number of stoneflies consumed in tanks containing either the light or dark substrate. Consumption of nymphs by trout was lower in tanks that contained the dark substrate (3 of 24 nymphs eaten) than in tanks that contained the light substrate (19 of 24 nymphs eaten). These data suggest that selection of dark resting spots by stoneflies has evolved, at least in part, as a means of reducing the risk of being found and eaten by visually hunting fish. The breakdown in substrate color selection during the hours of darkness also links visual predation to the distribution of nymphs. After all, animals need only be cryptic during the time when they are most vulnerable to predation by visual hunters (Endler, 1978). Choice of substrate by stoneflies may also function in concealing them from their own prey as has been shown for other aquatic insects (Moum and Baker, 1990).

Movement and Absence of Movement

Movement, and in some cases absence of movement, is an important component of crypticity. Once oriented correctly on the appropriate background, many cryptic animals remain immobile most of the time, and when they do move, they move so slowly that they attract as little attention as possible. In other cases, rapid motion followed by abrupt cessation of movement contributes to crypsis. The escape behavior of newborn northern water snakes (*Natrix sipedon*) illustrates this point. Although adults of this species are uniformly colored, the young have a crossbanded pattern and appear cryptic when lying motionless in vegetation. When disturbed, however, newborns zip along the ground and then come to a complete stop. The rapid movement blurs the crossbands, making the snake appear unicolored (Figure 13.11). This

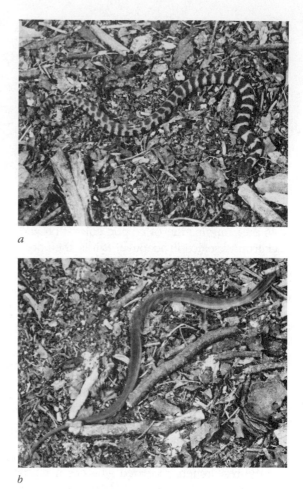

a

b

Figure 13.11 Newborn water snakes use a combination of cryptic coloration and movement to fool visually hunting predators. (*a*) Young snake with crossbanded pattern appears cryptic when lying still in vegetation. (*b*) When startled, the snake moves rapidly across the substrate, causing the crossbands to blur, and then comes to an immediate halt. A predator tracking the newborn's movement ends up searching for the snake way ahead of its actual position.

visual blending illusion and the abrupt transition from motion to stillness combine to lead the observer (and presumably a predator) to search for the snake ahead of its actual position, for one's eyes tend to follow the path of a moving object even after that object has stopped moving (Pough, 1976). Here, then, behavioral responses combined with cryptic coloration maximize camouflage.

Modification of the Environment

Some animals improve their chances of survival by modifying the background upon which they normally rest. The spider *Tetragnatha foliferens* hides under a leaf that it folds into a tube and fastens to the center of its web (Hingston, 1927a). From this vantage point, the spider is not only inconspicuous to predators, but it is in the best position to gain quick access to prey entangled in its web.

Insectivorous birds appear to use leaf damage as a clue to the location of cryptic caterpillars (Heinrich and Collins, 1983). Thus, it should come as no surprise that some species of palatable caterpillars avoid attracting birds by snipping off partially eaten leaves at the end of a feeding bout (Heinrich, 1979). Caterpillars of the underwing moth *Catocala cerogama* feed on the leaves of basswood trees and their dorsal coloration resembles the bark of basswood twigs. At the end of a nocturnal foraging bout, an individual of this species backs off the leaf on which it has been feeding, chews through the stalk of the leaf, turns around, and as the chewed leaf falls to the ground, crawls onto the twig where it will remain for the daylight hours. The branch on which the caterpillar was feeding now appears ungrazed. Although *Catocala cerogama* caterpillars are cryptically colored, they further enhance their chances of survival by removing evidence of their presence. (In contrast, unpalatable caterpillars—those with "hair," spines, or toxic glycosides—are quite blatant in their feeding behavior, foraging both day and night, and show no patterns of behavior associated with hiding leaf damage.) Finally, some animals achieve crypsis by carrying parts of their environment around with them. Many species of spider crabs (family Majidae) decorate themselves with algae, barnacles, sponges, and bot-

tom debris and become virtually indistinguishable from the ocean floor (Figure 13.12). According to Mary Wicksten (1980), who has studied camouflage in these crabs, even an experienced underwater naturalist might sit on a crab festooned with bottom debris before realizing that it is there.

OTHER FUNCTIONS OF COLOR

Evasion of predators is not the only function of color or pattern in animals. Color affects heat balance and thus plays a role in thermoregulation. Color and pattern are also important in many aspects of communication, including mate recognition, courtship, male–male competition, and territorial defense.

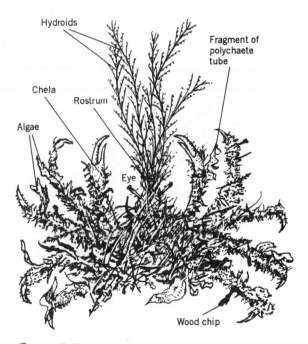

Figure 13.12 Camouflage through gaudy decoration. A crab has decorated its carapace, rostrum, and legs with strands of algae, strings of hydroids, and bottom debris. Such ornaments are picked up by the chelae or pincers, carried to the mouth for roughening of the edges, and then impaled on the shell. (Redrawn from Wicksten, 1980.)

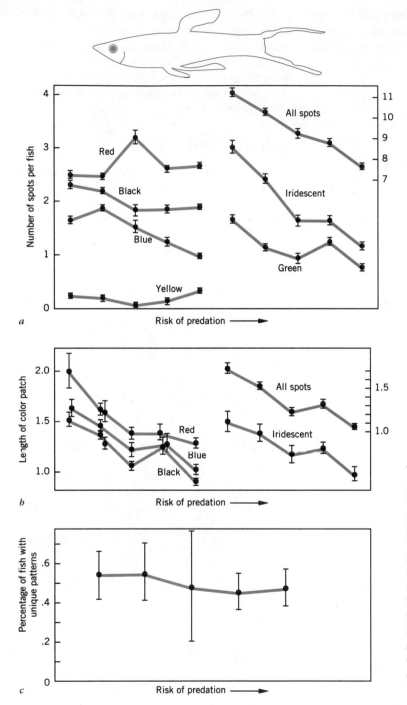

Figure 13.13 An animal's color is often a compromise solution to the problem of selection for conspicuousness in courtship displays and selection for inconspicuousness to visually hunting predators. Changes in the color and pattern of guppies as a function of predation pressure reflect the fine balance between these two selective forces. As predation risk increases, the number of spots (*a*), size of color patches (*b*), and diversity of patterns (*c*) decrease. (Modified from Endler, 1978.)

The various functions of animal color and pattern may act in concert or in opposition. Let us consider a case in which they act in opposition. Assuming that color and pattern are adjusted for thermoregulation, how can animals communicate effectively with mates and competitors, and at the same time be inconspicuous to visually hunting predators? Although some cryptically colored animals have evolved alternative means of exchanging information (e.g., relying on auditory or olfactory signals to communicate with conspecifics), many still rely on visual cues. As we will see, the color pattern displayed by a particular animal may be a compromise between factors that favor crypsis and those that favor conspicuousness.

John Endler's (1978) work with wild populations of guppies (*Poecilia reticulata*) in northeastern Venezuela and Trinidad provides an excellent example of how color patterns may represent a balance between mate acquisition and crypsis. Whereas female choice of mate and competition among males favor brighter colors and more visible patterns in guppies, selection by diurnal visual predators (at least six species of fish and one freshwater prawn) favors less colorful and less conspicuous patterns. It is interesting that as predation risk increases across communities, guppy colors and patterns become less obvious through (1) shifts to less conspicuous colors, (2) reductions in number of spots, (3) reductions in size of spots, and (4) slight reductions in color and pattern diversity (Figure 13.13). In areas in which guppies encounter low predation pressure, however, the balance shifts toward attracting mates, and colors and patterns become more conspicuous.

Polymorphism as Defense

Like most things, cryptic coloration is not foolproof. Although individuals of a cryptically colored species blend with their background, predators in a given area may develop a search image for that particular species (Chapter 12) and systematically search out and consume remaining individuals. If individuals of the prey species are widely spaced, however, predators will rarely encounter them and will soon forget the search image. Indeed, individuals of many cryptic species occur at widely spaced locations throughout their environment.

Other cryptic species get around the problem of predators forming search images by occurring in several different shapes and/or color forms, that is, by being polymorphic (*poly*, meaning many, and *morph* meaning form). We have already discussed some examples of polymorphism with respect to industrial melanism in peppered moths and fire melanism in fox squirrels and grasshoppers. In some cases, however, polymorphic species are not cryptically colored at all and rely solely on their diverse appearance to evade detection by predators. Whether cryptic or not, by being different, individuals of prey species can occur at higher densities without suffering increased mortality from predators searching for individuals with a specific appearance. Some species that occur at very high densities exhibit extreme polymorphism, making it almost impossible to find two individuals that look alike (Figure 13.14).

Gairdner Moment (1962) described the phenomenon in which members of a population look as little like each other as possible. In such populations the probability of an individual's having a certain appearance is inversely related to the number of other individuals in the population having that appearance. If one morph in a polymorphic population is much more common than another morph, then predators are likely to develop a search image for the more common, rather than the rare, morph. The end result is that predators take more of the common form relative to its frequency in the population. Thus, when two morphs are equally cryptic and are exposed to predators that use search images when hunting, the rare morph will have a selec-

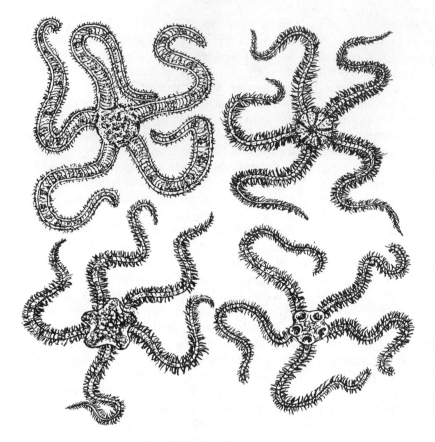

Figure 13.14 Although all the same species, these four brittle stars are dramatically different in appearance, thereby inhibiting the formation of search images in predators. (Drawn from photograph in Moment, 1962.)

tive advantage over the common morph. This form of frequency-dependent selection has been called apostatic selection (Clarke, 1969) or reflexive selection (Moment, 1962). The strength of apostatic selection varies as a function of factors such as density, palatability, and conspicuousness of prey (Allen, 1988). Furthermore, Greenwood (1984) indicates that predators need not hunt by search image to cause apostatic selection in prey. Some predators, for example, may simply have an aversion to prey that are rare or unfamiliar to them.

What experimental evidence do we have that being different pays off? Croze (1970), working on a sandy peninsula in England, placed 27 painted mussel shells with pieces of meat under them on the ground and exposed them to pre-

dation by carrion crows (*Corvus corone*). In some of the 14 trials, the shells were monomorphic (i.e., all the same color), while in others the shells were trimorphic (9 red, 9 yellow, and 9 black). The results, summarized in Table 13.2, show that crows took fewer of the trimorphic than the monomorphic prey. The percent survival for each of the three morphs in a trimorphic population was two to three times higher than that in monomorphic populations. Thus, a morph had a two to threefold selective advantage when occurring as part of a trimorphic population. Croze's results demonstrate that when prey populations occur at the same density, individuals in polymorphic populations experience less predation than those in monomorphic populations.

Table 13.2

SURVIVAL OF PAINTED MUSSEL SHELLS IN EITHER MONOMORPHIC OR TRIMORPHIC POPULATIONS WHEN EXPOSED TO PREDATION BY CARRION CROWS (IN PERCENTAGES)

	TYPE OF POPULATION	
SHELL COLOR	MONOMORPHIC	TRIMORPHIC
Yellow	10	31
Black	12	40
Red	19	45

Source: Data from Croze, 1970.

Before leaving the topic of polymorphism, we should make two points. First, throughout our discussions of cryptic and diverse coloration, we have focused on visually hunting predators and have ignored predators that detect prey through other means. Many animals detect prey using their sense of smell, and the formation of olfactory search images seems quite reasonable. In the case of prey hunted on the basis of olfactory cues, one could imagine animals that are cryptic by being either odorless or similar in odor to their background, or even populations in which odor polymorphisms exist (Edmunds, 1974).

Second, although being different to avoid being eaten may be the primary explanation of polymorphism in a population, it may be less important or totally unimportant in other populations. Populations of the banded snail *Cepaea nemoralis* are notoriously polymorphic in color and pattern, and explanations for their polymorphism have often focused on protection from avian predators through cryptic and diverse coloration. However, in some locations there are physiological differences relating to heat resistance among the various morphs, and such differences appear more important than predation in determining morph frequencies—just a reminder that there may be more to an animal's color and pattern than meets the eye.

Warning Coloration

Many animals that have dangerous or unpleasant attributes advertise this fact with bright colors and contrasting patterns. Bold markings, typically in black, white, red, or yellow, warn the predator of the prey's secondary defense mechanism and through this warning discourage an attack. The phenomenon is called aposematism and there are many familiar examples. The dramatic black-and-white markings of spotted and striped skunks (*Spilogale* and *Mephitis*) are truly exceptional amid an array of brown coat colors in mammals. The markings may serve, in part, to warn predators of the foul-smelling repellent that may, upon further harassment, be released from the skunks' anal scent glands. Many insects, such as the social wasps (*Vespula*), have a boldly patterned yellow and black body to warn of their painful sting, and the bright colors of several species of butterfly advertise their unpalatability. Frogs of the genus *Dendrobates*, and especially those in the genus *Phyllobates*, have toxic skin secretions. A single individual of the species *Phyllobates terribilis* has enough toxin in its skin to kill about 20,000 house mice or, in a more familiar currency, several adult humans. The Choco Indians of western Colombia make deadly weapons by simply wiping their blowgun darts across the back of one of these frogs (Myers and Daly, 1983). Not surprisingly, frogs that have toxic skin secretions are aposematically colored to warn predators that they are best left alone. In addition to conspicuous colors, characteristic noises (e.g., buzzes) and strong smells may also warn a predator. *Sternotherus odoratus*, indelicately but accurately called the "stinkpot," is a musk turtle of the eastern United States that ejects an odorous secretion when disturbed. The stink is thought to be an aposematic signal that warns predators of the turtle's bad-tasting flesh, pugnacious disposition, and unhesitating bite (Eisner et al., 1977).

Animals that are colored in this way often enhance their conspicuousness behaviorally. Many are active during the daytime, and individuals of some species form dense, obvious aggregations. Although rare forms in aposematic animals are typically selected against (predators will not be as familiar with the rare form as they are with the common form and may attack), they are at less of a disadvantage when they occur in clusters (Greenwood, Cotton, and Wilson, 1989). Thus, dense aggregations of aposematic prey not only emphasize the warning, but also function as areas in which rare forms may arise and survive.

The response of predators to aposematic coloration may be learned or innate. In the first case, predators sample some of the prey, discover their unpleasantness, and learn to avoid animals of similar appearance when searching for subsequent meals. Animals seem to learn to avoid prey that are conspicuously colored and distasteful more rapidly than they learn to avoid prey that are cryptically colored and distasteful (Gittleman and Harvey, 1980; Figure 13.15).

Sometimes two warningly colored species look alike. Apparently, two noxious species can benefit from a shared pattern because predators consume fewer of each species in the process of learning to avoid all animals of that general appearance. This phenomenon is called Mullerian mimicry. Although some predators learn through memorable experiences to avoid aposematic prey, others display innate avoidance. An innate response to warning coloration might be favored over a learned response, when the secondary defense of the prey has the potential of being fatal to the predator. Learning, at the moment of death, is of little value.

Sometimes, like advice, warning coloration is ignored. A predator that is starving might tackle a noxious prey that it would normally pass up during better times. Wolves will attack both skunks and porcupines when other prey is scarce. In addition, some predators are specialists and are able to eat certain aposematic ani-

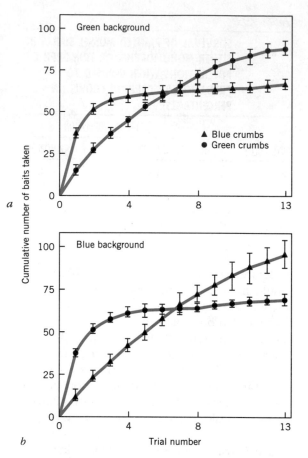

Figure 13.15 Predators learn more quickly to avoid distasteful prey that are conspicuous. Here, chicks were tested to determine the number of trials it would take to learn to avoid distasteful prey (crumbs covered with quinine sulfate and mustard powder) that was either cryptic or conspicuous. (*a*) When tested on a green background with blue (conspicuous) and green (cryptic) crumbs, chicks took fewer trials to learn to avoid the blue prey. (*b*) Similarly, when tested on a blue background with green (conspicuous) and blue (cryptic) crumbs, chicks took fewer trials to learn to avoid the green prey. (Modified from Gittleman and Harvey, 1980.)

mals, or the least noxious parts of them, as we saw for the black-headed grosbeak and black-backed oriole preying on the unpalatable monarch butterfly. Others (such as arthropods) are attracted to the bold patterns and movements of warningly colored animals but are unable to make the connection between color and inedibility, and they repeatedly attack. However, as long as an antipredator device confers a net advantage in terms of survival and reproduction, it will continue in the population.

Batesian Mimicry

Batesian mimicry is named after the nineteenth-century English naturalist Henry Walter Bates, and it refers to a palatable species adopting the warning characteristics of a noxious or harmful species. The harmless species is called the mimic and the noxious one, the model. By resembling a noxious species, the mimic gains protection from predators. The precise degree of protection experienced by the mimic varies as a function of numerous factors, including ratio of mimics to models, noxiousness of the model, memory of predators, availability of alternate prey, and whether mimics and models are encountered simultaneously or separately. Although in some instances the resemblance between model and mimic seems almost exact, the likeness typically does not have to be perfect because predators appear to generalize conspicuous features of noxious prey. Most known examples of mimicry are visual, probably reflecting the fact that we humans are visually oriented creatures. Other animals rely on smell and hearing more than vision, and thus olfactory and auditory mimicries may be quite common.

Many fascinating examples of Batesian mimicry can be found among insects and spiders. Some perfectly harmless flies mimic the yellow and black bands or buzzing sounds characteristic of bees and wasps. Predators all too familiar

with the painful stings of bees and wasps may leave these flies alone. Ants are typically avoided by insectivorous predators because of their sting or bite and unpleasant taste (formic acid lends ants their bad taste). It should come as no surprise, then, that ants have many mimics, and the likeness can include features of color, morphology, and behavior. Major R. W. G. Hingston (1927b) recorded several instances of spiders mimicking species of ants in India. One species of spider observed by Hingston closely resembled the large Indian black ant, *Camponotus compressus*. The spider was the size of a *Camponotus* worker and shared the uniform black color, elongate shape, and slender legs of the worker physique. Because ants have three pairs of legs and spiders four, the spider used its front pair of legs to simulate the antennae of the ant. These legs were thrust forward and the tips kept in continual motion to mimic the methodical movement characteristic of the ant's antennae. In a second species of ant studied by Hingston, individuals dragged workers of other species back to their nests, decapitated their victims, and threw the heads in a refuse heap. Hingston (1927b) reported that one species of spider curled into a tight motionless ball with its head and legs tucked underneath its pear-shaped abdomen (Figure 13.16) and sat amid the discarded heads. By mimicking a fragment of a noxious species (and associating with an even more formidable species), these spiders appear to gain protection from predators and may also be cryptic to their own prey that may wander by. Thus, mimicry can serve both defensive and foraging functions.

In a variation on the mimicry theme, some perfectly palatable individuals appear inedible by resembling inanimate objects. A caterpillar may take on the appearance of a broken twig on a limb and thus appear to be something of little interest to a predator. Certain tropical katydids have leaflike wings that are true to their model, even to the extent of showing "veins" and small "holes" from insect attacks. Some unabashed

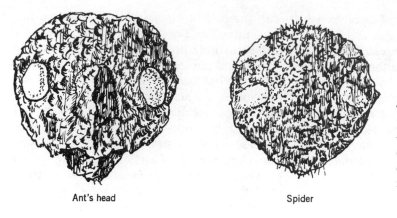

Ant's head   Spider

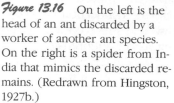

Figure 13.16 On the left is the head of an ant discarded by a worker of another ant species. On the right is a spider from India that mimics the discarded remains. (Redrawn from Hingston, 1927b.)

frogs and caterpillars crouch on leaves in imitations of bird droppings. Although some discussions of mimicry systems include resemblance to an inanimate object as part of Batesian mimicry (e.g., Edmunds, 1974), others consider such resemblance to be an example of crypsis (e.g., Pough, 1988).

Despite the diverse array of mimetic resemblances recounted in the literature, there are only a few studies that demonstrate that the purported mimics actually gain protection from their natural enemies. T. E. Reimchen (1989) first described a system of Batesian mimicry involving the juvenile stage of a snail (the mimic) and the tubes of a polychaete worm (the model) and then provided evidence that the resemblance actually conferred some degree of protection to the young snails. The snail, *Littorina mariae*, lives in the intertidal zone of the North Atlantic. The shells of some juveniles have a conspicuous white spiral, and the shells of others are yellow or brown. When adult, the snails are either yellow or brown and the white spiral possessed by some is visible only as a white apex on the shell. Egg masses of the snail are deposited directly on algal fronds and once the juveniles hatch, they disperse on the fronds. Snails with white-spiral coloration were observed only in habitats where the polychaete *Spirobis* was present. In these habitats, white-spiral phase juveniles are

virtually indistinguishable from the tubes of *Spirobis* that are cemented to the fronds (Figure 13.17). Reimchen collected the intertidal fish *Blennius pholis*, an important predator on juvenile snails, and conducted predation experiments in aquaria in the laboratory. Although the polychaete tubes are not noxious to the fish, they represent a substantial investment in time and energy as they are difficult to remove from the substrate, and once removed, they may

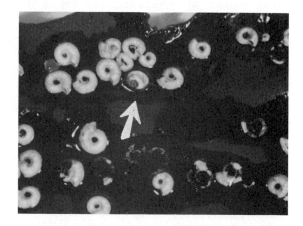

Figure 13.17 The white phase juvenile of a snail (shown by the arrow) is a Batesian mimic of the tubes of certain polychaetes. By resembling the tubes, young snails may gain protection from fish searching for food items on algal fronds.

prove to be unoccupied. In the experiments, blennies were housed one per aquarium and were presented with juvenile snails on either an algal frond with polychaete tubes or an algal frond without tubes. At each presentation, three juvenile snails (one white spiral, one yellow, one brown) were randomly positioned on the frond and the frond was lowered to the bottom of the tank. Once blennies detected a snail, they plucked it off the frond and swallowed the shell whole. Reimchen recorded the first snail taken at each trial. Overall, white-spiral snails suffered the lowest number of attacks and the reduction in attacks was greater on fronds with polychaete tubes (9.4%) than on fronds devoid of tubes (22.9%). Thus, in this unusual system of snail–polychaete mimicry, resemblance to the model does appear to confer a protective advantage to the mimic.

Diverting Coloration, Structures, and Behavior

A large number of animals have evolved colors, structures, and patterns of behavior that seem to serve in diverting a predator's attention, while they, and in some cases their offspring, escape with little or no damage. Whereas crypsis, diverse coloration, warning coloration, and Batesian mimicry function to help prey avoid an encounter with a predator, distraction devices come into play once a prey animal has been discovered, or when discovery seems all too imminent. Thus, we now begin our discussion of secondary defense mechanisms, that is, those mechanisms that operate during an encounter with a predator.

EYESPOTS

Eyespots can serve two defensive functions (Owen, 1980). First, if the spots are large, few in number, brightly colored, and suddenly flashed, they may startle or frighten the predator. Second, spots that are small and less gaudy may serve instead as targets to misdirect the predator. Such eyespots are typically located on nonvital portions of the body, and thus prey can often escape with less than fatal damage.

FALSE HEADS

Many predators direct their initial attack at the head of their prey. Some prey species have taken advantage of this tendency by evolving false heads that are located at their posterior end, a safe distance from their true heads. Lycaenid butterflies (Lepidoptera: Lycaenidae) display patterns of color, structure, and behavior that are consistent with deflecting predator attacks toward a false head (e.g., Robbins, 1981). Individuals of the species *Thecla togarna*, for example, possess a false head, complete with dummy antennae, at the tips of their hindwings (Figure 13.18). These butterflies enhance the structural illusion of a head at their hind end by performing two rather convincing behavioral displays. First, upon landing, the butterfly jerks its hindwings, thereby moving the dummy antennae up and down while keeping the true antennae motionless. *Thecla togarna*'s second ploy occurs at the instant of landing when the butterfly quickly turns so that its false head points in the direction of previous flight. An approaching predator is thus confronted with a prey that flutters off in the direction opposite to that expected. Experimental tests have demonstrated that markings associated with false heads misdirect the attacks of avian predators and, in particular, increase the possibility of escape if the prey is caught to begin with (Wourms and Wasserman, 1985).

AUTOTOMY

Rather than simply diverting a predator's attack toward a nonvital portion of the anatomy, some prey actually hand over a "disposable" body part to their attacker, almost as a consolation prize.

Figure 13.18 The false head of a butterfly. Note the pattern of markings that tends to focus attention on the posterior end of the butterfly and the dummy antennae and eyes at the tips of the hindwings. Markings and structure combine with behavior (e.g., movement of dummy, rather than true, antennae) to divert the attention of a predator away from the true head. (Redrawn from Wickler, 1968.)

Autotomy, the ability to break off a body part when attacked, has evolved as a predator defense mechanism in both vertebrates and invertebrates. Tail autotomy in lizards, for example, is commonly reported, as well as in some salamanders, a few snakes, and even some rodents. A more dramatic autotomy, however, is seen in sea cucumbers (members of the phylum Echinodermata), which, upon being attacked, forcefully expel their visceral organs (guts, in the vernacular) through a rupture in the cloacal region or body wall. The predator then may begin to feed on the sea cucumber's offering as it makes its slow escape. In most autotomy cases, the disposable body part is subsequently regenerated. As an example of the phenomenon, we will focus on tail autotomy in lizards.

Tail autotomy benefits the lizard in two ways: First, it allows the lizard to break away from its attacker; and second, if the detached tail continues to thrash and writhe, the attacker is distracted as the lizard runs for it (Arnold, 1988). (Although the vigor and duration of postautotomy tail movement varies among species, in some lizards the tail may thrash for as long as 5 minutes.) The effectiveness of tail autotomy is underscored by the presence of tails in the stomachs of predators, and the occurrence of tailless lizards and lizards with regenerated tails in natural populations.

Direct experimental evidence for the importance of tail autotomy as an antipredator device comes from a laboratory study by Benjamin Dial and Lloyd Fitzpatrick (1983). These researchers tested the effectiveness of tail autotomy and postautotomy tail movement in permitting the escape of lizards from mammalian and snake predators. In the first study, staged encounters were conducted between a feral cat (*Felis catus*) and two species of lizards, *Scincella lateralis* (a species with vigorous postautotomy tail thrashing) and *Anolis carolinensis* (a species with less vigorous postautotomy tail thrashing). Dial and Fitzpatrick recorded the cat's reaction to lizards of both species under two conditions: (1) thrashing tail trials—lizards and their autotomized tails were placed in front of the cat immediately after autotomy, and (2) exhausted tail trials—tails were allowed to thrash to exhaustion and then lizards and their autotomized tails were placed in front of the cat. In both types of trials autotomy was induced by the experimenters gripping the lizards' tails at the caudal fracture plane with forceps (in many species of lizards, tail breakage takes place at preformed areas of weakness). The results, summarized in Table 13.3, show that the dramatic postautotomy tail thrashing of *S. lateralis* is an effective escape tactic, whereas the more subdued tail movement of *A. carolinensis* is not. Note that in all of the thrashing tail trials with *S. lateralis*, the cat attacked the tail, rather than the lizard, and in all

Table 13.3

RESPONSES OF A FERAL CAT TO SIMULTANEOUS PRESENTATION OF AUTOTOMIZED TAILS (EITHER THRASHING OR EXHAUSTED) AND LIVE TAILLESS BODIES OF TWO SPECIES OF LIZARDS

	NUMBER OF RESPONSES		
TAIL	ATTACK TO TAIL	ATTACK TO BODY	ESCAPE OF LIZARD
	Anolis carolinensis		
Exhausted	0	8	3
Thrashing	0	6	1
	Scincella lateralis		
Exhausted	0	6	0
Thrashing	7	0	7

Source: From Dial and Fitzpatrick, 1983.

cases the lizard escaped. In 100 percent of the exhausted tail trials with this species, however, the cat attacked and captured the lizards. The results for *A. carolinensis* were quite different; the cat attacked the lizards and ignored the tails in all trials.

In the second experiment, Dial and Fitzpatrick (1983) examined whether postautotomy tail movement influenced predator handling time. The authors staged encounters between *S. lateralis* and the snake *Lampropeltis triangulum*, again using autotomized tails that were either thrashing or exhausted. On average, snakes required 37 seconds longer to handle thrashing tails than exhausted tails, providing the tailless lizard with more time to escape. Thus, for the lizard *Scincella lateralis*, postautotomy tail movement supplements the simple mechanism of breaking away from the predator's grasp and, depending on the type of predator, may either attract the predator's attention (as in the case of the cat) or increase the time required to handle the autotomized tail (as in the case of the snake). Either way, postautotomy tail movement enhances the opportunity for the lizard to escape.

Up until this point we have focused on the benefits of tail autotomy without mentioning po-

tential costs. Depending on the species of lizard, tail loss may lead to reductions in speed, balance, swimming or climbing ability, and, when the tail is used as a display, even to declines in social status (Fox and Rostker, 1982). Furthermore, regeneration of the tail must certainly entail costs in energy and materials. Many lizards, after all, have substantial fat deposits in their tails that are lost with the tail. One possible recourse to the cost of leaving behind energy reserves was suggested by Donald R. Clark (1971) after he observed that postautotomy tail movement in *S. lateralis* pushes the tail through leaf litter. Clark suggested that such movements propel the tail out of sight of the predator and facilitate later retrieval and eating by the original owner. There are some reports of lizards ingesting their own autotomized tails (—Judd, 1955, in describing the capture of a lizard that had escaped in his laboratory stated, "When an attempt was made to capture it, the skink snapped off the terminal one inch of its tail and wriggled free. However, it immediately turned around, and grasping the severed tail by its narrow end, gulped the whole thing down.") However, Dial and Fitzpatrick (1983) found that when snakes grabbed the tail of *S. lateralis* they never subsequently lost their hold. Thus, the question of whether lizards routinely lose their tail and eat it too remains unanswered.

FEIGNING INJURY OR DEATH

In ground-nesting birds such as the killdeer (*Charadrius vociferus*), a parent may feign injury in an elaborate effort to divert the attention of an approaching predator away from its nest and young, particularly soon after hatching when offspring are most vulnerable (Brunton, 1990). Upon spying a predator, an adult may suddenly begin dragging its wing, as it flutters away from its nest. The predator follows, and as it closes in, the killdeer suddenly recovers and flies away, giving a loud call. If all goes as planned, the predator will continue to wander off.

Some animals rely not on diverting the attention of a predator, but on causing the predator to lose interest. Because many predators kill only when their prey is moving, a prey animal that feigns death may fail to release the predator's killing behavior, and with any luck, the predator will lose interest and move along in search of a more lively victim. Perhaps the most familiar performer of death feigning is the opossum (*Didelphis virginiana*; Figure 13.19) and hence the phrase "playing possum" has come to be synonymous with "playing dead." Although their performance is less well publicized than that of the opossum, juvenile caimans (*Caiman crocodilus*) react aggressively toward humans when approached on land, but feign death when handled in water (Gorzula, 1978). The response of an individual to a particular predator may thus

Figure 13.19 An opossum playing dead.

vary as a function of context, and prey animals typically have several antipredator devices at their disposal.

Hognose snakes (*Heterodon platirhinos*) have a complex repertoire of antipredator mechanisms of which death feigning is one option. These fairly large nonvenomous or slightly venomous snakes occur in sandy habitats in the eastern United States. When first disturbed, the hognose opts for bluffing the predator—it flattens and expands the front third of its body and head, forming a hood and causing it to look larger. It then curls into an exaggerated S-coil and hisses, occasionally making false strikes at its tormentor. When further provoked, however, it drops the bluff and begins to writhe violently and to defecate. Then it rolls over, belly up, with its mouth open and tongue lolling. If the predator loses interest in the "corpse" and moves away, the snake slowly rights itself and crawls off.

The complete repertoire of antipredator mechanisms occurs in young hognose snakes, and Gordon Burghardt and Harry Greene (1988) have shown that newborn snakes are capable of making very subtle assessments of the degree of threat posed by a particular predator. The researchers conducted two experiments in which they monitored the recovery from death feigning (i.e., crawling away) of newly hatched snakes under various conditions. In experiment 1, the recovery of snakes was monitored in the presence or absence of a stuffed screech owl (*Otus asio*) mounted on a tripod 1 meter from the belly-up snake. In experiment 2, the snake recovered (1) in the presence of a human being staring at the snake from a distance of 1 meter, (2) in the presence of the same person in the same location but with eyes averted, and (3) in a control condition in which no human being was visible. Both the presence of the owl (experiment 1) and the direct human gaze (experiment 2) resulted in longer recovery times relative to the respective control conditions (Figure 13.20). When the human being averted his or

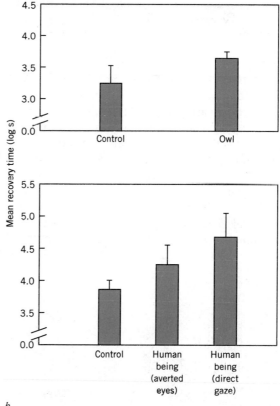

Figure 13.20 Mean time to recovery from feigning death in neonatal hognose snakes exposed to various recovery conditions. (*a*) In experiment 1, snakes recovered in the presence of a stuffed owl or in the absence of a stuffed owl (control condition). (*b*) In experiment 2, snakes recovered in the presence of a human being (eyes either staring at the snake or averted) or in the absence of a human being (control condition). Because the recovery times were skewed, the data were transformed into logarithms of seconds. (From Burghardt and Greene, 1988.)

her eyes, the recovery time was intermediate. Thus, young snakes are capable of using rather subtle cues to make adjustments in their antipredator behavior.

Pronouncement of Vigilance

Some prey appear to inform predators that they have been spotted. The advantage might be in discouraging the predator confronted with an alert and aware prey. Stalking predators, for example, may abandon the hunt once they receive the signal that they have been detected. Stotting, a stiff-legged jumping display performed by many species of deer, pronghorn, and antelope in which all four feet are off the ground simultaneously (Figure 13.21), appears to be just such a signal. The stotting display has attracted the attention of a number of investigators and arrival at prey detection as a plausible function has

Figure 13.21   Stotting by a Thomson's gazelle.

Table 13.4

HYPOTHESES TO EXPLAIN THE FUNCTION OF STOTTING

Benefits to the individual
  Signaling to the predator
    1. Pursuit invitation
    2. Predator detection
    3. Pursuit deterrence
    4. Prey is healthy
    5. Startle
    6. Confusion effect
  Signaling to conspecifics
    7. Social cohesion
    8. Attract mother's attention
  Signaling not involved
    9. Antiambush behavior
    10. Play
Benefits to other individuals
    11. Warn conspecifics

Source: After Caro, 1986a.

involved testing predictions from a diverse array of hypotheses.

At least 11 hypotheses have been proposed for the function of stotting (Caro, 1986a; Table 13.4). Although not mutually exclusive, the hypotheses range from the interpretation of stotting as a signal given by a hunted animal to either a predator or a conspecific to the interpretation that stotting has no signal value at all and is simply a form of play or alternatively a means to visually survey the flight path away from a predator. In the first true effort to distinguish among the hypotheses, Tim Caro (1986b) recorded the response of Thomson's gazelles (*Gazella thomsoni*) to naturally occurring predators, usually cheetahs (*Acinonyx jubatus*), in the Serengeti National Park of Tanzania. He analyzed prey behavior, cheetah behavior, and the outcome of hunts and found that cheetahs were more likely to abandon hunts when their prey stotted than when they did not stott (Table 13.5). These results, combined with other data that refuted many of the remaining hypotheses, sug-

Table 13.5

OUTCOME OF 31 CHEETAH HUNTS INVOLVING THOMSON'S GAZELLES THAT DID OR DID NOT STOTT

| | CHASE OCCURRED | | | |
	CHASE SUCCESSFUL	CHASE UNSUCCESSFUL	HUNT ABANDONED	TOTAL
Gazelle stotts	0	2	5	7
Gazelle does not stott	5	7	12	24

Source: Modified from Caro, 1986b.

gested that stotting typically functioned to inform the predator that it had been detected. Two other functions for stotting were supported by Caro's observations. First, mothers may stott to distract a predator from their fawn, a function much like the broken wing displays described for killdeer. Second, fawns appear to stott in order to inform their mother that they have been disturbed at their hiding place.

A more recent study suggests that the context of the cheetah–gazelle encounter and age of the performing gazelle are not the only factors to influence the function of stotting. Type of predator is another consideration. When hunted by coursing, rather than stalking, predators that rely on stamina to outrun their prey, gazelles appear to use stotting as an honest signal of their ability to outrun predators (FitzGibbon and Fanshawe, 1988). Coursing predators such as African wild dogs (*Lycaon pictus*) concentrate their chases on those individuals within a group that stott at lower rates and thus appear to use information conveyed in stotting to select their prey. In the study by FitzGibbon and Fanshawe, the mean rate of stotting of gazelles that were chased was 1.64 stotts per second and of those not chased was 1.86 stotts per second. By signaling their ability to escape at the start of a hunt, those gazelles with high stamina and/or running speeds may not have to prove their physical prowess by outrunning wild dogs in long, exhausting, and potentially dangerous chases. If the function of stotting varies with species of

predator, then we should not be surprised if future studies reveal that the function varies with species of prey as well. Finally, although often performed in the presence of predators, stotting also occurs during intraspecific encounters, and we can only guess what its function is under these circumstances.

Startle Mechanisms

Sometimes even an extra second or two is enough time for an animal to escape from what appears to be certain death. In some cases, a prey animal can escape if it can startle the predator into delaying for only an instant. Called deimatic displays by Edmunds (1974), startle mechanisms involve sudden and conspicuous changes in the appearance or behavior of prey that can produce confusion or alarm in a predator. The sudden presentation of a visual stimulus (such as large eyespots) or an auditory stimulus (such as squeaks, rattles, or screams) may startle the predator to the extent that it withdraws or hesitates just long enough for the prey to escape.

Many insects have deimatic displays that involve the sudden exposure of bold colors or patterns that are concealed when resting. Moths of the genus *Catocala* are palatable to avian predators and have cryptic barklike forewings that cover flamboyantly patterned hindwings (apparently, the word *Catocala* is derived from the Greek words *kato* and *kalos*, meaning "beautiful behinds"; Sargent, 1976). When crypsis fails and the moths are disturbed by predators, they suddenly flash their striking hindwings. Do the hindwings of *Catocala* serve as startle devices? Debra Schlenoff (1985) investigated this possibility by examining the response of blue jays (*Cyanocitta cristata*) maintained in indoor-outdoor aviaries to models of *Catocala* moths. She constructed artificial moths of gray cardboard, gave them pinon nut bodies (delicacies for blue jays), and attached brightly colored hindwings that popped into view when the model was removed by the bird from the presentation board (Figure 13.22). During the training phase, blue jays were taught to capture and eat artificial moths with uniform gray hindwings. During the test phase of the experiment, jays were presented with seven of the moths with gray hindwings and one randomly placed moth having a boldly patterned hindwing, that is, a *Catocala* hindwing. When jays picked up the moth with the *Catocala* pattern, they raised their crests and gave alarm calls, sometimes dropping or flying away from the moth model.

Auditory stimuli can also be startling to predators. Some species of arctiid moths click in response to touch or sound. These moths are usually distasteful and when harassed emit a repellent froth from their thorax. The disturbance clicks of arctiid moths often cause bats to abort their predatory attacks. Are the sudden clicks of these moths startle devices or simply warnings of bad taste? Big brown bats (*Eptesicus fuscus*) that were trained to fly to a platform where they sometimes received a mealworm reward, veered away from the platform when arctiid clicks were broadcast (Bates and Fenton, 1990). Individual bats, however, quickly habituated to the clicks and soon did not respond to the sound by avoiding the platform. In a second experiment, when broadcast clicks were paired with mealworms injected with quinine sulfate, bats rapidly

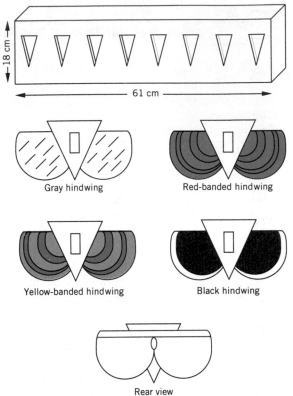

Figure 13.22 Presentation board and moth models. When the models were correctly positioned in the presentation board, the hindwings were folded back into the open triangles of the board and only the gray forewings were visible. Blue jays trained to capture models with gray hindwings exhibited startle responses when they captured randomly placed models with brightly colored hindwings. (From Schlenoff, 1985.)

learned to associate moth clicks with bad-tasting mealworms. Thus, in the case of bats and arctiid moths, the clicks may serve both as startle and warning devices, depending on the experience of the predator—inexperienced bats are startled by clicks, but experienced bats recognize clicks as a warning of distasteful prey.

Intimidation and Fighting Back

Prey animals have many ways of communicating "I am formidable" to a predator. Presumably, when a predator encounters a large, threatening, well-armed prey, it will continue on its way in search of a less challenging meal.

ENHANCEMENT OF BODY SIZE AND DISPLAY OF WEAPONRY

When dealing with potential predators, some animals employ the size-maximization principle. A cat hunches its back and erects its fur in the presence of a dog. Some toads and fishes inflate themselves when disturbed. In each case, the animal increases its size and appears more formidable, or unswallowable, as in the latter two examples. Several displays of intimidation through an increase in size are shown in Figure 13.23, but threat maximization need not always be visual. Loud calls, hisses, or growls emitted by prey may also cause a predator to look elsewhere for its next meal.

Sometimes eyes can be threatening. Several species of animals have utilized the relationship between eyes and threat and have developed eyespots as a means of repelling predators. Such eyes usually appear to be large, wide open, and staring straight ahead, although they may in ac-

Figure 13.23 Intimidation displays in several species of animals. The displays serve to make the animal appear larger and more formidable to predators. The animals shown here are (counterclockwise): frilled lizard, cat, short-eared owl, and spotted skunk. (Modified from Johnsgard, 1967.)

Figure 13.24 A toad directs its backside toward an attacker, revealing a pair of frightening eyespots.

tuality be sightless spots on the wings of a harmless insect or on the backside of a toad (Figure 13.24). Remember that eyespots may also function to startle a predator or to misdirect its attack. Finally, some animals display their weapons when confronting a predator. Ungulates often display their horns to a predator and paw at the ground, perhaps to draw attention to their dangerous hooves. Porcupines erect their spines and cats display their teeth. All these postures are probably meant to intimidate a predator.

CHEMICAL REPELLENTS

A wide variety of insects can discharge noxious chemicals when they are captured. Some of these chemicals are powerful toxins or irritants, and in some species they can be shot with considerable accuracy in several directions. The assassin bug (*Platymeris rhadamantus*) reacts to disturbance by spitting copious amounts of fluid in the direction of the attacker. The saliva is rich in enzymes and causes intense local pain when it comes in contact with membranes of the eyes or nose.

Other masters of chemical warfare are the bombardier beetles, which deter predators by emitting a defensive spray that contains substances stored in two glands that open at the tip of the abdomen (Dean et al., 1990; Eisner, 1958). Because the tip of the abdomen acts as a revolvable turret, the spray can be aimed in all directions (Figure 13.25). The chemical reactants from the two glands are mixed just before they are discharged, producing a sudden increase in temperature of the mixture. The hot spray is ejected, accompanied by audible pops, in quick pulses. The effect has been likened to that of the German V-1 "buzz" bomb of World War II (Dean et al., 1990).

Chemical deterrents are by no means limited to arthropods, as anyone who has had the misfortune of surprising a skunk or who owns a dog that has enjoyed the same experience must surely know. Although the defensive response of the horned toad (*Phrynosoma cornutum*) is perhaps less well known than that of the skunk, it is certainly no less spectacular. When disturbed, this small, spine-covered lizard of the southwestern United States may spatter its attacker with a stream of blood ejected from its eyes (Lambert and Ferguson, 1985). At the turn of the century, Charles Holder (1901) examined this behavior and suggested, on the basis of trials in which his fox terrier posed as a predator, that the ejected blood contained noxious components. Apparently, contact between the ejected blood and nasal membranes of the dog was particularly irritating, and only a single encounter was required to produce "a wholesome dread" in the lizards' canine tormentor. Whether the discharged blood actually contains noxious components, and just what these components might be, remains to be determined.

Group Defense

Up until now we have focused almost exclusively on strategies employed by individual animals to avoid being eaten. Some animals, however, are

Figure 13.25 The bombardier beetle ejects a hot irritating spray at its attackers. This beetle, tethered to a wire fastened to its back with wax, responds, with excellent aim, to a forceps pinch on its left foreleg.

social, and membership in a group makes accessible a host of antipredator tactics that are not available to solitary individuals. Generally, predators experience less success when hunting grouped rather than single prey because of the superior ability of groups to detect, confuse, and repel predators. In addition, an individual within a group has a lower probability of being selected during any given predator attack. We will now consider some examples of how social animals cope with predators. Keep in mind that group living has many advantages, including those totally unrelated to protection from predators (see Chapter 16).

ALARM SIGNALS

When a predator approaches a group of prey, one or more individuals within the aggregation may give a signal that alerts other members of the group to the predator's presence. Alarm signals may be visual, auditory, or chemical (see Chapter 17 for a general discussion of the means by which animals communicate alarm) and often

serve to either enlist support in confronting the attacker or inspire retreat to a safe location. In some cases the alarm may aid the signaler or its relatives; in other instances the alarm appears to benefit all those exposed to the signal by permitting members of the group to escape in a coordinated fashion. The proposed selective advantages of signaling alarm are covered in more detail in Chapter 19. We will focus our discussion here on the chemical alarm systems of some fishes and amphibians.

Some species of fish show escape responses to chemical stimuli from injured conspecifics. For example, if the skin of a minnow is broken, an alarm substance, called "Schreckstoff," is released from skin cells. Conspecifics that smell the chemical respond by rapid dashes followed by hiding and reduced activity. Although once thought to be peculiar to the minnows and their relatives, an analogous alarm system has been reported for other groups of fish, including the darters and gobies (Smith, 1982, 1989). In most cases, the alarm response is displayed by fishes that form schools.

Although the presumed function of releasing

Schreckstoff is to warn other fish within the school of the danger of attack, there is little experimental evidence for its effectiveness as an antipredator mechanism. Such evidence is available, however, for the alarm substance produced by injured tadpoles of the western toad, *Bufo boreas*. Individuals of this species live in ponds and lakes of western North America where the tadpoles form dense aggregations. Diana Hews (1988) first documented the response of toad tadpoles to release of the alarm substance and then tested whether tadpoles alerted by the substance had higher survival rates than those not exposed. Two natural predators of western toad tadpoles, giant water bugs (*Lethocerus americanus*) and dragonfly naiads (*Aeshna umbrosa*), were used in the experiments. When tested in aquaria, western toad tadpoles increased their activity and avoided the side of the tank that contained a giant water bug feeding on a conspecific tadpole (in a visually isolated but interconnected container). Tadpoles did not increase their activity or avoid the predator's side of the tank when the water bug was feeding on a tadpole of another species. It is important to note that toad tadpoles alerted by conspecific alarm substance were less vulnerable to predation. Dragonfly naiads had fewer captures per attack in tests with tadpoles exposed to the toad extract containing the alarm substance as compared to tests with tadpoles exposed to the control extract, water (Figure 13.26). In addition to warning conspecifics, the alarm substance of *B. boreas* may function directly in deterring predators. Many larval and adult toads are distasteful to predators because of a toxin in their skin, and this "bufotoxin" is a likely component of the alarm substance. Again, a given defensive mechanism can have more than one protective function.

IMPROVED DETECTION

Early detection of a predator can often translate into escape for prey, and groups typically are superior to lone animals in their ability to spot

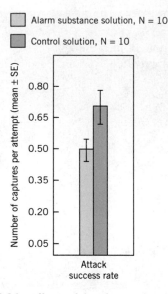

Figure 13.26 Effects of the alarm substance of toad tadpoles on attack success rate of dragonfly naiads. Tadpoles exposed to the alarm substance were less vulnerable to predation by naiads than those exposed to the control substance, water. (Modified from Hews, 1988.)

predators. Increases in the number of group members (and hence the number of eyes, ears, noses, etc.) often result in increases in the immediacy with which approaching predators are detected. Furthermore, as a result of the enhanced vigilance associated with groups, a given group member can often spend more time foraging and less time on the lookout for predators.

The benefits of increased predator-detecting ability may accrue to members of single-species or mixed-species groups. Florida scrub jays (*Aphelocoma coerulescens*) form single-species groupings that usually consist of from two to eight family members. Because these birds live in small, permanent groups of stable composition (see Chapter 19 for a discussion of communal breeding and helping at the nest in scrub jays), it is possible for individuals to coordinate their vigilance in a highly structured sentinel system (McGowan and Woolfenden, 1989). At any given time, only one family member typically sits on an exposed perch and continually scans the surroundings for predators. If a predator is spotted, the sentinel sounds the alarm, and family members respond by either mobbing a ground predator or by fleeing or monitoring the movements of an aerial attacker. Periodic exchanges among family members occur in order to relieve the sentinel bird of its duties. Sentinel systems have also been reported for mammals such as the dwarf mongoose (*Helogale undulata rufula*; Rasa, 1986) and the meerkat (*Suricata suricatta*; Moran, 1984), two species that live in family-based social groups.

The benefit of improved vigilance will apply to members of mixed-species groupings providing that they are on the lookout for the same species of predators and that they communicate detection to other group members. Additionally, some members of heterospecific groupings benefit if predators display a preference for individuals of the other prey species in their group. For example, Thomson's gazelles (*Gazella thomsoni*), a species familiar from our discussion of stotting, and Grant's gazelles (*G. granti*) often form mixed-species groups in the Serengeti National Park of Tanzania (FitzGibbon, 1990). When compared with those gazelles that remained as smaller groups of conspecifics, Thomson's gazelles that joined Grant's gazelles to form larger mixed-species flocks were less vulnerable to cheetahs. Grant's gazelles, on the other hand, benefited from the association because of the cheetahs' preference for the smaller Thomson's gazelles.

DILUTION EFFECT

Individuals within groups are safer not only because of their enhanced ability to detect predators, but also because each individual has a smaller chance of becoming the next victim. Called the "dilution effect," this advantage for grouped prey operates if predators encounter single individuals or small groups as often as large groups, and if there is a limit to the number of prey killed per encounter. As group size

increases, the dilution effect becomes more effective, and improved vigilance appears to provide relatively less benefit (Dehn, 1990).

Although this safety in numbers notion has intuitive appeal, in some cases, predators aggregate in areas where their prey are abundant. As a result of the gathering of predators, some grouped prey may actually suffer higher predation rates. In an examination of the balance between the forces of the dilution effect and the aggregating response of predators, Turchin and Kareiva (1989) studied grouping in aphids (*Aphis varians*). These small insects form dense clusters on the flowerheads of fireweed, and it is here that they are preyed upon by ladybird beetles, typically *Hippodamia convergens*. In one experiment, the researchers quantified per capita population growth rates (a measure of individual survivorship) for aphids living singly and for those living in colonies of over 1000 individuals and found that individual aphids benefited by forming groups. Grouping was only advantageous, however, in the presence of predators; when ladybird beetles were excluded from fireweed plants, individual survivorship of aphids did not increase with colony size.

The next question, then, is how do ladybird beetles respond to the grouping of their prey? Turchin and Kareiva found that beetles exhibited a strong aggregation response: More than 4 times as many beetles were found at aphid colonies of over 1000 individuals as compared to small and medium-sized colonies (Figure 13.27). In addition to gathering at large colonies, ladybird beetles also increased their feeding rate as aphid density increased. On average, beetles consumed 0.9 aphids per 10 minutes in colonies of 10 individuals as compared to 2.4 aphids per 10 minutes in colonies of 1000 individuals. Thus, group size in aphids appears to affect per capita growth rate of the aphid colony, number of predators attracted to the colony, and the rate at which predators feed. Given all these factors, does grouping reduce predation risk for aphids? Apparently so. When the researchers calculated the instantaneous risk of predation to an indi-

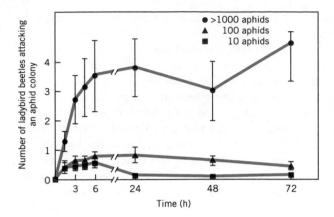

Figure 13.27 The tendency for prey individuals to form large groups and thereby dilute their chances of becoming the next victim may be countered by the tendency on the part of their predators to aggregate where prey are most common. Here, predatory ladybird beetles gather in larger numbers at aphid colonies that contain the most individuals. (Modified from Turchin and Kareiva, 1989.)

vidual aphid in a 10-minute period, they obtained values of 0.05 for colonies of 10 and 0.008 for colonies of 1000 or more. Thus, in the aphid–ladybird system, the dilution effect still occurs despite the strong tendency of predators to aggregate at large prey colonies. Turchin and Kareiva are quick to point out, however, that predators are not the only enemies of aphids. Parasitoids and pathogens may increase rapidly in large groups of aphids and may profoundly affect mortality, perhaps even eliminating the antipredator advantages of the dilution effect.

SELFISH HERD

In most groups, as we saw with the monarch butterfly, centrally located animals appear to be safer than those at the edges. By obtaining a central position, animals can decrease their chances of being attacked and increase the probability that one of their more peripheral colleagues will be eaten instead. This antipredator mechanism, often referred to as the "selfish herd" (Hamilton, 1971), emphasizes that although a given group appears to consist of members that coordinate their escape efforts, it is actually composed of selfish individuals, each trying to position as many others as possible between itself and the predator.

One might ask, then, are central locations within the group always the best? The answer is

no. In fact, a study on the antipredator advantages of schooling in fish suggests that the center is sometimes the most dangerous place to be. When in the company of a predatory seabass (*Centropristis striata*), silversides (*Menidia menidia*) at the center of a school suffered the most attacks (Parrish, 1989). Rather than assaulting the margins, seabass swim toward the center of the school, split the school into two groups, and then strike at the tail end of one of the groups, where individuals that were in the center now find themselves. The relative safety of a location within a group thus depends on the predator's method of attack. Because schools of fish undoubtedly cope with a number of predators, each potentially using a different attack strategy, the relative advantage of central versus peripheral locations may change. In addition, factors such as foraging efficiency (those in the front see the food first) and the energetics of locomotion (fish in the front of a school may experience more "drag" than those at the back) probably also influence optimal positions within the school.

CONFUSION EFFECT

Predators that direct their attacks at a single animal within a group may hesitate or become confused when confronted with several potential meals at once. No matter how brief, any delay in the attack will operate in favor of the prey. The so-called "confusion effect" was first described by Robert Miller (1922) for flocks of small birds in the presence of a hawk. He noted that upon detecting an approaching hawk, individual birds within the flock sat motionless in the foliage and all produced a shrill, quavering note, the rendition of which was known as the "confusion chorus." This particular call was hard to locate and Miller thought that it might function to distract attention from any particular individual in the group. Hawks apparently experienced difficulty in selecting a victim and were less successful in attacks on such groups than on a solitary bird. Miller described the dilemma of a hawk confronting a flock of prey in the following words, "... the more attention is divided, the greater is the possibility of failure."

The confusion effect is thought to be one of the primary antipredator advantages of schooling in fishes. Neill and Cullen (1974) examined the effects of size of prey school on the hunting success of two cephalopod predators (squid, *Loligo vulgaris*, and cuttlefish, *Sepia officinalis*) and two fish predators (pike, *Esox lucius*, and

perch, *Perca fluviatilis*). Whereas squid, cuttlefish, and pike are ambush predators, perch typically chase their intended victims. In most cases, predators were tested with fish of their natural prey species in schools of 1, 6, and 20 individuals. For all four predators, attack success per encounter with prey decreased as the size of the prey school increased (Figure 13.28). In the three ambush predators, increased size of prey school appeared to produce hesitation and behavior characteristic of conflict situations (such as alternating between approach and avoidance). Perch, on the other hand, switched targets more frequently as the size of prey school increased and, with each switch, reverted to an earlier stage of the hunting sequence. Under natural conditions, predators on fish may experience hunting success by restricting their attacks to individuals that have either strayed from the school or have a conspicuous appearance; in both cases, the predator can concentrate on the odd target. In some prey species, individuals within schools appear to segregate by size in order to reduce their conspicuousness to predators (Theodorakis, 1989).

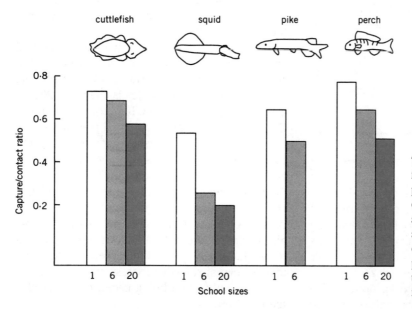

Figure 13.28 The confusion effect. As the size of the school of prey increases, four predators experience reduced hunting success due to either hesitation and conflict behavior (in the case of the cuttlefish, squid, and pike) or frequent switching of target prey (perch). (Modified from Neill and Cullen, 1974.)

MOBBING

Sometimes prey attack predators. Approaching and harassing one's enemies is termed mobbing, and this antipredator strategy typically involves visual and vocal displays, as well as frequent changes in position that culminate in swoops, runs, and direct hits on the predator. Mobbing is usually initiated by a single individual and then conspecifics, or members of another species, join in the fracas. When in hot pursuit of a predator, mobbers appear to have a greater chance of being preyed upon than nonmobbers, although there is some disagreement over whether mobbing actually entails a deadly risk to those that participate (Curio and Regelmann, 1986; Hennessy, 1986). Possible antipredator functions of mobbing include, but are not limited to, (1) confusing the predator, (2) discouraging the predator through either harassment or the announcement that it has been spotted early in its hunting sequence, (3) alerting others, particularly relatives, of the danger, and (4) providing an opportunity for others, again particularly relatives, to learn to recognize and fear the object that is being mobbed (Curio, 1978). Most evidence suggests that mobbing is not an act performed by a cooperative group of individuals attempting to protect the group as a whole, but rather the selfish act of individuals attempting to protect only those that will benefit them directly (that is, themselves and their mate, offspring, and relatives; Shields, 1984; Tamura, 1989).

Summary

Antipredator mechanisms may be classified as primary, secondary, and those characteristic of groups of prey. Primary defenses operate whether a predator is present or not and function to decrease the probability of an encounter with a potential predator. The prey may go undetected if it blends with its background (crypsis) or occurs in a wide variety of colors (diverse coloration). Alternatively, the prey may be detected by a predator and either be recognized as inedible (warning coloration) or go unrecognized as a potentially tasty meal (Batesian mimicry). Although many primary defenses involve color and pattern, the behavior of a prey animal is critical to the success of these mechanisms.

Secondary defenses operate during an encounter with a predator and function to increase an animal's chances of surviving the encounter. Amid the many options available, an individual may divert the predator's attention, inform the predator that it has been spotted early in the hunt, make the predator hesitate, or turn the tables and fight back.

Membership in a group makes available a number of protective devices that often combine primary and secondary defense mechanisms. Generally, predators experience less success when hunting grouped rather than solitary prey because of the superior ability of groups to detect, confuse, and discourage predators. In addition, during any given predator attack, an individual within a large group has a lower probability of being the one selected by the predator (dilution effect) and may use other group members as a shield between itself and the enemy (selfish herd).

Although for convenience we have discussed antipredator behavior as several distinct defensive mechanisms, our intent is not to imply that a given individual or species is characterized by a single protective strategy. Indeed, most animals face a large number of potential predators with a diverse array of methods for detecting and capturing prey, and thus a variety of deceptive and defensive tactics is crucial. The use of any particular device probably reflects the relative risks and energetic demands of the predator–prey encounter, and almost no defense works all the time. Finally, the behavior and color patterns of animals must be interpreted in the context of several selective forces; after all, animals must not only avoid being eaten, but must also feed and reproduce.

CHAPTER 14

Sexual Reproduction and Sexual Selection

Sexual Reproduction

On a warm rainy night in March upwards of 4000 wood frogs (*Rana sylvatica*) make their way toward a woodland pond. Breeding in this small amphibian, with its dark face patch resembling a robber's mask, is not only explosive but precisely synchronous (Berven, 1981). Within a 3-hour period almost all the wood frogs have arrived at the pond and the night's festivities begin; this is probably the one night of the year, and for most participants the one night of their lives, to reproduce. Males usually enter the pond first and begin emitting hoarse "clacks" from their sprawled, floating positions. Females may arrive somewhat later, and rather than bobbing at the surface with the males, they sit quietly at the bottom of the pond until they rise to the top to join the activities. Under crowded conditions, males actively search for females; but if few males are present in the pond, some may remain stationary and wait for females to pass (Woolbright, Greene, and Rapp, 1990). Once a female has been located, the male will try to clasp her in amplexus, the position that will allow him to fertilize her eggs. Attempts by a male to pair with a passing female often set off chain reactions as nearby males converge on the new couple. Amid this amphibian pileup, males strike each other with their forearms and clasp each other, as well as the amplexing pair. Males grabbing other males quickly realize their error from the warning sounds of their victim and because they have tested the firmness of their partner's body in a trial hug. Other males and females that are finished with egg laying are released because they lack the girth and firmness necessary to inspire a prolonged embrace (Noble and Farris, 1929). A male that has succeeded in locating a gravid (egg-filled) partner clasps her from behind, just under the forearms. Amplexus, however, does not ensure mating because of the boisterous intervention of other males. Typically,

rival males will attempt to dislodge the amplexing male by grabbing the female around the waist and prying him off by slowly shimmying up her back (Figure 14.1). The amplexing male's success in holding on depends to a large extent on his ability to hook his thumbs; small males are unable to reach around large females and are easily dislodged by more well-endowed suitors. Females, in fact, may show a preference for the larger males by physically dislodging weakly amplexing males from their backs. For male wood frogs, then, mating success hinges, in part, on being large enough to hook their thumbs around firm females.

Wood frogs often lay their eggs in enormous clusters in the same part of a pond year after year. Paired females, those with amplexing males hanging on, climb aggressively over other pairs and push toward the center of the mass of egg clusters. It is here that predation on eggs is likely to be lowest and temperature highest. Once in a suitable place, the female begins to release her eggs. As the eggs emerge, both she and the male use their hind legs to pack the eggs into a tight ball. During this process the male releases huge numbers of tiny sperm to fertilize the eggs. Fertilization complete, males quickly depart and resume their search for other females. Females, on the other hand, remain for some time with their legs clutching the egg mass in what has been described as a "deathlike stupor" (Berven, 1981). Upon recovering, they leave the pond, often with males in hot pursuit.

This brief sketch of procreation in the wood frog raises several questions about the reproductive process. Why, for example, do wood frogs go through all the antics associated with sexual reproduction instead of reproducing asexually, a far simpler process? Why do males fight for access to females? Why do females reject some suitors? Why do females produce larger gametes than do males, and why do they typically invest more in their offspring? These are some of the general questions associated with

Figure 14.1 Wood frogs exhibit explosive, synchronous breeding. On a single night in spring, thousands gather at woodland ponds where males compete intensely for females and females choose among available suitors. Here, one male (right) attempts to dislodge another male from the back of a female.

sexual reproduction in many species, questions that we will now consider.

THE EVOLUTION OF SEXUAL REPRODUCTION

Why bother with sex? Wouldn't life be much simpler without it? In the case of the wood frog, why don't females just produce exact genetic replicas of themselves and dispense with all the activities associated with mating on that damp night in spring? Scientists have pondered the question "why sex?" for nearly a century and have come up with several hypotheses and a plethora of mathematical models. Before getting into these, let's briefly review the basic mechanics of sex.

Sexual reproduction may have come into existence as early as 2.5 or 3.5 billion years ago (Bernstein, Byers, and Michod, 1981). Now widespread among plants and animals, sexual reproduction is defined as the process by which two parents donate genetic material to the creation of a new individual that differs genetically from both parents. Two basic steps comprise the sexual process: (1) production of gametes by meiosis and (2) production of a new individual by the fusion of two gametes to form the zygote. In step 1, female gametes (eggs) and male gametes (sperm) are produced by meiosis, a process of cell division that involves halving the number of chromosomes (Figure 14.2). Eggs and sperm, then, contain only a single set of chromosomes and thus are haploid. Most cells, though, are diploid—they contain two full sets of chromosomes. In step 2 of sexual reproduction, two haploid gametes, one from each parent, join at fertilization to form a diploid zygote, the offspring. (An exception to this general description occurs in insects such as wasps and bees—males are haploid, having been produced from unfertilized eggs, and females are diploid, having been produced by the joining of egg and sperm. We will discuss this phenomenon, known as haplodiploidy, in more detail in Chapter 19.) As we will see shortly, one important result of meiosis is the production of variable offspring. This variation occurs because (1) the chromo-

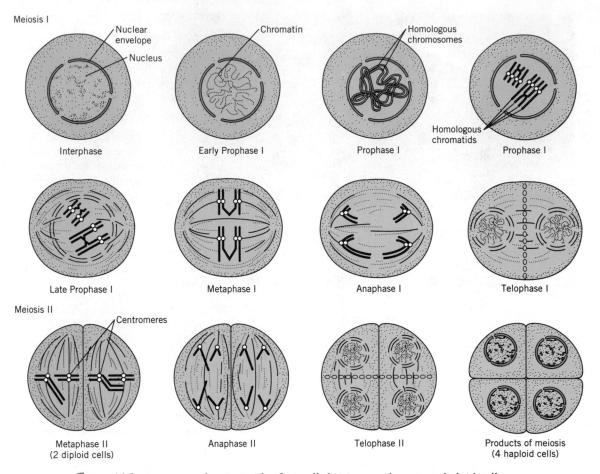

Figure 14.2 Summary of meiosis. The first cell division produces two diploid cells: each cell contains two full sets of chromosomes. The second meiotic division produces four haploid cells, each with one full set of chromosomes. Haploid gametes from two parents combine at fertilization to form a diploid zygote, the offspring. Meiosis helps guarantee production of variable progeny by sexually reproducing organisms because (1) the chromosome number of the gamete is reduced by one-half, necessitating its fusion with another haploid gamete carrying different genetic information and (2) exchange of genetic material can occur during the first meiotic division.

some number of the gamete is reduced by one-half, necessitating, in reproduction, its fusion with a second gamete carrying different genetic information, and (2) during the process of chromosome reduction, there is opportunity for exchange of genetic material between homologous chromosomes.

Sexual reproduction, then, is a complex process, and one with a number of costs, as we will see next. The question, why?, is underscored.

Why go through all this when asexual reproduction is so much simpler? Besides, sexual reproduction results in one individual's genes being diluted by those of the other parent, while in asexual reproduction, the offspring bears all the genes of only one parent.

COSTS OF SEXUAL REPRODUCTION

Although the subject of some debate, several costs are generally ascribed to sexual reproduction: (1) recombination, (2) meiosis, (3) production of sons, and (4) mating. The first cost, that of recombination, results from the fact that sexual reproduction involves the breakup of successful parental genotypes and the recombination of their parts. Because any given egg or sperm is likely to possess some deleterious recessive alleles, when two gametes unite there is a risk of producing offspring that are homozygous recessive for the trait in question. Barring new mutations, offspring of asexual parents are free of this risk because they are not formed by the fusion of two gametes.

The second cost, the cost of meiosis, is the result of halving one's genes. A sexual female propagates her genome only half as efficiently as an asexual female: A sexual female shares 50 percent of her genes with her offspring, but an asexual female shares 100 percent. This 50 percent reduction in the transmission of a mother's genes to the next generation is referred to as the cost of meiosis and inspired biologist George C. Williams (1975) to liken sexual reproduction to a game of roulette in which "the player throws away half his chips at each spin." Some scientists argue that meiotic cost is a mistaken concept. Treisman and Dawkins (1976) claim that the cost of sex does not lie in meiosis but in the wasting of resources in the production of males; after all, asexual females need not produce sons. Production of sons may thus be considered the third cost of sexual reproduction.

The fourth cost of sexual reproduction, that of mating, is also the subject of some controversy. This cost involves the expenditure of energy in developing mating structures, going through mating behavior, and avoiding unwanted attention from would-be mates. In 1978, Martin Daly announced that the costs of mating had been largely overlooked in theoretical discussions of the adaptive value of sexual reproduction. According to Daly, because sexual reproduction requires the careful orchestration of activities of two organisms whose interests are usually far from identical, it is more costly than asexual reproduction, which involves only a single organism.

Although costs of mating fall largely on the male in sexually reproducing species, only costs to females are considered when comparing costs incurred by sexual and asexual reproducers (because asexual reproducers are female). In their quest for genetic representation in future generations, sexually reproducing females may incur all three costs of mating, that is, the (1) production of elaborate reproductive organs, (2) performance of courtship and mating behavior, and (3) escape from unwanted sexual attentions. Additionally, during the breeding season sexual females may face increased risk of (1) predation due to the conspicuousness of competing and courting males, (2) disease transmission resulting from the close proximity required for mating, and (3) injury from males as reproductive males switch rapidly, and not always successfully, between courtship of females and aggressive interactions with rival males.

James Wittenberger (1981) claims that expenditures in time and energy associated with mating should not be considered when discussing the origin and maintenance of sexual reproduction because sex evolved long before such social interactions arose. Despite some disagreement over the precise costs of sexual reproduction, there is a consensus of opinion that sex is downright troublesome. Given its expense, sexual reproduction must confer some advantage to those individuals that practice it. What might the benefits of sex be?

ORIGIN AND MAINTENANCE OF SEXUAL REPRODUCTION

Most biologists agree that production of variable offspring is the key advantage to sexual reproduction: A parent that reproduces sexually has offspring with novel, or unique, genotypes, while a parent that reproduces without sex produces offspring that are genetically identical to each other. There are two hypotheses for the advantage of producing variable offspring: the long-term hypothesis and, not surprisingly, the short-term hypothesis.

The Long-Term Hypothesis

The long-term hypothesis is based on mathematical reasoning and states that populations with genetic recombination evolve more rapidly than those without, and thus sexual populations will survive environmental changes that cause the extinction of asexual populations (Fisher, 1930; Maynard Smith, 1971; Muller, 1932; Weismann, 1889). According to this interpretation, sexual reproduction accelerates evolution and thereby reduces the long-term risk of extinction.

But isn't this an argument based on group rather than individual selection? Yes—it suggests that individuals reproduce sexually because in the long run, sex is advantageous to the population. However, as discussed in Chapter 19, if such benevolent individuals did exist, they would soon be eliminated by selfish individuals behaving in a manner that maximizes their own fitness, irrespective of the effect of their actions on the population. Obviously, invoking group selection as a force to maintain sexual reproduction is controversial and constitutes the major objection to this hypothesis. Additionally, the elaborate mathematical models of the long-term hypothesis rely on the presence of extremely large populations, and critics claim that this re-

quirement means that the proposed benefit of sex is simply not available for a large segment of the earth's plants and animals. Yet many of them reproduce sexually just the same.

The Short-Term Hypothesis

In dramatic counterpoint to the long-term hypothesis, the short-term hypothesis suggests that sexual reproduction is advantageous to the individual in the short run, for among the variable offspring produced by a sexual parent, surely there are likely to be some of very high fitness. George C. Williams (1975), the major proponent of the short-term hypothesis, minimized reliance on mathematical models and compared life histories of plants and animals that reproduce sexually and asexually. As the evidence accumulated, a pattern emerged that highlighted potential advantages of sexual reproduction.

Organisms such as aphids, hydras, and some parasites alternate between sexual and asexual reproduction. In most cases, sexual reproduction is used in anticipation of dispersal into unpredictable environments because it is during these trying times that the production of novel genotypes is likely to be advantageous. In the case of aphids (Figure 14.3a), a group of soft-bodied insects whose feeding often causes curling or wilting of plant leaves, a foundress female that has successfully located a food source in spring and summer reproduces asexually. Her offspring live near her in an environment that is likely to be stable for several generations (with the exception of those individuals that overwinter, aphids typically live for only a few weeks). Sexual reproduction in aphids is reserved for descendants of foundress females that reproduce in late fall (Figure 14.3b). Because offspring of these fall parents will emerge in the spring in an environment different from that of their parents, a variety of genotypes is likely to include successful ones.

As do aphids that must overwinter and emerge in the spring, parasites that transfer from

Figure 14.3 Aphids alternate between sexual and asexual reproduction. (*a*) Scanning electron micrograph of a foundress female (largest individual on right) and her offspring on a leaf. Although this female and her offspring of spring, summer, and early fall reproduce asexually, descendants that reproduce in late fall switch to sexual reproduction. Production of genetically variable offspring in late fall is probably advantageous because these young must overwinter and face new conditions in the spring. (*b*) Annual cycle of an aphid showing the shift from asexual to sexual reproduction. (b; After Thornhill and Alcock, 1983.)

a

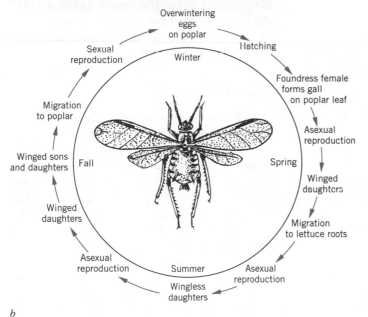

b

one host to another face tremendous changes in environment. Generally speaking, parasites reproduce asexually within a primary host (the first animal in which they live) and switch to sexual reproduction in the final host. Because final hosts are usually larger and more mobile than primary hosts, offspring of the parasite face unpredictable and widely varying environmental conditions. For example, the liver fluke (*Fasciola hepatica*) reproduces either sexually or asexually depending on the type of host in which it resides (Cheng, 1973). Eggs of this parasitic flatworm hatch in water, and the young penetrate the primary host, a snail, within 8 hours. Once inside the snail, individuals reproduce asexually and soon become free-swimming forms that attach to aquatic vegetation. Hungry sheep or cows munch on vegetation containing the tiny parasites and become final hosts. It is only in the livers of these large, mobile animals that *F. hepatica* reproduces sexually (Figure 14.4).

One further benefit to organisms that produce offspring through sexual reproduction is that genetically variable offspring may compete less with each other (Maynard Smith, 1976). The advantage of decreased sibling competition, however, applies only to species in which parents disperse before reproducing and then produce sedentary youngsters.

Williams (1975) summarized the short-term advantage of sexual reproduction with an analogy to the lottery. Imagine you are at the counter of your local drugstore preparing to purchase a stack of Megabucks tickets. Would you buy a group of tickets all with exactly the same numbers, or would you buy a stack of tickets in which each one had different numbers? In the interest of retiring at an early age, you would probably buy the tickets with different numbers. Similarly, in the face of unpredictable environmental conditions, parents that opt for the production of variable offspring are more likely to strike it rich (genetically speaking) and produce a winner

than parents that choose to produce redundant copies of the same old genotype.

According to the short-term hypothesis, sexually reproducing species have an advantage over species that reproduce asexually when multitudes of colonists are sent forth, among which a few will survive. But what of humans and other organisms that produce only a few offspring of which most are likely to survive? Williams suggests that sexual reproduction is a poor strategy indeed for organisms of low fecundity like ourselves. To make matters worse, our chances of switching to asexual reproduction and realizing true success at last appear quite dismal. In fact, the vast majority of vertebrates reproduce sexually, and it is noteworthy when asexual species such as whiptail lizards or ambystomatid salamanders evolve.

Critics of the short-term hypothesis point out that mathematical models demonstrate that sex will not create a reproductive advantage to the individual unless environmental conditions fluctuate dramatically from one generation to the next, and in the view of these critics, such conditions are highly unlikely.

Additional Hypotheses

Not all biologists are content to choose between only the short-term and long-term hypotheses. A variety of hypotheses, all concerned with advantages to the individual and not to the population, has arisen in the last two decades. Some researchers question the traditional focus on the production of variable progeny. Bernstein, Byers, and Michod (1981) argue that production of genetically variable offspring is only one consequence of recombination and that the generally accepted view of sexual reproduction is far too narrow. According to these authors, the sexual cycle functions primarily to repair errors in DNA; damage on one chromosome can be repaired by exchanging the damaged part with the homologous chromosome. From this perspec-

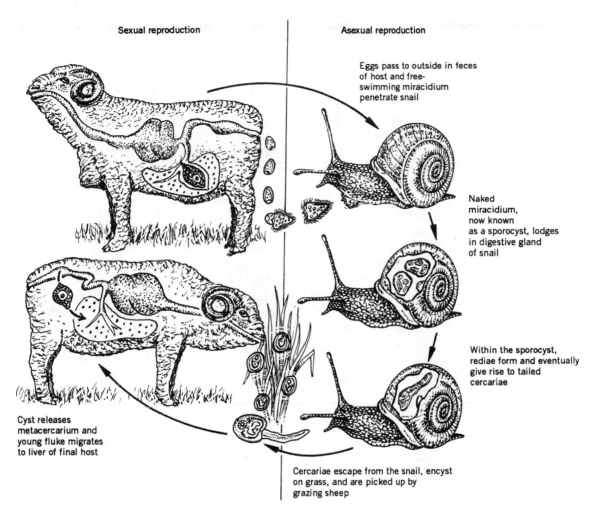

Sexual reproduction

Asexual reproduction

Eggs pass to outside in feces of host and free-swimming miracidium penetrate snail

Naked miracidium, now known as a sporocyst, lodges in digestive gland of snail

Within the sporocyst, rediae form and eventually give rise to tailed cercariae

Cercariae escape from the snail, encyst on grass, and are picked up by grazing sheep

Cyst releases metacercarium and young fluke migrates to liver of final host

Figure 14.4 Life cycle of the liver fluke. Although liver flukes reproduce asexually while in the digestive glands of snails, they switch to sexual reproduction upon transfer to final hosts such as sheep. Because final hosts are more mobile than primary hosts, the array of genotypes produced by sexual reproduction may come in handy when offspring face unpredictable environmental conditions. (Modified from Pantelouris, 1965.)

tive, then, the primary function of sex is the correction of genetic errors. Genetic variation, rather than being the key evolutionary advantage, is viewed as simply a by-product of the sexual process (Bernstein, Hopf, and Michod,

1988). Other researchers suggest that the advantage of sexual reproduction may lie in the process of genetic segregation and not necessarily in genetic recombination (Kirkpatrick and Jenkins, 1989). In diploid organisms, two genes for a trait segregate from each other during meiosis so that each gamete that is formed will end up with one or the other gene. The process of segregation promotes more rapid spread of new, beneficial mutations than does asexual reproduction in which two separate mutations are required for the beneficial trait to reach the homozygous state. Still others claim that sexual reproduction is favored in environments that are heterogeneous in space rather than in time as suggested by the short-term hypothesis (Bell 1982, 1988; Ghiselin, 1974), or that changes in the biotic environment, rather than the physical environment, generate the need for variable offspring (Glesner and Tilman, 1978).

Obviously, then, the answer to our question, why bother with sex?, remains, at least in the minds of biologists, elusive. In fact, the question has been called "maybe the biggest" unsolved problem in evolutionary biology (Michod and Levin, 1988). Perhaps the best approach to unraveling the mysteries of sexual reproduction is to keep in mind that the adaptiveness of sex may vary according to environmental conditions and the demographic characteristics of the organism in question. Indeed, there may not be a single selective advantage to sex, but rather a variety of factors involved in maintaining this mode of reproduction (Charlesworth, 1989).

Evolution of the Sexes

We usually think of sex as a process that involves males and females. However, this is not always the case. Bacteria, after all, reproduce sexually yet dispense with the requirement for two sexes. Usually, though, sexual reproduction does involve two sexes, so the question arises, what is meant by male and female?

Sexes are defined according to the amount of energy allocated to each kind of gamete. Females allocate more energy per gamete than do males: Females produce a small number of relatively large eggs, while males produce a large number of small, highly mobile sperm. Each egg comes equipped with informational molecules (DNA and RNA), a supply of nutrients, and molecules such as actin and tubulin that are required for proper cell division in the future zygote. In contrast, a sperm is little more than a streamlined package of genetic material powered by a flagellum that transports it to its rendezvous with the egg.

Although the genetic contributions of males and females to their offspring are equal, resource contributions are skewed toward the female parent. Take, for example, the female sea urchin whose eggs, as a result of their vast cytoplasmic stores, have a volume more than 10,000 times that of sea urchin sperm (Figure 14.5; Gilbert, 1988). From the moment of fertilization, the roles of the sexes are differentiated, thereby setting the stage for a conflict of interests that will last a lifetime.

One possible scenario for the evolution of the sexes was suggested by Parker, Baker, and Smith (1972). Imagine an ancestral population in which males and females did not exist, although gametes were produced meiotically. Perhaps individuals produced a range of gamete sizes and natural selection acted to favor both relatively large and relatively small gametes. Large gametes with their hefty energy stores would promote survival of the zygote after fertilization, while small, highly mobile gametes would fare better in the race to fertilization. An additional advantage to small gametes is that they could be produced rapidly and in large numbers. As the story goes, two selection pressures, one for large size and the other for large numbers, produced over generations a bimodal distribution of gamete sizes (Figure 14.6). One type of gamete, the egg, increased its girth by the accumulation of food stores for the upcoming zygote, and the second type, the sperm, unloaded the baggage of excess cellular material and became increasingly streamlined and specialized for the all-important race to the egg.

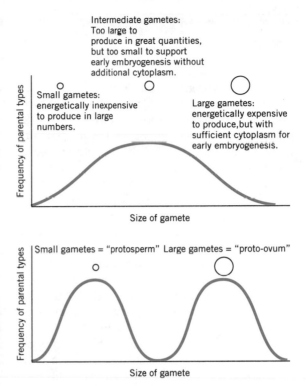

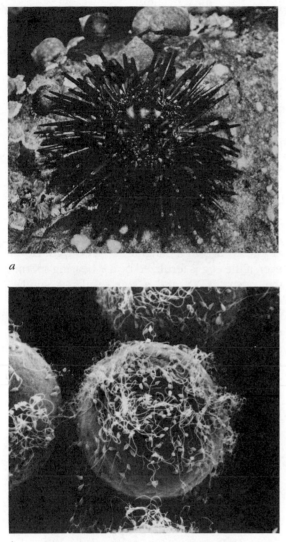

Figure 14.5 Sea urchins and the gametes they produce. (*a*) Sea urchins, the subject of many classic embryological studies, are reliable sources of gametes for developmental investigations. (*b*) Scanning electron micrograph of sea urchin sperm bound to the outer envelope of an egg. In sea urchins, as in most sexually reproducing organisms, eggs are vastly larger than sperm and this energetic difference serves to define the sexes.

Figure 14.6 One possible scenario for the evolution of the sexes was suggested by Parker, Baker, and Smith in 1972. Although an ancestral population of sexually reproducing organisms may have produced a range of gamete sizes, selection soon began to favor both relatively small and relatively large gametes: Large gametes, though expensive to produce, contained energy stores that would promote survival of the future zygote, while small gametes could be produced in large numbers. Over many generations, disruptive selection led to a bimodal distribution of gamete sizes with the ultimate result being production of large, energy rich eggs by females and large numbers of tiny sperm by males. (From Daly and Wilson, 1983; based on mathematical models of Parker, Baker and Smith, 1972.)

Sex Determination

Now that we see how gender is defined, we may ask, how is it determined? The mechanism by which an individual's sex is determined has puzzled embryologists for centuries. Aristotle claimed that sex was determined by the temperature of the male partner during intercourse—the more heated the passion, the greater the likelihood of producing sons (Gilbert, 1988). Aristotle even advised elderly men, who he assumed could not generate much heat, to have sex in the summer if they wished to be blessed with a male heir. Aristotle's notion is an example of environmental sex determination—external influences (conditions outside the cell) dictate the sex of the zygote. In the early 1900s, evidence began accumulating in favor of chromosomal sex determination—the sex of the zygote is determined by the chromosomes of the gametes. Today we know that the sex of an individual may be determined by either chromosomal or environmental means, depending on the species. (Human sex is determined chromosomally, so Aristotle's notion has been rejected.)

CHROMOSOMAL SEX DETERMINATION

In birds and mammals, as in the majority of animals studied to date, the sex of an individual is determined by chromosomes. Each cell of a female mammal carries two X chromosomes, while each cell of a male carries one X and one Y chromosome. It is the Y chromosome that contains the single gene for maleness. During meiosis each egg receives one X chromosome, and each sperm receives either an X or a Y. If an egg is fertilized by an X-bearing sperm, the zygote is XX and a daughter is produced. However, if the egg is fertilized by a Y-bearing sperm, the zygote is XY and a son results. A similar situation characterizes sex determination in birds except that it is the female that carries two different chromosomes (Z and W), while males carry two Z chromosomes. In birds and mammals then, the chance encounter between a certain egg and a certain sperm determines whether a son or daughter is produced. Sex

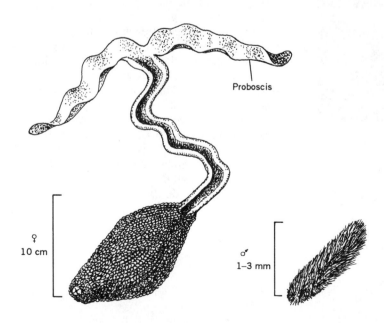

Figure 14.7 The echiuroid worm *Bonellia viridis* exhibits location-dependent sex determination; larvae that settle on or near the proboscis of an adult female are drawn into her mouth and down to her uterus where they develop into tiny parasitic males. Larvae that land on the rocky substrate become females. (After Barnes, 1968.)

determination in these two taxa occurs at fertilization and gender is a lifelong possession.

ENVIRONMENTAL SEX DETERMINATION

Location-dependent sex determination, one form of environmental sex determination, is characteristic of the rock-dwelling echiuroid worm *Bonellia viridis* (Figure 14.7; Agius, 1979; Baltzer, 1914). Although only 10 centimeters (4 inches) long, females of this marine invertebrate have a proboscis that can extend for over a meter. The rather striking proboscis functions not only to sweep food into the digestive tract of the female, but also to produce a compound that attracts and masculinizes larvae. If a larva lands on or near the proboscis, it enters the mouth of the female and migrates to the uterus, where it develops into a tiny parasitic male. The male spends the rest of his life locked within the body of the female fertilizing her eggs. However, if a larva settles on a rocky substrate away from the proboscis, it becomes a female. So, in *B. viridis*, then, distance from the proboscis of an adult female determines the sex of the larvae.

Another form of environmental sex determination involves temperature dependence. Many reptiles, for example, lack chromosomes involved in sex determination, and gender is determined after fertilization by the temperature at which eggs incubate (Bull, 1983; Raynaud and Pieau, 1985). Although organisms that display this second form of environmental sex determination are alike in lacking sex chromosomes, they differ markedly in the particular effect that temperature has on the sex of offspring. In the American alligator (*Alligator mississippiensis*; Figure 14.8a, Ferguson and Joanen, 1983) and in a terrestrial lizard, the leopard gecko (*Eublepharis macularius*; Bull, 1987), high incubation temperatures produce male hatchlings, while low temperatures result in female hatchlings. The opposite is true for many turtles, for which high temperatures produce females and low temperatures, males. The snapping turtle (*Che-*

lydra serpentina; Figure 14.8b), described uncharitably in one field guide (Conant, 1975) as "ugly both in appearance and disposition," displays a system of sex determination in which females develop at high and low extremes of temperature, while males predominate at intermediate temperatures (Yntema, 1981). In still other reptiles, incubation temperature has no effect on sex determination. Instead sex is determined chromosomally. Why is it that some species run hot and cold in producing sons or daughters while others rely on sex chromosomes? At present we do not know, but biologists are now examining the relationship between an individual's incubation temperature and later fitness in an attempt to discover the adaptive significance of environmental sex determination.

Sex Change

In sharp contrast to the permanence of gender in birds, mammals, and other animal groups, species in some groups change sex. In molluscs, for example, sex change occurs among species of gastropods (snails and slugs) and bivalves (clams and mussels; Wright, 1988). Sex change, however, is not limited to invertebrates. In fact, sexual flexibility is quite widespread among fish: At least 14 families contain species that exhibit sex change from female to male (protogyny) and 8 families contain species that change from male to female (protandry; Warner, 1984). Some suggestions for why sex changes occur in fish include (1) external fertilization and less complex sexual differentiation lower the physiological cost of change in fish relative to other vertebrates; and (2) mating in fish does not involve many complex or learned patterns of behavior, and thus it is easier than in other vertebrates to function quickly and successfully as the other sex (Warner, 1978). As we will see from the examples that follow, the primary determi-

a

b

Figure 14.8 Some egg-laying reptiles lack sex chromosomes and gender is determined by incubation temperature. (*a*) In the American alligator, eggs that incubate at high temperatures produce males while those that incubate at low temperatures result in females. (*b*) Snapping turtles display a system of sex determination in which females develop at high and low extremes of temperature, while males predominate at intermediate temperatures.

nants of sex change mechanisms are the social organization and mating system of the species in question (Ross, 1990).

PROTANDRY

The nosestripe anemonefish (*Amphiprion akallopisos*) exhibits protandry, the change from male to female (Fricke and Fricke, 1977). As the name of the fish suggests, individuals of this species live inside large sea anemones (Figure 14.9). Unlike many other organisms, anemonefish can move about the stinging tentacles of their host unharmed. Although not immune to the poison of the anemone, *A. akallopisos* possesses a substance in its mucus that suppresses

Figure 14.9 Sex changes with social environment in anemonefish of the genus *Amphiprion*. Breeding pairs live with unrelated subadults in sea anemones. If the female of a breeding pair dies, the male quickly changes sex, and a subadult of the social group matures to become the new male. The change from male to female is termed protandry and occurs in social systems in which large size is more beneficial to females than to males.

the release of the stinging cells. Nestled in the arms of their lethal host, anemonefishes live in male–female pairs and each pair normally resides with a small number of unrelated subadults and juveniles. Membership in a social group is somewhat exclusive and is limited by the size of the anemone. The largest and oldest member is always female, and she is quick to dominate her mate and the younger individuals inhabiting the anemone. If the female member of a breeding pair dies, the male quickly changes sex, and a subadult of the social group matures to become the new male. Social control of sex change enables recruitment of individuals independent of the sex of arriving youngsters and eliminates the need for leaving the anemone in search of mates. Because a small male can successfully fertilize and defend the eggs of a large female, size is not critical to the reproductive

success of male anemonefish. In contrast, large females produce many more eggs than small females, and thus female reproductive success is strongly influenced by body size. In the anemonefish, large size is more beneficial to females than to males, and protandry allows individuals to maximize reproductive success.

PROTOGYNY

The Pacific cleaner wrasse (*Labroides dimidiatus*) exhibits protogyny, the change of sex from female to male (Robertson, 1972). This light blue fish with a black racing stripe down its side lives in coral outcrops of the Great Barrier Reef. Amid the branching coral, it sets up cleaning stations and makes a living by feeding on ectoparasites and mucus removed from the skin, mouth, and gill chambers of other fish (Figure 14.10). As

Figure 14.10 The Pacific cleaner wrasse exhibits protogyny, the change from female to male. Male wrasses control harems of five or six females; and large males are more successful than small males at acquiring and defending mates. Protogyny appears to be an adaptation to mating systems in which older, larger males enjoy greater mating success than do smaller, younger ones. Cleaner wrasses make their living by removing ectoparasites from other fish. Here, a butterfly fish receives a grooming from the smaller wrasse.

customers wait in line for its services, this small fish drifts in and out of the mouths and over the bodies of much larger and normally pugnacious hosts. Typically, cleaner wrasses live in groups consisting of a single male and five or six females. A male actively defends his harem and mates with each member daily. However, if the male is removed from the harem, the largest female changes into a male. With sperm production underway within a week's time, this new male takes over as harem master. Unlike the anemonefish, being a small male has dire consequences in the mating system of the Pacific cleaner wrasse. Current theory suggests that protogyny in fish is an adaptation to mating systems in which older, larger males enjoy substantially greater mating success than do smaller, younger males (Charnov, 1982). Apparently, the best strategy for a cleaner wrasse is to function as a female when small and as a male when large enough to successfully compete for mates.

Sex Ratio

FISHER'S THEOREM OF THE SEX RATIO

Why in most species are the number of males and females approximately equal? In 1930 mathematician R. A. Fisher explained the prevalence of 1:1 sex ratios (the ratio of males to females) by stating that in a population in which individuals mate at random, the rarity of either sex will automatically set up selection pressure favoring production of the rare sex. Once the rare sex is favored, the sex ratio gradually moves back toward equality. For example, if males are rare in a population, parents should produce sons since on average, these scarce males will enjoy greater mating success and will produce more offspring than will females. Similarly, if females are in short supply, parents that invest in daughters will enjoy greater genetic success (i.e., have more grandchildren) than those that produce sons. As

originally stated, Fisher's theorem proposed that parents should produce sons and daughters with equal likelihood. A more general version of this theory is that parents should expend equal resources in the production of offspring of the two sexes, and thus, an equal sex ratio is the special case in which sons and daughters cost the same to produce. (For a review of theories concerning sex allocation, see Frank, 1990.)

MANIPULATION OF THE SEX RATIO

The Trivers-Willard Hypothesis

Fisher's theorem of the sex ratio and the fact that sex is chromosomally determined in birds and mammals appeared to provide quite reasonable explanations for the prevalence of 1:1 sex ratios in these groups. However, in 1973, Robert Trivers and Dan Willard shook the foundations of sex ratio theory by proposing that mammalian parents should have the ability to manipulate the sex ratio of their offspring before birth. Mothers should, for example, adjust the sex ratio of their young in accordance with their own ability to invest energy and resources. Because variance in reproductive success is expected to be greater among male than among female mammals (for reasons that will be discussed later in this chapter and in the next chapter, most females have a single mate while some males have several mates and other males none), a high-quality son will have greater reproductive success than a high-quality daughter. Conversely, low-quality offspring should be female, because regardless of quality, all females find mates. In short, Trivers and Willard predicted that mammalian mothers in good condition should produce sons, while mothers in poor condition should produce daughters. Could this radical idea be true? Are there instances in which mammalian parents produce male and female offspring in ratios that deviate from the standard 1:1? Also, if skewed sex ratios do exist, do they have any adaptive significance?

Figure 14.11 Females of the South American opossum appear to manipulate the sex ratio of offspring according to maternal condition: Females in good condition produce an excess of sons while those in poor condition favor daughters.

Examples of Apparent Sex Ratio Manipulation in Mammals

Over the last few years biologists have sexed red deer fawns (Clutton-Brock, Albon, and Guiness, 1986), hamster pups (Huck, Labor, and Lisk, 1986), baboon infants (Altmann, 1980), and a host of other young animals in search of evidence of skewed sex ratios. Slowly but surely they have found sex ratios that depart from 1:1. The numbers they have come up with, though, have not been shocking. In fact, most biases in the sex ratio have been subtle, along the lines of six of one sex to four of the other. We will consider two instances of apparent sex ratio manipulation, one involving a marsupial, the other a primate.

Steven Austad and Mel Sunquist (1986) examined whether opossum mothers manipulate the sex ratio of offspring in accordance with maternal condition (Figure 14.11). Because of the ease with which marsupial young can be monitored, opossums are ideal for sex ratio research. Young opossums are born in an embryonic state after only 13 days of gestation. At birth

they are deaf and blind and have only tiny buds where their tail and hind limbs will be. Despite a poorly developed hind end, newborn opossums have well-developed forelimbs, which they use to crawl the 4 or 5 centimeters from their mother's vagina to her pouch. Once in the pouch, each youngster attaches to a teat, where it firmly remains for the next 60 or 70 days as the nipples of the mother swell, creating an airtight seal with the newborn's rigid mouth. Young are weaned at approximately 100 days and have little contact with their mother after the cessation of nursing.

Austad and Sunquist radiocollared 40 females of the South American opossum, *Didelphis marsupialis*, in central Venezuela. Because the females wore radiocollars, they could be located in their dens during the day. Twenty females were left alone (controls), and sardines were left outside the dens of the other 20 females (experimentals) at dusk. Upon emerging from their dens, experimental females could enjoy a few appetizers before going off to search for the main meal. It was assumed that the experimental and control females were equally successful in finding other food. As expected, experimental females invested more in reproduction than did control females: Individual young of the experimentals—those given food supplements—were larger than young of controls at every stage of development. Although both male and female offspring of the experimentals were larger, males seemed to benefit most because more of them survived and the survivors themselves had greater reproductive success than did their sisters. Even more remarkable was the finding that food-supplemented mothers (i.e., mothers in good condition) produced an excess of sons. On average, experimentals produced 1.4 sons for each daughter, while controls produced an equal number of sons and daughters.

In line with the Trivers–Willard model, Austad and Sunquist reasoned that high-quality sons, able to outsurvive and outcompete other males, would have more opportunities to mate and pass along their mother's genes than would high-quality daughters whose reproductive output is limited by the number of young they themselves can bear. Although most opossums in the wild live for only one reproductive season, 19 second-year females were trapped during the course of the study. These elderly females suffered from cataracts, weight loss, and lack of coordination and produced nearly 1.8 daughters for each son. Presumably, excess daughters produced by mothers in poor condition are likely to find mates despite their less than ideal condition.

Although the mechanism of apparent sex ratio manipulation in *D. marsupialis* is unknown, Austad and Sunquist suggest three possibilities. First, because female opossums typically bear more young than can make their way to the pouch and find a teat, it is possible that males of food-supplemented mothers are simply better than females in their ability to reach the pouch. Second, well-fed mothers might distinguish between male and female offspring at birth and selectively pluck females off as they crawl toward the pouch. A third possibility is that manipula-

tion of the sex ratio occurs prior to birth, an intriguing notion indeed.

Manipulation of the sex ratio has been reported for several species of primates in which social rank is important in determining patterns of maternal investment. In baboons and macaques, daughters inherit the rank of their mother, while sons leave home upon reaching sexual maturity. Recent evidence suggests that high-ranking baboon and macaque mothers bias their investments toward daughters and low-ranking mothers toward sons (Altmann, 1980; Silk et al., 1981; Simpson and Simpson, 1985). These strategies of investment make sense in light of patterns of natal dispersal (dispersal of young animals from their place of birth); if low-ranking mothers produced female rather than male offspring, these offspring would remain at home, saddled with their mother's low status.

Spider monkeys (*Ateles paniscus*; Figure 14.12) exhibit the opposite pattern of natal dispersal: Females leave their natal groups upon reaching sexual maturity and males remain at home to breed. Thus, unlike baboon and macaque mothers whose sons leave the natal group, high-ranking mothers of *A. paniscus* are in a position to help their sons achieve high rank at home, with all the attendant sexual advantages. Meg Symington (1987) found that, generally speaking, high-ranking mothers of this species tended to invest in sons while low-ranking mothers produced daughters. Production of daughters by low-ranking females would seem to make sense as these daughters could leave their past (i.e., the social rank of their mother) behind and start fresh in a new social group. Similarly, investment in sons would seem to be the best option for mothers with social clout because their privileged male youngsters could remain at home and bask in the advantages of high rank. Although field data suggest that spider monkeys manipulate the sex ratio of offspring according to the predictions of the Trivers–Willard hypothesis, it remains to be seen whether benefits of high maternal rank in this species

Figure 14.12 In spider monkeys, females leave their natal groups upon reaching sexual maturity. Consistent with this pattern of natal dispersal is the tendency for low-ranking mothers to produce daughters and high-ranking mothers to produce sons. Investment in daughters by low-ranking females makes sense in light of the fact that these daughters will leave their past behind and start fresh in a new social group.

affect the reproductive success of sons over daughters.

Other studies suggest that a number of factors may affect the relative profitability of producing male and female offspring (Clutton-Brock and Iason, 1986). Although studies of opossums and spider monkeys indicate that variation in the comparative fitness of sons and daughters may favor change in the sex ratio, in other species, sibling competition for mates or resources may skew the sex ratio. If sons and daughters are equally costly, but siblings of one sex are more likely to compete with each other than siblings of the other sex, then parents would be wise to bias the sex ratio of their progeny toward the less competitive sex. In yet other species, parents tend to compete with offspring of one sex. In this case, it may be to the advantage of the parents to produce more offspring of the sex with which they do not compete. Alternatively, if offspring of one sex assist their parents in later reproductive attempts, parents might benefit by biasing the sex ratio toward the more cooperative sex, providing that the initial costs of sons and daughters are equal. We see, then, that a number of social and ecological circumstances can affect the relative benefits of producing sons or daughters.

Unanswered Questions

Many biologists have been swept up in the search for evidence of adaptive manipulation of the sex ratio. No one doubts that there are great advantages in being able to vary the sex ratio of progeny, and we have noted several theoretical advantages, but the question remains: Is such manipulation really possible? Some biologists remain unconvinced (Clutton-Brock and Iason, 1986). Perhaps the largest hurdle facing the area of sex ratio research is identification of the immediate mechanism(s) responsible for skewed sex ratios. Is differential mortality or motility of X- and Y-bearing sperm responsible for adjustments in the sex ratio prior to conception? Or is the sex ratio modified after fertilization through differential mortality of male and female embryos? Although it is clear that future research must address the question, how are adjustments in the sex ratio occurring?, the primary questions remain. Does such adjustment occur at all, and if it does, on what adaptive basis?

Sexual Selection

Why did the elaborate plumage of peacocks evolve (Figure 14.13)? Surely the long, shimmering train is energetically costly to produce.

Figure 14.13 A peacock displays his train. Perhaps the elaborate plumage evolved to attract mates.

Furthermore, the iridescent splendor is heavy, as well as conspicuous to predators. Equally striking are the enormous horns on some male scarab beetles (Figure 14.14). At what cost did they evolve? How can we reconcile the production of such bizarre structures with natural selection?

DARWIN'S VIEW OF SEXUAL SELECTION

Darwin (1871) was the first to suggest that spectacular structures such as the plumage of peacocks and the horns of male beetles could arise and be maintained through the process of sexual selection. In Darwin's view, there was a clear distinction between selection for traits that en-

hance survival and reproduction (natural selection) and selection solely for traits that increase an individual's success at acquiring mates (sexual selection): Although gaudy feathers and cumbersome horns were incompatible with his theory of natural selection, such structures persisted because they enabled their bearers to gain access to mates. (In contrast to Darwin's view of natural selection and sexual selection as two distinct processes, many researchers today regard sexual selection as a subset of natural selection. Because both processes have their evolutionary effects through the differential reproductive success of individuals, the separation of natural and sexual selection has all but disappeared in the literature. There are, however, some evolutionary biologists who still believe that Darwin's original distinction is a useful one [Arnold, 1983].)

According to Darwin, sexual selection arises when individuals of one sex, usually males, gain a mating advantage over other individuals of the same sex. Males could gain a competitive edge by fighting with each other, and winners could claim the spoils of victory—females. Intense fighting and competition for mates could lead to selection for increased size and elaborate weapons such as the horns of male beetles. Such competition between males for access to females, termed intrasexual selection, constitutes the first component of sexual selection.

Darwin proposed a second component, intersexual selection, to explain the existence of extravagant traits such as the train of a peacock. In intersexual selection (sometimes called epigamic selection or mate choice), individuals of the sex in demand, usually females, exert selection that favors individuals of the opposite sex with preferred attributes. Thus, males not only fight with each other for access to females, but they also compete to attract females through the elaboration of structures or behavior patterns. Darwin's concept of intersexual selection rested on the assumption that females prefer males ornamented or behaving in a particular way. Al-

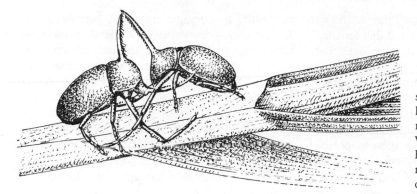

Figure 14.14 Males of the scarab beetle *Golofa porteri* have a head horn and enormously elongated front legs. Weapons such as these may have evolved in the context of male–male competition. (Redrawn from Eberhard, 1980.)

though the existence of female preferences was a virtual certainty to Darwin, questions as to how sexual preferences might be maintained by selection, or how preferences might arise in the first place, were left unanswered. For these reasons the scientific community, for a time, largely rejected Darwin's ideas on intersexual selection.

RUNAWAY SELECTION

In 1930 R. A. Fisher reinvigorated Darwin's theory of intersexual selection by tackling questions about how female preferences become established in the first place. Fisher suggested an initial stage in which males possessing a character that has come to be attractive to females have some other heritable advantage over competing males, one contributing to fitness. Once the character used as a basis for female mating preference becomes established through its link with greater male fitness, it will then become subjected to selection simply for its attractive properties. In response to the question of how female preferences would be maintained in a population, Fisher reasoned that (1) a female that mates with an attractive male will have attractive sons, providing that the attractive character of the male is inherited and (2) attractive sons will acquire more mates than other males and thus will leave more progeny. In short, the advantage to a female in choosing an attractive

mate is that she will reap the genetic benefits of many grandchildren.

Obviously, this series of selective events would constitute a runaway selection process, essentially producing a positive feedback loop, favoring ever more exaggerated male characteristics and females that find them attractive. So when does it all stop? When does a peacock tail become too long? According to Fisher, the process will be stabilized only when natural selection balances sexual selection. In other words, when the peacock's tail becomes too energetically costly to produce, or when it is so long and heavy that he cannot escape tigers, selection will no longer favor its increased length.

Lande (1981) and Kirkpatrick (1982) have developed mathematical models of Fisher's ideas for the evolution of female choice. These models demonstrate that runaway selection can indeed result in mate choice for characteristics that are arbitrary or even disadvantageous to the health and survival of individuals, providing that females prefer to mate with males that possess them.

ALTERNATIVE IDEAS FOR THE EVOLUTION OF FEMALE CHOICE: CHOOSING MALES WITH GOOD GENES

Amotz Zahavi (1975) suggested an alternative to Fisher's runaway selection model for the evolution of female choice. Termed the handicap principle, Zahavi's alternative stated that females prefer a male with a trait that reduces his chances of survival but announces his superior genetic quality precisely because he has managed to survive despite his "handicap." In short, male secondary sexual characteristics act as signals indicating high fitness, and females choose males with handicaps because their superior genes may help produce viable offspring. One problem with Zahavi's idea, of course, is that offspring inherit the handicap as well as the good genes.

Zahavi's suggestion that female choice of mate ought to have beneficial genetic effects in terms of offspring viability met with considerable debate (Maynard Smith, 1985). The runaway selection hypothesis, after all, does not require that females base their preferences on male traits that signal viability; furthermore, mathematical models show that Fisher's feedback process is sufficient to explain the development of extravagant traits in males. However, just because the runaway selection hypothesis can explain male secondary sexual characteristics, should we stop looking for other possible explanations? There is, in fact, a genetic model that provides support for the good genes hypothesis. This model shows that heritable differences in viability can provide a basis for mate choice providing that the elaborate ornaments and displays accurately represent a male's general health (Andersson, 1986). This brings us to an important question, do secondary sex traits develop in relation to the condition of the animals that possess them?

William D. Hamilton and Marlene Zuk (1982) tackled the question of whether the elaborate ornaments of males represent reliable signals of health and nutritional status by examining plumage coloration in North American birds. They suggested that only males in top physical condition would be able to maintain bright, showy plumage. Because bird species vary in their sus-

ceptibility to parasitic infection, Hamilton and Zuk predicted that the degree of male brightness would be correlated with the risk of attack by parasites. Accordingly, if males of a species vary substantially in their parasite load, then it would behoove a female to choose a male that honestly signals his good health, for the offspring of this male may experience increased viability if they inherit his resistance to parasites. In species in which the risk of infection by parasites is minimal, such information has little value to females in the market for a mate, and thus they should not display a preference for males with showy feathers—in these species, then, we would not expect males to be very brightly colored.

In testing their hypothesis, Hamilton and Zuk surveyed the literature on avian parasites and determined the risk of infection for each bird species. They then ranked each species from one (very dull) to six (very striking) on a plumage showiness scale. In support of their ideas, there was a significant association between showiness and the risk of parasitic infection—those species with the highest incidence of infection from blood parasites had the most showy males.

Although consistent with the good genes hypothesis, the Hamilton and Zuk (1982) results could also be interpreted as arising from the runaway selection process. A stronger test would entail demonstrating that females that choose males with super bright plumage produce offspring of higher viability than do other females.

In the last decade, the influence of parasites on sexual selection has been examined in many species of birds (e.g., Borgia, 1986; Møller, 1991), as well as in a host of other animals including gray treefrogs (*Hyla versicolor*; Hausfater, Gerhardt, and Klump, 1990), three-spined sticklebacks (*Gasterosteus aculeatus*; Milinski and Bakker, 1990), and humans (*Homo sapiens*; Low, 1990). Despite intense scrutiny of the idea that females choose mates using cues that will lead to more viable offspring, the debate between proponents of the runaway selection hypothesis and the good genes hypothesis continues today (Bradbury and Andersson, 1987; Maynard Smith, 1991).

BATEMAN'S PRINCIPLE AND TRIVERS'S THEORY OF PARENTAL INVESTMENT

Two ideas that have played critical roles in the development of sexual selection theory are Bateman's principle (1948) and Trivers's (1972) theory of parental investment. Although it was known for some time that males of most species vary more than females in the number of mates and surviving young produced, Bateman (1948) was the first to quantitatively document relative variance of the sexes in reproductive success. In an ingenious experiment, he placed from three to five adult fruit flies (*Drosophila melanogaster*) of each sex into laboratory containers used for breeding. As a result of these housing arrangements, each female had a choice of several suitors and each male had to contend with a number of rival males. Any offspring produced by matings could be traced to specific parents as a result of genetic markers, so Bateman was able to measure the reproductive success of all the males and females. He found that male reproductive success was far more variable than was female reproductive success: Many males failed to mate at all (a full 21% failed to mate even once), successful males usually mated with more than two females, but most females copulated with only one or two males. The relationship between number of matings and number of offspring produced also differed between the sexes. Typically, females gained nothing by mating with more than one male, while males, with each copulation, continued to accrue the genetic benefits of additional progeny.

Bateman explained his results in terms of the energetic investment of each sex in gametes: Although females produce a small number of large, energetically expensive eggs, males produce millions of small, relatively inexpensive sperm. As a result of differences in the number of gametes produced by the sexes, females (or more correctly, their eggs) become a limited resource for which males compete. According to Bateman, this gametic difference helps to explain the "undiscriminating eagerness" of male fruit flies and the "discriminating passivity" of females. Because males of almost all sexually reproducing organisms produce smaller gametes than do females, and because in many species males tend to have a certain proclivity for promiscuity, Bateman's principle has consequences far beyond the sex life of fruit flies in laboratory vials.

In 1972, Robert Trivers suggested that differential investment by the sexes in offspring, rather than in gametes, was responsible for competition and mate choice. In Trivers's view, the sex that provides more parental investment for offspring (usually the female) becomes a limiting resource for which the sex that invests less (usually the male) competes. The result is similar to that derived from Bateman's principle: Males compete for access to females and females have the luxury of choosing among available suitors. Trivers's theory is perhaps best tested by examining the select group of nontraditional species—those in which males invest more than females in the care of offspring. If Trivers is correct, we would predict that females of these species compete for males, who in turn are quite discriminating in their choice of mates. As we will see in the following chapter on parental care and mating systems, Trivers's hypothesis is often supported by these exceptions to the rule.

Support for Trivers's ideas also comes from the study of within-species variation in the courtship roles of some insects. Within certain species of katydids (relatives of the grasshoppers), courtship roles are far from fixed. Indeed, in one species (the genus and species of this particular katydid are undescribed and thus scientific names cannot be provided), whether males compete for females or females compete for

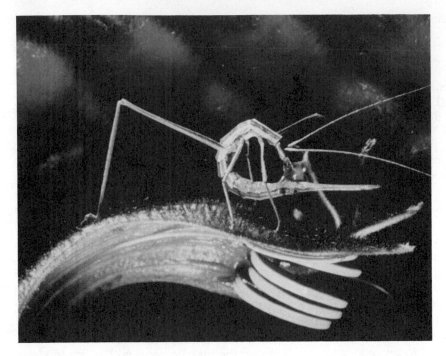

Figure 14.15   A female katydid eating part of the spermatophore deposited by her mate. This protein-rich gift constitutes the male katydid's parental investment. When food is scarce, the gift has relatively greater value, and rather than females choosing males, males choose females.

males varies with the relative importance of male parental investment. In katydids, male parental investment consists of a nuptial meal. Basically, at the time of mating, a male katydid transfers his spermatophore (a packet of sperm and fluids) to the female. Following separation of the couple, the female bends and eats part of the spermatophore (Figure 14.15). This protein-rich meal is important to successful reproduction in that both the number and fitness of her offspring are enhanced by the male's gift. The relative importance of the male's gift, however, varies with food availability. The gift is especially important to females when food is scarce, and it is during these times that females compete for males. In contrast, when food is plentiful, the relative value of the gift declines, and males compete for females (Gwynne and Simmons, 1990). These results are consistent with Trivers's theory on the importance of patterns of parental

investment in determining which sex competes for access to individuals of the other sex: Changing courtship roles within a species of katydid result from variation in the relative importance of male parental investment.

Despite the continuing accumulation of evidence in favor of Trivers's ideas, scientists have begun to challenge the notion that patterns of parental investment are prime determinants of the nature and strength of sexual selection. Rather than parental investment being the single, all-important determinant of sexual selection, there may be several factors, independent of parental care, that impact on mating effort and success. In other words, there are more conflicts between the sexes than simply who cares for the kids. Such conflicts might include which sex searches for mates, and once the sexes actually meet, which sex is more accepting of potential partners (Hammerstein and Parker, 1987).

MALE MATING COSTS REVISITED

In recent years some researchers have questioned the long-held notion that males incur trivial costs when producing gametes (Dewsbury, 1982b; Simmons, 1988). They are quick to point out that although gamete for gamete, sperm are vastly smaller and cheaper to produce than eggs, sperm are probably never passed along one at a time. Instead, millions of sperm are transferred in groups along with accessory fluids in ejaculates (unpackaged fluids) or spermatophores (packaged fluids). Although the cost of producing a single gamete may be minuscule, the costs of producing sperm groups and accessory fluids may limit the reproductive potentials of males. In field crickets (*Gryllus bimaculatus*), for example, the costs of spermatophore production (calculated by determining the percentage of the donor male's body weight comprised by his spermatophore) are relatively greater for small males than for larger ones. Small males appear to cope with these higher costs not by producing smaller spermatophores, but by increasing the refractory period between matings (time from when the male attaches a spermatophore to a female until the onset of the next attempt at courtship, when another spermatophore is ready). Thus, at least for small male field crickets, the costs of spermatophore production appear to limit their number of matings (Simmons, 1988).

In line with this notion of spermatophore and ejaculate costs, psychologist Donald Dewsbury (1982b) predicted that males should not distribute gametes among females in a hit-and-run fashion, but in a somewhat more prudent manner. Understanding male mating costs is essential to our understanding of the evolution of reproductive patterns, and males may, in fact, have higher costs than previously believed. In most cases, however, we can still expect females to be more selective than males as a result of their greater gametic and parental investment. With this introduction to some of the historical and theoretical background for sexual selection, we will now consider intrasexual selection, and then intersexual selection, in more detail.

A CLOSER LOOK AT INTRASEXUAL SELECTION

Intrasexual selection has led males to evolve a broad spectrum of attributes related to intense competition for mates. Some mechanisms operate prior to copulation while others have their effect once mating has occurred. We will now consider a few of the many tactics males use to gain that competitive edge in the mating game.

Adaptations that Help a Male Secure Copulations

Sexual Enthusiasm   Despite costs associated with mating, males of most species appear to follow a strategy of copulating with as many females as possible. To this end, males employ a variety of precopulatory mechanisms to ensure that few, if any, mating opportunities slip by. If we assume that males have a lower investment in gamete production and parental care, it is apparent that they would usually pay a small price for mating mistakes. Thus it behooves a male to take advantage of each mating opportunity that he encounters. The result is that males can be expected to have a low threshold for sexual arousal. Males may be so easily triggered, in fact, that they end up directing their sexual attentions to biologically inappropriate stimuli such as the wrong species, inanimate objects, and individuals of the same sex (Figure 14.16). We have already mentioned the fact that the movement of one male wood frog toward a female is sufficient to set off a chain reaction among neighboring males. The end result may be a pileup in which males, more often than not, end up clasping each other.

The Coolidge Effect is a good example of the seemingly boundless sexual enthusiasm of the males of many species. As the story goes (Ber-

Figure 14.16 Most males have a low threshold for sexual arousal and may direct their ardor to biologically inappropriate objects. This male Rio Grande leopard frog (on top) has mistaken both the species and sex of his partner, a male bull frog.

mant, 1976), President and Mrs. Coolidge were visiting a farm and Mrs. Coolidge was the first to tour the premises. She was shown a yard holding a number of hens and one rooster. When she was told that the one rooster was sufficient because he could copulate many times each day, she said, "Please tell that to the President." When the President came along, he was told the story, to which he asked, "Same hen every time?" No, he was told. After nodding slowly, the President said "Tell *that* to Mrs. Coolidge."

In fact, the phenomenon whereby sexually satiated males immediately resume copulation upon presentation of a new female has been reported for many rodents. In his studies of Norway rats (*Rattus norvegicus*), Alan Fisher (1962) found that the introduction of a novel female to a male soon after the termination of copulatory behavior resulted in reactivation of his mating behavior (Table 14.1). Whereas male rats left with the same female generally reached sexual satiation in approximately 1.5 hours, some males could be maintained in a state of cyclic sexual arousal for up to 8 hours simply by introducing new females at appropriate in-

tervals. Furthermore, it is not simply the case that "absence makes the heart grow fonder," for removal and replacement of the same female did not reactivate sexual behavior.

Dominance Behavior Males in some species may secure copulations by dominating other males and thereby excluding competitors from females. This behavior, of course, places a certain premium on greater male strength and more effective weaponry. Since the female makes no such investments, males tend to diverge from females in both appearance and aggressiveness. The degree to which this sexual dimorphism (difference in appearance of the sexes) occurs can be used as an index of role separation of the sexes in mating and reproduction.

In many birds and mammals, males are larger than females presumably because large body size improves the fighting ability and hence the reproductive success of males. It can be expected that the greatest levels of sexual dimorphism will be found in species in which the competition between males is most intense. This is, in fact, often the case, but sexual dimorphism

Table 14.1

	SEXUAL PERFORMANCE OF 16 MALE RATS		
TEST CONDITION	MEAN NUMBER OF INTROMISSIONS	MEAN NUMBER OF EJACULATIONS	MEAN TIME TO SATIATION (MIN)
Same female	43.0	6.9	93.8
Different female	85.5	12.4	257.7

Source: Fisher, 1962.

can result from other factors as well (Arak, 1988a; Hedrick and Temeles, 1989). Apparently, in many organisms the effect of female body size on female reproductive success is equally important in determining which sex is more "built to last" than the other. Additionally, although environmental factors usually affect males and females in a similar manner, sexual dimorphism in body size sometimes results from sex differences in niche utilization (e.g. food preference; Selander, 1966), or predation pressures (Bergmann, 1965).

In species in which males form dominance relationships, high-ranking males typically engage in more sexual activity than low-ranking males (Dewsbury, 1982a). In their attempts to monopolize sexual access to females, dominant males sometimes interfere directly with the copulations of subordinate males. Daniel Estep and colleagues (1988) examined interference by dominant males and consequent inhibition of sexual behavior among subordinate males of the stumptail macaque (*Macaca arctoides*). Long-term observations of a captive colony of this Old World primate revealed the existence of a stable, linear dominance hierarchy among adult males. Within the dominance hierarchy, copulation frequencies were rank related; the alpha, or highest ranking, male had almost as many copulations as the other 17 adult males combined (Table 14.2). Within this group, the three lowest ranking males failed to achieve a single copulation during the 34-week study. If low-ranking males mated at all, they usually did so surreptitiously,

out of view of dominant males. When tested in the absence of dominant males, low-ranking males copulated far more frequently than they did in the presence of dominant males. In short, dominant male stumptail macaques monopolize access to females by inhibiting subordinate males through aggressive threats and disruption of mating attempts.

Female Mimicry and Satellite Behavior
In keeping with the old adage "out of sight, out of mind," subordinate stumptail macaques cope with dominant males by copulating on the sly. The question arises, then, how do males of other species deal with dominant males? Two common tactics employed by subordinate males are female mimicry and satellite behavior. In the

Table 14.2

	PATTERNS OF COPULATION IN A CAPTIVE COLONY OF STUMPTAIL MACAQUES	
MALE RANK	TYPE OF COPULATION	TOTAL COPULATIONS[a]
Alpha male	Visible	424
	Surreptitious	—
14 other adult males[b]	Visible	69
	Surreptitious	393

[a] Per 1050 hours of observation.
[b] Three of 18 males in the colony failed to secure a copulation.
Source: Estep et al., 1988.

former, subordinate males mimic females and slip inside the territory of a dominant male. Once there, these pseudo females are able to steal copulations. In bluegill sunfish (*Lepomis macrochirus*), small males mimic the coloration and behavior of females and thereby succeed in entering the nests of larger males. Once inside the nest, the small male positions himself between a female and the larger male, all the time engaging in behavior typical of a gravid female (e.g., slow dipping movements and exaggerated rubbing of the side of the male). At the time of spawning, both the larger male and the female mimic release sperm (Gross, 1982).

The second strategy, satellite behavior, is characteristic of many amphibians in which males employ loud (and energetically expensive) calls to attract females. Although all males may be sexually mature and capable of calling, some remain silent and associate closely with a calling male, ready to intercept females attracted to the calls of the other male.

An example of satellite behavior is found in natterjack toads (*Bufo calamita*; Arak, 1988b). On spring evenings, males of this species gather at the edges of temporary ponds and call loudly to attract females. Callers typically adopt a head-up posture and energetically belt out their call (Figure 14.17). Satellites, on the other hand, keep a low profile and remain stationary and silent in a crouched position next to a calling male. If a female is attracted to the caller, a satellite male will make every effort to intercept and clasp her. In natterjack toads, body size and call intensity are highly correlated with mating success; in general, small males with weak calls are at a reproductive disadvantage. Not surprisingly, the consistency with which males in a population adopt the calling or satellite tactic is correlated with their body length: small males tend to be satellites; large males tend to be callers; and intermediate males are switch hitters, alternating opportunistically between the two strategies.

Figure 14.17 A male natterjack toad calls for females at the edge of a pond. Calling males compete with other males called satellite males that crouch silently nearby and intercept females on their way to the pond.

Adaptations Favoring Use of a Male's Sperm

Mate Guarding In addition to the option of adopting alternative reproductive strategies, males display a host of adaptations that increase the probability that their sperm, and not the sperm of a competitor, fertilize the eggs of a particular female. In some species males guard females during the time surrounding copulation. Precopulatory mate guarding is particularly common in Crustacea, whose mating is often restricted to a short period of time after the female molts (Dunham, Alexander, and Hurshman, 1986). The tiny amphipod *Gammarus lawrencianus* (Figure 14.18), for example, lives in estuaries along the coast of North America. Precopulatory mate guarding by males of this species consists of a highly stereotyped sequence of behavior and may be initiated at any time prior to the female molt. Typically, a free-swimming male grabs a passing female and draws her to his ventral surface. Once the male has a firm hold on the female, he moves her in such a way that the long axis of her body is at right angles to his. In this posture the male pal-

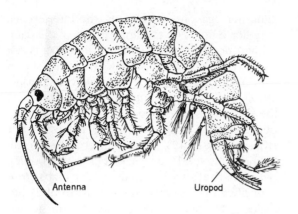

Figure 14.18 Male amphipods guard females during the precopulatory period, just before the female molts. By maintaining constant contact with a premolt female, the male ensures that his sperm rather than the sperm of a rival are used in fertilization.

pates the female with various appendages, and after a minute or so, rotates her into the precopula position in which her body is now underneath and parallel to his. It is during the period of palpation and inspection that mate guarding decisions are made. Males are more likely to guard females with an empty brood pouch than females with a brood pouch full of juveniles or recently fertilized eggs (Dunham, 1986), and although females early in their reproductive cycle are also typically rejected (presumably because of the greater time investment in guarding), males that have gone a long time without female social contact will accept such females (Dunham and Hurshman, 1990). Once in the precopula position, the two swim as a single unit until the female molts and sperm transfer is complete. Thus, the male has ensured that his sperm, rather than the sperm of a competitor, are used in fertilization.

In some amphibians, males maintain the vigil for long periods. *Atelopus oxyrhyncus*, a bright yellow frog of the Venezuelan cloud forest, ex-hibits a particularly prolonged amplexus (Dole and Durant, 1974). During the nonbreeding season, males and females of this species maintain individual home ranges on the leaf-covered forest floor. As the reproductive season approaches, males clasp neighboring females and the two move as a unit to nearby streams to spawn. Although most males initiate amplexus approximately 1 month prior to arrival at the spawning site, one notorious male clasped a female for a full 125 days prior to breeding. Prolonged mate guarding in *A. oxyrhyncus* is costly not only to females that are faced with carrying males on their backs, but also to males that are reduced to feeding only when insects chance to land on the head of their mate.

In some cases, a male may protect his reproductive interests by guarding his mate after copulation as well. If a female were to mate with more than one male, the first male to copulate with her would presumably experience reduced chances of fertilizing her eggs as a result of sperm competition with the second male (see later discussion). A male usually guards his mate during the postcopulatory period either by staying close to her and chasing away other males or by maintaining direct physical contact with her, thereby impeding access to her.

Rove beetles (*Leistotrophus versicolor*) employ a unique variation on the theme of postcopulatory mate guarding (Alcock and Forsyth, 1988). Males and females of this species aggregate at dung and carrion, because these substances attract flies that serve as their food. Male beetles compete for territories on the dung and thereby gain access to food and mates. Rather than guarding females after copulation, males attack and chase them. Such ungallant behavior probably serves to move still receptive females away from dung and rival males. Males then quickly return to the tasks at hand, namely catching flies and courting new females. Traditional postcopulatory mate guarding might be ineffective in rove beetles because several males are often present at a given patch of dung, and while one male was chasing off another male, a second rival could gain access to his mate.

Mechanisms to Displace or Inactivate Rival Sperm Although physical struggles among males for females are often conspicuous, perhaps the fiercest clashes related to mating occur more quietly—in the female reproductive tract. If the female has mated with more than one male, then their sperm must compete for the opportunity to fertilize her eggs. Commonly called sperm competition, the phenomenon may be defined as the competition among spermatozoa from different males to fertilize ova within a single female during a reproductive cycle. For example, the quality of a male's sperm—measured, perhaps, by their motility or the number of viable sperm per ejaculate—may determine the proportion of young that he sires. Another factor that may come into play is his position in the line of suitors to mate with a specific female. In some species, the advantage goes to the first male to mate with the female, in others the last male, and in still others, mating order does not seem to be an important factor in determining patterns of paternity. G. A. Parker (1970, 1984) pioneered research in the area of sperm competition and described the phenomenon as a "push–pull" relationship between two evolutionary forces—one that acts on males to displace previous ejaculates left by rivals, and the other that acts on early males to prevent such displacement.

Male insects are particularly expert at displacing or neutralizing sperm already present in the female reproductive tract. In most insects, eggs are fertilized at oviposition (egg laying) with sperm from previous matings stored by the female in a special storage organ called the spermatheca. She releases the sperm as the eggs pass through her reproductive passages. However, because males interfere with each other through sperm competition, the sperm of different males do not necessarily have equal chances of fertil-

izing her eggs. In some cases, the interference is rather crude. For example, males of the damselfly *Calopteryx maculata* use their penis not only to transfer sperm, but also to remove sperm previously deposited by competitors (Waage, 1979). Backward-pointing hairs on the horns of the damselfly penis appear to aid in scooping out clumps of entangled sperm left by earlier rivals (Figure 14.19). Some male crustaceans employ equally subtle tactics. Rather than scooping out their rival's sperm, male spider crabs (*Inachus phalangium*) push the ejaculates of earlier males to the top of the female's sperm storage receptacle, seal them off in this new location with a gel that hardens, and then place their own sperm near the female's oviduct, the prime location for fertilization to occur (Diesel, 1990).

Sperm competition in moths and butterflies (Lepidoptera) takes a somewhat different form. Male Lepidoptera produce two different types of sperm. One type, "eupyrene" sperm, is a typical insect spermatozoa and is used to fertilize eggs. The second type, called "apyrene" sperm, is smaller, completely lacking in nuclear (genetic) material, and usually comprises at least half of the sperm complement in any given ejaculate. Although apyrene sperm migrate to the spermatheca upon transfer to a female, they cannot fertilize the eggs because they lack genetic material. The function of apyrene sperm in moths and butterflies has been a mystery since their discovery in the early 1900s. Silberglied, Shepherd, and Dickinson (1984) invoked the idea of sperm competition to explain the widespread occurrence of "dud" sperm in the Lepidoptera. They suggest that apyrene sperm embark on a "seek and destroy" mission, the ultimate goal of which is to displace or inactivate eupyrene sperm from previous matings by other males. As a second alternative, they suggest that apyrene sperm may serve to prevent or delay further matings by the female. The functional equivalent of cheap filler, apyrene sperm may be used to stuff the spermatheca of a female. Distension of the sperm storage organ by these energetically

a

b

Figure 14.19   (*a*) A copulating pair of damselflies. (*b*) The penis of a male damselfly serves not only to transfer sperm to the female but also to remove sperm previously left by competitors; backward pointing hairs on the horns of the penis function to remove clumps of rival sperm.

low-cost duds could signal a successful insemination and thereby reduce the female's receptivity to further matings.

Mechanisms To Avoid Sperm Displacement

Since males have evolved ways to put their own sperm at an advantage at the cost of the sperm of other males, obviously natural selection would have evolved ways for those "other males" to keep this from happening. So, we will now consider some tactics utilized by male mammals to avoid or reduce the effects of matings by other males.

To begin with, male mammals may increase the proportion of their own sperm in the female reproductive tract through their mating behavior. For example, despite the fact that female Norway rats (*Rattus norvegicus*), golden hamsters (*Mesocricetus auratus*), and deer mice (*Peromyscus maniculatus*) can become pregnant after only one complete ejaculation series, males of these species typically attain multiple sets of ejaculations (Dewsbury, 1984). In this way, not only do they leave more ejaculate in the female tract, but they also tend to tie up receptive females longer, rendering them less available to competitors. Multiple and prolonged copulation may also increase fertility or accelerate pregnancy initiation. In some species, then, males that attain multiple ejaculations sire a greater proportion of pups in resulting litters than do males attaining just one ejaculation.

Like females of some promiscuous rodent species, females of the Rocky Mountain bighorn sheep (*Ovis canadensis canadensis*), usually mate with several males within a single period of estrus. Sperm competition in this species is intense because subordinate males (termed coursing rams) are remarkably successful in forcing copulations on ewes guarded by defending rams (Hogg, 1988). Defending and coursing rams copulate with estrous ewes at extremely high rates (0.83 and 0.90 copulations per hour of estrus, respectively; estrus typically lasts for 2

days). Similar to multiple ejaculations of rodents, frequent copulations by individual males of these bighorns presumably function to increase the proportion of their own sperm in the female reproductive tract. Defending rams copulate at especially high rates immediately after successful copulation by a coursing ram. Such "retaliatory" copulations by dominant males (Figure 14.20) may serve a function akin to that of the damselfly penis; immediate copulations in response to mating by a coursing ram may result in mechanical displacement of rival sperm. Alternatively, retaliation may be advantageous to the dominant male because sperm from the last matings are more likely to fertilize eggs. A phenomenon similar to retaliatory copulation has been reported for roof or brown rats (*Rattus rattus*). In this species, males exhibit reduced postejaculatory intervals (time from ejaculation

Figure 14.20 Competition for mates is intense among males of Rocky Mountain bighorn sheep. Although dominance and access to mates are established by fierce physical clashes, male competition continues in the reproductive tracts of females. Dominant males copulate at high rates immediately after successful copulation by a subordinate male. Such retaliatory copulations may displace the sperm of the subordinate male.

to next intromission) if a second male has copulated with the female in the interim (Estep, 1988).

Repellents and Copulatory Plugs   In an effort to reduce the likelihood of future matings by competitors, males of many taxa apply a repellent odor to their mates. In the case of the neotropical butterfly *Heliconius erato*, for example, males transfer an "antiaphrodisiac" pheromone to their mate during copulation, which makes her repulsive to future suitors (Gilbert, 1976).

In other species, males may deposit a copulatory plug, a thick, viscous material that tends to clog the female's reproductive tract. A wide variety of species deposit copulatory plugs, from garter snakes (Devine, 1977) to small mammals. It has been suggested that the copulatory plugs of rodents evolved in the context of "chastity enforcement" (Voss, 1979), although with the exception of guinea pigs (Martan and Shepherd, 1976), copulatory plugs do not appear to block subsequent inseminations.

Dewsbury (1988) examined the function of copulatory plugs in deer mice (*Peromyscus maniculatus*; Figure 14.21) by allowing females to mate with two males in succession. In the control group of females, the copulatory plug produced by the first male was left in place. In the experimental group of females, the first male's copulatory plug was removed before introduction to the second male. It was found that the percentages of pups sired by the first male did not differ between groups of females with or without copulatory plugs. Dewsbury suggests that rather than protecting against subsequent inseminations, copulatory plugs in deer mice function to ensure that sperm are retained within the female reproductive tract.

Although for the vast majority of species the precise function of copulatory plugs remains a mystery, R. Robin Baker and Mark A. Bellis (1988,

Figure 14.21   In deer mice, copulatory plugs deposited by males probably function in retention of sperm in the female reproductive tract rather than in prevention of inseminations by rival males.

1989) suggest an interesting hypothesis. They note that in mammalian ejaculates, there is a large proportion of abnormal sperm. In healthy humans, for example, up to 40 percent of the sperm are irregular in size or shape, perhaps having two heads or two tails. Baker and Bellis suggest that such irregular sperm assume a "kamikaze" role, sometimes staying behind and forming a kind of plug or sperm aggregation (Figure 14.22) that acts as a roadblock or barrier in the lower region of the female reproductive tract (e.g., the vagina or cervix). In this manner, kamikaze sperm may thus protect "egg-getters" (i.e., normal sperm) from being ousted by a second ejaculate. Perhaps, as Baker and Bellis suggest, the proportions of normal and kamikaze sperm vary according to whether plugging or fertilization is of prime importance; when a male copulates with the same female several times during a single mating session, it would be advantageous to have a preponderance of normal sperm in early ejaculates and larger

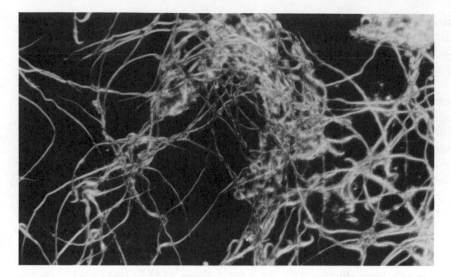

Figure 14.22 Intermeshed sperm, digested out from the copulatory plug of a rat, form the framework upon which seminal fluids coagulate. Baker and Bellis (1988) have suggested that the large numbers of apparently deformed sperm in mammalian ejaculates assume a "kamikaze" role and stay behind in a plug or sperm aggregation. By forming the framework of barriers in the female reproductive tract, kamikaze sperm could protect "egg-getter" sperm (i.e., sperm that are physiologically capable of fertilizing eggs) from ousting by a second ejaculate.

numbers of kamikaze sperm in the last ejaculate. The possibility that male mammals somehow manipulate ratios of kamikaze and egg-getter sperm in their ejaculates has exciting implications for future interpretation of patterns of reproductive behavior.

So far in our discussion of sperm competition, we have focused on interactions between males. Do females benefit by mating with more than one male? Multiple matings may, in fact, be to a female's advantage. Mating with several males may (1) increase the probability of fertilization, (2) increase the genetic diversity of offspring, (3) result in the accumulation of material benefits if males provide nutritional gifts at copulation, or (4) ensure that a female's sons are good at the game of sperm competition, providing that the trait is heritable. These and other

suggestions are reviewed in more detail by Møller and Birkhead (1989).

Sexual Interference: Decreasing the Reproductive Success of Rival Males

As we have seen, mating success is often measured by how well one advances one's own reproductive efforts, and how effectively one interferes with a competitor's efforts. We will now concentrate on the latter. Any behavior that reduces a rival's fitness by decreasing his mating success is called "sexual interference" (Arnold, 1976).

Perhaps the most effective animals when it comes to sexual interference are male salamanders. Adrianne Massey (1988) studied sexual interactions in field populations of red-spotted

a

b

Figure 14.23 There are few differences in heterosexual and homosexual courtship in salamanders. (*a*) A receptive female has straddled the tail of a male to stimulate the male to deposit a spermatophore. (*b*) A male mimics female behavior to cause the courting male to deposit a spermatophore. Pseudofemale behavior is one form of sexual interference used by male salamanders to decrease reproductive success of competitors.

newts (*Notophthalmus viridescens*) and noted three tactics of sexual interference: (1) spermatophore transfer interference, (2) pseudofemale behavior, and (3) amplexus interference. Although some of these terms may sound like behaviors in sporting events that deserve penalties, in the terminology of intrasexual selection, each represents a method by which male newts decrease the reproductive success of competing males.

Spermatophore transfer interference occurs when a rival male inserts himself between a female and the courting male that has just dismounted. The rival male not only induces spermatophore deposition by the first male, but also slips his own spermatophore into position so that the female picks up his spermatophore rather than that of the first male. This form of sexual interference depletes the first male's supply of spermatophores for future inseminations and prevents the first male from inseminating

the female (as well as permitting the intruding male to inseminate her). The second form of sexual interference, termed pseudofemale behavior, causes courting males to again waste spermatophores, but this time in the context of male–male pairings (Figure 14.23). Male red-spotted newts often clasp other males, although such pairings are usually brief because clasped males give a head-down display that elicits release. In some cases, however, the clasped male does not signal his maleness to the clasping male. Furthermore, once the clasping male dismounts, the clasped male may nudge him, in the manner of females, and get him to uselessly deposit his spermatophore. Finally, males of the red-spotted newt occasionally engage in amplexus interference. Here, an intruder simply inspects a pair in amplexus at close range, usually just before the mating male dismounts to deposit a spermatophore. The presence of a voyeur leads the amplexing male to pause. When he resumes his mating behavior, he usually picks it up at an earlier stage. The interference, then, causes the mating male to waste time and energy without increasing his probability of fertilizing the eggs.

Other species exhibit more dramatic forms of sexual interference, such as infanticide, the killing of a competitor's offspring (Figure 14.24; Hausfater and Hrdy, 1984). In some mammals, sexually selected infanticide often occurs when one or more males from outside the social group usurp the resident male. In Hanuman langurs (*Presbytis entellus*), the primate species in which infanticidal male intruders were first reported, infanticide occurs when individuals from all-male bands invade harems (Sugiyama, 1965). With her offspring gone, the female quickly returns to estrus; in Yukimaru Sugiyama's (1984) words, infants are attacked because they are "lit-

Figure 14.25   Nest sites are often in short supply for male tree swallows. In the absence of resident males, male "floaters" may kill unrelated young to gain access to nest sites and mating opportunities.

Figure 14.24   Although infanticide may serve several functions, it is a dramatic example of sexual interference. Here, a lion has killed the cub of a rival male.

tle more than an obstacle to activation of the mother's receptivity."

For male tree swallows (*Tachycineta bicolor*), nest sites are essential for acquiring a mate (Figure 14.25). Males of this species compete intensely among themselves for access to limited numbers of nest sites. Many individuals do not acquire nests, and thus there is a large population of male and female "floaters." When resident male tree swallows were experimentally removed from their nests shortly after the eggs hatched, some males from the floating population killed the nestlings and thereby acquired nest sites (Robertson and Stutchbury, 1988). In some cases, infanticidal males nested with the resident female, while in others they recruited females from the floating population. In tree swallows as well as langurs, then, killing of unrelated young serves not only to damage the male whose offspring are killed, but also to increase chances for mating with "widows" or other available females.

A CLOSER LOOK AT INTERSEXUAL SELECTION

Intersexual selection occurs when one sex is in the position of choosing individuals of the other sex as mates. This puts members of the sex being chosen in the position of competing among themselves for that privilege. Usually, as we have said, females do the choosing and males compete among themselves to be chosen. Since females must go through some form of pregnancy and then often must play the larger role in bringing up the offspring, they have a much greater investment—that is, they have more to lose. So, it is in their best interest to mix their genes with the best males possible. Thus, they are choosy. We will now consider just what male characteristics females might be scrutinizing.

Criteria by which Females Choose Mates

Characteristics used by females to select a mate should affect female fitness, be assessable, and vary among males (Searcy, 1979). Given these criteria, females are thought to choose mates on the basis of ability to provide sufficient sperm, useful resources, parental care, or good genes. Although some females may evaluate potential mates on only one characteristic, others may base their choice on multiple criteria. The following discussion considers evidence that females actually evaluate these qualities of males when searching for "Mr. Right."

Ability of Male to Provide Sufficient Sperm   Given the vast number of sperm produced by most males, is it possible that females in the market for a mate would take into consideration a male's sperm-producing ability? Indeed, females of some species appear to choose

"fresh" males over those that are "spent." Multiple copulations by male fruit flies (*D. melanogaster*) cause depletion of accessory gland substances and a temporary reduction in fertility. Not surprisingly, when given a choice, female fruit flies choose virgin males over sexually exhausted ones (Markow, Quaid, and Kerr, 1978). In species in which males can quickly regenerate sperm, such as some of the larger mammals—including humans, discrimination does not seem to be based on how many sperm a male is likely to be carrying at the time. (Thus we maintain the delicate integrity of our social fabric.)

Ability of Male to Provide High-Quality Resources or Parental Care   Females of some species may base their choice of a mate on the quality of resources provided by males. By so doing they could receive either immediate gains from gifts presented during the courtship period or more long-term benefits from access to valuable resources, such as food or nest sites, that are controlled by males. Females that exchange copulation for material goods could place themselves at an advantage. The increased commodities could obviously enhance reproductive output by enabling the female to live longer by being well fed or having access to a protected nest site. Not only would she have a competitive advantage over females without mates, but high-quality resources would likely improve the survivorship and competitive ability of her offspring. A female, then, must be able to assess the quality of a male's provisions not just at the moment, but in the future.

Females may also assess a male's parental abilities. This seems even more challenging than estimating value of future material resources, but it appears to occur, nonetheless. For example, females may use physical or behavioral features of males to predict parental quality. In fish such as the mottled sculpin (*Cottus bairdi*), large males are better able to guard eggs from predators and thus females probably use male size to assess parental ability (Downhower and

Figure 14.26 In red-winged blackbirds, females may use the size of the male's epaulette and his courtship intensity to estimate quality of care that he is likely to provide to her offspring.

Figure 14.27 A male European crossbill passes regurgitated seeds to a female. Female birds may judge the parental ability of a male by the quality and quantity of gifts provided during courtship feeding.

Brown, 1980). Several reliable predictors of male parental ability have also been identified for red-winged blackbirds (*Agelaius phoeniceus*; Figure 14.26). In this species, effort devoted to nest defense is correlated with the size of the epaulette (the patch of red feathers on the shoulder; Eckert and Weatherhead, 1987), and feeding effort is correlated with male courtship intensity (Yasukawa, 1981); therefore, females could use epaulette size and intensity of courtship to estimate quality of care that a male is likely to provide. In still other species of birds, females may judge male parental ability on the basis of the quality of nutritional gifts provided during the period of courtship (Figure 14.27). A large number of high-quality gifts may signal a male's superb foraging skill and willingness to feed his mate and offspring during incubation and post-hatching stages. Finally, some females may use

success of previous nesting attempts to judge the parental ability of males. Despite striking mate fidelity between the members of breeding pairs of some species of birds, bonds are often severed after unsuccessful breeding attempts (Coulson, 1966).

Ability of Male to Provide High Quality Genes   As we discussed earlier in the chapter, female choice may be based on the genetic characteristics of males. Such choice may occur at several levels. Because males are usually less selective than females and are prone to court individuals of the wrong species, discriminating females are often responsible for preventing matings between species. This most basic choice may relate to genetic compatibility—the bottom line being that matings between individuals of different species will not result in viable offspring. Females may take the process of mate selection one step further and discriminate against conspecifics from strains that are genetically different and restricted to specific habitats (such strains are called ecotypes). For example, assortative mating in fish occurs among ecotypes of the three-spined stickleback (*Gasterosteus aculeatus*; Hay and McPhail, 1975). One form of *G. aculeatus* is exclusively freshwater while the other is anadromous, spending most of its time in the ocean, but returning to streams to breed. In laboratory choice experiments, females prefer males of their own type, presumably because offspring of parents belonging to different ecotypes would be less well adapted to either habitat than offspring of parents of the same ecotype.

Females may evaluate potential mates on the basis of degree of genetic relatedness. Inbreeding caused by mating with close relatives increases homozygosity and hence the risk of producing offspring that are homozygous for deleterious or lethal recessive alleles. On the other hand, extreme outbreeding may cause the breakup of successful parental complexes of genes. Given the potential costs associated with mating with either very close relatives or complete strangers, females may strike a balance between extreme inbreeding and outbreeding when choosing a mate. Recall from Chapter 8 that female Japanese quail (*Coturnix coturnix japonica*) display a preference for individuals that are similar to, yet slightly different from, members of their own family. In an experiment, female quail spent more time near unfamiliar first cousins than near familiar siblings or unfamiliar unrelated individuals (Bateson, 1982).

Rather than choosing males according to genetic complementarity or degree of relatedness, the possibility also exists that females choose males solely on the basis of possession of good genes. Barring direct examination of a male's genotype, how could a female evaluate variation in genetic quality among suitors? Females might judge genetic quality by examining a male's (1) general physical well-being, (2) capacity to dominate rival males, or (3) capacity for prolonged survival. As an example of males' being assessed by some indicator of their potential fitness, consider the bowerbirds.

Male satin bowerbirds (*Ptilonorhyncus violaceus*) build unique structures called bowers in which they display to females in attempts to secure copulations (Figure 14.28). Males decorate their bowers with flowers and feathers and display strong preferences for inflorescences of certain colors; blue and purple flowers are relatively rare in the environment of satin bowerbirds and are preferred over yellow and white flowers, while orange, red, and pink inflorescences are completely unacceptable (Borgia, Kaatz, and Condit, 1987). The number of decorations on a bower is an important determinant

Figure 14.28 Males of the satin bowerbird decorate their bowers with flowers and feathers. The number of decorations on a bower is an important determinant of mating success, and males steal rare decorations from rivals with impunity. Female bowerbirds may judge genetic quality of a male by the attractiveness of his bower.

of male mating success, and males go to great lengths to steal rare decorations from the bowers of competitors. Bowers and their decorations appear to have no intrinsic value to either sex outside the context of sexual display and thus probably serve primarily as indicators of male quality. Females may favor males who exhibit exotic decorations because the ability to accumulate and hold these decorations indicates that a male is in top physical condition. Because the number of inflorescences on bowers is correlated with male age, the number of flowers, and specifically the number of rare decorations, could be used by females to assess the experience of a male. Thus, although female bowerbirds cannot directly examine the genotype of potential mates, they may evaluate a male's genetic quality on the basis of the attractiveness of his bower.

Consequences of Mate Choice for Female Fitness

Having explored those characteristics females might be assessing in potential mates, we come to the important question, do choosy females produce fitter offspring? A laboratory study of female choice in fruit flies (*D. melanogaster*) provided evidence that mate choice by females can indeed result in the production of fitter offspring (Partridge, 1980). Larvae produced by females that could choose their mates and females that could not were allowed to compete in laboratory vials with larvae from a standardized strain of flies. The proportion of offspring emerging as adults was slightly, but significantly, higher when females could choose their mates. These results suggest that in fruit flies, female choice improves the competitive ability of offspring. Similar results have been obtained for seaweed flies (*Coelopa frigida*). In these experiments, females that were allowed to exercise mate choice were not only more likely to mate, but also produced offspring with higher survival rates than did females given a single, randomly chosen male (Crocker and Day, 1987).

Strategies of Female Choice

Up until now we have focused on criteria used by females to choose mates, and we have said nothing about how females actually go about the process of selecting the best possible male. How much time should they devote to searching, and once located, should prospective mates be judged against a fixed standard or by some relative criterion?

Ideally, a female would have the opportunity to inspect every available suitor and choose the best from among them. In reality, however, females in search of mates probably operate under constraints of time, mobility, and memory (Janetos, 1980). A female whose reproduction is restricted to a particular season does not have all the time in the world to locate a mate and initiate breeding. Ecological conditions may limit her mobility and thereby prevent complete inspection of all males in the surrounding area. Also, the number of males that a female can remember at one time will certainly shape strategies of mate choice. In a theoretical analysis of female choice, Anthony Janetos (1980) compared five different strategies of mate selection and concluded that a strategy in which females rank and choose among available males on a relative basis (comparison to an external standard—that is, the choice is based on relative differences among males that are sampled) is better than random mating or alternative strategies based on a fixed threshold criterion (comparison to some internal standard). Although mate choice based on relative differences among available males may be the more common strategy, mate choice based on comparison to some internal standard has been reported for cockroaches (*Nauphoeta cinerea*; Moore and Moore, 1988) and scorpionflies (*Hylobittacus apicalis*; Thornhill, 1980). Different tactics of mate choice are likely to evolve under different circumstances, and knowledge of how females choose mates is just as essential to our understanding of intersexual selection as knowledge of why females choose mates.

Mate Choice by Males

In spite of our emphasis on female choosiness, some males may also be selective in their choice of mating partners. The more a male has to offer in terms of parental care or material resources, the more selective he is expected to be. Two other factors that are important in the evolution of male choice are limited ability of males to fertilize all available females and variation in female quality. As an example of this latter factor, consider the case of thirteen-lined ground squirrels (*Spermophilus tridecemlineatus*). Despite the fact that male ground squirrels provide neither parental care nor material resources, they often reject females willing to mate with them. As it turns out, male choice is favored in this species because of predictable variation in the quality of females—variation that can be attributed to sperm competition (Schwagmeyer and Parker, 1990). Each female, on average, mates with two males and the first male sires about 75 percent of the litter. This percentage, however, is not fixed. Factors such as the time between matings of first and second males (second males wait until first males depart to begin mating, and longer delays favor first males) and the duration of the second male's longest copulation (increases favor second males), influence fertilization rate. Using measurements such as these, the researchers developed a model to determine whether male choice in ground squirrels is consistent with what would be expected from the known effects of sperm competition on fertilization rate. They then examined whether field data on male response to previously mated females conformed to the model. Consistent with their predictions, male ground squirrels were more likely to reject previously mated females the longer the time since the female's copulation with another male.

Methodological Problems Associated with Studying Mate Choice

Several methodological problems are commonly encountered when studying mate choice (Halliday, 1983). Aside from the fact that mate choice may be very subtle, it is often masked or confounded by extraneous effects, such as motivation of the female at the time of assessment. If a female does not display a preference for a male during a choice test, is it because she has rejected him as a potential mate, or is it simply because she is not receptive? Studies of female choice face the difficult task of unraveling the ties between motivational state and choice of mating partner. Additionally, female choice in the laboratory is often measured by orientation toward, or time spent near, a particular male; yet in some cases, these behavioral responses may not relate to actual mating inclination.

Perhaps the most difficult problem faced by students of mate choice is determining the precise roles of male competition and female choice within a single mating system. In the plethodontid salamander *Desmognathus ochrophaeus*, staged laboratory encounters between one female and two males (one large and one small) always resulted in courtship between the female and large male. Although one might be tempted to claim that females of this species display a clear preference for large males, Lynn Houck (1988) has shown that pairing between females and large males results mostly from the fact that large males attack and threaten small males and effectively exclude them from courtship activity. Further evidence against female choice for large males in *D. ochrophaeus* comes from observations that in the presence of only one potential mate, females mated with small males just as often as with large males. Houck is careful to point out, however, that female choice of mates cannot be ruled out completely for this species; female salamanders in the field are not restricted physically or socially and probably have the option of examining several males before engaging in courtship behavior.

Pregnancy block, also called the Bruce Effect, is an example of a phenomenon that has been interpreted both as a product of male–male competition and as a mechanism of female choice. In the Bruce Effect, the pregnancy of a recently inseminated female is terminated upon exposure to an unfamiliar male. Although first reported in house mice (*Mus musculus*; Bruce, 1959), pregnancy block has also been documented in a variety of rodents, including deer mice and voles. Unlike house mice in which pregnancy terminations only occur during the first few days following insemination, female voles that are exposed to unfamiliar males may resorb or abort fetuses up to day 17 of a 21-day gestation period (Stehn and Richmond, 1975).

Initial interpretations of pregnancy block focused on advantages to unfamiliar males (e.g., Trivers, 1972). A male, new to the neighborhood, that encountered a pregnant female could not only disrupt the previous insemination of a rival male, but also mate with the female once she became receptive again. Incidence of pregnancy termination in mice, however, is dependent upon factors such as presence of the first mate or other females, age of the female, and length of exposure to the unfamiliar male (Bruce, 1961, 1963; Chipman and Fox, 1966; Parkes and Bruce, 1961), suggesting that females have some control over whether or not termination of pregnancy occurs. Perhaps pregnancy block is not simply a product of male–male competition, but rather a mechanism by which females can select better mates or better mating environments (Schwagmeyer, 1979). Ability to terminate pregnancy might be beneficial under conditions in which a female, deserted by her initial mate, has little chance of successfully rearing the upcoming litter (Dawkins, 1976). Upon encountering a new male, her best option might be to terminate investment in the current pregnancy and mate with the newcomer in order to improve her chances of receiving parental assistance. Additionally, if the female carried the litter to term, there is always the possibility that the new male would commit infanticide; thus, it may be in the best interest of the female to cut her losses and resorb the doomed youngsters (Labov, 1981). Indeed, in a more recent study, newly pregnant house mice were more likely to exhibit

pregnancy block when exposed to infanticidal males than when exposed to noninfanticidal males (Elwood and Kennedy, 1990). These data suggest that females can somehow assess the risk to their future offspring and adjust their reproductive response in accordance with that risk. Although the functional significance of pregnancy block is far from resolved, the possibility that female rodents exercise mate choice even after copulation has exciting implications.

In summary, then, intrasexual and intersexual selection may operate within a single mating system. We should thus be very cautious about attributing any single characteristic of males to one form of sexual selection or the other. Perhaps the elaborate train of peacocks, often thought of as an example of a character that probably resulted from female choice, results instead from male–male competition, or some combination of the two processes. Although a recent study supports a role for female choice in the peafowl mating system (Petrie, Halliday, and Sanders, 1991), male–male competition cannot be excluded as an important factor affecting mating success in this species. From this classic example we see that determining the roles of intersexual and intrasexual selection in the lives of animals remains an enormous challenge.

Summary

Although females that reproduce sexually incur costs associated with recombination, production of sons, and mating, they apparently benefit through production of variable progeny. Among sexually reproducing organisms, males and females are defined according to the amount of energy allocated to each gamete; females produce small numbers of relatively large eggs while males produce large numbers of small, highly mobile sperm. The sexes are further differentiated on the basis of parental investment; whereas females often invest time and energy in the care of offspring, males typically invest little in parental effort and focus instead on mating effort. Sex differences in parental investment and energetic contributions per gamete create a conflict of interest that has far-reaching implications for reproductive behavior.

The sex of an individual may be determined through environmental or chromosomal means. In species that exhibit environmental sex determination, sex is determined by (1) temperature at which eggs incubate (e.g., some turtles and lizards) or (2) location at which an individual settles (e.g., an echiuroid worm). Although many instances of environmental sex determination have been reported, the advantages to having sex determined by features of the environment rather than by a specific combination of sex chromosomes remain a mystery. Equally intriguing are deviations from the standard 1:1 sex ratio that have been noted in species with chromosomal sex determination. The profitability of producing sons or daughters may vary according to factors such as maternal condition, maternal rank, or which offspring are likely to be competitive or cooperative. Although many examples of sex ratios skewed in a direction consistent with anticipated gains for parents have been reported, the immediate mechanism responsible for unequal sex ratios has yet to be identified. At present, the issue of adaptive manipulation of the sex ratio remains somewhat controversial.

Sex differences in gametic and parental investment are important determinants of sexual selection. Sexual selection results from (1) competition within one sex for mates (intrasexual selection) and (2) preferences exhibited by one sex for certain traits in the opposite sex (intersexual selection). Because females invest more in gametes and parental care, they are usually a limited resource for which males compete. Competition among males for access to females appears responsible for the evolution in males of a broad spectrum of physical and behavioral attributes that enhance fighting prowess. Weapons, large size, and physical strength have evolved in males of many species. Additionally, males devote much time and energy to ensuring that their sperm and not the sperm of a competitor are used to fertilize a female's eggs. In some species this may take the form of dominance behavior, mate guarding, or alternative reproductive strategies, while in others, repellents or copulatory plugs are used. Because reproductive success is measured in relative terms, males may also enhance their position by employing tactics of sexual interference that decrease the sexual success of other males.

As a result of their greater investment in gametes and parental care, females are usually more discriminating than males in their choice of mates. Intersexual selection generated by females that prefer males with certain characteristics could be based on a male's ability to provide sufficient sperm, material resources, parental care, or good genes. Discrimination purely on the basis of male genetic quality is the most problematic. Two hypotheses have been put forth to explain the evolution of female choice when males provide neither parental care nor material resources. The good genes hypothesis suggests that females choose males whose genes will enhance the viability of offspring (that is, male secondary sex characteristics act as signals of high fitness). The runaway selection hypothesis, however, does not require that females base their preferences on male traits that signal viability. Instead, runaway selection can yield mate choice for traits that are arbitrary, or even disadvantageous, providing that females prefer to mate with males that possess them. When selecting a mate, females probably operate under constraints of time, mobility, and energy.

CHAPTER
15

Parental Care and Mating Systems

Parental Care

Inside silken nest chambers *Stegodyphus mimosarum* spiderlings sit on the abdomen of an adult female and begin to devour her. Chances are the victim is their mother. *Stegodyphus mimosarum*, a social spider found in Africa, is only one of about 10 species of spiders in which young eat their mother at the end of the brood care period (Seibt and Wickler, 1987). Unlike species of solitary spiders in which gerontophagy (the term coined by Seibt and Wickler to describe the consumption of elderly individuals) is confined to the mother and her offspring, social *Stegodyphus* have a more varied menu in that they can choose between eating their own mother or someone else's. However, in the interests of seeing her genes passed on through the survival and reproduction of her relatives, the designated victim should "prefer" to be eaten by her own offspring (particularly if she cannot produce another brood) rather than by a nonrelative. Because spiderlings do not cannibalize each other or individuals of different spider species, gerontophagy in *Stegodyphus mimosarum* appears to represent a mother's final act of parental care as she sends her offspring into the world on a full stomach. This bizarre example illustrates the often unexpected adaptive relationships between parents and offspring. In this chapter, we will focus first on parent–offspring relationships and then on mating systems in an effort to better understand these reproduction-enabling mechanisms across the spectrum of animal life.

SOME MODELS OF THE PARENT–OFFSPRING RELATIONSHIP

In recent years, several different views on the nature of the parent–offspring relationship have emerged. We will consider the following three: (1) parental provision, (2) conflict, and (3) symbiosis. Summarized by Alberts (1986), these three models represent different perspectives on the organization of parent–young interactions. In some instances, the initial research that led to the development of these perspectives was conducted with the laboratory strain of the Norway rat (*Rattus norvegicus*). Here we will consider each model in turn, using examples from the rat, and in some cases, from other animals.

The Parental Provision Model

Perhaps the most traditional view of the parent–offspring relationship is presented in the parental provision model. This model emphasizes the one-way flow of resources from parent to offspring and thus focuses on the parent as the provider for dependent young. For example, through their milk supply, rat mothers provide their young not only with nutrition, but also with antibodies to fight infection (pups cannot produce their own antibodies until 2 to 3 weeks of age). Through their nest-building and brooding activities, mother rats supply the warmth necessary for pup growth and development. Furthermore, young rats do not spontaneously urinate or defecate, and licking by the mother is essential to stimulate these processes. Finally, because pups have fairly limited abilities when it comes to moving from one place to another, mother rats also provide transportation, typically by grasping infants at the back of the neck and carrying them to another location. From these few highlights of parent–young interactions in rats, we see the steady flow of resources—food, antibodies, warmth, mechanical stimulation, and transportation—from parent to offspring.

Mother rats, of course, are not the only parents in the animal world that provide for their offspring, and in fact, when it comes to early nutrition, there are some interesting methods of parental provision. Our description of the mother rat's supplying food to her young focused on the period after birth (before birth, nutrition is supplied via the placenta); our next

a

b

Figure 15.1 Mother caecilians provide nourishment to their developing young before birth. (*a*) The caecilian, *Dermophis mexicanus*. Embryonic young of this species feed by using their tiny teeth to scrape the lining of their mother's oviduct. (*b*). Scanning electron micrograph of fetal teeth.

examples concern the provision of food prior to birth. Consider, for example, the way in which limbless amphibians, called caecilians, sometimes provide nourishment to their developing offspring. Females of the species *Dermophis mexicanus* (Figure 15.1a) supply food to their offspring, which typically number between 1 and

12, during their year-long gestation period. Prior to birth, embryonic caecilians gain nourishment from their mother by using their tiny teeth to scrape the lining of her oviducts (Wake, 1977, 1980; Figure 15.1b). Female sand sharks (*Carcharias taurus*) also provide their developing embryos with an interesting source of nourishment, their own excess eggs. Springer (1948) studied oophagy (egg eating) in sand sharks and described embryonic sharks moving about inside the oviducts of the mother, snapping up extra eggs. Not surprisingly, each oviduct of females of this species typically contains only one advanced embryo.

As a final example of parental provision—specifically the provision of transportation—female scorpions come to mind (Figure 15.2). Females of the species (*Euscorpius flavicaudis*) carry their 30 or so young on their back from birth to the first instar (an instar is a stage of development), a period of about 7 days. This apparently monumental effort (the brood weighs almost half of what the mother weighs after having given birth) serves to protect the defenseless young from predation (Benton, 1991). Although later instars and adults are quite capable of defending themselves (most of the 1500 or so species of scorpions have a sting that is somewhat less painful than that of a honey bee, but about 25 species produce venom that can kill a human being; Polis, 1990), first instar young are slow moving, lack claws, and have a soft cuticle and blunt sting. Female scorpions thus provide a critical resource, in this case, a safe ride, to their dependent young.

The Conflict Model

The conflict model is based on parental investment theory put forth by Robert Trivers. Trivers (1972) defined parental investment as "any investment by the parent in an individual offspring that increases that offspring's chance of surviving (and hence reproductive success) at the cost of the parent's ability to invest in other offspring"

Figure 15.2 A mother scorpion provides transportation to her young.

(p. 139). Whereas the parental provision model portrays offspring as somewhat passive recipients of parental resources, Trivers's model predicts a fundamental conflict between parents and offspring, one in which the offspring are active participants, behaving so as to extract more investment than their parents are willing to provide.

At the heart of parent–offspring conflict is the basic fact that in species that reproduce by sexual reproduction, parents and offspring are not genetically identical. Because in outbred, diploid animals, parents and offspring share only one-half of their genes, maximization of a particular offspring's fitness may not always be achieved in the same manner that parental fitness is maximized (Trivers, 1974). Thus, as both parents and offspring behave to increase their own inclusive fitness, they are destined to disagree over the amount of parental care given and over how long the period of parental investment should last.

Trivers's theory of parental investment is based on costs and benefits to the individuals involved. The ratio of cost to mother/benefit to offspring of a particular act changes across the

period of offspring development (Figure 15.3). In the period directly after birth, cost to the mother is small and benefit to the offspring large, and mother and offspring can continue the dependent relationship without conflict. However, as the youngster grows, it becomes more and more expensive to care for, and soon

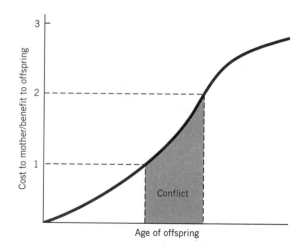

Figure 15.3 Trivers's (1974) model of parent–offspring conflict. (From Wilson, 1975.)

the cost to the mother exceeds the benefit to the offspring (that is, the cost/benefit ratio exceeds 1; see Figure 15.3). Parent–offspring conflict begins at this time, a time when the mother's behavior reduces her fitness, but not the inclusive fitness of her offspring. The conflict ends when continued investment costs the mother twice the benefit received by the offspring (in other words, the cost/benefit ratio exceeds 2, again see Figure 15.3). At this point, independence is desirable for the offspring because its inclusive fitness is best served not by continuing to extract parental investment, but by the mother's production of new siblings. The basic point, then, is that offspring will tend to favor a level of investment by their parents that is higher than the level deemed appropriate by their parents, and conflict results.

The changing cost/benefit ratio of parental care can be seen in the interactions between cats and their kittens. Female cats initiate almost all nursing bouts from the time their kittens are born until the young are approximately 20 days old (Schneirla, Rosenblatt, and Tobach, 1963). Between 20 and 30 days after birth, kittens and mothers initiate feedings with equal frequencies. However, soon after day 30, suckling is initiated almost entirely by the kittens, and mothers begin to avoid and actively discourage nursing by their offspring (Figure 15.4).

Do offspring stand a chance against their larger, more experienced parents in the competitive struggle to maximize inclusive fitness? Trivers (1974) argued that selection should favor parental attentiveness to signals from offspring because offspring are best at evaluating their own needs. As adults become increasingly proficient at assessing the needs of their young, he predicted that successive offspring would have to employ more convincing tactics to elicit care from their experienced parents. Richard Alexander (1974) disagreed with Trivers's description of conflict between parents and offspring, claiming that parents are heavily favored to win the parent–offspring contest. According to Al-

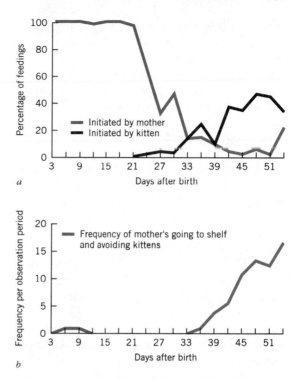

Figure 15.4 Weaning conflict in cats as illustrated by changes in mother–young interactions during the postnatal period. (*a*) The frequency of mother-initiated nursing bouts decreases over time, and the frequency of feedings initiated by kittens increases. (*b*) During the period of parent–offspring conflict, mothers begin to avoid their offspring and spend increasingly more time on a shelf away from the kittens. (*a*; Modified from Schneirla, and Rosenblatt, 1961; *b*; Modified from Schneirla, Rosenblatt, and Tobach, 1963.)

exander, offspring are at the mercy of their parents, tolerated and manipulated for the sole purpose of increasing parental fitness. Alexander believed that any adaptation that facilitated success on the part of the offspring could not evolve if it was harmful to the inclusive fitness of the parent. His reasoning centered on the fact that gene(s) responsible for extorting extra parental investment would be passed on to the young or that same offspring, and thus the offspring's inclusive fitness, would suffer in the same manner as that of its parents. Daly and Wilson (1978) succinctly summarized Alexander's point by stating that, "an offspring who 'cheats' its parent becomes a parent of offspring who will cheat in turn" (p. 158).

Although Alexander's ideas seem plausible given that parents are physically superior to their offspring and thus seem to occupy the position of power in the relationship, the prediction of routine parental victory has not stood up to examination with mathematical models. In a series of papers, Parker and Macnair (1978, 1979; Macnair and Parker, 1978, 1979; Parker, 1985) used game theory models (see Chapter 4) to show that a "conflictor" gene (that is, a gene that causes its bearer to extract more care from a parent than the parent is willing to give) could invade populations of "nonconflictors." In essence, these models demonstrated that there was no reason to expect that a parental victory would be more likely than an offspring victory. In fact, the most likely outcome turns out to be a level of investment intermediate between the best situation for parents and the best one for offspring. Finally, although Trivers's (1974) original models were based on a scenario in which the mother produced one offspring per year and received no help from the father in caring for them, and all subsequent offspring were full siblings of the offspring in question (that is, they had the same father and thus their coefficient of relatedness was 0.5), the models of Macnair and Parker explore parent–offspring conflict in a variety of different situations, including when two parents invest equally in offspring, and when offspring have more than one father, a situation known as multiple paternity. In view of the evidence presented in these and other mathematical models of parent–offspring conflict, Alexander (1979) has modified his views predicting parental victory.

Although the notion of conflict between parents and offspring has provided a fascinating perspective on parent–offspring interactions, in practice, the models are often difficult to test (Clutton-Brock, 1991). It is often difficult to know, for example, just what the best level of investment is for parents and for offspring. How do we go about assessing such a level? Furthermore, measures of overt behavior, such as rates of begging by nestling birds or tantrums by young mammals that are not permitted to nurse, may not accurately represent the intensity of conflict between parents and offspring. Although obvious, these patterns of behavior may not be the best measures of conflict. Sometimes the problem is of a different nature. In the collared peccary (*Tayassu tajacu*), a relative of the domestic swine that lives in the desert and scrub areas of the southwestern United States, Mexico, and Central and South America, there is little sign of overt behavioral conflict at the time of weaning. Peccary mothers, it turns out, rarely reject suckling attempts by their young (Babbitt and Packard, 1990). The question arises, then, is the relationship between a peccary mother and her young perfectly harmonious, or is it simply that we need a way other than recording rejections of offspring by the mother to measure potential conflict? From these few examples, we see that testing models of parent–offspring conflict can be a difficult and complex matter, indeed.

The Symbiosis Model

Our final model of the parent–offspring relationship involves a metaphor from the biological world. Jeffrey Alberts and his colleague David Gubernick (1983) likened the interaction between a mother rat and her litter to a symbiotic association, that is, an association that benefits both participants. (Typically, the term symbiosis is used to describe the interactions between two different species. Thus, Alberts and Gubernick use the term in a different sense, that is, to describe the relationship between parents and offspring in a single species.)

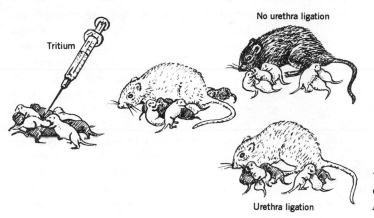

No urethra ligation

Tritium

Urethra ligation

Figure 15.5   Design of experiments by Gubernick and Alberts (1983). (From Alberts, 1986.)

We have already described how the parental provision model emphasizes that mother rats provide their young with resources such as milk, antibodies, and warmth, and that licking by the mother provides the mechanical stimulation necessary for inducing urination by the pups. The symbiosis model emphasizes that, at least in some cases, the flow of resources is more complex than simply from mother to offspring. In fact, resources may flow from mother to offspring and from offspring to mother—a situation called "bidirectional exchange" by Alberts and Gubernick. It is this two-way exchange of resources, an exchange that benefits both mother and young, that inspired the symbiosis metaphor. As an example of symbiosis in parent–offspring interactions, let us take a closer look at what actually happens when a mother rat licks her pups.

At first glance, the act of maternal licking of young would seem to largely benefit the offspring. When the mother rat grooms the anogenital region of a pup, she not only cleans the pup, but stimulates urination. Resources (mechanical stimulation in this particular case) seem once again to be flowing from mother to pup. The key to the symbiosis model, however, is that the mother rat ingests the urine obtained from licking her offspring. Ingestion of pup urine by the mother raises the possibility of a more com-

plex path of resource exchange. Could the licking of pups provide the mother with important resources?

That resources flow from offspring to mother was demonstrated in a series of experiments by Gubernick and Alberts (1983) in which radioactively labeled water, called tritium, was injected into a few pups in a litter. The next day, the radioactive label could be detected in the mother. Water was obviously moving from pup to mother, but how? The next experiment answered this question. When urination was prevented in labeled pups by tying off the urethra (a procedure termed urethra ligation), the label was not detected in the mother, demonstrating that pup urine was the source of water transferred to the mother (Figure 15.5). As it turns out, by licking the anogenital region of her pups and consuming their urine, a rat mother can reclaim about two-thirds of the water passed to her litter in her milk (water is the main component of the milk of most species, and in the case of rats, a mother's milk is about 73% water). The neatness of this system of recycling is illustrated by the fact that the peak level of water transfer from pups to mother occurs when the pups are 15 days old, which is exactly the time of peak milk production (Figure 15.6).

Even if mother rats use both water and pup urine as fluid sources, how important a source

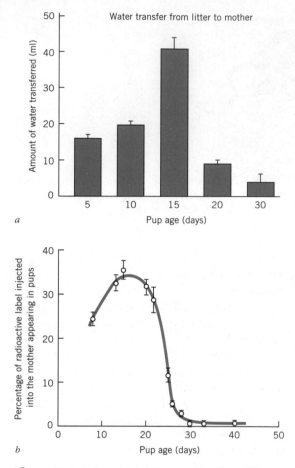

a

b

Figure 15.6 (*a*) Amount of water transferred from a litter of eight rat pups to their mother at several points after birth. The peak level of water transfer from pups to mother occurs when the pups are 15 days old; this is also the time of peak milk production. (*b*) Amount of milk flow from a lactating rat to her litter of eight pups as measured by the rate of transfer of a radioactively labeled substance injected into the mother and found in the pups 48 hours later. (a; From Gubernick and Alberts, 1983; b; Babicky et al., 1970.)

is pup urine? Apparently, very important. Friedman, Bruno, and Alberts, (1981) found that mothers deprived of both pup urine and water become twice as dehydrated as mothers deprived only of water. Additional experiments suggested that pup urine also serves as a source of electrolytes (salts), sodium in particular. We see, then, that rather than a one-way flow of resources from mother to pup, there is an exchange of resources between them. Water, electrolytes, and nutrients move from mother to pup in the form of milk, and then the mother reclaims water and electrolytes by ingesting pup urine. This exchange of resources is beneficial to both mother and pup, and herein lies the resemblance to symbiotic associations.

Evaluation of the Models

Now that we are familiar with some models of parent–offspring interactions, the question arises, which model is best? We agree with Alberts's (1986) analysis, that these models and metaphors were put forth to stimulate research by providing new ways to view the relationship between parents and offspring. Each model provides a conceptual framework within which specific hypotheses can be formulated and tested. The value of these models, then, is that they open our eyes and our minds to different approaches to the study of parent–offspring interactions. The models need not be mutually exclusive, and their usefulness in describing parent–young interactions probably varies from species to species.

PROVIDING THE CARE

Maternal and Paternal Care

Male versus Female Care and Mode of Fertilization   In species with parental care, why do parental duties fall to the female in some species and to the male in others? Mode of fertilization (internal versus external) appears to be an important variable in determining which parent cares for the offspring (Ridley, 1978). Specifically, female care is associated with internal fertilization and male care with external fertilization. Three alternative hypotheses—the certainty of paternity hypothesis, the gamete order hypothesis, and the association hypothesis—have been proposed to explain the evolution of male versus female care as it relates to mode of fertilization. We will consider each hypothesis in turn.

The certainty of paternity hypothesis is based on the idea that parental solicitude toward young is correlated with likelihood of genetic relatedness (Trivers, 1972). Females, regardless of mode of fertilization, can be absolutely certain that they are related to their offspring. Certainty of maternity guarantees that 50 percent of a mother's genes are present in her progeny. Males, especially of species with internal fertilization, cannot be so confident. Even though a male copulates with a female, he has no guarantee that his sperm, rather than the sperm of a competitor, will fertilize her eggs. In short, because males of internally fertilizing species run the risk of investing time and energy in raising another male's offspring, the odds run against the evolution of paternal behavior. Reliability of paternity is assumed to be greater when eggs are fertilized externally instead of inside the female, and thus external fertilization opens the way for paternal care. Although appealing to our sense of intrigue, the certainty of paternity hypothesis has not held up under scrutiny. Theoretical models developed by Maynard Smith (1977) and Werren, Gross, and Shine (1980) raise questions as to the usefulness of the paternity hypothesis as a general explanation for the evolution of patterns of parental care. Briefly stated, these authors argue that if paternity is similar for all matings regardless of whether or not there is paternal care (that is, within a population, the reliability of paternity does not differ between males that are parental and those that are nonparental), then paternity itself cannot influence selection for parental behavior by males.

In essence, any effect of reliability of paternity on the evolution of male care should cancel out.

The gamete order hypothesis suggests that patterns of parental care result from differences in opportunities of males and females to desert offspring. Natural selection should favor desertion of offspring by whichever parent has the earliest opportunity, thereby forcing the remaining partner to provide care (Dawkins and Carlisle, 1976). According to this hypothesis, the partner releasing gametes last gets saddled with parental responsibilities. In species with internal fertilization, females are likely to care for young because they are unable to desert the embryo(s). By contrast, in externally fertilizing species, females usually release their eggs before males release sperm, and thus females, rather than males, have the first opportunity to leave the scene.

The association hypothesis, perhaps not as flashy as the first two hypotheses, relies only on the proximity of adults and offspring to explain the evolution of male versus female care (Williams, 1975). With internal fertilization, it is usually the female that carries the embryos, and thus she is in the best position to care for the young when they enter the world. Paternal behavior is less feasible because fathers may not be in the vicinity at the time the eggs are laid or the young are born. According to this hypothesis, external fertilization, particularly when it occurs in a territory defended by a male, would be associated with parental care by males. Because territorial males are almost always in the neighborhood of the eggs they fertilize, paternal behavior is likely to evolve.

These three hypotheses, of course, cannot cover all the cases. The evolution of male versus female care in some species remains somewhat of a mystery. In an attempt to test the hypotheses in certain species, Gross and Shine (1981) examined reams of data for amphibians (35 families) and teleost fish (182 families), two groups that show both male and female care and internal and external fertilization. On the basis of their survey, the authors rejected the gamete order hypothesis, discovered some instances where the certainty of paternity hypothesis failed to predict the pattern of parental care, and concluded that the association hypothesis was most consistent with available data.

Paternal investment in fish and amphibians usually takes the form of solitary male care rather than shared male and female responsibilities. In contrast, paternal investment in birds and mammals almost always occurs in addition to maternal care. Can the association hypothesis explain this basic dichotomy in patterns of parental behavior? As Gross and Shine (1981) point out, the association hypothesis should not be invoked to explain this basic difference in the parental care of ectothermic (fish and amphibians) and endothermic vertebrates (birds and mammals), but it is valuable in interpreting the distribution of male versus female care within each group. Instead, the higher frequency of biparental care in birds and mammals with paternal investment probably reflects the fact that parental care in these groups usually involves both the feeding and guarding of young. Two parents are better than one when it comes to feeding offspring and thus biparental care evolves (Maynard Smith, 1977). Fish and amphibian parents, on the other hand, rarely feed their offspring and parental duties consist largely of guarding, a task that may be performed almost as well by one parent as by two (Emlen, 1973).

Having stated the broad generalization of the adequacy of solitary male care in fish and amphibians, we should point out that there are many exceptions to this rule, including some cases of biparental care in fish. For example, cichlids (family Cichlidae; approximately 1800 species worldwide) are an important lineage of freshwater fishes found in Africa, South and Central America, and India. Most cichlids breed in areas rich in potential predators, and unlike many fish, they display elaborate parental behavior that includes solitary male care, maternal care, and biparental care (Keenleyside, 1979).

What is it about some species of cichlids that led to the evolution of biparental care? In studying several species of biparental cichlids in Lake Jiloa, Nicaragua, McKaye (1977) discovered that less than 10 percent of breeding pairs were able to raise their offspring to the stage at which they left their natal territory. If two parents have such dismal luck in protecting their offspring, how successful could a single parent be? Experimental evidence verifies the severe handicap suffered by single parents of biparental cichlid species: In the presence of predators, removal of either member of a brooding pair resulted in increased loss of young compared to control situations in which both parents remained with the brood (Keenleyside, 1978). Thus, under certain ecological conditions—in this case, extremely intense predator pressure—two parents may be necessary to successfully rear the young.

A second example of biparental care among fish involves the unusual case of parental feeding of young—consider the elaborate parental behavior of another cichlid, the discus (*Symphysodon discus*). Two days after fertilization, the fry of this species hatch, aided by both parents who chew open the egg cases. The two parents then deposit their youngsters on aquatic vegetation where the wrigglers dangle from threads as they live off their yolk supply. Within 2 or 3 more days, however, the brood can swim freely and they attach themselves to their parents and begin feeding on parental skin secretions (Skipper and Skipper, 1957; Figure 15.7). Quite possibly, young that can feed off two parents may grow and develop more rapidly than those that have only a single parent. In some fish, whether or not they live in predator-rich waters or their pattern of care involves the feeding of young, biparental care may be necessary to ensure the survival and health of offspring.

Patterns of Parental Care and Phylogenetic History   What role might phylogenetic history play in predicting patterns of parental care? The relationship between phylogeny

Figure 15.7 Young cichlids of the biparental species *Symphysodon discus* graze on parental skin secretions.

and strategies of reproduction is perhaps best illustrated by a broad comparison between patterns of parental care in birds and mammals. In mammals, internal gestation and lactation necessitate a major parental role for the female and restrict the ability of the male to aid offspring during early development. Male mammals cannot take over the duties of pregnancy and lactation for their mates, and rather than hanging around during the period of early development, males, in most cases, seek mating opportunities

elsewhere. There are, however, some mammalian species in which the male remains with the female during the rearing of offspring. This raises the interesting question of why male lactation has not evolved in such species, given that the physiological hurdles to such an occurrence seem somewhat surmountable (Daly, 1979). Although we do not have the answer to this question, two possibilities have been suggested. First, the reproductive success of male–female pairs may not be limited by the lactational abilities of the female at all, but rather by other factors such as food availability (Daly, 1979). According to this argument, then, males have evolved other forms of parental investment and the addition of lactation to the list of paternal activities may simply not be useful. The second possibility concerns potential limitations on a male's time and energy. If a male remains with a female in order to prevent other males from mating with her, then perhaps the amount of time and energy that the male devotes to guarding either his mate or their territory makes the evolution of lactation in males impossible (Clutton-Brock, 1991).

In contrast to the extended period of internal development in mammals, birds develop outside the mother's body. Embryos, along with food in the form of yolk, are packed in eggs that develop largely in the external environment. Because male birds are just as capable as their mates at providing care, parental duties such as incubation, feeding, and guarding are usually partitioned somewhat equally between the sexes. We see from this general comparison of birds and mammals that the basic biological attributes of a lineage constrain evolutionary possibilities—phylogenetic history, indeed, is important in determining patterns of parental care.

Sex Role Reversals

In the vast majority of internally fertilizing species, sperm is the male's only contribution to the survival of his offspring. As previously mentioned, Trivers (1972) argued that differential parental investment by the sexes governs the operation of sexual selection: The sex with greater parental investment becomes a limiting resource, essentially an object of competition among individuals of the sex investing less. The end result of the traditional asymmetry in patterns of parental investment is that males compete among themselves for access to females, and females, in turn, are selective in their choice of mates. (Recall, however, from our discussion of sexual selection in Chapter 14, that patterns of parental investment are probably not the sole determinants of sexual selection.) Are there exceptions to the general rule of predominantly female care in internally fertilizing species? Although in the minority, there are some species in which the burden of parental care falls largely or entirely on the male. These nontraditional species provide an excellent test of the importance of relative parental investment in controlling sexual selection. Let's consider two well-known examples of species that exhibit the phe-

nomenon of sex role reversal in parental care—the giant water bug and the Northern jacana.

Female giant water bugs (*Abedus herberti*) lay their eggs on the backs of males. Oviposition represents the end of female investment and male water bugs are left with the task of ensuring development of their offspring. Males aerate the eggs by exposing them to the air–water interface and through gentle rocking motions just below the surface. They assist nymphs during hatching and avoid eating their carefully nurtured investment by simply not feeding while nymphs on their back are emerging (Figure 15.8). Trivers's (1972) theory predicts that because male water bugs show greater parental investment than do females, they should be a limited resource for which females compete. We would also expect females to actively pursue males during courtship, and males, in turn, should be quite discriminating. Robert Smith (1979) tested these predictions in the field and in the laboratory by carrying a large kitchen strainer with which to

Figure 15.8 A male giant water bug broods eggs glued to his back. A single nymph has just hatched and the male, as part of his complete takeover of parental duties, suppresses his predatory urges while the young assume independence.

catch these insects while stalking Arizona streams. Smith's data indicated that space on the backs of males was a limiting resource during periods of peak egg production (from April to August). In staged encounters in the laboratory, females always approached males first and although males usually responded with a display called "pumping" (a rapid rocking movement that sometimes reached a crescendo of 300 pumps per minute), males occasionally opted not to display and thereby gracefully declined the female's invitation to mate. Trivers's predictions were further met by Smith's observations that males exercised considerable control over a female's success in courtship, mating, and egg laying. At the conclusion of these male-orchestrated activities, females departed, the weight of parental care now resting squarely with the male.

The Northern jacana (*Jacana spinosa*), previously called the American jacana, a small bird of tropical marshes best identified by its spindly greenish legs and toes, provides a second example of sex role reversal in parental care. Donald Jenni and Gerald Collier (1972) conducted an extensive study of the population dynamics, behavior, and social organization of individually marked jacanas in Costa Rica. They found that jacanas have a polyandrous mating system in which a given female is simultaneously paired with several males. At their study site, Northern jacanas breed year-round and a female defends a large territory that may encompass from one to four male territories. In addition to helping her mates defend their individual territories, the female independently repels intruders from her entire territory. A female's critical role in territorial defense is matched by her dominant role in courtship. Parental activities, however, are the province of the male (Jenni and Betts, 1978). Nest building and incubation are male duties exclusively, a division of labor underscored by the fact that only male jacanas have incubation patches (highly vascularized bare patches of skin

on the belly). Once the precocial young have hatched, males brood and defend their chicks (Figure 15.9). A female aids her mate in care of the brood only in the sense that she helps repel intruders from his territory, a task that she performs at all times regardless of whether or not young are present. Finally, morphological correlates appear to accompany the female jacana's malelike role in territorial defense and courtship: Breeding females weigh approximately 145 grams; adult males typically weigh in at a trim 89 grams.

Giant water bugs and Northern jacanas are not the only species in which males assume primary responsibility for care of offspring. With the notable exception of mammalian species, some level of sex role reversal has been reported in other species of insects and birds, as well as in crustaceans, fishes, and amphibians. The courtship behavior of males and females in species exhibiting sex role reversal in patterns of parental care is strong support for Trivers's hypothesis relating differential parental investment to sexual selection. As is often the case in biology, exceptions to the rule provide the clearest insight.

Alloparental Care and Adoption

Care of offspring need not be performed only by parents. In well over 150 species of birds and 120 species of mammals, some form of care by individuals other than the genetic parents, a pattern of behavior called alloparental care, has been reported (reviewed by Riedman, 1982). Individuals that provide such care for conspecific young are called alloparents (Wilson, 1975).

Although sometimes alloparental care may involve the actual adoption of another individual's offspring (particularly following the disappearance of the young animal's genetic parents), more typically such care involves helping the genetic parents rear their litter or brood. In birds, older offspring, and sometimes unrelated individuals, help parents by participating in ter-

Figure 15.9 Male Northern jacana and his two chicks confront an intruder, a purple gallinule. In this polyandrous species, males assume all parental responsibilities and females defend areas that encompass the smaller territories of their mates.

ritory and nest defense, feeding of nestlings and fledglings, and removal of fecal sacs. In 1987 Michael Tarburton and Edward Minot added incubation to the list of possible activities performed by alloparents. In an Australian population of white-rumped swiftlets (*Aerodramus spodiopygius*), the single nestling of a breeding pair's first clutch incubates the egg of the second clutch before fledging. Mammalian alloparents also display most—and sometimes all (see later discussion)—of the patterns of behavior exhibited by true parents. In the prairie vole (*Microtus ochrogaster*), a small rodent that inhabits grasslands in the central and western United States, most individuals in older litters remain at the natal nest where they brood, groom, and retrieve pups in their parents' subsequent litters (Wang and Novak, 1992). In some mammals, alloparental care may even involve nursing. Female bighorn sheep (*Ovis canadensis*) that have lost their own lambs sometimes permit the lambs of other females in their herd to nurse from them (Hass, 1990). In fact, some females

were observed to nurse up to seven different lambs (Table 15.1). Because bighorn females tend to remain within their natal group, allomothers are probably suckling relatives.

From these examples we see that alloparents are often related to the young that they care for, and quite often, they are older offspring that remain at home and care for their younger siblings. Why would older offspring postpone departure from the natal territory and remain, at least for a while, as nonbreeders within the parental group? Why not leave home and attempt independent reproduction? At least four factors probably enter into such a decision: (1) risks associated with dispersal, (2) chances of finding a suitable territory, (3) chances of finding a mate, and (4) chances of successful reproduction once established (Emlen, 1982). As we will see, in our more detailed discussion of helpers at the nest in Chapter 19, severe ecological conditions, such as extremely high costs of reproduction or shortages of territories or mates, may limit the option of independent breeding and favor remaining at

Table 15.1

NUMBER OF SUCKLING BOUTS, AND PERCENT OF BOUTS WITH OWN LAMB (AS OPPOSED TO LAMBS OF OTHER FEMALES IN THE HERD) FOR BIGHORN SHEEP ON THE NATIONAL BISON RANGE, MONTANA, 1983–1984[a]

YEAR	FEMALE	AGE	TOTAL BOUTS	PERCENTAGE OWN	NUMBER LAMBS NURSED
1983	01	3+	69	38	7
	02	6	46	52	7
	03	3	73	97	2
	04	2	82	100	1
	05	2	16	94	2
	06	7	88	11	7
	07	8+	66	97	2
	08	8+	75	0	7
	09	8+	67	0	7
1984	10	5	63	100	1
	02	7	36	69	2
	11	7	41	34	2
	01	9+	28	14	3
	07	9+	57	0	1
	06	8	30	0	2

[a] Only females that nursed alien lambs >5 percent of bouts were considered allomothers. A " + " indicates females estimated to be > 4 years old during 1980.
Source: Hass, 1990.

home with parents (Emlen, 1982; Emlen and Vehrencamp, 1983). However, even if older offspring remain at home during periods of harsh environmental conditions or territory or mate shortages, why should they help? The possible advantages associated with helping will also be discussed in Chapter 19, and natal dispersal is covered in more detail in Chapter 16.

Brood Parasitism

Intraspecific Brood Parasitism Waterfowl exhibit two patterns of behavior that may broadly be characterized as alloparental care in that an individual other than the genetic parent provides care for conspecific young (Eadie, Kehoe, and Nudds, 1988). In several species of waterfowl, some individuals lay their eggs in the nests of other individuals of their species, and from that point on, the foster parent incubates these eggs along with her own and then cares for all the young once they hatch (although especially common in ducks, geese, and swans, this phenomenon has also been reported in other species of birds, including house sparrows, swallows, and starlings). In another situation involving waterfowl, the young of some individuals mix with the young of other conspecifics soon after hatching, and then adults provide all subsequent care for these mixed broods. Obviously, these patterns of behavior in waterfowl are not what we usually think of when we consider alloparents as helpers at the nest. Rather than the genetic parent's staying with its offspring and simply receiving assistance from an alloparent, in waterfowl, the genetic parent permanently leaves its offspring in the care of an alloparent.

Does the genetic parent (the "donor") profit from leaving its young in the care of another individual (the "recipient")? Do the recipients of foreign young suffer reduced reproductive success? In some instances, the repercussions for both the donors and recipients of eggs are known. In bar-headed geese (*Anser indicus*), for example, some females lay their eggs in the nests of other females in their colony, and although the hatching success of their donated eggs may be only 5–6 percent, the gain in reproductive success is higher than if such females had not produced any eggs (Weigmann and Lamprecht, 1991). In other words, donors benefit. The hatching success of the eggs of recipients, on the other hand, is substantially less (29%) than that of females incubating only their own eggs (67%). Recipients are thus harmed by the presence of eggs that are not their own in their nests. In bar-headed geese, then, the term intraspecific brood parasitism—a term that implies an advantage for the donor and a disadvantage for the recipient—accurately describes the phenomenon of females laying their eggs in the nests of other conspecifics. In many other cases, however, the benefits to donors and costs to recipients have not been documented in sufficient detail to justify using the term parasitism. In situations such as these, until the precise costs and benefits to participants have been quantified, the neutral term brood amalgamation (amalgamation meaning mixing or commingling) may be more appropriate for describing these phenomena in waterfowl (Eadie, Kehoe, and Nudds, 1988).

Interspecific Brood Parasitism   Some animals exploit the parental behavior of other species. Approximately 1 percent of all species of birds lay their eggs in the nest of another species, thus leaving host parents with the task of raising foster young. The phenomenon has been named interspecific brood parasitism because of the apparent benefit experienced by the true parents who dispense with all parental responsibilities, and the harm that befalls the heterospecific host parents who typically experience a reduction in reproductive success (see later discussion). Interspecific brood parasites include several species of honeyguides, cuckoos, finches, and cowbirds, as well as one species of duck (Payne, 1977).

Damage to the host or its young may be directly inflicted by either the parasitic adult or its offspring (reviewed by Payne, 1977). The adult, for example, must place its egg in the host's nest when the host is beginning to incubate its own eggs. Upon discovering a nest after incubation or hatching has begun, a female cuckoo may eat the eggs or kill the young, causing the potential host to nest again. If the cuckoo is on time, however, she may throw out a host egg before laying her own in the nest. To add insult to injury, cuckoos often lay their eggs from a perch above the nest and their thick-shelled eggs break the host eggs when they strike them. Nestling cuckoos are renowned for methodically evicting eggs or young from the nest of their foster par-

Figure 15.10   Nestling parasitic cuckoo evicting a host egg from the nest.

ents. By positioning itself under a nearby egg or nestling, a young cuckoo may lift a nestmate onto its back, slowly work its way to the edge of the nest, and nudge the host egg or nestling to its death below (Figure 15.10). Nestling honeyguides employ an even more gruesome tactic to ensure full attention from their foster parents. At hatching, young honeyguides use their hooked bills to kill their nestmates. An alternative strategy to the outright killing of foster siblings is simply to monopolize parental care. Brood parasites usually mature more rapidly than host young, thus gaining a critical head start in growth and development. Additionally, their huge mouths, brightly colored gapes, and persistent begging often elicit preferential and prolonged feeding by foster parents (Figure 15.11). In England, a young cuckoo was so successful at begging that adults of three different species were observed feeding the youngster (Hardcastle, 1925). Host young are usually no match for their larger, more aggressive foster sibling and often die from starvation, crowding, or trampling.

In response to the devastating effects of brood parasitism, host species have developed ways to avoid being parasitized. The most common defenses are those used to reduce predation—the host species simply conceal their nests and defend them when they are discovered. Some hosts identify and remove the eggs or young of parasites from mixed clutches or broods. In response, many aspects of the laying behavior of brood parasites are designed for better deception of hosts. Cuckoos (*Cuculus canorus*) time their laying for the late afternoon when hosts are less attentive (e.g., Davies and Brooke, 1988), and while some passerines spend at least 20 minutes laying an egg in their own nests, female cuckoos can deposit an egg in a host nest in less than 10 seconds (Wyllie, 1981). Parasitic eggs or young often resemble those of the host species. Some species of parasitic finches mimic mouth color and pattern of host nestlings while others go so far as to imitate begging calls and head-waving behavior of their nestmates. Such co-evolution between brood parasite and host may produce adaptations and counteradaptations of increasing complexity (Davies and Brooke, 1988).

Figure 15.11 Parasitic young are often larger than their foster parents. Here a European cuckoo begs for food from its smaller foster parent, a hedge sparrow.

Before leaving the topic of interspecific brood parasitism, we should note that a somewhat similar phenomenon has been reported for three species of mouth-brooding fish in Africa (Ribbink, 1977). Like other mouth-brooding cichlids, *Haplochromis polystigma, Haplochromis macrostoma,* and *Serranochromis robustus* females pick their eggs up soon after laying and brood them in their mouth until the fry are ready to emerge. At the time of emergence, females release their offspring at a suitable site and remain to guard them. In the event of danger, the young retreat to the safety of their mother's mouth. Females of these three predaceous cichlids have been observed tending mixed broods consisting of their own fry and the fry of *H. chrysonatus,* a plankton-feeding cichlid. Although *H. chrysonatus* apparently broods its own eggs, it turns the duties associated with guarding the fry over to foster parents of the other cichlid species. Is the phenomenon in which some parental duties are carried out by foster parents of other fish species analogous to that of interspecific brood parasitism in birds?

In reviewing the evidence for what had initially been described as "cuckoo-like" behavior on the part of *H. chrysonatus,* Kenneth McKaye (1981) noted several differences between the phenomena of mixed-species broods in fish and interspecific brood parasitism in birds. For example, in contrast to the situation in birds, in fish, (1) the energetic cost of rearing foster young is minimal, (2) brood sizes are large, and when a predator attacks, it does not take the entire brood, and (3) foster young or their parents pose no threat to host young, and most important, (4) unlike the devastating effects of brood parasitism on the reproductive fitness of avian hosts, caring for the young of another species may actually increase the reproductive success of foster parents. In fish, it is possible that the host may actually "adopt" the young of other species. Such adoptions might be advantageous to the foster parent if the addition of foreign fry to a brood reduces the probability that the par-

ent's own young will be taken by a predator (see Chapter 13 for a discussion of the benefits of being part of a group when predators are nearby). Although this explanation for the occurrence of mixed-species broods in fish is possible, and indeed intriguing given that the three cichlid species seen guarding broods that contained their own young and those of *H. chrysonatus* inhabit waters rich in potential predators, more data are needed on the benefits and costs to fish foster parents of rearing mixed-species broods. Clearly, however, if guarding the young of other species is beneficial to these cichlid parents, then the term parasitism is inappropriate—if both parties benefit, the relationship may actually be mutualistic. Until the precise details of the relationship are known, we will use the more neutral term interspecific brood amalgamation to describe the formation of mixed-species broods in cichlids. Table 15.2 contrasts characteristics of interspecific brood parasitism in birds and brood amalgamation in fish.

Mating Systems

Earlier in this chapter we described how parents and offspring often disagree over the details of their interactions. As we will see in the pages that follow, such conflict is certainly not restricted to the parent–offspring relationship. In fact, adult males and females are often at odds over what constitutes an ideal mating relationship. Generally speaking, males focus on mating effort while females tend to emphasize parental effort. However, this disparity is not the only variable influencing the mating strategies of the sexes. Ecological factors, such as predation, resource quality and distribution, and the availability of receptive mates, influence the ability of males to monopolize females and the ability of females to choose among potential suitors (Em-

Table 15.2

COMPARISON OF INTERSPECIFIC BROOD CARE IN BIRDS AND FISH

CHARACTERISTIC	BIRDS	FISH
Size of brood	Small	Large
Effect of predation on brood	Total loss	Partial loss
Parental feeding of young	Yes	No
Energy devoted by foster parents to feed alien young	High	None or low
Competition among members of brood	High	Probably low
Behavior of alien young or their natural parent toward host young	May kill or remove	No removal or replacement
Active recruitment of alien young into brood	No	Yes
Apparent nature of relationship	Parasitic	Mutualistic
Ultimate "adaptive" response of host to alien young	Rejection	Acceptance and search for foster young

Source: Modified from McKaye, 1981.

len and Oring, 1977). Because such ecological conditions often vary both within and between locations, considerable flexibility is usually associated with the mating patterns of a given spe-

cies. In the sections that follow we (1) address the problems associated with categorizing mating relationships, (2) present a simple classification of mating systems, and (3) provide a detailed look at how some animals solve the dilemma of reproducing.

PROBLEMS ASSOCIATED WITH CATEGORIZING MATING SYSTEMS

Attempts to categorize mating systems have been hampered by ambiguous terminology, and as a result of this problem, there is no generally accepted system of classification. Mating systems are usually defined on the basis of criteria such as number of individuals of a species that a male or female copulates with, whether or not males and females form a bond and cooperate in care of offspring, and finally, duration of the pair bond. Although some biologists define mating systems solely on number of copulatory partners per individual (e.g., Thornhill and Alcock, 1983; Table 15.3), others use all three criteria to differentiate mating relationships (e.g., Wittenberger, 1981; Table 15.4). For example, according to Thornhill and Alcock's (1983) scheme, monogamy is defined as the mating system that re-

Table 15.3

CLASSIFICATION OF MATING SYSTEMS BASED ONLY ON THE NUMBER OF COPULATORY PARTNERS PER INDIVIDUAL

Monogamy:	An individual male or female mates with only one partner per breeding season
Polygyny:	Individual males may mate with more than one female per breeding season
Polyandry:	Individual females may mate with more than one male per breeding season

Source: Thornhill and Alcock, 1983.

Table 15.4

CLASSIFICATION OF MATING SYSTEMS BASED ON THE NUMBER OF COPULATORY PARTNERS PER INDIVIDUAL AND NATURE OF THE PAIR BOND

Monogamy:	Prolonged association and essentially exclusive mating relationship between one male and one female
Polygyny:	Prolonged association and essentially exclusive mating relationship between one male and two or more females at a time
Polyandry:	Prolonged association and essentially exclusive mating relationship between one female and two or more males at a time
Promiscuity:	No prolonged association between the sexes and multiple matings by members of at least one sex

Source: Wittenberger, 1979.

sults when "an individual male or female mates with only one partner per breeding season." Wittenberger (1981), however, defines monogamy as "a prolonged association and essentially exclusive mating relationship between one male and one female at a time," thus using characteristics of the pair bond as well as number of partners to define the mating system.

Determination of Mating Exclusivity

Determination of the degree of mating exclusivity practiced by males and females is often difficult to assess under natural conditions. In some small mammals, copulation takes place in runways or underground burrows, and as one can well imagine, the mating activities of nocturnal species are especially difficult to observe. Against such odds, Jan Randall (1991) was able to identify 26 matings over a 13-month period by banner-tailed kangaroo rats (*Dipodomys spectabilis*), nocturnal rodents that inhabit deserts of the southwestern United States. Only 5 of these 26 matings occurred above ground, however. The remainder took place in the underground burrow system and were thus inferred from patterns of male behavior and the presence of copulatory plugs in the reproductive tracts of females (see Chapter 14). Even the mating habits of highly visible animals, such as birds, may be difficult to decipher. Val Nolan (1978) studied the ecology and behavior during the breeding season of prairie warblers (*Dendroica discolor*) from 1952 to 1965 and during 5524 hours of observation saw only nine copulations. The difficulty in seeing copulations has led to the common practice of inferring mating relationships from noncopulatory social behavior. Nest cohabitation, extensive overlap of male and female home ranges, parental care by both males and females, and repeated sightings or trappings of male–female pairs are often cited as evidence of monogamy. However, as pointed out by Wickler and Seibt (1983), indirect evidence, such as the physical association between adult males and females, may be misleading and ultimately we will need more precise information to determine the mating system of a species.

Determination of Paternity

Even if we are lucky enough to observe copulation by kangaroo rats or prairie warblers, we still need more information in order to ascertain the mating strategies of males and females in these and other species. We cannot assume, for example, that all copulations result in the fertilization of eggs or that the number of eggs fertilized at different matings is identical (Thornhill and Alcock, 1983). In fact, determination of paternity is perhaps the most important step needed for the study of mating systems. At present the following three techniques are available for assessing paternity: (1) genetic markers, (2) protein electrophoresis, and (3) DNA "finger-printing." The use of genetic markers (pieces of genetic material that bear or produce a distinctive feature) is of limited value in the determination of paternity of most species under natural conditions. This technique requires knowledge of the inheritance pattern of the trait in question (e.g., coat color or plumage pattern) and information such as this is usually obtained through captive breeding experiments.

Electrophoresis, the process by which proteins in samples of blood, saliva, or other tissue (e.g., liver) are separated according to charge or molecular weight through placement in an electric field, has been used to resolve human paternity cases for many years. Once proteins have been separated on a gel, each individual from which a sample was taken can be assigned a genotype according to the resulting pattern of bands. Typically, the genotype of the mother is compared to that of the offspring in order to reconstruct the genetic contribution of the father. Once the paternal genotype has been determined, the banding patterns of all "potential" fathers are compared to this genotype and males with genotypes that do not match the paternal banding pattern are excluded. Because protein electrophoresis determines patterns of paternity through an exclusion procedure, it provides information on who is not the father, rather than who is the father. Electrophoretic techniques are not suitable for paternity determination in species that exhibit low heterozygosity (variation) in proteins (sufficient variation in alleles coding for proteins is necessary in order to contrast combinations of alleles in the mother and offspring to determine alleles contributed by the father) or those in which the identity of the mother is uncertain. The problem of uncertain maternity often occurs in field studies of small mammals in which young are captured only after they have left the nest and thus cannot be linked to a specific female.

Recent advances in the technology for manipulating and analyzing DNA have led to DNA fingerprinting as a means of determining maternity and paternity (Jeffreys, Wilson, and Thein, 1985). Some elements of DNA, known as "minisatellites," are highly variable in both structure and extent of repetition within the genome. It is now possible to extract DNA from a sample of blood or other tissue, separate DNA fragments on the basis of molecular weight using electrophoresis, and label these fragments with a radioactive probe. Parentage can be determined by comparing the resulting DNA fingerprint (banding pattern of labeled minisatellites) of potentially related individuals. DNA fingerprinting differs from electrophoresis in that it provides positive determination of maternity and paternity rather than relying on exclusion procedures.

When studying mating relationships in field populations, it is usually desirable to sample tissues nondestructively, and unfortunately, neither electrophoresis nor DNA fingerprinting always allow this. In addition, although the technique of DNA fingerprinting has already begun to revolutionize the study of animal behavior, it has the drawback of being very expensive. Weatherhead and Montgomerie (1991) use DNA fingerprinting in their studies of avian behavior and estimate that even after all the laboratory facilities have been set up, the costs of fingerprinting may run in the neighborhood of $100–$200 per individual bird.

Defining the Pair Bond

Imprecise terminology poses problems for classifications of mating systems that are based, in part, on the nature of the bond between mating partners. Although the term "pair bond" consistently appears in the literature on mating systems, there is still no universally accepted definition. The term also downplays the conflict between male and female reproductive strategies by implying that mating partners function as a cooperative unit. Douglas Mock (1983), in his review of the study of avian mating systems, emphasized that cooperative tendencies between mating partners are expected only when such tendencies fit both of the individuals' genetic interests, and he concluded by describing the pair bond as "a grudging truce in which acts of generosity are regarded with mild skepticism." Finally, although some biologists have set definite time limits on what is meant by "prolonged association" (e.g., 20–25% of the breeding season according to Wittenberger and Tilson, 1980), ambiguity is often associated with attempts to categorize mating systems on the basis of duration of the pair bond.

There are, indeed, a number of problems associated with defining mating systems, but nonetheless, a fundamental understanding of the nature of how animals get together (and stay together) is critical to the larger questions regarding social organization.

A CLASSIFICATION OF MATING SYSTEMS

Randy Thornhill and John Alcock (1983) presented a classification of mating systems based solely on the number of copulatory partners per individual (see Table 15.3), a system attractive for its simplicity. According to them, polygyny is a mating system in which some males copulate with more than one female during the breeding season. Monogamy is a mating system in which a male and female have only a single mating partner per breeding season, and polyandry, the least common mating system, is one in which some females mate with more than one male during the breeding season.

Defining mating systems in this way eliminates the need to define pair bond or prolonged association. In addition, promiscuity, a fourth category of mating system included in the classification scheme of Wittenberger (1981; see Table 15.4), is redefined under polygyny or polyandry. (Throughout much of the literature on mating systems, polygyny and polyandry are grouped under the general category of polyg-

amy ["many marriages"] as opposed to monogamy ["single marriage"].)

Polygyny

According to theory, a male's reproductive "strategy" will involve devoting time and energy to parental effort or to mating effort. In many cases because males are physically capable of fertilizing a large number of females, they will favor the option of mating rather than parenting. As males maximize their reproductive output through multiple matings, polygyny results, and we find that this is the most common mating system among vertebrates, including humans (Flinn and Low, 1986). Polygyny, itself, may be subdivided into female defense, resource defense, and lek polygyny.

Female Defense Polygyny   Female defense polygyny occurs when females live in groups that can be defended by a male. Although in some species female gregariousness may be related to cooperative hunting or increased predator detection, in other species clumping is more directly related to reproduction. For example, female elephant seals (*Mirounga angustirostris*) become sexually receptive less than 1 month after giving birth. Each year pregnant females haul out onto remote beaches of Año Nuevo Island, California. Female gregariousness, a shortage of suitable birth sites, and a tendency to return annually to traditional locations result in the formation of dense aggregations of receptive females. Under these crowded conditions, a single dominant male, weighing in the vicinity of 8000 pounds and sporting an enormous overhanging proboscis (Figure 15.12), can monopolize sexual access to 40 or more females (LeBoeuf, 1974). This male defends his harem against all other male intruders in bloody, and sometimes lethal, fighting.

Resource Defense Polygyny   Resource defense polygyny occurs when males defend re-

Figure 15.12 A male elephant seal. Males of this species compete for access to and defend large numbers of females at birthing sites. This is an example of female defense polygyny.

sources essential to females, rather than defending females themselves. Those resources can be nest sites or food. If such resources are unevenly distributed in an area, then it is possible to defend areas that have more or better resources than do others (Emlen and Oring, 1977). In some cases there is good evidence that female choice is often based on the quality of resources controlled by a male. Thus, even though receptive females do not live together, a male can monopolize a number of mates by controlling critical resources. Male scorpion flies (Mecoptera: Panorpidae, Figure 15.13) fiercely defend the area around a dead arthropod (Thornhill, 1981). Standing next to his nutritious find, a male will disperse a sex attractant and display with rapid wing movements and abdominal vibrations. If a female approaches, copulation is the fee that she must pay in order to gain access to this food resource. Thus, while males enjoy the benefits of mating with several females, females enjoy a hearty meal that may reduce the amount of time that they need to spend foraging.

Resource defense polygyny produces a number of significant problems for females. For example, males in such cases usually do not help in rearing the young. Also, the activity around these areas may attract predators, and other receptive females may increase the competition for commodities such as food. Polygynous matings will be advantageous to females only when the benefits achieved by mating with a high-quality male and gaining access to his resources more than compensate for these costs. In other words, a female may reproduce more successfully as a secondary mate on a high-quality territory than as a monogamous mate on a lower quality territory. Jared Verner and Mary Willson (1966) coined the phrase "polygyny threshold" to describe the difference in territory quality needed to make secondary status a better reproductive option for females than primary status.

Figure 15.13 A male scorpionfly. Scorpionflies display resource defense polygyny in that a male will defend the area around a dead insect and females must mate with him to gain access to the meal.

Gordon Orians (1969) elaborated on Verner and Willson's ideas and reasoned that if polygyny is always to a male's advantage and yet it does not always occur, then the circumstances under which it does occur must be those in which there is some advantage to the female. He reasoned that females should join a harem when this decision confers greater reproductive success than monogamous alternatives. According to this argument, then, the average reproductive success of females should not decrease as harem size increases. Figure 15.14 illustrates the polygyny threshold model's way of relating differences in territory quality to female choice of already mated versus unmated males.

The polygyny threshold model has been extremely important in shaping our views of mating systems, especially with respect to the mating relationships of birds. Predictions of the model have been subjected to numerous tests in the field, and most studies confirm the expected relationship between number of mates and environmental factors believed to reflect territory quality. However, the critical question is whether unmated females do better by accepting secondary rather than primary status. Although this point can be tested by comparing the success of

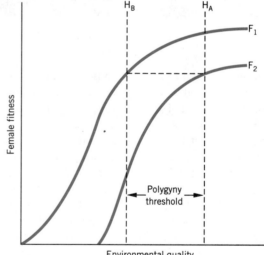

Figure 15.14 The polygyny threshold model. The polygyny threshold is the difference in quality between H_a and H_b that would favor polygynous matings.

F_1 = fitness curve for monogamous females
F_2 = fitness curve for secondary females
H_a = highest quality breeding habitat
H_b = marginal breeding habitat

(From Wittenberger, 1979; modified from Orians, 1969.)

monogamous and secondary females, procuring the necessary data on reproductive success is no easy task.

Theoretically, harem size is a function of the benefits this kind of organization bestows. Therefore, harems of different sizes should impart equal levels of advantages, as seems to be true in birds. However, in mammals this may not be the case at all. Data collected on yellow-bellied marmots (*Marmota flaviventris*), for example, do not agree with the prediction of equal fitness for females in harems of different sizes (Downhower and Armitage, 1971). Yellow-bellied marmots are polygynous rodents that inhabit forest clearings and alpine meadows of the western United States (Figure 15.15). Active from early May to mid-September, they spend the rest of the year in hibernation. For over 20 years, Kenneth Armitage has studied the social behavior of marmots in Colorado. Through a combination of livetrapping, direct observation, and genetic analysis of blood samples, he and his colleagues have discovered that although some individuals are solitary, others live in social groups consisting of one adult male, one or more adult females, and a number of yearlings and juveniles. Matrilines, consisting of from one to five related females, often occupy a common home range and a male's harem may consist of one or more matrilines. Not surprisingly, a male marmot's reproductive output, measured in terms of the number of yearlings produced, increases as harem size increases. However, in apparent contradiction to Orian's notion that polygyny occurs when it is advantageous to females, the reproductive success of females decreases with increases in harem size (Figure 15.16).

In light of the finding that female reproductive success varied as a function of harem size, Downhower and Armitage (1971) suggested that the typical harem size of two in yellow-bellied marmots is a compromise between the polygynous inclinations of the male and the preference for monogamy on the part of the female. Elliott (1975) and Wittenberger (1979) criticized this

Figure 15.15 Yellow-bellied marmot. Results from studies on the social behavior of this polygynous mammal do not support the polygyny threshold model.

interpretation and suggested that loss in annual reproductive output of females living in large harems could be compensated by increased survival and hence greater *lifetime* reproductive success. According to their argument, female marmots in large harems may lose a little on an annual basis but surely they would make up for this loss by longer survival and increased opportunity for reproduction. The problem with this argument is that survivorship of females does not vary as a function of harem size. Al-

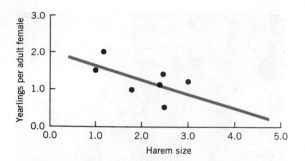

Figure 15.16 Contrary to predictions of the polygyny threshold model, female reproductive success, measured in terms of the number of yearlings produced per adult female, decreases as harem size increases in yellow-bellied marmots. (Modified from Downhower and Armitage, 1971.)

though Armitage (1986) revised the earlier Downhower and Armitage model (1971), the polygyny threshold model still does not apply to yellow-bellied marmots. The model is based on the assumption that polygyny evolves when already mated males attract additional females to their territory. In yellow-bellied marmots, however, males are attracted to females that reside in suitable habitat. In marmots then, females appear to determine the spacing patterns that make polygyny possible. Furthermore, the organization of marmots into matrilines does not meet the implicit assumption of the polygyny threshold model that females act independently of other females. Results from the study of the behavior and ecology of yellow-bellied marmots have served not only to question the universality of the polygyny threshold model, but perhaps more importantly, to emphasize the conflicting nature of male and female reproductive interests.

Lek Polygyny The third category of polygyny, lek polygyny, occurs when males defend "symbolic" territories that are often located at traditional display sites termed leks. Males of lek species do not provide parental care and defend only their small territory on the lek, not groups of females that happen to be living together nor resources associated with specific areas. Females visit these display arenas, select a mate, copulate, and leave. This extreme form of polygyny occurs when environmental factors make it difficult for males to monopolize females either directly as in female defense polygyny or indirectly as in resource defense polygyny.

A well-known example of a species that displays lek polygyny is the hammer-headed bat (*Hypsignathus monstrosus*; Figure 15.17), an inhabitant of lowland rain forests of west and central Africa. Adult males are nearly twice the weight of adult females and not only have an enormous muzzle that terminates in flaring lip flaps, but also possess a larynx that fills more

Figure 15.17 Male hammer-headed bat at his calling site on the lek.

than one-half of their body cavity. In comparison to adult males, mature females have a much less dramatic muzzle and a larynx approximately one-third the size. Jack Bradbury (1977), intrigued by the extreme sexual dimorphism, modifications of the male larynx, and reports of aggregations of calling males, observed the bats from platforms in the forest canopy, netted animals at the leks where they assembled, and tracked individual males and females by radiotelemetry.

As with other lek species, male hammer-headed bats show no parental care and have no significant resources within their display territories. Adult males select sites that are at least 10 meters apart along streams and rivers. These linear arrays of displaying males may contain over 100 individuals and occur at traditional sites used for at least 60 years. Individual males are quite particular about their choice of calling branch, selecting the exact same place night after night and often making from 5 to 10 passes by the location to ensure that they land at just the right spot. Once in position, they begin wing flapping and loud calling (a noise described by Bradbury as sounding like "a glass being rapped hard on a porcelain sink"). These displays often continue without interruption for hours. Interested females fly along the calling assembly and make short hovering inspections of males. The mere pause of a female near a male inspires him to accelerate his calling rate, and should she move closer, he pulls his wings tightly against his body and emits several long buzz notes. If the female turns and leaves, the male will extend his wings and resume his display. However, if the female lands on his calling branch, signaling that he has been chosen to donate his sperm, he copulates rapidly and typically resumes calling within 1 minute of mating. Copulation appears to occur exclusively on these display territories and some males have far greater success than others. In fact, at the major assembly site studied in 1974, 6 percent of the males achieved 79 percent of the copulations.

How might lek behavior such as that seen in the hammer-headed bat have evolved? Several hypotheses have been suggested. These have been divided into two groups based on potential benefits for males or females.

From the male perspective, group, rather than individual, displays may increase signal range or the amount of time that signals are emitted (e.g., Lack, 1939; Snow, 1963). Alternatively, males may aggregate because they require specific display habitats that are limited and patchily distributed (e.g., Snow, 1974). Leks may provide protection from predators through increased vigilance—with more eyes watching, a predator would have a harder time sneaking up on them (e.g., Wiley, 1974). Leks may also serve as information centers where males exchange the latest news on good foraging sites (e.g., Vos, 1979). Males may gather near "hot spots," areas through which a large number of females are likely to pass (e.g., Bradbury, Gibson, and Tsai, 1986; Payne and Payne, 1977). Finally, leks may arise because less successful males generally have better mating chances in the vicinity of highly successful males. Because certain males are extremely successful at attracting females, other less successful males gather around such "hotshots" and obtain more copulations than they would have had they displayed by themselves (Beehler and Foster, 1988).

From the female perspective, large groups of males may facilitate mate choice (Alexander, 1975; Bradbury, 1981). After all, it might be easier to distinguish between superior and inferior males when comparison shopping is possible. Mating within a group of males may reduce vulnerability of females to predation since any predator might be distracted by so many displaying individuals (Wittenberger, 1978), or if males aggregate in less desirable habitat, lek mating may reduce competition between the sexes for resources (Wrangham, 1980). All of these hypotheses could be useful as explanations for the development of leks, but for now they are best regarded as possible explanations.

Monogamy

Monogamy occurs when a male and female have only a single mating partner per breeding season. Because sperm from one male is often sufficient to fertilize a female's limited number of eggs, monogamy is sufficient from the female perspective, but for males, confining copulation to a single female is a rather conservative means of ensuring genetic representation in the next generation. What ecological circumstances might favor monogamy over polygyny?

Monogamy is likely to occur when males are unable to monopolize females. For example, when receptive females are scarce or widely distributed, it might be more beneficial for a male to remain with a given female rather than to search endlessly for additional mates. In other cases, male monogamy may be enhanced by female reproductive synchrony (Knowlton, 1979). When females within a population are receptive at the same time, a male's chances of copulating with several females in succession are exceedingly small. For example, synchrony in the availability of receptive females appears to force monogamy on male termites. Pair formation in these insects takes place during the nuptial flight when vast numbers of individuals from neighboring colonies become airborne simultaneously (Wilson, 1971). During the flight, each male attempts to sequester a single female for himself. The short duration of the flight, the intense competition for mates, and each female's steadfast refusal to actually mate until well established in a safe burrow all favor a male termite's guarding the sole female he has managed to sequester. In fact, males of most termite species remain with their mate for life (Figure 15.18).

Although paternal care is more likely to be a side effect rather than a prerequisite for monogamy (Wickler and Seibt, 1983), monogamous males are often in the position of providing useful materials to their mate and offspring. Hornbills (Bucerotidae) are large-bodied monogamous birds of the Old World tropics and

Figure 15.18 Male termite closely pursuing a female he has met on the nuptial flight. Synchrony in the availability of receptive females forces monogamy on male termites.

subtropics. All 45 species display the peculiar habit of sealing their nestholes. After a female enters a nesthole to begin egg laying, she carefully plasters mud around the edge of the hole until only a small crack remains to connect her to the outside world. Once sealed within her nesthole, a female relies entirely on her mate to bring food for herself and her nestlings (Figure 15.19). Although variable across species, the period of time that a female remains ensconced in her nest may extend to over 4 months. Nesthole sealing, apparently an adaptation to reduce predation on females and nestlings (Kemp, 1971, 1976), obviously means that females will be monogamous, and the need for male care ensures that he, too, will not take another mate. A male hornbill, constrained by the enormous task of feeding his mate and offspring, simply cannot

Figure 15.19 A male red-billed hornbill feeding his mate that is sealed inside the nest hole. Male hornbills are monogamous because they cannot simultaneously provision two nests.

care for a second nest (Leighton, 1986). Thus, monogamy in hornbills is linked to their adaptation to reduce nest predation, the success of which rests on the male's ability to provide parental care.

Monogamous mating systems, although common among birds (approximately 91% of bird species are monogamous; Lack, 1968), are relatively rare among mammals (about 3% of mammal species display monogamy; Kleiman, 1977). (It should be noted, however, that for the most part, the values 91% and 3% have been arrived at on the basis of individual accounts of patterns of association between males and females and not on assessment of mating exclusivity; the percentages might be very different, probably much lower in the case of birds, if monogamy was defined according to the number of individuals with which males and females mate.) This striking phylogenetic difference is said to result from the difference between avian and mammalian parents discussed earlier in the chapter, namely that whereas male birds are equally adept as females at most aspects of parental care, male mammals typically can provide little useful help during the female-dominated activities of gesta-

tion and lactation (Trivers, 1972). This broad generalization rests on the assumption that male parental investment positively affects female reproductive success and has led to the idea that, in some species, monogamy has evolved because male parental care is essential to female fitness. Silver-backed jackals (*Canis mesomelas*; also called black-backed jackals) are monogamous canids that inhabit the brush woodland of the Serengeti Plain of Tanzania. For over a decade, Patricia Moehlman (1986) has studied this species and yet has never observed a single mate change among breeding pairs. Adult members of a breeding pair appear to mate exclusively with their partner and exhibit a high degree of behavioral synchrony that includes cooperative hunting and tandem marking of territory boundaries. Male silver-backed jackals participate in parental activities such as guarding the den and regurgitating food for the young. In this species, male parental investment seems critical to the reproductive success of the female; in the two instances in which the male of a pair disappeared during the whelping season, the female and pups all died within a week. However, as we will see next, absence of the male from a monogamous pair does not always result in such a dramatic outcome.

Patricia Gowaty (1983) removed males of pairs of the eastern bluebird (*Sialia sialis*) to examine the effects of male parental care on female reproductive success in this species. Males of this "apparently monogamous" species (a phrase used by Gowaty to acknowledge that mating habits were described on the basis of observed associations among breeding adults and not on knowledge of genetically effective matings) were removed from the territories of breeding females after eggs had been laid. These experimentally deserted females were compared to paired females with respect to several measures of reproductive success (e.g., clutch size, number of eggs hatched, number of nestlings fledged). Contrary to general expectation, lone and paired females did not differ for any measure of reproductive success considered, indicating that female fitness is not always dependent on male parental care, even in monogamous species. Gowaty suggests that rather than male parental care, dispersion of nest sites is the important variable maintaining monogamy in eastern bluebirds. So, if parental care by monogamous males does have positive effects on female fitness, these effects may be subtle and much more difficult to observe in some species than originally expected. It is also apparent that long-term field studies that permit determination of lifetime reproductive success are critical to evaluating the role of paternal behavior in monogamous mating systems.

Polyandry

Polyandry, defined as the mating relationship that occurs when some females have more than one mate during the breeding season, is the least common of mating systems. Because most males offer only sperm in return for copulation and because the sperm from one male is often sufficient to fertilize all the eggs of a given female, most females have little to gain by mating with more than one male. However, if copulation opens the door to critical resources or male parental assistance, then mating with several males may result in reproductive benefits for females. From a male perspective, however, polyandry seems to be the most unsatisfactory mating option. If, in exchange for copulation, a male must commit himself to the provision of resources or parental care, then the last thing he should want is to share his mate with other males. The rarity of polyandry results from the basic fact that under most circumstances, polyandrous matings are disadvantageous to both males and females.

Polyandry occurs to varying degrees in different species. Females of some species mate with more than one male in order to supplement sperm received from previous matings (e.g., fruit flies; Pyle and Gromko, 1978), while others copulate with several males to gain access to resources that males exchange for copulation (e.g., hummingbirds; Wolf, 1975). These mating relationships do not involve sex role reversal (that is, females competing for males that provide the bulk of parental care and are selective in their choice of mate) and are best described as combinations of female polyandry and male polygyny. "True" polyandry is always associated with sex role reversal like that described for the Northern jacana in the section on parental care. Thus, in its most striking form, true polyandry is characterized by males assuming most, if not all, parental responsibilities, thereby freeing females to pursue mating opportunities.

Two hypotheses have been proposed to account for the evolution of sex role reversal and polyandry. Graul, Derrickson, and Mock (1977) suggested that sex role reversal and polyandry evolved in response to dramatic fluctuations in food availability. During periods of food scarcity, a male's takeover of parental duties may be necessary if the energy reserves of his mate have been drastically depleted by egg production. Faced with an exhausted mate, the best option for a male may be to increase parental investment. Should food availability increase, the female can lay additional eggs in a second nest that she might tend (double clutching) or leave to another male. The second hypothesis suggests that high predation pressure, rather than food availability, is the ecological factor of importance in the evolution of role reversal and polyandry (Maxson and Oring, 1980). Intense predation on nests may force males into parental duties so that females can lay a second clutch and thereby increase the chances for both partners of fledging young.

The spotted sandpiper (*Actitis macularia*; Figure 15.20) provides an example of a polyandrous species subjected to intense nest predation and at the same time demonstrates the flexibility associated with strategies of mating and parental care (Oring and Lank, 1986). For several years, Lewis Oring and colleagues have studied a population of spotted sandpipers on Little Pelican Island in Leech Lake, Minnesota (a name that fails to conjure up images of a swimmer's paradise). Female spotted sandpipers lay separate clutches with one or more males on all-purpose territories that they defend against other females. Although females typically acquire mates sequentially, an occasional female will have two males on her territory at the same time. Males engage in territorial defense and provide most of the parental care. Female spotted sandpipers engage in a strategy of bet hedging, vigorously pursuing mating opportunities when they arise, and in their absence, investing in parental care. In essence, if a new male appears, she will mate with him; if no new males appear, she will tend to her young. Variations in social behavior result from a combination of changing environmental circumstances and social conditions. As is characteristic of other species that display sex role reversal, female spotted sandpipers are larger and more striking in their coloration (that is, they have a denser spot pattern) than males.

Nest loss plays an important role in determining the incidence of polyandry in spotted sandpiper populations. Intense predation on nests leads to concentrations of individuals in more protected areas and thereby increases the potential for females to acquire multiple mates. However, even in relatively safe sites such as Little Pelican Island (the small size of the island makes it unsuitable for year-round occupation by most mammalian predators), nest losses to predation are quite high. For example, despite the removal of some mammalian predators from the island, only 37 percent of the 1289 eggs laid

Figure 15.20 Polyandry is practiced by spotted sandpipers. Here two females fight over prime nesting habitat on which the winner will likely pair with several males in succession.

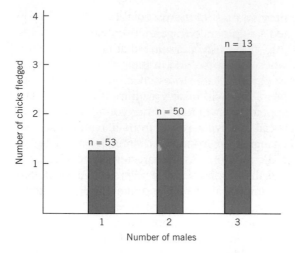

Figure 15.21   In spotted sandpipers, female reproductive success, measured in terms of number of chicks fledged, increases with the number of males monopolized by each female. (From Oring and Lank, 1986.)

between 1973 and 1982 hatched. Although deer mice (*Peromyscus maniculatus*) are the most consistent source of egg destruction, a single mink on the island in 1975 caused total breeding failure. Females counter high predation rates with a remarkable ability to rapidly lay replacement clutches. Interclutch intervals may be as short as 3 days and one female laid as many as 20 eggs (totaling more than 400% of her body weight) over a period of 42 days (unpublished data cited in Emlen and Oring, 1977). In summary, intense predation on spotted sandpiper nests leads to clumping of individuals in more protected areas and increases potential benefits associated with laying several clutches (Figure 15.21).

Summary

Parental care is one component of the overall life history pattern of a species. Because animals have limited time, energy, and resources to devote to reproduction, evolutionary "decisions" must be made concerning the amount of care and who should assume parental responsibilities. The fact that providing parental care entails trade-offs is reflected in Trivers's (1972) definition of the term parental investment: "any investment by the parent in an individual offspring that increases that offspring's chance of surviving (and hence reproductive success) at the cost of the parent's ability to invest in other offspring." The issue of who provides the care appears related to factors such as mode of fertilization (male care being associated with external fertilization and female care with internal fertilization) and phylogenetic history (in mammals, parental care usually rests with the female, while in birds, parental duties are typically partitioned somewhat equally between the sexes). The phenomenon of conspecifics other than the genetic parents caring for young is called alloparental care.

In the past several years, at least three different views, or models, of the parent–offspring relationship have emerged. The parental provision model represents the most traditional view of parent–offspring interactions in that it emphasizes the one-way flow of resources such as food and warmth from parent to offspring. Rather than treating offspring as passive recipients of parental resources, the conflict model portrays offspring as active participants in a somewhat contentious relationship. According to this model, because parents and offspring are not genetically identical, they are destined to disagree over the amount of care given and over how long the period of offspring dependence should last. The third model uses a metaphor from the biological world to describe the parent–offspring relationship. In taking into account the bidirectional exchange of resources that sometimes occurs between parents and their offspring, the symbiosis model views the parent–offspring relationship as an association that benefits both participants. (Recall that this is an unusual use of the term symbiosis. Rather than describing interactions between two different species [the traditional use of the term], it is used here to describe interactions between parents and young within a species.)

Conflicts of interest are not restricted to parent–young interactions, and in fact, such conflicts characterize most social behavior. In the case of mating relationships, males usually produce more offspring by seeking additional mates. In contrast, females tend to emphasize parental effort rather than mating effort and can usually produce more offspring by gaining male parental investment. The disparity between the sexes in parental investment interacts with ecological factors such as predation, resource quality and distribution, and the availability of receptive mates to shape the mating system of a species.

Mating systems are usually defined on the basis of one or more of the following factors: (1) number of individuals that a male or female copulates with, (2) whether or not males and females form a bond and cooperate in care of offspring, and (3) duration of the pair bond. Attempts to categorize mating systems, however, are often hampered by problems associated with defining the pair bond and determining mating exclusivity and paternity. We favor a system of classification based solely on the number of copulatory partners per individual. According to this system, polygyny occurs when some males copulate with more than one female during the breeding season. Monogamy is a mating system in which a male and female have only a single mating partner per breeding season, and polyandry, the least common mating system, occurs when some females mate with more than one male during the breeding season.

PART
Three

Behavior of Groups

CHAPTER

16

Sociality and Dispersion

iny *Holocnemus pluchei* spiderlings hatch from the eggs that their mother has held close to her body for some time (Figure 16.1). For the next 4 or 5 days, the spiderlings remain on the web of their mother, but at the end of this grace period, each spiderling is faced with a choice— to build its own web or to move onto the web of a conspecific, typically that of a larger individual. (Although up to 15 conspecifics may share a single web, most groups consist of just two individuals.) The choice is by no means an easy one—there are costs as well as benefits associated with group living in *H. pluchei* (Jakob, 1991). Spiderlings that move onto a web that is already occupied catch fewer prey items than do spiderlings that opt for a solitary existence, because the larger occupants of the web outcompete them. Some group-living spiderlings may actually be eaten by their "hosts." Why do some

spiderlings opt for sharing a web, given that they lose food (and possibly their lives), rather than constructing their own web? It seems that webs are quite costly structures to build. One estimate suggests that the average spiderling must forage for 5 days in order to recoup the energy used in web building (in the life of a spiderling, 5 days is a substantial amount of time). When spiderlings move onto a web that is already occupied, they make few or no home improvements. Thus, because they add little or no silk, they essentially dispense with the costs of web building. In short, group living in this species entails losing food, but saving silk (Jakob, 1991).

This brief sketch of the options available to *H. pluchei* spiderlings reveals that there are both pros and cons associated with living socially. In the sections that follow we consider the variety of costs and benefits experienced by group-living animals. We then continue with this type of cost/benefit analysis in our consideration of patterns of dispersion among animals, examining first the option of whether to leave home upon reaching maturity, and second, whether to maintain exclusive use of a particular area.

Figure 16.1 Holocnemus pluchei female with eggs. Soon after hatching, spiderlings have the option of building their own web or moving onto the web of a conspecific. Although living socially saves the spiderling the time and energy needed to build a web, it also results in decreased food due to competition from the original owner of the web.

Living in Groups

Jane Goodall, the famed African researcher, is reported to have said, "One chimpanzee is no chimpanzee at all." In her efforts to encourage research institutes using chimpanzees to offer them a more natural social environment, Goodall illustrates the fact that many animals are highly social and into the very fabric of their lives is woven the need to be with others of their kind. In fact, if we think about animals in general, we are likely to come up with case after case of animals that live in groups.

We should quickly point out that not all group-living animals are technically social. For example, some arthropods tend to gather in places that have a certain moistness and temperature. They, therefore, incidentally end up in the same vicinity through what is called passive grouping. Other species, though, are actually social, that is, they seek out each other's company. Being social, however, has both costs and benefits (Alexander, 1974).

THE COSTS OF SOCIALITY

The costs to individuals that live in groups may include increased competition for mates, nest sites, or food resources. Consider the case of the humbug damselfish (*Dascyllus aruanus*; Figure 16.2). This small fish lives in fairly stable social groups of about 5 to 35 individuals amid branching corals in lagoons off the Great Barrier Reef of Australia. The groups are organized into strict dominance hierarchies in which each individual has a rank. Each hierarchy is organized according to a simple rule—larger fish are always dominant over smaller fish—and this hierarchical organization influences patterns of feeding within groups. Humbugs feed on plankton by orienting into the current and intercept-

ing prey items that flow past their home coral. As it turns out, the larger (more dominant) fish within a group feed farther upstream than do the smaller fish and thereby gain first access to food arriving in the current. Smaller, lower quality prey eventually reach the smaller group members farther back. This method of feeding within the group probably explains the suppressed growth observed for juveniles living in groups with many large adults (Forrester, 1991). We see, then, an example in which social living entails the cost of competing with other group members for food.

Other costs associated with social living include increased exposure to parasites or disease, increased conspicuousness due to the increased level of activity in groups, and a greater possibility of providing parental care to young that are not one's own or spending time trying to find one's own young from amid the clamoring throng. Mexican free-tailed bats (*Tadarida brasiliensis mexicana*), for example, live in aggregations, sometimes with millions of individuals, where mismatches between nursing mothers and their offspring sometimes occur (McCracken, 1984). Although it may seem surprising

Figure 16.2 Group living may entail the cost of competing with other group members for food. Humbug damselfish, for example, live in groups amid staghorn corals off the Great Barrier Reef. In this species, the larger, more dominant group members gain first access to the best food floating by in the current. As a result of strategic positioning by larger group members, the growth of smaller individuals is often suppressed.

Figure 16.3 A crèche of Mexican free-tailed bat young. Mothers returning from a night's foraging sometimes mistakenly nurse the offspring of another female. In this species, then, group living sometimes entails the cost of providing parental care to another group member's young.

that about 17 percent of mothers suckle young that are not their own, imagine the task of a mother when she returns to the cave after foraging for insects and tries to locate her own pup amid a dense mass of wriggling youngsters (Figure 16.3). Social living may also provide opportunities for exploitation of some individuals by others. In large groups, a male may have difficulty guarding his mate from other males, and hence he may wind up caring for another male's offspring. Females may also be duped into caring for young that are not their own. Amid the seemingly frantic activity of a nesting bird colony, a female may fail to notice the arrival of another female at her nest and the quick deposit of an egg by the intruder (for more extensive discussion of intraspecific brood parasitism, see Chapter 15).

THE BENEFITS OF SOCIALITY

Obviously, since so many species are social, there must be certain distinct advantages to group living. Recall from our discussion of antipredator behavior in Chapter 13 that with membership in a group comes a variety of antipredator tactics that are not available to solitary individuals. Predators are usually less successful when hunting grouped rather than solitary prey because of the superior ability of groups to detect, confuse, and repel predators. Furthermore, during an attack by a predator, an individual within a group has a smaller chance of becoming the next victim.

Individuals living in groups may also have greater foraging success. In some species, individuals in groups cooperatively hunt together with increased success, on average, for each individual. Cooperative hunting sometimes enables predators to kill animals larger than themselves. This seems to be the case for Harris' hawk (*Parabuteo unicinctus*; Bednarz, 1988), a bird of prey that lives (and hunts) in family groups in the southwestern United States. During the early morning hours, family members typically gather at one perch site, from which the group then splits into smaller subgroups of from one to three individuals, each subgroup making alternate short flights throughout the family's home area. This type of movement has been described as "leapfrogging." Upon discovering a rabbit, one of three hunting tactics, or more typically a mixture of the three, may be employed by the hawks. The most common, the "surprise pounce," occurs when several hawks arrive from different directions and converge on a cottontail or jackrabbit unfortunate enough to be out in the open. Even if the rabbit escapes under vegetation, however, the safety may only be temporary. At this point, the hawks employ their "flush-and-ambush" tactic, a strategy in which one or two hawks flush the rabbit from the cover, and then family members perching nearby pounce on it. "Relay attack" is the third, and least common, hunting tactic used by family members. Here, they engage in a constant chase of a rabbit, with the position of "lead" bird changing each time the lead attempts a kill and misses. Once killed, the prey is shared among all members of the hunting party.

Using this combination of hunting techniques, Harris' hawks probably confuse and ex-

haust their prey. In fact, success in hunting rabbits is correlated with group size—hunting parties of five to six individuals do better than smaller parties (Figure 16.4a). The average energy intake available per individual from rabbit kills is also higher in groups of five or six members than in smaller groups (Figure 16.4b). As noted previously, cooperative hunting makes possible the killing of prey that may be much larger than the hawks themselves. Jackrabbits, for instance, are two to three times heavier than female and male hawks, respectively.

Unlike the situation for Harris' hawk in which members of hunting parties participate in all aspects of foraging from searching for prey to eating it, ants of the species *Formica schaufussi* forage individually, but form groups to bring oversized prey back to the nest. Once a scout has found a relatively large food item (typically a dead insect), it then lays a chemical trail back to the nest. Recruits eventually assemble at the prey item and then either drag or carry the prey

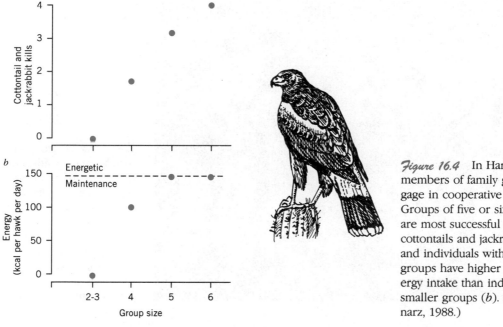

Figure 16.4 In Harris' hawks, members of family groups engage in cooperative hunting. Groups of five or six individuals are most successful at killing cottontails and jackrabbits (*a*) and individuals within such groups have higher average energy intake than individuals in smaller groups (*b*). (From Bednarz, 1988.)

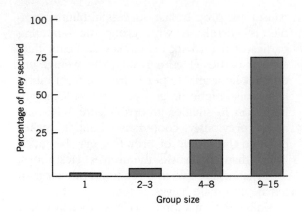

Figure 16.5 In the ant *Formica schaufussi*, workers forage individually but retrieve heavy prey items in groups. Such group retrieval is important in defense of prey against workers of other ant species. Here, in staged encounters with a competitor *Myrmica americana*, larger retrieval groups of *Formica schaufussi* are more successful than smaller groups at securing prey. (Data from Traniello and Beshers, 1991.)

back. In this way, *F. schaufussi* can include a larger range of prey items in its diet. The retrieval groups also are likely to be able to defend the prey against other ant species in the area. In staged contests in the field, larger retrieval groups of *F. schaufussi* were more successful in securing prey from their competitor *Myrmica americana* than were smaller groups (Figure 16.5). Here, then, groups are effective not only in prey retrieval, but also in defense of the prey (Traniello and Beshers, 1991).

Another species in which more than one factor influences the formation of groups is the lion (*Panthera leo*). Although ideas on patterns of grouping in lions initially focused on hunting success (Figure 16.6), more recent data suggest that this factor alone cannot explain the formation of prides in this species (Packer, 1986; Packer, Scheel, and Pusey, 1990). First of all, scavenging may be a more efficient means of obtaining food than hunting, and under conditions such as these, groups may be necessary to de-

Figure 16.6 A group of lions sharing a wildebeest kill. Studies of group living in lions have often focused on hunting success. As it turns out, some of the major functions of grouping in lions relate to defense of cubs, space, and scavenged food. (Photo from Willy Bemis.)

fend carcasses against lions from other prides. Group living also appears important in protecting cubs from nomadic males that commit infanticide, and in the defense of the pride's home area against intrusion by neighboring prides. Thus, for lions, the benefits of group living include defense of food, young, and living areas against conspecifics.

Next we will see that in spite of the advantages of living in one's social group, some animals leave those benefits, never to return.

Natal Philopatry and Natal Dispersal

NATAL PHILOPATRY, NATAL DISPERSAL, AND BREEDING DISPERSAL

Some animals are born in one place and then move to another place where they breed, never to return to their birthplace. This general behavior has been called natal dispersal, and it has come to be defined in many ways. Here, we will use Walter E. Howard's (1960) definition of natal dispersal as "the movement an individual makes from its point of origin to the place where it reproduces or would have reproduced had it survived and found a mate." This definition refers to the permanent movement of individuals away from their birth site. (A second type of dispersal, called breeding dispersal, refers to movements between consecutive breeding sites. For example, upon weaning their young, some female rodents abandon the nest and offspring and move to another site where they will give birth to their next litter. Our present discussion focuses on natal, rather than breeding, dispersal.)

Natal dispersal can be contrasted with natal philopatry. In natal philopatry, offspring remain at their natal area and share the home range or territory with their parents (Waser and Jones,

1983). Some authors who are more interested in the genetic than in the behavioral consequences of dispersal have defined natal philopatry as limited dispersal and include cases in which young animals move no more than 10 home ranges from their natal nest (Shields, 1987). Here, because our focus is on the behavioral interactions between parents and their adult progeny, we will use the definition of natal philopatry as absolute nondispersal of young. (In a broad sense, philopatry can also refer to the tendency of some migratory species to return seasonally to their place of birth, but we are not concerned with such behavior here.)

COSTS AND BENEFITS OF NATAL PHILOPATRY VERSUS NATAL DISPERSAL

The costs and benefits of philopatry and natal dispersal may be categorized as either genetic or somatic (Shields, 1987). Genetic aspects include two opposing forces, reduction of inbreeding effects and conservation of successful genomes (see following discussion). Somatic considerations, on the other hand, involve survival and reproductive success. In order to fully understand the selective advantages of philopatry and natal dispersal, a consideration of both genetic and somatic aspects is necessary.

Potential genetic costs of philopatry involve those of inbreeding. Inbreeding, or mating between relatives, may be extreme and involve mating between members of a nuclear family (e.g., matings between parents and offspring, or between siblings; Shields, 1982). From the standpoint of genetics, inbreeding can be costly because it increases homozygosity and hence (1) increases the risk of production of offspring that are homozygous for deleterious (harmful) or lethal recessive alleles and (2) reduces variation among offspring. The somatic costs associated with philopatry might involve increased competition. If conditions are crowded at home, then philopatric young may have to compete with relatives for food, nest sites, or mates. Limited access to critical resources could result in lower survival and/or fecundity relative to those offspring that disperse (Shields, 1987).

What benefits might be associated with philopatry? Genetically speaking, inbreeding may not be all bad. A certain level of inbreeding, for example, could maintain complexes of genes that are particularly well adapted to local conditions. If animals that dispersed from the natal site settled a short distance away (in most cases, dispersal distances are surprisingly short; Shields, 1982) and mated with nonrelatives, such complexes of genes would be diluted, or perhaps weeded out of the population. Thus, we see that there may be an optimal level of inbreeding (Shields, 1982).

Somatic benefits of philopatry include the advantages of being familiar with the local physical and social setting. Such familiarity may enable philopatric young to be efficient not only at finding and controlling food, but also in escaping from predators. Familiarity with family and neighbors is also likely to reduce levels of aggression and stress associated with social interactions. In short, philopatric young may live longer and leave more offspring because of the relatively low risks and energy use associated with living in familiar surroundings (Shields, 1982). Dispersers, in contrast, are thought to face

high energy costs and risks of predation as a result of increased movement and lack of familiarity with the terrain (e.g., Ambrose, 1972; Metzgar, 1967). In some cases, dispersers may encounter high levels of aggression when attempting to establish residence in an existing population (e.g., Joule and Cameron, 1975). Despite such costs, recall that dispersers may avoid crowded conditions at the natal area and thereby benefit through increased access to mates and critical resources.

NATAL PHILOPATRY AND NATAL DISPERSAL IN BIRDS AND MAMMALS—SOME HYPOTHESES

Males and females of a particular species often differ in whether or not they disperse from the natal nest. Even more striking is the observation that the direction of the sex bias in natal dispersal differs between birds and mammals (Table 16.1). In the majority of bird species that have been studied, females are more likely to disperse than are males. In mammals, however, just the reverse is true—males are more likely to disperse than are females. We can ask why this is so—why the difference?

Paul J. Greenwood (1980) suggested a combination of two hypotheses to explain the observed patterns of sex-biased dispersal in birds and mammals. First, although acknowledging that the benefits associated with familiarity with local physical and social conditions would often favor philopatric tendencies in offspring, he also recognized that the genetic costs associated with close inbreeding would favor some level of dispersal. A sex bias in dispersal seems to be the perfect compromise—extreme inbreeding is prevented because members of one sex disperse, and individuals of the other sex experience the somatic benefits of philopatry. However, this first hypothesis, although explaining why sex biases in dispersal tendencies occur, does not explain which sex leaves home, and why the direction of the bias differs in birds and mammals.

Table 16.1

NUMBER OF SPECIES OF MAMMALS AND BIRDS IN WHICH NATAL DISPERSAL IS MALE-BIASED, FEMALE-BIASED, OR IN WHICH OFFSPRING OF BOTH SEXES DISPERSE

	PREDOMINANT DISPERSING SEX		
	MALE	FEMALE	BOTH
Mammals	45	5	15
Birds	3	21	6

Source: Data from Greenwood, 1980.

Greenwood's (1980) second hypothesis addresses the issue of which sex disperses in birds and mammals and why. He suggests that sex differences in dispersal are related to differences in the ways in which male and female birds and mammals compete for critical resources. First, he notes, most mammals are polygynous in that a single male defends a group of females. Young or subordinate males, unable to gain access to females, may disperse in order to increase their chances of mating. Additionally, female mammals often live in matrilineal social groups (groups of mothers, daughters, and granddaughters) in which the benefits of living with kin may be quite high. Faced with such a social system, males may disperse and thereby avoid the genetic costs of extreme inbreeding. Greenwood thus linked male dispersal in mammals to their typical mate-defense mating systems.

Birds differ from mammals in that most appear to be monogamous, but more important, most species have resource-defense mating systems. In other words, rather than competing for females directly, male birds usually compete for territories that attract females. Under conditions such as these, familiarity with a particular area might be more important to males than to females. Female birds might disperse in order to avoid the genetic costs of close inbreeding and to choose territories with the best resources. Therefore, Greenwood (1980) linked male-biased dispersal in mammals to mate-defense mating systems and female-biased dispersal in birds to resource-defense mating systems. However, Greenwood's hypotheses fail to explain patterns of dispersal in territorial mammals. Although female-biased dispersal would be predicted, as with birds, this pattern was observed in only one of 13 species of territorial mammals examined (Liberg and von Schantz, 1985).

F. Stephen Dobson (1982), focusing strictly on dispersal data for mammals, suggested that predominant dispersal by male mammals was related, in part, to competition for mates. He reasoned that because most mammals are poly-

gynous, competition for mates would be more intense among males than among females, and thus dispersal should be more common in males. Furthermore, Dobson said that in species with monogamous mating systems, levels of competition for mates would be more equal between the sexes, and males and females should disperse in similar proportions. When Dobson examined dispersal data for species of mammals with different mating systems, he found remarkable agreement with his hypothesis (Table 16.2). Although this hypothesis highlighted the links in mammals between (1) male-biased dispersal and polygynous mating systems and (2) equal dispersal by males and females in monogamous species, it failed to explain female-biased dispersal in either monogamous birds or in species of polygynous mammals that have resource-defense mating systems. Finally, although Dobson introduced the idea that differences between males and females in levels of competition for mates might be involved in sex differences in dispersal tendency in mammals, he also noted that avoidance of inbreeding was probably a significant influence on natal dispersal in many species.

Olof Liberg and Torbjorn von Schantz (1985) presented the "Oedipus" hypothesis to explain sex-biased dispersal in birds and mammals. This hypothesis, based on reproductive competition

Table 16.2

NUMBER OF SPECIES OF MAMMALS IN WHICH NATAL DISPERSAL IS MALE-BIASED, FEMALE-BIASED, OR IN WHICH OFFSPRING OF BOTH SEXES DISPERSE, AS A FUNCTION OF TYPE OF MATING SYSTEM

MATING SYSTEM	PREDOMINANT DISPERSING SEX		
	MALE	FEMALE	BOTH
Monogamous	0	1	11
Polygynous or promiscuous	46	2	9

Source: Data from Dobson, 1982.

between parents and offspring, differs from the hypotheses of Greenwood (1980) and Dobson (1982) in that the question of dispersal is viewed from the standpoint of the parents, not the offspring. Liberg and von Schantz assume that it is usually in the offspring's best interest to stay at home. Because parents occupy the superior position in the parent–offspring relationship, it is they that make the "decisions" about those that can stay and those that should leave. Furthermore, if parents benefit by letting some offspring stay at home but cannot afford to have all offspring stay, then they should encourage those young that exact the highest toll to leave.

According to Liberg and von Schantz (1985), levels of reproductive competition between parents and male and female offspring are related to mating systems and modes of reproduction in birds and mammals. With respect to polygynous or promiscuous mating systems, male offspring that remain at home are predicted to compete with their fathers for mates, but female offspring do not compete with either parent. In monogamous mating systems, neither sons nor daughters pose a competitive threat to either parent. Furthermore, there may be differences between mammals and birds based on their different modes of reproduction—specifically, egg laying versus gestation and birth. In birds, a daughter that is allowed to stay at home can cheat her mother, or both parents in the case of monogamous species, by laying eggs in the family nest and thereby imposing her own breeding costs on her parents (recall our general discussion of intraspecific nest parasitism in Chapter 15). In mammals, however, daughters cannot hide their reproductive efforts because pregnancy and birth are obvious events, and parents cannot be fooled into caring for their daughter's offspring. According to Liberg and von Schantz, for sons, there are no corresponding ways of cheating parents in either birds or mammals.

Using this information on reproductive competition between parents and offspring as a func-

tion of mating system and mode of reproduction, Liberg and von Schantz (1985) make the following predictions regarding dispersal and philopatry in birds and mammals: In monogamous birds, sons cannot steal copulations from their fathers, but daughters can cheat parents by laying eggs in the family nest. In such cases, parents should force daughters to disperse. In polygynous and promiscuous birds, sons can steal copulations from their fathers and daughters can cheat parents, so both sexes should be driven away from the nest. In the case of monogamous mammals, sons cannot steal matings from their fathers, and daughters cannot trick parents into caring for offspring, so the prediction is that both sexes should be allowed to remain at home. Finally, in polygynous and promiscuous mammals, sons can cheat their fathers out of copulations, but daughters cannot fool parents into caring for offspring, so daughters should be allowed to stay home and sons should be forced to disperse. The predictions of the Oedipus hypothesis are summarized in Table 16.3.

Liberg and von Schantz's (1985) hypothesis provides explanations for some of the areas of disagreement between the hypotheses of Greenwood (1980) and Dobson (1982). For example, Dobson (1982) argued that in monogamous mammals, competition for mates should be fairly equal between the sexes, and there should be no sex bias in dispersal. As pointed out by Greenwood (1980), however, most bird species are monogamous and most are characterized by female-biased dispersal. The Oedipus hypothesis provides the explanation that irrespective of mating system, because female birds can cheat their parents by laying eggs in the family nest, they should be forced to disperse. A drawback of the Oedipus hypothesis is that its predictions are valid only when it is in the best interests of both male and female offspring to remain at home. The hypothesis does not apply to situations in which young animals disperse on their

Table 16.3

PREDICTIONS OF THE OEDIPUS HYPOTHESIS

MATING SYSTEM	BIRDS (DAUGHTER CAN CHEAT PARENTS BY LAYING EGGS IN THE FAMILY NEST)	MAMMALS (DAUGHTER CANNOT TRICK PARENTS INTO CARING FOR HER OWN YOUNG)
Monogamous (son cannot steal matings from father)	Daughters forced to disperse	Equal proportions of both sexes disperse
Polygynous or promiscuous (son can steal matings from father)	Daughters and sons forced to disperse	Daughters allowed to stay; sons forced to disperse

Source: Modified from Liberg and von Schantz, 1985.

own initiative, perhaps to avoid inbreeding or to gain accesss to better resources.

We see, then, that none of the three hypotheses alone can explain patterns of dispersal in all of the species that have been studied. It is probably true, however, that there are multiple influences on dispersal (Dobson and Jones, 1985). Factors that influence dispersal probably differ in their importance between species, sexes, and individuals. Thus, each time that we ask why animals disperse, we should consider the hypotheses that such movement reduces (1) competition for resources, (2) the probability of inbreeding, and (3) competition for mates. Finally, we should also examine the relative roles that parents and offspring play in dispersal.

Territoriality

In much the same way that patterns of dispersal or philopatry can be examined through an analysis of the costs and benefits, we can also consider patterns of space use by animals. For example, we can ask, should an animal maintain exclusive use of an area or should it share the area with others? As we will see, such "deci-

sions," like those involving dispersal, depend to a large extent on access to resources and mates.

HOME RANGES, CORE AREAS, AND TERRITORIES

We begin with some definitions. The home range of an individual animal is that area in which it carries out its normal activities. Within the home range there is often an area in which most activities are concentrated—this location is called the core area. In some cases, the core area may be the area immediately surrounding the nest site, or perhaps a food or water source. Although the home ranges of individuals may overlap, such overlap is less likely in core areas. If overlap of the core area, or even the entire home range, is prevented by agonistic behavior by the resident(s) (individuals, pairs, or groups may actively defend a given area), then the defended area is called a territory (Noble, 1939). Although this definition of territory emphasizes active defense of an area, other definitions downplay defense and emphasize instead the exclusive use of space (Schoener, 1968). This latter definition is often more practical for describing the space use patterns of animals whose territorial behavior is difficult to observe in the field. Because of the secretive habits of many small mammals, for example, it is virtually impossible to state with any certainty that the exclusive use of an area is maintained by active defense (Ostfeld, 1990). Territories may be used solely for feeding, mating (recall our discussion of leks in Chapter 15), or raising young. Alternatively, they may be used for a variety of purposes, in which case they are called multipurpose territories.

COSTS AND BENEFITS OF TERRITORIALITY

Recall from our discussion of how optimality models have been used to study territorial behavior (Chapter 4) that there are both costs and benefits associated with defense of an area (Brown, 1964). Depending on the type of territory maintained, some benefits might include (1) a longer-lasting food supply given that fewer individuals have access to it, (2) increased access to mates, if individuals of the opposite sex prefer to mate with territory holders, or (3) a greater opportunity to rear offspring at a high-quality site. On the other hand, acquiring or defending a territory can be downright costly when one considers the energy needed to patrol territory boundaries, to display to potential intruders, and to forcibly evict trespassers or a previous owner. Territory acquisition and defense also take time away from other essential activities such as foraging. Consider, for example, the dilemma faced by great tits (*Parus major*), small perching birds that live in England and feed on seeds and insects. In this species, activities related to feeding and territorial defense occur in different microhabitats and are completely incompatible (Ydenberg and Krebs, 1987). Whereas feeding occurs within 3 meters of the ground, territory defense, particularly the singing component, takes place high in trees, about 10 meters above the ground. Obviously, then, there are trade-offs involved in territorial behavior—time spent in defense cannot be spent foraging.

How do great tits "decide" how to allocate time between territorial defense and feeding?

One factor that seems to influence time allocation in these small birds is food availability. In one field experiment, supplemental food was placed on tables and provided to some males on their territories—these males were designated "owners." Owners, however, were not the only birds to make use of the extra food. Males on neighboring territories also visited the food tables and were called "visitors." A third group of males—"control" males—did not visit the feeding tables. After 5 days of supplemental feeding on certain territories, each male was exposed to a single simulated intrusion that consisted of exposure to a stuffed male great tit mounted on a stake. At the same time that the stuffed male was placed on each male's territory, 1 minute of taped great tit song was played in the background. The responses of the three types of males were scored. Males that had access to supplemental food (both owners and visitors) responded more strongly to the stuffed intruder than did control males (Figure 16.7). Owners and visitors exhibited shorter latencies to approach the stuffed intruder, spent more time in its vicinity, and were more likely to attack. Control males, on the other hand, were more likely to respond to the intruder with song, a less energetically expensive display than outright approach and attack. These results suggest that access to the supplementary food allowed owners and visitors to meet their food requirements rapidly and thereby to spend more time in territorial activity, particularly in energetically expensive activity (Ydenberg, 1984). That visitors were similar to owners in their level of response to the intruder suggests that the response was not simply a result of birds defending a rich food resource on their own territory.

Economically speaking, then, territoriality is expected to occur only when the benefits outweigh the costs. A net benefit associated with territoriality translates into increased fitness for the individual territory holder. Factors such as abundance, spatial distribution, and renewability are thought to influence whether or not a par-

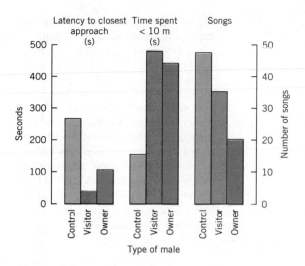

Figure 16.7 Great tits, like many animals, cannot feed and defend their territory at the same time. Thus, there are trade-offs between foraging and territorial defense. One factor that seems to influence the allocation of time between the two activities is food availability. Here, male great tits that had access to supplemental food (owners and visitors; see text for details) responded more strongly to a stuffed "intruder" than did control males that did not make use of the extra food. Although owners and visitors engaged in energetically expensive methods of territorial defense (approach and attack), control males engaged in less expensive displays such as singing. Thus, access to high levels of food appeared to allow males to devote more time and energy to territory defense. (From Ydenberg and Krebs, 1987.)

ticular resource is economically defendable (Davies and Houston, 1984). Generally speaking, territoriality is favored when resources are moderately abundant. When resources are overly abundant, extreme pressure from intruders may make territory defense too costly. At the opposite extreme, low resource levels may necessitate defense of such an enormous area, just to ensure access to adequate supplies, that patrolling costs become prohibitive. With respect to the spatial distribution of resources, defending resources that occur in discrete patches should be less costly than defending those that are widely dispersed. Finally, it should be beneficial to defend resources that are renewed slowly, but not resources that renew themselves more rapidly than they are consumed. Quickly renewable resources, after all, should not be strongly affected by exploitation by intruders, but those resources that are renewed slowly could be significantly reduced by intruders.

THE ECOLOGY OF TERRITORIALITY IN MICROTINE RODENTS

Let us now consider territoriality in the microtine rodents, a group that includes the voles, lemmings, and muskrats. These small mammals exhibit amazing variation in the degree of territoriality both among species and between sexes within species. In some species, only females are territorial; in other species only males are territorial; and in still others, territoriality is exhibited by both male and female members of a breeding pair. Hypotheses put forth to explain the different patterns of territoriality suggest that whether or not females are territorial depends on characteristics of the food supply, and whether or not males are territorial depends on the availability of reproductive females (reviewed by Ostfeld, 1990). The important point here is that males and females differ in terms of which resource is most critical to their reproductive success, that is, which resource is worth defending—food for females, and mates for males. The two major hypotheses that we will discuss—the "females in space" hypothesis (Ostfeld, 1985) and the "females in space and time" hypothesis (Cockburn, 1988; Ims, 1987a)—are similar in their explanations for female territoriality, but differ somewhat in their explanations for male territoriality.

The Females in Space (FIS) Hypothesis

As previously mentioned, for females, food appears to be the critical resource influencing re-

(a) Before food added   (b) After food added

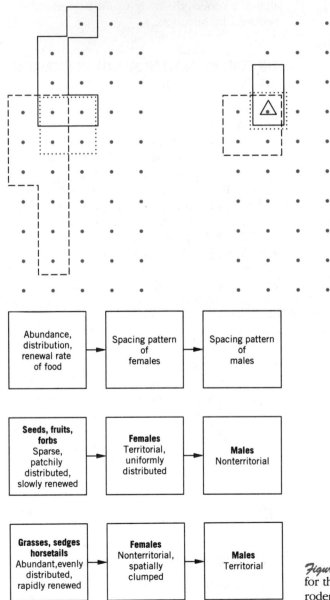

Figure 16.8 In microtine rodents, females are predicted to be territorial when food resources are scarce. In this field experiment with gray-sided voles, the home ranges of specific females were measured both before and after supplemental food was added (triangle denotes location of supplemental food). Addition of food resulted in smaller home ranges and a greater degree of home range overlap. In other words, abundance of food resulted in a relaxation of territorial behavior among neighboring females. (Modified from Ims, 1987b.)

Abundance, distribution, renewal rate of food	→	Spacing pattern of females	→	Spacing pattern of males

Seeds, fruits, forbs Sparse, patchily distributed, slowly renewed	→	**Females** Territorial, uniformly distributed	→	**Males** Nonterritorial

Grasses, sedges horsetails Abundant, evenly distributed, rapidly renewed	→	**Females** Nonterritorial, spatially clumped	→	**Males** Territorial

Figure 16.9 The females in space (FIS) hypothesis for the occurrence of territoriality among microtine rodents. (Modified from Ostfeld, 1990.)

productive success. Rick Ostfeld (1985) proposed that characteristics of the food supply, specifically abundance, distribution, and renewal rate, ultimately determine the occurrence of territoriality in microtine rodent females. Some species of voles feed primarily on seeds, fruits, and forbs (called the SFF diet by Ostfeld). Such foods are sparse, patchy in distribution, and slowly renewed. Ostfeld thus predicted that females of species feeding primarily on foods from the SFF diet would be territorial. Other species of voles feed predominantly on grasses, sedges, and horsetails (the GSH diet). Because these plants are abundant, widely distributed, and rapidly renewed, Ostfeld predicted that females of species utilizing the GSH diet would be nonterritorial. Although there are some exceptions, comparisons across species generally support these predictions (Table 16.4). In addition, two single-species studies in which the amount of available food was either changed or experimentally manipulated also support the idea that characteristics of the food supply determine spacing patterns of females. Bank voles (*Clethrionomys glareolus*) and gray-sided voles (*Clethrionomys rufocanus*) feed mostly on fruits and seeds, and as predicted by Ostfeld, females are territorial. However, when food was unusually abundant for a population of bank voles (Ylonen, Kojola, and Viitala, 1988), or when supplemental food in the form of oats, corn, and sunflower seeds was provided to a population of gray-sided voles (Ims, 1987b), the degree of territoriality exhibited by females decreased. In the study with gray-sided voles, females that exploited the same supplemental food store became more clumped in distribution, and the degree of overlap in their home ranges increased considerably (Figure 16.8).

Most male mammals do not help in the rearing of offspring, and as a result, access to mates is thought to be a more critical determinant of male reproductive success than is access to food. Ostfeld (1985) predicted that the spatial distribution of females determines whether or not

males are territorial. Thus, food resources determine the spatial organization of females, which, in turn, affects the spacing pattern of males (Figure 16.9). According to this idea, when females are nonterritorial, they occur in clumps and can be more easily defended by males (that is, males can be territorial). When females are territorial, however, they are widely distributed, making defense by males much more difficult. Under these latter conditions, males are not expected to be territorial but are predicted to roam widely in search of receptive females. An inevitable result of such wandering is that the ranges of males will overlap. According to the FIS hypothesis, then, territoriality in males is expected to be associated with lack of territoriality in females, and conversely, lack of territoriality (home range overlap) in males is expected when females are territorial. Interspecific comparisons generally support these predictions (refer again to Table 16.4).

Is there evidence from single-species studies to support the idea that the spatial distribution of females determines the spatial distribution of males? In the study previously described in which supplemental food was given to populations of gray-sided voles, a reduction in the overlap of male home ranges (that is, an apparent increase in territorial behavior) coincided with the decreased territoriality and spatial clumping of females (Ims, 1987b). Although consistent with the FIS hypothesis, these results might also be explained by other factors such as breeding synchrony among females (see later discussion). Finally, results contrary to the predictions of the FIS hypothesis were obtained by Rolf Ims (1988). He placed female gray-sided voles in wire mesh cages and arranged them in either a clumped or a dispersed distribution, and over a 10-day period, he recorded patterns of space use by males. The home ranges of males overlapped more (that is, males were less territorial) when females were clumped than when they were widely dispersed. Ims suggested that the clumped arrangement of females induced in-

Table 16.4

TYPE OF DIET AND THE OCCURRENCE OF TERRITORIALITY IN MICROTINE RODENTS: TESTING THE FEMALES IN SPACE HYPOTHESIS

SPECIES	DIET[a]	FEMALE TERRITORIALITY	MALE TERRITORIALITY
Clethrionomys gapperi	SFF	yes	no
Clethrionomys glareolus	SFF	yes	no
Clethrionomys rufocanus	SFF	yes	no
Microtus arvalis	SFF	yes	no
Microtus ochrogaster	SFF	yes	yes
Microtus pinetorum	SFF	yes	yes
Microtus pennsylvanicus	SFF	yes	no
Microtus agrestis	GSH	no	yes
Microtus californicus	GSH	no	yes
Microtus xanthognathus	GSH	no	yes

[a] SFF = seeds, fruits, and forbs; such foods are scarce, patchy in distribution, and slowly renewed. GSH = grasses, sedges, and horsetails; such foods are abundant, widely distributed, and rapidly renewed.
Source: Modified from Ostfeld, 1985.

tense competition among the males, making territorial behavior too costly.

It is important to note that the predictions of the FIS hypothesis pertain only to the breeding season, the time when food is especially important to females because of the high costs of reproduction and the time when males compete for copulations. During the nonbreeding season, Ostfeld (1985) predicted that the patterns of sex-specific territoriality exhibited by microtine rodents during the breeding season should break down, and mixed-sex aggregations should form. In support of this prediction, communal groups comprised of males and females of different ages have been found during late fall and winter in several species of voles.

The Females in Space and Time (FIST) Hypothesis

Andrew Cockburn (1988) and Rolf Ims (1987a), although agreeing with Ostfeld (1985) on the causes of female territoriality in microtine rodents, proposed a modification of factors responsible for male territoriality. They suggested, based on ideas outlined earlier by Emlen and

Oring (1977), that the distribution of fertilizable females not only has a spatial component, but also a temporal component. In other words, the availability of females varies in both space and time. Not surprisingly, then, this has been called the "females in space and time" hypothesis.

The temporal component focuses on the degree of breeding synchrony in the population. Female voles and lemmings typically produce several litters within a breeding season. Although the first period of estrus in a female's life appears to be induced by exposure to a male (this first estrus is called male-induced estrus), subsequent estrus periods occur just after giving birth. Within hours of parturition, a female becomes sexually receptive for about 1 day. This period of receptivity is known as postpartum estrus. A female that is impregnated during postpartum estrus does not become receptive again until she gives birth to her litter about 3 weeks later. Most microtine females are thus sexually receptive for fairly short periods of time (about 1 day) at regular intervals (every 3 weeks) throughout the breeding season.

Do most females within a population become

sexually receptive at the same time (that is, synchronously) or at widely dispersed points in time (that is, asynchronously)? According to Ims (1987a), when females become receptive asynchronously, males should search for females, stay with them only during their brief period of sexual receptivity, and then move on in search of other mates. Regardless of the spatial distribution of females, the time taken for a male to search for new mates is probably less than the 3 weeks it would take for a particular female to enter estrus again. When breeding is asynchronous, then, males should not be territorial but should roam widely in search of receptive females (Figure 16.10). Ims further predicts that these patterns of movement and mating will most likely result in a promiscuous mating system. Because males have highly overlapping home ranges, a single female will probably mate with more than one male over the course of the breeding season, and males, constantly on the move in search of receptive females, will probably mate with more than one female.

The male spacing strategy is predicted to change dramatically when breeding occurs synchronously. Because resources, in general, are more easily defended through territorial behavior when they have a clumped distribution, males can monopolize females that become receptive at the same time. If such females have a clumped distribution or fairly small home ranges, then a single male can monopolize a group of females, and the mating system is likely to be polygamous (see Figure 16.10). If, however, such females have a dispersed distribution or extremely large home ranges, then a single male may monopolize only a single female, and the mating system is likely to be monogamous (again see Figure 16.10). Thus, although type of mating system depends on the spatial distribution of females, the spacing system of males can be predicted by the temporal component of female distribution alone.

It turns out, unfortunately, that it is extremely difficult to disentangle the effects on male ter-

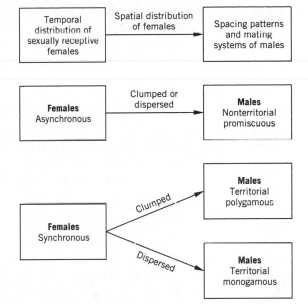

Figure 16.10   The females in space and time (FIST) hypothesis for the occurrence of territoriality among microtine rodents. (Drawn from Ims, 1987a.)

ritorial behavior of spatial and temporal clumping of females because the likelihood of synchronous breeding is probably tied to the spatial distribution of females. Because social interactions, olfactory cues, and shared food resources may be involved in achieving breeding synchrony, such synchrony is probably more common among females that are clumped. Thus, clumping by females may result in territoriality by males, but the question remains, is male territorial behavior due to the spatial distribution of females, breeding synchrony among females, or both?

At least one study has supported the females in space and time hypothesis, and one has not. When a population of meadow voles (*Microtus pennsylvanicus*) was monitored, females were found to breed asynchronously, and as predicted by the FIST hypothesis, males were not territorial (McShea, 1989). However, in another study, this time involving the gray-sided vole, females in the population were found to breed synchronously, but contrary to the predictions of the FIST hypothesis, males were nonterritorial (Kawata, 1985).

Unanswered Questions

We see, then, that there are studies that support and studies that refute both the females in space hypothesis and the females in space and time hypothesis. There are also certain territorial systems that cannot be completely explained by either of the hypotheses. Pine voles (*Microtus pinetorum*) and prairie voles (*Microtus ochrogaster*; Figure 16.11), for example, eat seeds, fruits, and forbs. According to the females in space hypothesis, because these foods are sparse, patchy, and slowly renewed, females should be territorial—and they are. However, contrary to the predictions of the FIS hypothesis, males of both species are also territorial. Rather than roaming widely in order to come in contact with as many females as possible, male pine voles and prairie voles appear to mate monogamously (FitzGerald and Madison, 1983; Getz et al., in press). Although there is some variation within populations, in both species, a given male and female share a nest and territory. The FIST hypothesis predicts a monogamous mating system only when breeding occurs synchronously, and females have widely dispersed or exceptionally large home ranges. To date, the degree of breeding synchrony has not been studied in natural populations of pine voles or prairie voles, but female home ranges do not appear particularly large or widely spaced.

INTERSPECIFIC TERRITORIALITY

In general, individuals of the same species exploit the same resources, and thus territories are established and defended against conspecifics. Additionally, as we have seen, certain resources may be particularly important to one sex, and not as important to the other sex, and this can lead to sex-specific territoriality within a species.

Figure 16.11 A pair of prairie voles. Understanding the spacing patterns of these small, apparently monogamous rodents has proven to be a challenge for those who study territoriality.

Thus we see that female gray-sided voles maintain exclusive areas to gain access to food resources, but that the home ranges of males overlap with those of other males and with those of females. Sometimes, however, members of different species exploit the same resources, and this can lead to interspecific territoriality.

Consider the case of the Beau Gregory (*Eupomacentris leucostictus*), a territorial reef fish common to the shallow waters of the West Indies. Throughout the year, males and females of this pugnacious species defend their individual territories against conspecifics as well as against certain heterospecifics (individuals of other species). As it turns out, the level of aggression exhibited by *E. leucostictus* toward other species is closely tied to the amount of overlap in their diets (Ebersole, 1977). Territory holders are most aggressive to trespassing species that have diets very similar to their own. We see, then, that interspecific territoriality is indeed linked to exploitation of the same resources by different species.

Summary

Living in groups has both costs and benefits. In some cases, group members face increased competition for mates, nest sites, or food resources. Other costs associated with living socially include increased exposure to parasites or disease, increased conspicuousness to predators or prey, and a greater possibility of providing care to young that are not one's own.

Group living also has many benefits, particularly those relating to eating and not being eaten. Some predators, for example, engage in cooperative hunting and as a result have higher capture success and average energy intake per individual than do solitary hunters. Cooperative hunting may also allow predators to kill animals much larger than themselves. Because groups are usually more successful than solitary individuals at detecting, confusing, and repelling predators, there are also antipredator advantages associated with grouping. In addition, during an attack by a predator, an individual within a group has a smaller chance of becoming the next meal. Finally, living in groups may aid in defending young, food, or space against conspecifics.

Natal dispersal may be defined as "the movement an individual makes from its point of origin to the place where it reproduces or would have reproduced had it survived and found a mate" (Howard, 1960). Thus, natal dispersal involves permanent movement away from the birth site. Breeding dispersal, on the other hand, refers to movements between consecutive breeding sites. Natal philopatry occurs when offspring remain at home until death.

As with group living, there are costs and benefits associated with philopatry and natal dispersal, and in this case they are categorized as either somatic (involving survival and reproductive success) or genetic. Potential genetic costs of philopatry focus on the costs of inbreeding. Inbreeding increases homozygosity, and thus reduces variation among progeny, and increases the risk of producing offspring that are homozygous for deleterious or lethal recessive alleles. If conditions are sufficiently crowded at home, then philopatric young may face intense competition from relatives for food, nest sites, or mates. Limited access to critical resources could result in reduced survival or fertility of philopatric young relative to those offspring that disperse. Competition with relatives thus constitutes a somatic cost of philopatry.

Because a certain level of inbreeding may actually be beneficial due to the maintenance of gene complexes that are particularly well suited to local conditions, there are also genetic benefits associated with philopatry. Somatic benefits of philopatry center on the advantages of being familiar with the local physical and social setting. Such familiarity typically involves low risks and energy use and may result in increased survival and fecundity for philopatric young. Dispersers, on the other hand, are thought to face high energy costs and risks of predation as a result of increased movement and lack of familiarity with the physical and social environment.

The direction of sex differences in dispersal differs between birds and mammals. In most birds, females are more likely to disperse than are males. In most mammals, however, males are more likely to disperse than are females. At least three hypotheses have attempted to explain patterns of dispersal in birds and mammals. Greenwood's (1980) hypothesis focuses on differences in the ways in which male and female

birds and mammals compete for critical resources (in birds, mating systems are largely
resource-defense, and in mammals mate-defense). The Oedipus hypothesis of Liberg and
von Schantz (1985) focuses on the abilities of
male and female offspring to cheat their parents.
Dobson (1982) proposed that competition for
mates was responsible, in part, for patterns of
dispersal in mammals.

The home range of an animal is the area in
which it carries out most of its normal activities.
Some animals may defend all or a certain portion of their home range. The defended area, or
area maintained for their exclusive use, is called
a territory. In some instances, territories are
used solely for feeding, mating, or raising young,
while in others they serve a variety of purposes.

Some benefits associated with territoriality include a longer lasting food supply because fewer
individuals have access to it, increased access to
mates, and a greater opportunity to rear young
on a high-quality site. Territoriality can be quite
costly, however, given the energy needed to advertise ownership and to keep out and evict
intruders. Territory acquisition and defense also
take time away from other essential activities
such as foraging. From the standpoint of economics, territoriality is expected to occur only
when the benefits outweigh the costs. Factors
such as abundance, spatial distribution, and renewability are thought to influence whether a
particular resource is economically defendable.
Although it is usually individuals of the same
species that exploit the same resources and thus
territories are usually defended against conspecifics, sometimes individuals of different species
use the same resources and interspecific territoriality results.

CHAPTER
17

Maintaining Group Cohesion: Description and Evolution of Communication

*D*ogs rarely, if ever, run with wolves. One reason is because wolves often view dogs as prey, and that ends their socializing right there. Another reason is because, even if a dog were allowed into the group, it would not be too long until there was a mix-up in communication and the dog would suffer. Wolves have a range of behaviors that enable them to communicate with each other in complex ways. The signals must not be misread in such a dangerous and able species. The unfortunate dog is likely to misunderstand such signals, or send an inappropriate signal itself, resulting in its loss of standing in the wolves' social register.

Wolves use a multi-media approach to communication, conveying messages in a variety of sensory modalities. An erect tail is a dominant wolf's flag of confidence. Licking a muzzle may be a sign of subordinance or the means by which a pup begs for food. Territories are marked with scents in urine. Growls serve as threats. Howling wards off intruders. Wolves, then, say it with sights, sounds, and scents.

Communication is a diverse phenomenon and, as we will see, it is widespread in the animal kingdom. But before we become involved with the hows and whys of communication, let us focus our thoughts on the nature of communication and the means by which communication systems may have evolved.

Defining Communications

You might think that since we spend most of our day communicating with others, a definition of communication would roll glibly from our tongues. However, the more involved we become in the study of communication, the more difficult it seems to define the process.

The broadest definition of communication is the transfer of information (Batteau, 1968). Many people, however, choose to narrow this defini-

tion with one or more of the following restrictions: (1) The sender must intend to communicate, (2) the interacting individuals must be members of the same species, and (3) the receiver must respond to the message. Let's consider each of these restrictions.

First, must the sender *intend* to communicate? Does a student in the back of the room whose head periodically falls to his chest intend to communicate that he is sleeping? Is he that bold? Clearly, messages can be sent unintentionally.

Another problem lies in assessing intent. How would you determine whether a female gypsy moth intends to attract a mate by releasing her powerful sex attractant? If you decide that the chemical is released by the female for the purpose of communicating her presence to a male, are you also willing to accept that a yucca plant, which produces an odor attractive to yucca moths only at the time of day that the moths are active, is communicating with the moths? Narrowing the definition to information transfer among animals at least eliminates the problem of deciding whether communication takes place between plants and animals (Green and Marler, 1979).

This leads us to the second restriction, that communication must be between individuals of the same species (Diebold, 1968; Frings and Frings, 1964). If a sparrow detects an owl and emits a warning call that is heeded by nearby sparrows, it is clearly communicating. However, if it alerts a chickadee as well, is that communication? When foragers of some ant species find food, they leave a chemical trail that recruits and guides other workers to the site. The snake *Leptotyphlops* follows such trails to the ants' nest and devours the brood (Burghardt, 1970). Did the ant that laid the trail communicate with the snake as well as its nestmates? The snake may have been "eavesdropping" on the ant's message to other ants. However, some messages, for instance the bright markings on some prey species that warn a potential predator of unpalatability, are specifically directed at members of other species. Obviously, there may be problems with this caveat.

The third restriction stipulates that the receiver must respond. For example, one definition of behavior is "a process by which the behavior of one individual affects the behavior of others" (Altmann, 1962). Although this definition eliminates some of the problems we have discussed, it introduces others. Suppose one individual sends a message but it is ignored. If a hungry peahen continues to eat while a peacock struts his stuff (Figure 17.1), has he failed to communicate or is she responding "no"? There are other situations in which the receiver's appropriate response would be *not* to do something. Consider, for example, a male red-winged blackbird that sees another male displaying at the territory boundary and heeds the message by *not* entering the foreign turf. How could one ever determine what he *would have done* if he had not received that message? Then there are responses that are gradual. For example, the song of a male budgerigar stimulates the ovaries of the female to develop and this, in turn, eventually results in changes in the female's behavior. How long must an observer wait before deciding whether or not there has been a response?

Furthermore, perhaps not all situations in which the behavior of one animal alters that of another involve communication. For instance, is an animal that causes another to respond by pushing or shoving communicating? Many people believe that communication is not involved when brute force is used to achieve ends (Cullen, 1972; Dawkins, 1986).

We see, then, that the harder we try, the more difficult it becomes to define communication. However, a definition often helps to focus thoughts. So, recognizing that any definition will be somewhat controversial, we will say that communication occurs when a signal that evolved for the role is sent to a receiver and the recipient responds.

Reasons for Communication

The questions regarding communication have an interesting twist in that some are easily answered, but one has proven exceedingly thorny. Specifically, we can often answer the who, what, where, and when of the process, but the *why* must be inferred. Furthermore, observations are often consistent with different explanations. As a result, there has been a lively controversy regarding the evolution of animal communication systems. So let's ask then, why communicate?

SHARING INFORMATION

It has been suggested that communication results in shared information and that this sharing

Figure 17.1   A peacock courting a disinterested female. Does communication take place in the absence of a response by the receiver?

is adaptive. Communication makes private information, such as an animal's internal state or what it is likely to do next, available to others who may then respond appropriately (Cullen, 1966; Smith, 1977).

Consider how important it is for a courting male orb-web spider (Argiopidae) to properly identify himself and his intentions to a female by plucking on her web with a species-specific rhythm. The females of these species are usually larger than the males and earn their living by eating small arthropods similar to the would-be suitor (Figure 17.2). If the male fails to provide the signals by which the predatory female can distinguish his approach from that of a prey species, the male may quickly turn from suitor to snack (Bristowe, 1958). So the male signals the female and communicates both his identity and intentions.

The sharing of information may also enhance

Figure 17.2 The female *Argiope* spider is much larger than the male (upper right). If a courting male fails to identify himself, he could easily be mistaken for prey.

survival by coordinating group activities. For example, when a honeybee informs her fellow workers of the location of a rich nectar source, the food stores of the entire hive, upon which she depends, are increased.

The question immediately arises, however, what guarantees that animals will tell the truth? In the previous examples, both the sender and receiver would be expected to gain from honest signals, so we might predict honesty. However, there are some situations in which the sender might gain by lying. Consider, for instance, a female choosing from among several courting males. She would gain the most by choosing the one of highest quality. Therefore, each male would gain by advertising that he would be the best father of her young. Each might try to exaggerate these traits. Also consider two opponents at a territorial border. The one that displayed greater strength or willingness to fight might have an advantage if the opponent used the display to decide how to react. We might expect, then, that a "cheater," one that made false claims about its qualities, resources, or future behavior, would have an advantage. In a population of honest signalers, a mutant cheater would, therefore, be more successful and the cheating mutation would become more common. In other words, honesty might not be an evolutionarily stable strategy (ESS). (You may recall from Chapter 4 that an ESS is a strategy that cannot be replaced by an alternative one if it is adopted by most members of the population.) It has been argued, therefore, that displaying easily bluffed information, such as motivation or future behavior, is unlikely to be an ESS (Maynard Smith, 1972, 1974, 1982). If this argument is pushed to its conclusion, the function of communication must not be to share information.

MANIPULATING OTHERS

What, then, is an alternative explanation for communication? Richard Dawkins and John Krebs

(1978) suggest that animals communicate not to convey information, but to manipulate the behavior of others to their own advantage. This idea stems from Dawkins's (1976) presentation of an animal as a machine designed to ensure the preservation and propagation of its genes. Simply put, if a particular action by another animal would enhance the spread of one's kinds of genes, then it is beneficial to cause the companion to behave in that way. Although this could be accomplished either by brute force or by persuasion, the latter is often energetically more efficient. As we have seen, animals may be genetically programmed to respond in a standard way to a specific stimulus. The response has been selected through evolution because it enhances survival in most situations. The specific stimulus has been selected from all those available because it is usually a reliable indication of the appropriate situation for that response. In communication, the individual sending a message might be viewed as providing the stimulus to cause the receiver to perform the behavior when it is beneficial to the sender, regardless of its consequences for the receiver.

However, the receiver would be under a different set of selective pressures: to be a mind reader and predict the true future action of the signaler (Krebs and Dawkins, 1984). In this view, the selective pressures on receivers are similar to those suggested by the hypothesis that communication functions in sharing information.

TESTING THE HYPOTHESES?

We should ask, then, how might we determine whether animals are communicating to share information or to manipulate the other individual? Keep in mind that when we say "to" here, it is used simply as a shorthand way of referring to adaptiveness, the result of natural selection.

Signals and the Predictability of Future Behavior

It would seem that these two hypotheses regarding the function of communication make differ-

ent predictions about whether the signal can be used to predict the subsequent behavior of the signaler. If the animal is sharing information about its motivation or intentions, we might expect the signal to be predictive. However, if the role of the signal is to manipulate the recipient, there would be no reason to expect it to indicate future behavior. So, let us see whether the predictiveness of signals can be used to test the hypotheses.

Some studies do indicate that certain signals are predictive of future behavior. In some species a threat display is reliably followed by attack. For example, stomatopods (*Gonodactylus bredini*), commonly known as mantis shrimps, are marine crustaceans that ferociously defend the burrows and cavities in which they live. Stomatopods have two large forelimbs, called raptorial appendages, that can unfold and shoot forward in a manner similar to that of the praying mantis (Figure 17.3*a*). The raptorial appendages, which are adapted either for spearing or smashing, are used both in prey capture and territorial defense. Therefore, combatants may be seriously injured or even killed during the battles over burrows. Readiness to attack is signaled by a threat display, called a meral spread, in which the raptorial appendages are spread out, thereby exposing a depression called a meral spot, while less dangerous appendages, the maxillipeds, are extended in a circular pattern (Figure 17.3*b*). The meral spread is significantly correlated with an ensuing attack (Caldwell and Dingle 1976; Dingle and Caldwell, 1969).

Studies of other species have shown that an attack signal is only a weak indicator of future behavior (Caryl, 1979). However, since we might expect that the signaler's behavior will depend on the reaction of the receiver, should we really expect a signal to predict subsequent behavior? If, for example, an opponent signals that it will attack and the other individual retreats, no attack would be made. Can we say, then, that the signal did not predict attack (Waas, 1990)? Because of such questions, we often find that the predictive

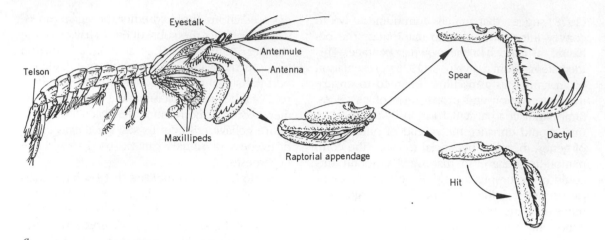

a

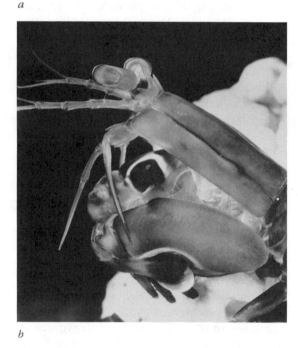

b

Figure 17.3 (*a*) The mantid shrimp, a stomatopod, has two raptorial appendages that can be used in prey capture or combat. (*b*) The threat display of a stomatopod, called a meral spread, is a good predictor of attack. Here the display is seen from the side.

value of signals may not be a good test of the hypotheses regarding why animals communicate.

Signals and Honesty

Another way to test the hypotheses might be to ask whether the signals are honest and reliable: The information sharing hypothesis predicts honesty; the manipulation hypothesis predicts dishonesty. Unfortunately, this approach does not provide a clear answer either because we can find examples of both truth and deceit in different animal signals. The courtship display of a male smooth newt (*Triturus vulgaris*), for example, is an honest signal. The amount of sperm a male newt can provide can vary greatly, depending on how long it has been since his last mating. Females would benefit by choosing a male with a large sperm offering. The courting male reveals his suitability as a mate by "fanning" his tail at a rate that is highly correlated with the amount of sperm he eventually deposits (Halliday, 1976).

On the other hand, male pied flycatchers (*Ficedula hypoleuca*) may be deceitful when attracting females. In this species, males may mate with more than one female. When they do, the females have a lower reproductive success than do monogamously mated females because they receive less help in providing food for the young. We would expect, then, that if a female could distinguish between a mated and unmated male, she would avoid one that was already mated. Males, in contrast, would increase their reproductive success by attracting a second female. It has been suggested that a female that chooses to mate with a previously mated male may do so because she is unaware that he has mated. One way a male may mislead a female about his mating status is to hold two widely separated territories. The second female may then be unaware that he is already mated until she has laid her eggs and it is too late in the season to start over with a new male (Alatalo et al., 1981; Alatalo and Lundberg, 1984; Alatalo, Lundberg, and Stahlbrandt, 1982). Females may also judge a male's mating status incorrectly because, although unmated males spend more time on their territory and sing more frequently than mated males when no female is present, their behavior is similar when a female is present, making it difficult for a female to distinguish between them (Searcy, Eriksson, and Lundberg, 1991). Thus, we see that males of some species are honest when attracting females, but those of other species are not.

Since we can find examples of both honesty and deceit in animal communication, it may be more fruitful to consider factors that might favor one or the other.

Tight Association with a Physical Attribute Although a signaler might gain if it could lie, we often find that signals closely tied to a physical or physiological condition simply cannot be faked. Since size is usually a good predictor of fighting success and an attribute that is difficult to fake, many displays allow opponents to judge one another's size. For instance, in the threat display, male stalk-eyed flies (Diopsidae) face one another head-to-head with their forelegs spread outward parallel to the eyestalks (Figure 17.4). This pose allows each competitor to compare the length of its eyestalks to the length of its rival's. Eyestalk length increases with body size, and males with shorter eyestalks usually retreat without a fight (Burkhardt and de la Motte, 1983; de la Motte and Burkhardt, 1983).

Stable Social Unit A stable social unit also favors honest communication. When some degree of cooperation is essential for an individual's survival, honesty should be favored over dishonesty (Markl, 1985; van Rhinjn and Vodegel, 1980). One reason we should expect honesty in

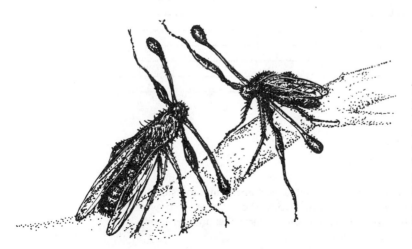

Figure 17.4 An aggressive display of male stalk-eyed flies. In this pose, each male can determine the distance between the eyes of its opponent and thereby the rival's size. Size is correlated with fighting success. Since there is little a combatant can do to alter its body size, this is an example of honest signaling.

a stable social environment is that individuals will both send and receive signals at different times. Thus, the advantages of sending dishonest signals would be reversed when the animal is the receiver. Therefore, the advantages of receiving honest signals might outweigh the advantages of sending dishonest ones, and honesty might come to predominate in the population.

Another reason that honesty should be expected in stable social units is that the members often recognize one another and remember previous interactions. Working with vervet monkeys (*Cercopithecus aethiops*) in Kenya's Amboseli National Park, Dorothy Cheney and Robert Seyfarth (1988) tested the idea that members of a social group would cease believing a known liar. Vervet monkeys utter different vocal signals, or calls, during different situations. They have two acoustically different calls that warn of the encroachment of another group of monkeys on their territory. One call, the "wrr," is given when another group is spotted in the distance, perhaps as far away as 200 meters. The other, a "chutter," is given if the other group comes so close that there are threats, chases, or actual contact (Figure 17.5). Vervet monkeys also have a stable

social group in which the animals recognize each individual and its calls. The question was, what would happen if one individual lied and falsely alarmed the group of the approach of another group? Would the other monkeys believe the liar again?

To answer this question, Cheney and Seyfarth broadcast tape recordings of the calls of a member of the group to create the illusion that it was lying. The calls were broadcast from a speaker that was concealed in a place where the group might expect that individual to be. First, an individual's "chutter" call was played to see what the baseline response of the others would be. The next day, when no other group was in sight, they played that individual's "wrr" call eight times, at approximately 20-minute intervals. They found that the other monkeys gradually stopped responding to that individual's warning that another group was in sight. The monkeys no longer believed the liar. However, if another monkey uttered the "wrr" call, the group still believed the warning and responded appropriately. So we see that in a stable social group in which individuals are recognized, others may soon learn not to believe a dishonest signaler.

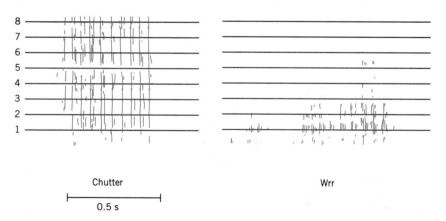

Chutter

|---------------|
0.5 s

Wrr

Figure 17.5 Vervet monkey alarm calls that alert group members of the approach of a neighboring troop are of two types. The "wrr" is given when another group is spotted in the distance. The "chutter" warns that another group is nearby. (From Cheney and Seyfarth, 1991.)

Cheating Costs Outweighing Benefits

Although it may not be intuitively apparent, honesty will also be favored in a population when an honest signal is costly to the sender (Zahavi, 1975, 1977). A signal, such as a loud call or a vigorous display, costs energy and may also incur risk. The louder the call and the more vigorous the display, the greater the risk. A signal that says "I am very strong" might cost more than a signal that says "I am strong." However, the cost can be borne by a truly strong individual more easily than by a weaker one. Thus the signaler proves its quality by its ability to withstand its costly handicap. A mathematical model has been developed that shows that if honesty were costly, it could be an ESS (Grafen, 1990).

The aggressive displays of the little blue penguin (*Eudyptula minor*) show a clear relationship between cost and honesty. Little blue penguins have a repertoire of aggressive displays that differ in cost (risk of injury), effectiveness in deterring an opponent, and ability to predict attack. Although two displays may involve the same ritualized posture, one that involves moving within the rival's striking distance is riskier to perform than one that is performed while stationary and out of the opponent's reach. By its choice of display, a penguin conveys information about both its willingness to sustain injury while performing the display and its willingness to fight. It chooses a display with costs that represent the value it places on the resource. Roughly 10 percent of penguin interactions are not settled by displays and end in fighting. During battles, penguins commonly suffer flesh wounds and sometimes eye loss (Figure 17.6). These injuries could make it more difficult to obtain sufficient food or to breed successfully. Thus, attempting to intimidate an opponent into retreat by bluffing a strong motivation to attack could be quite costly if the rival called the bluff and a fight ensued (Waas, 1991). The encounters invariably begin with low-risk displays and escalate until one opponent retreats or a fight occurs. The process is somewhat analogous to human actions at an auction—the bids begin low and gradually increase until bidders unwilling to pay the price drop out of the process. The price that little blue penguins must pay is risk of injury. As a territorial contest escalates and the price of the property increases, one "bidder" usually decides that the territory is not worth that great a risk. In the case of these penguins, the signals remain honest because they are costly.

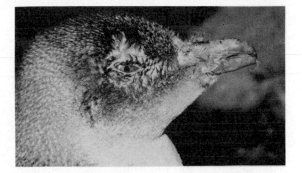

Figure 17.6 Eye and head injuries in a little blue penguin resulting from fights. Little blue penguins have an extensive repertoire of aggressive displays. Displays vary in cost (the risk of injury). By its choice of display, a penguin indicates its willingness to fight. Signals remain honest because they are costly.

It has also be argued, however, that receivers of signals pay costs as well and that this will corrupt honest signals (Dawkins and Guilford, 1991). In some signals, such as the aggressive displays of little blue penguins, the receiver pays for an honest signal by replying with an honest signal, which can be costly to the receiver. In other cases, the cost of assessing the signal may be time. In some species, females use the duration of a male's display to assess his quality. Female field crickets (*Gryllus integer*), for instance, judge a suitor's quality by the length of his song (Hedrich, 1986). The time she spends listening to male songs could be spent on other activities. Furthermore, calling often attracts parasites. By remaining close to a calling male to assess his quality, the female places herself at risk. A receiver may pay dearly for challenging the honesty of the signal. If receivers are unwilling to accept the cost of a test, honest signals may be corrupted by occasional dishonest bluffs. For example, a newly molted stomatopod is virtually defenseless with its soft body exposed. In this condition, it may dishonestly use the meral spread described earlier and deceive an opponent, causing it to retreat. It can get away with the bluff because the receiver might pay dearly if it chose to test the honesty of the signal and, in fact, the signaler was *not* newly molted (Adams and Caldwell, 1990). We often find, then, that the same characteristics of a signal that make it costly to send also make it costly to receive. As a result, the receiver may settle for a signal that does not guarantee honesty but is less costly to receive.

The Nature of Animal Communicative Signals

Animals communicate by sending a signal—a particular posture, movement, sound, or chemical—that has a specific meaning. The familiar golden arches of McDonald's is a good example of the use of an arbitrary symbol given an invariable and distinct meaning among humans. For someone familiar with the symbol, the golden arches does more than just signify the location of a restaurant. The type of restaurant and the menu are standard across the country, so a traveler knows exactly what to expect.

Stereotypy is an important aspect of signals, and we find that throughout an animal population, every member sends the same signal in a remarkably similar way (Wilson, 1980). This contrasts with human language, in which the same thought can be expressed in remarkably different words by different individuals. For example,

if you think back to recent elections, you may remember how the diverse speeches of the candidates made almost identical promises.

Finally, animal "language" is remarkably economical. Each species has a small number of signals and messages. In general, a species has between 15 and 45 signals or displays (including postures, movements, sounds, and chemicals)

Figure 17.7 A discrete display. The courtship posture of the male cutthroat finch has evolved to a typical intensity. Notice the small difference in the form of the display when it is shown at low intensity (*a*) and at high intensity (*b*). The similarity in these two forms of the display is even more apparent when the displaying males in (*a*) and (*b*) are compared to the noncourting male in the lower left in (*b*). The birds on the right of both drawings are females. (Drawn from Morris, 1957.)

that are used to convey about a dozen messages (Smith, 1969).

Increasing the Amount of Information Conveyed

Among some invertebrates and many vertebrates, the meaning of a communication signal can be varied by alterations in the manner in which it is presented or the context in which it is presented.

DISCRETE VERSUS GRADED DISPLAYS

Animal signals may be discrete or graded. Discrete displays are those that are either present or absent regardless of the strength of the stimulus, just as a telephone rings with the same loudness and at the same rate no matter how urgent the call is. A discrete display is performed at a "typical intensity" that is very similar from one time to the next. For example, a male cutthroat finch (*Amadina fasciata*) ruffles his feathers and assumes a characteristic courtship posture that is very similar from one time to the next. As you can see in Figure 17.7, there is only a minute difference between the low- and the high-intensity display. If this posture appears at all, it is nearly in its maximal form (Morris, 1957).

In contrast, graded displays are more variable—that is, they may be performed with various levels of intensity. In general, the intensity or duration of a given signal increases with the animal's motivation. The aggressive display of the rhesus monkey (*Macaca mulatta*) provides an example. At low intensities, the display is simply a hard stare. As aggression rises, the monkey rises to a standing position, still staring. At highest intensities, it stares with an open mouth while bobbing its head and slapping the ground with its hands (Figure 17.8; Altmann, 1962).

COMPOSITE SIGNALS

Sometimes two or more signals are combined to produce a new meaning. For example, a zebra mare has a threat face that is used in aggressive encounters. However, when this display is presented while the mare has her hindquarters pre-

Figure 17.8 A graded display. The intensity of the aggressive display of the rhesus monkey is graduated. At first (left), the animal threatens with a hard stare. The stare continues during higher intensity displays, but the animal stands (middle). The display may escalate to its highest form in which the animal stands and stares with an open mouth while bobbing its head and slapping the ground with its hands. (Modified from Wilson, 1972.)

sented to the stallion and her tail moved to the side, the meaning of the threat face is changed and it now indicates a readiness to mate (Trumler, 1956).

METACOMMUNICATION

In metacommunication, one signal defines or sets the meaning of those that follow. An example is the invitation to play. Many of the actions used in play are the same as those employed in adult aggressive and sexual behaviors. To ensure that the playmate will interpret the actions correctly, they are prefaced with a signal. The most familiar example of a play invitation is the play bow of puppies and other canines (Figure 17.9a). Chimpanzees, baboons, Old World monkeys, and sometimes even human children elicit

play with a relaxed open-mouthed expression called a play face (Figure 17.9b). The play bow and play face signal that the actions that follow are not aggressive or sexual but are instead performed in the spirit of "fun."

CONTEXT

The meaning of a particular signal may vary with the context in which it is given. This is the case with a chemical signal sent by the queen honeybee (*Apis* spp.). The chemical trans-9-keto-2-decenoic acid is picked up from the queen as the workers groom her and is distributed throughout the hive along with the food that is shared among workers. When attained in this manner, the chemical prevents the development of any additional queens. However, the queen

a

b

Figure 17.9   Metacommunication—an invitation to play. The play bow (*a*) used by dogs and other canines and the play face (*b*) are used by many primates to initiate play. These signals invite play and inform the playmate that the actions that follow are for fun even though they may be similar to the aggressive and/or reproductive actions of adults.

also exudes the chemical as she soars skyward on her nuptial flight. In this context, the same chemical causes males to gather around her. In other words, it serves as a queen inhibitor or as a sex attractant, depending on the context.

Ritualization—The Evolution of a Communicative Signal

How and from what did the signals used in communication evolve? Julian Huxley (1923) noted that in the course of the evolution of communicative signals, certain behaviors lost their flexibility and original function and assumed a stereotyped pattern. He applied the term ritualization to this process.

Ritualization is a fascinating concept and one that directly relates the principles of evolution to changes in behavioral patterns. We will first consider the hypotheses for the advantages of stereotypy in displays, then types of behaviors that served as the starting point for evolution of communicative signals, and finally the nature of the modifications made on those acts as they evolved into displays.

ADVANTAGES OF STEREOTYPY IN DISPLAYS

There is general agreement that displays become increasingly stereotyped as a result of ritualization. However, there is disagreement regarding why this should be so.

Reduction of Ambiguity

An increase in the stereotypy of a signal accomplishes two things: It increases the display's attention-getting properties, and it makes it distinguishable from others carrying different messages (Cullen, 1966). A clear example is courtship rituals. Because a mating between members of different species generally produces nonviable or sterile offspring, the reproductive effort is not rewarded with the perpetuation of the

parents' genes in future generations. Therefore, a female must be careful to choose a mate of her own species. This discrimination is easier if the male provides unambiguous signals identifying his species. If the courtship display of a male varied from one performance to the next or was executed differently by each male in the population, a female might confuse it with another display and not be attracted. Worse yet, since differences in the courtship displays of species living close to one another usually prevent the costly error of mating with the wrong species, flexibility in the displays of the suitors of these species might make all males look alike to a female. As a result, she might squander her genes on a male of an incorrect species. A stereotyped display reduces the possibility of confusion because, as Desmond Morris (1957) explained, "a signal that is constant in form cannot be mistaken." Some people believe that, although information may be lost as the signal becomes stereotyped, the increase in the clarity of the signal may compensate for this.

Concealment of Conflicting Tendencies

Other people believe that the loss of information that accompanies stereotypy might be advantageous instead of detrimental (Krebs and Davies, 1978). This suggestion is consistent with the hypothesis that communication occurs to manipulate the actions of others. Following this reasoning, it is often better for the sender to minimize information about its internal state. The value of hiding conflicting tendencies to fight or flee is easily seen in threat signals. Because a competitor could easily take advantage of any hesitation or indecision, it would be better to bluff one's way through an aggressive encounter rather than revealing conflicting tendencies.

Comparison of Performances For Assessment of Qualities

Stereotypy may have been selected to allow the receiver to accurately assess some quality in the sender (Zahavi, 1980). The receiver uses displays as reliable indicators of some quality such as strength, size, aggressiveness, or parental ability. If all the senders are performing the same display, then comparisons among them are easier. We are familiar with this procedure from sports such as figure skating. Each skater is asked to trace certain patterns in the ice and the exactitude of the performance is an indication of skill. Because each skater must perform the same task, it is possible to decide who is the best. Similarly, a female animal may be better able to judge the qualities of her potential mate or a male may be better able to determine the qualities of his competitor if the display carrying a certain message is stereotyped so that all males must perform the same actions.

THE RAW MATERIAL FOR EVOLUTION

Even Michaelangelo's magnificent *David* began as a block of marble and the sculptor had to start somewhere. How did the sometimes elaborate communicative displays sculptured by natural selection begin? Three sources of raw material are traditionally recognized: intention movements, displacement activities, and autonomic responses.

Intention Movements

Animals may begin functional behavior patterns with some characteristic action. This behavior may be the incomplete initial motions of an entire pattern or may represent the preparatory phase of the action. By themselves, these actions have little adaptive value. Because it is often possible to judge from these activities just what the animal intends to do, these have been named intention movements (Heinroth, 1910). Through evolution, intention movements may be formalized into a display. For example, in species such as the ring-billed gull (*Larus delawarensis*) that bite during an attack, the threat display often emphasizes a widely gaping mouth. However, although jabbing is a movement associated with this gull's attack, in the threat display, the bird does not attempt to make contact with its opponent. The gulls' threat display may have originated with the intention movement of opening

Figure 17.10 "Sky pointing" by a blue-footed booby. This display is part of courtship. It probably evolved from the intention movements of flight. (Photo by John D. Palmer.)

the mouth, an obvious prerequisite for biting. Initially, an opponent witnessing this intention movement may have anticipated the attack in time to flee. Because injury was avoided, natural selection may have favored this reaction to the open mouth. Then selection could act on the aggressor's signaling and the gaping mouth became a stereotyped display.

The threat display of the ring-billed gull also illustrates another common process in the ritualization of an action: the development of anatomical features that emphasize the signal. In this case, the widely open mouth reveals a vermilion interior (Moynihan, 1958).

Numerous avian displays originated with intention movements for flight or walking (Daanje, 1950). A bird about to take flight goes through a sequence of preparatory motions. It will usually begin by crouching, pointing its beak upward, raising its tail, and spreading its wings just slightly. One or more components of the takeoff leap have been ritualized into communicative signals in different species. As shown in Figure 17.10, the part of the courtship display of the blue-footed booby (*Sula nebouxii*) aptly named "sky pointing" is a ritualized version of flight intention movements. Notice that the wings are spread and the the tip of the beak and the tail point upward. Although this display had its origins in flight intention movements, it is difficult to see how a bird could ever take off from this ritualized pose. Obviously, the movement has changed during the ritualization process. In some birds, such as bitterns (Daanje, 1950) and green herons (Meyerriecks, 1960), the movements associated with walking or hopping have been ritualized into a protective display. The intention movements that precede a jump include a bow followed by stretching so that the head and anterior end of the body are elevated while the hind end is lowered. The second phase of these intention movements, stretching, is the part that has been ritualized in bitterns. During its protective display, the bittern sits motionless while lifting its head and stretching its long neck. This posture helps the bird blend in

Figure 17.11 The cryptic posture of a bittern. This display, which is believed to have originated as intention movements for walking, illustrates how the rate of performance of an action may be altered during ritualization. In this case, the action has become frozen into a posture.

with the reeds of its habitat, thereby making it difficult for a predator to spot it (Figure 17.11).

Displacement Activities

Displacement activities are actions performed by animals in conflict situations that are irrelevant to either of the conflicting tendencies. When an animal cannot "decide" which of two opposing actions to perform, for instance to fight or to flee, it may perform displacement activities such as preening, nesting, or eating. Similar to intention movements, displacement activities are often incomplete actions.

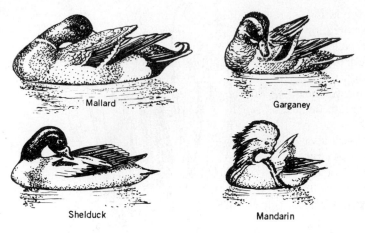

Mallard

Garganey

Shelduck

Mandarin

Figure 17.12 Mock preening by courting male ducks. This display is thought
to have evolved from displacement
preening. The movements emphasize
the bright markings on the wings.
(Modified from Tinbergen, 1951.)

Courtship is a time of conflicting tendencies. Sexual partners must come together to mate in spite of the aggressive tendencies that tend to keep them apart. Since there are conflicting tendencies to approach and fight during courtship, displacement activities would be expected. It has been suggested that the mock preening of the courtship displays of males of many duck species (Figure 17.12), including the familiar mallard (*Anas platyrhynchos*), had their origins in displacement preening (Lorenz, 1972). In a similar manner, displacement nesting behavior may have evolved into the portion of the courtship display of the great crested grebe (*Podiceps cristatus*) that was named the "penguin dance" by Julian Huxley (1914). During this display the partners first shake their heads in a stereotyped motion and then dive for weeds as they would when building a nest. But the weeds are not used to set up housekeeping. Instead, the mates present them to one another (Figure 17.13) in a ritualized behavior.

Autonomic Responses

The autonomic nervous system regulates many of the automatic body functions—digestion, circulatory activities such as heart rate and diameter of the blood vessels, certain hormone levels, and thermoregulation to name a few. It is thought that many displays originated with autonomic functions. These may be divided into four categories (Morris, 1956).

Urination and Defecation   If you are a dog owner, you know about the elaborate territorial urination of dogs. A walk with a male dog is always slow paced; there is not a tree on the block that he fails to mark. What is the evolutionary origin of this ceremony? Desmond Morris (1956) maintains that it has its roots in the tendency of animals to lose control of their bladder and bowel functions at times of stress. If you have raised a puppy, you probably had to clean up puddles whenever it was excited. A similar response, elicited by the threat of a territorial intruder, may have been the origin of ritualized territorial marking of territories with urine.

Vasodilation   The dilation of specific blood vessels may cause certain structures to become enlarged, or if the vessels are close to the surface, dilation may increase the coloration of particular areas. A change in the distribution of blood is a common action of the autonomic nervous system at times of stress or conflict. The blushing of embarrassed humans provides a familiar example. Similar circulatory changes, called a sex flush, occur during sexual excite-

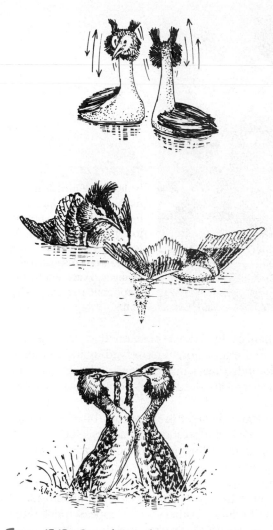

Figure 17.13 Courtship in the great crested grebe. This part of the courtship ceremony, the "penguin dance," evolved from displacement nest building. The mates dive for weeds to present to one another.

ment. On the other hand, we may turn "white as a ghost" when frightened. It has been suggested that similar circulatory responses have become ritualized into communicative signals in some species. In birds such as the turkey, jungle fowl, and bateleur eagle, the naked head or neck skin may flush and fleshy appendages may swell due to vasodilation. These changes now have

signal value during courtship and aggressive encounters.

Respiratory Changes   Perhaps the best examples of animal signals that developed from respiratory changes are the inflation displays of birds during which the males fill pouches on their body with air to attract mates. For instance, the male frigate bird (*Fregata minor*) has a pouch on its throat that is inconspicuous when deflated, but when inflated, its enormous size and brilliant scarlet color attract passing females (Figure 17.14).

The origin of bubble nest construction in the Siamese fighting fish (*Betta splendens*) may also be from respiratory movements (Figure 17.15). A male will gulp a mouthful of air at the surface of the water and blow saliva-coated bubbles. These bubbles form a cluster at the surface of the water into which the fertilized eggs are placed. This provides a well-oxygenated nest for the developing embryos. It has been suggested that this behavior evolved from the fish's habit of taking mouthfuls of air at the surface when the oxygen content of the water was low. When the fish gulps air, a few air bubbles generally escape from its mouth and float at the surface (Braddock and Braddock, 1959).

Thermoregulatory Responses   In birds and mammals the erection of feathers or hair helps to adjust body temperature. When a bird fluffs its feathers, it raises each feather just to the point where the tip falls back on the next most posterior feather. This increases the thickness of the insulating layer. When feathers are raised even further, they are described as ruffled. These feathers no longer overlap and entrap air, allowing the bird to cool itself rapidly. Alternatively, the bird can hold its feathers flat against the body surface, giving it a sleeked appearance. This is the typical position of the feathers in an active bird since it reduces resistance to air during flight while still allowing heat to dissipate because the insulating layer of feathers is small. It has been suggested that the thermoregulatory

Figure 17.14 A courting male frigate bird. The male inflates his huge brilliant red throat pouch to capture the interest of females. This display originated from respiratory changes controlled by the autonomic nervous system.

Figure 17.15 A male Siamese fighting fish building a bubble nest, an action that is thought to have originated with respiratory movements.

changes in feather position that accompanied social interactions first served as indications of the individual's internal state and then gradually became ritualized into elaborate signals (Morris, 1956).

Many displays in numerous bird species are based on the position of feathers. Sleeking, fluffing, and ruffling are common in aggressive and appeasement displays (Smith, 1977). For example, a zebra finch (*Poephila guttata*) that is prevented from escaping from a dominant individual fluffs its feathers as an appeasement signal (Morris, 1954). Feather position is also important in courtship displays such as the one of the male bicheno finch (*Poephila bichenovi*). The fluffing of his feathers during courtship accentuates his species-specific black and white markings (Morris, 1956).

Sometimes only a specialized region of feathers is erected. Perhaps the most familiar example of this is the courtship display of the peacock, or the turkey, in which the exquisite tail feathers are erected in a beautiful fan shape. Other species have specialized feathers on other parts of the body that serve as a courtship signal, when erected. These feathers may form a crest as in the sulfur-crested cockatoo (*Kakatoe galerita*), an ear tuft as in several species of owls, a chin growth as in the capercaillie (*Tetrao urogallus*), a throat plume as in some herons, or eye tufts such as the shiny green patch between the eyes of the bird of paradise (*Lophorina superba*).

In mammals, the erection of hair, which creates an insulating layer just as the erection of feathers does in birds, may also be ritualized into communicative displays. Who can mistake the meaning of the hair-on-end posture of a frightened cat? In some species, such as the rufous-naped tamarin (*Saguinus geoffroyi*), a squirrel-sized South American monkey, the meaning of the message varies with the part of the body on which hair is erected. When all the hair is erect, the individual is likely to attack or behave indecisively. However, when only the tail hair is erected, it will probably flee (Moynihan, 1970).

Sweating is another means by which the autonomic nervous system of many mammals regulates body temperature. Deodorant ads remind us that perspiration is increased during stress. This response is so common that we often describe our condition when we are apprehensive, such as waiting for exam scores to be posted, as "sweating it out." This autonomic response is thought to have provided the raw material for many animal signals. It has been suggested that the scent glands of mammals may have evolved from sweat glands and that the evolutionary origin of territorial marking may have been the sweating that occurs in frustrating situations (Morris, 1956).

Although these three classes of behaviors have traditionally been considered as sources of displays, E. O. Wilson (1980) points out that "ritualization is a pervasive, highly opportunistic evolutionary process that can be launched from almost any convenient behavior pattern, anatomical structure or physiological change." For example, the patterns involved in prey catching have been ritualized in the male gray heron (*Ardea cinerea*). During courtship, he erects his crest and certain other body feathers and points his head downward as if to strike at an object and snaps his mandibles closed, movements similar to those used during fishing (Verwey, 1930). Food exchange has also been ritualized. Billing, the touching of bills, is derived from the parental feeding of young and has taken on a variety of symbolic meanings in different species of birds. It is common in courtship and appeasement displays in which it functions to establish or maintain bonds. Mated pairs of masked lovebirds (*Agapornis personata*) bill to reassure one another during greetings and after a spat (Figure 17.16). The appeasement display of the Canada jay (*Perisoreus canadensis*) also includes billing.

Figure 17.16 Lovebirds billing. This display is derived from parental feeding behavior and now signals nonaggression during greetings and after conflicts.

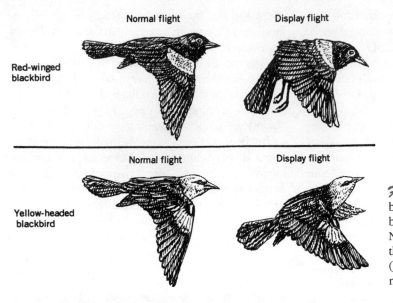

Normal flight   Display flight

Red-winged
blackbird

Normal flight   Display flight

Yellow-headed
blackbird

Figure 17.17   Ritualized flight in blackbirds. Almost any behavior can be a starting point for ritualization. Notice the exaggeration of motion that developed during ritualization. (Modified from Orians and Christman, 1968.)

The subordinate bills while squatting and quivering its wings, the posture of a young bird begging for food (Wickler, 1972). Flight is another common behavior that has been ritualized for communication. The yellow-headed blackbird (*Xanthocephalus xanthocephalus*) and the red-winged blackbird (*Agelaius phoeniceus*) perform an aerial display involving ritualized flight patterns to entice females. As you can see in Figure 17.17, ritualized flight is more conspicuous because certain movements are accentuated and because the movements reveal special plumage patterns (Orians and Christman, 1968). Portions of certain fiddler crab courtship patterns may have developed from the movement of the male entering his burrow. It is speculated that this may be the origin of the courtship display of a male *Uca beebei*. When a female approaches him and his burrow, he raises his body and flexed cheliped to an almost vertical position, an exaggeration of his movements into his burrow during the final stages of courtship. This exposes his dark underside, which probably functions to guide a female to his burrow (Christy and Salmon, 1991).

THE RITUALIZATION PROCESS

In ritualized patterns, the communicative signal can be distinguished from the behaviors from which it evolved. If one is reasonably certain of the action from which a social signal evolved, a comparison of the initial and present versions of the behavior elucidates the changes that the ancestral action has undergone. The specific changes involved in this behavioral metamorphosis have been categorized in a variety of ways. The scheme presented here is basically that of Niko Tinbergen (1952), elaborated with details and examples from a variety of other workers (Blest, 1961; Daanje, 1950; Eibl-Eibesfeldt, 1975). Keep in mind that the categories may overlap.

Formalization of the Movements

Formalization often results in the exaggeration of all or portions of the original pattern in one or more of five basic ways.

Change in Intensity   The intensity of a behavior refers to features such as the duration of the action or the extent of movement during the performance of a component of the display. As a result, one part of the action is emphasized.

Change in the Rate of Performance

Sometimes a change in the rate at which an action is performed makes it more noticeable. You have probably seen how this works in the theater. Pantomimists often slow down an action so that its meaning is more easily perceived by the audience. On the other hand, a film can be shown at a faster than normal speed to convey increased excitement or confusion. The same basic techniques are apparently also effective in the animal world since changes in the rate of performance are common as a behavior becomes ritualized to function in communication. For example, the rhythmic protective displays of some Saturnid moths involve the same wing movements used in flight but the wings beat at a slower frequency (Blest, 1957*a*).

An extreme degree of the change in rate of performance is the freezing of activity into a sustained posture. In this context, recall the protective posture of the bittern (Daanje, 1950). The display originated with a series of movements that preceded walking or hopping, but the movements have been ritualized into a display in which the bird sits motionless with its bill directed upward. Saturnid moths also have displays in which they remain stationary and expose their eyespots to frighten predators. It has been suggested that all the wing displays of the Saturnid moths evolved from flight movements; these stationary postures are furthest from the original pattern (Blest, 1957*b*).

Development of Rhythmic Repetition

Some displays may have originated as a discrete act but evolved into a repeated performance. The rhythmic waving of the large claw of a male fiddler crab, used to entice females into his burrow, probably began as a single thrusting of the claw in aggressive encounters. Likewise, the rhythmic bioluminescent flashing of male fire-

flies that now communicates his sex, reproductive intention, and species may have originated as a single flash serving to illuminate an area in the search for a suitable spot to land. Males may have occasionally flashed as they alighted near females and this evolved into a display (Lloyd, 1968). When woodpeckers are searching for food, the sounds produced by their beaks beating against a tree vary greatly; however, during their courtship drumming they have a unique species-specific pattern.

Change in Components of Original Behavior Pattern

Portions of the original behavior pattern may be deleted, combined with new actions, or performed in a new sequence. For instance, ritualized beak wiping in some bird species consists of little more than the bow that normally precedes the actual cleaning of the beak (Morris, 1957). The full forward display of the green heron, during which the bird crouches forward and flips its tail back and forth, is derived from flight intention movements. During actual preflight motions, the tail is never lowered while the bird is crouched (Meyerriecks, 1960).

Change in Orientation

Sometimes the display action is not directed toward the stimulus that initiates it as it was before the behavior became ritualized. As an example consider the inciting behavior of certain species of ducks in which a female urges her mate to act aggressively toward an intruder. In the behavior's original form, a female that felt threatened would run to her drake for protection. She would stop in front of him and look back over her shoulder at the approaching threat. The male would protect the territory and the female by attacking the intruder. Now, in ritualized inciting she still runs to her drake and stares over her shoulder but the angle at which she holds her neck is always the same. Sometimes this will result in her looking at the interloper but at other times she is

568 CHAPTER 17/MAINTAINING GROUP COHESION: DESCRIPTION AND EVOLUTION OF COMMUNICATION

looking at something totally irrelevant to the situation.

Emancipation

During ritualization, behaviors become freed from the internal and external factors that originally mediated them. As a result, the ritualized action may appear in a different context than the ancestral form. In other words, it may no longer be elicited by the stimuli that were previously effective. Furthermore, fluctuations in the tendency to execute the ancestral form do not influence the likelihood of performing the ritualized version. For example, a drake need not be thirsty to engage in display drinking during courtship.

Development of Conspicuous Structures

Sometimes the display behavior is accentuated by anatomical specializations such as brightly colored areas on the body, enlarged claws, antlers, manes, sailfins, and tumescent bodies.

Channels for Communication

Communication can involve any of a variety of sensory channels—vision, audition, chemical, touch, and electric fields. Each channel has both advantages and limitations (Table 17.1). As we will see, the channel employed for a particular signal will depend on the biology and habitat of the species, as well as the function of the signal.

VISION

There are two obvious properties of visual signals. The first is ease of localization. If the signal can be seen, the location of the sender is known. For example, when a male is displaying to attract a mate, there is never any doubt as to his precise location. The receiver can see him and, therefore, respond to him in terms of his exact location as well as his general presence. The second property is rapid transmission and fade-out time.

The message is sent literally at the speed of light and as soon as the sender stops displaying, the signal is gone. For instance, if a displaying bird suddenly spots a hawk and flees, its position will not be revealed by any lingering images.

In addition, visual systems provide a potential for a rich variety of signals. The diversity is possible because of the number of stimulus variables available that can be perceived by most animals. These include brightness and color as well as spatial and temporal pattern (altered by the animal's movements and posturing).

Visual signals have their disadvantages, of course. The most obvious is, quite simply, that if the sender cannot be seen, its signals are useless; and vision is easily blocked by all sorts of obstructions from mountains to fog. Also, visual signals are useless at night or in dark places (including the depth of the sea), except for light-producing species. Furthermore, visual powers weaken with distance, so visual signals are not usually employed for long-distance communication. As distance increases, visual signals generally become simpler and bolder and necessarily carry less information. A male iguanid lizard (*Anolis auratus*) uses head bobs, up and down movements of the head, in two displays. One is a "challenge" display used in territorial defense. The defender gets within a few centimeters of the intruder, turns sideways, and bobs its head. At such close range, the display is easy to see. The "assertion" display, given from conspicuous places within the territory, is thought to attract females. The intended receiver may be several meters away and unattentive. Although this display also consists of head bobs, unlike the challenge display, its beginning movements are at high acceleration and velocity, serving to get the attention of the intended viewer. The movements of the assertion display are also more exaggerated than those of the challenge display, which increases the distance over which the display can be detected (Fleishman, 1988).

Some visual signals are badges, aspects of an animal's appearance that have been modified to convey information. Some badges, such as the

Table 17.1

CHARACTERISTICS OF DIFFERENT SENSORY CHANNELS FOR COMMUNICATION

FEATURE	TYPE OF SIGNAL				
	VISUAL	AUDITORY	CHEMICAL	TACTILE	ELECTRIC
Effective distance	Medium	Long	Long	Short	Short
Localization	High	Medium	Variable	High	High
Ability to go around obstacles	Poor	Good	Good	Good	Good
Rapid exchange	Fast	Fast	Slow	Fast	Fast
Complexity	High	High	Low	Medium	Low
Durability	Variable	Low	High	Low	Low

white dot on the tip of the wings of a male black-winged damselfly (Coenagrionidae; Figure 17.18), are permanent. Others are inconspicuous when the animal is at rest but can be flashed to convey information. An example of this type of badge is the dewlap (a loose fold of skin hanging from the throat) of a male anole, a type of lizard (Figure 17.19). Most of the time the dewlap is out of sight, but it is extended like a broad shield when the male is attracting a female or defending a territory. Other badges, such as the color of an animal, may change with physiological state or as a form of display behavior. A male scarlet tanager, for example, is only scarlet during the breeding season. Squid and octopus, as well as many species of fish, change color as a display.

AUDITION

Sound signals have a number of advantages. They can be transmitted over long distances, especially in water. Although sound is transmitted at a slower speed than light, it is still a rapid means of transmitting a message. The transient nature of sound makes possible rapid exchange

Figure 17.18 The white dot on the tip of a male black-winged damselfly's wings is a permanent visual badge, an aspect of anatomy that serves as a communicative signal. In this case, the badge identifies sex. The female lacks this mark. (Photo by Stephen Goodenough.)

Figure 17.19 An anole with his dewlap extended. The dewlap is an example of a visual badge, an aspect of anatomy that serves as a communicative signal. This badge can be hidden most of the time and extended to court a female or defend a territory.

and immediate modification, but it does not permit the signal to linger. After the message has been sent, the signal disappears without a trace. Sound signals have an additional advantage of being able to convey a message when there is limited visibility, such as at night, in water, or in dense vegetation.

Every music lover is aware of the tremendous variety of sounds that can be produced by the temporal variation of just two parameters of sound: frequency (pitch) and amplitude (loudness). Depending on the species, animals may vary either or both of these aspects of sound, or they may, like a drummer, simply alter the pattern of presentation. Thus, some species, such as birds, sing the melody and others, such as crickets, provide the beat.

The type of sounds used by a particular species in communication will be determined by how the animal produces them. Sounds may be generated by respiratory structures, beating on objects in the environment, or rubbing appendages together. Many structures specialized for sound production have evolved in association with respiratory structures. Mammals have a larynx and birds a syrinx. You are probably aware that frogs and toads use air sacs or chambers as resonators, but do you know that similar structures are used by some birds such as frigate birds (*Fregata*), swans (*Cygnus*), and various grouse and even by primates such as howler monkeys (*Alouatta*) and orangutans (*Pongo pygmaeus*)? The environment may also be used to produce auditory signals. Humans often tap their toes when listening to music, but for some animals foot stamping itself is the signal. Rabbits, birds, and some hoofed mammals stamp their feet on the ground as a means of communication. This musical theme has several variations—beavers (*Castor*) slap the water with their tails and woodpeckers drum on trees. Arthropods most commonly produce their auditory signals by rubbing together various parts of their exoskeleton. Crickets, for example, sing by opening and closing their wings. Each wing has a thickened edge, called a scraper, that rubs against a row of ridges, the file on the underside of the other wing cover. Whenever the scraper is moved across the file, a pulse of sound is produced. This method of generating sounds is called stridulation.

CHEMICAL

The chemical senses, smell and taste, are a third channel for communication. Information may be carried by chemicals over long distances, especially when assisted by currents of air or water. The rate of transmission and fadeout time are slower than for visual or auditory signals. Depending on the function of the signal, this may be an advantage rather than a drawback. Sometimes, for example in the delineation of territory boundaries, a durable signal is more efficient because it remains after the signaler has gone. Furthermore, chemical signals can be used in situations in which visibility is limited. The ease with which the sender of a chemical signal can be located varies with the chemical emitted, but it is usually more difficult to locate a signaler using chemicals than one using visual or auditory signals.

Chemical signals can be varied to serve different functions by the evolutionary choice of the chemical used. We might expect, for instance, that chemical signals used to attract mates would indicate the sender's species so that only suitable mates would respond. Larger molecules offer greater possibilities for diversity. The more atoms in a molecule, the more ways those constituents can be arranged. By combining atoms in different ways, distinctive molecules can serve as sex attractants identifying different species.

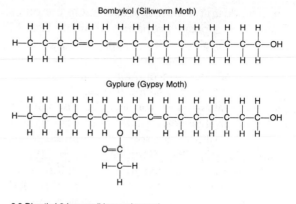

Bombykol (Silkworm Moth)

Gyplure (Gypsy Moth)

2,2-Dimethyl-3-isopropylidenecyclopropyl propionate (American Cockroach)

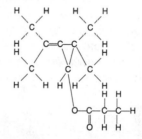

Figure 17.20 Sex attractants of three different insects. Notice that each molecule has between 10 and 17 carbons. The size of the molecule has been adjusted through evolution so that it is large enough to allow for species specificity but not so large that it is difficult to synthesize or that its volatility is severely reduced. (Modified from Wilson, 1963.)

We do indeed find that sex attractants usually have 10 to 17 carbon molecules and molecular weights ranging from 180 to 300 (Figure 17.20), a size range that permits species specificity. An added benefit of increased molecular size in sex attractants is that higher molecular weight molecules are more effective in attracting mates, at least for insects. However, limits are placed on the size of the chemicals used as signals. Larger molecules are energetically expensive to make and store, and they are only effective over shorter distances. Territorial markers also tend to be high molecular weight compounds. These less volatile chemicals persist longer, a characteristic desirable in this situation. On the other hand, the opposite properties would be favored for a chemical signaling alarm. The high volatility conferred by low molecular weight would create a steeper concentration gradient around the caller, thereby making it easier for recruits to locate the signaler. Furthermore, fast fade-out time is advantageous for an alarm call so that the signal disappears soon after the danger is gone (Wilson, 1965; Wilson and Bossert, 1963).

Chemicals produced to convey information to other members of the same species are called pheromones. Some of these, releaser pheromones, have an immediate effect on the recipient's behavior. A good example of a releaser pheromone is a sex attractant, a chemical that attracts suitable mates and guides their approach. The most famous sex attractant is probably that of the female silk moth, *Bombyx mori*.

She emits a minuscule amount, only about 0.01 micrograms, of her powerful sex attractant, bombykol, from a small sac at the tip of her abdomen (Figure 17.21). This pheromone is carried by the wind to any males in the vicinity. As few as 200 molecules of bombykol reacting with the receptor hairs on the male's antennae have an immediate effect on the male's behavior—he turns and flies upwind in search of the emitting female (Schneider, 1974). Other examples of releaser pheromones are trail pheromones, which direct the foraging efforts of others, and alarm substances, which warn others of danger.

Primer pheromones exert their effect more slowly, by altering the physiology and subsequent behavior of the recipient. In insect societies, queens control the reproductive activities of nestmates largely through primer pheromones. For example, a queen honeybee (*Apis*

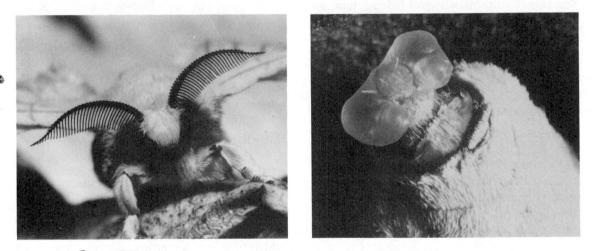

Figure 17.21   The antennae of a male silk moth (left) and the tip of the abdomen of a female silk moth (right). Each antennae is about a quarter of an inch long. The feathery appearance is due to numerous branches, each containing many odor-receptor hairs. Half of the odor receptors on the male's antennae are tuned to the female sex attractant, bombykol. This is why the male is sensitive to such minute quantities of the sex attractant. The tip of the female silk moth's abdomen has a pair of glands, shown in an expanded active state here. Within these glands is a small quantity of the sex attractant. This sex attractant is an example of a releaser pheromone because it causes an immediate effect on the behavior of the male. When he detects the pheromone, he turns and flies upwind in search of the female.

mellifera L.) produces several compounds from her mandibular gland that ensure that she will remain the only reproductive individual in the colony (Winston and Slessor, 1992). This pheromone coats the queen's entire body surface but is most concentrated on her head and feet. Most of the pheromone spreads through the colony by the activities of the workers attending the queen, but some is spread through the wax of the comb (Naumann et al., 1991). It prevents the development of rival queens within a hive by preventing the workers from feeding larvae the special diet that would cause them to develop into queens and by inhibiting the development of the workers' ovaries so that they are incapable of producing eggs. When the queen dies, the inhibiting substance is no longer produced, and new queens can develop (Wilson, 1968). In ants, as in honeybees, the queens inhibit reproductive processes of workers. It has long been suspected that the queen's control is exerted through primer pheromones and evidence is now emerging to support this suspicion (Vargo and Passera, 1991).

Vertebrates also produce primer pheromones that influence reproductive activity in various ways. These primer pheromones may help regulate reproductive activities so that they occur in the appropriate social or physical setting. In species in which there is cooperative care of the young, it may be advantageous for females to give birth synchronously. In others, in which dominant individuals control access to the resources needed to raise offspring, a subordinate female may raise more young if she gives birth at a time when she may have the least competition for the needed resources (McClintock, 1983).

In many mammalian species, primer pheromones from other females and males are important in regulating such events as the onset of puberty, the time of ovulation, or the time of estrus. Mice (*Mus musculus*), for example, produce an estrus inducer, an estrus inhibitor, and an adrenocortical activator. Odors produced by

female mice produce disturbances in the reproductive cycle if several females are housed together. Females living in groups of four are more likely than a female caged alone to enter a hormonal condition that mimics pregnancy (Lee and Boot, 1956). When even more females are crowded together, their estrous cycles become irregular and sometimes even cease. These effects are eliminated if the olfactory bulbs of the mice are removed, preventing them from detecting the odors of other females. The odor of a male's urine is also effective in overriding the effects of all-female grouping. For example, the odor of a male mouse can initiate and synchronize the estrous cycles and cause females to ovulate within a few days of smelling it (Whitten, Bronson, and Greenstein, 1968). The effect of the odor of an unfamiliar male on the pregnancy of a newly impregnated female mouse is even more dramatic: It blocks the pregnancy.* The embryo fails to implant in the uterus and the female returns to estrus. The molecule in the male's urine that causes these effects is species-specific and depends on the presence of androgens (male sex hormones). Castration removes the molecule from the urine, but testosterone replacement returns it to castrates (Bronson, 1971, 1974; Bruce, 1959; Parkes and Bruce, 1961).

TOUCH

Many kinds of animals communicate by physically touching. These tactile signals are obviously only effective over short distances and are not effective around barriers. Tactile messages can be sent quickly and it is easy to locate the sender, even in the dark. The message may be varied by how and where the recipient is touched. Sometimes the message is simple, perhaps an expression of dominance. For example, a dominant

---

* The pregnancy block caused by an unfamiliar male, called the Bruce effect, was also discussed in Chapter 14.

dog may place its leg on the back of a subordinate. However, complex messages may also be communicated by touch. Honeybee scouts inform nestmates of the location of a food source by dancing. The recruits cannot see the choreography because the hive is so dark, but they follow the dancers' movements with their antennae.*

Grooming among certain primates, such as rhesus monkeys (*M. mulatta*), is a special form of tactile communication. Assuming that self-grooming has skin care as its primary function, Maria Boccia compared several aspects of social grooming and self-grooming: body site preference, duration of grooming (both overall and to specific areas of the body), and the method of grooming (stroking and/or picking). She reasoned that if the primary function of both social grooming and self-grooming were hygiene, then these physical aspects of grooming would be the same in both. However, social grooming was found to be different from self-grooming in each of these respects. Therefore, she concluded that skin care is not the most important factor molding the form of social grooming (Boccia, 1983). Furthermore, she showed that the message of the tactile signal varies depending on the body site being groomed (Boccia, 1986). In other words, the monkey's response depended on which part of its body was groomed. The animal being groomed was likely to move away from the groomer when the posterior part of the body was groomed. The responses to the five body regions typically groomed may reflect a continuum from a tendency to maintain an affiliation at one extreme to a tendency to terminate the interaction at the other extreme.

SUBSTRATE VIBRATIONS

Some signals are encoded in the pattern of vibrations of the environmental substrate, such as the ground or the water surface. These are called seismic signals. For example, a female white-lipped frog (*Leptodactylus albilabris*) of the dense Puerto Rican rain forest literally "feels the earth move" when she is courted, since the male sits on the ground or some other solid substrate and thumps his courtship message, causing the substrate to reverberate beneath his foot (Lewis and Narins, 1985). Seismic signals are also used by Cape mole-rats (*Georychus capensis*), animals that usually live alone in underground burrows. A male attracts a mate from neighboring burrows by drumming with his hindlegs, generating a specific pattern of vibration (Narins et al., 1992).

Seismic signals travel farther than auditory ones. They are, therefore, well suited for long-distance communication, particularly in areas in which vision is obstructed (Narins et al., 1992). (In watching old Westerns, you may have seen the hero place an ear to the ground to determine whether help was on its way by feeling the vibrations caused by distant running horses.) In the examples of the white-lipped frog and the Cape mole-rat, you can see that both live in places in which vision would be impeded.

Tapping on the surface of water creates seismic signals of a slightly different sort—ripples, or surface waves (Figure 17.22). For example, a female water strider (*Rhagadotarsus*) must receive "good vibes" from a courting male before she agrees to mate. Using his legs, the male taps out a patterned sequence of ripples in the water surface. If a female approaches a calling male, he switches to a courtship vibratory serenade that stimulates her to copulate and release her eggs. A male may defend his signaling site by producing other patterns of surface vibrations that serve as aggressive signals. If an intruder does not retreat when suitably warned, the defender will attack him (Wilcox, 1972).

Rhagadotarsus is not the only type of water strider to tap out messages on the water surface. Ripple signals also play a role in mating and territorial defense of other species of water strid-

---

* The bee dances will be discussed in more detail in the section of the next chapter dealing with recruitment.

Figure 17.22 Water striders communicate by tapping out messages on the surface of the water, thereby creating surface waves. The pattern of the wave determines the message. Here, a male taps out a repel signal.

ers, including *Limnoporus* spp. (Vepsalainen and Nummelin, 1985; Wilcox and Spence, 1986) and *Gerris* spp. (Hayashi, 1985; Wilcox, 1979; Wilcox and Ruckdeschel, 1982).

ELECTRIC FIELDS

Two distantly related groups of tropical freshwater fish produce electric signals used in both orientation (electrolocation; see Chapter 11) and communication. These groups are the knife (gymnotid) fish of South America and the elephant-nose (mormyrid) fish of Africa.

The electric signals are generated by electric organs that are derived from muscle in most species, but in one family of gymnotids they are derived from nerve (Hopkins, 1977, 1988). When a normal muscle cell contracts, or when a nerve cell generates an impulse, a weak electric current is generated. The modified muscle (or nerve) cells in an electric organ also generate a weak electric current and because they are arranged in stacks, their currents are added, resulting in a stronger current. When an electric organ discharges, the tail end of the fish, where

the electric organ is located, becomes momentarily negative with respect to the head. Thus, an electric field is created around the fish (Figure 17.23a).

This electric field is the basis of the signal. A diversity of signals is created by varying the shape of the electric field, the waveform of the electric discharge, the discharge frequency, the timing patterns between signals from the sender and receiver and by stopping the electric discharge.

There are two general categories of electric signals produced by the weakly electric fish. Mormyrids and some gymnotids produce brief pulses at low frequencies. Other electric fish produce signals continuously at high frequencies (Figure 17.23b). Electric signals are perceived by specialized electroreceptor organs embedded in the fish's skin (Hopkins, 1974).

What are the characteristics of electric signals? When the electric organ discharges, an electric field is created instantaneously. It also disappears at the instant the discharge stops. As a result, electrical signals are ideally suited for transmitting information that fluctuates quickly, such as aggressive tendencies. An electric signal does not propagate away from the sender, but instead it exists as an electric field around the sender. Because an electric signal is not propagated, its waveform is not distorted during transmission. As a result, the waveform of the electric signal may be a reliable cue to the sender's identity, and different waveforms may be used for different signals (Hopkins, 1986a). Indeed, we find that, although the waveform is generally constant for a particular individual, it is different in males and females and among different species (Hagedorn, 1986; Hopkins, 1986b). Electric signals are well suited for communication in the environment in which these fish live. Both groups of electric fish are active at night and generally live in muddy tropical rivers and streams where visibility is poor. Electric signals can move around obstacles and are undisturbed by the suspended matter that creates murky

a

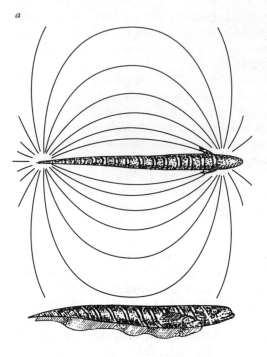

b

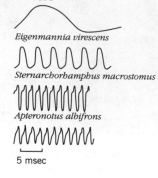

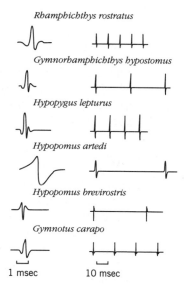

Figure 17.23 (*a*) Discharges from an electric organ create an electric field around a weakly electric fish that is used as a communication signal. The signal can be varied by altering the shape of the electric field, the waveform of the electric discharge, the discharge frequency, and the timing patterns between signals from the sender and receiver and by stopping the electric discharge. (*b*) Some species of weakly electric fishes produce electric signals with a delay between each pulse and others produce a continuous "buzz" of signals. (From Hopkins, 1974.)

water. However, electric signals are only effective over short distances, about 1 to 2 meters, depending on the depth of the water and the relative positions of the sender and receiver (Hopkins, 1986b). The shortness of the effective distance of electric signals may actually be an advantage. These fish often live in groups sometimes consisting of several species. Since many individuals will be signaling at once, the short effective distance of the signal may reduce electric "noise."

The weakly electric fish use electric signals to convey the same messages that other organisms send via alternative channels. Using electrical discharges, these fish can threaten one another or proclaim submission. Males of some species not only advertise their sex and species on the basis of electric signals but also court females by "singing" an electrical courtship song (Hagedorn, 1986; Hopkins, 1972, 1974, 1986b; Hopkins and Bass, 1981). In one species, *Eigenmannia virescens*, interruptions in the electrical discharge, called chirps, are used by males both to defend their territory and to court females (Kramer, 1990). In addition, electric signals are part of the mechanism promoting schooling in at least one mormyriform fish, *Marcusenius cyprinoides* (Moller, 1976).

Selective Forces Acting on the Form of the Signal

Why do birds sing and fireflies flash? Why isn't it the other way around? These questions sound absurdly simple, but in fact they have no simple answers. Trying to give a simple answer to why a signal has a certain form is somewhat like attempting to name the single ingredient responsible for the flavor of a stew. It cannot be done because, in both cases, the final product results from an intermingling of many elements. In communication there are many factors, including the anatomy, physiology, and behavior of the species as well as features of the habitat, that are important in determining whether or not any given signal is possible and, even more important, which signals can be successful in conveying a message. Nonetheless, it is possible to name the ingredients that blend in a stew, so it should be possible to label the selective forces that may have had a hand in molding a signal. Throughout the discussion of selective forces that follows, try to keep in mind that the pigeonholing is artificial because the factors that have been teased apart are, in reality, constantly interacting in the shaping of the signal.

SPECIES CHARACTERISTICS

Anatomy and Physiology

The anatomy and physiology of the sender and receiver will dictate which sensory channels can be most easily exploited for communication. An animal must have the anatomical structures necessary to produce the signal. For example, although it is not the sole factor, ant anatomy is undoubtedly one reason why ants do not rely on sound for communication. Ants lack an anatomical structure that could produce sounds loud enough to be heard at the distances over which they forage. Anatomy is also the reason why whales have no displays involving fluffed feathers.

Consider how an attribute as simple as body size might influence the form of a display. First, a small body size may limit the usefulness of visual signals and favor the use of another sensory channel. Small individuals have difficulty seeing and being seen over distances. Is it any wonder then that semaphoring plays no role in an ant's long-distance communication? Since the anatomy of ants makes it unlikely that auditory and visual signals can be effective, natural selection has favored other channels of communication, for example, chemicals.

Second, if visual signals are employed by a species, body size may influence the form of the signal. For example, in addition to whole body movements and posturing, there are two forms of visual display common among the primates of Central and South America: facial expression and the erection of hair on parts of the body. It has been suggested that these two forms of display are functionally equivalent. The evolutionary choice of form may depend on body size since the raising of long hair on a small body is more visible at greater distances than is a change in expression on a diminutive face (Moynihan, 1967). The small species of New World primates have displays involving the erection of hair but

a

b

Figure 17.24 A marmoset (*a*) and a woolly monkey (*b*). Among the New World monkeys, small species, such as the squirrel-sized marmoset, communicate more frequently with displays involving the
erection of hair, while larger species, such as the
woolly monkey, rely more heavily on facial expressions. This is probably because differences in facial
expression would be difficult to see in small species.

have little variation in facial expression, while
just the opposite is true among the larger primates. Tamarins and marmosets are squirrel
sized, so subtleties of facial expression would
be difficult to see at a distance. However, the
erection of tufts and ruffs of hair would increase
visibility and, both tamarins and marmosets have
long silky fur that makes their displays more
effective (Figure 17.24*a*). Although the New
World monkeys tend to be poker-faced compared to their Old World cousins, larger species
such as the capuchins (*Cebus*), best known for
soliciting coins for organ-grinders, the common
spider monkey (*Ateles*), and the woolly monkey

(*Lagothrix*) have a richer variety of facial expressions (Figure 17.24*b*) than do the smaller New World primates such as the tamarins and marmosets. The faces of the larger monkeys are big enough for a visual signal as subtle as a facial expression to be seen and deciphered.

Body size may also affect the form of visual displays that involve movements of the entire body. For example, it may affect the kinds of movements used in the display. Smaller individuals are often more agile. You may have noticed a relationship between body size and agility among human athletes: Gymnasts are usually more petite than weight lifters. Likewise, in the animal kingdom, smaller species are often better acrobats. Among herons (Meyerriecks, 1960) and gulls (Moynihan, 1956), for example, the larger species have fewer and less elaborate aerial displays than do the smaller species.

The anatomy and physiology of the receiver are also important forces acting on the design of a signal (Guilford and Dawkins, 1991, 1992). Natural selection should favor signals that are easy for a receiver to detect and discriminate, and this will depend on its sensory abilities. For example, it is not surprising that bird song is more melodious than the courtship songs of insects when the responsiveness of their auditory receptors is considered. The tympanal membrane of insects, unlike the ears of birds and mammals, cannot detect differences in pitch (Marler and Hamilton, 1966). An insect suitor must, therefore, identify himself not by his tune but by his rhythm. In the symphony of animal sound, the insect is a percussionist. In addition, the low frequencies in the "chuck" call of male frogs (*Physalaemus pustulosus*) used to attract females is thought to have been selected because the female's sensory system is most sensitive to low-frequency sounds. The tuning of the female's ears is thought to have evolved before the advent of the chuck call since female sensory system of two related frog species (*Physalaemus pustulosus* and *P. coloradorum*) have similar biases toward lower than average frequencies in the chuck call. However, the chuck call has evolved only in male *P. pustulosus*. Thus, the chuck call could not have been responsible for the evolution of the female bias (Ryan et al., 1990).

Furthermore, some signals are easier for a receiver to discriminate, that is, to recognize that they belong to a particular category. Warning colors, for instance, help a predator recognize prey as unpalatable (see Chapter 13).

Finally, signals should be easy for the receiver to remember, especially if they are used to assess the qualities of the sender. This is because some quality of the signal from several individuals must be remembered and compared if the "best" is to be chosen.

Behavior

The behavioral repertoire of a species may also influence the form of communication signals. Some behaviors are too important to stop during communication. It is obvious that a brachiator, such as a gibbon that swings from branch to branch, could not evolve a display such as the chest pounding of the gorilla or it would be constantly plummeting to the ground. Although there may be occasions when communication is so important that it warrants the cessation of other activities, there are times when it is advantageous to continue another behavior during signal emission. Some sensory channels, vision for example, minimize an animal's freedom for other activities. The sender must employ a major part of its motor equipment and the receiver must turn its eyes toward the signal and thereby impair the performance of other activities (Marler, 1967). Producing and receiving sound or chemical signals interfere much less with other activities.

The relationship between the type of foraging of ants and the nature of their recruitment signal provides a specific example of the way in which the behavioral characteristics of a species may influence the form of a signal. The recruitment

Figure 17.25 Tandem running in ants. This means of recruitment is efficient when a single companion is all that is necessary to exploit a food source. The recruited ant maintains contact with the caller by continuously patting the leader's posterior parts.

signal is nicely suited for the food source exploited by each species. Some species of *Leptothorax* feed on large stationary items such as dead beetles. For this type of prey, the assistance of a single companion is all that is necessary. A volunteer is solicited by "tandem calling". The successful scouting forager returns to the nest and regurgitates a sample of food to several nestmates. Then it turns around and exudes a droplet of a calling pheromone from the poison gland in its elevated abdomen. The nestmate that touches the caller's hindlegs or abdomen with its antennae is led to the food source. The recruit keeps in contact by continually patting the leader on the posterior parts (Figure 17.25; Möglich, Maschwitz, and Hölldobler, 1974).

Other species, including the fire ant (*Solenopsis*), relish live insects much larger than themselves. Because of the size of the prey, many workers must be recruited; and because the prey moves about, each new position of the quarry must be indicated. A fire ant scout that has been successful in locating food lays a chemical trail during its return to the nest. Other workers are attracted by this odor and follow the trail to the food source. If they are rewarded, they too leave a chemical trail on their return trip. Several features of this recruitment policy are well suited to the type of food source exploited. First, because workers leave a trail when, and only when,

food is found, the intensity of the chemical provides an index of the richness of the food source. The pheromone is volatile—within 2 minutes after the food is gone, the trail disappears. Because the trail self-destructs, the confusing remains of old trails are eliminated. Furthermore, the trail can change position to track the moving prey (Wilson, 1971).

Finally, the leaf cutters (*Atta*) and the seed eaters (*Pogonomyrmex*) exploit more permanent or renewing food sources. Their highways require durable road signs. Again, the recruitment signal is well suited for its purpose. These species lay odor trails with long-lasting chemicals. In addition, they mark their paths with persistent visual cues by cutting the vegetation (Hölldobler, 1976; Figure 17.26).

ENVIRONMENTAL CHARACTERISTICS

Determination of the Sensory Channel Employed for a Signal

It is not surprising that the characteristics of a species' habitat may also mold the form of its signals. An obvious variable of great importance is the amount of light available. Darkness severely limits visual displaying unless the animal sports a biotic lantern. Among bioluminescent organisms, the color of emitted light may be an adaptation increasing the efficiency of the signal in environments with different qualities of prevailing light. For example, among North American fireflies most (21 of 23) species active during twilight emit yellow light, while the majority (23 of 32) of dark-active species produce green light. The emission of yellow light at dusk is thought to make the signal easier to distinguish from the green light that is still reflected from foliage (Lall et al., 1980).

Night is not the only cause of darkness; vision is also restricted in many aquatic habitats, favoring communication via the other sensory channels. Low-frequency sounds are transmitted farther than light in deep sea waters (Payne, 1972).

companions over an entire ocean basin (Payne and Webb, 1971).

Audition is not the only sensory channel exploited by animals in dark aquatic environments. Both chemical and electric signals have been selected for communication. The bullhead catfish, for example, has a sophisticated social system in the murky waters it calls home. Each individual's identity and status is identified by chemical cues (Todd, 1971). A description of the experiments that demonstrated this ability can be found in the next chapter in the section on individual recognition. As previously mentioned, electric signals are another solution to the problem of communicating in perpetually murky waters, but this capability is limited to a few organisms, the gymnotid fish of South America and the mormyrid fish of Africa (Hopkins, 1974, 1980).

Vision may be impeded by obstacles rather than darkness. Dense vegetation, tall grasses, or even thick mists and mountain top clouds may keep mated pairs from keeping an eye on each other. However, in some species of birds, out of sight is not out of mind. Duetting and antiphonal singing (in which a male and female alternate phrases in a song) occur primarily in species that inhabit areas in which vision may be limited. Their songs are often elaborate. In some antiphonal songs, the mates weave their alternate syllables together so precisely that to a human eavesdropper, it sounds as if the melody is being created by a single songster (Thorpe, 1972). Although there may be additional forces active in the evolution of this mutual displaying (Smith, 1977), the nature of the habitat is an obvious contributing factor.

Influence on the Specific Form of a Signal

Dialects in the Language of Bees The physical characteristics of the environment may do more than simply favor a particular sensory channel for communication; they may mold the form of the signal within any given channel. The

Figure 17.26 Trails cut by ants through the vegetation leading from the nest to food-collecting areas. These trails are marked with durable chemicals. This type of signal is useful when the food source exploited is permanent or renewable.

Therefore, the eerie songs of the finback whale are well designed for long-distance communication. They are extremely low pitched, loud, and pure in tone. In more precise terms, the frequencies in their calls are restricted to a band only about 3–4 Hz wide centered at 20 Hz. It has been estimated that before competition from the roar of ships' propellers, these tones could be heard over distances of 1000 to 6500 kilometers. At one time, therefore, the whales' wails may have enabled them to keep in touch with dialects in the language of bees has been interpreted in this light (Gould, 1982). When a honeybee scout (*Apis* spp.) discovers a rich food source, she returns to the hive and informs her fellow workers. If the food is more than a certain distance from the hive, she performs a waggle dance (see the section in the next chapter on recruitment for a detailed discussion). Her gyrations inform the recruits of the distance and direction to the food source. Distance information is encoded in the number of waggles during the straight run of the dance. However, among different races of bees, the flight distance represented by each waggle varies (Figure 17.27). Why should these differences, or dialects, in the bee language exist?

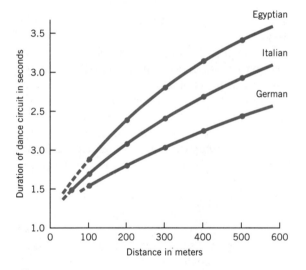

Figure 17.27 Dialects in the dance of three races of honeybees. A scout bee informs other workers of the location of a food source by a waggle dance. The number of waggles during the straight run of the dance signifies the distance to the food. However, the number of meters designated by each waggle varies among the races. These dialects may be related to differences in the environment. Bees inhabiting colder habitats have larger foraging ranges. In their dance, each waggle denotes a greater travel distance than does a waggle in the dances of races with small foraging ranges. (Modified from Gould, 1982.)

One reason for differences in bee language might be that the climate influences foraging distances and that this in turn was a selective pressure for divergence of the bee language into dialects (Gould, 1982; Towne and Gould, 1988). Frigidity is a key factor determining winter survival of a honeybee colony. The nestmates survive the freeze by metabolizing stores of honey to produce heat and by sacrificing outer layers of bees in the colony to provide insulation. Larger colonies would be expected in icier climates because additional members would be required to gather larger larders and to increase the thickness of the living comforter insulating the colony. The greater demand for food would increase the average foraging distances for those races living in frigid environs. To reduce the time and energy a worker would invest in waggling the location of a food source to recruits, selection might sacrifice some precision in the language of these races. The dance was made more efficient by having a waggle signify a greater distance. The suggestion that the distance signified by a waggle is influenced by the average distance traveled during foraging is supported by the observation that artificial swarms of German bees (*A. m. carnica*), a race that generally experiences severe winters causing them to forage over longer distances and dwell in larger colonies, choose larger and more distant nest boxes than their cousins from sunny Italy (*A. m. ligustica*).

It is puzzling to note, however, that the behavior of honeybee species in Thailand does not seem to support the suggestion that differences in climate and foraging range lead to dance dialects. Although the three species studied (*Apis florea, A. cerana,* and *A. dorsata*) forage over strikingly different distances, their dance languages indicate distances in much the same way. So, perhaps there is no universal relationship among climate, foraging distance, and dance dialects. Alternatively, it may be that other factors, such as nesting behavior and the cues used to follow the dances, are also important. Although two of the honeybee species in Thailand (*A. florea* and *A. dorsata*) live in open nests and visual signals play a role in their dance communication, those in Europe nest in cavities and use only sound and touch to follow the dances. It may be easier to follow long waggling runs by watching than by sound and touch. Thus, open-nesting Thailand species may be able to use the same dialect no matter how far they usually travel for nectar; greater distance is simply signified by longer waggle runs (Dyer and Seeley, 1991).

The Structure of Bird Song   Researchers studying bird song have also attempted to correlate the structure of the signal with characteristics of the environment. Comparative studies have demonstrated that differences in bird song characteristics such as the range of frequencies employed in the song and the occurrence of pure tones or trills (rapid sequences notes) can be correlated with aspects of the species' habitat. For example, an interspecific study in Panama revealed that species singing below the canopy in tropical forests declared their presence with low-pitched whistles but others inhabiting grasslands used rapid trills centered on higher frequencies (Morton, 1975). Even within a species, the type of song has been shown to vary with the habitat of the singer. The song of the great tit (*Parus major*) is ideal for this type of analysis because the species has such a large geographic distribution. Forest dwellers from England, Poland, Sweden, Norway, and Morocco were found to have songs with a lower pitch, a narrower range of frequencies, and fewer notes per phrase than open woodland birds from England, Iran, Greece, Spain, or Morocco (Figure 17.28). In fact, birds from similar habitats in Oxfordshire, England, and Iran, separated by 5000 kilometers, sing more similar melodies than two English populations occupying different habitats (Hunter and Krebs, 1979).

Why should these correlations between song structure and type of environment exist? Differences in the transmission of sounds in various habitats may provide a clue. Two processes counteract the transmission of sound. One, attenuation (weakening), influences how far the sound will carry. The second, degradation, determines how distorted the signal becomes during transmission. Exactly how these factors interact to modify the structure of bird song is not certain, but several researchers have presented reasonable hypotheses.

Some investigators emphasize the role of attenuation (Hunter and Krebs, 1979; Morton, 1975). Eugene Morton measured the attenuation of various frequencies in the tropical rain forest and in the grasslands of Panama by broadcasting sounds on a loudspeaker and recording them on tape recorders positioned at increasing distances from the speaker. Higher pitched tones attenuated more rapidly than lower pitched tones in both habitats. A major difference between the habitats in sound attenuation was that in the forest, tones of about 2 kHz carried better than expected. Forest species seem to take advantage of this. Unlike the birds of the grasslands, those in the forest sing in frequencies of 1.5–2.5 kHz and rarely employ frequencies above 3.5 kHz. Thus, as seen in Figure 17.29, the distribution of frequencies in the songs of forest birds seems to be ideally suited for maximum projection with minimum effort.

However, other investigators (Marten and Marler, 1977; Marten, Quine, and Marler, 1977; Wiley and Richards, 1978) do not find consistent differences in the attenuation of sounds in grassland and forest environment. How then can song differences be explained? Mac Hunter and John Krebs (1979) still invoke the frequency-dependent pattern of attenuation (i.e., low frequencies carry farther than high-pitched tones) to explain the song structure of great tits, but they suggest that other factors, such as predation, may explain why low notes are not used in all habitats. The forest populations employ low frequencies and thus their songs seem designed for long-range communication. Other factors, such as visibility

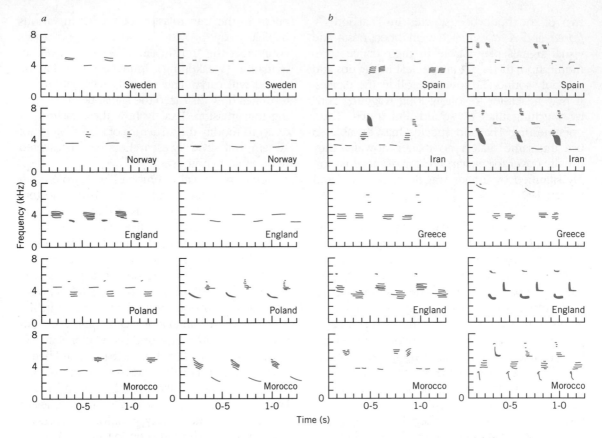

Figure 17.28 Song structure of populations of great tits in many countries: (*a*) forest dwellers, (*b*) woodland inhabitants. The lefthand column in each section shows songs that are "typical" of each area, and the righthand column shows songs that emphasize the characteristic features of the songs of that area. Notice that the songs of the species living in forests have a lower pitch, fewer frequencies, and fewer notes per phrase than the woodland species. (From Hunter and Krebs, 1979.)

to predators in the open environment, may have selected for higher frequency songs that carry shorter distances in grassland species.

On the other hand, R. Haven Wiley and Douglas Richards (1978) argue that since attenuation is similar in both environments, one should consider how the other factor determining how far a sound can be transmitted, degradation, might mold bird song. A major source of distortion that is present in forests but not grasslands is reverberation from foliage. This is greater for high-frequency sounds because they have more cycles per second and, therefore, have a greater chance of striking small objects such as leaves and branches. Reverberation of high frequencies may deflect them away from the receiver, making them more difficult to detect or may cause echoes that would scramble complicated song patterns. Therefore, a song designed for maximum detection within a forest should emphasize

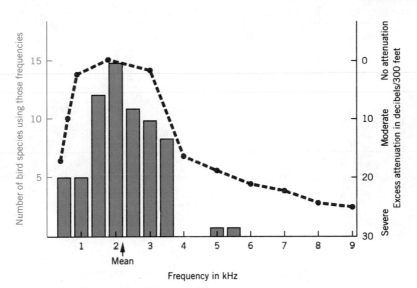

Figure 17.29 The relationship between frequencies used in the songs of birds from lowland forest areas of Central America and the attenuation of various frequency sounds in that environment. The histograms indicate the number of bird species using various frequencies in their songs. The curve indicates the degree of attenuation of sounds of various frequencies in the same habitat. The most common frequencies used in bird songs in the area are around 2 kHz, the frequencies that attenuate the least. (Modified from Morton, 1975.)

low frequencies and have few trills or trills with widely spaced notes. The major source of degradation in grasslands would be unpredictable gusts of wind. Since part or all of a song might be "gone with the wind" at any moment, redundancy would be favored in the songs of grassland species; the repetition of notes such as is found in trills would be expected. The use of higher frequencies might also be predicted since they would increase the chance of completing the tune between gusts of wind that would distort the song. These predictions, that higher frequencies and more trills would be common in grassland species, are partially borne out by the observations of Fernando Nottebohm (1975) on the South American rufous-collared sparrows (*Zonotrichia capensis*). Notice in Figure 17.30 that the grassland populations do employ faster trills. Selection to avoid deg-

radation could also explain the difference in the average frequency of songs found by Hunter and Krebs (1979) and Morton (1975). Recall that in both studies the bird living in grasslands sang higher pitched tunes.

Language and Apes

Although animals do not normally use language in the sense that we do, several investigators have studied the ability of apes—chimpanzees (*Pan troglodytes*) and gorillas (*Gorilla gorilla*)—to learn language. Perhaps these studies have been prompted by a human desire to communicate with apes or to better understand what it means to be human. Regardless of the motivation for the studies, the results have philo-

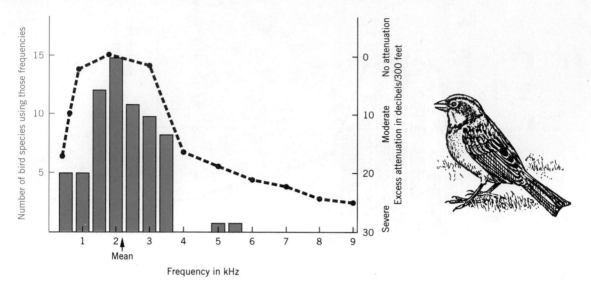

Figure 17.30   Songs of grassland and forest dwelling rufous-collared sparrows in South America. The songs of grassland species have more and faster trills. This difference might be explained by selection to avoid degradation of the song by gusts of wind typical of open country. (Modified from Nottebohm, 1975.)

sophical and practical implications for many areas of study, including linguistics, anthropology, zoology, sociology, and neurobiology (Savage-Rumbaugh, Rumbaugh, and Boysen, 1980).

Early studies were designed to teach chimpanzees to talk. In the longest and most thorough of these attempts, Keith and Cathy Hayes gave a chimp named Viki formal speech training using techniques resembling modern behavior modification. They would mold Viki's lips into the correct shape for the pronunciation of a particular sound when necessary and would reward her for successively closer approximations to the desired pronunciation. Eventually she was able to say three words—mama, papa, and cup. However, her pronunciation was not perfect. In fact, it was a voiceless aspiration. Furthermore, she never achieved consistently correct usage and often seemed to have trouble remembering the words (Hayes and Hayes, 1951).

The failure to train chimps to talk is probably due to the differences in the anatomy of the speech-producing organs between humans and apes. The shape and position of the skull is different in apes. Accompanying these changes are alterations in the position and shape of the larynx, tongue, epiglottis, and palate. Such structural changes alter the resonating properties of the vocal cavities. In addition, differences in the musculature around the mouth, the shape of the lips, and the arrangement of teeth make it possible for humans to pronounce many sounds that a chimp cannot (Lenneberg, 1967).

A more fruitful approach to demonstrating the ability of apes to acquire language skills has been to use nonverbal languages. For example, the Gardners (1969) trained Washoe, a young chimpanzee, to communicate using the American Sign Language for the Deaf, ASL (Figure 17.31). The Gardners believed that an interesting

Figure 17.31 Chatting with a chimp in American Sign Language.

and intellectually stimulating environment would assist the development of language skills. For this reason Washoe and the chimpanzees they have trained since then are reared as much as possible like human children. However, spoken English was not permitted around Washoe because it was feared that it might encourage her to ignore signs. After 4 years of training, Washoe had a reported vocabulary of 132 signs. Her signs were not restricted to requests. Washoe used the signs to refer to more than just the original referent; she applied them correctly to a wide variety of referents. For example, Washoe extended the use of the sign for dog from the particular picture of a dog from which she learned it to all pictures of dogs, living dogs, and even the barking of an unseen dog (Gardner and Gardner, 1969). She also apparently invented combinations of signs to denote objects for which she had no name. Classic examples are her signing "water bird" for a swan on a lake and "rock berry" for a Brazil nut (Fouts, 1974). By the time she knew 8–10 signs, Washoe had begun to string them together. Examples of typical early combinations are "please tickle," "gimme food," and "go in" (Gardner and Gardner, 1969).

Once the Gardners had demonstrated that chimpanzees could be taught to use a gestural language, other ape signing projects were begun. Roger Fouts (1973) and Herbert Terrace and coworkers (1979) continued working with chimpanzees while Francine Patterson (1978) extended the studies to a gorilla. The techniques employed in these signing studies were similar to those originally used by the Gardners, except that spoken English was permitted in the presence of the apes.

At the same time that the Gardners were working with Washoe, David Premack was training another chimpanzee, Sarah, to use plastic chips of various shapes and colors as words

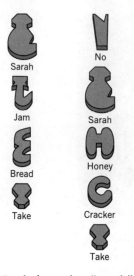

such as if–then and same–different (Premack, 1976).

The more successful aspects of the pioneering studies with Washoe and Sarah were combined in the LANA Project (Rumbaugh, 1977; Rumbaugh, Gill, and von Glaserfeld, 1973). The chimpanzee Lana was trained to use a computer to communicate in Yerkish, a symbolic language invented for the purpose. Yerkish words, called lexigrams, are geometric figures built from combinations of nine simple design elements such as lines, circles, and dots (Figure 17.33). Lana chose words by pressing a computer key labeled with the lexigram. When a key was depressed, it lit up and the lexigram simultaneously ap-

Figure 17.32 Symbols used as "words" by Sarah, a chimpanzee. Sarah learned to communicate using these plastic shapes.

(Figure 17.32). Unlike Washoe, Sarah lived in a cage in a laboratory. She did have frequent human contact and was taken outside to play. Not only were Sarah's language medium and learning environment different from Washoe's, but the conditions for her demonstration of language abilities were also different. Most of her use of language consisted of using one word out of a choice of several to complete a preformed statement or arranging four to five words into a sentence of a specific word order. Premack established certain criteria for accepting that Sarah was using a particular chip as a word. She had to be able to use the plastic chip to request the object it stood for, to select the proper chip when asked the name of the referent, and to "describe" the referent of a particular chip using other chips. Premack's strategy was to break down linguistic rules into simple units and to teach them to the chimp one at a time. In this way Sarah was taught not only to name many objects, but also more complicated relationships

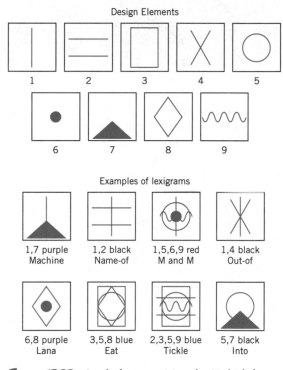

Figure 17.33 Symbols comprising the Yerkish language. Each "word" is a combination of a few geometric shapes. They were embossed on keys of a computer keyboard. Chimpanzees have learned to communicate by pressing the appropriate keys.

peared on a projection screen. It is significant to note that Lana was required to use lexigrams in an appropriate order. In other words, she had to learn syntax, the rules governing word order in a sentence. For example, she learned that pressing lexigram keys to spell out "Please machine give . . ." might be rewarded but that pressing "Please give machine . . ." was not an acceptable way to make a request. Unlike Sarah, who was given a limited choice of words to use at one time, Lana always had her complete vocabulary available to her. This computerized language system eliminated the problems of social cuing and the difficulty of interpreting the symbols that plague the sign language studies.

Lana developed a large vocabulary and mastered Yerkish grammar. She also coined new words just as Washoe did. For example, she called an orange soda a "Coke-which-is-orange" (Rumbaugh, 1977).

In summary, the ape language projects using Washoe, Sarah, and Lana demonstrated that apes (1) are relatively adept at learning words, (2) readily string words together in short sequences so that the strings adhere to rules of grammar if that is required, and (3) have the ability to coin new words.

These skills are obviously necessary for language, but do they demonstrate the use of true language? Before that question can be answered, a set of criteria defining true language must be established. Although definitions vary, most linguists would agree on two essential elements. First, words or signs must be used as true symbols that can stand for, or take the place of, a real object, event, person, action, or relationship. Symbols should permit reference to objects or events that are not present. Second, words or signs should be combined to form novel phrases or sentences that are understandable to others. This necessitates a knowledge of syntax because a change in the order of symbols can alter the meaning of the message.

The ability of apes to master these two language skills has been questioned. First let's ex-

amine the evidence that apes use words as symbols. When a chimp uses a "word" (sign, plastic chip, or lexigram) to name an object, is it used as a symbol that stands for the object, one that can be used to refer to the object even when it is not present, or is the animal producing the "word" because it has been associated with the object through reward? When a pigeon pecks a red key for food and a green key for water, no one assumes that the animal is using the key to represent the item. However, when an animal as intelligent as an ape presses a key that is labeled, not with a color but with a "word," the assumption is often made that it is using language and not simply performing a complicated behavior for a reward. This conclusion is especially tempting when the animal strings words together in a sequence for a reward. How could it be determined whether the animals were using the words as symbols or just mastering a complex conditioned response? It has been argued (Savage-Rumbaugh et al., 1980) that in the studies just described, the apes were not required to do anything that eliminates the possibility that they were simply using words as labels rather than as symbols.

However, other experiments might be interpreted as a demonstration of a chimpanzee's ability to use "words" as true symbols. These tests were done with two chimpanzees, Sherman and Austin (Figure 17.34). They were trained, as Lana was, to communicate by pressing computer keys embossed with lexigrams. The emphasis in this program was on interanimal communication, so the animals were not taught to produce strings of lexigrams or to adhere to grammatical rules. In addition, they were raised in a social preschool setting. The animals were first taught to name foods by pressing lexigram keys. It is important to note that the animals were taught to distinguish between the use of a food name as its name and the use of the food name as a request because they were never allowed to eat the same item that they named.

Following this training, the animals' ability to

Figure 17.34 Sherman and Austin, two chimps, have shown that apes can cooperate with one another to solve problems using the symbolic language, Yerkish. They could only communicate when they had access to the computer keyboard.

communicate with one another symbolically was investigated. In one test, Sherman and Austin were able to specify foods to one another using lexigrams. One of the chimpanzees was taken to a different room where he watched the experimenter bait a container with one of 11 different foods or drinks. That animal was then led back to the keyboard and asked to name the food in the container. Using the information gained by observing the response of the first animal, the second chimp was permitted to use the keyboard to request the food. If both animals were correct, they were given the food or drink. Sherman and Austin were able to communicate with one another in this manner whether or not they used the same keyboard and even if the experimenter was ignorant of the identity of the food in the container. The animals communicated regardless of which of them was the observer. The animal that did the requesting based on the information provided by his knowledgeable pal could demonstrate that he knew which item he was asking for by selecting its picture from a group. However, when the chimp who knew the identity of the food was prevented from using the keyboard to describe the contents of the container, he could not transmit the information to his buddy (Savage-Rumbaugh, Rumbaugh, and Boysen, 1978a).

The chimps also passed the next test—using symbols to inform one another of the appropriate tool to use to solve a problem. The animals were kept in separate rooms. One chimp had to decide which one of six tools he needed to obtain hidden food and then ask the other one for that tool via the keyboard. They could successfully cooperate in this manner only when the keyboard was turned on (Savage-Rumbaugh, Rumbaugh, and Boysen, 1978b). Clearly Sherman and Austin were using words as symbols and not simply labeling objects.

Do the language abilities of apes meet the second criterion for true language? Do the animals spontaneously produce novel sentences that follow grammatical rules? The answer to this question has been controversial. Terrace and coworkers (1979) argue that chimpanzees have not demonstrated that they can create a sentence. Most of their analysis was done on the utterances of Neam Chimpsky, a young male chimp usually called Nim. Videotapes and films of the signing of other chimpanzees were also studied. Nim was trained to sign in ASL by a method modeled after the Gardners'. He acquired a reasonable vocabulary of signs and would string them together spontaneously. Nim was not rewarded any differently for sequences that followed traditional English usage, but his teachers usually signed to him in combinations with stereotyped orders.

When all of Nim's multisign combinations were recorded and analyzed, Terrace and coworkers (1979; 893) concluded that Nim did not spontaneously produce sentences characteristic of human language. Instead, Nim's utterances were "unstructured combinations of signs, in which each sign is separately appropriate to the situation at hand." Since Nim created a tremendous variety of combinations of signs even though he was not required to do so, these signs could not have been learned by rote memorization. However, Nim's utterances were often not only initiated by the trainer's signing, they also tended to imitate, either partially or fully, the teacher's preceding signs.

Furthermore, Nim's language development was different from that of a human child; his phrases did not grow in complexity or length as a child's do. His two sign combinations did show a preferred word order. For example, he would sign "more + X" more frequently than "X + more." However, important to Terrace's argument is that Nim did not add new information to his utterance as he added a third word, the way a child does when it learns longer sentences. Nim expanded his most frequent two-sign combination "play me" to his most frequent three-sign sequence "play me Nim." The third word, Nim, provides information already present in the two-word utterance. This is not the way a child usually increases its phrase length. The third word added by a child supplies new information. A youngster might elaborate "sit chair" to "sit daddy chair." Although children are known for making a short story long, their tales are not usually as redundant as Nim's 16-sign utterance "give orange me give eat orange me eat orange give me eat orange give me you." Not only did Nim's phrases remain at the same level of complexity, they also seemed to be stuck at an average length of 1.5 signs. In contrast, a child's sentences continue to get longer.

Needless to say, Terrace's conclusion that an ape cannot create a sentence was challenged by others doing ape language research. Fouts, Patterson, and the Gardners argue that Nim's language abilities were stunted by the operant conditioning procedures used in his training. Allen Gardner backs his claim "that you can turn it [imitation] on and off, depending on the type of training you give" with a videotape of a chimp who shows little or no imitation of his trainer's signs until the last third of the tape when operant conditioning techniques were begun. During this last section of the tape, 70 percent of the chimp's signs were imitative. The Gardners claim that Washoe's signs are not only novel, but are also spontaneous. Washoe would often initiate signing and would sign to other animals. The Rumbaughs argue that because of the way in which Nim was trained, he never understood the meaning of words and that is why he was unable to create a sentence. In addition, Nim's trainers changed so often that he may not have had the opportunity to form the relationships claimed to be essential for language development. In defense of Washoe's language abilities, the Gardners argue that Terrace could not appreciate the conversations between chimp and trainer because he analyzed isolated segments of the film. This prevented him from seeing the execution of a complete sign, causing him to misinterpret it (Marx, 1980).

The controversy about whether apes possess language stifled interest in this research for awhile, but it has been reinstilled by the remarkable abilities of Kanzi, a bonobo or pygmy chimpanzee (*Pan paniscus*). Kanzi was born in

1980, the son of Matata, who was part of a language study by Sue Savage-Rumbaugh. Kanzi was enlisted in the program when Savage-Rumbaugh realized that he had begun to learn communication skills simply by observing his mother's training. Kanzi was never trained to use lexigrams. Instead, his constant human companions used lexigrams, gestures, and speech to communicate with each other and with him. In this way they served as communicative models. Kanzi picked up the use of language in much the same way as a child would. When Kanzi was 18 months old, he spontaneously began to use nonritualized gestures to indicate the direction in which he would like to travel or actions he would like to have performed. Then, a year later, he spontaneously began to use lexigrams to communicate. Once he learned to use lexigrams, he began to use them to refer to items such as food or objects, or to locations that were not in sight (Savage-Rumbaugh, 1986). Kanzi is now a language star and communicates on a board with 256 lexigrams (Figure 17.35). He has the grammatical skills of a two-and-a-half-year-old human child. There may still be those who doubt that apes can acquire language skills, but Kanzi's abilities have certainly convinced critics that the issue deserves serious consideration (Greenfield and Savage-Rumbaugh, 1990; Linden, 1992).

Figure 17.35 Kanzi, the pygmy chimpanzee that has demonstrated the most advanced language skills so far. He communicates with a computer keyboard that has over 250 lexigrams. He was not trained by operant conditioning techniques. Instead, he observed and interacted with humans who used gestures and lexigrams to communicate.

Communication and Animal Cognition

Many people have wondered what it is like to be an animal—whether nonhuman animals have thoughts or subjective feelings. Such musings have led some investigators to seriously consider whether nonhuman animals are cognitive, conscious, aware beings.

Donald Griffin (1978, 1981, 1984) has suggested that tapping animal communication lines is a way that we might find out whether animals have conscious thoughts or feelings. After all, the only way we know about the thoughts or feelings of other people is when they *tell* us either through verbal or nonverbal communication. So, if nonhuman animals also have thoughts and feelings, they probably communicate them to others via their communication signals. If we could learn to speak their language, we could eavesdrop and thereby glimpse into the animal mind.

Most people agree that one sign of cognition is the ability to form mental representations of objects or events that are out of sight. You may recall that we discussed this in relation to problem solving in Chapter 5. Another way to look for cognition is to ask whether animal signals are symbolic, that is, whether they refer to things that are not present (Smith, 1991). We have seen that certain apes can learn a language that uses symbols (Savage-Rumbaugh, 1986) and Alex, an African Grey Parrot is able to vocally request more than 80 different items, even if they are out of sight. In addition, he can quantify and categorize those objects. He has shown an understanding of the concepts of color, shape, and same versus different on both familiar and novel objects (Pepperberg, 1991). However, these animals have been taught to use language.

We might next consider whether the signals that animals use in nature to communicate with each other are symbolic. The apparent simplicity of this question is deceptive because observations are often open to alternative interpretations. For example, the waggle dance of the honeybee that was described earlier is symbolic in that it contains information on the distance and direction to a distant food source. For this reason, Donald Griffin (1984) has suggested that the waggle dance is an indication of thinking or awareness. However, James and Carol Gould (1986) argue that it is not evidence of thought since the dances are genetically preprogrammed; bees can perform and understand dances without previous experience.

Another sign of cognition might be whether animals adjust signals according to conditions at the time. One way we might see such an adjustment in signaling is if an individual determined whether or not to signal by the composition of its audience. In other words, if an individual sees a predator, does it always sound an alarm or does this depend on its present company? The company does seem to affect the likelihood of alarm calling among domestic chickens (*Gallus gallus*). A cock is more likely to call in the presence of an unalerted companion (Karakashian, Gyger, and Marler, 1988). This has been interpreted as an indication that the cock chooses whether or not to call (Marler, Karakashian, and Gyger, 1991). The choice of calling or withholding the call would, then, be taken as evidence of cognition.

However, an alternative interpretation of this observation is that the call is not one that warns a companion of a predator, but rather one that indicates conflicting tendencies to flee or freeze (Smith, 1991). To avoid detection by a predator, a lone cock might decide to flee for cover if the predator were still at a distance, but to freeze if the predator were close at hand. When a cock is alone, then, the only factor it needs to consider when deciding how to avoid detection is the distance of the predator. Since there are no conflicting tendencies, no call is given. However, if a companion is nearby, there is another factor to consider when deciding how to avoid detection—the behavior of the companion. Consider a situation in which a predator has approached to a distance at which the best way for a lone cock to avoid detection would be to remain motionless. However, if it is with a companion and the companion moves, this might draw the attention of the predator. In that case, the safest response might be to flee for cover. As a result, there may be a moment of indecision and the call might be a response to the vacillating tendencies to flee or freeze. The indecision would only occur in the presence of another chicken, so the call would only occur when others were present (Smith, 1991). Again we find that the question of animal awareness is difficult to explore scientifically.

Summary

There are many possible definitions of communication. The broadest definition, a transfer of information, may be qualified by requiring that there be an intent to communicate, that it occur between animals, that those animals be conspecifics, and/or that the message result in a change in the recipient's behavior. One workable definition is "a transfer of information by a signal that evolved for the purpose that modifies the recipient's actions."

Communication is a keystone of sociality. One hypothesis for the selective advantage of communication is that it is beneficial for group members to share information. Interactions can be more appropriate when individuals inform one another of their inner state or their future course of action. Another hypothesis views communication as selfish. A display can be seen as a way to manipulate others to do one's bidding without resort to brute force.

Approaches to testing the hypotheses might be to determine whether signals are good predictors of future behavior and whether the signals are honest. Neither approach yields conclusive answers. Some signals are good predictors of behavior and others are not. One reason that a signal may not accurately predict future behavior is that the receiver may respond in a way that makes it unnecessary for the sender to behave as the signal predicted it would.

Whereas the information sharing hypothesis predicts that signals should be honest, the manipulation hypothesis predicts dishonesty. Examples of both honesty and of deceit in animal signaling are common. We might expect honesty when the signal is tied to a physical attribute and cannot be faked, in stable social units in which members are individually recognized, and when the costs of cheating outweigh the benefits.

Animal communication systems differ from human language in many ways. Animal signals are more specific; the meaning of a display is usually constant. Furthermore, animal displays are stereotyped. In other words, each performance of a display is the same, even if different individuals are displaying. In addition, animals have far fewer signals and messages than we have. Each species usually has between 15 and 45 signals with which to send approximately 12 messages.

Although animal communication systems are more rigid than ours, some animals, particularly the higher vertebrates, have ways of increasing the amount of information that can be conveyed with a limited number of signals. An example is the use of graded, rather than discrete, displays. Inflexible signals that have evolved to a "typical intensity" are called discrete displays. Some displays are graded, however, so that their intensity reflects fluctuations in motivation. Sometimes signals are combined to form a composite display that has a new meaning. In metacommunication one signal alters the meaning of those that follow. For example, some species set the scene for play with a characteristic signal. This ensures that the actions that follow will be considered as fun even though they are often the same behaviors used by adults during aggression or reproduction. Finally, the meaning of the signal may be changed by the context in which it is shown.

The signals used in communication are stereotyped modifications of ancestral behaviors but not everyone agrees on why stereotypy evolved. One idea is that it increases the attention-getting properties of the signal and makes it clearly different from signals conveying different messages. Another suggestion is that the loss of variability helps to hide conflicting tendencies in the sender. Alternatively, stereotypy may have evolved to facilitate the comparison of performances by different individuals. If all those who display must do the same actions, it is possible to determine who does it best.

Traditionally, three evolutionary sources for displays have been recognized. The first category is intention movements, those behavior fragments that may precede a functional action. Displacement activities are a second category. The last group, autonomic responses, includes urination and defecation, vasodilation, respiratory changes, and thermoregulatory responses. Although these are the most common sources for signals, almost any behavior can be ritualized.

Ritualization, the evolution of a display, changes the form of the flexible ancestral action into a stereotyped signal. If one is reasonably sure of the precursor behavior, it can be compared to the current display to determine the changes in the original form. Typically, all or a part of the ancestral behavior pattern is exaggerated by changes in the duration or extent of movement, by alterations in the rate of performance, or by repetition. In addition, the original actions may be combined in a new order or some parts may be deleted. The display might also be directed toward a new stimulus. As the signal becomes "emancipated" from the factors that originally mediated it, the behavior may be shown in a new context and/or be motivated by different factors. Frequently, these changes in behavior are accentuated by anatomical modifications such as bright colors, antlers, manes, and the like.

Any sensory channel may be used for communication. Signals within each channel have characteristic properties that make them more or less useful depending on the species, the environment, and the function of the signal.

Visual signals are easy to locate, are transmitted quickly, and disappear just as fast. However, visual signals must be seen and are, therefore, only useful when there is enough light and where there are few obstacles to obscure the signal. Visual signals include badges, anatomical specializations that convey a message, as well as movements and postures.

Auditory signals can be transmitted long distances. The rate of transmission and fade-out is rapid. They do not require light and, in fact, work well under water. The sounds may be generated by respiratory structures, the rubbing of appendages, or beating on parts of the environment.

Chemicals can convey messages over great distances, particularly when assisted by currents of air or water. They are transmitted more slowly, are more durable, and are usually more difficult to locate than visual or auditory signals. Furthermore, they are effective in environments with limited visibility.

Chemicals used to convey information to conspecifics are called pheromones. When the chemical has an immediate effect on the behavior of the recipient, as occurs with sex attractants and trail and alarm substances, it is called a releaser pheromone. Primer pheromones act slowly. They exert their effect by altering the physiology of the recipient.

Tactile signals are only effective over short distances. It is easy to locate the sender. They are transmitted and disappear rapidly and are effective in areas with limited visibility. Grooming is a special form of tactile communication. It is practiced by many species but is especially prominent among primates.

Seismic signals are those caused by vibration of the environment. They are well suited to communication over long distances, particularly when vision is limited.

Electric fields are used for communication among the mormyriform fish of Africa and the gymnotid fish of South America. Transmission and fade out are almost instantaneous but the signals do not travel far. The signals are effective when visibility is limited.

The channel used for communication and even the detailed structure of a signal may be influenced by a variety of factors. Obviously, the anatomy and physiology of the species is important in directing the evolution of their signals. The sender must be capable of generating the signal and the recipient must be able to detect it. The species' behavior is also a factor. Some signals can be sent while continuing other behaviors, a benefit to active species. Often, as occurs in ants, a specific behavior, such as the type of prey preferred, will determine the desirable characteristics of a signal.

The physical habitat may also be important in

the evolutionary choice of a signal. Visibility is of prime importance. If it is limited by darkness or by obstacles, channels other than vision are favored.

The structure of a signal in any sensory channel may be influenced by the environment. For example, differences in the average winter temperature may have resulted in the dialects in the language of honeybees. Those races that inhabit colder areas must forage over longer distances. In this situation, some precision in the distance information provided by the waggle dance may have been sacrificed for the sake of efficiency. A single waggle would signify a greater flight distance.

Among other species of honeybees, the nesting habits may be more important than foraging range in shaping the dance language. Although three species in Thailand have very different foraging distances, their dances indicate distance in similar ways. Perhaps long waggling runs can be followed more easily by open nesting species of Thailand, which can use vision, than by European species that nest in cavities and follow dances using touch and sound.

Likewise, the lower pitch and scarcity of trills in the songs of birds living in the forest instead of grasslands may be attributed to characteristics of the environment. The two factors that work against the detection of an auditory signal are attenuation, influencing how far the sound will carry, and degradation, determining how distorted the signal becomes during transmission. Both of these factors have been invoked by various researchers to explain the differences in the songs of birds in different habitats.

Animal communication signals are not true language because animals do not use signals as symbols that can take the place of their referent and because they do not string signals together to form novel sentences. However, many workers have been interested in elucidating the language abilities of apes and have, therefore, attempted to teach them language. The early studies designed to train chimpanzees to speak failed because an ape's anatomy does not allow it to make the sounds of words. However, later studies used languages that do not require voice. Chimpanzees and gorillas have learned to communicate with their trainers in American Sign Language for the Deaf. Other chimps communicate using plastic tokens of different shapes or via a computer keyboard displaying geometric configurations that stand for words. These studies have shown that apes can learn "words" and may even coin new ones. They can also learn to string words together into short phrases. If it is required, they may even follow some grammatical rules in constructing "sentences."

In spite of these abilities, the language capacity of apes has been challenged. It has been argued that the chimps have not demonstrated an ability to use the "words" as symbols. However, even if this is a true criticism of the earlier studies with Washoe, Sarah, Lana, and Nim, more recent work with the chimps Sherman and Austin has shown that the apes can communicate with one another to solve problems if, and only if, they have access to their computer keyboards. Since they were able to conduct cognitive operations with only symbolic information available, true symbolization seems to be possible for apes. However, whether apes can achieve the second hallmark of language, the ability to construct novel sentences that follow grammatical rules, is still being debated.

Some people have suggested that knowledge about the communication systems of animals may provide an insight into the question of animal cognition. Although many people accept that some species, primarily apes and Alex, the African Grey Parrot, have been taught to use signals that represent items that are out of sight, there is less agreement that the signals animals use to communicate with one another in nature are symbolic. Another sign of cognition might be the adjustment of signals according to conditions at the time. Deciding whether animal communication meets this criterion has also been controversial.

CHAPTER
18

Maintaining Group Cohesion: Functions of Communication and Contact

*W*hether we walk across the Arctic tundra, the African plains, or the Amazon basin, we see animals in groups. The question arises, then, what holds such groups together? As we will see, the answers are often deceptively simple and interrelated. We will find that social bonds are cemented by both communication and physical contact. We will begin by considering the functions of communication.

Functions of Communication

RECOGNITION OF SPECIES

"To each his own" is more than a cliché in the animal world. Some behaviors, particularly aggressive and reproductive acts, are only adaptive when they are directed toward members of the same species. It is disadvantageous to waste time and energy in mating if the production of viable fertile offspring is unlikely. Likewise, time and energy should not be squandered defending a territory from an individual that is not competing for resources or mates. Such competition is usually most intense from members of the same species. Only these individuals would have identical requirements. Therefore, it is often impor-

tant for an animal to be able to recognize individuals of its own species.

Species recognition is particularly important when there are many closely related species living in the same area. For instance, on one small beach measuring about 56 square meters, an area slightly smaller than a quarter of a tennis court, there may be 12 species of fiddler crabs (*Uca*) courting (Crane, 1941). In spite of this density of species, a female fiddler crab still chooses the correct species for a mate. Her discrimination is probably based largely on the distinctive courtship display of the males of each species in which the large claw, or cheliped, is waved about in a specific pattern. You can see a few of the differences in Figure 18.1. One species of fiddler crab, *Uca rhizophorae*, courts by moving his cheliped up and down. Another fiddler crab, *U. annulipes* waves his claw in large circles. *U. pugilator* holds his cheliped out to the side of his body and waves it in tight circles. These differences are easily spotted, even by humans who are not as attuned to subtle differences in cheliped gyrations as female fiddler crabs.

The relationship between sexual behavior and the maintenance of species identity is also obvious among frogs and toads (Figure 18.2). Male frogs usually attract their mates by produc-

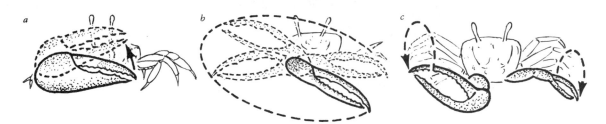

Figure 18.1   Courtship displays of three species of fiddler crabs. The difference in the courtship displays is an important factor that allows females to choose males of their own species. (*a*) *Uca rhizophorae* waves his cheliped vertically. (*b*) *Uca annulipes* beckons females by waving his cheliped in a large circular pattern. (*c*) *Uca pugilator* extends his claw to the side of his body and waves it in circles during courtship. (Modified from Schöne and Schöne, 1963.)

Figure 18.2 A male American toad (*Bufo americanus*) calling. Male frogs and toads attract females by croaking a species-specific call. In response, interested females approach the suitor. The male amplifies the sounds produced by his vocal cords by inflating the large vocal sac beneath his chin.

ing species-specific calls that advertise species identity, sex, reproductive state, and location. Judging by the response of nearby females, these calls seem to be the amphibian equivalent of, "I'm over here. Come and get me." Often the males of several species will serenade together in a chorus, and a female must choose one of her "own kind" from the variety of callers at the local pond. To make the proper choice even more difficult for a female, this discrimination must often be made in the dark. As a result, selection has favored clear species differences in croaks. The power of selection is most obvious in areas where the ranges of closely related species overlap. For example, in southeast Australia, two such species of treefrogs, *Hyla ewingi* and *H. verreaux*, inhabit some of the same areas. In both species, the male's call is a trill, a rapidly repeated series of sound impulses. The calls are so similar in the areas where only one of these species is found that females are not very good at distinguishing between them. Where the species coexist, however, the calls

have become different: The *H. ewingi* call is slower and the *H. verreaux* call is faster than the respective calls in areas occupied by only one of the species. Females from the overlap zone easily detect the differences in call rate and preferentially approach a male whose call has the right tempo (Littlejohn and Loftus-Hills, 1968).

Birds also use song as an important diagnostic clue to species membership. Flycatchers use song to identify members of their species and aggressively defend their territories from only these individuals. This was demonstrated by placing stuffed mounts of various bird species in pairs within the territory of a flycatcher. These models could be made to "sing" through loudspeakers placed under their perches. In one series of experiments, a flycatcher (*Myiarchus crinitus*) male tolerated models of a kingbird, a yellow-bellied flycatcher, a yellow-browed tyrant, a Baltimore oriole, and a red-eyed vireo. That is, he tolerated the intruder unless the dummy sang a *M. crinitus* song; the territory holder attacked any model that dared to sing the owner's song. That the birds rely on song, rather than appearance, for species recognition was demonstrated by a territory holder's reaction to mounts of males of his own species. The living and stuffed males could peacefully coexist as long as the model sang the song of a different species. When the flycatcher model sang the appropriate species song, it was attacked until the tape was switched making him change his tune (Lanyon, 1963).

RECOGNITION OF CASTE

In social insects, the duties associated with the maintenance and perpetuation of the commune are divided among colony members. A group of individuals in a colony that is anatomically and behaviorally specialized to perform specific duties is called a caste. Each colony member identifies its caste membership, causing others to respond to it appropriately. You may recall from the previous chapter that a queen honeybee communicates her royal status with pheromones. In addition to preventing the develop-

ment of additional queens, these chemicals elicit the appropriate solicitude from the nurse workers who provide her with food and cleanse her body by constant licking (Figure 18.3). The queen's chemical dominance is conferred by three unique substances: trans-9-keto-2-decenoic acid and trans-9-hydroxy-2-decenoic acid from the mandibular glands and another substance produced by a gland at the base of the sting (Koschevnikov's gland; Wilson, 1971). The pheromones are picked up as the nurse workers lick the queen's body and are rapidly transmitted throughout the hive as the workers exchange food with one another.

Among the social Hymenoptera (e.g., ants, bees, and wasps), the males, or drones, can also be identified as a group. Among honeybees (*Apis* spp.), a drone's only contribution to the hive is to mate with the queen during her nuptial flight. During this flight, the local males provide the queen with enough sperm to last her entire lifetime. Before the drones make this genetic contribution to the next generation, they must be kept by the female workers since the males

Figure 18.3 Worker honeybees caring for their queen. The queen (in the center) maintains her dominion by pheromones. The nurse workers respond to the queen's chemical signals by grooming and feeding her. Pheromones from the queen also prevent rival queens from developing within the hive.

cannot even feed themselves. It is, therefore, helpful for these insects to be able to distinguish males from females. When food becomes scarce, as it does in the fall, the females drive away or simply starve the drones, thereby reserving the available resources for those who make an active contribution to the colony (Gould and Gould, 1988).

Furthermore, some social insects are able to distinguish individuals of different life stages. This is important so that proper care is given to the stages that require it. Almost all ant species fastidiously organize the members of their colony so that eggs, larvae, and pupae are each placed in separate piles. In fire ants (*Solenopsis saevissma*) a chemical is produced by the larvae that identifies their life stage. If this substance is extracted from the larvae and placed on corn cob grits, the workers will carry these morsels to the larval piles, groom them, rub them with antennae and palpi, and care for them as they do their brood. If they are alarmed, the workers will pick up the treated dummies and run with them as they would their larvae (Glancey et al., 1970).

RECOGNITION OF POPULATION

Sometimes animals must not only be able to discriminate their own species, but also their own population from among others of that species. This ability is not universal, but it is widespread throughout the social animals.

Protection of Resources

Many social insects can distinguish nestmates from intruders of the same species (conspecifics). Their response to conspecific invaders falls somewhere on a gradient of hostility. At one end of the scale, the alien is tolerated but offered less food until it takes on the colony odor. More frequently, however, the response is violent. The intruder is attacked and killed or driven from the nest (Wilson, 1971). Often, invaders are un-

welcome because they are in search of booty. For example, honeybee workers, sometimes called "robber bees," trespass in other colonies to steal food. In other species, such as the acacia-ant (*Pseudomyrmex ferruginea*), intruders are treated with hostility because the queen may be at stake. The queen, the only reproductive individual in the invaded colony, is usually killed by intruders (Mintzer, 1982).

Colony membership is believed to be communicated primarily by odor. This has been demonstrated in army ants (*Eciton* spp.) by exposing workers to pieces of blotting paper on which members from either the home colony or an unfamiliar colony had been active. Workers responded to the treated paper as they would to the ants that left the scents (Schneirla, 1971).

The source of the odor differences among colonies of social insects may be (1) the diet or nesting environment, as it seems to be in honeybees (Kalmus and Ribbands, 1952) or (2) a colony recognition pheromone, as it appears to be in the acacia-ant (Mintzer, 1982).

Some vertebrates, particularly those that hold group territories, shun neighboring populations. The maintenance of a territory has several functions, one of which is the protection of resources. For example, a wolf pack defends a territory that is probably no larger than is necessary to provide enough food for the pack. Although territory size is usually stable, it sometimes fluctuates with prey density (Peterson, 1977). Nonetheless, the range covered by a single pack may be 125 to 555 square kilometers (Mech, 1977), so the pack cannot rely simply on its presence to deter invasion of its territory. Ownership is announced by both auditory and chemical means—one reason wolves howl is to discourage invasion of their territory. Howling can be heard by human ears for several miles and probably over even greater distances by wolf ears. It has been estimated that howling may advertise a pack's presence over an area of about 127 square kilometers (Joslin, 1966).

When an intruder howls from any spot within a pack's territory, the pack members usually howl a warning in return from wherever they happen to be (Harrington and Mech, 1983). Wolves reinforce this long-range communication with long-lasting, but more localized scent marks. These marks convey information not contained in the vocalizations, for example, the boundaries of the territory and the length of time since the pack passed by. Together the invisible chemical fenceposts and the pack sing-alongs help neighboring packs avoid one another and the possible harm that accompanies such an encounter. When a wolf pack spots one or more unfamiliar wolves on its territory, the trespassers are chased and attacked. Most of these flee. Many of those that do not will be killed (Mech, 1970).

Mate Choice

In other species, communicating population identity might facilitate the choice of a mate possessing adaptations to the local environment (Nottebohm, 1969, 1970, 1972). This suggestion arose as a possible explanation for the observation of song traditions, or dialects, within populations of some bird species. If you are interested in bird identification, you have probably listened to records of the songs of various species so that you could learn to identify the bird by its song in the field. This is possible because, as we have discussed, males sing species-specific songs. However, sometimes there is a population of birds whose songs are similar to their immediate neighbors, but consistently different in some characteristics from the songs of males of other populations. In spite of the differences in rendition among populations, the species song is still identifiable. This is analogous to our ability to recognize the English language in spite of differences in pronunciation by individuals from different parts of the country. Likewise, the dialects in bird song may allow a human trained ear to identify the population of the singer just as we are often able to identify a Bostonian and a Southerner by their accents.

Regional differences in the song of male white-crowned sparrows (*Zonotrichia leucophrys*) in the San Francisco Bay area of California can be seen in the spectrograms (plots of sound frequency versus time) of the songs of 18 males shown in Figure 18.4. Notice that the structure of the second part of the song is consistent within each region but varies between populations. The dialect is learned from older males during the first month or two of life (Marler and Tamura, 1964). Dialects have been found in many other species as well, including chaffinches (Marler, 1952; Promptoff, 1930), cardinals (Lemon, 1966), the plain titmouse (Dixon, 1969), short-toed treecreepers (Thielcke, 1965), the Argentinian chingolo sparrow (Nottebohm, 1969), and the montane white-crowned sparrow (Baptista and King, 1980).

If a female can recognize a male as belonging to her population by the dialect he sings, then his song might serve as a means of advertising unseen genes for subtle, perhaps physiological, adaptations to the local habitat. Choosing this "home town boy" as a mate might increase the chance of her offspring possessing the same adaptations (Nottebohm, 1969, 1970, 1972).

There are several other hypotheses for the function of dialects (see Baker and Cunningham, 1985; Kroodsma and Byers, 1991; Payne, 1981; Rothstein and Fleischer, 1987). However, to gain some insight into how researchers might test a hypothesis regarding the function of behavior, we will consider further the hypothesis that female birds use dialects as a cue indicating whether her suitor is adapted to the local environment.

The local adaptation hypothesis requires that the female prefers a mate that sings in the dialect of the region where she was raised. We might begin by asking whether or not there is evidence of such preferences in any species. Indeed, such a preference has been demonstrated in certain amphibians, such as the cricket frog (*Acris crepitans*), which also have regional dialects in their mating call. Female cricket frogs show a clear preference for the mating call of males from

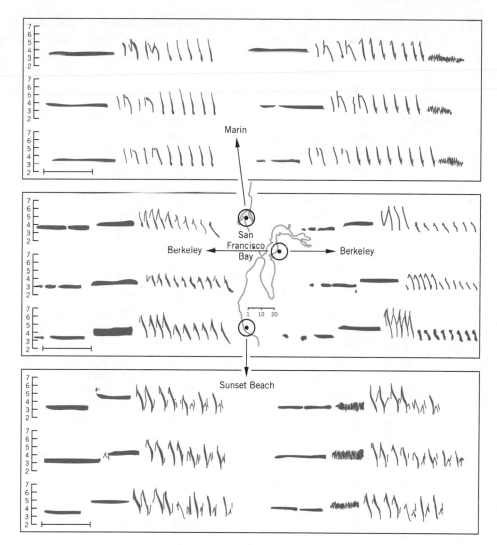

Figure 18.4 Dialects in the songs of white-crowned sparrows in the San Francisco Bay area. Notice that although the details of the second part of the song are similar within an area, they differ between populations. These learned song traditions may provide a means by which one bird can identify another from an alien population. (From Marler and Tamura, 1964.)

their natal area over the calls of males from distant places. This preference results from the tuning of the female's auditory nervous system to the characteristics of the local call (Capranica, Frishkopf, and Nevo, 1973).

Admittedly, the determination of preference is sometimes difficult to assess, but some experiments on birds have lent support to this idea. For instance, fledgling (10–25-day-old) female white-crowned sparrows were captured and listened to tapes of songs sung by males from their home region. The females were serenaded in this fashion for 1½ hours each day until the end of the period during which these birds learn songs (50 days old). They were then hastened into reproductive readiness by alterations in the photoperiod and implants of estradiol, a female sex hormone. Once they were in reproductive condition, the females were observed again during taped serenades by males. On the first day, the taped song was in a dialect unfamiliar to the females, and on the second day the recording was in the dialect they had previously heard in the laboratory (that of their home area). The females responded to the familiar dialect with significantly more perch hopping and by giving the display which, in nature, solicits copulation (Baker, Spitler-Nabors, and Bradley, 1982). A possible conclusion based on these data is that the females were sensitized by their early song learning experience so that they were more responsive to songs sung in their home dialect.

If these laboratory tests reflect what happens in nature, one might predict that a male singing in a female's natal dialect would be chosen as a mate over a male singing in an alien dialect. The next step might be to determine how females behave in nature. Some studies have addressed this issue and indicate that in nature females do not seem to show a preference for their natal dialect. One study, for instance, focused on mated pairs in the zone where the Presidio and San Franscisco dialects meet. The members of each pair were marked, the male song recorded, and the pair's breeding success measured. After the breeding season, the females were trapped and injected with the male hormone testosterone, a technique that will induce them to sing. The members of most pairs sang different dialects. Thus, the female did not choose a male that sang in her native dialect. Furthermore, matching song types made no difference in the reproductive success of the pair. So, there does not seem to be a reproductive advantage to choosing a male with the native dialect (Petrinovich and Baptista, 1984). Another study also indicates that dialect is unimportant in mate

choice by females in nature. This study revealed that female white-crowned sparrows living in a mixed-dialect population do not show a preference for either song type. Again we see that an individual female does not prefer a male that sings the same dialect as her father. In fact, a male's dialect may not play any role in a female's mate choice. If a particular female chooses a mate with a certain dialect during one breeding season, during the next breeding season, she is just as likely to pick a male with a different dialect as she is to choose one with the same dialect (Chilton, Lein, and Baptista, 1990).

The hypothesis that dialects help females choose a male with genes adapted to the local environment would also predict genetic differences among populations with different dialects. The genetic differences would be expected because the dialects would limit dispersal and intermixing of neighboring populations. Indeed, several studies have shown genetic differences among six white-crowned sparrow populations that sing in different dialects (Baker, 1974, 1975; Baker, Thompson, and Sherman, 1982). However, not everyone agrees that these genetic differences are caused by female preferences for natal dialects (Hafner and Peterson, 1985; Zink and Barrowclough, 1984). Therefore, a decision on whether dialects, in fact, influence mate choice will await future experimentation.

RECOGNITION OF KIN

Among many species of animals, relatives are treated differently from others, but this is not always because the animals can identify their kin. In some cases, juveniles may remain in their natal area, and neighbors and kin are then one and the same. Thus, an animal may simply behave differently toward its neighbors or toward familiar animals. In other situations, kin-biased behavior may result from individual or species recognition. There are some situations, however, in which animals seem to discriminate relatives that they have never seen before. When the interaction between two individuals *is* influenced by their genetic relatedness, we may describe it as kin discrimination (Waldman, 1987; Waldman, Frumhoff, and Sherman, 1988). We will discuss some of the methods that might be used in kin discrimination in Chapter 19.

The ability to discriminate kin may have at least two selective advantages. First, it may be a method of preventing inbreeding (Hoogland, 1982) or at least optimizing the degree of inbreeding (Bateson, 1980). Second, kin discrimination is a necessary thread in the fabric of nepotism—favoritism shown to relatives. The message of the old saying "Blood is thicker than water"—that family members deserve preferential treatment—is indeed heeded by many animal species. For example, in Belding's ground squirrels (*Spermophilus beldingi*), mothers and daughters, as well as pairs of sisters, cooperate by giving warning calls when a terrestrial predator approaches and by defending one another's nursing pups from other members of their species that would kill them. This cooperation enhances survival but is limited to close relatives (Sherman, 1981). Because discrimination of kin is so pertinent to a discussion of altruism, a more complete discussion of this topic is found in Chapter 19.

RECOGNITION OF INDIVIDUALS

We rather routinely see animals recognizing each other as individuals. We need look no further than two old dogs lounging in the shade of a tree who have known each other all their lives; or we can watch a coral fish guard its territory rather cursorily against its familiar neighbors but vigorously attack a stranger that happens to swim by.

Is it Communication?

Although examples of individual recognition are widespread through the animal kingdom, it is not clear whether they are examples of com-

munication. You may recall from the previous chapter that many investigators insist that the signaler *intend* to send a message and others believe that the signal must have evolved for the purpose of communication. In many cases, the cues used to identify individuals are aspects of individual variation. The dogs beneath a tree may identify one another by their appearance or by odors. Although these characteristics are distinctive enough for individuals to recognize one another, they are not intentional, nor have they evolved for the purpose of individual recognition. For these reasons, some people do not consider individual recognition to be communication. However, whether or not it is considered communication, individual recognition is important to many social interactions, so we will consider it here.

Importance

There are many situations in which individual recognition plays a role. For instance, it is often important for mates and for parents and offspring to recognize one another. In addition, territorial males will conserve energy if they can distinguish neighbors from strangers. A stranger at a territorial border might be more likely to attempt to usurp a portion of territory than a neighbor that happens to be sitting at the boundary. Furthermore, in close-knit social groups, such as higher primates, the effect of a signal often depends as much on the status of the signaler as it does on the signal's intrinsic properties (Marler, 1965).

Methods of Recognition

Sometimes an individual is identified by the way in which the signal is delivered. Many bird species respond differently to the song of a stranger than to the song of a neighbor. The experiments of Ken Yasukawa and colleagues (1982) illustrate the techniques used to demonstrate song-based discrimination among male red-winged blackbirds (*Agelaius phoeniceus*). In playback experiments, a recording of a male's own song, the song of a male whose territory borders his, or the song of a distant male was broadcast through a speaker placed within the subject's territory but near the boundary with the neighbor whose song was used in the neighbor presentation. The following aggressive responses of the territory holder were recorded: (1) the number of songs per minute, (2) the proportion of songs accompanied by the song spread display (an aggressive display in which the red epaulets on the singer's wings are made more obvious), (3) the number of flights per minute, and (4) how close the male came to the speaker. The responses to the tapes of the three singers were significantly different in three of these measures (song rate, intensity of the song spread display, and approach to speaker). These differences in response to tapes of the male's own song versus that of a neighbor and a stranger suggest that individuals can be recognized by differences in their songs.

Another technique employed to determine whether or not the males can discriminate among songs of certain individuals is "speaker replacement," that is, replacing the individual with a speaker that broadcasts its song. Male territory holders were captured and replaced by speakers positioned within their territory. The territory was defended only by recordings of songs. The songs played were either one of the original, now captive, territory holder or those of unfamiliar males. The number of times the territory was invaded by other male red-winged blackbirds and the duration of their trespass were recorded. Often the intruder was a male whose territory bordered that of the removed male. In these cases, the songs of the missing neighbor (the removed territory owner) were much more effective than those of a stranger in discouraging flights through the speaker territory, further evidence that birds distinguish neighbors from strangers by their songs.

In some instances, individuals are distinguished by variations in the expression of some characteristic other than a communicative signal.

The faces of many primate species are so variable that individuals are easily distinguished by humans. George Schaller (1963) worked with mountain gorillas and reported that he could distinguish individuals by their noses. Jane van Lawick-Goodall (1971) lived among the wild chimpanzees in Tanzania for 10 years. She also identified individuals by morphological or behavioral differences. Ancient Flo had a "deformed, bulbous nose and ragged ears" while William "had long upper lips that wobbled" and David Graybeard had a "handsome face and well marked silvery beard." Both workers surmised that the apes also used differences in appearance for individual recognition.

A characteristic odor may be the variable that serves as a personal signature. You are probably aware that dogs can identify one another by the trail of characteristic odors each leaves behind. Also, bullhead catfish live in communities in which each fish recognizes the others as individuals by their smell. In experiments, a yellow bullhead (*Ictalurus natalis*) demonstrated its ability to distinguish odors from two individuals by responding differently when the odor of one or the other test fish was introduced into its aquarium. When aquarium water from the tank of the bullhead that had been designated as the "positive" animal was introduced, the bullhead being tested would receive a food reward if it rose to the surface, stuck its head out of the water, and opened its mouth within 5 seconds. However, if the water from the aquarium of the bullhead that had been designated as the "negative" subject was introduced into the tank of the bullhead being tested, the required response was to retreat to a shelter that had been provided. By responding appropriately, the fish would avoid an electric shock. After approximately 25 training trials, the bullheads were able to discriminate the odors correctly 95 percent of the time. If the fish's odor receptors were destroyed experimentally, it lost this ability (Todd, 1971; Todd, Atema, and Bardach, 1967).

Some Say it With Scents, Sounds, or Simply Touch.

The feathery antennae of this male luna moth provide a large and exquisitely sensitive surface for receiving chemical signals.

A klipspringer in South Africa is marking the boundary of its territory by rubbing objects with a secretion from its preorbital gland.

A tiger in India is investigating the chemical messages left on a tree by other tigers.

Dolphins "talk" to one another using whistles, squeaks, groans, and clicks. They are quite "talkative" and chatter to one another almost continuously. However, because their calls are rapid and intricate, deciphering their language has proven to be challenging.

Ants communicate, not only with chemicals, but also with touch. Here, two ants (*Paraponera* spp.) in Bolivia communicate by touching one another with their antennae.

The Tugs and Ties of Society.

A red deer stag holds his harem by controlling the behavior of others through a variety of forms of communication. A stag assesses his opponent carefully, during roaring duels and visual inspection of size during a parallel walk, and fights only in cases where he is likely to win. He herds wandering females back into the group by walking beside them with a characteristic stance and gait. Chemical messages are also important. A stag marks landmarks with secretions from his preorbital glands. After sniffing a spot where a female has urinated, he may make a characteristic facial expression involving curling back his lip and raising his head, a response that may play a role in detecting the female's sex pheromones.

Touch can be a social tie. Here bull elephants entwine trunks in a friendly test of strength.

This open-mouthed threat by a lowland gorilla displays his formidable canines. Many threat displays provide cues, such as weapon size, that allow rivals to assess one another's fighting ability.

Courtship, a Language of Friendly Persuasion.

In some cases, courtship helps species identification. The croaks of this painted reed frog (*Hyperolius marmoratus*) declare his sex, species, and mating condition.

Courtship among king penguins begins when the male performs an ecstatic display in which he stretches up and throws his head back so that his bill points skyward. When an unmated hen is attracted by his antics, they begin a bout of antiphonal calling. While the bond is being formed, the male and female stand opposite each other, facing away with their bills. Then, the male parades in front of the female, advertising himself with his neck extended, his chest thrust forward, and his back arched. If the female accepts, she will follow him.

The Raggiana's bird of paradise (*Paradisea raggiana*) of New Guinea courts a female by spreading his beautiful red feathers.

The courtship dance of scorpions, such as the striped scorpions (*Centruroides vittatus*) shown here, may last for hours or even days. The male and female face one another, each with its abdomen raised high in the air, as they move about in circles. After these initial formalities, the male grasps the female with his claws (pedipalps) and they begin to dance backward and forward. During the dance, he pulls the female over his package of sperm and special hooks on her body cause the sperm to be thrust into her reproductive tract.

The intimidation display of the frilled lizard (*Chlamydosaurus*) make it look larger and more intimidating to its predator, causing the predator to make another choice.

When its wings are folded, the leaf katydid (*Pterachroza ocellata*) blends with its background. However, when alarmed, it stands on its head and suddenly exposes the brightly colored undersurface of its wings, thereby startling its predator.

The bright colors of this poison arrow frog (*Dendrobates punilio*) warn away potential predators.

SEXUAL ATTRACTION

Besides being species-specific, the signals that function to attract a mate must be effective over long distances so that males and females can locate one another even if the species members are widely distributed. For this reason, chemical and auditory signals are used commonly, but not exclusively, for attraction. The sex attractant pheromone of the female silk moth (*Bombyx mori*) was described in the previous chapter. If the wind favors the distribution of the pheromone, males from miles away may be attracted. The sex attractant pheromone of another species, *Actias selene*, is also potent. In one experiment, males displaced 46 kilometers relocated newly hatched females (Immelmann, 1980).

Auditory signals also carry well, especially when amplified by communal displaying or by special anatomical or environmental structures. The courtship songs of crickets (Hoy 1978; Moiseff, Pollack, and Hoy, 1978; Popov and Shuvalov, 1977) and the calls of frogs and toads (Littlejohn, 1977; Rheinlaender et al., 1979; Wells, 1977) have been shown to attract females. As you can see in Figure 18.5, female crickets will approach and congregate on a loudspeaker that broadcasts the courtship song of a male of its species.

COURTSHIP

Once the individuals are close enough to interact, they court before committing themselves to mating. There are several functions of courtship: identification of species and sex, reduction of aggressive tendencies, coordination of the mates' behavior and physiology (Etkin, 1964), and assessment of male qualities.

Identification

In his courtship display, a male signals his species, sex, reproductive condition, and location. It is necessary for a male to identify himself as belonging to a certain species so that any errors

Figure 18.5 Female crickets aggregating on a loudspeaker that is broadcasting the courtship song of the male of that species. One of the functions of the courtship song is to attract appropriate mates.

made during the initial attraction can be corrected before reproductive effort is wasted on a member of the wrong species. The advertising male must also make clear his sex because it is not always easy to tell. Although it is easy to distinguish the sexes of some species, such as dogs, in other species the sexes are remarkably similar, as we see in sparrows. The male also announces his location. In some cases, the location is obvious, as with a strutting peacock. In other cases, it must be deduced, as when a toad croaks in the blackness of night.

Reduction of Aggression

Since members of the opposite sex are not always friendly when they first meet, a second function of courtship is to reduce aggressive tendencies so that the couple can approach each other safely. Among carnivorous invertebrates, including the orb-weaving spiders we discussed in the previous chapter, it is usually the male that is in danger. The female is larger and he is about the size of her usual prey. He reduces her

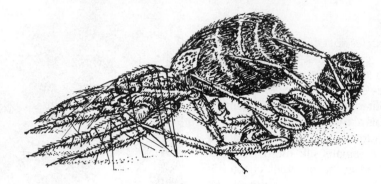

Figure 18.6   Mating in the crab spider, *Xysticus cristatus*. A function of courtship is to reduce aggressive tendencies. In predatory invertebrates such as the crab spider the female is usually larger than the male. A male crab spider must court the female properly or he will be mistaken for prey. The threat of becoming a meal is countered by the male crab spider by restraining the female with silken threads. The female is strong enough to break these ties but remains immobile while the male injects his sperm.

aggressive tendencies by tapping his message on her web. The male crab spider (*Xysticus cristatus*) ties the dangerous female down with silken threads before he attempts copulation (Figure 18.6). Evidently, the act of tying her down triggers her reproductive behavior because the threads are not strong enough to restrain her physically (Bristowe, 1958).

Among certain vertebrates, the female must enter the male's domain. To do so, she must overcome his aggressive tendencies. When a female stickleback (*Gasterosteus aculeatus*) enters a male's territory, for instance, she reduces the probability of attack by assuming a "head-up" position that displays her egg-swollen abdomen and distinguishes her from an intruding male. The subsequent zigzag courtship dance of the male consists of swimming toward the female, a movement of incipient attack, followed by swimming away, as he leads the female toward the nest (Tinbergen, 1952).

Coordination of Behavior and Physiology

The third function of courtship is to coordinate the couple's behavior and physiology. Among

certain species of birds, the male's song may stimulate a female into reproductive condition (see Kroodsma and Byers, 1991) and among salamanders female receptivity may be increased by male courtship pheromones (reviewed in Verrell, 1988). For example, during courtship, a male mountain dusky salamander (*Desmognathus ochrophaeus*) applies a courtship pheromone to the female by pulling his lower jaw across the female's back and angling his snout so that his premaxillary teeth scrape the female's skin, thereby "injecting" the pheromone into the female's circulatory system. The female then indicates her receptivity by assuming a tail-straddling position and the male deposits a spermatophore (Figure 18.7a). Lynne Houck and Nancy Reagan (1990) demonstrated that the male's courtship pheromone makes the female more receptive. They staged a total of 200 courtship encounters between 50 pairs of salamanders on each of 4 nights. By removing the mental gland, which produces the pheromone, from each male, they prevented the males from delivering any pheromone during courtship. These glands were then used to create an elixir containing the courtship pheromone. Thirty minutes before

a

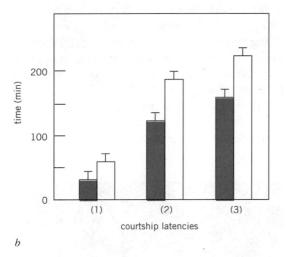

b

Figure 18.7 During courtship a male mountain dusky salamander injects a female with his courtship pheromone, which serves to make her more receptive. (*a*) The male (on the left) alternately scrapes the female's back with his teeth and swabs the area with courtship pheromone, which is produced by a gland beneath his chin. The female (on the right) indicates her readiness to mate by assuming a tail-straddling position. (*b*) Courtship behaviors and mating occur more quickly when a female has been treated with courtship pheromone than when she has been treated with saline. Pheromone treatment did not significantly shorten the average time from the beginning of courtship to active courtship. However, after pheromone treatment, the average time until the female is in a receptive tail-straddling position is 43 minutes (26%) shorter, and the average time until mating (spermatophore transfer) was 59 minutes (28%) shorter. (From Houck and Reagan, 1990.)

some encounters, each female was treated with this pheromone-containing elixir. Before other encounters, the same female was treated with a saline solution. After a female received a pheromone treatment, she assumed a tail-straddle position indicating receptivity 43 minutes (26 percent) sooner and mated 59 minutes (28 percent) sooner than she did after receiving a saline injection (Figure 18.7b).

Assessment

Finally, courtship may allow a female to judge the qualities of her suitor so that she can choose the one most likely to enhance her own reproductive success. Most commonly a female is required to invest more time and energy than a male to raise offspring successfully. It is, therefore, important that she choose a mate with the best genes possible. This means that males must advertise and females must discriminate from among them. Some aspects of courtship suggest that it provides a means for evaluating the suitor's qualities, including his physical prowess, ability to provide food for the offspring, or even the extent of his commitment. These suggestions stem from several studies, including those on the courtship displays of falcons and eagles.

Their aerial acrobatics may allow a female to judge her suitor's speed and coordination, skills important in hunting and, therefore, in his ability to provide food for his offspring. In addition, the courtship of the common tern (*Sterna hirundo*) requires males to catch fish and offer them to the female. She compares the quantity of fish provided by her various suitors and usually chooses the best fisherman. The number of fish a male provides during courtship, the quantity provided to his chicks, and the fledgling success of the clutch are significantly correlated. The quality of the courtship offering, then, is a reliable indicator of the male's ability to provide for the pair's offspring (Wiggins and Morris, 1986). Additional discussions on the use of courtship signals in assessing the characteristics of the mate are found in Chapter 14.

SYNCHRONIZATION OF HATCHING

The eggs of precocial birds such as ducks and geese are laid in large groups, or clutches. If they are separated, hatching occurs over several days. However, when they are incubated together, they hatch within a short time of one another. The obvious question is, how do chicks still entombed within their shell know when their siblings are emerging? The answer is quite simple: They call to one another. The chicks begin to pipe several days before hatching, and just before a chick bursts forth into the outside world, its piping peaks. The hatchlings hear the rising chorus and they prepare to emerge from their eggs. Synchronization of hatching is adaptive because within hours after the first chick emerges, the mother leads her brood away from the nest, thereby reducing the likelihood of predation. Stragglers are likely to perish (Vince, 1969).

BEGGING AND OFFERING FOOD

Individuals of many species of social insects beg for food. When signaled in a specific manner,

individuals will regurgitate food to other members of the colony, a process called trophallaxis. The larvae of ant species such as *Formica sanguinea* or *Solenopsis invicta* beg for food by rocking their heads back and forth, flexing their mandibles, and "swallowing" rapidly (Hölldobler and Wilson, 1990). In Figure 18.8 you can see a wood ant (*Formica rufo*) elicit trophallaxis from another worker by touching her head with her antennae and forelegs.

The parents of many vertebrate species tend the offspring by feeding them, and communication can involve signals that prompt the parent to offer food. Nestling birds often beg for food by opening their mouths widely, sometimes uttering begging calls. Patterns or colors inside the mouth act as releasers that cause the parent to provide a meal either by poking pieces of food into the gaping mouth or by regurgitation to the youngster (Figure 18.9). As the nestlings get older, they may beg for food with conspicuous wing movements. A juvenile may gesture,

Figure 18.8 Individuals of social insect species may beg for food in various ways. Here, wood ants exchange food. One worker causes another to regurgitate food (a process called trophallaxis) by tapping her forelegs and antennae to the nestmate's head and mouth parts.

Figure 18.9 The gaping mouths of nestling birds begging for food. Nestlings of many bird species solicit by opening their mouths widely, thereby revealing species-specific markings inside the mouth that cause the parent to provide food.

as the bald ibis (*Geronticus eremita*) does, by waving its wings slowly, or it may signal, as many songbird nestlings do, by quivering its wings.

Mammals invite their youngsters to nurse with a characteristic stance. An ungulate mother that commonly has only one or two offspring at a time, such as a deer or an antelope, spreads her stiffened legs apart so that the young can reach the teats on her underside. Other mother ungulates, such as pigs, have too many young to suckle while standing. Instead, they lie on their sides to communicate a willingness to nurse (Figure 18.10; Fraser, 1968).

ALARM

One of the most fascinating facets of animal communication involves alarms—when one animal warns another of danger. The dangers in the wild are of many stripes. Among the most common are predators. Individuals may even have to be on guard against other members of their species (conspecifics) bent on either infanticide or the takeover of their territory. In many spe-

cies, the individual that detects the danger alerts others in the group with an alarm signal. Sometimes, as in ants, the signal serves to enlist support in confronting the danger. In other instances, as in certain passerine birds, alarm calls cause others to flee to a protective area or if they are already in cover, to freeze. (A discussion of the hypotheses for both the selective advantages of emitting alarm calls and the mechanisms for their evolution can be found in Chapter 19.)

Methods of Giving Alarms

Alarm may be communicated in a variety of ways. Some methods, such as the rump patches of many ungulates, are visual. When herd-dwelling species such as deer or antelope are home on the range and one animal senses danger, it indicates this by making the patch more conspicuous with the erection of rump patch hairs or the elevation of its tail (Figure 18.11; Smith, 1977).

The use of chemicals to alert others is widespread among animals, particularly among the

Figure 18.10 A mother pig nursing her young. Many ungulate mothers must signal a willingness to allow the young to nurse. The mother pig does this by lying on her side.

Figure 18.11 The alarm signal in white-tailed deer. An animal that senses danger will raise its tail while running away. This makes the white rump markings more visible and warns others of the threat.

social insects (Maschwitz, 1966). For instance, when a bee stings and pulls away, she leaves her stinger and the attached innards in the victim's body (Ghent and Gary, 1962). Her mutilated flesh liberates a chemical alarm substance, isoamyl acetate, that attracts other workers to assist in defending the hive against the threat. This chemical, which may be familiar to you as the paint solvent banana oil, is produced by the cells lining the sting pouch. She can also release the alarm by simply pointing the tip of her abdomen upward and opening the sting chamber to dispense the pheromone. At the same time she beats her wings, thereby hastening the diffusion of the substance toward her nestmates. Although accessory pheromones are probably liberated simultaneously, isoamyl acetate is sufficient to alarm and attract worker bees. This can be demonstrated by their response to cotton balls soaked in pure isoamyl acetate (Boch, Shearer, and Stone, 1962).

Chemical alarm substances are also produced by some vertebrates. For example, if the skin of a minnow or any other fish in the order Ostariophysi is broken, a chemical is released from club-shaped epidermal cells. This chemical causes other fish, particularly those of the same species, to swim away and avoid the area for hours or even days. Escape responses have also been elicited by a substance liberated from the skin of toad tadpoles *Bufo* (Schneider, 1966).

Alarmed black-tailed deer (*Odocoileus hemionus columbianus*) release a chemical from their metatarsal glands, located on the outside of the hindlegs. To humans the scent is similar to that of garlic. Normally the release of this chemical is accompanied by visual and auditory alarm signals, so it is difficult to isolate the effects of the pheromone alone. Under experimental conditions, a deer presented with the pheromone becomes quietly alert, seemingly searching for some danger (Müller-Schwarze, 1971).

A number of species use sounds to communicate danger. We see this when a gerbil repeatedly thumps its back feet on the ground or when a beaver slaps its tail on the water. In other animals, special vocalizations have evolved. Certain pronghorn (*Antilocapra americana*), for instance, emit a blow sound when they encounter suspected danger that may serve to alert nearby individuals (Waring, 1969). The alarm calls of Belding's ground squirrel and of the small passerine birds will be discussed in detail in the next chapter.

Specificity Among Alarms

Most species use the same signal to indicate any source of danger, but some species use specific calls to designate the type of threat. For example, vervet monkeys (*Cercopithecus aethiops*) classify their most common predators into one of at least three groups—snakes (e.g., python), mammals (e.g., leopard), or birds (e.g., eagle). Considering the hunting strategy of each type of predator, the characteristics of the alarm call and the response of conspecifics within hearing distance seem to be adaptive. The low amplitude alarm call emitted when a snake predator is encountered captures the attention of individuals near the caller that might be in danger from the slow-moving reptile without attracting other predators in the area. The response of others is to look at the ground, the most likely place to find a snake. However, when a major mammalian predator such as a leopard is seen, very loud, low-pitched, and abrupt chirps are emitted. These properties make the call audible from a great distance and make the caller easy to locate by its fellows. The most common response to the chirp is to scatter and run for cover in the trees, a relatively safe haven from the ambush style of attack characteristic of a leopard. The staccato grunts of the threat–alarm bark that are sounded when an avian predator is spotted are also loud and low pitched. These features allow the grunts to be easily located and to be transmitted over long distances, thereby broadcasting the position of the predator. When the threat–alarm bark is heard, others will look up and/or hastily retreat to thickets. The dense brush makes it difficult for a swooping eagle to catch them (Struhsaker, 1967). The responses just described were also typical of those to playback tapes of these three types of call. The responses, then, are specific to the nature of the alarm call, and not to the appearance of the particular type of predator (Seyfarth, Cheney, and Marler, 1980).

Similarity Among Species

Alarm signals, particularly of species living in the same area, are often remarkably similar. As an example, note the striking resemblance of the alarm calls of several species of passerine

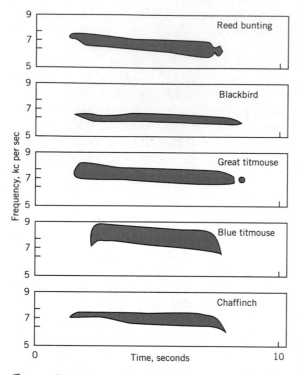

Figure 18.12 Alarm calls of several species of small songbirds. Notice the structural similarity of the calls. As a result, an alarm call will alert birds of many species (Modified from Marler, 1959.)

birds shown in Figure 18.12. A possible explanation for this similarity is that when species have common predators, it is advantageous to be alerted to danger regardless of who sounds the alarm. It should not matter to a small bird feeding on the ground whether or not it is a conspecific that alerts it to an eagle overhead. If the divergence among alarm calls of species living in the same area is slight, the chance of a predator striking unnoticed is decreased. Thus there is selection for convergence of alarm signals (Marler, 1973; Wilson, 1974).

A second possible reason for similarity of alarm calls is that they have the same functions. The source of an alarm pheromone used to recruit assistance should be easily located over a

reasonable distance; workers who are too far away to detect the danger themselves should be able to travel to the site in time to be of service. Molecules with a low molecular weight generally meet these requirements well. They are more volatile and, therefore, disperse rapidly over a reasonable area. This creates a steep concentration gradient that facilitates locating the caller. Natural selection has, therefore, chosen chemicals with low molecular weights to serve as alarm substances (Wilson, 1971).

Likewise, it has been suggested that, ideally, an alarm that triggers dispersal to safer areas should be difficult to locate (Marler, 1957). In that way, the caller would be in less personal danger. Armed with a knowledge of the cues known to assist the localization of a sound source, Peter Marler was able to predict the type of bird calls that would be best suited for alarm. The alarm calls typical of small passerine birds shown in Figure 18.12 have the characteristics that would be expected to make them difficult to locate—they begin and end gradually and employ only a few wavelengths centered on 8 kHz.

It is interesting that there are some birds that take advantage of the similarity among alarm calls to foil competitors. For instance, in one study (Moller, 1988), 63 percent of the alarm calls issued by great tits (*Parus major*) were false alarms. These were given when other individuals, particularly sparrows, were in the caller's feeding area, but when no predator was in sight. The other individuals generally heed the warning and flee. The result of the false alarm call is that the feeding area is cleared of competitors and the caller can feed undisturbed.

DISTRESS

Besides alarm calls, which are emitted when an individual senses danger but is not being attacked, some species also have distress calls that are reserved for the moments when injury seems imminent. These calls are often quite loud and may serve to startle and confuse the predator while simultaneously warning conspecifics.

In some species, distress calls will cause aggregation of conspecifics. When jackdaws in Europe or crows or robins in the United States hear distress calls, they approach the source of the call and attack the predator en masse, a response called mobbing. Likewise, when an elk calf in the Jackson Hole Wildlife Preserve fell into a ditch and began emitting distress calls, other elk gathered at the ditch (Frings and Frings, 1977).

In some cases, as in vervet monkeys (*Cercopithecus aethiops*), it is primarily the parent who responds to the call of a distressed youngster. Among primates, distress calls often take the form of screams. When Dorothy Cheney and Robert Seyfarth (1980) played the screams of an infant vervet monkey to a group of three mothers, all of which were separated from their offspring at the time of the test, the mother of the infant whose screams were heard looked toward the source of the sound more quickly and for a longer time than the others. In addition, the mother of the distressed youngster was also significantly more likely to approach the speaker broadcasting the screams. It is interesting that the other mothers hearing the tape were likely to look at the *mother* of the taped juvenile, as if expecting her to respond.

CALMING

Some displays have the opposite effect of alarm. One example is the departing signal given by water birds, such as gannets, boobies, anhingas, and cormorants. If a group member takes off without an appropriate signal, it will alarm the others. An unalarmed individual that is going to leave the scene will perform a "pretake-off" display (van Tets, 1965). Another type of calming signal serves as an indication that all is clear. For example, a black-tailed prairie dog that has been startled will later perform a jump–yip display that seems to indicate that the danger is past.

After the jump–yip display, it begins another activity (Smith, 1977).

SIGNALS USED IN AGONISTIC ENCOUNTERS

In many species, social interaction involves agonistic behavior, that group of actions involved in conflicts that includes aggressive behaviors, such as threats, and submissive behaviors, such as appeasement or avoidance. Since such behavior is usually the result of competition, it is most common between members of the same species and, furthermore, members of the same sex. Once dominance is attained by a particular individual, it is often flaunted with a badge or characteristic stature so that in an established social group, each individual knows its place without continuous testing.

There is a variety of signals used to threaten another of the same species. As was mentioned in the discussion of ritualization in Chapter 17, these displays often originated from portions of attack behavior such as biting and rushing. Typically an animal makes itself appear as large and intimidating as possible, often while showing its weapons (Eibl-Eibesfeldt, 1975; Figure 18.13). Although the displays are well known, their interpretation has been controversial.

As we saw in the previous chapter, the traditional ethological view is that the displays convey information about the intentions of the signaler so that the receiver can respond appropriately. Many of the threat signals seem to reflect incompatible tendencies to attack or flee. Given these conflicting tendencies, and differences in the internal and external stimuli, it is not surprising that a given threat display may be followed by either attack or escape. Nonetheless, the display still carries some information, albeit imprecise, about what the signaler will do next (Hinde, 1981). The general idea is that one animal signals its intention to behave in a particular manner and the other uses that information to respond appropriately.

However, you may recall that it has been ar-

Figure 18.13 Threat signal in a baboon. Notice that the animal's weapons, its teeth, are displayed.

gued that this system of signaling should be vulnerable to lying. Assume that disputes are settled by displaying. Each contestant signals how likely it is to attack or how long it is willing to display. The individual that announces the stronger resolve wins and the loser retreats. Intentions, however, are easy to bluff. If a mutant appeared that was prone to lying, it could easily fake a strong resolve because others would assess its determination as greater than theirs. As a result, the liar should win every dispute. A successful liar would be favored by natural selection and soon so much of the population would be prone to prevarication that it would pay to ignore displays and require each opponent to prove its claims (Maynard Smith, 1979). Consider this analogy to humans. If you and a rival bet $50 on who could run faster, would you be willing to take your opponent's word that he or she can run a 3-minute mile or would you insist that the claim of fleet feet be demonstrated by action?

Because animals might lie, an alternative

Figure 18.14 A threat display by a male Siamese fighting fish (*Betta splendens*). Individuals of many species make themselves appear larger when threatening another. Here the fish flares its opercula (gill covers) outward and spreads its fins, actions that help to intimidate its rival.

interpretation of threat signals has become popular. Displays used in threatening an opponent should provide cues that can be used to assess the rival's true abilities. Therefore, displays are related to fighting ability and difficult to fake (Dawkins and Krebs, 1984). Only then is the neighborhood weakling prevented from bluffing its way to triumph.

Some displays emphasize size. Since the bigger animal often wins, size is a reliable cue to the probability of fight success and it is difficult to fake. Birds emphasize size by fluffing feathers. Fishes often erect their fins and flare their gill covers outward (Figure 18.14), and some even puff themselves up.

Sometimes an indirect cue provides an indication of size. For example, the pitch of a call is often related to body size. The lower the pitch of the call, the larger one would predict the caller to be. This relationship between body size and pitch of the call has been clearly demonstrated for the toad *Bufo bufo*. Furthermore, since physical strength is important in contests over possession of females, larger males are rarely replaced by smaller ones. In these disputes, males clasp onto a female's back, sometimes for several days, until she lays her eggs that he then fertilizes. Since males outnumber females in the ponds where they mate, there is intense competition for mates. The males will try to dislodge rivals from a female's back by pushing and wrestling. When a male is attacked, he calls. In laboratory studies where taped croaks were played to male toads, the pitch of the defender's call influenced the probability of attack. Challengers were less likely to attack following a low-pitched call (Davies and Halliday, 1978; Figure 18.15).

Other displays are indirect trials of strength or endurance. For example, red deer stags, *Cervus elaphus*, roar at one another while defending their harem (Figure 18.16). The roaring contest seems to follow rules. One stag roars several times and then listens while another answers

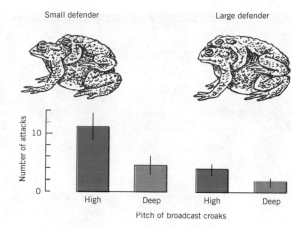

Figure 18.15 The results of an experiment demonstrating that male toads (*Bufo bufo*) use the pitch of their rival's croak to assess his size and fighting ability. When a male toad finds a suitable female, he will ride on her back until she lays eggs that he can fertilize. Rival males may try to dislodge him. When attacked, the defending male calls. In this experiment, medium-sized males attacked either small or large males that were clasping females. The defending male was prevented from croaking by a rubber band through his mouth. During each attack, tape-recorded croaks were played to simulate the call of a defending male. Regardless of the actual size of the defender, he was less likely to be attacked when the pitch of the broadcast croak was deep. The pitch of the call is only one factor used to assess size. Notice that large defenders are attacked less frequently than small males. The strength of the defender's kick may be another cue. (Modified from Davies and Halliday, 1978.)

with his best bellows. Frequently the contest involves several stags. As would be expected if they were using roaring for assessment, rivals' roars rarely overlap. As the contest progresses, the rate of roaring increases until one competitor gives up (Figure 18.17). Because roaring is so exhausting, the stag's ability to continue the vocal duel is an indication of strength. At the peak of the rut, harem holders often roar day

Figure 18.16 A red deer stag (*Cervus elaphus*) roaring at a rival. Red deer stags challenge one another by bellowing. In this vocal duel each male takes a turn roaring at the other. The pace of the bellowing increases until one contestant gives up. Because roaring is strenuous, it is a reliable cue to use for the assessment of a rival's fighting ability.

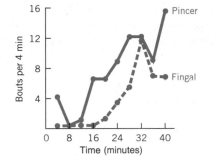

Figure 18.17 A roaring contest between two red deer stags. The males defend their harems by roaring at rivals. As the males take turns roaring, the tempo of the contest accelerates. Finally one male gives up and leaves. (Modified from Clutton-Brock and Albon, 1979.)

and night, causing them to lose as much as 20 percent of their body weight. Those that become so exhausted that they can no longer roar lose their females to the challenger.

Several aspects of the roaring duels are con-sistent with the idea that they are used for as-sessment. First, roaring ability requires both strength and endurance and is, therefore, a cue that is difficult to fake. In addition, the maximum and mean roaring rates of an individual stag are significantly correlated with his fighting ability. Furthermore, on the rare occasions when fights do occur, the combatants have equal roaring skills more often than is expected by chance (Clutton-Brock and Albon, 1979; Clutton-Brock et al., 1979).

CONTACT

Just as there are signals that maintain distance between groups, there are others that maintain contact within a group. These are most common in habitats in which visibility is poor. For example, the South American tapir *Tapirus terrestris*, a hoglike mammal, forages in the dense vegetation of the rain forest. As these animals move through the forest, they produce a "sliding squeal" that helps them keep in touch, thus enabling them to reassociate at any time (Hunsaker and Hahn, 1965).

RECRUITMENT

Reasons

Recruitment is a specific type of assembly in which the members of the group are directed to a specific place to assist in a particular job such as food retrieval, nest construction or defense, or migration. Recruitment is common among social insects (Wilson, 1980). Some species have several recruitment systems, each for a different purpose. The African weaver ant (*Oecophylla longinoda*), for instance, has different signals for recruiting others to food, to new terrain, for emigration, for attacking intruders that are nearby, and for attacking intruders that are distant (Hölldobler and Wilson, 1978).

Methods

As we will see, social insects may employ a variety of methods to recruit others.

Tandem Running or Odor Trails   Tandem running, in which an individual maintains physical contact with the leader, and odor trails in ants have been described in the previous chapter. Odor trails are also left by termites, even the primitive species that never leave the nest to forage such as *Zootermopsis nevadensis*. Alistair Stuart (1963a,b; 1967) has shown that when a *Zootermopsis* nymph detects a change in light intensity or air currents that might indicate a break in the nest wall, it writes a trail in a substance produced by the sternal gland on its fifth abdominal segment as it runs from the site of damage. The nymph gets its nestmates' attention by bumping them while jittering from side to side. This bumping alarms the others that then follow the trail back to the breach in the wall to repair it.

Bee Dancing   The most remarkable and controversial example of recruitment communication is in honeybees (*Apis* spp.). Although it may take several days before a new food source is noticed, soon after one honeybee discovers the prize, many recruits join the harvest. The dramatic increase in the number of bees utilizing a new food source is evident in Figure 18.18. We know that the initial discoverer does not lead the others to the food because recruitment is observed even if she is captured as she leaves the hive on a return trip (Maeterlinck, 1901).

How does the pioneer scout enlist the assistance of her sisters? The first step in scientifically addressing a question such as this is to observe the bees' behavior. Observations have revealed that as a scout bee sips the sugary solution provided by either a newly blossoming patch of flowers or an experimental feeding dish, she marks the spot with a chemical from the Nasonoff's gland at the tip of her abdomen. Traces of

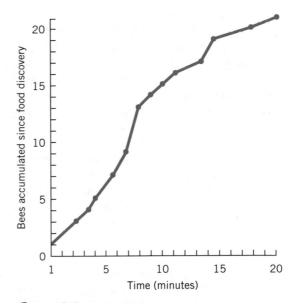

Figure 18.18   The number of bees arriving at a newly discovered food site over time. Notice how quickly workers are recruited. (Modified from Wenner, 1971.)

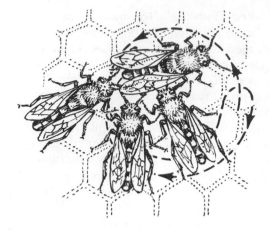

Figure 18.19 The round dance of a honeybee scout. After finding food close to the hive, a scout returns and does a round dance on the vertical surface of the comb. This dance consists of circling alternately to the right and left. It informs the recruits that food can be found within a certain radius of the hive.

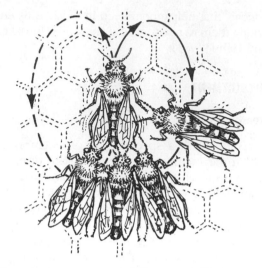

Figure 18.20 The waggle dance of a honeybee scout. When a scout finds food at some distance from the hive, she returns and does a waggle dance. Aspects of the dance correlate with the distance and direction to the food source.

the nectar or a dusting of pollen adhere to her body. The scout then returns to the hive and mounts the vertical comb. When other workers gather around her, she regurgitates a sample of the nectar to these nestmates. If the food source was relatively close to the hive, she begins to dance in circles, once to the left and once to the right. This circling, or round dance (Figure 18.19), is repeated many times. However, if the bonanza was far from the hive, she performs a waggle dance, during which she moves in the pattern of a figure 8. It is a simple dance; straight ahead, circle to the right, then up the center again and loop to the left (Figure 18.20). During the central straight run of the dance, she vibrates or waggles her abdomen and produces a low frequency buzz at about 250 Hz. Her sisters, unable to see in the darkness of the hive, crowd around and follow her movements with their antennae. Eventually there is a rapid exodus of recruits that later show up at the food site.

Several aspects of the waggle dance are correlated with the location of the food source. Direction to the food find is indicated by the angle of the waggle run of the dance. There is close agreement between the angle of this run with respect to gravity and the angle that the recruits must assume with the sun as they fly to the food source. On the vertical comb within the dark hive, bees substitute straight up for the position of the sun (gravity tells them which end is up), and then the scout waggles at the angle of the flight path to the food relative to the sun. So when the food is to be found in the direction of the sun, the waggle run of the dance is oriented straight up. If the recruits should fly off with the sun directly behind them, the waggle run would point straight down on the comb. Likewise, a food source 20 degrees to the right of the sun would be indicated by a straight run directed 20 degrees to the right of vertical (Figure 18.21). The angle of the dance is adjusted

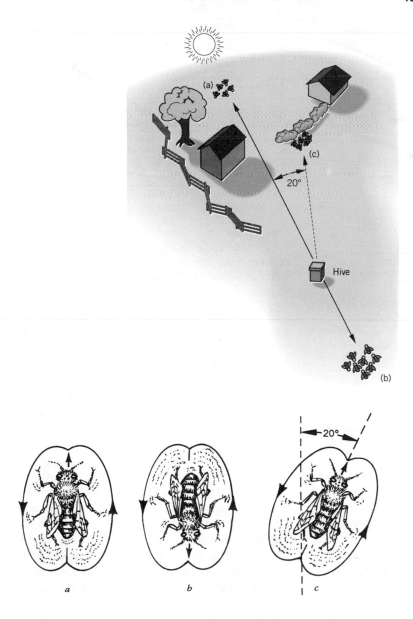

Figure 18.21 Direction correlations of the waggle dance of honeybees. The dancer indicates the direction to the food find by the orientation of the waggle run of her dance relative to gravity. When the bee is dancing on the vertical surface of the comb within the dark hive, the direction of the sun is straight up. The angle that recruits should assume with the sun as they fly to the food source is equal to the angle of the waggle run relative to vertical. (*a*) The food is in the direction of the sun so the dancer waggles straight up. (*b*) When the recruits should fly directly away from the sun to reach the food, the waggle run is oriented straight down on the comb. (*c*) A food source located 20 degrees to the right of the sun would be indicated by a dance oriented so that the waggle run was 20 degrees to the right of vertical.

by approximately 15 degrees an hour to compensate for the apparent movement of the sun across the sky.

Distance to the food source, or more accurately the energy that was expended during the scout's return from the goal, is correlated with

two other features of the straight run. The longer the duration of the straight run, which is emphasized by the dancer's waggling, the greater the distance to the food source (von Frisch, 1971). This relationship is shown in Figure 18.22. The distance to the food is also correlated

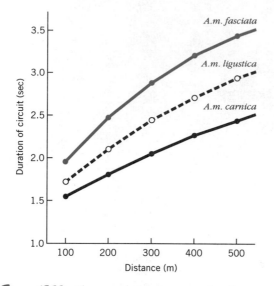

Figure 18.22 The correlation between the distance to a food source and the duration of waggling in the dance of honeybees. The number of waggles increases with the distance to the food source. (Data from von Frisch, 1971.)

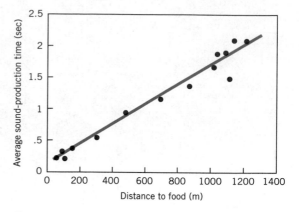

Figure 18.23 Correlation between distance to the food source and buzzing during the waggle dance of honeybees. The duration of buzzing increases with the distance to the food source. (Modified from Wenner, 1964.)

with the duration of buzzing during the straight run (Figure 18.23). As the distance to the food increases, the sound trains of buzzes are longer (Wenner, 1964).

Most of what we know about the bee dance comes from experiments performed by Karl von Frisch. Most biologists today believe that honeybees communicate the distance and direction to the food source in the waggle dance. Stated more precisely, von Frisch's dance-language hypothesis is that the information in the dance guides recruits to the general area of the food source and then olfactory cues, odor molecules adhering to the body of the scout bee and the bee odor left to mark the find, help them locate the exact site. Von Frisch provided the following evidence for his hypothesis:

1. **The behavior of the recruits depends on the scout bee's dance.** When food is located near the hive, the returning scout bee does

a round dance. The recruited foragers randomly search for the source in the vicinity of the hive. However, when the location of the food site exceeds a certain distance, the scout performs a waggle dance and the recruits begin to search preferentially in the direction of the food source.

These statements are based on the results of several of Karl von Frisch's experiments. His procedure, which is still used by some investigators, was to establish a feeding station where bees can imbibe a weak unscented sugar solution. Because the solution is weak, the returning foragers do not dance. On the test day a stronger sucrose solution is set out at the feeding station and a scent, perhaps lavender oil, is added to it. The forager discovering the bounty returns to the hive and dances. During this performance, the hive is closed to prevent bees from exiting. Meanwhile, the food is removed from the feeding station and plates emitting lavender fragrance, but containing no reward, are strategically placed. When the hive is reopened, recruits leave and search for the food. When they detect the food odor emanating from a scent plate, they are attracted to it. The direction in which the bees are searching is determined by counting

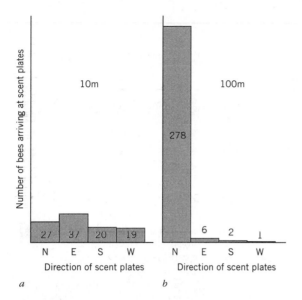

Number of bees arriving at scent plates

10m

278

N E S W

Direction of scent plates

2/ 3/ 20 19

100m

6 2 1

N E S W

Direction of scent plates

a

b

Figure 18.24 Number of bees arriving at scent plates positioned to the north, south, east, and west of the hive when the training station was 10 meters to the east of the hive (*a*) and 100 meters to the north of the hive (*b*). When the food source is close to the hive, the recruits search randomly within a certain radius of the hive. However, when the food source is distant, they look preferentially in the direction of the training station. (Modified from von Frisch, 1971.)

the number of bees arriving at each scent plate.

In one experiment, bees were trained to a feeding station 10 meters to the east of the hive. After sipping the strong scented sucrose on the test day, the scout returned to the hive and did a round dance. While the hive was closed, the feeding station was emptied and scent plates were positioned to the north, south, east, and west, each at a distance of 25 meters. As you can see in Figure 18.24a, the recruits arrived at each scent plate in almost equal numbers. Apparently they did not know the direction to the nectar. However, when the same feeding station was moved to a position 100 meters north of the hive, the returning discoverer did a waggle dance. This time almost all the bees arrived at

the northern scent plate, indicating that they had direction information (Figure 18.24b).

The accuracy of the direction information was further demonstrated in fan experiments. Following his usual procedures, von Frisch trained bees to a feeding station and then placed scent plates so that they were arrayed like a fan. Each plate was 550 meters from the hive and separated from its neighbors by 150 meters. The overwhelming majority of recruits appeared at the scent plate nearest to the original feeding station (Figure 18.25).

To examine the accuracy of distance information, von Frisch performed a step experiment. After training bees to a feeding station, scent plates were placed in line with the empty feeding station at intervals closer or farther away.

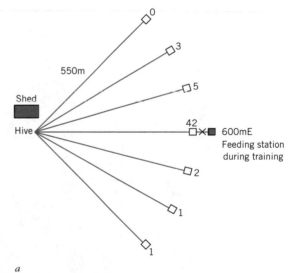

0

3

5

550m

Shed

Hive

42

600mE
Feeding station
during training

2

1

1

a

Figure 18.25 Results of a fan experiment showing that after following a waggle dance, the recruits search preferentially in the direction of the training station. Bees were trained to a feeding station 600 meters east of the hive. The greatest number of recruits arrived at the food-free scent plate in the direction of the training station. An "X" denotes the position of the feeding station during training and squares show the positions of the scent plates. (Modified from von Frisch, 1967.)

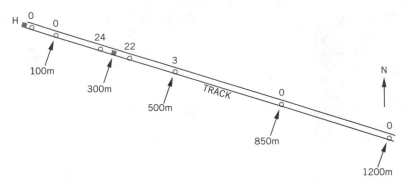

Figure 18.26 The results of an experiment showing that after following a waggle dance, the recruits preferentially search for food at the distance of the training station from the hive. Most of the recruits arrived at the scent plates closest to the original food station. The solid blue square at 300 meters shows the position of the feeding station during training and circles indicate the positions of the food-free scent plates. (Modified from von Frisch, 1967).

Figure 18.27 A mechanical bee that can "talk" to live bees. The mechanical bee is constructed of brass covered in beeswax. On its back is a single wing made from part of a razor blade. An electromagnet causes the wing to vibrate so that it mimics the sound patterns produced by real dancing bees. A tiny plastic tube connected to a syringe releases droplets of scented sugar water to recruit bees, thus simulating regurgitation by live scout bees. Computer software choreographed the robot's dance so that it directed the bees to a food source in a direction chosen by the experimenter.

Again, the recruits appeared at the scent plates closest to the original feeding station (Figure 18.26).

2. The recruits search for the food in all directions from the hive unless they have attended a waggle dance. Normal waggling can be disoriented by placing the hive in a horizontal position and depriving the bees of a view of the sun or a patch of blue sky. (When dancing on a horizontal surface, the bees orient the dance relative to the sun and if the sun is not directly visible, they determine its position from the pattern of polarization in the sky.) After attending a disoriented dance, recruits search randomly. However, if the hive is restored to its vertical position, both the dance and the search become oriented.

3. When faced with detours, recruits act as if they have direction information. In his detour experiment, von Frisch placed the feeding station so that the scout had to fly around a building to reach it. When she danced, she indicated the direction to the food as a straight line, the bee line. Some of the recruits were observed to fly up and over the building, suggesting that they were using direction information from the dance.

It has even been possible to "talk" to bees in their own language using a mechanical model of a dancing bee (Figure 18.27). A computer controlled the model's movements and sound production. The dancing model bee fed recruits scented sugar water through a syringe, mimicking the manner in which a real bee would feed the recruits upon returning to the hive. The model danced indicating a food source 250 meters to the south of the hive. Scent plates without food were placed 250 meters to the north, south, southeast, and southwest of the hive. Although the model's dance was not as effective as a live bee's dance in recruiting others to look for food or in accurately directing them to the correct location, most of the recruits showed up at the feeding station indicated by the dance (Figure

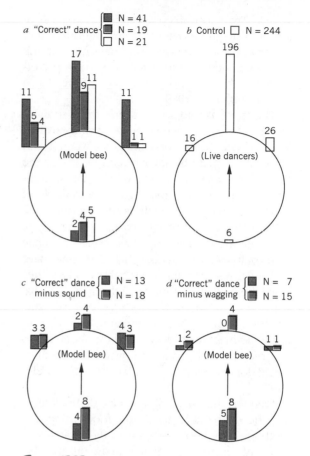

Figure 18.28 The dance of a mechanical bee can direct recruits to a distant food source. The figure shows the results of experiments in which the model bee danced by both waggling and buzzing in a manner that mimicked a dance of a live bee directing recruits to a food station located 250 meters to the south of the hive. Scent plates were placed 250 meters to the north, south, southeast, and southwest of the hive. Most of the recruits showed up at the scent plate signaled by the model bee's dance (*a*). The number of bees recruited and the ability of the recruits to locate the correct scent plate are somewhat less than that of bees recruited by a live dancer (*b*). When model bees' dances lack accompanying sound patterns (*c*) or waggling (*d*) the recruits show no preference for the direction indicated by the dance. (From Michelsen et al., 1989.)

18.28). Two critical components of the dance are the wagging movements and the buzzing. If either of these components is missing from the model's dance, the recruits showed no preference for the direction indicated by the dance (Michelsen et al., 1989).

The story, as told so far, seems quite straightforward, but the dance-language hypothesis has, in fact, been controversial. Adrian Wenner and associates explain honeybee recruitment on the basis of odor cues. Olfactory information available to bees includes the scent of both the food and of the locality as well as the bee odor left at the site. Odors brought back on the body of the scout are detected by the recruits' antennae as they follow the dancer. If the recruits have had previous experience with those odors, they leave the hive and monitor all the sites with that scent at which they had been previously successful. Their response to unfamiliar odors is to leave the hive, drop downwind for about 200 to 300 meters, and progress upwind in search of that odor (Wenner, 1971; Wenner and Wells, 1990).

Wenner has experimentally demonstrated that bees are able to locate food using this method. They know how far to fly because distance information is available in the buzzing that accompanies waggling. Furthermore, bees can be conditioned to visit a food dish in response to an injection of the food scent into the hive. Thus they can learn to associate a particular odor with food in a certain location (Johnson and Wenner, 1966; Wenner and Johnson, 1966). In addition, one can manipulate the number of new recruits by varying the odor of either the food or its location (Wenner, 1971).

It should also be mentioned that Wenner believes that the results of von Frisch's fan experiments are consistent with the odor hypothesis for recruitment. He believes that the array of scent plates might produce a continuous gradient of odor. If one assumes that each bee is searching for the center of the odor field, then the probability that a bee will arrive at a partic-

ular plate depends on the distance of that plate from the center of the odor field. Following this reasoning, the expected distribution of bees would not be significantly different from what von Frisch found (Wenner, 1971).

So, with both dance and olfactory information available simultaneously, how does one determine which cues are actually being used by the bees? If the odor information and the dance directions send the bees to different sites, then one could tell which cue the bees were using by observing where they arrived.

One way to create conflicting information is to get the dancers to "lie." Bees do not normally deceive one another, but they can be made to prevaricate by certain experimental manipulations. If the sun, or a suitable bright light, is seen by a dancing bee, she will orient her waggle run to the light rather than to gravity. Normally, nestmates attending this dance also see the light and interpret the movements without confusion. However, because the bees can dance relative to either light or gravity, the experimenter can manipulate the information in the dance at will, simply by changing the bees' sensitivity to light. A bee's ability to detect light is influenced by its three ocelli, non–image-forming light detectors. When the ocelli are covered with black paint, the bee is 6 times less sensitive to light. In spite of this dimness, she can still forage and dance.

In a clever experiment (Gould, 1975), the bees were presented with a light that was bright enough to be seen by a normal bee, but not by an ocelli-painted one. The ocelli of scouts were painted, and they danced indicating the flight direction relative to gravity. However, the normal recruits interpreted the dance relative to the artificial sun. Since the frame of reference of the dancer was different from that of the recruits attending the dance, the message sent was different from the one received. The dance-language hypothesis predicts that depending on cues present in the dance, the misinformation provided by the dance should cause recruits to arrive at a scent plate as far off from the foraging

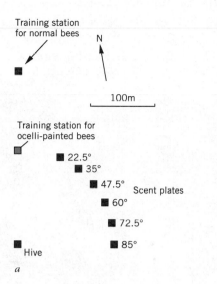

Training station for normal bees

N

100m

Training station for ocelli-painted bees

■ 22.5°
■ 35°
■ 47.5°   Scent plates
■ 60°
■ 72.5°
■ 85°

Hive

a

Figure 18.29 The results of misdirection experiments using ocelli-painted foragers. The experiment was designed to determine whether bees utilized olfactory or dance information by creating a situation in which these two cues would send the recruits in different directions. When a bright light is presented to bees, they interpret it as the sun and orient or interpret dances relative to it. The light sensitivity of some bees was reduced by covering their ocelli with black paint. The ocelli-treated foragers did not see the artificial sun and they danced indicating the direction to the food source relative to gravity. The untreated recruits interpreted the dance relative to the "sun." (*a*) The experimental set-up. By shining a light on the hive from different directions, the experimenters could reorient the dancers. However, the ocelli-painted recruits attending the dances would not see the light and would, therefore, misinterpret the dance information. (*b*) The results of the experiment when the bees were trained on diluted unscented sucrose (von Frisch's techniques). The direction indicated by the dance is signified with an arrow. Notice that the greatest number of bees arrived here. (*c*) The results of the experiment when the bees were trained on concentrated scented sucrose (Wenner's techniques). The direction indicated in the dance is denoted with an arrow. Notice that the bees did not show a preference for this station. (Gould, 1975.)

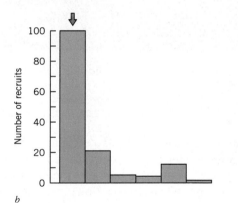

b

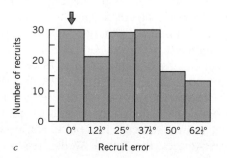

c   Recruit error

station as the light deviated from the vertical. However, the odor search hypothesis predicts that bees would not be misdirected.

When this experiment was performed using von Frisch's training techniques, the bees were misdirected as predicted. However, when Wenner's methods were used, the bees were not deceived (Figure 18.29). The difference in the procedures is that Wenner trains his bees on concentrated sucrose containing the scent that will be used in later tests. Unlike von Frisch's methods, this causes foragers to dance during training and it allows food odor to accumulate in the hive (Gould, 1975, 1976).

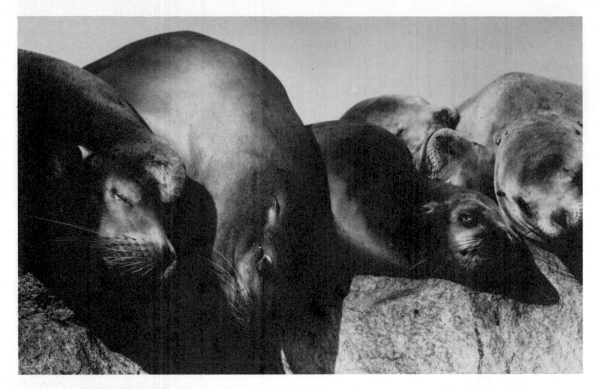

Figure 18.30 Contact animals, sea lions. Some animals maintain social bonds through bodily contact.

Can these conflicting results be reconciled to answer our original question regarding the cues actually used by the recruited bees? Perhaps they can, if the problem is considered from the bee's point of view. For a bee, collecting nectar is a matter of life and death. She may die if she runs out of energy while searching for food. Therefore, one would expect that a forager would use all information available—both dance and odor. Certain circumstances might make one of these cues easier to use. For example, James Gould (1975) suggests that the different training techniques of von Frisch and Wenner may be sampling different stages in the use of a food source. In early spring, when the frigid siege of winter is ending and bees begin to venture out of the hive in search of nourishment, flower patches are scarce. When a patch first comes into bloom, scouts may dance to guide recruits to the site. However, after the blossoms have been sipped for a few days, their fragrance begins to accumulate in the hive. At that time, odor alone may be sufficient to recruit bees to the site.

Although Gould's suggestion offered a compromise between opposing viewpoints and many consider the issue settled (Robinson, 1986), the voice of skepticism about the dance-language hypothesis, although small, has yet to be completely silenced. Gould's misdirection experiments have been criticized as lacking adequate controls (Rosin, 1978, 1980a,b, 1984). Furthermore, the results of some experiments of proponents of the dance-language hypothesis do not seem to fit neatly into place. For instance, Adrian Wenner and Patrick Wells (1987) point out that in some experiments cited as indicating that the dance contains information about the direction to the food source (Gould, Henerey, and MacLeod, 1970), 87 percent of the marked bees that were seen to follow the dance and leave the hive failed to appear at any feeding dish. Those that did arrive at the predicted site took 30 times longer than would have been needed for a direct flight. If nothing else, the odor hypothesis remains a "thorn in the foot" of the dance-language hypothesis, causing an occasional limp in its otherwise self-confident gait.

In any event, the dance-language controversy should serve as a useful reminder that an answer to a question often depends on how it is asked. We generally find that it is best to design experiments with the natural history of the organism in mind.

Maintenance of Social Bonds by Physical Contact

It is easy for humans to understand the significance of physical contact as a signal of social bonding, given our penchant for hugging or otherwise holding each other. Social bonds between certain other animals are also cemented by physical contact. These species are called contact animals (Figure 18.30), as opposed to distance animals that live in groups but prefer to remain at a specified distance from one another. Among contact animals, cuddling, nuzzling, and touching in general tend to firm social bonds (Eibl-Eibesfeldt, 1975).

Many of the greeting signals exchanged by some animals as they encounter each other serve as an assurance of nonaggression. Chimpanzees often greet by touching hands or sometimes by placing a hand on the companion's thigh (Goodall, 1965). Sea lions rub noses and lions rub cheeks. African wild dogs greet one another by pushing their muzzles into the corners of each other's mouths (Schaller, 1972).

A special type of contact behavior, grooming, is exhibited by both invertebrates and verte-

brates. The workers of most social insects groom their nestmates. This is usually done by licking, although they occasionally pick up extraneous material with their mandibles. Grooming probably plays a role in cleanliness as well as providing a means for disseminating colony odor and pheromones. You may recall that the substance produced by the queen honeybee that inhibits the development of other queens is distributed around the colony in this manner.

Grooming is also found in a variety of vertebrates. Birds who prefer bodily contact, such as waxbills (*Estrildidae*), babblers (*Timalidae*), white eyes (*Zosteropidae*), and parrots (*Psittacidae*), groom or preen one another. Rodents groom each other by gently nibbling on each other's fur. But allogrooming (grooming another individual) is most prominent among primates. They usually spend large parts of their day grooming one another (Figure 18.31).

Grooming probably originated as a means of hygiene, at least among vertebrates. However, allogrooming has become important in signaling conciliation and bonding. Although grooming may still be essential for the care of the skin, these social functions are often more important than hygiene (Boccia, 1983).

One social function of grooming is to suppress aggression. When a bird is threatened, for instance, it can often halt the attack if it acts as if it is about to be preened. We also find that allogrooming among rodents is most common in conflict situations. Likewise, mountain sheep dwelling in the area south of the Mediterranean, the mouflon (*Ovis ammon*), employ licking, a form of grooming, in a ritualized appeasement ceremony (Pfeffer, 1967). For primates, grooming appears to be a very relaxing experience, one that reduces tension and restores relationships.

The importance of contact in the formation of bonds, especially between parents and offspring, has already been discussed so it should not be surprising that a second function of primate grooming behavior is the formation and expression of social bonds. Most grooming oc-

Figure 18.31 Tactile communication—social grooming among crab-eating macaques. Although the grooming originally functioned only for skin care, its social functions have now become more important. Primates spend a major portion of their day grooming. It helps form and maintain social bonds.

curs within close relationships, and in species in which grooming is rare, it occurs most commonly between mothers and their offspring. Furthermore, there is an increase in grooming between males and females when the females are in estrus. This increase reflects consort bonds (Carpenter, 1942).

A third function of social grooming among primates, enlisting support for agonistic encounters, stems from the two functions just described, the reduction of aggression and maintenance of social bonds. Robert Seyfarth (1976, 1977) has constructed a model that is useful in predicting which individuals are most likely to groom one another. One of the most important factors he lists is the ability of the monkey to support the groomer in future aggressive encounters. Of course, every individual would want to groom the most dominant animals so there would be competition for access to these animals. Low-ranking animals would lose the competition for

access to the higher ranking animals. Therefore, the model predicts that most grooming will occur between individuals of similar rank. Indeed, a relationship between grooming and agonistic support has sometimes been observed, especially in chimpanzees (DeWaal, 1978).

Summary

There are numerous functions for communication. Recognition of species, population, caste, kin, or individuals is often possible from their displays. Species recognition is important so that reproductive efforts are not wasted on members of the wrong species and so that aggression is directed toward those individuals that are competing for the same resources. Among some species, groups share resources, and in these cases, it is important to recognize the members of one's population so that outsiders can be prevented from utilizing one's space, food, and mates. The social insects share the duties of perpetuating and maintaining the nest. A caste is a group of individuals specialized to perform certain responsibilities. So that each member of the colony will receive the care it requires, its caste must be communicated. The recognition of kin may help to reduce inbreeding and is essential for nepotism. Individual recognition has obvious advantages—parents and offspring may have to interact, mates may have to locate one another, and males may distinguish strange and familiar competitors.

Sexual reproduction is often dependent upon communication. First a mate must be attracted and then courted. During courtship the male advertises himself and the female assesses his qualities, aggression is reduced, and the behavior and physiology of the mates is coordinated. Offspring may communicate with one another to synchronize their hatching. An offspring may have to signal its desire to be fed or the parent may have to indicate its willingness to feed the young.

Since animals in nature are surrounded by potential dangers, an individual may alert others when it is alarmed. Usually all sources of danger are signified by the same signal, but sometimes, as in vervet monkeys, a different call is used for each class of predators. The alarm signals of species living in the same environment are usually very similar. One reason might be that the more eyes looking for danger, the more likely it is that the threat will be detected. Alternatively, the similarity may result from the shaping of the signal to suit its purpose. When danger seems imminent, it may be communicated with a distress call. On the other hand, other signals have evolved to indicate that all is clear so that others are not needlessly alarmed.

Communication is also important in aggressive interactions. Many animal disputes are settled without fighting. The animals threaten one another with stereotyped displays. The traditional interpretation of these displays is that they indicate what the signaler is likely to do next. More recently it has been suggested that the threat displays allow competitors to assess one another's abilities. This way the weaker individual can accept its loss and leave without risking injury.

Other signals allow animals to keep in touch with one another. Sometimes it is only important to remain in contact so that future association is possible. At other times, signals can facilitate assembly. When individuals are brought together to perform a specific duty, as occurs in social insects, it is called recruitment. Tandem running and odor trails are two means of recruitment, but the most elaborate recruitment signals are those of the honeybees. When a scout bee finds food, she recruits nestmates to help in the harvest by dancing. A round dance is performed when the food is close to the hive. It informs recruits that food is within a certain distance of the hive but does not specify the location. The waggle dance is performed when food is farther from the hive. Aspects of this dance correlate with the direction and distance to the food source. The angle that the recruits should assume with the sun as they fly to the food is the same as the angle of the waggle run of the dance with respect to gravity. The recruits know how far to fly by the duration of both the waggling and the buzzing that accompanies it. Because it is so important for bees to find food, they also use olfactory information to locate it. The spot is marked with a chemical from the Nasonoff's gland, and odors of both the locale and the food substance are brought back on the scout's body. The relative importance of dance language and odor in honeybee recruitment has been controversial.

Physical contact is another means of maintaining and reaffirming social bonds. In fact, many greeting signals used to reassure individuals when they reunite involve touching. Individuals of many species groom one another, but grooming is probably most obvious among primates. Grooming plays an important role in cementing social bonds.

CHAPTER 19

Altruism

What Is Altruism?

Nice guys finish last.—or do they? In recent years the concept of "nature, red in tooth and claw" has been changing with the realization that some animals cooperate with other members of their species. For example, an individual Belding's ground squirrel (*Spermophilus beldingi*) risks being spotted by the approaching predator when it sounds an alarm. Nevertheless, it barks at the sight of a badger and all those in the area scurry to safety. In other species, such as wild turkeys (*Meleagris gallopavo*), a male attracts a mate by displaying on common mating areas where both dominant and subordinate males aggregate. These communal displays surely increase the chance of being seen by a predator, yet the subordinate has little hope of immediate fatherhood. Members of still other species cooperate in rearing offspring that do not belong to them. One can see why it is adaptive for a parent to care for its own offspring until they can survive on their own, but in some species, for example the dwarf mongoose (*Helogale parvula*), even unrelated immigrants may forgo reproducing themselves and assist another female in raising offspring by bringing food to the young and by guarding the den from predators (Figure 19.1). Perhaps the best example of cooperation among animals involves the social insects. The workers in groups of social insects toil tirelessly to care for their colony. They may even die in the defense of the nest. However, the young they help rear are not their own; the workers are sterile.

When we observe animals cooperating with or helping one another, such as in these examples, we often describe their behavior as altruistic. We might identify an action as altruistic when it appears to be costly or detrimental to the altruist but is beneficial to another member of its species. Although it is more difficult to measure, altruism is more precisely defined in terms of fitness: It is the behavior of an individual that raises the fitness (number of offspring produced

that live to breed) of another individual at the expense of the altruist's individual fitness (Hamilton, 1964).

Our discussion of cooperative behavior will begin with some of the hypotheses for how aid giving may have evolved. This should set the stage for later sections in which the various hypotheses will be evaluated as possible explanations for the evolution of specific types of aid giving behavior. However, as we will see, questions of the evolution of altruism are only part of the fascinating puzzle surrounding such behavior.

Hypotheses for the Evolution of Altruism

At first, it may not be obvious just why the evolution of altruism has been difficult to explain. Part of the reason seems to lie in the philosophy

Figure 19.1 Dwarf mongooses are small carnivores that live in groups of about 10 individuals consisting of a breeding pair and helpers. Some helpers are unrelated to the group. Important jobs of helpers are to watch for predators and to bring food to the young.

and values that many of us have been taught. We may believe that good or altruism *should* triumph over evil or selfishness. However, as morally satisfying as it may seem to humans, a "good deed" costs the altruist. The cost may be time, energy, or increased danger—factors that may reduce the altruist's reproductive potential. Unless the benefits to the altruist outweigh these costs, it is difficult to see how the behavior would have been selected because natural selection does not give points for being nice. Natural selection gives points for reproductive success, period. Evolution occurs when individuals carrying certain inherited traits leave more reproductively capable offspring than do individuals that lack the alleles* for those characteristics. By this reasoning, a population of altruists would be vulnerable to a selfish mutant, one that would accept the altruists' service and never reciprocate. A selfish mutant would gain at the altruists' expense. As a result, alleles promoting selfish behavior would multiply more quickly in the population, always pressing out alleles for altruism. It would be predicted, therefore, that selfish individuals would inherit the earth. However, we do see examples of helpful or cooperative behavior in many species of animals. So, let's look at some of the explanations offered to resolve this enigma.

INDIVIDUAL SELECTION

As we saw in Chapter 4, natural selection responds to the fitness costs and benefits of a behavior and selects the behavioral alternative that shows the most favorable balance sheet. We would expect, then, that if animals tend to perform some behavior, even a cooperative one, the benefits of the deed must outweigh the costs. When we see helpful or cooperative behavior,

we might expect that the donor must be gaining more than it is losing.

When the behavior seems to benefit others, at some cost to the actor, it is not always easy to see what the benefit to the actor may be. However, there are several ways that an individual may gain through an action that appears to be costly. Consider alarm calling by birds. It seems at first that a bird that sounds an alarm at the sight of the predator risks drawing the predator's attention to itself while warning other birds in the flock. It has been suggested, however, that the caller may, in fact, benefit by the act. The call may cause the predator to leave, since its presence has obviously been discovered and the flock is on the alert. What about animals that help others raise young? In some species of cichlid fishes, adults adopt unrelated young into their own brood, caring for them and defending them from predators as if they were family (Figure 19.2). Although this may at first seem to be quite a generous act, on closer inspection it seems that the parents gain because the adopted young reduce the predation risk of their own

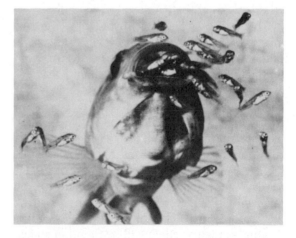

Figure 19.2 Cichlids caring for fry. Some cichlids adopt fry of other pairs and care for them as if they were their own. This seemingly altruistic behavior is actually quite selfish. By adopting unrelated young, the parents reduce predation on their own offspring.

---

* You may recall from the detailed discussion in Chapter 3 that an allele is one of the alternative forms of a gene. For instance, fruit flies have several alleles that influence the length of their wings: one for normal wings, one for short wings (apterous), and one for vestigial wings.

young (McKaye and McKaye, 1977). So, there may be personal gains that outweigh the costs of a helpful action.

KIN SELECTION

We have said that fitness is a measure of reproductive success. So far we have defined individual fitness, which is measured by the number of offspring an individual leaves that survive and reproduce. Offspring carry copies of some of an individual's alleles into future generations, and if a particular allele survives, so does the trait for which it codes. Now we can expand the definition a bit. We can remind ourselves that family members other than offspring also possess copies of some of that individual's alleles. An allele is an allele, regardless of its bearer. Therefore, if *family members* are assisted in a way that increases their reproductive success, the alleles that the altruist has in common with them are also duplicated, just as they would be if the altruist reproduced personally. Phrased another way, because common descent makes it likely that a certain percentage of the alleles of family members are identical, helping kin is another way to perpetuate one's own alleles.

The fraction of alleles shared between two individuals as a result of common descent is called their coefficient of relatedness, r. You can see that r also indicates the probability that two individuals will have inherited a particular allele from a common ancestor. In fact, it is possible to calculate the percentage of alleles you are likely to share with particular relatives. You may recall from our discussions in Chapter 3 that with few exceptions, animals have two alleles for each gene and that these separate during the formation of gametes (eggs or sperm). There is, therefore, a 50–50 chance (a probability of 0.5) that any particular allele will be found in an egg or sperm produced by the parent. Thus, an individual is likely to have 50 percent of its alleles in common with a parent, sibling, or an offspring ($r = 0.5$), and 25 percent in common with a

half sibling or a grandparent ($r = 0.25$), but only 12.5 percent (one-eighth) in common with a first cousin ($r = 0.125$). In contrast to individual selection, which considers only the survival of alleles through one's own offspring, kin selection also includes the survival of alleles via the offspring of relatives (Figure 19.3).

Kin selection is based on the idea of inclusive fitness, which considers all adult offspring, an individual's own or those of its relatives, that are alive because of the actions of that individual. Since an individual shares more alleles by common descent with certain relatives, the possibility of genetic gain increases with the closeness of the relationship. Helping a cousin that shares an average of only one-eighth of the same alleles is less productive than assisting a brother or sister that is likely to have half of its alleles in common with the altruist. In other words, the fitness gained through family members must be devalued in proportion to their genetic distance (diminished relatedness); the allele for an altruistic act would survive if a small number of close relatives or a large number of distant relatives were assisted. We can calculate inclusive fitness by counting offspring, but not all offspring are counted, and a relative's offspring do not count as much as one's own.

Which offspring are included when determining inclusive fitness? According to W. D. Hamilton (1964), when calculating inclusive fitness we count the offspring that would have existed without help or harm from others and some fraction (determined by r) of the offspring of relatives that the individual helped or harmed. Simply put, to calculate an individual's inclusive fitness we add (1) the individual's offspring and (2) the extra offspring that a relative was able to raise because of that individual's efforts, devalued by the genetic distance between them (e.g., multiplied by r) and subtract any of the individual's offspring that resulted from the help or harm of others (Grafen, 1982, 1984).

What does all this have to do with the evolution of altruism? It gives us a broader view of

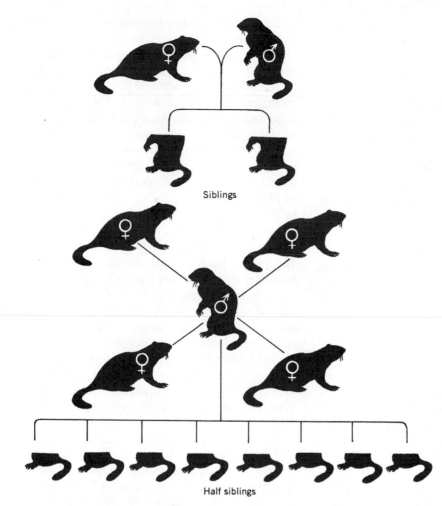

Siblings

Half siblings

Figure 19.3 Relatives share a portion of their alleles by common descent. Parents and offspring share 50 percent of their alleles, as do siblings. Half siblings share only 25 percent.

the costs and benefits of a particular helpful or cooperative behavior so that we see how the allele promoting that action may have spread through the population. We will find that an allele promoting a helpful behavior will spread if the fitness benefits are greater than the fitness costs. (Notice that fitness benefits and costs also tell us the probability that the allele[s] favoring the helpful behavior will be shared by the offspring. So, phrased slightly differently, the allele favoring the helpful behavior will spread if helping results in leaving more copies of the allele promoting helping than would be left if the in-

dividual did not help.) The benefits, the fitness gains to the altruist, are calculated as the extra offspring that were raised by relatives because of the actions of the altruist, multiplied by r, to take into account the genetic distance between the altruist and the offspring. The cost, or the fitness loss to the altruist, is the difference in the animal's lifetime number of offspring that results from choosing to be helpful rather than behaving in some other way. It is, of course, difficult to know "what would have been." However, we might estimate the number of offspring produced by animals that followed one course of action with the number produced by those that did not. There are reasons that this estimate may be inaccurate (Grafen, 1984), but it is a beginning. In the end we would expect the allele to spread if the benefit exceeds the cost. If, for example, a male animal helped his brother raise four more nieces than he could have without assistance, the altruist's genetic gain would be 1 (r × number of extra nieces = 0.25 × 4 = 1). If he raised one fewer offspring of his own by helping his brother raise nieces than he could have if he had devoted himself entirely to raising his own young, then his genetic cost would have been 0.5 (r × number fewer offspring = 0.5 × 1 = 0.5). In this case, because the benefit (1.0) exceeds the cost (0.5), the allele promoting helping brothers would spread through the population. So we see that altruism, a behavior that lowers one's individual fitness, could evolve if it increased the fitness of enough relatives that also carry alleles for altruism.

Kin-Biased Behavior

If altruism evolves by kin selection, then we might expect animals to assist relatives more often than they do others. Kin-biased behavior such as this may be based on a variety of mechanisms, some of which might involve the ability to discriminate kin from nonkin. We might wonder, then, how an animal might determine who its relatives are. We will consider four possible mechanisms of kin discrimination (Blaustein, Bekoff, and Daniels, 1987; Holmes and Sherman, 1983).

Location   When relatives are distributed throughout the habitat in a predictable way, kin selection can work if the altruistic deeds are directed toward those individuals found in areas where relatives are most likely to be found. For instance, we might see location as a mechanism underlying kin-biased behavior in species in which there is limited dispersal from the nest or burrow. As you may recall from Chapter 16,

in many mammal species, the males disperse and the females remain in their natal area. Among many bird species, however, the females tend to disperse and the males remain at home. In any event, those animals that remain in the natal area tend to be related. Therefore, any good deed directed toward a neighbor will usually be received by a relative.

The most obvious example of how location might work as a mechanism underlying kin-biased behavior occurs when a parent identifies its offspring as the young in its nest or burrow. In many species of birds, parents will feed any young they find in their nest. We see this among bank swallows (*Riparia riparia*), for instance. The parent bank swallow learns its nest hole locations and feeds any chicks in these holes, including any neighbor's chicks placed inside by experimenters. However, a parent will ignore its own chicks if they are moved to a nearby nest hole. After about 2 weeks, at the time the young begin to leave the nest and fledglings unrelated to the parent might enter the nest, the parent begins to reject foreign young experimentally placed in the nest (Beecher, Beecher, and Hahn, 1981).

Familiarity   Kin may also be treated differently from nonkin because they are familiar. In some species, the young learn to recognize the individuals with which they are raised through their experiences during early development, and then, later in life, they treat familiar animals differently from unfamiliar ones. The ideal setting for this learning is a rearing environment such as a nest or burrow that excludes unrelated individuals.

Familiarity is apparently a mechanism used by a young spiny mouse (*Acomys cahirinus*) to distinguish its siblings. When weanling pups are released into a test arena, they often huddle together in pairs and the members of the pair are generally siblings. However, the spiny mice do not identify their siblings per se, but rather they prefer to huddle with familiar pups, their littermates. We know this because siblings separated soon after birth and raised apart treat one another as nonsiblings. However, if unrelated young are raised together, they respond to one another as siblings. So we see that kin-biased behavior among spiny mice seems to develop as a result of familiarity (Porter, Tepper, and White, 1981).

Strangers might be recognized as kin by this method if they are associated with a mutual relative such as a mother. For example, sisters born in different litters might identify each other as siblings because their common mother treats them both as offspring.

Phenotype Matching In some species, animals are able to identify their kin even if they have never met before. It has been suggested that this is possible by "phenotype matching" (Alexander, 1979; Holmes and Sherman, 1982; Lacy and Sherman, 1983). Phenotype, as you may recall from Chapter 3, is the physical, behavioral, and physiological appearance of an animal. Because one's genetic inheritance has so much to do with appearance, family members often resemble one another in one or more characteristics. Furthermore, family members may have similar characteristics that are not inherited, but instead come from the environment. One way this occurs is through maternal "labeling." Individuals could acquire a specific characteristic, such as a scent, while developing in the uterus or within the same group of eggs. Alternatively relatives might be labeled later in life. Social insects often use nest odors, which are at least partially due to food odors, as a cue to identify nestmates, and nestmates are usually relatives. As a result of interactions with known kin, family members may learn to pick out distinguishing characteristics. Later, when a stranger is encountered, its phenotype may be matched or compared to the learned family image to determine whether the stranger is related.

There are several examples of kin discrimination among animals that appear to be me-

diated by phenotype matching. Sweat bees (*Lasioglossum zephyrum*) normally guard the entrance of their nest so that only nestmates, usually their sisters, are permitted to enter. In one experiment, the breeding of sweat bees was controlled so that the resulting bees were genetically related to varying degrees. The probability that guard bees, which had never before met any of the inbred bees, would permit the intruders to enter their nest increased with the degree of genetic relatedness between the guard and the trespasser (Figure 19.4; Greenberg, 1979). The possibility that the discrimination at the nest entrance was based on phenotype matching was explored by raising some guards with only their sisters, others only with nonsisters, and still others in nests that had both sisters and nonsisters (Buckle and Greenberg, 1981).

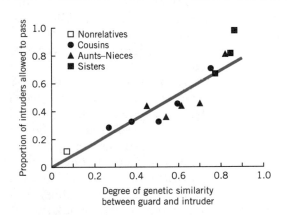

Figure 19.4 The relationship between the percentage of intruders accepted by a guard sweat bee and the degree of genetic similarity between the guard and intruder. Guard sweat bees usually allow only their nestmates, which are usually their sisters, to enter the nest. In this experiment, the guards were presented with bees that they had never met before but that, as a result of controlled breeding experiments, were related to them in degrees. The proportion of the trespassers permitted to enter increased with the closeness of the genetic relationship between the guard and the stranger. (Modified from Greenberg, 1979.)

In this experiment, the guard bee allowed intruders to enter if they were related to its own nestmates, even if those nestmates were not its own relatives. These results are consistent with the hypothesis that the guard compares some aspect of the unfamiliar intruder's phenotype (probably an odor) with the "phenotype template" that it has learned characterizes a nestmate. If the stranger matches the template, it is allowed to enter.

In contrast to the sweat bees who use known relatives (their nestmates) to form an image of what kin should look like, tadpoles of the Cascades frog (*Rana cascadae*) behave as if they match their *own* phenotype with that of a stranger to identify kin. This was shown in a series of experiments by Andrew Blaustein and colleagues (summarized in Blaustein, Bekoff, and Daniels, 1987; Blaustein and Waldman, 1992; Blaustein and O'Hara, 1986). All tadpoles that hatch from a single egg mass are siblings, and in nature, Cascades frog tadpoles remain near their hatching site and form small cohesive schools. At first, then, one might suspect that location or familiarity might be the basis of their preference to associate with kin. However, when these hypotheses were tested, location cues and interaction with siblings were shown to be unnecessary. In one series of experiments, tadpoles were reared for about 4 weeks in one of 4 conditions: (1) exposure only to their siblings, (2) exposure to both sibling and nonsibling tadpoles, (3) exposure only to nonsiblings, and (4) total isolation. Each tadpole was then given the choice of associating with siblings or with nonsiblings. The test aquarium was a long tank with mesh barriers forming two end compartments. About 2 dozen siblings were placed in one end compartment and the same number of nonsiblings at the other. The tadpole being tested swam freely within the central chamber. The amount of time a tadpole spent near each end was recorded and used to determine whether the tadpoles preferred to associate with kin. Indeed, most of the tadpoles for all four rearing conditions did spend more time swimming near their siblings. Even the tadpoles raised with nonsiblings preferred to swim near their own siblings that they had never met before. So we see that Cascades frog tadpoles can identify their brothers and sisters, even if they did not grow up with them (O'Hara and Blaustein, 1981). Since the siblings develop within a common jelly matrix, it was then suspected that the developing tadpoles may learn to identify some cue that comes from the jelly, but this too was shown to be unimportant. Tadpoles reared without any jelly mass or with the jelly mass of nonsiblings substituted for their own still preferred to associate with their siblings. Furthermore, tadpoles raised in isolation preferred to associate with full siblings over half siblings, maternal half siblings over paternal half siblings, and paternal half siblings over nonsiblings (Blaustein and O'Hara, 1982). A possible explanation for this ability is that the tadpoles, even those raised alone, learn their own cues, and when they later make an association choice, they match their own phenotype with those of other tadpoles.

Recognition Alleles It has also been suggested that animals might inherit the ability to recognize kin, instead of having to learn the important cues (Hamilton, 1964). This would be possible if an individual inherited a recognition allele or group of alleles that enabled it to recognize others with the same allele(s). The postulated allele(s) would have three simultaneous effects: It would (1) endow its bearer with a recognizable label and (2) the sensory ability to perceive that label in others, and (3) it would cause the bearer to behave appropriately toward others with the label. This recognition system has been named the "green beard effect" to indicate that the label could be any conspicuous trait, such as a green beard, as long as the allele responsible for it also causes its owner to behave appropriately to other labeled individuals (Dawkins, 1976, 1982; Hamilton, 1964). Demonstrating the existence of recognition alleles has proven quite difficult, primarily because it is so hard to eliminate all the possible opportunities for learning recognition cues during an animal's lifetime.

Some kin-biased behaviors do seem to be based on recognition alleles, however. The larvae of the sea squirt *Botryllus schlosseri* settle near their siblings. These larvae, which superficially resemble a frog tadpole, are planktonic for a short time and then settle, attach to the sea bottom, and develop to the adult form. When groups of siblings settle, they tend to clump together, but groups of unrelated larvae settle randomly. In this case, sibling recognition seems to be based on the alleles of the region of DNA that determines whether grafted tissue will be associated with the host, called the histocompatibility locus. If the siblings do not share an allele in this locus, they do not settle together. Unrelated larvae that happen to share an allele are just as likely to settle together as are siblings sharing an allele (Grosberg and Quinn, 1986).

Experiments in house mice (*Mus musculus*) also indicate that social preference may be influenced by a small genetic difference. Mice were repeatedly inbred so that all their alleles were identical except for a single gene, the same region that was shown to be important in kin discrimination in sea squirts. Specifically, it was a gene in the region of H-2, the major histocompatibility complex. In mating preference tests, these inbred male mice could choose among females that carried different alleles of the gene just described. Surprisingly, this allele made a difference in their preference. Males from most strains chose females whose allele differed from theirs, but males from a few strains preferred mates whose allele matched theirs. It is thought that this small amount of genetic material might alter mating choices by influencing a female's odor (Yamazaki et al., 1980). A gene linked to H-2 is concerned with odor production and the quantity and chemistry of the odor produced depend on which alleles are present. Furthermore, house mice can detect these odor differ-

ences (Boyse et al., 1982). Individuals of outbred populations of mice maintained in enclosures large enough to allow normal patterns of social competition and reproductive behavior, also prefer mates whose histocompatibility genes differ from their own (Potts, Manning, and Wakeland, 1991). So, a small region of genetic material may be able to both label an individual and cause a particular response by another animal.

RECIPROCAL ALTRUISM

Reciprocal altruism is evolution's version of "you scratch my back and I'll scratch yours" or "one good deed deserves another." Reciprocal altruism can be selected if the favor is returned with a net gain to the altruist. Although evolution's tally sheet records gains and losses in terms of inclusive fitness, reciprocal altruism does not require that the cooperating animals be related (Trivers, 1971).

At first it is difficult to understand how reciprocal altruism evolved when there is a time lag between the deed and repayment because delayed restitution opens the possibility of cheating. A cheater is an individual that received a service but failed to repay the altruist at a later time.

In reciprocal altruism, the costs and benefits to the altruist depend on whether the recipient returns the favor. You may recall from Chapter 4 that evolutionary game theory is designed to handle situations, such as this, in which the best course of action depends on what others are doing.

The Prisoner's dilemma is a mathematical game that has been applied to assist our understanding of the evolution of reciprocal altruism. The name of the game comes from an imaginary story in which two suspects for a crime are arrested and kept in separate jail cells to prevent them from communicating. Certain that one of them is guilty, but lacking sufficient evidence for a conviction, the district attorney offers each a deal. Each prisoner is told that there is enough incriminating evidence to guarantee a short jail term, but that freedom might be obtained by "squealing" and providing enough evidence to send the other to jail for a long time. However, if each informs on the other, then they both go to jail for an intermediate length of time.

Given these payoffs for selfishness (informing) and altruism (concealing evidence), selfishness is the best strategy. A selfish prisoner cannot lose. If the partner is altruistic, the selfish prisoner goes free. Even if the partner turns out to be selfish, too, the first prisoner is better off being selfish than taking the rap for both of them. We see, then, that neither prisoner would be expected to behave altruistically. The cost, if the other does not reciprocate, would be too great. If we translate jail sentences into fitness payoffs, the solution to the Prisoner's dilemma seems to imply that reciprocal altruism cannot evolve. Indeed, this may be true *if* the prisoners will never meet again.

In real life, however, individuals interact repeatedly and when they do, reciprocal altruism might evolve. Assume, for instance that the prisoners are Bonnie and Clyde, long-term partners in crime, who are repeatedly arrested. In this case, Bonnie would succeed if she altruistically concealed evidence against Clyde each time they were arrested, *as long as* Clyde reciprocated. However, if Clyde informed on Bonnie, she should do likewise at her first opportunity. If, in a subsequent encounter, he concealed evidence, she should also do so.

This strategy, called tit-for-tat, can be a winner in the game of Prisoner's dilemma if there are repeated encounters between the prisoners. The strategy of tit-for-tat might be paraphrased, "Start out nice and then do unto others as they did unto you." In this strategy, an individual begins by being cooperative and in all subsequent interactions matches the other party's previous action. This strategy, then, involves both retaliation (it follows defection, or cheating, with defection) and forgiveness (it "forgets" a defection and cooperates if the partner later cooperates). If a

population of individuals adopts this strategy, it cannot be overrun by a selfish mutant (Axelrod, 1984). So, when the individuals have repeated encounters, reciprocal altruism can be an evolutionarily stable strategy (ESS), one that cannot be invaded by another strategy (see Chapter 4).

We see then that game theory modeling of the evolutionary process informs us that a population that is genetically programmed to unfailingly perform some service for others can be invaded and overrun by any mutants that cheat because a swindler's gain is always greater than a sucker's. However, alleles creating altruists that retaliate against cheaters by refusing to assist them on the next encounter will survive and prosper if there are enough individuals carrying the allele in the initial population to make meetings between them frequent enough. In other words, reciprocal altruism can evolve if there is discrimination against cheaters (Axelrod and Hamilton, 1981; Dawkins, 1976).

Besides discrimination against cheaters, another condition that is required for the evolution of reciprocal altruism is a fair return on the altruist's investment. There are certain factors that lead to a profitable balance sheet and will, therefore, favor the evolution of reciprocal altruism: (1) The opportunity for repayment is likely to occur, (2) the altruist and the recipient are able to recognize one another, and (3) the benefit of the act to the recipient is greater than the cost to the actor. These factors are most likely to occur in a highly social species with a good memory, long life span, and low dispersal rate (Trivers, 1971).

If these characteristics bring our species to mind, it is not surprising. Robert Trivers points out that reciprocal altruism is particularly common among humans. Not only do humans help the infirm through various programs (with the expectation that they, too, might someday benefit from such a program), they also help one another in times of danger and they share food, tools, and knowledge. Trivers even argues that our feelings of envy, guilt, gratitude, and sym-

Figure 19.5 Vampire bats are reciprocal altruists. A vampire bat that has just fed will regurgitate blood to a hungry roostmate, thereby saving it from starvation. Only bats that have a prior association will regurgitate blood to one another.

pathy have evolved to affect our ability to cheat, spot cheaters, or avoid being thought of as a cheater.

Reciprocal altruism is not limited to humans, however. Vampire bats (*Desmodus rotundus*),

for instance, share food with needy familiar roostmates even if they are not related (Wilkinson, 1984, 1990; Figure 19.5). This generosity may make the difference between life and death for the recipient. If a vampire bat fails to find food on two successive nights, it will starve to death unless a bat that has successfully fed regurgitates part of its blood meal. The hungry bat begs for food by first grooming, which involves licking the roostmate under the wings, and then by licking the donor's lips. A receptive donor will then regurgitate blood (Figure 19.6). The regurgitated food must be enough to sustain the bat until the next night, when it may find its own meal. Although the benefit to the recipient is great, the cost to the donor is small. Since a bat's body weight decays exponentially following a meal, the recipient may gain 12 hours of life and, therefore, another chance to find food. However, the donor loses less than 12 hours of time until starvation and usually has about 36 hours, another 2 nights of hunting, before it would starve (Figure 19.7).

Vampire bats roost in somewhat stable groups of both related and unrelated members. A typical group consists of 8 to 12 adult females and their

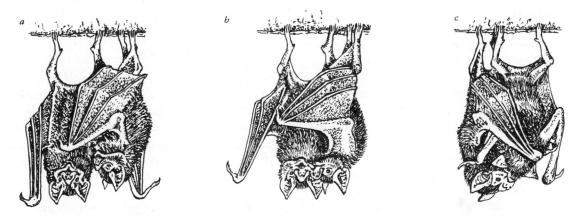

Figure 19.6 A vampire bat that was unsuccessful in obtaining a meal during a night's hunt begs for food from a roostmate. First it grooms the roostmate by licking it under the wings (*a*), then it licks it on the lips (*b*). If receptive, a well-fed roostmate will respond by regurgitating blood to the hungry partner (*c*). (From Wilkinson, 1990.)

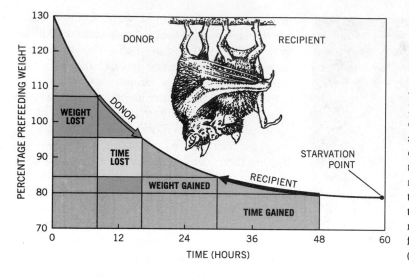

Figure 19.7 Blood sharing by vampire bats costs the donor less than the benefit to the recipient because the weight loss after a meal is exponential. By donating 5 milliliters of blood to a hungry roostmate, the donor loses only about 6 hours of the time it has left until starvation. However, the recipient may gain 18 hours, and therefore a chance to hunt again. (From Wilkinson, 1990.)

pups, a dominant male, and perhaps a few subordinate males. Males leave their mothers when they are about 12 to 18 months old, but females usually remain well past reproductive maturity. Thus, many of the individuals in a roost are related, but perhaps not as closely as might be expected. According to biochemical analysis of blood samples, only half of the individuals in a cluster share the same father. Furthermore, about once every 2 years, a new female will join the group.

Although the groups may change slightly over time, there are numerous opportunities to share food. Females may live as long as 18 years, and in one study, two tagged females shared the same roost for more than 12 years. Additionally, a blood meal is not always easy to obtain. On any given night, roughly 33 percent of the juveniles less than 2 years old and 7 percent of older bats fail to feed.

Generally, only individuals who have had a prior association share food. In an experiment, a group of bats was formed from two natural clusters in different areas and maintained in the laboratory. Aside from a grandmother and granddaughter, all the bats were unrelated. The bats were fed nightly from plastic measuring bottles, so that the amount of blood consumed

by each bat could be determined. Then, every night one bat was chosen at random, removed from the cage, and deprived of food. When it was reunited with its cagemates the following morning, the hungry bat would beg for food. In almost every instance, blood was shared by a bat that came from the starving bat's population in nature. Furthermore, there seemed to be pairs of unrelated bats that regurgitated almost exclusively to one another, suggesting a system of reciprocal exchange.

PARENTAL MANIPULATION

Parental manipulation, as proposed by Richard Alexander (1974), is a means for the evolution of altruism in which parents are selected to produce offspring that behave altruistically toward their siblings. This selection occurs because the parents' fitness is increased in spite of a reduction in the personal fitness of a particular youngster.

Alexander's reasoning goes as follows: If an allele were to cause an offspring to maximize its personal fitness by grabbing more than its fair share of parental favors from siblings, it would be beneficial to that individual as a juvenile. However, this behavior would decrease the parents' fitness because each and every offspring parents produce, even the one that loses sibling disputes, bears copies of the parents' alleles. The greedy juvenile will mature and someday become a parent. When this happens, it will pass this allele to at least some of its own offspring and their selfishness would reduce the parent's fitness. Thus, as a parent, the selfish sibling would become the victim of its genetic inheritance. In the end, an allele favoring selfish behavior toward siblings would reduce its bearer's overall reproductive success (i.e., there would be selection against such selfish behavior). The bottom line of Alexander's argument is that parents will always win when their fitness conflicts with that of their offspring.

Richard Dawkins (1989) later pointed out that this argument is flawed because it assumes a genetic asymmetry between the parent and offspring that does not exist. The genetic relationship between a parent and offspring is 50 percent, whether you look at it from the parent's or from the offspring's point of view. Thus, the argument could just as easily be made with the parent and offspring actions reversed and then the opposite conclusion would be reached. If a parent had an allele that improved its fitness as a parent because one of the offspring was favored, then that allele would be expected to have lowered its fitness when it was a juvenile. By this line of reasoning, the child should always win. We see, then, that there is no blanket answer to the question of whether parent or offspring will be favored by natural selection in the battle of the generations. In the end, we might expect the parent–offspring conflict to end in compromise.

Examples of Cooperation Among Animals

As we consider various forms of cooperation among animals, we will note many similarities among distantly related groups. We will also no-

tice that the selective forces leading to similar forms of cooperation may be quite different.

ALARM CALLING

Individuals of many social species emit an alarm call when a predator or other source of danger is discovered. As a result, others in the area are alerted to the danger, enabling them to take appropriate action to protect themselves.

Alarm Calls of Birds

Is Avian Alarm Calling Altruistic? In other words, does the bird who calls to alert its neighbors assume risk? The answer has been controversial. If an acoustic engineer were to design a sound to be difficult for humans to localize, its structure would be similar to that of the alarm call of certain birds (Marler, 1955). Because of this, it has sometimes been inferred that the caller is difficult to locate and may not be endangering itself. However, laboratory experiments on raptors such as barn owls (*Tyto alba*; Konishi, 1973; Shalter and Schleidt, 1977), pygmy owls (*Glaucidium perlatum* and *Glaucidium brasilianum*), and goshawks (*Accipiter gentilis*; Shalter, 1978) showed that these predators are able to locate the source of alarm calls. Although avian alarm calls are usually almost pure tones (i.e., they consist of few frequencies) and pure tones are difficult for a barn owl to locate, the calls are usually centered between 6–8 kHz, wavelengths among those most easily located by barn owls (Figure 19.8). This suggests that the caller might indeed attract the attention of the predator and thus be behaving altruistically.

Hypotheses for Evolution of Avian Alarm Calling Although alarm calling is quite common among birds, the selective forces that have led to it are still not clear.

Individual Selection There are several suggestions for how alarm calling may have evolved via individual selection, but a common theme among them is that calling reduces the caller's

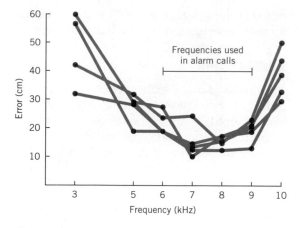

Figure 19.8 The overlap between the frequencies employed in alarm calls and those that a barn owl locates most easily. (Modified from Konishi, 1973.)

risk of predation. An alarm call informs a predator that it has been detected, and since many predators prefer to sneak up on unwary prey, the call may discourage an attack. For instance, in one study, although the goshawks and pygmy owls tested could orient to alarm calls, the predators were less likely to attack the prey if an alarm was sounded (Shalter, 1978). In fact, the one goshawk tested that had been captured as a mature hunter failed to respond to 9 of the 10 alarm calls. (It can be speculated that she was particularly unresponsive because she had experienced more unprofitable hunting experiences with alarm-calling birds than did the other birds tested.) In this way, calling may avert a predator's attack.

A slightly different variation on this theme is that the caller gains by warning others because they flee and no longer lure the predator to the area. A group of birds feeding in an open area is likely to be noticed by a predator. Once in the area, a predator will be a hazard to the would-be caller as well. However, a warning causes the entire flock to take flight. Thus, the caller's chances of survival are increased above what they would have been had it not warned

the others (Dawkins, 1976) and caused them to clear out, thereby reducing the predator's interest in that area.

A corollary to this hypothesis is that alarm calls evolved because they indirectly reduce the chance of the predator developing a preference for the caller's particular species. If a hunt is successful, the predator will be more likely to return to the area on future hunting expeditions. In time, then, the predator will become more familiar with the general area in which the prey was obtained and with the prey's habits. An individual bird may be safer if the predator never learns to prefer its species (Trivers, 1971).

It has even been argued that the caller is completely selfish. It has seen the predator and knows its location. Calling informs others that a predator is near but does not tell them where it is. When a predator draws near, it is advantageous to get out of sight, but it is even better to take the flock with you as cover. Because only the caller knows the predator's location, it can place itself within the flock in such a way that it is less likely than the others to be the victim (Charnov and Krebs, 1975).

Kin Selection If kin selection were the basis of giving warning calls, then the caller would be saving others that bear alleles in common with it. There may, in fact, be a risk in calling, but if the risk is outweighed by the benefit to family members, the allele for calling could be maintained in the population. Since this hinges on the presence of relatives in the rescued flock, John Maynard Smith (1965), who originally made this suggestion, added that alarm calling could only have evolved during the breeding season, when it would enhance the survival of the caller's offspring.

There are several lines of evidence suggesting that kin selection is not the force that led to alarm calling among birds. For one thing, alarm calls *are* given throughout the year, rather than being restricted to the breeding season. In addition, since many flocks consist of several species, one must wonder how many of the caller's relatives could be listening, since the caller's alleles would be diluted in any such flock. Furthermore, there are data for several bird species indicating that relatives usually disperse and thus simply would not be present to heed the warning (Trivers, 1971).

Reciprocal Altruism   It seems reasonable at first that reciprocal altruism could account for the evolution of warning calls; the risk assumed by one caller would be repaid when others return the favor. However, even Robert Trivers (1971), an originator of the concept of reciprocal altruism, believes that this is an unlikely explanation. Recall that altruism can only evolve by this means if cheaters are discriminated against. There is no evidence that warning calls are ever withheld because of the past behavior of a companion. In addition, the members of many flocks come and go, making it difficult to identify cheaters.

Alarm Calls of Ground Squirrels

Although the basis for the evolution of avian alarm calls is still disputed, Paul Sherman (1977) has built a convincing argument that some alarm calls of Belding's ground squirrels (*Spermophilus beldingi*) evolved via kin selection (Figure 19.9). These rodents are often victims of aerial predators, such as hawks, or of terrestrial predators, such as coyotes, long-tailed weasels, badgers, and pine martens. The alarm calls for these classes of predators are different. If the villain approaches on the ground, the alarm is a series of short sounds, whereas the warning of an attack from the air is broadcast as a high-pitched whistle. Interestingly, the selective forces behind the evolution of alarm calls to aerial predators appear to be different from those behind the evolution of alarm calls to terrestrial predators.

Individual Selection   Although the alarm calls to aerial predators appear to promote self-preservation, those to terrestrial predators do not. When a hawk is spotted overhead or when

Figure 19.9   A female Belding's ground squirrel emitting an alarm call. When a terrestrial predator is spotted, females with living relatives nearby are more likely to call than are either females without kin as neighbors or males that rarely have kin in the vicinity. Observations such as these support the kin selection hypothesis for the evolution of alarm calling to terrestrial predators.

an alarm whistle is heard, near pandemonium breaks out in a Belding's ground squirrel colony. Following the first warning, others also whistle an alarm and all scurry to shelter. As a result, a hawk is rarely successful. However, when it is, the victim is most likely to be a noncaller. In one study, only 2 percent of the callers were

Table 19.1

ALARM CALLING AND SURVIVAL IN BELDING'S GROUND SQUIRRELS AT TIOGA PASS, CALIFORNIA.[a]

CATEGORY	NUMBER OF GROUND SQUIRRELS			
	CAPTURED	ESCAPED	PERCENT CAPTURED	P (χ^2 TEST)
Aerial predators				
Callers	1	41	2%	
Non-callers	11	28	28%	<0.01
Total	12	69	15%	
Terrestrial predators				
Callers	12	141	8%	
Non-callers	6	143	4%	<0.05
Total	18	284	6%	

[a] All data are from observations made during attacks by hawks (n = 58) and predatory mammals (n = 198) that occurred naturally during 1974–1982.
Source: Sherman, 1985.

captured, but 28 percent of the noncallers were caught (Table 19.1). The most frequent callers were those that were in exposed positions and close to the hawk, regardless of their sex or relationship to those around them. Thus it seems that the alarm whistles given at the sight of a predatory bird directly benefit the caller by increasing its chances of escaping predation (Sherman, 1985).

In contrast, individual selection does not seem to be behind the evolution of ground squirrel alarm trills, which are issued in response to terrestrial predators. In this case, the caller is truly assuming risk; we know this because significantly more callers than noncallers are attacked. As can be seen in Table 19.1, 8 percent of the ground squirrels that called in response to terrestrial predators were captured, while only 4 percent of the noncallers were caught. The predators, even coyotes whose hunting success often relies on the element of surprise, did not give up when an alarm call was

sounded. Furthermore, the caller was not manipulating its neighbors to its own advantage. Generally, the reactions of other ground squirrels were to sit up and look in the direction of the predator or to run to a rock. Their reaction did not create pandemonium that might confuse a predator. Nor did the caller seek anonymity in the midst of aggregating conspecifics (Sherman, 1977).

Kin Selection   The evidence for kin selection's being the basis for alarm trills to terrestrial predators in ground squirrels is strengthened

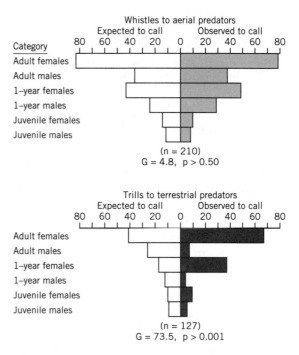

Figure 19.10   Expected and observed frequencies of alarm calls by Belding's ground squirrels to aerial and terrestrial predators. Expected frequencies are those that would be predicted if the animals called randomly. The calls to aerial predators are close to the expected frequencies. However, the calls to terrestrial predators are more likely to be given by females with relatives nearby, than would be predicted if the animals called randomly. (From Sherman, 1985.)

by information on the structure of ground squirrel society. The squirrels in an area are usually strongly related because of the stability of females in the population. Because daughters tend to settle and breed near their birthplace, the females within any small area are usually genetically related to one another. The sons, on the other hand, cut the proverbial apron strings before the first winter hibernation and set off independently, never to return to their natal burrow.

The population of Belding's ground squirrels studied by Paul Sherman lives in Tioga Pass Meadow high in the Sierra Nevada Mountains of California (Sherman, 1977, 1980a,b, 1985). Members of this population have been individually marked since 1969 and their genealogies are known so Sherman is able to keep records of which individuals called and when. His data, which are summarized in Figure 19.10, suggest that the ground squirrels practice nepotism, favoritism to family members. Notice in Figure 19.10 that when a terrestrial predator appears, females are more likely than are males to sound an alarm. This is consistent with kinship theory because it is females that are more likely to have nearby relatives that would benefit from the warning. In addition, reproductive females are more likely than are nonreproductive females to call (Figure 19.11). An even finer distinction can be made: Reproductive females with living relatives call more frequently than do reproductive females with no living family!

Alarm calling in the round-tailed ground squirrel, *Spermophilus tereticaudus*, also appears to be a result of kin selection (Dunford, 1977). The females tend to settle near their mothers and the males emigrate, just as Belding's ground squirrels do. It is the females, whose neighbors are likely to be family, that are most likely to call. An interesting difference between this species and Belding's ground squirrel

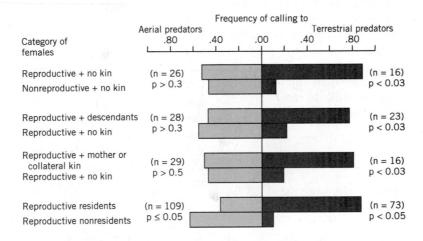

Figure 19.11 The effects of residency and genetic relatedness on the frequency of alarm calling to terrestrial and aerial predators in Belding's ground squirrels. Notice that when a terrestrial predator approached, reproductive females called more frequently than nonreproductive females. Furthermore, reproductive females with kin nearby called more than reproductive females with no kin and residents called more frequently than nonresidents. Kinship and residency did not affect the frequency of calling when an aerial predator approached. (From Sherman, 1985.)

is the incidence of male alarm calling. During the spring, when the males in an area are not likely to have kin nearby, they rarely warn others of impending danger. However, by July, a few juvenile males have settled near their mothers and sisters, and these males do issue warnings. If there were individual gain from calling, one would predict that males would emit alarm calls throughout the year. Instead, as predicted by the kin selection hypothesis, they only call during the times of the year when kin are nearby.

COOPERATIVE BREEDING AND HELPING

Cooperative breeding, in which individuals (helpers) assist in the care and rearing of another's young rather than producing offspring of their own, seems to be an ideal example of altruism. It was first described among birds (Skutch, 1935). Indeed, Alexander Skutch, (1961) has defined a helper as "a bird which assists in the nesting of an individual other than its mate, or feeds or otherwise attends a bird of whatever age which is neither its mate nor its dependent offspring." Cooperative breeding has, in fact, been described in more than 222 bird species, but it has also been reported in approximately 120 species of mammals (Emlen, 1991) and several species of fish. In any species, though, the helper facilitates the genetic legacy of others.

Helpers' Duties

Basically helpers give parental care to offspring that are not their own. A more precise term for helping is alloparental care (Wilson, 1975). Most commonly helpers assist in one of two ways, in providing food for or protecting the offspring of others, but there are other ways to help as well.

Providing Food   Most helpers bring food to another individual's offspring. Florida scrub jay (*Aphelocoma coerulescens*) helpers, for example, deliver about 30 percent of the food con-

sumed by the nestlings. Although this help does not increase the total amount of food brought to the nest, it does reduce the parents' share of the job and enables the parents to enjoy better health, as is suggested by the improvement in survivorship of breeders with helpers. In one study, 87 percent of the breeders with helpers lived to the next year, but only 80 percent of the breeders without helpers made it (Stallcup and Woolfenden, 1978).

Among some mammals, the helper may bring food to the mothers as well as the offspring. Blackbacked jackals* (*Canis mesomelas*) deliver food by regurgitation (Figure 19.12). Helpers contribute between 18 and 32 percent of all regurgitations to pups. In addition, they contribute to the nourishment of the lactating mother that might remain with the pups while the others hunt (Moehlman, 1979).

Protection of Offspring   One advantage of group living is increased vigilance and defense against predators. Therefore, it is not surprising that helpers provide extra protection for the young. Besides issuing alarm calls to warn the chicks, Florida scrub jay helpers actively defend them by mobbing predators such as snakes (Woolfenden, 1975). Jackal families with helpers always have an adult on guard to drive away predators, whereas groups lacking helpers may have to leave the pups unattended while the others hunt. Among fish, the helpers mainly contribute by offspring protection; they can contribute little to the nourishment of the young.

Other Activities   Depending on the species, however, helpers may engage in a variety of other activities. In certain bird species, for instance, helpers may build and clean nests or incubate and brood the nestlings (Skutch, 1987). Although most primate helpers serve primarily

_____
* Also known as silverbacked jackals.

a

b

Figure 19.12 A jackal helper about to regurgitate food to a pup (*a*) and chasing away a predator (*b*).

as babysitters, certain ringtailed lemur helpers are an exception. These helpers may nurse the infants in addition to assisting in their care (Pereira and Izard, 1989). In another primate, the saddle-backed tamarin (*Saguinus fuscicollis*), there are two types of helpers. One type consists of extra males whose job it is to carry the heavy juveniles. At birth a tamarin is almost 20 percent of the adult weight and typical litters consist of twins. Thus, carrying these youngsters is burdensome, and if the duty were not shared, the mother might not be able to obtain enough nourishment for herself and to ensure a milk supply (Terborgh and Goldizen, 1985).

Are Helpers Altruistic?

Helpers appear to help, then, but do they? Also, if they do, is there some disadvantage to themselves? If the helper is an altruist, then there should be benefits for the assisted breeder and costs to the helper. Let's first take a closer look at the benefits to the breeders, then examine the costs to helpers.

Fitness Benefits to Those Helped   Although there are exceptions, many studies do indicate that the helped reap fitness benefits from the helper's assistance. These benefits may take two forms—increased survival of young and increased survival of the breeders. We have mentioned the Florida scrub jay, and this species is one of the most thoroughly studied to date (Woolfenden and Fitzpatrick, 1984, 1990). These birds live in territories containing one breeding pair and from 0 to 6 helpers. The breeding success for pairs with helpers clearly exceeds that of pairs without helpers (Woolfenden, 1975). Figure 19.13 shows the breeding success of experienced pairs with and without helpers during one 5-season study. It can be seen that the presence of helpers has no effect on the number of eggs laid but does increase the chance that the young will hatch, leave the nest, and become independent birds.

Similarly, breeding success increases with the presence of helpers in some mammalian and fish species. Blackbacked jackals form monogamous breeding pairs. Between one and three of the young from previous litters remain with their parents and help them rear the next pups. As seen in Figure 19.14, the breeding success of a pair of blackbacked jackals increases with the number of helpers (Moehlman, 1979). A similar relationship is found in the Princess of Burundi cichlid fish (*Lamprologus brichardi*). More young are successfully raised when the young from one brood help their parents guard the eggs and larvae of subsequent broods from predators (Taborsky and Limburger, 1981).

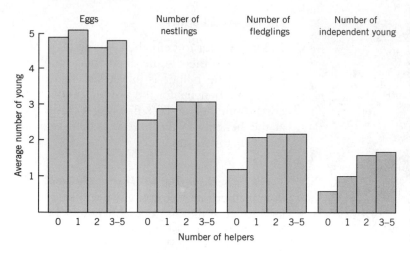

Figure 19.13 The relationship between the number of Florida scrub jay helpers and the breeding success of the experienced parents. Helpers do not increase the number of eggs laid. They do, however, increase the chances that the eggs will hatch and that the young will survive to become independent. (Data from Woolfenden, 1975.)

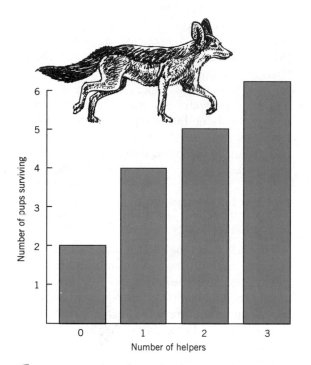

Figure 19.14 The relationship between the number of blackbacked jackal helpers and the number of the pups surviving. (Modified from Moehlman, 1979.)

But increased breeding success in groups with helpers is not by itself sufficient to demonstrate that breeders are actually benefiting from the assistance of helpers because correlation does not show cause and effect. To get a feeling for why this is so, consider the following example: The number of bars is usually greater in towns with more churches. This is not because the fear of God causes people to drink, but rather because the frequency of both enterprises increases as the population does.

In the case of helpers, some other factor, territory quality for example, could be responsible for the increase in both reproductivity and group size. If reproductive success is higher among individuals on good territories, and the young stay with their parents, then we would expect groups on good territories to have more helpers and higher reproductive success, even if the helpers did nothing at all.

One way to examine whether territory quality, not the presence of helpers, is responsible for the enhanced reproductive success of groups with helpers is to remove the helpers and look at the effect, if any, on reproductive success. If helpers really do help, we would expect that their removal would lower reproductive success. On the other hand, if they do not help, their removal should have no effect. In one experiment, helpers were removed from the nests of gray-crowned babblers (*Pomatostomus temporalis*), a bird that lives in year-round territorial groups of 1–13 birds in the open woodland of Queensland, Australia. Gray-crowned babbler parents are usually assisted by a variable number of their offspring from previous broods, but in this experiment, only one helper was permitted to remain in 9 of the breeding groups. These groups then raised an average of 0.8 young, less than half the number of fledglings produced by the 11 control groups that had more assistance. Therefore, the positive relationship between breeding success and the number of helpers found in gray-crowned babblers does seem to be a result of the presence of helpers (Brown et al., 1982).

However, not all data are consistent with the idea that helpers increase the fitness of the breeders; Amotz Zahavi (1974) reached the opposite conclusion from his studies of the Arabian babbler (*Turdoides squamiceps*). These birds nest communally in the deserts of Israel. As in the previous examples, the young from previous broods remain to help their parents raise the next brood, but Zahavi did not find that helpers increased the reproductive success of the group. Furthermore, he observed agonistic behavior within the breeding groups. One nestling died as the result of a wound probably caused by a peck. Zahavi suggests that helpers may do more harm than good by reducing the amount of food available for nestlings and by increasing the chances that a predator will be attracted by activity around the nest. He argues that the helpers are only tolerated because they are offspring of the breeders and because they help to defend the nest when a predator approaches. His conclusions remain controversial, however, because Jerram Brown's (1975) reanalysis of Zahavi's data did show a positive relationship between the breeding success and the presence of helpers.

When considering the balance of benefits and costs of harboring a helper from the breeder's point of view, it is interesting to note that in at least one species, the pied kingfisher (*Ceryle rudis*), helpers are only tolerated when their services are needed. These birds usually have primary helpers, which are older offspring, and may also have secondary helpers, which are unrelated. Heinz-Ulrich Reyer (1980) compared two colonies of pied kingfishers in East Africa. Breeding pairs at Lake Naivasha typically have only one primary helper. When males apply for a job as a secondary helper, they are persistently chased away by the male territory holder. In contrast, at Lake Victoria, secondary helpers are eventually tolerated and permitted to stay and feed the young. Why? The answer is that the services of secondary helpers are needed to raise offspring at Lake Victoria whereas they are not at Lake Naivasha. These birds fish for a living and Lake Victoria is a harder lake to fish. Victoria's rougher waters increase the length of time it takes to catch a fish and the fish are smaller. Furthermore, the fishing grounds are farther from the colony. With the additional fish provided by secondary helpers, the breeding pair can raise more offspring (Table 19.2).

Cost to the Helper   Remember, for helping to involve altruism it must be somehow costly to the helper. Unfortunately, data for evaluating this prediction are scarce. In the case of the Florida scrub jay, however, it seems that young birds would do better if they set up housekeep-

Table 19.2

THE EFFECT OF HELPERS ON THE REPRODUCTIVE SUCCESS OF PAIRS OF PIED KINGFISHERS

	Lake Victoria			Lake Naivasha		
	MEAN	SD	n	MEAN	SD	n
Clutch size	4.9	0.6	22	5.0	0.6	8
Young hatched	4.6	0.5	14	4.5	0.7	2
Young fledged						
No helpers	1.8	0.6	14	3.7	0.9	9
1 Helper	3.6	0.5	12	4.3	0.5	4
2 Helpers	4.7	1.0	6	—	—	—

Source: Reyer, 1980.

ing for themselves (Woolfenden and Fitzpatrick, 1990). Remember that Florida scrub jay helpers are raising younger siblings, as are most helpers. Since an individual is likely to have as many alleles in common with a sibling as with an offspring, the helper is making some genetic profit by increasing the number of offspring its parents raise. Groups with a helper raise an average of 0.32 more offspring than do unassisted parents. However, a pair breeding for the first time raises an average of 1.36 offspring. Although these calculations are oversimplified because they neglect many factors that would alter the predicted outcome, the general conclusion that a young scrub jay would make more genetic profit by breeding rather than helping seems valid. So why does the helper remain?

Ecological Conditions Favoring Helping

Older offspring must either leave home and attempt independent breeding or postpone departure from the natal territory and remain, at least for a while, within the parental group. In some species, the young may remain at home because they benefit so greatly from group living. Juveniles remaining with a group may enjoy access to some critical resource (either now or in the future), protection from predators, more food as a result of cooperative hunting, or more effective territorial defense (Stacey and Ligon, 1987). The costs of dispersing may also favor juveniles remaining at home. At least four factors probably enter into the decision: (1) risk of dispersal, (2) chances of finding a suitable territory, (3) chances of finding a mate, and (4) chances of successful reproduction once established. As we will see, the costs and benefits of helping are influenced by certain ecological conditions. Severe ecological conditions, such as the prohibitive costs of reproduction or shortages of territories or mates, limit the options of independent breeding and favor staying with the parents (Emlen, 1982; Emlen and Vehrencamp, 1983).

Habitat Saturation in Stable Environments   A young male scrub jay probably stays with his parents because he has little choice. It is very unlikely that he could find a suitable area for a territory because the available space is filled. A bird that wins a territory keeps it for life (Woolfenden, 1975). The most common way for a male to acquire a territory is by inheriting a portion of his parents' property, either by replacing his father after his death or by subdivision of his father's territory. If there is more than one son helping, the most dominant one is favored in the property settlement (Woolfenden and Fitzpatrick, 1978).

The scrub jay's problem is typical of that encountered by animals living in stable environments. When ecological conditions are predictable, population numbers increase and suitable habitat becomes saturated. The result is severe competition for territories. An individual has the option of leaving the parental nest and breeding independently only if a territory can be established by challenging and defeating a breeder, successfully competing for any vacancies resulting from the death of nearby breeders, or by inheriting or budding off a portion of the parental territory. If an individual must postpone breeding until a territory is available, it is best to wait at home because this is an area of proven quality, a factor that will increase the chances of surviving until the next year. Meanwhile, alliances can be formed that may enable the takeover of other territories, including a portion of the parents'. The helper's waiting time is best spent increasing his inclusive fitness by helping to raise siblings (Emlen, 1982).

This contention is supported by a comparison of the extent of helping in populations of the acorn woodpecker (*Melanerpes formicivorus*) residing in habitats with varying degrees of saturation. These gregarious birds, best known for their meticulous habit of individually caching thousands of acorns in storage trees called "granaries" (Figure 19.15), have been studied in several locations in the United States. Study sites

Figure 19.15 An acorn woodpecker places an acorn in a previously drilled hole in a dead tree. Food stores such as these must last the communal group through the winter.

of southeastern Arizona seem free of the shortage of available housing. In contrast to their California and New Mexico counterparts, acorn woodpeckers in Arizona tended to disperse or migrate between seasons (Stacey and Bock, 1978). Average group size at the Arizona study site was 2.2 and only 16 percent of the breeding units had helpers. In short, the frequency of occurrence of helping in populations of acorn woodpeckers varies directly with the scarcity of open territories (Figure 19.16).

As usual, however, there are some parts of these stories that do not fit as neatly into the interpretation of helping as a consequence of habitat saturation. In populations of both the Florida scrub jay and the acorn woodpecker, we occasionally find available territories remaining

in California, New Mexico, and Arizona vary with respect to woodpecker density, territory turnover rate, and territory fidelity. Interestingly enough, the occurrence of helping at these three locations parallels the gradient in difficulty associated with territory establishment (Emlen, 1982). Michael and Barbara MacRoberts (1976) studied acorn woodpeckers in coastal California and noted that not a single territory became vacant during their 3-year study. In this extremely saturated habitat, birds were permanently territorial and lived in family groups that consisted, on average, of 5.1 adults plus the young of the year. Forty-nine percent of the juveniles remained at home and 70 percent of the groups had helpers. Young acorn woodpeckers in the Magalena Mountains of New Mexico face somewhat better odds in their quest for suitable territories than do those on the West Coast: Nineteen percent of the territories in New Mexico became vacant over a 3-year study by Peter Stacey (1979). Average group size at this locality was 3.0 and 29 percent of all youngsters stayed at home. Helpers were present in only 59 percent of breeding groups. As a final comparison, acorn woodpeckers in the Huachuca Mountains

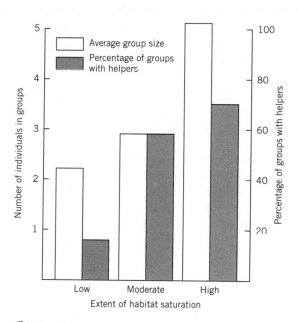

Figure 19.16 The relationship between the extent of habitat saturation and both the average group size and the frequency of helping behavior in acorn woodpeckers. In environments with few available territories, the group size and the percentage of groups with helpers is much greater than in areas of low habitat saturation. (Data from Emlen, 1982.)

vacant. These inconsistencies certainly deserve further attention, but we should not be too quick to abandon the hypothesis completely because of them. There is some evidence, for example, that the Florida scrub jay helpers that decline to establish their own territories are somehow biologically inferior and would be unable to establish and maintain a territory of their own. Instead, they survive as nonbreeders in the relative safety of their parents' territory and gain some indirect fitness by helping to raise siblings (Woolfenden and Fitzpatrick, 1990).

It has also been suggested that the benefits of group living, rather than the inability to win and maintain a territory, may explain why territories remain vacant. A major benefit gained by acorn woodpeckers that remain on their natal territory is present and future access to the trees needed for acorn storage. A juvenile does better remaining on its parents' high-quality territory as a nonbreeder and postponing breeding than it would if it attempted to breed on a low-quality territory (Koenig and Stacey, 1990).

Biased Sex Ratios Sexual partners, rather than territories, may be in short supply for some cooperatively breeding birds. Typically involving an excess of males, this demographic constraint limits the option of becoming established as an independent breeder (Emlen and Vehrencamp, 1983). Male-biased sex ratios characterize the splendid fairy-wren (*Malurus splendens*), an inhabitant of heathlands and scrubs of western Australia. This small passerine whose tail makes up nearly half of its body length (Figure 19.17) has been studied since 1973 by Ian Rowley and colleagues. Females of this species suffer much greater annual mortality than do males (57% and 29% respectively; Rowley, 1981) and thus are frequently in short supply. As would be predicted from the scarcity of females in the populations of splendid wrens, helpers tend to be males awaiting an available mate. For some individuals, the wait can be as long as 5 years. As patterns of mortality and the resultant sex ratios

Figure 19.17 A male splendid wren carries an insect to feed its young. When adult females are in short supply in the population, this breeding male can count on his sons to help rear the next brood.

vary, so does the percent of groups with helpers: When females are scarce, male helpers are a dime a dozen. Breeding females with helpers survived better than those without helpers (Rowley and Russell, 1990; Russell and Rowley, 1988). Furthermore, helpers, combined with breeding experience, lead to more females renesting after a first brood had been raised.

Cost of Reproduction in Unstable Environments Some populations of cooperative breeders live in environments in which conditions fluctuate. Breeding attempts among novices are largely unsuccessful for species inhab-

Figure 19.18 White-fronted bee-eater. When environmental conditions are harsh, young bee-eaters forsake independent reproduction and remain at home to rear their younger siblings.

iting variable, unpredictable habitats. Faced with erratic changes in environmental conditions, juveniles often remain at home and help rear relatives.

Support for the idea that helping should increase with the degree of breeding difficulty comes from studies on white-fronted bee-eaters (*Merops bullockoides*) (Figure 19.18). The life history of this bird makes it a good species in which to look for a relationship between helper frequency and environmental harshness. It lives in the savannahs and scrub-grasslands of the Rift Valley in eastern and southern Africa. The breeding of these insectivores is most successful when it is timed to coincide with periods of insect abundance. However, synchronization of these events is difficult to achieve because there is extreme variation in both the timing and amount of rainfall and the response of insects to the rain. As a result, the ease of raising offspring and, therefore, the cost of breeding independently vary from year to year.

The frequency of helping does indeed increase during hard times. During periods of lit-

tle rainfall and hence low food availability, bee-eaters suffer high losses from nestling starvation. It is during these periods of extremely harsh environmental conditions, when chances of successful reproduction are slim, that older offspring remain in their natal groups as helpers. Once conditions become more favorable, however, young bee-eaters are more likely to disperse and initiate breeding on their own. Notice in Figure 19.19 that helping increases with decreases in either rainfall or the availability of food, conditions that make it difficult to raise young (Emlen, 1982).

The Evolution of Cooperative Breeding

In this discussion, and throughout most of this book, we have assumed, as most people do, that behavior is a result of natural selection. However, the view that helping is an adaptation shaped by natural selection has been challenged (Jamieson, 1986, 1989, 1991; Jamieson and Craig, 1987), and this has led to considerable debate

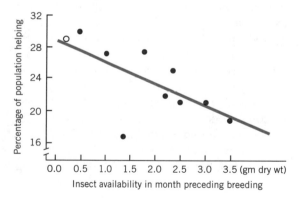

Figure 19.19 Incidence of helping in white-fronted bee-eaters varies as a function of food availability; when insect abundance increases, fewer young remain at home to help care for younger siblings. (The closed circles indicate insect availability during the month before breeding. The open circle indicates the insect availability for one colony during the month following breeding.) (From Emlen, 1982.)

(Emlen et al., 1991; Ligon and Stacey, 1991; White et al., 1991). Put simply, Ian Jamieson has suggested that helping is a by-product of the evolution of parental care and communal breeding. As we have seen, for some species dispersal may be difficult. Additionally, there may be advantages to group living, including increased vigilance against predators and cooperative foraging. As a result, natural selection may favor group breeding in some species. Natural selection may also favor individuals that feed begging youngsters wherever they are found. Communal breeding brings nondispersing older animals, usually older siblings, in contact with begging youngsters. Helping, then, is seen as an unselected consequence of group living.

If helping is an unselected by-product of communal breeding, we might expect all adults to follow the general rule, "If it begs, feed it." However, this is not always the case. Whether or not an immigrant acorn woodpecker feeds nestlings depends on when it joined the group. If the immigrant arrived before the eggs were laid, it will feed the youngsters. But if it arrived after the eggs were laid, it will ignore the begging (Stacey and Ligon, 1991). Furthermore, among white-fronted bee-eaters, only about half of the nonbreeding members of the group become helpers, even though all are exposed to the same begging stimuli (Emlen and Wrege, 1989).

It has also been argued that helping, at least in some species, is costly to the helper and, therefore, we might anticipate natural selection to refine parental responses, causing helping to be weeded out (Emlen et al., 1991). Among white-winged choughs (*Corcorax melanorhamphos*), for instance, helpers have a variety of duties, one of which is incubating the eggs. Although all the helpers develop a brood patch, not all incubate the eggs. Those that do incubate lose weight in proportion to the time they spend on the nest (Figure 19.20; Heinsohn, Cockburn, and Mulder, 1990). It would be surprising that such a costly behavior would be maintained

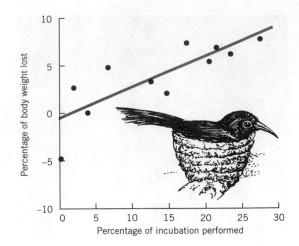

Figure 19.20   Some white-winged chough helpers incubate the eggs. Those that do lose weight steadily. It would be expected that such a costly behavior would be selected against if there were no benefits to helping that outweighed the costs. (From Heinsohn, Cockburn, and Mulder, 1990.)

even if it originated as an unselected by-product of selection for communal breeding and parental care.

Jamieson and Craig's suggestion that helping has not been selected has been useful in focusing attention on the untested assumptions that can pervade the study of any behavior. As a result, it has stimulated research in many new directions. Keeping in mind that any direct or indirect benefits that result from helping are consistent with the hypothesis that helping is a result of natural selection, but not proof of it, we will next consider some of these benefits.

Individual Selection   Although in many species older offspring remain at home during periods of either harsh environmental conditions or territory or mate shortages, why should they help? Furthermore, not all helpers are relatives. We have already discussed one example of unrelated helpers, the pied kingfisher. Others, including mammals such as the dwarf mongoose (*Helogale parvula*) can be listed. The inclusive fitness of helpers cannot be increased by the additional offspring they help to raise.

The question is, then, what might they be gaining? Many different hypotheses have been offered to answer this question (Brown, 1987; Emlen, 1991).

Enhanced Survivorship As we have already noted, some ecological conditions make it likely that an individual will live longer if it stays with a group. Helping, then, may be "paying" the breeders for permission to stay on a territory of known quality until it can win one of its own. In addition, helping may allow the group to grow larger, resulting in an increase in the benefits due to group living.

Increased Reproduction in the Future There are also many ways that helping might increase the chances of successfully raising offspring in the future. We have seen that helping may be a way of gaining a territory, as it is for the Florida scrub jay.

When mates are scarce, the helper may gain an opportunity to acquire the original breeder as a mate in a later year. Frequently helpers who are unrelated to the breeder are lacking a mate of their own. Sometimes, this helper assists a female breeder one year and mates with her the following year. This is the case in the pied kingfisher. You may recall that the secondary helpers are unrelated to the breeders. All secondary helpers are unmated. More than half of them return the following year and of these, half succeed in mating with the female they assisted (Reyer, 1980, 1984, 1986).

If the helper finally becomes a breeder, it may gain assistance in raising its own young from the offspring it helped to rear. The white-winged chough, a bird that cannot breed without helpers, provides an excellent example of this. Groups with less than seven members cannot support even a single youngster through the first winter. The main reason seems to be that foraging is a skill that takes years for a chough to master. Their diet consists of small invertebrates that are found by digging in the soil. Finding sufficient food is a difficult and time-consuming activity. Juveniles must be fed for up to 8 months, and even then they are poor foragers (Heinsohn, Cockburn, and Mulder, 1990). Thus, helpers are a necessity and the group grows slowly, usually only through reproduction. The young remain in the group and help raise the next generation of choughs (Heinsohn, 1991).

Increased Reproductive Success as Breeder In addition, the helper may benefit if its activities help it to learn or practice skills that will be necessary when it must raise its own young. Although there may be examples of this among certain birds, the best illustration might be among primates. Immature female vervet monkeys (*Cercopithecus aethiops sabaeus*) often babysit for the infants of others. Juvenile females that initiate caretaking more often and have more experience carrying infants are more successful in raising their own firstborn when they become mature than are females with less babysitting experience (Fairbanks, 1990).

Kin Selection Most helpers are previous offspring of a breeding pair that assist in the rearing of their siblings. Since an individual is as closely related to a sibling as to an offspring, it seems reasonable that kin selection might play a role in the evolution of cooperative breeding. Furthermore, most cooperatively breeding species live in habitats that either prevent the individual from breeding independently or that make it difficult to raise young successfully without assistance.

Another observation suggesting that kin selection may play a role in the evolution of helping is that males are more likely than are females to remain and assist their parents. It has been suggested that the "decision" to become a helper may be influenced by how certain the individual is of its genetic relationship to the young (Charnov, 1981). Differences in the degree of genetic closeness to the young result from the occasional unfaithfulness of a female to her mate. In these cases, the young are not all fathered by the same male. A female can always count on her offspring having 50 percent of her alleles. However, half siblings share only a quarter of their alleles. So, if some of her siblings had a different father, they will share less than 50 percent of their alleles with one another. However, the possibility of cuckoldry presents a worse problem for a male. If an intruder fertilizes his mate's eggs, then he is genetically unrelated to the young. However, his genetic relationship to his siblings must be at least 25 percent, even if their fathers are different. This guarantee may predispose males to become helpers and rear siblings rather than offspring. The observation that helpers are usually male has also been explained by individual selection. Recall that a male's opportunity for mating independently is often limited by a scarcity of suitable territories. He may have to wait to inherit at least a portion of his father's territory (Woolfenden and Fitzpatrick, 1978.)

There is still much controversy regarding the importance of kin selection in helping. The hypotheses we have discussed are not mutually exclusive. Instead, any one of them, or any combination, might be valid for a particular species. The question arises, then, how might we determine, for any given species, whether any of the proposed hypotheses for the evolution of helping by individual or kin selection are correct? The answer is simple—we make predictions that are consistent with each hypothesis and then evaluate the predictions. The evaluation may be based on observation or experimentation.

Stephen Emlen and Peter Wrege (1989) used 5 years of data on white-fronted bee-eaters to test alternate hypotheses for helping behavior

(Table 19.3). The birds in the study were individually marked and their family relationships were known. Each of the hypotheses we have discussed makes certain predictions that can be tested by examining the fitness benefits and the characteristics of helpers.

If helping increases an animal's chances of living to the next breeding season, we would predict that survival would increase with group size and helpers should have higher survival than nonhelpers. These predictions were not fulfilled in bee-eaters. Individuals in small groups survived as well as those in large groups. Besides, breeders, helpers, and nonhelpers all survived equally well. Thus, enhanced survival does not seem to be an advantage of helping in bee-eaters.

The hypothesis that helping increases future reproductive success predicts that animals that help should have a better chance of breeding than nonhelpers and that animals that have never bred should be more likely to help than those that have offspring. Neither of these predictions was met in bee-eaters. Forty percent of all helpers are former breeders and half of the birds that hatched in one season became breeders the following year, without having helped others at all.

If the hypothesis that helping increases reproductive success applies, we would predict that animals that had been helpers would raise more offspring than those that had never helped. Again, this prediction is not fulfilled in bee-eaters. There were no significant differences in the number of young fledged between helpers and nonhelpers.

Finally, the hypothesis that helping increases the helper's inclusive fitness through the extra offspring that relatives produce makes several predictions. First, breeding units with helpers should produce more young than units lacking helpers and/or the number of surviving young should be greater for units with helpers. Second, helpers should be closely related to those that

Table 19.3

PREDICTIONS CONCERNING (1) FITNESS GAINS ACCRUING TO HELPERS AND (2) PERSONAL CHARACTERISTICS OF BIRDS THAT BECOME HELPERS

HYPOTHESIS	PREDICTION 1	PREDICTION 2
Helping results in direct fitness gains through:		
Increased survival to next breeding season	Survival probability should increase with group size; birds that have served as helpers should have a higher survival than those that have not	No prediction
Increased future opportunity to breed	Birds that have served as helpers should have a greater probability of breeding than those that have not	Birds that have not bred before should be more likely to help than individuals that have already achieved breeding status
Increased reproductive success as breeder	Reproductive success for first-time breeders should increase as some function of the amount of previous helping experience —and/or— The likelihood of first-time breeders having helpers of their own should increase with previous helping experience	Birds that have no prior experience should be more likely to help than birds that have extensive prior experience
Helping results in indirect fitness gains due to the extra related offspring reared	The presence of helpers should significantly increase the production of young at helped nests —and/or— The presence of helpers should increase the survival of recipient breeders —and— Helpers should, on average, be closely related to the nestling beneficiaries	Birds should be more likely to help when the recipients are close kin than when they are distant kin or unrelated

Source: Modified from Emlen and Wrege, 1989.

are helped. Bee-eater helpers did, indeed, benefit by increasing the production of nondescendant kin. Although they did not help the breeders live longer, they did dramatically increase nestling survival. Helpers and breeders are closely related, so the extra nestlings that were raised due to the helper's efforts increased its inclusive fitness greatly. It was calculated, in fact, that most of the benefit a bee-eater derives from helping is through the extra nondescendant kin that survive. So we see that in bee-eaters, kin selection is the major factor leading to the evolution of helping.

COOPERATION IN ACQUIRING A MATE

Males of some species cooperate in attracting a mate. Some even relinquish the opportunity to pass their alleles into the future generation personally, at least temporarily. Instead, they concentrate their efforts on making another male more attractive to females. This presents a problem that should now be familiar: Since behaviors persist over time only when the alleles for them are perpetuated, how could this form of cooperation have evolved?

Kin Selection

When the cooperating males are related, kin selection seems to be a reasonable explanation for the evolution of this collaboration. Two examples, one from birds and the other from mammals, are presented to support this claim.

Wild Turkeys Strangely, most male wild turkeys (*Meleagris gallopavo*) in some Texas populations never mate. Toward the end of a young cock's first autumn, when he is about 6 to 7 months old, he and his brothers forsake the others in their family and form a sibling group that will be an inseparable unit until death. This sibling group and all other juvenile male sibling units in the area flock together for the winter. During the first winter each male's status within this "fraternity" is determined. Only one of the males will mate, that individual being determined through competition.

Each male's reproductive fate is decided by the outcome of two contests. One competition is for dominance within the sibling group. Brothers battle by wrestling, spurring, striking with the wings, and pecking at the head and neck. Endurance is the key to success. The turkeys fight until they are exhausted. When one remains able to do battle, however weakly, he is

the winner. The second contest is between rival sibling groups. The groups challenge and fight one another until a dominance hierarchy is established. The sibling group with the most members is usually victorious. Renegotiation of rank is rare; the dominance hierarchy within and between sibling groups is stable.

When the breeding season begins, females interested in mating visit the open meadows where the males congregate. Although the male winter flocks have disbanded, individual sibling groups remain together. The brothers of each unit court the hens by strutting in unison (Figure 19.21), even though only the dominant male in the highest ranking sibling group will mate. Of 170 tagged males displaying at 4 grounds, not more than 6 males accounted for all 59 observed matings. If a subordinate male is presumptuous enough to attempt a mating, the dominant male chases him away and he then mates with the hen.

A subordinate male gains inclusive fitness by helping his brother to perpetuate his alleles. On the other hand, without his assistance, the brother could not be successful; the cooperative efforts of siblings are necessary for their unit to become dominant and the synchronous strutting of siblings makes the dominant male more attractive to the hens. Thus, the subordinate brother reproduces "by proxy" (Watts and Stokes, 1971).

Lions Male lions (*Panthera leo*) also cooperate in attaining mates. Although their genetic relationship may not be as close as the turkeys', the males of a pride are usually related. The group of males, called a coalition, generally consists largely of brothers, half brothers, and cousins that left their natal pride as a group. Some calculations indicate that the males of a coalition are almost as closely related as are half siblings (Bertram, 1976). They remain together, and after 1 to 3 years of traveling nomadically, they challenge the males of other prides. The coalition may take over a pride by slowly driving

Figure 19.21 Communal displaying in wild turkeys. Groups of brothers strut together to attract mates. In some populations, only the most dominant male in the most dominant sibling group will have the opportunity to copulate. Since the cooperating males are closely related, kin selection is a reasonable hypothesis for the evolution of this behavior.

out the resident males, or it may be a hostile takeover involving serious fighting (Figure 19.22). In such contests, the larger coalition usually wins. The reward for the victors is a harem of lionesses. When the females come into reproductive condition, which is sometimes hastened by the new males killing any cubs that are present, they often do so simultaneously. During the 2- to 4-day period when a female is in reproductive condition, she mates about every 15 minutes around the clock. Any of the males in the coalition may be the first to find her, mate with her, and keep others away by his presence. A female may change mates during this period, but generally not more than once a day (Bertram, 1975). Thus the male who mates may gain fitness directly. When another male takes over, it is likely to be a relative. Then the first male still may gain fitness indirectly.

Individual Selection

As you must now be aware, similar behaviors can evolve from different mechanisms and more than one mechanism may shape a behavior. Although kin selection may be responsible for the

Figure 19.22 Two male lions fighting for control of a pride.

evolution of assistance in mate acquisition among wild turkeys in Texas, it is only part of the story among lions and is not involved in other cases of cooperative courting. As we will see, it may be to a male's advantage to assist another's reproductive efforts if this service increases his chances of personal reproduction, now or in the future.

Lions It is now known that roughly half of male coalitions contain at least one unrelated male (Packer, 1986; Packer and Pusey, 1982). Immediately the question arises, then, why would an unrelated male be accepted in a coalition? The answer turns out to be quite simple— the larger the coalition, the greater a male's reproductive success (Figure 19.23). Larger coalitions have a better chance of ousting the current coalition in a pride, maintaining control of that pride, and perhaps even gaining residence in a succession of prides. A solitary male has little chance of reproducing and, therefore, much to gain by joining another coalition. A small coalition may also gain by accepting an unrelated male because the extra member may help it take over prides. Indeed, coalitions accept unrelated companions only while they are not yet resident in a pride (Packer and Pussey, 1987).

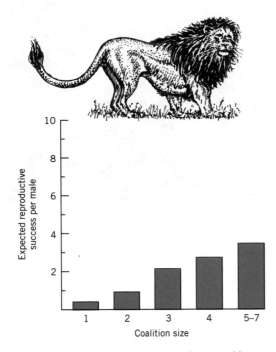

Figure 19.23 The size of male coalitions of lions is related to each male's reproductive success. As the coalition size increases, so does reproductive success. (Data from Packer et al., 1988.)

Larger coalitions also remain in control of a pride for a longer time than do smaller ones. A coalition of three to six males may remain in control as long as 2 to 3 years. A coalition of two might be in possession for over a year. If a lone male manages to gain control of a pride, which happens infrequently, his tenure generally lasts only a few months (Bertram, 1975). As a result, the lifetime success of a male lion increases by cooperating with other males in taking over a pride, even if all the males are not related (Packer et al., 1988).

Long-Tailed Manakins Among male lions cooperation in acquiring a mate may have evolved due to both direct and indirect fitness gains. However, cooperating long-tailed manakins (*Chiroxiphia linearis*) are not related. Therefore, there are no indirect fitness gains and kin selection cannot be playing a role. It takes two, or sometimes three, of these beautiful male birds to court a female. First, they must attract a female, and two males will perch together and call synchronously. The call, which sounds similar to the word *toledo*, may be emitted as many as 19 times a minute and 5000 times a day. If these vocalizations are successful in attracting a female, the males move to a display perch. The next step is for one male to stimulate the other to perform the acrobatic display that will induce the female to mate. The male is put "in the right mood" by the solicitation display executed by his masculine partner. When the males are ready, they proceed to the jump display that will stimulate the female. This courtship display cannot be performed alone. In one of the most common variations of the jump display, called the up–down variant, the movements of the males are somewhat like those of two children on a seesaw, but the birds do not use any apparatus. One male jumps into the air emitting a wheezy *buzzee* call and hangs there momentarily; just as he lands, the other male jumps up. In this way they alternately jump up and down. Another common variant of the jump display is the cartwheel variant. In this form of the courtship display, the males make a continuous moving circle around one another. The pattern of their movements resembles that of objects being juggled. They begin by perching next to one another on the display branch. The front male jumps up and backward just far enough to land on the spot where the second male had previously perched. Collision is avoided because while the first male is in midair, the partner moves anteriorly to occupy the spot where the other male had perched. Now this male takes a turn at jumping up and moving to the rear. The courtship sequence may be repeated only once or as many as a hundred times in succession. When the display bout is over, one male leaves the display branch and watches while the remaining male does a solo performance of the precopulatory display. If his gymnastics have impressed the female, she mates with him.

Although the males take turns jumping to court a female, it is always the same male who mates. The benefit to this male is obvious but what is the advantage to the other male? It is not indirect fitness because it seems unlikely that the two males are related. A typical brood consists of only one or two offspring and there is no reason to assume that the siblings are necessarily the same sex. Furthermore, just before and after each breeding season, the young, particularly the subadult males, disperse. It seems unlikely therefore that male relatives would stay in proximity for the 3–4 years it takes them to acquire adult plumage. If the partners are not genetically related, then the nonbreeder is not increasing his indirect fitness. So why doesn't he compete in attracting females?

A possible answer to how the cooperative courting of the long-tailed manakin evolved emerges when the choices of the subordinate bird are considered. Although he will not mate while he is a member of his current male–male alliance, his chances of mating would not be increased by deserting his partner. Solitary males cannot mate. They cannot even perform the courtship display. If he cannot dominate this partner, the chances of his becoming the dominant member of another pair would be low. However, if the subordinate male outlives his partner, it is likely that a younger male, one that can be dominated, will become his new associate. Then it will be his turn to mate and raise his personal fitness (Foster, 1977).

Reciprocal Altruism

Cooperation in mate acquisition may also evolve through reciprocal altruism. An example of this is the coalitions formed among male olive baboons (*Papio anubis*; Figure 19.24). A male who lacks a female consort sometimes enlists the help of a friend in order to win another male's mate. The following scenario is typical of what often happens. Male A is consorting with an estrous female. Male B covets this female since he has none of his own, so he solicits the help of Male C and the two form an alliance and challenge Male A. While the battle is in progress, Male B gets away with the female. Male C has acted altruistically; he risked injury while assisting another to acquire a mate. However, at some time in the future, the shoe will be on the other foot and he will enlist the help of Male B in winning a consort of his own (Packer, 1977).

EUSOCIALITY

The pinnacle of altruism is eusociality. This term describes groups that possess three characteristics: sterile workers that labor on behalf of other individuals that are engaged in reproduction, cooperation in the care of young, and an overlap of at least two generations capable of sharing in the colony labor (Wilson, 1971).

The Social Insects

The members of a group of social insects labor together almost as if they are parts of a superorganism. Reproduction is the sole responsibility of the queen. The sterile castes perform several

Figure 19.24 An alliance between two male olive baboons. The two males on the right are cooperating to challenge the male on the left. Alliances such as this one are formed to win the lone male's consort for one of the challengers. At a later time, the male that was assisted will have to reciprocate.

Figure 19.25 One of the jobs of sterile worker honeybees is to guard the nest. They check the odors on all incoming bees. Here, a guard bee inspects an approaching drone at the hive entrance. If the intruder is not a nestmate, it is expelled from the hive.

altruistic services. One is defense. Soldier ants and worker bees provide examples. Some members of each ant colony, the soldiers, specialize in colony defense. They place themselves between the threat and the colony and fight intruders to the death. Even when they are injured, they appear to behave in the best interest of the colony. Wounded soldiers leave the nest, or if they are already outside, they refuse to enter. This is adaptive because dead bodies within the colony would pose a sanitation problem. Honeybees also have nest guards (Figure 19.25). As honeybee workers defend the nest, they commit suicide. When a bee stings an intruder, the barbs on the stinger anchor it in the victim's body, causing it to be torn from the bee's body. The sterile workers do more than just defend the colony: They also provide food for nestmates. In almost all eusocial species, the workers indiscriminately share food with colony members. Finally, the amazing aspect of eusocial behavior is that the workers are sterile. They toil tirelessly caring for offspring that are not their own, without the hope of ever having offspring of their own.

The sterility of workers raises an interesting question. How can the alleles for the altruistic deeds be perpetuated when the individuals displaying the behaviors do not produce any offspring? This observation mystified even Darwin (1859). In his *Origin of Species by Means of Natural Selection*, he wrote: ". . . I can see no real difficulty in any character having become correlated with the sterile condition of certain members of insect communities: the difficulty lies in understanding how such correlated modifications could have been accumulated by natural selection."

Kin Selection and the Proposed Role of Haplodiploidy

The hymenopterans (ants, bees, and wasps) have a near monopoly on eusociality. Although it originated at least 11 times within the single order Hymenoptera, it arose only twice among the myriad of other arthropods: in the termites (Wilson, 1980) and in the aphids (Aoki, 1977, 1979, 1982). The preponderance of eusociality among the hymenopterans provided a valuable clue to W. D. Hamilton (1964) as he puzzled over the evolution of the behavior. It drew his attention to another trait of the hymenopterans that is rare outside that order, their means of sex determination, called haplodiploidy. Only a few other arthropod groups, including the mites, thrips, and whiteflies, display haplodiploidy.

In haplodiploidy, unfertilized eggs usually develop into males and fertilized eggs typically develop into females. The single queen in a hymenopteran nest has a mating flight and stores the sperm obtained at this time for the rest of her 10 or more years of life. The queen carefully doles out the sperm. Some, but not all of the eggs are fertilized. An unfertilized egg develops into a male that has only the single set of chromosomes donated by the mother. A male is, therefore, haploid. A fertilized egg, on the other hand, develops into a female. This means that a female hymenopteran has two sets of chromosomes, one from her mother and the other from her father; she is diploid.

Hamilton was able to make sense of the link between eusociality and haplodiploidy. He noticed that haplodiploidy can change the rules for determining the degree of relationship among the kin within a colony and results in peculiar asymmetries in the closeness of relationships among colony members. For example, a surprising and important outcome of haplodiploidy is that the sister workers within a colony could be, on the average, more closely related to one another and to the siblings they help to raise than they would be to their own offspring! How can this be? Since the queen is diploid, her eggs are formed by meiosis and are not identical. Each egg contains replicates of 50 percent of the mother's chromosomes, one of her two copies of each chromosome. In other words, the coefficient of relationship of the mother to her offspring is 0.5. The mother's chromosomes are only half of the genetic legacy of each daughter and, as you see, the sisters share an average of half of the maternal alleles. This means that they have an average of 25 percent of their alleles in common through their mother. If only one drone fertilizes the queen, then the father's contribution to each of his daughters is identical and accounts for half of her genetic inheritance. Since the father is haploid, identical copies of each of his chromosomes are in each sperm cell. Therefore, the sisters would have 50 percent of their alleles in common by virtue of their father's input and an average of 25 percent in common because of their mother's donation (Figure 19.26). As a result, sisters could share an average of 75 percent of their alleles by descent, a larger fraction than the 50 percent that a mother shares with her offspring. Thus, the sister workers increase their inclusive fitness more by rearing reproductively capable siblings than they would if they produced their own offspring. The greater genetic profit would favor daughters re-

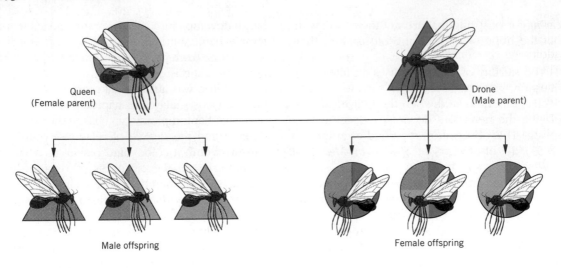

Queen
(Female parent)

Drone
(Male parent)

Male offspring

Female offspring

Degrees of relatedness in haplodiploid species

	Daughter	Son	Mother	Father	Sister	Brother
Female	0.5	0.5	0.5	0.5	0.75	0.25
Male	1	0	1	0	0.5	0.5

Figure 19.26 The genetic contributions of a male and female of a haplodiploid species to their female offspring. The possible combinations are indicated by colors. Female offspring have a 50 percent chance of sharing a gene from their diploid mother (A or a). The maternal contribution is only half of the female offspring's genes. As a result, sisters have 25 percent of their genes in common through their mother's gametes. The father is haploid so his sperm are all identical. In other words, sisters of haplodiploid species share 25 percent of their genes because of their mother's contribution and 50 percent of their genes because of their father's contribution, or a total of 75 percent of their genes by common descent. Male offspring are haploid. Each male's entire set of genes comes from his mother. Thus, the average degree of relationship among brothers is 50 percent.

maining at home to augment their mother's reproductive efforts. In this way, then, the evolution of eusociality might be explained.

Sex Ratio Data as Support for the Kin Selection Hypothesis Another asymmetry in closeness of relationships paved the way for Robert Trivers and Hope Hare's (1976) test of the kin selection hypothesis. Because of haplodiploidy, males are more closely related to their mother than they are to their sisters. A result of this asymmetry is that the interests of the royal figure

are at odds with those of her daughters. The queen shares an equal number of her alleles with her sons and daughters, so from her perspective, the optimal way to perpetuate her alleles is with an equal investment in sons and daughters. In contrast, a female worker is more closely related to her sisters than to her brothers. Sisters, we know, may share 75 percent of their alleles by descent. However, they share only 25 percent of their alleles with their brothers. (Each egg formed by the mother contains copies of half of her alleles. These are the only alleles a son inherits, but a daughter receives an additional set of chromosomes from the father. Therefore, half of the maternal half, or 25 percent, of a female worker's alleles are likely to be present in her brother.) This asymmetry in relationship led Trivers and Hare to expect a sex ratio skewed toward reproductive females and away from males; a female offspring would enjoy a threefold increase in inclusive fitness if she raised sisters rather than brothers. They calculated that from the female worker's point of view, the optimal sex ratio is 3:1 in favor of females, but from the queen's point of view, it is 1:1. Thus there is a conflict of interest between the queen and the workers.

Although it might seem, at first, that the queen must be the winner in this conflict because she could determine the sex of her offspring by choosing whether or not to fertilize the egg, it turns out that the workers have many ways to alter the sex ratio. Among honeybees, for instance, drones must develop in wider cells than workers do. A queen lays a drone egg only in wide cells. Since the workers build the cells, they have the last say in the number of drones produced. These cells are always on the periphery of the comb, and the workers can keep the queen away from this region of the comb. Workers may even fill the drone cells with honey, thereby preventing the queen from laying eggs in them. If a drone egg is laid, the workers can abort its development by eating the egg or the larva that develops from it. In many hymenopteran species, male offspring must be cared for and nourished by their sisters. Consequently, the workers have the opportunity to modify the sex ratio by feeding their brothers less than their sisters or even by starving them. Actually, with the focus on the evolution of eusociality, it is only the reproductively capable siblings, those that can perpetuate copies of the worker's alleles, that are of interest. Because it is only the future queens that should be counted as females in the sex ratio, the workers have another means

of altering the ratio. All diploid eggs will develop into females but only those that are fed an appropriate diet will become reproductively capable. Since it is the workers that feed the larvae, they have control over the number of potential queens. Furthermore, the male and female reproductives are often different sizes and this necessitates an unequal energy investment in raising them. Therefore, the quantity of investment in these reproductives is more important than the actual numbers of individuals of each sex. If eusociality among the Hymenoptera truly evolved via kin selection and the sterile workers are actually behaving in ways that increase the frequency of their own alleles, then the combined weight of reproductive females within a hive should be three times greater than the combined weight of reproductive males.

This prediction was confirmed. Trivers and Hare accumulated data on the sex ratios (investment ratios) of 21 ant species. Although there is a good deal of scatter in the data, the investment in reproductively capable females was found to exceed that in reproductive males by an amount close to 3:1 (Figure 19.27).

A remarkable exception to the usual sex ratio among ants also supports the kin selection hypothesis. It has been argued that sterile castes evolved because the workers increased their inclusive fitness more by raising siblings than by producing their own offspring. As evidence, it was noted that sex ratios are often biased toward females, the colony members that share a greater proportion of alleles with the workers that care for them. However, in some ant species the workers that attend to the daily routine of running the nest and caring for the brood are slaves that are genetically unrelated to the colony members. The soldiers of slave-making species wage war against colonies of other ant species and drag back the pupae or larvae that later hatch to serve their colony as slave workers. The slaves work diligently, performing all the duties they are genetically programmed to perform in their home colony. Because the nursemaids car-

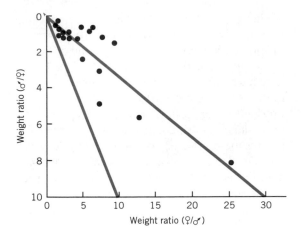

Figure 19.27 The investment ratio (male/female) measured as the dry weight of reproductive adults of each sex for 21 species of ants. The ratio is biased toward females by an amount close to 3:1. This observation is consistent with the kin selection hypothesis for the evolution of eusociality. (Modified from Trivers and Hare, 1976.)

ing for the brood are unrelated to the colony, their fitness is not increased by altering the 1:1 sex ratio that is optimal for the queen. Although sex ratio data were only available for two slave-making species (*Leptothorax duloticus* and *Harpagoxenus sublaevis*), Trivers and Hare noted that the dry weight of female reproductives approximately equaled the dry weight of male reproductives. The slaves cannot revolt because if a mutation leading to rebellious behavior did occur, it would not be passed on. The future reproductives being nurtured by the slaves bear the queen's alleles, not the slaves'. The 1:1 investment ratio in the two sexes of slave-making ants is expected because it is optimal for the queen mother, and it is her offspring's alleles, not those of the slaves, that are passed into future generations.

Problems with the Kin Selection Hypothesis and Sex Ratio Data Hamilton's kin selection

hypothesis relates the high frequency and the extreme degree of altruism among the Hymenoptera to the unusually close genetic relationship among the sister workers in the nest. However, since he formulated his ideas, there have been repeated reports of multiple matings and insemination in many hymenopteran species. For example, a queen honeybee (*Apis mellifera*) makes the best of her mating opportunity by mating with an average of 17.25 males (Adams et al., 1977) and by storing most of the 6 million sperm she receives from each male (Kerr et al., 1962; Woyke, 1964). If the sperm from each male remains clumped during storage, then workers of the same age within a colony are likely to have the same father. The rules for calculating relatedness then would remain virtually untarnished (Trivers and Hare, 1976). However, biochemical studies show that the sperm of different drones mix during storage in the spermatheca of honeybee queens. In one study it was shown that 8 of 12 queens were utilizing the sperm of more than one male in significant proportions. Five of these queens were fertilizing eggs with the sperm of three males (Page and Metcalf, 1982). In another experiment, queens were artificially inseminated with sperm from six males that had genetic markers that identified the father of the offspring. Again, the results indicate that the sperm of all mates mix while they are stored within the queen and are, therefore, used in fluctuating, but representative proportions (Laidlaw and Page, 1984). Likewise, there is evidence in two species of social wasps that the queens mate more than once and use the sperm from different males in relatively constant proportions through time (Ross, 1986).

If the sperm of different males mix during storage, then the female workers within a colony may have different fathers. Workers fathered by different drones share only 25 percent, rather than 75 percent, of their alleles by descent. In this extreme case, the workers would be no more closely related than are half siblings in diploid species such as the ones discussed in previous sections. More realistically, if the queen mates twice and the sperm mix randomly, the workers will share 50 percent of their alleles by common descent. They will have an average of 25 percent of their alleles in common from their mother, and they will have a 50 percent chance of having the paternal half of their alleles donated by the same father. Remember that, because males are haploid, all the sperm from a single father are identical $(0.25 + 0.25 = 0.5)$. Obviously, the exact genetic relationship among the sister workers of a colony will depend on the number of males that mated with their mother, the extent of mixing of sperm during storage, the utilization of the sperm by the queen, and the outcome of any competition between sperm of different males. Unfortunately, data on these issues are scarce.

The reality of multiple insemination among some hymenopteran species obviously has ramifications for the kin selection hypothesis. Hamilton (1964) pointed out that multiple insemination would reduce the tendency to evolve sterile castes because the workers would be expected to increase their fitness more by rearing their own offspring that would share a greater proportion of their alleles than would their reproductive siblings. Although kin selection could still operate, multiple insemination weakens the argument that haplodiploidy was a major factor in the evolution of eusociality.

Furthermore, the sex ratio data of Trivers and Hare (1976), which has been heralded by some biologists as decisive support for the kin selection hypothesis, has been criticized by others. Richard Alexander and Paul Sherman (1977) are particularly negative in their evaluation of the sex ratio data. First, they argue that the 3:1 female-biased sex ratio predicted by Trivers and Hare is wrong. You recall that this ratio was based on the idea that a worker's investment in a particular sex should reflect the degree of genetic closeness to it. A worker's sister is assumed to share 75 percent of her alleles by descent while her brother carries only 25 per-

cent. The predicted result is that workers should invest 3 times as much effort in female reproductives as in males. The critics point out that these genetic relationships are wrong. One reason for the error, multiple insemination, has already been discussed. Another cause for error is that workers of some eusocial species, including a host of ant species, sometimes lay unfertilized eggs. These eggs develop into males. A worker is more closely related to her son than she is to her brother. Because a son shares an average of 50 percent of her alleles by descent while a brother shares only 25 percent, a worker has more to gain by rearing sons. The other workers (her sisters) would be rearing nephews. Thus, worker-laid eggs play havoc with the calculations of the genetic relationships that form the basis for predicting the optimal sex ratio from the worker's perspective.

As a second point in their argument that a 3:1 female-biased sex ratio would not be predicted, Alexander and Sherman suggest that the statistics used to determine how closely the observed ant sex ratio data fit the prediction are inappropriate.

A third major point made by Alexander and Sherman is that the ant sex ratio data are consistent with an alternative hypothesis, that of local mate competition. According to this hypothesis, which by the way is another of Hamilton's (1967) contributions, a female-biased sex ratio will result if close relatives (brothers) are mating rivals. In a situation in which sons compete with one another for local mates, one son's gain is another's loss. If mating opportunities are limited, and one son could fertilize all the available females, then additional sons are superfluous. Under these conditions a mother disseminates her alleles just as well by raising only one son. In other words, a mother should produce only as many sons as are needed to maximize the number of her grandchildren. As a result, when there is mating rivalry among close relatives, the optimal sex ratio for a parent is biased toward females. If there is local mate competition among eusocial species, particularly among those ant species studied by Trivers and Hare, then the queens' optimal sex ratio, as well as the workers', should be shifted toward females. However, because data on matings within the family are scarce, Alexander and Sherman could point to only one ant species (*Myrmica schencki*) included in Trivers and Hare's analysis that is known to have local mate competition.

So where does all this leave us? We still have no definitive answer to how important kin selection may have been in the evolution of eusociality, but we have seen the way in which science progresses. Hamilton hypothesized that haplodiploidy predisposed the hymenopterans to eusociality because with that means of sex determination, females increase their inclusive fitness more by raising reproductively capable siblings than by producing their own offspring. Because this idea was new and logical, it stimulated much discussion and research. Data from many areas were marshaled to evaluate the hypothesis. The ratio of investment in reproductive males and females by several ant species is biased toward females, the sex that shares the greatest percentage of alleles with the workers. This is consistent with Hamilton's ideas because it is another instance of workers behaving in a way that will increase the frequency of copies of their own alleles in the next generation. Kin selection rapidly became the most popular explanation for the evolution of eusociality. Then some people began to have reservations about the kin selection hypothesis and the sex ratio data that support it. One important issue has been determining the actual genetic relationships among nestmates since the queens of many species seem to mate more than once. The genetic relationship among workers is even more difficult to calculate when worker-laid eggs, which will develop into males, are considered. Finally, the prediction of an optimal sex ratio biased 3:1 in favor of females and the analysis of the sex ratio data have been criticized. This leaves us with a hypothesis based on an extraordinarily close relationship between sister

workers that may not exist and is supported by controversial data.

Nonetheless, although the closeness of genetic relationship among social insects cannot entirely explain the evolution of eusociality, it does seem to be a factor that predisposes the group toward eusociality. Eusociality is common among the hymenoptera, which are haplodiploid. Other eusocial species also seem to be closely related.

There are two groups of insects, termites and certain aphids, that, like the hymenoptera, have sterile castes. Among termites, a high degree of relatedness, perhaps resulting from extreme inbreeding, may predispose the group to eusociality (Bartz, 1979; Reilly, 1987). Aphids, small insects that obtain nourishment by sucking the juices from plants, may be almost genetically identical to other colony members. There are two types of one developmental stage of certain aphids of the family Pemphigidae. One of these, the secondary type of the first instar larvae, serves as a soldier acting in the colony's defense (Foster, 1990). When larvae of a syrphid fly, a predator of aphids, enters the colony, the aphid soldiers insert their piercing mouthparts into the predator, killing it. This action could not have evolved by enhancing the soldier's reproductive success since the soldier is preprogrammed to die. Instead, it is thought to be a result of an usually high genetic relatedness among the members of an aphid colony. Indeed, because the individuals develop from unfertilized eggs of a single female, they are nearly identical genetically (reviewed by Ito, 1989).

As we saw in the evolution of helping, additional factors to be considered are those that influence the costs and benefits of dispersing or remaining with the natal group.

Individual Selection   So far in this chapter we have seen that there may be several factors, acting in unison or independently, that promote altruism in various species. Therefore, it is wise for us to examine some of the other factors that might favor eusociality.

Special Needs   Social grouping may be necessary for the survival of some species. Consider, for instance, the termite. In termites, both sexes are diploid, so the assumed advantages of haplodiploidy do not apply to these insects. Nonetheless, as we have seen, individuals within a termite colony are highly related to one another (Reilly, 1987). The high degree of genetic relatedness presumably sets the stage for the evolution of eusociality, but it is not a complete explanation. We may wonder, then, what other factors may have played a role in the evolution of eusociality among termites.

You may also be aware that termites are unable to digest wood, the primary component of their diet. The termite's gut contains symbiotic protozoans that break down the cellulose from the wood in the termite's diet to a soluble carbohydrate. A termite must be supplied with these protozoans after hatching and again after each molt. The protozoans are obtained by licking the anal secretions of other termites in the colony. Since the cellulose-digesting intestinal flagellates are essential and they must be obtained from other termites, it is adaptive for these insects to live in groups. Given this restriction and the fact that a nonreproductive is at least as closely related to a sibling as it would be to its own offspring, the termite path was paved for eusociality (Lin and Michener, 1972).

A species need not have the requirement for transmitting intestinal symbionts for eusociality to be advantageous. Earlier in this chapter, in our discussions of helping and assistance in mate acquisition, it was suggested that any situation in which there is a low probability of successfully raising offspring alone makes it advantageous for individuals to assist the reproductive efforts of others, especially if those that benefit are relatives. The same contention may be valid when applied to the evolution of eusociality. Let's review some ecological factors related to nest building that could send some species toward eusociality.

Ecological Factors Favoring Eusociality
Eusociality in some insects may have developed

as a result of some of the same ecological pressures that contribute toward social behavior in other species.

1. THE NEED TO DEFEND THE NEST   The nest may have to be protected against predators or parasites. An example of the importance of assistance in protecting the nest is provided by the social wasp (*Mischocyttarus mexicanus*). The nests of this species may be founded by one to several females. When there are several foundresses, they are usually closely related and only one female actually lays eggs, although each of them is capable of doing so. The variation in the number of females establishing a nest makes it possible to determine whether lone or group nesting strategies are more successful and to isolate some of the factors responsible for success.

Marcia Litte's (1977) comparison of one-foundress nests and multiple-foundress nests demonstrated the importance of a larger group for nest defense. The two important predators of these wasps are birds and ants. Because the sting of this wasp is mild, multifoundress nests were no more effective in driving away birds than were smaller nests. However, the wasps are very effective in defending against ants. A lone foraging ant was rarely permitted to approach within 4 centimeters of the nest and those that succeeded in mounting the nest were thrown off. When they are home, all adult female wasps actively defend the nest against predators. However these females must also forage. While a single foundress is foraging, her nest is left unattended, but in a multifoundress nest, there is always someone available to guard the home front. Undefended nests fail more often than those protected by continuous vigilance. Since a female would make greater genetic gain by successfully rearing relatives than by failing in her attempt to raise her own offspring, eusociality is favored. Other people, however, have argued that parasite or predator pressure favored the development of eusociality among bees (Lin and Michener, 1972) and wasps (Evans, 1977).

Sometimes the nest must be protected from takeover by other females of the same species. Conspecific pressure, not predation or parasitism, has been shown to provide an advantage to cooperating for the paper wasp (*Polistes metricus*). George Gamboa (1978), for example, did not detect any predation when he studied *Polistes*. Furthermore, he observed that the presence of additional foundresses had no effect on the prevalence or severity of parasitism. However, he did notice that single-foundress nests were usurped by a challenging foundress significantly more frequently than were multiple-foundress nests—when a single foundress leaves her nest unattended during her foraging trips, there is a greater opportunity for a coup. In addition, a lone foundress may be unable to defend her nest without assistance even if she is at home at the time of assault. Interestingly, a coup did not have the same devastating effect on multiple-foundress nests. When a takeover was successful in a nest with several foundresses, the queen was not dethroned; rather, a subordinate foundress was replaced.

2. THE NEED FOR ASSISTANCE IN NEST BUILDING During colonial times in the United States, it was hard work to build a house. As a result, children often remained in the home of their parents even when they were married and about to begin their own families. When the dwelling became crowded, an addition was built.

For a eusocial insect, nest building is no less of a chore. The nests are often intricate and take a great deal of time and energy to construct (Figure 19.28). Therefore, in the early stages of sociality, a newly matured adult might have been better off using the mother's nest as a safe haven for her eggs, even if this required enlarging it. Eventually this cooperation may have evolved toward rearing younger siblings rather than one's own offspring (Andersson, 1984).

The problem of defending and building a nest makes it difficult for a single parent to raise offspring successfully. In addition, a female that emerges from hibernation late in the season may

Figure 19.28 A paper wasp nest illustrates the elaborate nests of social insects. The difficulty of defending against takeover by conspecifics may be a factor favoring eusociality in some species.

find it difficult to find a suitable spot for a nest (West-Eberhard, 1975). Under these circumstances, she may be better off staying with a group and assisting someone else's reproductive efforts so that she has a chance to inherit an established productive nest later in the season.

When it is difficult to found a new nest, a wasp may be better off accepting small but guaranteed fitness returns than she would be gambling on her ability to raise a brood to independence. Consider the fitness consequences to a female if she died before the brood was independent. If she is a worker, she would still derive some fitness, because other workers would continue to care for the brood. However, if she is a solitary foundress, she loses everything. Thus, the "assured fitness returns" may favor eusociality (Gadagkar, 1990a,b).

Increased Probability of Successful Reproduction in the Future   You may recall that earlier in this chapter it was argued that helping by some vertebrates could be considered a form of payment for permission to remain on the breeding territory while waiting to take over. A similar situation may hold for some eusocial invertebrates.

Eusociality among some wasps may have been favored by a low probability of reproducing alone coupled with the possibility of taking over a productive nest in the future. Females of various wasp species become workers at established colonies. Sometimes, but not always, these so-called joiners are helping to produce relatives. Marcia Litte (1977) reports that there is a dominance hierarchy among the foundress and the joiners in the wasp species *Mischocyttarus mexicanus*. This hierarchy determines who is next in line to ascend the "throne." Litte examined the process of queen replacement by experimentally removing queens from established colonies. In every case, the right of egg laying was assumed by the female just under the former queen in the dominance hierarchy.

Parental Manipulation   For some biologists the explanation for the evolution of eusociality among insects lies in Alexander's hypothesis of parental manipulation. A parent's concern must be to maximize its overall fitness through all of its reproductive offspring, whereas the interest of any particular offspring is to maximize its own fitness. The reasoning behind Alexander's conclusion that the parent will always win in conflicts over fitness was given earlier in this chapter. The proponents of the parental manipulation hypothesis believe that eusociality is evidence that the queen has won this parent–offspring conflict. The queen's fitness is elevated more if her daughters remain at the nest and help raise additional siblings than if they leave and rear their own offspring. This is because the queen shares an average of 50 percent of her alleles with her offspring but only 25 percent of her alleles with grandchildren. Therefore, selection should favor her turning some of her offspring into workers (Charnov, 1978).

How might the queen manipulate her offspring into increasing her fitness at the expense of their own? One way would be by the queen behaviorally dominating her offspring. This is one interpretation of the actions of the queen in the primitively eusocial bee *Lasioglossum zephyrum*. The workers do the labor, but the queen must act as the foreman of the job and direct their activities. When a worker returns from a pollen-collecting trip, the queen usually meets her subject in a tunnel and then backs away. This stimulates the worker to follow her. The queen stops at the entrance of the cell that should be provisioned with the pollen. Without her direction, the workers are unable to locate an appropriate cell for their pollen and will deposit it in the tunnel. In addition, the presence of the queen stimulates activity within the hive, much as an office is busier when the boss is around. In an experiment, colony activity was measured before and after removing the queen. Sometimes one of workers immediately ascended the vacant throne. However, when no worker took that role, the activity level within the hive began to decline within 30 minutes after the queen was removed, and it remained significantly lower than it was while the queen was in place (Breed and Gamboa, 1977).

Alternatively, the queen could "force" her offspring to assist her reproductive activities by restricting the options available to them through controlling their diet. After all, it is a daughter's diet, not her genes, that determines whether she will be reproductively capable. The queen's domination might have originated with unequal food distribution among her daughters. Because smaller females are less likely to reproduce successfully, selection would favor their staying at home to rear siblings. The queens of many eusocial species, for example, honeybee queens, produce pheromones that inhibit the development of their worker daughters' ovaries. With the possibility of reproducing on her own restricted or eliminated, it would be to a daughter's advantage to enjoy whatever fitness she

Figure 19.29   A queen naked mole-rat, the only reproductive female of the colony, resting on the workers that feed her and help care for the young.

gained by helping to rear her reproductively capable siblings. At the same time, the mother would be maximizing her personal fitness.

Mole-Rats

Mole-rats are so named because they behave like moles, but look like rats. These rodents are particularly intriguing because two species, the naked mole-rat (*Heterocephalus glaber*; Jarvis, 1981) and the Damara mole-rat (*Cryptomys damarensis*; Bennett and Jarvis, 1988) appear to be eusocial. The social structure of these burrowing animals fits the classical definition of eusociality (Wilson, 1971) that we applied to social insects earlier in this chapter. First, breeding is restricted to a single female, aptly named the queen, even in groups with almost 300 members (Figure 19.29). Other adult females are smaller than the queen and neither ovulate nor breed. One to three males breed with the queen, although most adults males do produce sperm. Second, mole-rat colonies contain overlapping generations of offspring. Third, there is differentiation of labor among individuals within the colony (Figure 19.30). This specialization is much less rigid than among the castes of social insects. The duties assumed by the nonbreeding members seem to depend more on their size and age (Figure 19.31). Smaller members' duties generally include gathering food and transporting nest material. As they grow, they begin to clear the elaborate tunnel system of obstructions and debris. Larger members dig tunnels and defend the colony (Honeycutt, 1992; Lacey and Sherman, 1991; Lovegrove, 1991; Sherman, Jarvis and Braude, 1992).

Kin Selection   A close genetic relationship among mole-rats seems to have paved the way for the evolution of eusociality, just as it did among the social insects. Free-living colonies of naked mole-rats were sampled and the genetic similarities of colony members determined using the technique of DNA fingerprinting. There is an unusually high degree of genetic relatedness among the members of any single colony. Even between colonies, there is more genetic similarity than is found among nonkin of other free-living vertebrates. This similarity is thought to be a consequence of the extreme inbreeding within a colony (Faulkes, Abbott, and Mellnor, 1990; Reeve et al., 1990).

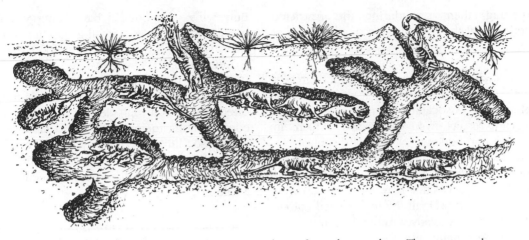

Figure 19.30 Division of labor among members of a mole-rat colony. The queen and perhaps one to three males breed. Others care for the young, gather food, dig tunnels, and defend the colony.

The inbreeding is promoted by their population structure. Naked mole-rats live for a long time and are prolific breeders. Two females caught in the wild have survived for 16 years in captivity, and they still breed. In nature litters, which may range in size up to 12, are born year-round every 70–80 days. When the queen dies, one of the larger females assumes her reign, after violent in-fighting among the other eligible females of the colony (Brett, 1991). Furthermore, there is little mixing of genes between colonies. Members of different colonies are quite aggressive toward one another (Lacey and Sherman, 1991). Intruders may even be killed. New colonies are thought to be formed by "budding." A group may leave the parental colony and seal off the intervening tunnels (Brett, 1991). High genetic similarity among mole-rats is not

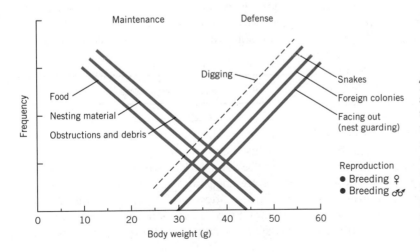

Figure 19.31 The duties assumed by mole-rats vary with their body size. Smaller individuals gather food and nesting materials. Slightly larger ones clear and dig tunnels. The largest members defend the colony from invasion by members of other colonies and from predatory snakes. (Data from Lacey and Sherman, 1991.)

surprising, then, considering the restricted breeding and the lack of intercolony mixing of genes.

Individual Selection The high degree of genetic relatedness cannot, by itself, explain the evolution of eusociality in mole-rats. Genetic similarity results from inbreeding, which, in turn, is a consequence of limited dispersal. So, for a fuller understanding of eusociality and inbreeding, we must probe deeper. We should consider the factors that may act in concert with genetics to favor remaining in the natal colony and helping raise relatives rather than dispersing to breed personally.

One predisposing factor may be the subterranean life-style of the naked mole-rat. Within their underground tunnels, naked mole-rats are fairly safe from predators. The dry regions inhabited by mole-rats have many plants with subterranean roots, tubers, and bulbs. Since naked mole-rats feed primarily on these, they need not leave the safety of the tunnel system to forage. Furthermore, the tunnel system can be expanded easily as the colony grows (Alexander, Noonan, and Crespi, 1991). Thus the benefits resulting from this subterranean life-style would seem to favor remaining with the natal group to raise related offspring, even if it means forgoing personal breeding.

These benefits cannot be the entire explanation for eusociality, since there are noneusocial mole-rat species that also live in underground tunnels. The question arises, then, what other ecological features have played a role in the evolution of eusociality in the naked mole-rat? The answer may involve the hazards of dispersal (Lovegrove, 1991).

One factor that makes dispersal a risky business for a mole-rat is that the tubers and bulbs they prefer to eat are distributed unevenly throughout the habitat. The natal colony may have access to a patch of food, but a group that sets off on its own may have to burrow extensively before encountering another rich area.

Burrowing uses quite a bit of energy. Consequently, members of a small group might die of starvation before locating a new food resource (Lovegrove, 1991; Lovegrove and Wissel, 1988).

We see, then, that the explanation for the evolution of eusociality in mole-rats is multifaceted. There is a high degree of genetic relatedness among members of naked mole-rat colonies, so kin selection may be a partial explanation. However, the benefits of group living and the costs of dispersal also seem to play a role.

Conclusions

There is no explanation for the evolution of altruism that applies to every example. Different life histories and ecological conditions may alter the relative importance of a particular evolutionary mechanism. Furthermore, the mechanisms suggested are not mutually exclusive and may be working simultaneously. To complicate matters, the same data may be consistent with more than one hypothesis. It is difficult to demonstrate conclusively that one particular hypothesis is correct because evolution is the central concern and behaviors are not fossilized. Therefore, there is no behavioral record of the evolutionary steps toward altruism. We must reconstruct what may have occurred by studying living species that display different degrees of cooperation.

Summary

Altruism is the performance of a service that benefits a conspecific at a cost to the one that does the deed. Strictly speaking, the benefits and costs are measured in units of fitness (the reproductive success of a gene, organism, or behavior). Since changes in fitness are nearly impossible to ascertain, however, the gains and losses are usually arbitrarily defined as certain goods or services that seem to influence the participants' chances of survival.

The prevalence of altruism among animals has puzzled evolutionary biologists. Evolution involves changes in the frequency of certain alleles in the gene pool of a population. If aiding a conspecific costs the altruist, then it should be less successful in leaving offspring bearing copies of its alleles than are the recipients of its services. As a result, the alleles for altruism would be expected to decrease in the population. Therefore, the existence of altruism seems at first to contradict evolutionary theory.

The hypotheses for the evolution of altruism can be arbitrarily classified into four overlapping classes.

1. **Individual selection** The general thrust of these hypotheses is that when the interaction is examined closely enough, the altruist will be found to be gaining, rather than losing, by its actions. The benefit may not be immediate; sometimes the gain is in the individual's future reproductive potential.

2. **Kin selection** When the beneficiaries of the good deeds are genetically related to the altruist, the enhanced reproductive success they enjoy will perpetuate the alleles that the altruist shares with them by virtue of their common descent. The altruist's relatives are more likely than nonrelatives to carry the alleles leading to altruism. Therefore, assisting relatives may increase the altruist's inclusive fitness, even if it decreases its personal fitness.

How can relatives be identified? There are several possibilities. One way might be to use location as a cue: The individuals that share one's home are likely to be kin. Alternatively, individuals might be identified as kin because they are recognized from prior social contact or as a result of their association with a known relative. Another possible way to recognize a relative is by certain traits that characterize fam-

ily members. In other words, an image of a family member may be developed that is matched or compared to the appearance of a stranger to determine whether it is related. Finally, recognition may be genetically programmed. Perhaps there are alleles that in addition to labeling relatives with a noticeable characteristic cause the altruist to assist others that bear the label.

3. **Reciprocal Altruism** Altruism might also evolve, in spite of the initial cost to the altruist, if the service is repaid with interest. In other words, altruism will be favored if the final gain to the altruist exceeds its initial cost. However, for reciprocal altruism to work, individuals that fail to make restitution must be discriminated against. Because of this requirement, certain factors make reciprocal altruism more likely. First, there should be a good chance that an opportunity for future repayment will arise. Second, the individuals must be able to recognize one another.

4. **Parental Manipulation** According to this hypothesis, there is selection for altruism toward siblings when it maximizes the parents' fitness. Any allele that causes an offspring to behave in a manner that reduces the parents' fitness would lower the selfish sibling's fitness when it becomes a parent and passes the trait on to its own offspring.

No single hypothesis applies to every example of altruistic behavior. In addition, these evolutionary mechanisms are not mutually exclusive; more than one may be responsible for a single example of altruism. To confuse matters even further, similar behaviors may evolve by different mechanisms in different species.

Members of some social species emit alarm calls to warn their neighbors of a predator's approach. For example, many small passerine birds do this. The suggested explanations for this behavior that invoke individual selection are numerous. The caller may benefit if the call discourages the predator's attack. The predator might look for easier victims because the call is notification that it has been detected and the prey are wary. It might also be discouraged because an alarm call causes the prey to scatter, making them more difficult to catch. Another suggestion is that the caller benefits by causing conspecifics to flee to a safer place where, in the anonymity of the flock, it can avoid being spotted by the predator. Kin selection and reciprocal altruism have also been invoked to account for avian alarm calling. However, given the natural history of the species that call, these hypotheses seem unlikely.

Ground squirrels emit two types of alarm calls, one to terrestrial predators and one to aerial predators. The calls seem to have been selected in different ways. Individual selection seems to be the best explanation for the evolution of alarm calls to aerial predators. However, kin selection seems to be the most likely mechanism for the evolution of ground squirrel alarm calls to terrestrial predators.

Another form of altruism is helping. A helper is an individual that assists in the rearing of offspring that are not its own, usually by providing food or by protecting the young. In most species helpers are previous offspring that are helping their parents raise their siblings so kin selection seems to be a reasonable explanation for these occurrences of helping. The helpers are not always relatives, however. In these cases, helping may be a means of maximizing individual fitness in the future. Helping commonly accompanies ecological conditions that make reproduction difficult or costly. Under such conditions, helping may be a means of obtaining permission to remain on a high-quality territory, of maintaining group or territory cohesiveness, of earning the future assistance of those helped, of obtaining a mate, or of protecting young from predators.

Another apparently altruistic behavior is helping another individual to acquire a mate. Kin selection is thought to have been important in the evolution of this behavior in some popula-

tions of wild turkeys and in lions because the cooperating males are related. However, some lion coalitions include unrelated males and in other species, such as the long-tailed manakin, the males that display together are unrelated. In such cases, it seems that cooperation increases the nonmating male's future chances of reproduction.

Finally, reciprocal altruism seems to explain the alliances formed by male olive baboons. One male may assist another to win a consort away from a third male. However, at some time in the future, the male that was assisted will have to repay the favor in kind.

Eusocial species are those that have sterile workers, cooperative care of the young, and an overlap of generations so that the colony labor is a family affair. The eusocial insects behave altruistically in several ways: Food is shared; the colony members specialized for defense often die performing their duty; and some members of the colony are sterile but care for the young of the colony's royalty.

Eusociality is common among hymenopterans (e.g., ants, bees, and wasps) but almost nonexistent outside that order. Hamilton noted that haplodiploidy, a term that describes a means of sex determination in which fertilized eggs develop into females and nonfertilized eggs develop into males, is also common in the order Hymenoptera but rare outside it. He suggested that haplodiploidy may have predisposed the hymenopterans to eusociality because it results in a closer relationship between sister workers and their siblings than the relationship that would exist between the workers and their own offspring. The female workers are likely to have 75 percent of their alleles in common with the reproductively capable siblings they help to raise, but they would have only 50 percent of their alleles in common with offspring they produced. As a result, a female worker makes greater gains in inclusive fitness by raising siblings than she would by producing offspring.

Trivers and Hare have supported Hamilton's

hypothesis with an analysis of the ratio of investment in males and females among various ant species. An optimal sex ratio is one that maximizes inclusive fitness. Since a queen has 50 percent of her alleles in common with her sons and daughters, the optimal ratio of investment in her reproductive offspring is 1 male:1 female. On the other hand, since the female workers share 3 times the numbers of alleles by common descent with their sisters as with their brothers, the optimal sex ratio from their point of view is 3 reproductively capable females:1 reproductively capable male. In the 21 species of ants analyzed, the ratio of investment in the sexes was biased toward females by an amount close to 3:1.

However, it is now known that a queen may mate with more than one male. If the sperm of more than one male mix during storage, then the genetic relationships among colony members are changed. Half sisters share fewer than 75 percent of their alleles by common descent. This observation weakens the argument that haplodiploidy strongly favored the evolution of eusociality.

As a result, the role of individual selection has been given more attention in recent years. Termites are eusocial but both males and females are diploid. The individual termite presumably gains by its close association with others because it must obtain intestinal symbionts from its colony members immediately after hatching and again after each molt. These symbionts are necessary for the insect to be able to break down the cellulose in the wood that forms the main portion of its diet. In addition, certain ecological factors may make eusociality a means of increasing fitness. Cooperative action may be needed to defend a nest against predators, parasites, or even conspecifics. The combined efforts of several individuals may make the chore of nest building easier. In these cases, an individual may maximize its lifetime fitness by postponing immediate reproduction in favor of assisting another individual in order to increase the likelihood of successful reproduction in the future.

Parental manipulation is a third hypothesis for the evolution of eusociality. It has been suggested that the queen forces her offspring to assist her reproductive efforts because the production of offspring raises her fitness more than the production of grandchildren. She may accomplish this by behaviorally dominating her workers or by restricting their reproductive options.

The naked mole-rat is a eusocial mammal. The colony members have a high degree of genetic relatedness as a result of inbreeding and this predisposes them to eusociality. This subterranean life-style and the hazards of dispersal are other factors that may have favored the evolution of eusociality in the naked mole-rat.

Appendix

Magnetoreception

Within the last few decades, magnetic sensitivity has moved out of the realm of mysticism into reality. Effects of magnetism on behavior have been demonstrated in many organisms—from mud-dwelling bacteria to birds and bees and beyond. Many scientists have now, in fact, turned their attention to the nature of the detector and several pieces of the puzzle are beginning to fit together. Some scientists even speculate that more than one type of receptor may be found in a single organism!

We considered evidence for the effects of magnetism on the behavior of organisms and the way that magnetic cues might be used for orientation and navigation in Chapter 11. Here, we are concerned with the hypothesized mechanisms for magnetoreception, including electroreceptors, magnetite deposits, photoreceptors, and the pineal gland.

ELECTRORECEPTORS

One mechanism by which a magnetic field might be detected is induction—that is, by the gener-

ation of an electric potential by the movement of an electrical conductor through a magnetic field. Although this mechanism does not seem to be common, presumably because the potentials generated on land or in freshwater would be too small, induction is the mechanism used by the electroreceptive cartilaginous fishes of the sea. The motion of a shark, ray, or skate swimming through a magnetic field induces an electric field that is well within the range of sensitivity of its electroreceptors. What makes these electric fields of possible use in orientation is that the potentials generated at the surface of the skin depend on the direction in which the fish is swimming relative to the earth's magnetic field. Notice in Figure A.1 that when a fish is swimming eastward, the potentials on its belly surface would be positive relative to those on its back surface. When the fish is swimming westward, the polarity of the induced field would be reversed. Thus, the potentials on its back surface would be more positive than those on its belly. However, no voltage difference between the back and belly would be created by a fish's swimming north or south. So, just by swimming, the fish gains information that could reveal its compass bearing. But the system may be even more sophisticated than this. It might also indi-

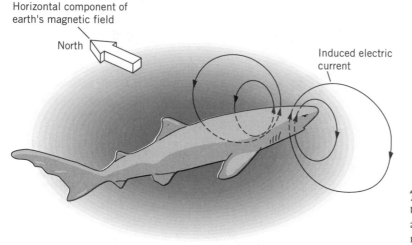

Horizontal component of earth's magnetic field

North

Induced electric current

Figure A.1 The orientation of the electrical current induced as a shark swims through a magnetic field.

cate the animal's latitude, or north–south position on the earth. This information would be encoded in the intensity of the electric field generated by the swimming movements. Intensity changes with latitude because of the regular variation in the vertical component of the earth's magnetic field (Kalmijn, 1978, 1984).

MAGNETITE DEPOSITS

Most organisms with a magnetic sense lack electroreceptors. Many of these have deposits of magnetic material, magnetite, which often forms chains or clumps and in vertebrates are commonly found in the head or skull.

There are two types of magnetic behavior of magnetite that could make it useful in a magnetoreceptor. First, the magnetite could be a permanent magnet if the crystal is large enough and of the right shape. It could then act as a compass needle aligning itself with magnetic lines of force. The smallest magnetically stable unit, called a single-domain, would rotate as it tracks the external field (Figure A.2). This turning might be detected by the nervous system. One suggested mechanism that might inform the nervous system of the movement of magnetite involves mechanoreception. When the miniature magnets align with a magnetic field, they might stimulate hair cells or pressure receptors. The pattern of stimulation might inform the animal about the strength and direction of the magnetic field. The effect of magnetite rotating would be amplified if the single-domains were connected in a chain. Indeed, chains of permanently magnetic grains of magnetite have been discovered in several organisms (Figure A.3).

Alternatively, a magnetite grain could be a superparamagnet. When a grain of magnetite is too small to be magnetically stable, it is still magnetic, but the direction of its magnetic field changes with that of an external field. The directions of the magnetic field of individual molecules within the grain are unstable and align themselves with the external field. In this case, the direction of the magnetic field of the magnetite grain could change without rotation of the grain itself.

Perhaps the following analogy will help clarify the difference in the behavior of a permanent magnet and a superparamagnet. Imagine the possible responses of a line of people instructed to always face a beacon of light when the light moved 90 degrees to the right. If the people form a line so that each person stands behind another with his or her hands on the shoulders of the person in front and never releases this grip, the entire column would have to rotate as a unit 90 degrees to the right. This would be analogous to the movement of a permanent magnet. However, if each person in line releases his or her hold on the person ahead, then each

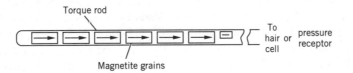

Torque rod

Magnetite grains

To hair or pressure cell receptor

Figure A.2 One hypothesis for a mechanism that might link the response of single-domain magnetite particles to changes in the magnetic field with the nervous system. The magnetite grains, which might form a chain, are permanent magnets that would turn like a compass needle in response to alterations in the direction of the magnetic field. This turning might stimulate a mechanoreceptor in a hair cell (Kirschvink and Gould, 1981).

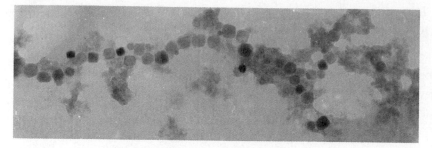

Figure A.3   A chain of single-domain magnetite particles from the skull of a sockeye salmon (Mann et al, 1988).

would be free to turn 90 degrees to the right. The individuals would then be standing shoulder to shoulder. The column would not move, but the people within it would rotate. This would be analogous to the behavior of a superparamagnet.

There are several suggestions for mechanisms that might link the nervous system to the behavior of supermagnetic grains. One proposal is that a series of such grains might be embedded in a rodlike stretch receptor (Figure A.4). When the grains aligned their fields with an external magnetic field perpendicular to the rod, the grains would repel one another and the rod would lengthen and stimulate a stretch receptor. A field parallel to the rod would align the magnetic fields of the grains so that they attracted one another, thereby shortening the rod (Kirschvink and Gould, 1981; Towne and Gould, 1985).

Some interesting research has been done on magnetoreception in certain mud-dwelling bacteria. The bacterium *Aquaspirillum magnetotactum* apparently has a chain of magnetite particles within its body that acts as a permanent magnet and aligns itself with the external field like a compass needle. The bacterium then swims in that direction. It has, in fact, been learned that the bacteria in the Northern Hemisphere move toward magnetic north and those in the Southern Hemisphere seek magnetic south. The reason for this difference is that the magnets in these bacteria point in opposite directions. In the Northern Hemisphere, the south pole of the bacterial compass needle is at the organism's

anterior end. However, the north pole of the bacteria of the Southern Hemisphere is at its anterior end. The internal magnet of these bacteria is so strong that even when they are dead, they passively align with a magnetic field. This response is appropriate for bottom-seeking organisms. Since the earth's magnetic field dips downward at the poles, becoming almost vertical, the magnetotactic bacteria of both hemispheres always swim downward, toward the muddy bottom they call home (Blakemore, 1975; Blakemore and Frankel, 1981; Frankel, 1981; Frankel and Blakemore, 1980; Frankel, Blakemore, and Wolf, 1979).

Magnetite deposits have also been found in many other organisms including honeybees (*Apis mellifera*; Gould, Kirschvink, and Deffeyes, 1978), homing pigeons (*Columba livia*; Walcott, Gould, and Kirschvink, 1979), the common Pacific dolphin (Zoeger, Dunn, and Fuller, 1981), yellowfin tuna (*Thunnus albacares*; Walker et al., 1984), the American eel (*Anguilla anguilla*; Hanson et al., 1984), and chinook salmon (*Oncorhynchus tshawytscha*; Kirschvink et al., 1985).

If the magnetite deposits function as magnetoreceptors in larger organisms, the information they provide would have to be transmitted to the nervous system. Therefore, associations between magnetite and the nervous system are of particular interest. Magnetite deposits in the abdomen of honeybees are within cells closely associated with the nervous system (Kutterbach et al., 1982). In addition, magnetite has been located in the particles that stimulate hair cells

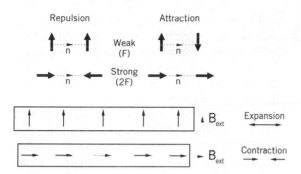

Figure A.4 One hypothesis for a mechanism that might link the response of superparamagnetic magnetite particles to changes in the magnetic field with the nervous system. The directions of the magnetic moments of individual molecules of magnetite within the deposit would track that of the imposed field. As a result, some directions of magnetic field might cause the magnetite molecules to repel one another. This might cause the expansion of a receptor structure that would in turn stimulate a stretch receptor. When exposed to a different orientation of magnetic field, the molecules would be attracted to one another and the receptor structure would contract (Kirschvink and Gould, 1981).

within the gravity proprioceptive structures of the guitar fish (O'Leary et al., 1981). Although it has not been found in every bird examined, magnetite has been reported near nerves in the heads of some pigeons (Walcott, Gould, and Kirschvink, 1979). In the bobolink (*Dolichonyx oryzivorus*), a bird that migrates across the equator, magnetite is consistently found in the sheaths around the olfactory nerve and bulb. It is also found in bristles that project into the nasal cavity (Beason and Nichols, 1984). (Might these bristles function as a hair-cell mechanoreceptor?) Interestingly, branches of the bobolink's trigeminal nerve appear to innervate the region in which magnetite deposits are found. These branches respond to earth-strength changes in the direction of the magnetic field (Beason and Semm, 1987; Semm and Beason, 1990). Thus, it is tempting to speculate that magnetite particles are magnetoreceptors.

It is important to keep in mind that, except in the mud-dwelling bacteria, the evidence that magnetite deposits are magnetoreceptors is circumstantial. Although magnetite grains are often large enough, may be connected in chains, and may be associated with structures in the nervous system, there is no clear demonstration that these deposits are magnetic detectors.

MAGNETORECEPTION AND LIGHT

It is interesting that in some organisms the reception of light and magnetic fields seem to be related. Such interactions have been shown in various responses of planaria (Brown and Park, 1967), honeybees (Leucht, 1984), crayfish (Sadauskas and Shuranova, 1984), turtles (Raybourn, 1983), pigeons (Semm and Demaine, 1986), quail and chickens (Krause, Cremer-Bartels, and Mitoskas, 1985), and rats (Reuss and Oclese, 1986).

Some of the pieces of the puzzle of this interaction between light and magnetoreception in vertebrates are falling into place (Figure A.5). The first step in the general scheme occurs in the retina, where some of the photoreceptors also serve as magnetoreceptors. Although several molecular mechanisms by which the photoreceptor molecules might detect magnetic fields have been suggested, the correct one is still not known (Leask, 1977a,b; Schulten, 1982; Schulten and Windmuth, 1986). The basic idea, however, is that photoreceptor molecules absorb light better under certain magnetic conditions. Thus, the amount of light absorption also provides information about the local magnetic field.

The suggestion that animals might somehow "see" the earth's magnetic field seems bizarre, so we might ask whether or not there is support for the idea. Indeed, several observations have suggested to researchers that the visual system is involved in magnetodetection. For example, the electrical activity of certain photoreceptors

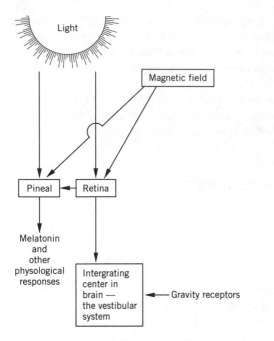

Figure A.5 An overview of the interaction between photoreception and magnetoreception.

in the eye of a blowfly (*Calliphora vicina*) is changed by alterations in the magnetic field (Phillips, 1987). Also, the retinas of certain vertebrates respond to light differently when the magnetic field is altered, even when they are removed from the animal (Krause, Cremer-Bartels, and Mitoskas, 1985; Raybourn, 1983). Furthermore, certain cells in the brain of a pigeon respond both to light and to the orientation of a magnetic field. However, both responses are substantially reduced when the bird is blinded (Demaine and Semm, 1985; Semm et al., 1984). Similarly, in the rat, the effect of magnetic fields on the synthesis of the pineal hormone melatonin is eliminated in blind animals (Oclese, Reuss, and Vollrath, 1985).

Information about magnetic fields from the photoreceptors would then be sent to various regions of the brain, where certain centers integrate information on light, magnetic fields, and gravity. This integration might tell the organism

the degree of inclination in the lines of force of the earth's magnetic field (see Chapter 11). Since the dip in the geomagnetic field varies regularly with latitude, this information might allow the animal to determine its north–south location. Thus, these integrative brain centers could be the neural basis of the magnetic compass.

The integration of information from the retina regarding light and magnetic fields with information from gravity receptors takes place in the brain's vestibular system. If the vestibular system is stimulated by tilting the animal, certain neurons in this brain region will respond to changes in the magnetic field. Visual and magnetic information are also integrated. Certain cells in the accessory optic system (the nucleus of the basal optic root) respond both to light and to directional changes in the natural magnetic field. However, these cells only respond to magnetic fields if the visual system is stimulated and intact (Semm et al., 1984).

The pineal gland in the brain may also play a role in magnetoreception. Not only does the pineal get information about magnetic conditions from the retina, it may also be able to pick up its own information (Demaine and Semm, 1985). The pineal then causes changes in the animal's behavior through its neural and endocrine activities. Alterations in the magnetic field cause changes in the electrical activity of certain cells in the pineal (Reuss, Semm, and Vollrath, 1983; Semm, Schneider, and Vollrath, 1980), changes in the synthesis of the pineal hormone melatonin (Cremer-Bartels et al., 1984; Stehle et al., 1988; Welker et al., 1983), and changes in the amount of cyclic AMP, a substance that regulates many cellular activities (Rudolph et al., 1988).

TWO MAGNETORECEPTOR SYSTEMS

It now seems that at least two migratory species, the eastern spotted newt* (*Notophthalmus viri-*

---

* Also known as the red spotted newt.

descens; Phillips, 1986) and the bobolink (Beason and Semm, 1987) have two independent systems for detecting magnetic fields. We do not know, however, how the animals use either system.

It is tempting to speculate that the pineal and retinal photoreceptor system of birds is part of their magnetic compass sense. An observation in favor of this speculation is that inexperienced homing pigeons are disoriented if they are transported to their release site in complete darkness. Thus, they seem to need light to use their magnetic compass (Wiltschko and Wiltschko, 1981).

To use the geomagnetic field as a map, an animal might merely compare the local intensity of the field with that at the goal. A receptor system used in a map sense, then, would not have to respond to the direction of the field, but it would be expected to respond to slight variations, less than 0.1 percent, in the magnetic field experienced (Beason and Semm, 1991). Although we have no idea whether magnetite deposits serve as a magnetoreceptor that is part of a map sense (or even if there *is* a magnetic map sense), it is thought that the amount of material typically found in pigeons could make such as receptor sensitive enough to small differences in magnetic field to fit the bill (Yorke, 1981).

References

ABLE, K. P. 1980. Mechanisms of orientation, navigation, and homing. In: S. A. Gauthreaux Jr., ed., *Animal Migration, Orientation, and Navigation*. Academic Press. New York, pp. 283–373.

ABLE, K. P. 1982. Skylight polarization patterns at dusk influence migratory orientation in birds. Nature 299:550–551.

ABLE, K. P. 1991. Common themes and variations in animal orientation systems. Am. Zool. 31:157–167.

ABLE, K. P. AND M. A. ABLE. 1990. Ontogeny of migratory orientation in the savannah sparrow, *Passerculus sandwichensis*: Calibration of the magnetic compass. Anim. Behav. 39:905–913.

ABU-GIDEIRI, Y. B. 1966. The behaviour and neuro-anatomy of some developing teleost fishes. J. Zool. London 149:215–241.

ACEVES-PINA, E. O. AND W. G. QUINN. 1979. Learning in normal and mutant *Drosophila* larvae. Science 206:93–96.

ADAMS, E. S. AND R. L. CALDWELL. 1990. Deceptive communication in asymmetric fights of the stomatopod crustacean *Gonodactylus bredini*. Anim. Behav. 39:706–717

ADAMS, J., E. D. ROTHMAN, W. E. KERR, AND Z. L. PAULINO. 1977. Estimation of the number of sex alleles and queen mating from diploid male frequencies in a population of *Apis mellifera*. Genetics 86:583-596.

ADLER, K. 1976. Extraocular photoreception in amphibians. Photochem. and Photobiol. 23:275–298.

ADLER, K. AND C. R. PELKIE. 1985. Human homing orientation: critique and alternative hypotheses. In: J. L. Kirschvink, D. S. Jones, and B. J. MacFadden, eds., *Magnetite Biomineralization and Magnetoreception in Organisms: A New Biomagnetism*. Plenum Press. New York, pp. 573–593.

ADLER, K. AND D. H. TAYLOR. 1973. Extraocular perception of polarized light by orienting salamanders. J. Comp. Physiol. 87:203–212.

AGIUS, L. 1979. Larval settlement in the echiuran worm *Bonellia viridis*: Settlement on both the adult proboscis and body trunk. Mar. Biol. 53:125–129.

AIDLEY, D. J. 1981. Questions about migration. In: D. J. Aidley, ed., *Animal Migration*. Cambridge University Press, Cambridge, N.Y. pp. 1–8.

ALATALO, R. V. AND A. LUNDBERG. 1984. Polyterritorial polygyny in the pied flycatcher *Ficedula hypoleuca*—evidence for the deception hypothesis. Ann. Zool. Fenn. 21:217–228.

ALATALO, R. V., A. CARLSON, A. LUNDBERG, AND S. ULFSTRAND. 1981. The conflict between male polygamy and female mongamy: The case of the pied flycatcher *Ficedula hypoleuca*. Am. Nat. 117:738–743.

ALATALO, R. V., A. LUNDBERG, AND K. STAHLBRANDT. 1982. Why do pied flycatcher females mate with already-mated males? Anim. Behav. 30:585–593.

ALBERTS, J. R. 1986. New views of parent–offspring relationships. In: W. T. Greenough and J. M. Juraska, eds., *Developmental Neuropsychobiology*. Academic Press. New York, pp. 450–478.

ALBERTS, J. R. AND D. J. GUBERNICK. 1983. Reciprocity and resource exchange: A symbiotic model of parent–offspring relations. In: L. A. Rosenblum and H. Moltz, eds., *Symbiosis in Parent-Offspring Interactions*. Plenum Press. New York, pp. 7–44.

ALCOCK, J. AND A. FORSYTH. 1988. Post-copulatory aggression towards their mates by males of the rove beetle *Leistotrophus versicolor* (Coleoptera: Staphylinidae). Behav. Ecol. Sociobiol. 22:303–308.

ALEXANDER, R. D. 1974. The evolution of social behavior. Ann. Rev. Ecol. Syst. 5:325–383.

ALEXANDER, R. D. 1975. Natural selection and specialized chorusing behavior in acoustical insects. In: D. Pimental, ed., *Insects, Science, and Society*. Academic Press. New York, pp. 35–77.

ALEXANDER, R. D. 1979. *Darwinism and Human Affairs*. University of Washington Press. Seattle. 317 pp.

ALEXANDER, R. D. AND P. W. SHERMAN. 1977. Local mate competition and parental investment in social insects. Science 196:494–500.

ALEXANDER, R. D., K. M. NOONAN, AND B. J. CRESPI. 1991. The evolution of eusociality. In: P. W. Sherman, J. U. M. Jarvis, and R. D. Alexander, eds., *The Biology of the Naked Mole-Rat*. Princeton University Press. Princeton, NJ, pp. 3–44.

ALLEN, J. A. 1988. Frequency-dependent selection by predators. Phil. Trans. Roy. Soc. London B 319:485–503.

ALTMANN, J. 1980. *Baboon Mothers and Infants*. Harvard University Press. Cambridge, MA.

ALTMANN, S. A. 1962. A field study of the sociobiology of rhesus monkeys (*Macaca mulatta*). Ann. N.Y. Acad. Sci. 102:338–435.

AMADON, D. 1948. Continental drift and bird migration. Science 108:705–707.

AMBROSE, H. W. 1972. Effect of habitat familiarity and toe-clipping on rate of owl predation in *Microtus pennsylvanicus*. J. Mamm. 53:909–912.

ANDERSSON, M. 1984. The evolution of eusociality. Ann. Rev. Ecol. Syst. 15:165–189.

ANDERSSON, M. 1986. Evolution of condition-dependent sex ornaments and mating preferences: Sexual selection based on viability differences. Evolution 40:804–816.

ANDREW, R. J. 1962. Evolution of intelligence and vocal mimicking. Science 137:585–589.

AOKI, S. 1977. *Colophina clematis* (Homoptera, Pemphigidae) an aphid species with "soldiers." Kontyu 45:276–282.

AOKI, S. 1979. Further observations on *Astegopteryx styracicola* (Homoptera, Pemphigidae) an aphid species with soldiers biting man. Kontyu 47:99–104.

AOKI, S. 1982. Soldiers and altruistic dispersal in aphids. In: M. D. Breed, C. D. Michener, and H. E. Evans, eds., *The Biology of Social Insects*. Westview Press. Boulder, CO, pp. 154–158.

ARAK, A. 1988a. Sexual dimorphism in body size: A model and a test. Evolution 42:820–825.

ARAK, A. 1988b. Callers and satellites in the natterjack toad: Evolutionarily stable decision rules. Anim. Behav. 36:416–432.

ARMITAGE, K. B. 1986. Marmot polygyny revisited: Determinants of male and female reproductive strategies. In: D. I. Rubenstein and R. W. Wrangham, eds., *Ecological Aspects of Social Evolution*. Princeton University Press. Princeton, NJ, pp. 303–331.

ARMSTRONG, D. P. 1991. Levels of cause and effect as organizing principles for research in animal behavior. Can. J. Zool. 69:823–829.

ARNOLD, A. P. AND S. M. BREEDLOVE. 1985. Organizational and activational effects of sex steroids on brain and behavior: A reanalysis. Horm. Behav. 19:469–498.

ARNOLD, E. N. 1988. Caudal autotomy as a defense. In: C. Gans and R. B. Huey, eds., *Biology of the Reptilia*, vol. 16. Alan Liss, Inc. New York, pp. 235–273.

ARNOLD, S. J. 1976. Sexual behavior, sexual interference and sexual defense in the salamanders *Ambystoma maculatum, Ambystoma tigrinum* and *Plethodon jordani*. Z. Tierpsychol. 42:247–300.

ARNOLD, S. J. 1983. Sexual selection: The interface of theory and empiricism. In: P. Bateson, ed., *Mate Choice*. Cambridge University Press. Cambridge, pp. 67–108.

ARONSON, L. R. 1971. Further studies on orientation and jumping behavior in the gobiid fish, *Bathygobius soporator*. Ann. N.Y. Acad. Sci. 188:378–407.

ASCHOFF, J. 1965. Circadian rhythms in man. Science 148:1427–1432.

ASCHOFF, J. 1967. Circadian rhythms. In: H. Brown and F. Favorite, eds., *Life Sciences and Space Research*. North-Holland, Publ. Amsterdam. pp. 159–173.

AUSTAD, S. N. 1983. A game theoretical interpretation of male combat in the bowl and doily spider, *Frontinella pyramitela*. Anim. Behav. 31: 59–73.

AUSTAD, S. N. AND M. E. SUNQUIST. 1986. Sex-ratio manipulation in the common opossum. Nature 324:58–60.

AXELROD, R. 1984. *The Evolution of Cooperation*. Basic Books. New York. 241 pp.

AXELROD, R. AND W. D. HAMILTON. 1981. The evolution of cooperation. Science 211:1390–1396.

AYALA, F. J. AND C. A. CAMPBELL. 1974. Frequency-dependent selection. Ann. Rev. Ecol. Syst. 5:115–138.

BABBITT, K. J. AND J. M. PACKARD. 1990. Parent–offspring conflict relative to phase of lactation. Anim. Behav. 40:765–773.

BABICKY, A., I. OSTADALOVA, J. PARIZEK, J. KOLAR, AND B. BIBR. 1970. Use of radioistope techniques for determining the weaning period in experimental animals. Physiol. Bohemoslov. 19:457–467.

BACON, J. AND B. MÖHL. 1983. The tritocerebral commissure giant (TCG) wind-sensitive interneurone in the locust. I. Its activity in straight flight. J. Comp. Physiol. 150:439–452.

BAERENDS, G. P. 1985. Do the dummy experiments with sticklebacks support the IRM concept? Behaviour 93:258–277.

BAGGERMAN, B. 1962. Some endocrine aspects of fish migration. Gen. and Comp. Endocrinol. Suppl. 1:188–205.

BAKER, M. C. 1974. Genetic structure of two populations of white-crowned sparrows with different song dialects. Condor 76:351–356.

BAKER, M. C. 1975. Song dialects and genetic differences in white-crowned sparrows (*Zonotrichia leucophrys*). Evolution 29:226–241.

BAKER, M. C. AND M. A. CUNNINGHAM. 1985. The biology of bird-song dialects. Behav. Brain Sci. 8:85–133.

BAKER, M. C., D. B. THOMPSON, AND C. L. SHERMAN. 1982. Allozyme frequencies in a linear series of song dialect populations. Evolution 36:1020–1029.

BAKER, M. C., K. J. SPITLER-NABORS, AND D. C. BRADLEY. 1982. Response of female mountain white-crowned sparrows to songs from their natal dialect and an alien dialect. Behav. Ecol. Sociobiol. 10:175–179.

BAKER, R. R. 1978. *The Evolutionary Ecology of Animal Migration*. Holmes & Meier Publishers. New York. 1012 pp.

BAKER, R. R. 1980. Goal orientation by blindfolded humans after long-distance displacement: Possible involvement of a magnetic sense. Science 210:555–557.

BAKER, R. R. 1981. *Human Navigation and the Sixth Sense*. Simon & Schuster. New York. 138 pp.

BAKER, R. R. 1985. Magnetoreception by man and other primates. In: J. L. Kirschvink, D. S. Jones, and B. J. MacFadden, eds., *Magnetite Biomineralization and Magnetoreception in Or-*

ganisms: A New Biomagnetism. Plenum Press. New York, pp 537–561.

BAKER, R. R. 1987. Human navigation and magnetoreception: The Manchester experiments do replicate. Anim. Behav. 35:691–704.

BAKER, R. R. AND M. A. BELLIS. 1988. 'Kamikaze' sperm in mammals? Anim. Behav. 36:936–939.

BAKER, R. R. AND M. A. BELLIS. 1989. Elaboration of the kamikaze sperm hypothesis: A reply to Harcourt. Anim. Behav. 37:865–867.

BALDA, R. P. 1980. Recovery of cached seeds by a captive *Nucifraga caryocatactes*. Z. Tierpsychol. 52:331–346.

BALDA, R. P. AND A. C. KAMIL. 1989. A comparative study of cache recovery by three corvid species. Anim. Behav. 38:486–495.

BALDACCINI, E. N., S. BENVENUTI, V. FIASCHI, AND F. PAPI. 1975. Pigeon navigation: Effects of wind deflection at the home cage on homing behavior. J. Comp. Physiol. 99:177–196.

BALTZER, F. 1914. Die Bestimmung und der Dimorphismus des Geschlechtes bei *Bonellia*. Sber. Phys. Med. Ges. Wurzb. 43:1–4.

BANDURA, A. 1962. Social learning through imitation. In: M. R. Jones, ed., *Nebraska Symposium on Motivation.* University of Nebraska Press, Lincoln.

BAPTISTA, L. F. AND J. R. KING. 1980. Geographical variation and song dialects of montane white-crowned sparrows (*Zonotrichia leucophrys oriantha*). Condor 82:267–284.

BAPTISTA, L. F. AND L. PETRINOVICH. 1984. Social interaction, sensitive phases and the song template hypothesis in the white-crowned sparrow. Anim. Behav. 32:172–181.

BAPTISTA, L. F. AND L. PETRINOVICH. 1986. Song development in the white-crowned sparrow: Social factors and sex differences. Anim. Behav. 34:1359–1371.

BARASH, D. P. 1977. *Sociobiology and Behavior.* Elsevier. Scientific Publishing Co. New York. 378 pp.

BARASH, D. P. 1982. *Sociobiology and Behavior,* 2nd ed. Elsevier, New York. 426 pp.

BARGIELLO, T. A., F. R. JACKSON, AND M. W. YOUNG. 1984. Restoration of circadian behavioural rhythms by gene transfer in *Drosophila*. Nature 312:752–754.

BARGIELLO, T. A., L. SAEZ, M. K. BAYLIES, G. GASIC M. W. YOUNG, AND D. C. SPRAY. 1987. The *Drosophila* clock gene *per* affects intercellular junctional communication. Nature (London) 328:686–691.

BARLOW, G. W. 1968. Ethological units of behavior. In: D. Ingle, ed., *The Central Nervous System and Fish Behavior.* The University of Chicago Press, Chicago, pp. 217–232.

BARLOW, G. W. 1989. Has sociobiology killed ethology or revitalized it? In: P. P. G. Bateson and P. H. Klopfer, eds., *Perspectives in Ethology,* vol. 8, Whither ethology? Plenum Press, New York, pp. 1–45.

BARLOW, G. W. 1991. Nature–nurture and the debates surrounding ethology and sociobiology. Am. Zool. 31:286–296.

BARNARD, C. J. AND C. A. J. BROWN. 1985. Risk-sensitive foraging in common shrews (*Sorex araneus* L.). Behav. Ecol. Sociobiol. 16:161–164.

BARNES, R. D. 1968. *Invertebrate Zoology.* Saunders, Philadelphia, PA.

BARTZ, S. H. 1979. Evolution of eusociality in termites. Proc. Nat. Acad. Sci. USA 76:5764–5768.

BASTIAN, J. 1981. Electrolocation. I. An analysis of the effects of moving objects and other electrical stimuli on the electroreceptor activity of *Apteronotus albifrons*. J. Comp. Physiol. A 144:465–479.

BATEMAN, A. J. 1948. Intra-sexual selection in *Drosophila*. Heredity 2:349–368.

BATES, D. L. AND M. BROCK FENTON. 1990. Aposematism or startle? Predators learn their responses to the defenses of prey. Can. J. Zool. 68:49–52.

BATESON, P. 1976. Specificity and the origins of behavior. In: J. Rosenblatt, R. A. Hinde, and C. Beer, eds., *Advances in the Study of Behavior*, vol. 6. Academic Press. New York, pp. 1–20.

BATESON, P. 1979. How do sensitive periods arise and what are they for? Anim. Behav. 27:470–486.

BATESON, P. 1982. Preferences for cousins in Japanese quail. Nature 295:236–237.

BATESON, P. 1983. Genes, environment and the development of behaviour. In: T. R. Halliday and P. J. B. Slater, eds., *Animal Behaviour*, vol. 3, Genes, development and learning. W. H. Freeman, New York, pp. 52–81.

BATESON, P. 1983. Optimal outbreeding. In: P. Bateson, ed., *Mate Choice*. Cambridge University Press. Cambridge, Massachusetts pp. 257–277.

BATESON, P. 1987. Imprinting as a process of competitive exclusion. In: J. P. Rauschecker, and P. Marler, eds., *Imprinting and Cortical Plasticity*. John Wiley & Sons. New York, pp. 151–168.

BATESON, P. 1990. Is imprinting such a special case? Phil. Trans. Roy. Soc. London B. 329:125–131.

BATESON, P. P. G. 1980. Optimal outbreeding and the development of sexual preferences in Japanese quail. Z. Tierpsychol. 53:231–244.

BATESON, P. P. G. AND P. H. KLOPFER, EDS. 1989. Preface to *Perspectives in Ethology*, vol. 8, Whither ethology? Plenum Press, New York. v–viii

BATESON, P. P. G. AND P. H. KLOPFER, EDS. 1991. *Perspectives in Ethology*, vol. 9. Human understanding and animal awareness. Plenum Press. New York. 314 pp.

BATTEAU, D. W. 1968. The world as a source; the world as a sink. In: S. J. Freedman, ed., *The Neuropsychology of Spatially Oriented Behavior*. Dorsey Press. Homewood, IL, pp. 197–203.

BEACH, F. A. 1950. The snark was a boojum. Am. Psychol. 5:115–124.

BEACH, F. A. 1976. Sexual attractivity, proceptivity, and receptivity in female mammals. Horm. Behav. 7:105–138.

BEACH, F. A. 1981. Historical origins of modern research on hormones and behavior. Horm. Behav. 15:325–376.

BEACH, F. A. AND B. J. LeBOEUF. 1967. Coital behavior in dogs. 1. Preferential mating in the bitch. Anim. Behav. 15:546–558.

BEASON, R. C. AND J. E. NICHOLS. 1984. Magnetic orientation and magnetically sensitive material in a transequatorial migratory bird. Nature 309:151–153.

BEASON, R. C. AND P. SEMM. 1987. Magnetic responses in the trigeminal nerve system of the bobolink (*Dolichonyx oryzivorus*). Neurosci. Lett. 80:229–234.

BEASON, R. C. AND P. SEMM. 1991. Neuroethological aspects of avian orientation. In: P. Berthold, ed., *Orientation in Birds*. Birkhäuser Verlag. Basel, Switzerland, pp. 106–127.

BECK, W. AND W. WILTSCHKO. 1982. The magnetic field as reference for the genetically encoded migratory direction in Pied Flycatchers (*Ficedula hypoleuca* PALLAS). Z. Tierpsychol. 60:40–46.

BEKOFF, M. AND J. A. BYERS. 1981. A critical reanalysis of the ontogeny and phylogeny of mammalian social and locomotor play: An ethological hornet's nest. In: K. Immelmann, G. W. Barlow, L. Petrinovich, and M. Main, eds., *Behavioral Development. The Bielefeld Interdisciplinary Project*. Cambridge University Press. Cambridge. pp. 296–337.

BEDNARZ, J. C. 1988. Cooperative hunting in Harris' hawks (*Parabuteo unicinctus*). Science 239:1525–1527.

BEECHER, M. D. 1990. The evolution of parent–offspring recognition in swallows. In: D. A. Dewsbury, ed., *Contemporary Issues in Com-*

parative Psychology. Sinauer Associates Inc., Sunderland, Massachusetts. pp. 360–380.

BEECHER, M. D., I. M. BEECHER, AND S. HAHN. 1981. Parent–offspring recognition in bank swallows (*Riparia riparia*): Development and acoustic basis. Anim. Behav. 29:95–101.

BEEHLER, M. N. AND M. S. FOSTER. 1988. Hotshots, hotspots, and female preference in the organization of lek mating systems. Am. Nat. 131:203–219.

BEKOFF, A. AND J. A. KAUER. 1984. Neural control of hatching: Fate of the pattern generator for the leg movements of hatching in post-hatching chicks. J. Neurosci. 4:2659–2666.

BELETSKY L. D. AND G. H. ORIANS. 1987. Territoriality among male red-winged blackbirds II. Removal experiments and site dominance. Behav. Ecol. Sociobiol. 20:339–349.

BELETSKY L. D. AND G. H. ORIANS. 1989. Territoriality among male red-winged blackbirds III. Testing hypotheses of territorial dominance. Behav. Ecol. Sociobiol. 24:333–339.

BELING, I. 1929. Über das Zeitgedächtnis der Bienen. Z. Vergl. Physiol. 9:259–338.

BELL, G. 1982. *The Masterpiece of Nature.* University of California Press. Berkeley. 635 pp.

BELL, G. 1988. Uniformity and diversity in the evolution of sex. In: R. E. Michod and B. R. Levin, eds., *The Evolution of Sex: An Examination of Current Ideas.* Sinauer Associates, Inc. Sunderland, MA, pp. 126–138.

BELOVSKY, G. E. 1978. Diet optimization in a generalist herbivore: The moose. Theoret. Pop. Biol. 14:105–134.

BELTZ, B. S. 1990. New dimensions in neuroanatomy: Visualizing the morphology, physiology and chemistry of neurons. Am. Zool. 30:513–529.

BENHAMOU, S. 1989. An olfactory orientation model for mammal's movements in their home ranges. J. Theor. Biol. 139:379–388.

BENHAMOU, S. AND P. BOVET. 1989. How animals use their environment: A new look at kinesis. Anim. Behav. 38:375–383.

BENHAMOU, S. AND P. BOVET. 1992. Distinguishing between elementary orientation mechanisms by means of path analysis. Anim. Behav. 43:371–377.

BENNETT, N. C. AND J. U. M. JARVIS. 1988. The social substructure and reproductive biology of colonies of the mole-rat, *Cryptomys damarensis* (Rodentia, Bathyergidae). J. Mamm. 69:293–302.

BENTLEY, D. R. AND R. R. HOY. 1972. Genetic control of the neuronal network generating cricket *Teleogryllus gryllus* song patterns. Anim. Behav. 20:478–492.

BENTLEY, P. J. 1982. *Comparative Vertebrate Endocrinology.* Cambridge University Press. Cambridge. 485 pp.

BENTON, T. G. 1991. Reproduction and parental care in the scorpion, *Euscorpius flavicaudis.* Behaviour 117:20–28.

BENVENUTTI, S., V. FIASCHI, L. FIORE, AND F. PAPI. 1973. Homing performances of inexperienced and directionally trained pigeons subjected to olfactory nerve section. J. Comp. Physiol. 83:81–92.

BENVENUTTI, S., A. J. BROWN, A. GAGLIARDO, AND M. NOZZOLINI. 1990a. Are American homing pigeons genetically different from Italian ones? J. Exp. Biol. 148:235–243.

BENVENUTTI, S., V. FIASCHI, A. GAGLIARDO, AND P. LUSCHI. 1990b. Pigeon homing: A comparison between groups raised under different conditions. Behav. Ecol. Sociobiol. 27:93–98.

BERGMANN, G. 1965. Der sexuelle Grössendimorphismus der Anatiden als Anpassung an das Höhlenbrüten. Commentat. Biol. 28:1–10.

BERMANT, G. 1976. Sexual behavior: hard times with the Coolidge Effect. In: M. H. Siegel and H. P. Zeigler, eds., *Psychological Research: The Inside Story,* Harper and Row. New York, pp. 76–103.

BERNSTEIN, H., F. A. HOPF, AND R. E. MICHOD. 1988. Is meiotic recombination an adaptation for repairing DNA, producing genetic variation, or both? In: R. E. Michod and B. R. Levin, eds., *The Evolution of Sex: An Examination of Current Ideas*. Sinauer Associates, Inc. Sunderland, MA, pp. 139–160.

BERNSTEIN, H., G. S. BYERS, AND R. E. MICHOD. 1981. Evolution of sexual reproduction: Importance of DNA repair, complementation, and variation. Am. Nat. 117:537–549.

BERTHOLD, P. 1991. Spatiotemporal programmes and genetics of orientation. In: P. Berthold, ed., *Orientation in Birds*. Birkhäuser Verlag, Basel. pp. 86–105.

BERTRAM, B. C. R. 1975. Social factors influencing reproduction in wild lions. J. Zool. London 177:463–482.

BERTRAM, B. C. R. 1976. Kin selection in lions and evolution. In: P. P. G. Bateson and R. A. Hinde, eds., *Growing Points in Ethology*. Cambridge University Press. New York. pp. 281–301.

BERVEN, K. A. 1981. Mate choice in the wood frog, *Rana sylvatica*. Evolution 35:707–722.

BINGMAN, V. P. 1981. Savannah sparrows have a magnetic compass. Anim. Behav. 29:962–963.

BINGMAN, V. P. 1984. Night-sky orientation of migratory Pied Flycatchers raised in different magnetic fields. Behav. Ecol. Sociobiol. 15:77–80.

BINGMAN, V. P. AND K. P. ABLE. 1979. The sun as a cue in the orientation of the white-throated sparrow, a nocturnal migrant bird. Anim. Behav. 27:621–622.

BINKLEY, S. 1988. *The Pineal: Endocrine and Nonendocrine Function*. Prentice Hall. Englewood Cliffs, NJ. 304 pp.

BINKLEY, S. 1990. *The Clockwork Sparrow*. Prentice Hall. Englewood Cliffs. N.J. 262 pp.

BITTERMAN, M. E. 1979. Historial introduction. In: M. E. Bitterman, V. M. Lolord, J. B. Overmier, and M. E. Rashotte, eds., *Animal Learning Survey and Analysis*. Plenum Press, New York.

BLACK, G. A. AND J. B. DEMPSEY. 1986. A test for the hypothesis of pheromone attraction in salmonid migration. Environ. Biol. Fishes 15:229–235.

BLAKEMORE, R. P. 1975. Magnetotactic bacteria. Science 190:377–379.

BLAKEMORE, R. P. AND R. B. FRANKEL. 1981. Magnetic navigation in bacteria. Sci. Am. 245 (Dec):58–65.

BLAUSTEIN, A. R. AND D. S. WALDMAN. 1992. Kin recognition in anuran amphibians. Anim. Behav. 44:207–221.

BLAUSTEIN, A. R. AND R. K. O'HARA. 1982. Kin recognition in *Rana cascadae* tadpoles: Maternal and paternal effects. Anim. Behav. 30:1151–1157.

BLAUSTEIN, A. R. AND R. K. O'HARA. 1986. Kin recognition in tadpoles. Sci. Am. 254 (Jan):108–116.

BLAUSTEIN, A. R., M. BEKOFF, AND T. J. DANIELS. 1987. Kin recognition in vertebrates (Excluding primates): Empirical evidence. In: D. J. C. Fletcher and C. D. Michener, eds., *Kin Recognition in Animals*. John Wiley & Sons. New York, pp. 287–357.

BLEST, A. D. 1957a. The evolution of protective displays in the Saturnioidea and Sphingidae (Lepidoptera). Behaviour 11:257–309.

BLEST, A. D. 1957b. The function of eyespot patterns in the Lepidoptera. Behaviour 11:209–257.

BLEST, A. D. 1961. The concept of ritualization. In: W. H. Thorpe and O. L. Zangwill, eds., *Current Problems in Animal Behavior*. Cambridge University Press. London, pp. 102–124.

BLOUGH, P. M. 1989. Attentional priming and visual search in pigeons. J. Exp. Psych. An. Behav. Proc. 15:358–365.

BLOUGH, P. M. 1991. Selective attention and search images in pigeons. J. Exp. Psych. An. Behav. Proc. 17:292–298.

BOCCIA, M. L. 1983. A functional analysis of social grooming patterns through direct comparison with self grooming in rhesus monkeys. Int. J. Primatol. 4:399–418.

BOCCIA, M. L. 1986. Grooming site preferences as a form of tactile communication and their role in the social relations of rhesus monkeys. In: D. M. Taub and F. A. King, eds., *Current Perspectives in Primate Soical Dynamics*. Van Nostrand Reinhold. New York, pp. 505–518.

BOCH, R., D. A. SHEARER, AND B. C. STONE. 1962. Identification of isoamyl acetate as an active component in the sting pheromone of the honey bee. Nature 195:1018–1020.

BOHNER, J. 1983. Song learning in the zebra finch (*Taeniopygia guttata*): Selectivity in the choice of a tutor and accuracy of song copies. Anim. Behav. 31:231–237.

BOHNER, J. 1990. Early acquisition of song in the zebra finch, *Taeniopygia guttata*. Anim. Behav. 39:369–374.

BOLLENBACHER, W. E., S. L. SMITH, W. GOODMAN, AND L. I. GILBERT. 1981. Ecdysteroid titer during larval-pupal-adult development of the tobacco hornworm, *Manduca sexta*. Gen. Comp. Endocrin. 44:302–306.

BOLLES, R. C. 1969. Avoidance and escape learning: Simultaneous acquisition of different responses. J. Comp. Physiol. Psychol. 68:355–358.

BOND, A. B. AND D. A. RILEY. 1991. Searching image in the pigeon: A test of three hypothetical mechanisms. Ethology 87:203–224.

BORGIA, G. 1986. Satin bowerbird parasites: A test of the bright male hypothesis. Behav. Ecol. Sociobiol. 19:355–358.

BORGIA, G., I. M. KAATZ, AND R. CONDIT. 1987. Flower choice and bower decoration in the satin bowerbird *Ptilonorhyncus violaceus*: A test of hypotheses for the evolution of male display. Anim. Behav. 35:1129–1139.

BOVET, D. 1977. Strain differences in learning in the mouse. In: A. Oliverio, ed., *Genetics, Environment and Intelligence*. North-Holland Publishing Co., Amsterdam, pp. 79–92.

BOYSE, E. A., G. K. BEAUCHAMP, K. YAMAKAZI, J. BARD, AND L. THOMAS. 1982. A new aspect of the major histocompatibility complex and other genes in the mouse. Oncodevelopmental Biol. Med. 4:101–116.

BRADBURY, J. W. 1977. Lek mating behavior in the hammer-headed bat. Z. Tierpsychol. 45:225–255.

BRADBURY, J. W. 1981. The evolution of leks. In: R. D. Alexander and D. W. Tinkle, eds., *Natural Selection and Social Behavior*. Chiron Press. New York, pp. 138–172.

BRADBURY, J. W. AND M. B. ANDERSSON, EDS. 1987. *Sexual Selection: Testing the Alternatives*. John Wiley & Sons. New York. 308 pp.

BRADBURY, J. W., R. M. GIBSON, AND I. M. TSAI. 1986. Hotspots and the evolution of leks. Anim. Behav. 34:1694–1709.

BRADDOCK, J. C. AND Z. I. BRADDOCK. 1959. Development of nesting behavior in the Siamese fighting fish (*Betta splendens*). Anim. Behav. 7:222–232.

BRANNON, E. L. 1982. Orientation mechanisms of homing salmonids In: E. L. Brannon and E. O. Salo, eds., *Salmon and Trout Migratory Behavior Symposium*. School of Fisheries. University of Washington, Seattle, pp. 219–227.

BRANNON, E. L. AND T. P. QUINN. 1990. A field test of the pheromone hypothesis for homing by Pacific salmon. J. Chem. Ecol. 16:603–609.

BRANNON, E. L., T. P. QUINN, G. L. LUCCHETTI, AND B. D. ROSS. 1981. Compass orientation of sockeye salmon fry from a complex river system. Can. J. Zool. 59:1548–1553.

BREED, M. D. AND G. J. GAMBOA. 1977. Behavioral control of workers by queens in primitively eusocial insects. Science 195:694–696.

BREEDLOVE, S. M. AND A. P. ARNOLD. 1983. Hormonal control of a developing neuromuscular system. II. Sensitive periods for the androgen-induced masculinization of the rat spinal nucleus of the bulbocavernosus. J. Neurosci. 3:424–432.

BRELAND, K. AND M. BRELAND. 1961. The misbehavior of organisms. Am. Psychol. 16:681–684.

BRELAND, K. AND M. BRELAND. 1966. *Animal Behavior*. Macmillan. New York.

BRETT, R. A. 1991. The population structure of naked mole-rat colonies. In: P. W. Sherman, J. U. M. Jarvis, and R. D. Alexander, eds., *The Biology of the Naked Mole-Rat*. Princeton University Press. Princeton, NJ, pp. 97–136.

BRINES, M. L. 1980. Dynamic patterns of skylight polarization as clock and compass. J. Theor. Biol. 86:507–512.

BRISTOWE, W. S. 1958. *The World of Spiders*. Collins, London.

BROCK, M. A. 1974. Circannual rhythms. I. Free-running rhythms in growth and development of the marine cnidarian *Campanularia flexosa*. Comp. Biochem. Physiol. A 51:377–383.

BROCKMANN, J., A. GRAFEN, AND R. DAWKINS. 1979. Evolutionarily stable nesting strategy in a digger wasp. J. Theor. Biol. 77:473–496.

BRODY, S. AND S. A. MARTINS. 1979. Circadian rhythms in *Neurospora crassa:* Effects of unsaturated fatty acids. J. Bacteriol. 137:912–915.

BROIDA, J. AND B. SVARE. 1982. Strain-typical patterns of pregnancy-induced nestbuilding in mice: Maternal and experiential influences. Physiol. Behav. 25:153–157.

BROIDA, J. AND B. SVARE. 1983. Mice: Progesterone and the regulation of strain differences in pregnancy-induced nestbuilding. J. Behav. Neurosci. 97:994–1004.

BRONSON, F. H. 1971. Rodent pheromones. Biol. Reprod. 4:344–357.

BRONSON, F. H. 1974. Pheromonal influences on reproductive activities in rodents. In: M. C. Birch, ed., *Pheromones*. North Holland Publishing Co. Amsterdam, pp. 344–365.

BROWER, L. P. AND W. H. CALVERT. 1985. Foraging dynamics of bird predators on overwintering monarch butterflies in Mexico. Evolution 39:852–868.

BROWER, L. P., W. N. RYERSON, L. L. COPPINGER, AND S. C. GLAZIER. 1968. Ecological chemistry and the palatability spectrum. Science 161:1349–1351.

BROWN, F. A., JR. 1971. Some orientational influences of nonvisual terrestrial electromagnetic fields. Ann. N.Y. Acad. Sci. 188:224–241.

BROWN, F. A., JR. AND Y. H. PARK. 1967. Association between photic and subtle geophysical stimulus patterns—a new biological concept. Biol. Bull. 132:311–319.

BROWN, F. A., JR. AND Y. H. PARK. 1975. A persistent monthly variation in the responses of planarians to light and its annual modulation. Int. J. Chronobiol. 3:57–62.

BROWN, F. A., JR., J. W. HASTINGS AND J. D. PALMER. 1970. *The Biological Clock: Two Views*. Academic Press, New York.

BROWN, J. L. 1964. The evolution of diversity in avian territorial systems. Wilson Bull. 76:160–169.

BROWN, J. L. 1975. Helpers among Arabian babblers *Turdoides squamiceps*. Ibis 117:243–244.

BROWN, J. L. 1982. The adaptationist program. Science 217:884–886.

BROWN, J. L. 1987. *Helping and Communal Breeding in Birds*. Princeton University Press. Princeton, NJ. 354 pp.

BROWN, J. L., E. R. BROWN, S. D. BROWN, AND D. D. DOW. 1982. Helpers—Effects of experimental removal on reproductive success. Science 215:421–422.

BROWNELL, P. H. 1984. Prey detection by the scorpion. Sci. Am. 251 (Dec):86–98.

BRUCE, H. M. 1959. An exteroceptive block to pregnancy in the mouse. Nature (London) 184:105.

BRUCE, H. M. 1961. Time relations in the pregnancy block induced in mice by strange males. J. Reprod. Fert. 2:138–142.

BRUCE, H. M. 1963. Olfactory block to pregnancy among grouped mice. J. Reprod. Fert. 6:451–460.

BRUCE, V. G. 1972. Mutants of the biological clock in *Chlamydomonas reinhardi*. Genetics 70:537–548.

BRUCE, V. G. 1974. Recombinants between clock mutants of *Chlamydomonas reinhardi*. Genetics 77:221–230.

BRUNTON, D. H. 1990. The effects of nesting stage, sex, and type of predator on parental defense by killdeer (*Charadrius vociferus*): Testing models of parental defense. Behav. Ecol. Sociobiol. 26:181–190.

BUCKLE, G. R. AND L. GREENBERG. 1981. Nestmate recognition in sweat bees (*Lasioglossum zephyrum*): Does an individual recognize its own odor or only odours of its nestmates? Anim. Behav. 29:802–809.

BULL, J. J. 1983. Evolution of sex determining mechanisms. Benjamin Cummings Publishing Co. Menlo Park, CA.

BULL, J. J. 1987. Temperature-sensitive periods of sex determination in a lizard: Similarities with turtles and crocodilians. J. Exp. Zool. 241:143–148.

BULLOCK, T. H. 1973. Seeing the world through a new sense: Electroreception in fish. Am. Sci. 61:316–325.

BULLOCK, T. H. AND T. SZABO. 1986. Introduction. In: T. H. Bullock and W. Heiligenberg, eds., *Electroreception*. John Wiley and Sons. New York, pp. 1–12.

BURGHARDT, G. M. 1970. Defining "communication." In: J. W. Johnston, D. G. Moulton, and A. Turk, eds., *Communication by Chemical Signals*. Appleton-Century-Crofts. New York, pp. 5–18.

BURGHARDT, G. M. AND H. W. GREENE. 1988. Predator simulation and duration of death feigning in neonate hognose snakes. Anim. Behav. 36:1842–1844.

BURKHARDT, D. AND I. DE LA MOTTE. 1983. How stalk-eyed flies view stalk-eyed flies: Observations and measurements of the eyes of *Cyrtodiopsis whitei* (Diopsidae, Diptera). J. Comp. Physiol. A 151:407–421.

BURROWS, M. T. AND R. N. HUGHES. 1990. Variation in growth and consumption among individuals and populations of dogwhelks, *Nucella lapillus*: A link between foraging behavior and fitness. J. Anim. Ecol. 59:723–742.

BURROWS, W. 1945. Periodic spawning of the Palolo worms in Pacific waters. Nature 155:47–48.

BYERS, D. R., L. DAVIS, AND J. A. KIGER. 1981. Defect in cyclic AMP phosphodiesterase due to the *dunce* mutation of learning in *Drosophila melanogaster*. Nature 289:79–81.

CAHALANE, V. H. 1961. *Mammals of North America*. Macmillan. New York, 682 pp.

CALDWELL, R. L. AND H. DINGLE. 1976. Stomatopods. Sci. Am. 234 (Jan):80–89.

CALVERT, W. H. AND L. P. BROWER. 1986. The location of the monarch butterfly (*Danaus plexippus* L.) overwintering colonies in Mexico in relation to topography and climate. J. Lepidopterists' Soc. 40:164–187.

CALVERT, W. H., L. E. HEDRICK, AND L. P. BROWER. 1979. Mortality of the monarch butterfly (*Danaus plexippus* L.): Avian predation at five overwintering sites in Mexico. Science 204:847–851.

CAMHI, J. M. 1980. The escape system of the cockroach. Sci. Am. 248 (Dec):158–172.

CAMHI, J. M. 1984. *Neuroethology*. Sinauer Associates Inc. Sunderland, MA. 416 pp.

CAMHI, J. M. 1988. Escape behavior in the cockroach: Distributed neural processing. Experientia 44:401–408.

CAMHI, J. M., W. TOM, AND S. VOLMAN. 1978. The escape behavior of the cockroach *Periplaneta americana*. II. Detection of natural predators by air displacement. J. Comp. Physiol. 128:203–212.

CAPLAN, A. L., ED. 1978. *The Sociobiology Debate*. Harper & Row Publishers, New York. 514 pp.

CAPRANICA R. R., L. S. FRISHKOPF, AND E. NEVO. 1973. Encoding of geographic dialects in the auditory system of the cricket frog. Science 182:1272–1275.

CARACO, T. 1981. Energy budgets, risk and foraging preferences in dark-eyed juncos (*Junco hyemalis*). Behav. Ecol. Sociobiol. 8:213–217.

CARACO, T. 1983. White-crowed sparrows (*Zonotrichia leucophrys*): Foraging preferences in a risky environment. Behav. Ecol. Sociobiol. 12:63–69.

CARO, T. M. 1986a. The functions of stotting: A review of the hypotheses. Anim. Behav. 34:649–662.

CARO, T. M. 1986b. The functions of stotting in Thomson's gazelles: Some tests of the predictions. Anim. Behav. 34:663–684.

CARO, T. M. 1988. Adaptive significance of play: Are we getting closer? Trends Ecol. Evol. 3:50–54.

CARPENTER, C. R. 1942. Sexual behavior of free-ranging rhesus monkeys (*Macaca mulatta*). I. Specimens, procedures and behavioral characteristics of estrus. J. Comp. Phys. Psychol. 33:113–142.

CARPENTER, F. L., D. C. PATON AND M. A. HIXON. 1983. Weight gain and adjustment of feeding territory size in migrant hummingbirds. Proc. Nat. Acad. Sci. USA 80:7259–7263.

CARTER, R. V. AND L. M. DILL. 1990. Why are bumble bees risk-sensitive foragers? Behav. Ecol. Sociobiol. 26:121–127.

CARTWRIGHT, B. A. AND T. A. COLLETT. 1983. Landmark learning in bees. J. Comp. Physiol. 151:521–543.

CARYL, P. G. 1979. Communication by agonistic displays: What can games theory contribute to ethology? Behaviour 68:136–169.

CAVINESS, V. S., JR. AND P. RAKIC. 1978. Mechanisms of cortical development: A view from mutations in mice. Ann. Rev. Neurosci. 1:297–326.

CAVINESS, V. S., JR. AND R. L. SIDMAN. 1972. Olfactory studies of the forebrain in the reeler mutant mouse. J. Comp. Neurol. 145:85–104.

CHAMPALBERT, A. AND J.-P. LACHAUD. 1990. Existence of a sensitive period during the ontogenesis of social behaviour in a primitive ant. Anim. Behav. 39:850–859.

CHANCE, M. R. A. 1962. Social behavior and primate evolution. In: M. F. Ashley Montague, ed., *Culture and the Evolution of Man*. Oxford University Press. New York. 376 pp.

CHARLESWORTH, B. 1989. The evolution of sex and recombination. Trends Ecol. Evol. 4:264–267.

CHARNOV, E. L. 1976. Optimal foraging, the marginal value theorem. Theoret. Pop. Biol. 9:129–136.

CHARNOV, E. L. 1978. Evolution of eusocial behavior: Offspring choice or parental parasitism. J. Theoret. Biol. 75:451–466.

CHARNOV, E. L. 1981. Kin selection and helpers at the nest: effects of paternity and biparental care. Anim. Behav. 29:631–632.

CHARNOV, E. L. 1982. *The Theory of Sex Allocation*. Princeton University Press. Princeton, NJ.

CHARNOV, E. L. AND J. R. KREBS. 1975. The evolution of alarm calls. Am. Nat. 109:107–112.

CHENEY, D. L. AND R. M. SEYFARTH. 1980. Vocal recognition in free-ranging vervet monkeys. Anim. Behav. 28:362–367.

CHENEY, D. L. AND R. M. SEYFARTH. 1988. Assessment of meaning and the detection of unreliable signals by vervet monkeys. Anim. Behav. 36:477–486.

CHENEY, D. L. AND R. M. SEYFARTH. 1991. Truth and deception in animal communication. In:

C. A. Ristau, ed., *Cognitive Ethology: The minds of other animals*. Lawrence Erlbaum, Assoc. Hillsdale. N. J. pp. 127–151.

CHENG, T. C. 1973. *General parasitology*. Academic Press. New York. 965 pp.

CHERRY, J. D. 1984. Migratory orientation of tree sparrows. MS thesis, State University of New York at Albany.

CHICHIBU, S. 1981. EOD time series analysis of *Gnathonemus petersii*. Adv. Physiol. Sci. 30:165–178.

CHILTON, G., M. R. LEIN, AND L. F. BAPTISTA. 1990. Mate choice by female white-crowned sparrows in a mixed-dialect population. Behav. Ecol. Sociobiol. 27:223–227.

CHIPMAN, R. K. AND K. A. FOX. 1966. Factors in pregnancy blocking: Age and reproductive background of females: Numbers of strange males. J. Reprod. Fert. 12:399–403.

CHRISTY, J. H. AND M. SALMON. 1991. Comparative studies of reproductive behavior in mantis shrimps and fiddler crabs. Am. Zool. 31:329–337.

CLARK, D. R. 1971. The strategy of tail-autotomy in the ground skink, *Lygosoma laterale*. J. Exp. Zool. 176:295–302.

CLARK, R. B. 1960. Habituation of the polychaete *Nereis* to sudden stimuli. 1. General properties of the habituation process. Anim. Behav. 8:82–91.

CLARKE, A. M. AND A. D. B. CLARKE. 1976. Early experience: Myth and evidence. Open Books. London. 314 pp.

CLARKE, B. 1969. The evidence for apostatic selection. Heredity 24:347–352.

CLAYTON, F. L. AND R. A. HINDE. 1968. Habituation and recovery of aggressive display in *Betta splendens*. Behaviour 30:96–106.

CLAYTON, N. S. 1987. Song tutor choice in zebra finches. Anim. Behav. 35:714–721.

CLUTTON-BROCK, T. H. 1991. *The Evolution of Parental Care*. Princeton University Press. Princeton, NJ.

CLUTTON-BROCK, T. H. AND G. R. IASON. 1986. Sex ratio variation in mammals. Quart. Rev. Biol. 61:339–374.

CLUTTON-BROCK, T. H. AND P. H. HARVEY. 1979. Comparison and adaptation. Proc. Roy. Soc. London B. 205:547–565.

CLUTTON-BROCK, T. H. AND P. H. HARVEY. 1984. Comparative approaches to studying adaptation. In: J. R. Krebs and N. B. Davies, eds., *Behavioural Ecology: An Evolutionary Approach*. Sinauer Assoc., Sunderland, MA, pp. 7–29.

CLUTTON-BROCK, T. H. AND S. D. ALBON. 1979. The roaring of red deer and the evolution of honest advertisement. Behaviour 69:145–170.

CLUTTON-BROCK, T. H., F. E. GUINESS, AND S. D. ALBON. 1982. *Red Deer: The Behavior and Ecology of Two Sexes*. University of Chicago Press, Chicago. 378 pp.

CLUTTON-BROCK, T. H., S. D. ALBON, AND F. E. GUINESS. 1986. Great expectations: Dominance, breeding success, and offspring sex ratios in red deer. Anim. Behav. 34:460–471.

CLUTTON-BROCK, T. H., S. D. ALBON, R. M. GIBSON, AND F. E. GUINESS. 1979. The logical stag: Adaptive aspects of fighting in red deer (*Cervus elaphus* L.). Anim. Behav. 27:211–225.

COCKBURN, A. 1988. *Social Behaviour in Fluctuating Populations*. Croom Helm. New York.

COLE, S., F. R. HAINSWORTH, A. C. KAMIL, T. MERCIER, AND L. L. WOLF. 1982. Spatial learning as an adaptation in hummingbirds. Science 217:655–657.

COLLETT, T. S. AND M. F. LAND. 1975. Visual spatial memory in a hoverfly. J. Comp. Physiol. 100:59–84.

CONANT, R. 1975. *A Field Guide to Reptiles and Amphibians of Eastern and Central North America*. Houghton Mifflin Co. Boston.

COOPER, J. B. 1985. Comparative psychology and

ethology. In: G. A. Kimble, and K. Schlesinger, eds., *Topics in the History of Psychology.* Lawrence Erlbaum Associates Inc., Hillsdale, New Jersey, pp. 135–164.

COOPER, J. C., A. T. SCHOLZ, R. M. HORRAL, A. D. HASLER, AND D. M. MADISON. 1976. Experimental confirmation of the olfactory hypothesis with homing, artificially imprinted coho salmon (*Oncorhynchus kisutch*). J. Fish. Res. Bd. 28:703–710.

COOPER, J. R., F. E. BLOOM, AND R. H. ROTH. 1978. *The Biochemical Basis of Neuropharmacology*, 3rd ed. Oxford University Press. New York.

COOPER, R. M. AND J. P. ZUBEK. 1958. Effects of enriched and restricted early environments on the learning ability of bright and dull rats. Can. J. Psychol. 12:159–164.

COTT, H. B. 1940. *Adaptive Colouration in Animals.* Methuen, London. 508 pp.

COULSON, J. C. 1966. The influence of the pair bond and age on the breeding biology of the kittiwake gull (*Rissa tridactyla*). J. Anim. Ecol. 35:269–279.

COX, C. R. AND B. J. LEBOEUF. 1977. Female incitation of male competition: A mechanism of mate selection. Am. Nat. 111:317–335.

COX, G. W. 1985. The evolution of avian migration systems between temperate and tropical regions of the world. Am. Nat. 126:451–474.

CRAIG, C. L. AND G. D. BERNARD. 1990. Insect attraction to ultra-violet reflecting spider webs and web decorations. Ecology 71:616–623.

CRAIG, W. 1918. Appetites and aversions as constituents of instincts. Biol. Bull. 34:91–107.

CRANE, J. 1941. Eastern Pacific Expeditions of the New York Zoological Soc XXVI: Crabs of the genus *Uca* from the west coast of Central Am. Zoologica 26:145–203.

CREMER–BARTELS, G. K. KRAUSE, G. MITOSKAS, AND D. BRODERSEN. 1984. Magnetic field strength as an additional zeitgeiber for endogenous rhythms? Naturwiss. 71:567–574.

CREWS, D. 1974. Effects of castration and subsequent androgen replacement therapy on male courtship behavior and environmentally induced ovarian recrudescence in the lizard, *Anolis carolinensis.* J. Comp. Physiol. Psychol. 87:963–969.

CREWS, D. 1979. Neuroendocrinology of lizard reproduction. Biol. Reprod. 20:51–73.

CREWS, D. 1983. Control of male sexual behaviour in the Canadian red-sided garter snake. In: J. Balthazart, E. Prove, and R. Gilles, eds., *Hormones and Behaviour in Higher Vertebrates.* Springer-Verlag. New York, pp. 398–406.

CREWS, D. 1984. Gamete production, sex hormone secretion, and mating behavior uncoupled. Horm. Behav. 18:22–28.

CREWS, D. 1987. Diversity and evolution of behavioral controlling mechanisms. In: D. Crews, ed., *Psychobiology of Reproductive Behavior.* Prentice Hall. Englewood Cliffs, NJ, pp. 88–119.

CREWS, D., V. TRAINA, F. T. WETZEL, AND C. MULLER. 1978. Hormone control of male reproductive behavior in the lizard, *Anolis carolinensis*: Role of testosterone, dihydrotestosterone, and estradiol. Endocrinol. 103:1814–1821

CREWS, S. T., J. B. THOMAS, AND C. S. GOODMAN. 1988. The *Drosophila* single-minded gene encodes a nuclear protein with sequence similarity to the *per* gene product. Cell 52:143–152.

CROCKER, G. AND T. DAY. 1987. An advantage to mate choice in the seaweed fly, *Coelopa frigida.* Behav. Ecol. Sociobiol. 20:295–301.

CROOK, J. H. 1964. The evolution of social organisation and visual communication in the weaver birds (Ploceinae). Behaviour Suppl. 10:1–178.

CROOK, J. H. 1970. The socio-ecology of primates. In: J. H. Crook, ed., *Social Behavior in Birds and Mammals*. Academic Press, London. pp. 103–166.

CROWLEY, P. H., D. R. DEVRIES, AND A. SIH. 1990. Inadvertent errors and error-constrained optimization: Fallible foraging by blue-gill sunfish. Behav. Ecol. Sociobiol. 27:135–144.

CROZE, H. 1970. Searching image in carrion crows. Z. Tierpsychol. Suppl. 5:1–86.

CULLEN, E. 1957a. Adaptations in the kittiwake to cliff nesting. Ibis 99:275–302.

CULLEN, E. 1957b. Adaptations in the kittiwake in the British Isles. Bird Study 10:147–179.

CULLEN, J. M. 1966. Ritualization of animal activities in relation to phylogeny, speciation and ecology. Reduction of ambiguity through ritualization. Phil. Trans. Roy. Soc. Ser. B. 241:363–374.

CULLEN, J. M. 1972. Some principles of animal communication. In: R. A. Hinde, ed., *Nonverbal Communication*. Cambridge University Press, Cambridge, England pp. 101–122.

CURIO, E. 1976. *The Ethology or Predation*. Springer-Verlag, Berlin. 250 pp.

CURIO, E. 1978. The adaptive significance of avian mobbing. I. Teleonomic hypotheses and predictions. Z. Tierpsychol. 48:175–183.

CURIO, E. AND K. REGELMANN. 1986. Predator harassment implies a deadly risk: A reply to Hennessy. Ethology 72:75–78.

DAAN, S. AND J. ASCHOFF. 1982. Circadian contributions to survival. In: J. Aschoff, S. Daan, and G. A. Groos, eds., *Vertebrate Circadian Systems, Structure and Function*. Springer-Verlag. Berlin, pp. 305–321.

DAANJE, A. 1950. On locomotory movements in birds and the intention movement derived from them. Behaviour 3:48–98.

DALY, M. 1978. The cost of mating. Am. Nat. 112:771–774.

DALY, M. 1979. Why don't male mammals lactate? J. Theor. Biol. 78:325–345.

DALY, M. AND M. WILSON. 1978. *Sex, Evolution and Behavior*. Duxbury Press. North Scituate, MA. 387 pp.

DALY, M. AND M. WILSON. 1983. *Sex, Evolution and Behavior*. 2nd ed. PWS-Kent Publishers, Boston MA. 402 pp.

DARLING, F. F. 1937. *A Herd of Red Deer*. Oxford University Press. New York.

DARWIN, C. 1859. *On the Origin of Species*. Murray, London.

DARWIN, C. 1871. *The Descent of Man and Selection in Relation to Sex*. Murray, London.

DARWIN, C. 1873. *Expression of the Emotions in Man and Animals*. Murray, London.

DAVIDSON, J. M., C. A. CAMARGO, AND E. R. SMITH. 1979. Effects of androgen on sexual behavior in hypogonadal men. J. Clin. Endocrinol. Metab. 48:955–958.

DAVIES, N. B. 1978. Territorial defense in the speckled wood butterfly (*Pararge aegeria*): The resident always wins. Anim. Behav. 26:138–147.

DAVIES, N. B. AND A. I. HOUSTON. 1984. Territory economics. In: J. R. Krebs and N. B. Davies, eds., *Behavioural Ecology: An Evolutionary Approach*. Sinauer Associates, Inc. Sunderland, MA, pp. 148–169.

DAVIES, N. B. AND M. DE L. BROOKE. 1988. Cuckoo versus reed warblers: Adaptations and counteradaptations. Anim. Behav. 36:262–284.

DAVIES, N. B. AND T. R. HALLIDAY. 1978. Deep croaks and fighting assessment in toads (*Bufo bufo*). Nature (London) 274:683–685.

DAVIS, R. L. AND J. A. KIGER, JR. 1981. Dunce mutants of *Drosophila melanogaster*: Mutants defective in the cyclic AMP phosphodiesterase enzyme system. J. Cell Bio. 90:101–107.

DAWKINS, M. 1971. Perceptual changes in chicks: Another look at the "search image" concept. Anim. Behav. 19:556–574.

DAWKINS, M. S. 1986. *Unravelling Animal Behavior*. Longman Group Ltd. Essex, England. 159 pp.

DAWKINS, M. S. 1989. The future of ethology: How many legs are we standing on? In: P. P. G. Bateson and P. H. Klopfer, eds., *Perspectives in Ethology*, vol. 8, Whither ethology? Plenum Press, New York, pp. 47–54.

DAWKINS, M. S. AND T. GUILFORD. 1991. The corruption of honest signaling. Anim. Behav. 41:865–873.

DAWKINS, R. 1976. *The Selfish Gene*. Oxford University Press. New York. 224 pp.

DAWKINS, R. 1980. Good strategy or evolutionarily stable strategy? In: G. Barlow and J. Silverberg, eds., *Sociobiology: Beyond Nature/Nurture*. Westview, Boulder, CO.

DAWKINS, R. 1982. *The Extended Phenotype*. Freeman, Oxford. 307 pp.

DAWKINS, R. 1989. *The Selfish Gene*, 2nd ed. Oxford University Press. New York. 352 pp.

DAWKINS, R. AND J. R. KREBS. 1978. Animal signals: Information or manipulation? In: J. R. Krebs and N. B. Davies, eds., *Behavioural Ecology—an Evolutionary Approach*. Sinauer Associates, Inc. Sunderland, MA, pp. 282–309.

DAWKINS, R. AND J. R. KREBS. 1984. Animal signals: Mind-reading and manipulation? In: J. R. Krebs and N. B. Davies, eds., *Behavioural Ecology—an Evolutionary Approach*. Second ed. Sinauer Associates, Inc. Sunderland, MA, pp. 380–402.

DAWKINS, R. AND T. R. CARLISLE. 1976. Parental investment, mate desertion and a fallacy. Nature 262:131–133.

DAWSON, B. V. AND B. M. FOSS. 1965. Observational learning in budgerigars. Anim. Behav. 13:470–474.

DAYTON, T. 1985. Statistical and methodological critique of Baker's chapter. In: J. L. Kirschvink, D. S. Jones, and B. J. MacFadden, eds., *Magnetite Biomineralization and Magnetorecep-tion in Organisms: A New Biomagnetism*. Plenum Press. New York, pp. 563–572.

DE LA MOTTE, I. AND D. BURKHARDT. 1983. Portrait of an Asian stalk-eyed fly. Naturwiss. 70:451–461.

DEAN, J., D. J. ANESHANSLEY, H. E. EDGERTON, AND T. EISNER. 1990. Defensive spray of the bombardier beetle: A biological pulse jet. Science 248:1219–1221.

DECKARD, B. S., B. LIEFF, K. SCHLESINGER, AND J. C. DeFRIES. 1976. Developmental patterns of seizure susceptibility in inbred strains of mice. Developmental Psychobiol. 9:17–24.

DeFRIES, J. C., E. A. THOMAS, J. P. HEGMANN, AND M. W. WEIR. 1967. Open-field behavior in mice. Analysis of maternal effects by means of ovarian transplantation. Psychonomic Science 8:207–208.

DEGUCHI, T. 1979. Circadian rhythm of serotonin N-acetyltransferase activity in organ culture chicken pineal gland. Science 203:1245–1247.

DEHN, M. M. 1990. Vigilance for predators: Detection and dilution effects. Behav. Ecol. Sociobiol. 26:337–342.

DELCOMYN, F. 1980. Neural basis of rhythmic behavior in animals. Science 210:492–498.

DEMAINE, C. AND P. SEMM. 1985. The avian pineal gland as an independent magnetic sensor. Neurosci. Lett. 62:119–122.

DeVALOIS, R. L. AND G. H. JACOBS. 1968. Primate color vision. Science 162:533–540.

DEVINE, M. C. 1977. Copulatory plugs, restricted mating opportunities and reproductive competition among male garter snakes. Nature 267:345–346.

DeVOOGD, T. J. 1990. Recent findings on the development of dimorphic anatomy in the avian song system. J. Exp. Zool. Suppl. 4:183–186.

DeVRIES, D. R., R. A. STEIN, AND P. L. CHESSON. 1989. Sunfish foraging among patches: The patch departure decision. Anim. Behav. 37:455–467.

DEWAAL, F. B. M. 1978. Exploitative and familiarity-dependent support strategies in a colony of semi-free-living chimpanzees. Behaviour 66:268–312.

DEWSBURY, D. A. 1978. What is (was?) the 'Fixed Action Pattern'? Anim. Behav. 26:310–311.

DEWSBURY, D. A. 1982a. Dominance rank, copulatory behavior, and differential reproduction. Quart. Rev. Biol. 57:135–159.

DEWSBURY, D. A. 1982b. Ejaculate cost and male choice. Am. Nat. 119:601–610.

DEWSBURY, D. A. 1984. Comparative Psychology in the Twentieth Century. Hutchison Ross, Stroudsburg, Pennsylvania.

DEWSBURY, D. A. 1984. Sperm competition in muroid rodents. In: R. L. Smith, ed., *Sperm Competition and the Evolution of Animal Mating Systems.* Academic Press. Orlando, FL, pp. 547–571.

DEWSBURY, D. A. 1988. A test of the role of copulatory plugs in sperm competition in deer mice. (*Peromyscus maniculatus*) J. Mamm. 69:854:857.

DEWSBURY, D. A. 1989. A brief history of the study of animal behavior in North America. In P. P. G. Bateson and P. H. Klopfer eds., *Perspectives in Ethology,* vol. 8, Whither ethology? Plenum Press, New York, pp. 85–122.

DIAL, B. E. AND L. C. FITZPATRICK. 1983. Lizard tail autotomy: Function and energetics of postautotomy tail movement in *Scinella lateralis.* Science 219:391–393.

DIAMOND, M., L. DIAMOND, AND M. MAST. 1972. Visual sensitivity and sexual arousal levels during the menstrual cycle. J. Nervous and Mental Disease 155:170–176.

DIEBOLD, A. R. 1968. Anthropological perspectives. In: T. A. Sebeok, ed., *Animal Communication.* Indiana University Press. Bloomington, pp. 525–571.

DIECKMANN, C. AND S. BRODY. 1980. Circadian rhythms in *Neurospora crassa*: oligomycin resistant mutations affect periodicity. Science 207:896–898.

DIESEL, R. 1990. Sperm competition and reproductive success in the decapod *Inachus phalangium* (Majiidae): A male ghost spider crab that seals off rivals' sperm. J. Zool. London 220:213–223.

DIJKGRAAF, S. AND A. J. KALMIJN. 1962. Verhalt ungeversuche zur Funktion der Lorenzinischen Ampullem. Naturiwiss. 49:400.

DILL, L. M. AND A. H. G. FRASER. 1984. Risk of predation and the feeding behavior of juvenile coho salmon (*Oncorhynchus kisutch*). Behav. Ecol. Sociobiol. 16:65–71.

DILL, P. A. 1971. Perception of polarized light by yearling sockeye salmon (*Oncorhynchus nerka*). J. Fish. Res. Bd. Canada 28:1319–1322.

DILL, P. A. 1977. Development of behaviour in alevins of Atlantic salmon, *Salmo salar* and rainbow trout, *S. gairdneri.* Anim. Behav. 25:116–121.

DINGLE, H. 1980. Ecology and evolution of migration. In: S. A. Gauthreaux, Jr., ed., *Animal Migration, Orientation, and Navigation.* Academic Press. New York, pp. 1–101.

DINGLE, H. AND R. CALDWELL. 1969. The aggressive and territorial behavior of the mantis shrimp *Gonodactylus bredini* Manning (Crustacea: Stomatopoda). Behaviour 33:115–136.

DIXON, K. L. 1969. Patterns of singing in a population of the Plain titmouse. Condor 71:94–101.

DOBSON, F. S. 1982. Competition for mates and predominant juvenile male dispersal in mammals. Anim. Behav. 30:1183–1192.

DOBSON, F. S. AND W. T. JONES. 1985. Multiple causes of dispersal. Am. Nat. 126:855–858.

DOLE, J. W. AND P. DURANT. 1974. Movements and seasonal activity of *Atelopus oxyrhynchus* (Anura: Atelopodidae) in a Venezuelan cloud forest. Copeia 1974:230–235.

DORST, J. 1962. *The Migration of Birds*. Houghton Mifflin. Boston.

DOWNHOWER, J. F. AND K. B. ARMITAGE. 1971. The yellow-bellied marmot and the evolution of polygyny. Am. Nat. 105:355–370.

DOWNHOWER, J. F. AND L. BROWN. 1980. Mate preference of female mottled sculpins, *Cottus bairdi*. Anim. Behav. 28:728–734.

DOWSE, H. B., J. C. HALL, AND J. M. RINGO. 1987. Circadian and ultradian rhythms in *period* mutants of *Drosophila melangaster*. Behav. Gen. 17:19–35.

DRESLER, A. 1940. Über eine jahreszeithliche Schwankung der spektralen Hellempfindlichkeit. Licht. 10:79–82.

DRYDEN, G. L. AND J. N. ANDERSON. 1977. Ovarian hormones: Lack of effect on reproductive structures of female Asian musk shrews. Science 197:782–784.

DUDAI, Y. 1977. Properties of learning and memory in *Drosophila melanogaster*. J. Comp. Physiol. 114:69–89.

DUDAI, Y. 1979. Behavioral plasticity in a *Drosophila* mutant, dunce DB276. J. Comp. Physiol. 130:271–275.

DUDAI, Y., Y.-N. JAN, D. BYERS, W. G. QUINN, AND S. BENZER. 1976. Dunce, a mutant of *Drosophila* deficient in learning. Proc. Nat. Acad. Sci. USA 73:1684–1688.

DUFTY, A. M. 1989. Testosterone and survival: A cost of aggressiveness? Horm. Behav. 23:185–193.

DUNBRACK, R. I. AND L. M. DILL. 1983. A model of size dependent surface feeding in a stream dwelling salmonid. Environ. Biol. Fish 8:203–216.

DUNFORD, C. 1977. Kin selection for ground squirrel alarm calls. Am. Nat. 111:782–785.

DUNHAM, P. J. 1986. Mate guarding in amphipods: A role for brood pouch stimuli. Biol. Bull. 170:526–531.

DUNHAM, P. J. AND A. HURSHMAN. 1990. Precopulatory mate guarding in the amphipod, *Gammarus lawrencianus:* Effects of social stimulation during the post-copulation interval. Anim. Behav. 39:976–979.

DUNHAM, P. J., T. ALEXANDER, AND A. HURSHMAN. 1986. Precopulatory mate guarding in an amphipod, *Gammarus lawrencianus* Bousfield. Anim. Behav. 34:1680–1686.

DUNLAP, J. C. 1990. Closely watched clocks. Trends in Genetics 6:159–165.

DYER, A. B., R. LICKLITER, AND G. GOTTLIEB. 1989. Maternal and peer imprinting in mallard ducklings under experimentally simulated natural social conditions. Dev. Psychobiol. 22:463–475.

DYER, F. C. 1991. Bees acquire route-based memories but not cognitive maps in a familiar landscape. Anim. Behav. 41:239–246.

DYER, F. C. AND T. D. SEELEY, 1991. Dance dialects and foraging range in three Asian honey bee species. Behav. Ecol. Sociobiol. 28:227–233.

EADIE, J. McA., F. P. KEHOE, AND T. D. NUDDS. 1988. Pre-hatch and post-hatch brood amalgamation in North American Anatidae: A review of hypotheses. Can. J. Zool. 66:1709–1721.

EALES, L. A. 1985. Song learning in zebra finches: Some effects of song model availability on what is learnt and when. Anim. Behav. 33:1293–1300.

EARLE, S. A. 1979. The gentle whales. National Geographic 155(1):2–17.

EBERHARD, W. G. 1980. Horned beetles. Sci. Am. 242 (Mar): 166–182.

EBERSOLE, J. P. 1977. The adaptive significance of interspecific territoriality in the reef fish *Eupomacentrus leucostictus*. Ecology 58:914–920.

EBIHARA, S. AND H. KAWAMURA. 1981. The role of the pineal organ and the suprachiasmatic nucleus in the control of circadian locomotor rhythms in the Java Sparrow, *Padda oryzivora*. J. Comp. Physiol. 141:207–214.

EBIHARA, S., K. UCHIYAMA, AND I. OSHIMA. 1984. Circadian organization in the pigeon, *Columba livia*, the role of the pineal and the eye. J. Comp. Physiol. 154:59–69.

ECKERT, C. G. AND P. J. WEATHERHEAD. 1987. Male characteristics, parental care and the study of mate choice in the red-winged blackbird (*Agelaius phoeniceus*). Behav. Ecol. Sociobiol. 20:35–42.

ECKERT, R. 1972. Bioelectric control of ciliary activity. Science 176:473–481.

EDMUNDS, M. 1974. *Defence in Animals*. Longman. New York. 357 pp.

EDNEY, E. B. 1954. Woodlice and the land habitat. Biol. Rev. 29:185–219.

EIBL-EIBESFELDT, I. 1956. Einige Bemerkungen über den ursprung von Ausdrucksbewegungen bei Säugetieren. Z. Säugetier. 21:29–43.

EIBL-EIBESFELDT, I. 1966. Das Verteidigen der Eiablageplätze bei der Hood-Meerechse (*Amblyrhynchus cristatatus venustissimus*) Z. Tierpsychol. 23:627–631.

EIBL-EIBESFELDT, I. 1975. *Ethology: The Biology of Behavior*, 2nd ed. Holt, Rinehart and Winston. New York. 625 pp.

EISENBERG, J. F., N. A. MUCKENHIRN, AND R. RUDRAN. 1972. The relation between ecology and social structure in primates. Science 176:863–874.

EISNER, T. 1958. The protective role of the spray mechanism of the bombardier beetle, *Brachynus ballistarius* Lec. J. Insect Physiol. 2:215–220.

EISNER, T., W. E. CONNER, K. HICKS, K. R. DODGE, H. I. ROSENBERG, T. H. JONES, M. COHEN, AND J. MEINWALD. 1977. Stink of the stinkpot turtle identified: W-phenylalkanoic acids. Science 196:1347–1349.

ELLIOTT, P. F. 1975. Longevity and the evolution of polygamy. Am. Nat. 109:281–287.

ELWOOD, R. W. AND H. F. KENNEDY. 1990. The relationship between infanticide and pregnancy block in mice. Behav. Neur. Biol. 53:277–283.

EMLEN, J. M. 1973. *Ecology: An evolutionary approach*. Addison-Wesley Publishing Co., Reading, MA. 493 pp.

EMLEN, S.T. 1967a. Migratory orientation in the indigo bunting, *Passerina cyanea*. Part I. Evidence for the use of celestial cues. Auk 84:309–342.

EMLEN, S. T. 1967b. Migratory orientation in the indigo bunting *Passerina cyanea*. Part II. Mechanisms of celestial orientation. Auk 84:463–489.

EMLEN, S. T. 1969. The development of migratory orientation in young Indigo Buntings. Living Bird 8:113–126.

EMLEN, S. T. 1970a. Celestial rotation: Its importance in the development of migratory orientation. Science 170:1198–1201.

EMLEN, S. T. 1972b. Exploring the mysteries of migration. In: *The Marvels of Animal Behavior*. National Geographic Society. Washington DC, pp. 271–284.

EMLEN, S. T. 1972. The ontogenetic development of orientation capabilities. NASA Spec. Publ. NASA SP-262, pp. 191–210.

EMLEN, S. T. 1975. Migration: Orientation and navigation. In: D. Farner, ed., *Avian Biology*, vol. V. Academic Press, New York. pp. 129–219.

EMLEN, S. T. 1976. Altruism in mountain bluebirds? Science 191:808–809.

EMLEN, S. T. 1982. The evolution of helping. I. An ecological constraints model. Am. Nat. 119: 29–39.

EMLEN, S. T. 1991. Evolution of cooperative breeding in birds and mammals. In: J. R. Krebs and N. B. Davies, eds., *Behavioural Ecology, An Evolutionary Approach,* 3rd ed. Blackwell Scientific Publications. Oxford, pp. 301–337.

EMLEN, S. T. AND L. W. ORING. 1977. Ecology, sexual selection, and the evolution of mating systems. Science 197:215–223.

EMLEN, S. T. AND P. H. WREGE. 1989. A test of alternative hypotheses for helping behavior in white-fronted bee-eaters of Kenya. Behav. Ecol. Sociobiol. 25:303–319.

EMLEN, S. T. AND S. L. VEHRENCAMP. 1983. Cooperative breeding strategies among birds. In: A. H. Brush and G. A. Clark, Jr., eds., *Perspectives in Ornithology.* Cambridge University Press. Cambridge, pp. 93–133.

EMLEN, S. T., H. K. REEVE, P. W. SHERMAN, P. H. WREGE, F. L. W. RATNIEKS, AND J. SHELLMAN-REEVE. 1991. Adaptive versus nonadaptive explanations of behavior: The case of alloparental helping. Am. Nat. 133:259–270.

ENDLER, J. A. 1978. A predator's view of animal color patterns. Evol. Biol. 11:319–364.

ENQUIST, M. AND O. LEIMAR. 1990. The evolution of fatal fighting. Anim. Behav. 39:1–9.

ENRIGHT, J. T. 1970. Ecological aspects of endogenous rhythmicity. Ann. Rev. Ecol. Syst. 1:221–238.

ENRIGHT, J. T. 1972a. A virtuoso isopod. Circalunar rhythms and their tidal fine structure. J. Comp. Physiol. 77:141–162.

ENRIGHT, J. T. 1972b. When the Beachhopper looks at the moon: The moon-compass hypothesis. In: S. R Galler, K. Schmidt-Koenig, G. J. Jacobs, and R. E. Belleville, eds., *Animal Orientation and Navigation.* NASA Spec. Publ. NASA SP-262, pp. 523–555.

EPSTEIN, R., C. E. KIRSHNIT, R. P. LANZA, AND L. C. RUBIN. 1984. "Insight" in the pigeon: Antecedents and determinants of an intelligent performance. Nature 308:61–62.

ESTEP, D. Q. 1988. Copulations by other males shorten the post-ejaculatory intervals of pairs of roof rats, *Rattus rattus.* Anim. Behav. 36:299–300.

ESTEP, D. Q., K. NIEUWENHUIJSEN, K. E. M. BRUCE, K. J. DE NEEF, P. A. WALTERS, S. C. BAKER, AND A. K. SLOB. 1988. Inhibition of sexual behavior among subordinate stumptail macaques, *Macaca arctoides.* Anim. Behav. 36:854–864.

ETKIN, W. 1964. *Social Behavior and Organization Among Vertebrates.* University of Chicago Press. Chicago. 307 pp.

EVANS, H. E. 1977. Extrinsic and intrinsic factors in the evolution of insect sociality. BioScience 27:613–617.

EWER, J., M. ROSBASH, AND J. C. HALL. 1988. An inducible promoter in *Drosophila* conditionally rescues adult *per-* mutant arrhythmicity. Nature 333(6168):82–84.

EWERT, J.-P. 1967. Aktivierung der verhaltensfolge beim beutefang der erdkröte (*Bufo bufo* L.) durch elektrische mittelhirn-reizung. Z. vergl. Physiol. 54:455–481.

EWERT, J.-P. 1969. Quantitative analyze von Reiz-Reaktionsbeziehungen bei visuellen Auslosen der Beutefang-Wendereaktion der Erdkröte (*Bufo bufo* L.). Pflugers Archiv 308:225–243.

EWERT, J.-P. 1987. Neuroethology of releasing mechanisms: Prey-catching in toads. Behav. Brain Sci. 10:337–405.

EWERT, J.-P., E. M. FRAMING, E. SCHURG-PFEIFFER, AND A. WEERASURIYA. 1990. Responses of medullary neurons to moving visual stimuli in the common toad. I. Characterization of medial reticular neurons by extracellular recording. J. Comp. Physiol. A 167:495–508.

EWING, A. W. 1963. Attempts to select for spontaneous activity in *Drosophila melanogaster.* Anim. Behav. 11:369–378.

FAGAN, R. 1981. *Animal Play Behavior.* Oxford University Press. New York. 684 pp.

FAIRBANKS, L. A. 1990. Reciprocal benefits of allomothering for female vervet monkeys. Anim. Behav. 40:553–562.

FANCY, S. G., L. F. PANK, D. C. DOUGLAS, C. H. CURBY, G. W. GARNER, S. C. AMSTRUP, AND L. REGELIN. 1988. Satellite telemetry: A new tool for wildlife research and management. U.S. Fish and Wildlife Service Resource Publication 172:1–54.

FAULKES, C. G., D. H. ABBOTT, AND A. L. MELLNOR. 1990. Investigation of genetic diversity in wild colonies of naked mole-rats (*Heterocephalus glaber*) by DNA fingerprinting. J. Zool. London 221:87–97.

FELDMAN, J. F. 1989. Circadian rhythms In: K. C. Smith, ed., *The Science of Photobiology*, 2nd ed. Plenum Press. New York, pp. 193–213.

FELDMAN, J. F. AND J. DUNLAP. 1983. *Neurospora* in the study of circadian rhythms. Photochem. Photobiol. Rev. 7:319–368.

FELDMAN, J. F. AND M. N. HOYLE. 1973. Isolation of circadian clock mutants of *Neurospora crassa*. Genetics 75:605–613.

FELDMAN, J. F. AND N. WASSER. 1971. New mutation affecting circadian rhythmicity in *Neurospora* In: M. Menaker, ed., *Biochronometry*. National Academy of Science. Washington, DC, pp. 652–656.

FELTMATE, B. W. AND D. DUDLEY WILLIAMS. 1989. A test of crypsis and predator avoidance in the stonefly *Paragnetina media* (Plecoptera: Perlidae). Anim. Behav. 37:992–999.

FENG, A. S. 1977. The role of the electrosensory system in postural control in the weakly electric fish *Eigenmannia virescens*. J. Exp. Biol. 66:141–158.

FENTON, M. B. 1983. *Just Bats*. University of Toronto Press, Toronto. 165 pp.

FERGUSON, D. E., H. F. LANDRETH, AND M. R. TURNIPSEED. 1965. Astronomical orientation of the southern cricket frog, *Acris gryllus*. Copeia. 1965:58–66.

FERGUSON, M. W. J. AND T. JOANEN. 1983. Temperature-dependent sex determination in *Alligator mississippiensis*. J. Zool. London 200:143–177.

FERSTER, C. B. AND B. F. SKINNER. 1957. *Schedules of Reinforcement*. Appleton-Century-Crofts. New York.

FILDES B. N., B. J. O'LOUGHLIN, J. L. BRADSHAW, AND W. J. EWENS. 1984. Human orientation with restricted sensory information: No evidence for magnetic sensitivity. Perception 13:229–236.

FINK, L. S. AND L. P. BROWER. 1981. Birds can overcome the cardenolide defence of monarch butterflies in Mexico. Nature 291:67–70.

FINLEY, J., D. IRETON, W. M. SCHLEIDT, AND T. A. THOMPSON. 1983. A new look at the features of mallard courtship displays. Anim. Behav. 31:348–354.

FISHER, A. C., JR. 1979. Mysteries of bird migration. National Geographic 156(2):154–193.

FISHER, A. E. 1962. Effects of stimulus variation on sexual satiation in the male rat. J. Comp. Physiol. Psychol. 55:614–620.

FISHER, J. AND R. A. HINDE. 1949. The opening of milk bottles by birds. Brit. Birds 42:347–357.

FISHER, R. A. 1930. *The Genetical Theory of Natural Selection*. Oxford University Press. Oxford.

FITZGERALD, R. W. AND D. M. MADISON. 1983. Social organization of a free-ranging population of pine voles, *Microtus pinetorum*. Behav. Ecol. Sociobiol. 13: 183–187.

FITZGIBBON, C. D. 1990. Mixed-species grouping in Thomson's and Grant's gazelles: The antipredator benefits. Anim. Behav. 39:1116–1126.

FITZGIBBON, C. D. AND J. H. FANSHAWE. 1988. Stotting in Thomson's gazelles: An honest signal of condition. Behav. Ecol. Sociobiol. 23:69–74.

FLEISHMAN, L. J. 1988. Sensory influences on physical design of a visual display. Anim. Behav. 36:1420–1424.

FLINN, M. V. AND B. S. LOW. 1986. Resource distribution, social competition, and mating patterns in human societies. In: D. I. Rubenstein and R. W. Wrangham, eds., *Ecological Aspects of Social Evolution*. Princeton University Press. Princeton, NJ, pp. 217–243.

FORMANOWICZ, D. R., JR., M. S. BOBKA, AND E. D. BRODIE. 1982. The effect of prey density on ambush-site changes in an extreme ambush type predator. Am. Mid. Nat. 108:250–255.

FORRESTER, G. E. 1991. Social rank, individual size and group composition as determinants of food consumption by humbug damselfish, *Dascyllus aruanus*. Anim. Behav. 42:701–711.

FOSTER, M. S. 1977. Odd couples in manakins: A study of social organization and cooperative breeding in *Chiroxiphia linearis*. Am. Nat. 111:845–853.

FOSTER, W. A. 1990. Experimental evidence for effective and altruistic colony defence against natural predators by soldiers of the gall–forming aphid *Pemphis spyrothecae* (Hemiptera: Pemphigidae). Behav. Ecol. Sociobiol. 27:421–430.

FOUTS, R. S. 1973. Acquisition and testing of gestural signs in four young chimpanzees. Science 180:978–980.

FOUTS, R. S. 1974. Language: Origins, definition and chimpanzees. J. Human Evol. 3:475–482.

FOX, S. F. AND M. A. ROSTKER. 1982. Social cost of tail loss in *Uta stansburiana*. Science 218:692–693.

FRAENKEL, G. S. AND D. L. GUNN. 1961. *The Orientation of Animals—Kineses, Taxes and Compass Reactions*. Dover Publications. New York.

FRAME, G. W. AND L. H. FRAME. 1980. Cheetahs: In a Race for Survival. National Geographic 157(5):712–728.

FRANK, S. A. 1990. Sex allocation theory for birds and mammals. Ann. Rev. Ecol. Syst. 21:13–55.

FRANKEL, R. B. 1981. Bacteria magnetotaxis vs. geotaxis. EOS 62:850.

FRANKEL, R. B. AND R. P. BLAKEMORE. 1980. Navigational compass in magnetic bacteria. J. Magn. Mater. 18:1562–1564.

FRANKEL, R. B., R. P. BLAKEMORE, AND R. S. WOLF. 1979. Magnetite in magnetotactic bacteria. Science 203:1355.

FRASER, A. F. 1968. Reproductive behaviour in ungulates. Academic Press. New York. 202 pp.

FREI, U. 1982. Homing pigeon's behavior in the irregular magnetic field of western Switzerland. In: F. Papi and H. G. Walraff, eds., *Avian Navigation*. Springer-Verlag. Berlin, pp. 129–139.

FREI, U. AND G. WAGNER. 1976. Die Anfangsorientierung von Brieftauben im erdmagnetisch gestörten Gebiet des Mont Jorat. Rev. Suisse Zool. 83:891–897.

FRICKE, H. AND S. FRICKE. 1977. Monogamy and sex change by aggressive dominance in coral reef fish. Nature 266:830–832.

FRIEDMAN, M. I., J. P. BRUNO, AND J. R. ALBERTS. 1981. Physiological and behavioral consequences in rats of water recycling during lactation. J. Comp. Physiol. Psychol. 95:26–35.

FRINGS, H. AND M. FRINGS. 1964. *Animal Communication*. Blaisdell Publishing Co. New York. 204 pp.

FRINGS, H. AND M. FRINGS. 1977. *Animal Communication*, 2nd ed. University of Oklahoma Press. Norman, OK.

FROMME, H. G. 1961. Untersuchungen über das Orientierungsvermögen nachlich ziehender Kleinvögel (*Erithacus rubecula, Sylvia communis*). Z. Tierpsychol. 18:205–220.

FRYXELL, J. M., J. GREEVER, AND A. R. E. SINCLAIR. 1988. Why are migratory ungulates so abundant? Am. Nat. 131:781–798.

GADAGKAR, R. 1990a. Origin and evolution of eusociality: A perspective from studying primitively eusocial wasps. J. Genetics 69:113–125.

GADAGKAR, R. 1990b. Evolution of eusociality: The advantage of assured fitness returns. Phil. Trans. Roy. Soc. London B. 329:17–25.

GALEF, B. G., JR. 1976. Social transmission of acquired behavior. In: J. S. Rosenblatt, R. H. Hinde, E. Shaw, and C. Beer, eds., *Advances in the Study of Behavior*, vol. 6. Academic Press. New York. pp. 77–100.

GALEF, B. G., JR. 1990. An adaptationist perspective on social learning, social feeding, and social foraging in Norway rats. In: D. A. Dewsbury, ed., *Contemporary Issues in Comparative Psychology*. Sinauer Associates. Inc. Sunderland, MA, pp. 55–79.

GALEF, B. G., JR., L. A. MANZIG, AND R. M. FIELD. 1986. Imitation learning by budgerigars: Dawson and Foss (1965) revisited. Behavioral Proc. 13:191–202.

GALEF, B. G. AND S. W. WIGMORE. 1983. Transfer of information concerning distant foods: A laboratory investigation of the "information-centre" hypothesis. Anim. Behav. 31:748–758.

GAMBOA, G. J. 1978. Intraspecific defense: Advantage of social cooperation among paper wasp foundresses. Science 199:1463–1465.

GAMOW, R. I. AND J. F. HARRIS. 1973. The infrared receptors of snakes. Sci. Am. 228(May):94–100.

GARCIA, J. AND R. A. KOELLING. 1966. Relation of cue to consequence in avoidance learning. Psych. Sci. 4:123–124.

GARDNER, G. F. AND J. F. FELDMAN. 1980. The *frq* locus in *Neurospora crassa*: A key element in circadian clock organization. Genetics 96:877–886.

GARDNER, R. A. AND B. T. GARDNER. 1969. Teaching sign language to a chimpanzee. Science 165:664–672.

GASTON, S. AND M. MENAKER. 1968. Pineal function. The biological clock in the sparrow? Science 160:1125–1127.

GEIST, V. 1971. *Mountain Sheep*. University of Chicago Press, Chicago.

GENDRON, R. P. 1986. Searching for cryptic prey: Evidence for optimal search rates and the formation of search images in quail. Anim. Behav. 34:898–912.

GENDRON, R. P. AND J. E. R. STADDON. 1983. Searching for cryptic prey: The effect of search rate. Am. Nat. 121:172–186.

GENDRON, R. P. AND J. E. R. STADDON. 1984. A laboratory simulation of foraging behavior: The effect of search rate on the probability of detecting prey. Am. Nat. 124:407–415.

GETZ, L. L., B. McGUIRE, T. PIZZUTO, J. E. HOFMANN, AND B. FRASE. In press. Social organization of the prairie vole (*Microtus ochrogaster*). J. Mamm.

GHENT, A. W. AND N. E. GARY. 1962. A chemical alarm releaser in honey bee stings (*Apis mellifera* L.). Psyche 69:1–6.

GHISELIN, M. T. 1974. *The Economy of Nature and the Evolution of Sex*. University of California Press. Berkeley.

GIBB, J. A. 1960. Populations of tits and goldcrests and their food supply in pine plantations. Ibis 102:163–208.

GILBERT, L. E. 1976. Postmating female odor in *Heliconius* butterflies: A male-contributed antiaphrodisiac? Science 193:419–420.

GILBERT, S. F. 1988. *Developmental Biology*. Sinauer Associates, Inc. Sunderland, MA. 843 pp.

GILL, F. B. AND L. L. WOLF. 1975. Economics of feeding territoriality in the golden-winged sunbird. Ecology 56:33–45.

GILLETTE, R., M. U. GILLETTE, D. J. GREEN, AND R. HUANG. 1989. The neuromodulatory response: Integrating second messenger pathways. Am. Zool. 29:1275–1286.

GIRALDEAU, L.-A. AND D. L. KRAMER. 1982. The marginal value theorem: A quantitative test using load size variation in a central place forager, the eastern chipmunk, *Tamias striatus*. Anim. Behav. 30:1036–1042.

GITTLEMAN, J. L. 1989. The comparative approach in ethology: Aims and limitations. In: P. P. G. Bateson and P. H. Klopfer, eds., *Perspectives in Ethology*, vol. 8. Whither Ethology? Plenum Press, New York, pp. 55–83.

GITTLEMAN, J. L. AND P. H. HARVEY. 1980. Why are distasteful prey not cryptic? Nature 286:149–150.

GLANCEY, B. M., C. E. STRINGER, C. H. CRAIG, P. M. BISHOP, AND B. B. MARTIN. 1970. Pheromone may induce brood tending in the fire ant (*Solenopsis saevissma*). Nature (London) 226:863–864.

GLESNER, R. R. AND D. TILMAN. 1978. Sexuality and the components of environmental uncertainty: Clues from geographic parthenogenesis in terrestrial animals. Am. Nat. 112:659–673.

GODDARD, S. M. AND R. B. FORWARD, JR. 1991. The role of the underwater polarized light pattern, in the sun compass navigation of the grass shrimp, *Palaemonetes vulgaris*. J. Comp. Physiol. 169:479–491.

GOODALL, J. 1965. Chimpanzees of the Gombe Stream Reserve. In: I. DeVore, ed., *Primate Behavior*. Holt, Rinehart and Winston. New York, pp. 425–473.

GOODENOUGH, J. E. 1978. The lack of effect of deuterium oxide on the period or phase of the monthly orientation rhythm of planarians. Int. J. Chronobiol. 5:465–476.

GOODENOUGH, J. E. 1980. The monthly orientation rhythm of planarians is not generated by the interaction of solar- and lunar-day rhythms J. Interdiscipl. Cycle Res. 11:117–124.

GORDON, D. A. 1948. Sensitivity of homing pigeons to the magnetic field of the earth. Science 108:710–711.

GORZULA, S. J. 1978. An ecological study of *Caiman crocodilus* inhabiting savanna lagoons in the Venezuelan Guayana. Oecologia 35:21–34.

GOSS, R. J. 1969. Photoperiodic control of antler cycles in deer. I. Phase shift and frequency changes. J. Exp. Zool. 170:311–324.

GOSS-CUSTARD, J. D. 1977. Optimal foraging and the size selection of worms by redshank, *Tringa totanus*, in the field. Anim. Behav. 25:10–29.

GOTO, K., D. L. LAVAL-MARTIN, AND L. N. EDMUNDS. 1985. Biochemical modeling of an autonomously oscillatory circadian clock in *Euglena*. Science 228:1284–1288.

GOTTLIEB, G. 1965. Imprinting in relation to parental and species identification by avian neonates. J. Comp. Physiol. Psychol. 59:345–356.

GOTTLIEB, G. 1968. Prenatal behavior of birds. Quart. Rev. Biol. 43:148–174.

GOTTLIEB, G. 1971. *Development of Species Identification in Birds*. University of Chicago Press, Chicago.

GOTTLIEB, G. 1976. Early development of species-specific auditory perception in birds. In: G. Gottlieb, ed., *Neural and Behavioral Specificity: Studies in the Development of Behavior and the Nervous System*. Academic Press, New York, pp. 237–280.

GOTTLIEB, G. 1978. Development of species identification in ducklings. IV. Change in species-specific perception caused by auditory deprivation. J. Comp. Physiol. Psychol. 92:375–387.

GOTTLIEB, G. 1985. Development of species identification in ducklings. XI. Embryonic critical period for species-typical perception in the hatchling. Anim. Behav. 33:225–233.

GOULD, J. L. AND C. G. GOULD. 1986. Invertebrate intelligence. In: R. J. Hoage and L. Goldman, eds., *Animal Intelligence, Insights into the Animal Mind*. Smithsonian Institution Press. Washington DC, pp. 21–36.

GOULD, J. L. AND K. P. ABLE. 1981. Human homing: An elusive phenomenon. Science 212: 1061–1063.

GOULD, J. L. 1982. Why do honeybees have dialects? Behav. Ecol. Sociobiol. 10:53–56.

GOULD, J. L. 1975. Honey bee recruitment: The dance-language controversy. Science 189:685–692.

GOULD, J. L. 1976. The dance language controversy. Quart. Rev. Biol. 51:211–244.

GOULD, J. L. 1980. The case for magnetic sensitivity in birds and bees (such as it is). Am. Sci. 68:256–267.

GOULD, J. L. 1982. The map sense of pigeons. Nature. 296:205–211.

GOULD, J. L. 1985. Absence of human homing ability as measured by displacement experiments. In: J. L. Kirschvink, D. S. Jones, and B. J. MacFadden, eds., *Magnetite Biomineralization and Magnetoreception in Organisms: A New Biomagnetism*. Plenum Press. New York, pp. 595–600.

GOULD, J. L. 1986. The local map of honeybees: Do insects have cognitive maps? Science 232:861–863.

GOULD, J. L. AND C. G. GOULD. 1988. *The Honey Bee*. Scientific American Library. W. H. Freeman and Co. New York. 239 pp.

GOULD, J. L., J. L. KIRSCHVINK, AND K. S. DEFFEYES. 1978. Bees have magnetic resonance. Science 201:1026–1028.

GOULD, J. L., M. HENEREY, AND M. C. MACLEOD. 1970. Communication of direction by the honey bee. Science 169:544–554.

GOULD, S. J. AND R. C. LEWONTIN. 1979. The spandrels of San Marco and the Panglossian paradigm: A critique of the adaptationist programme. Proc. Royal Soc. London 205:581–598.

GOWATY, P. A. 1983. Male parental care and apparent monogamy among eastern bluebirds (*Sialia sialis*). Am. Nat. 121:149–157.

GRAFEN, A. 1982. How not to measure inclusive fitness. Nature (London) 298:425–426.

GRAFEN, A. 1984. Natural selection, kin selection and group selection. In: J. R. Krebs and N. B. Davies, eds., *Behavioural Ecology*, 2nd ed. Sinauer Associates, Inc. Sunderland, MA, pp. 62–84.

GRAFEN, A. 1990. Biological signals as handicaps. J. Theoret. Biol. 144:517–546.

GRANT, B., G. A. SNYDER, AND S. F. GLESSER. 1974. Frequency-dependent mate selection in *Mormoniella vitripennis*. Evolution 28:259–264.

GRANT, B., S. BURTON, C. CONTOREGGI, AND M. ROTHSTEIN. 1980. Outbreeding via frequency-dependent mate selection in the parasitoid wasp. *Nasonia* (*Mormoniella*) *vitripennis* Walker. Evolution 34:983–992.

GRAUL, W. D., S. R. DERRICKSON, AND D. W. MOCK. 1977. The evolution of avian polyandry. Am. Nat. 111:812–816.

GRAYSON, J. AND M. EDMUNDS. 1989. The causes of colour and colour change in caterpillars of the poplar and eyed hawkmoths (*Laothoe populi* and *Smerinthus ocellata*). Biol. J. Linn. Soc. 37:263–279.

GREEN, S. 1975. Dialects in Japanese monkeys. Z. Tierpsychol. 38:305–314.

GREEN, S. AND P. MARLER. 1979. The analyses of animal communication. In: P. Marler and J. G. Vandenbergh, eds., *Handbook of Behavioral Neurobiology*, vol. 3, Social behavior and communication. Plenum Press. New York, pp. 73–158.

GREENBERG, L. 1979. Genetic component of bee odor in kin recognition. Science 206:1095–1097.

GREENFIELD, P. M. AND E. S. SAVAGE-RUMBAUGH. 1990. Grammatical combination in *Pan paniscus*: Processes of learning and invention in the evolution and development of language. In: S. T. Parker and K. R. Gibson, eds. *"Language" and intelligence in monkeys and apes: Comparative developmental perspectives*. Cambridge University Press. Cambridge, England. pp. 540–578.

GREENGARD, P. 1976. Possible role for cyclic nucleotides and phosphorylated membrane proteins in postsynaptic actions of neurotransmitters. Nature 260:101–108.

GREENWOOD, J. J. D. 1984. The functional basis of frequency-dependent food selection. Biol. J. Linn. Soc. 23:177–199.

GREENWOOD, J. J. D., P. A. COTTON, AND D. M. WILSON. 1989. Frequency-dependent selection on aposematic prey: Some experiments. Biol. J. Linn. Soc. 36:213–226.

GREENWOOD, P. J. 1980. Mating systems, philopatry and dispersal in birds and mammals. Anim. Behav. 28:1140–1162.

GRIFFIN, D. R. 1955. Bird navigation. In: A. Wolfson, ed., *Recent Studies in Avian Biology*. University of Illinois Press. pp. 154–197.

GRIFFIN, D. R. 1978. Prospects for a cognitive ethology. Behav. Brain Sci. 1:527–538.

GRIFFIN, D. R. 1981. *The Question of Animal Awareness*, 2nd ed. Rockefeller University Press. New York. 209 pp.

GRIFFIN, D. R. (ED.). 1982. *Animal Mind—Human Mind*. Springer-Verlag. Berlin. 427 pp.

GRIFFIN, D. R. 1984. *Animal Thinking*. Harvard University Press. Cambridge, MA. 237 pp.

GRIFFIN, D. R. 1991. Progress toward a cognitive ethology. In: C. A. Ristau, ed., *Cognitive Ethology, the Minds of Other Animals*. Lawrence Erlbaum. Hillsdale, NJ, pp. 3–17.

GRIFFIN, D. R., F. A. WEBSTER, AND C. R. MICHAEL. 1960. The echolocation of flying insects by bats. Anim. Behav. 8:141–154.

GRIFFITHS, M. 1988. The platypus. Sci. Am. 258 (May):84–91.

GROSBERG, R. K. AND J. F. QUINN. 1986. The genetic control and consequences of kin recognition by the larvae of a colonial marine invertebrate. Nature 322:456–459.

GROSS, M. R. 1982. Sneakers, satellites and parentals: Polymorphic mating strategies in North American sunfishes. Z. Tierpsychol. 60:1–26.

GROSS, M. R. AND R. SHINE. 1981. Parental care and mode of fertilization in ectothermic vertebrates. Evolution 35:775–793.

GUBERNICK, D. J. 1980. Maternal 'imprinting' or maternal 'labelling' in goats? Anim. Behav. 28:124–129.

GUBERNICK, D. J. 1981. Parent and infant attachment in mammals. In: D. J. Gubernick and P. H. Klopfer, eds., *Parental Care in Mammals*. Plenum Press. New York, pp. 243–306.

GUBERNICK, D. J. AND J. R. ALBERTS. 1983. Maternal licking of young: Resource exchange and proximate controls. Physiol. Behav. 31:593–601.

GUBERNICK, D. J. AND R. J. NELSON. 1989. Prolactin and paternal behavior in the biparental California mouse, *Peromyscus californicus*. Horm. Behav. 23:203–210.

GUBERNICK, D. J., K. C. JONES, AND P. H. KLOPFER. 1979. Maternal "imprinting" in goats? Anim. Behav. 27:314–315.

GUILFORD, T. AND M. S. DAWKINS. 1987. Search images not proven: A reappraisal of recent evidence. Anim. Behav. 35:1838–1845.

GUILFORD, T. AND M. S. DAWKINS. 1991. Receiver psychology and the evolution of animal signals. Anim. Behav. 42:1–14.

GUILFORD, T. AND M. S. DAWKINS. 1992. Understanding signal design: a reply to Blumberg & Alberts. Anim. Behav. 44:384–385.

GUNN, D. L. 1937. The humidity reactions of the woodlouse, *Porcellio scaber*. J. Exp. Biol. 14:178–186.

GUNN, D. L. AND J. S. KENNEDY. 1936. Apparatus for investigating the reactions of land arthropods to humidity. J. Exp. Biol. 13:450–459.

GURNEY, M. E. 1981. Hormonal control of cell form and number in the zebra finch song system. J. Neurosci. 1:658–673.

GURNEY, M. E. AND M. KONISHI. 1980. Hormone induced sexual differentiation in brain and behavior in zebra finches. Science 208:1380–1382.

GWADZ, R. 1970. Monofactorial inheritance of early sexual receptivity in the mosquito *Aedes atropalus*. Anim. Behav. 18:358–361.

GWINNER, E. 1971. A comparative study of circannual rhythms in warblers. In: M. Menaker, ed., *Biochronometry*. National Academy of Science. Washington, DC, pp. 405–427.

GWINNER, E. 1978. Effects of pinealectomy on circadian locomotor activity rhythms in European starlings, *Sturnus vulgaris*. J. Comp. Physiol. 126:123–129.

GWINNER, E. AND I. BENZINGER. 1978. Synchronization of a circadian rhythm in European starlings by daily injections of melatonin. J. Comp. Physiol. 127:209–213.

GWINNER, E. AND W. WILTSCHKO. 1978. Endogenously controlled changes in migratory direction of the garden warbler, *Sylvia borin*. J. Comp. Physiol. 125:267–273.

GWYNNE, D. T. AND L. W. SIMMONS. 1990. Experimental reversal of courtship roles in an insect. Nature 346:172–174.

HAFNER, D. J. AND K. E. PETERSON. 1985. Song dialects and gene flow in the white-crowned sparrow *Zonotrichia leucophrys nuttalli*. Evolution 39:687–694.

HAGEDORN, M. 1986. The ecology, courtship and mating of gymnotiform electric fish. In: T. H. Bullock and W. Heiligenberg, eds., *Electroreception*. John Wiley & Sons. New York, pp. 497–525.

HAILMAN, J. P. 1967. Cliff-nesting adaptations of the Galapagos swallow-tailed gull. Wilson Bull. 77:346–362.

HAILMAN, J. P. 1967. The ontogeny of an instinct: The pecking response in chicks of the Laughing Gull (*Larus atricilla* L.) and related species. Behaviour Suppl. 15:1–159.

HAILMAN, J. P. 1969. How an instinct is learned. Sci. Am. 221 (Dec):98–106.

HALL, J. 1979. Control of male reproductive behavior by the central nervous system of *Drosophila*: Dissection of a courtship pathway by genetic mosaics. Genetics 92:437–457.

HALLIDAY, T. R. 1976. The libidinous newt. An analysis of variations in the sexual behavior of the male smooth newt, *Triturus vulgaris*. Anim. Behav. 24:398–414.

HALLIDAY, T. R. 1983. The study of mate choice. In: P. Bateson, ed., *Mate Choice*. Cambridge University Press. Cambridge, pp. 3–32.

HALPIN, Z. T. 1991. Introduction to the symposium: Animal behavior: Past, present and future. Am. Zool. 31:283–285.

HAMILTON, W. D. 1964. The genetical evolution of social behaviour. J. Theoret. Biol. 7:1–52.

HAMILTON, W. D. 1967. Extraordinary sex ratios. Science 156:477–488.

HAMILTON, W. D. 1971. Geometry for the selfish herd. J. Theor. Biol. 31:295–311.

HAMILTON, W. D. AND M. ZUK. 1982. Heritable true fitness and bright birds: A role for parasites? Science 218:384–387.

HAMMERSTEIN, P. AND G. A. PARKER. 1987. Sexual selection: Games between the sexes. In: J. W. Bradbury, and M. B. Andersson, eds., *Sexual Selection: Testing the Alternatives*. John Wiley & Sons. New York, pp. 119–142.

HAMNER, W. M., M. SMYTH, AND E. D. MOLFORD, JR. 1968. Orientation of the sand-beach isopod *Tylos punctatus*. Anim. Behav. 16:405–409.

HANDLER, A. N. AND R. J. KONOPKA. 1979. Transplantation of a circadian pacemaker in *Drosophila*. Nature (London) 279:236–238.

HANSON, M., G. WIRMARK, M. OBLAD, AND L. STRID. 1984. Iron–rich particles in European eel (*Anguilla anguilla* L.). Comp. Biochem. Physiol. A 79:221–224.

HARDCASTLE, A. 1925. Young cuckoo fed by several birds. Brit. Birds 19:100.

HARDEN-JONES, F. R. 1968. *Fish Migration*. St. Martin's Press. New York. 325 pp.

HARDER, L. D. AND L. A. REAL. 1987. Why are bumble bees risk averse? Ecology 68:1104–1108.

HARDIN, P. E., J. C. HALL, AND M. ROSBASH. 1990. Feedback of the *Drosophila period* product on circadian cycling of its messenger RNA levels. Nature 343:536–540.

HARLOW, H. F. AND M. K. HARLOW. 1962. Social deprivation in monkeys. Sci. Am. 207 (Nov):136–146.

HARLOW, H. F., M. K. HARLOW, AND S. J. SUOMI. 1971. From thought to therapy: Lessons from a primate laboratory. Am. Sci. 59:538–549.

HARLOW, H. F. 1949. The formation of learning sets. Psychol. Rev. 56:51–65.

HARRINGTON, F. H. AND L. D. MECH. 1983. Wolf pack spacing: Howling as a territory-independent spacing mechanism in a territorial population. Behav. Ecol. Sociobiol. 12:161–168.

HARRIS, W. A. 1980. The effects of eliminating impulse activity on the development of the retinotectal projection in salamanders. J. Comp. Neurol. 194:303–317.

HART, B. 1980. Neonatal spinal transection in male rats: Differential effects on penile reflexes and other reflexes. Brain Res. 185:423–428.

HASLER, A. D. AND A. T. SCHOLZ. 1983. *Olfactory Imprinting in Homing Salmon*. Springer-Verlag. Berlin. 134 pp.

HASLER, A. D. AND W. J. WISBY. 1951. Discrimination of stream odors by fishes in relation to parent stream behavior. Am. Nat. 85:223–238.

HASLER, A. D., A. T. SCHOLZ, AND R. M. HORRAL. 1978. Olfactory imprinting and homing in salmon. Am. Sci. 66:347–355.

HASS, C. C. 1990. Alternative maternal-care patterns in two herds of bighorn sheep. J. Mamm. 71:24–35.

HASTINGS, J. W. 1970. Cellular-biochemical clock hypotheses. In: *The Biological Clock*. F. A. Brown, Jr., J. W. Hastings, J. D. Palmer. Academic Press. New York, pp. 63–91.

HASTINGS, J. W., B. RUSAK, AND Z. BOULOS. 1991. The physiology of biological timing. In: C. L. Prosser, ed., *Neural and Integrative Animal Physiology*. Wiley-Liss. New York, pp. 435–546.

HAUSFATER, G. AND S. B. HRDY, EDS. 1984. *Infanticide: Comparative and Evolutionary Perspectives*. Aldine. New York. 598 pp.

HAUSFATER, G., H. C. GERHARDT, AND G. M. KLUMP. 1990. Parasites and mate choice in gray treefrogs, *Hyla versicolor*. Am. Zool. 30:299–312.

HAVERKAMP, L. J. 1986. Anatomical and physiological development of the *Xenopus* embryonic motor system in the absence of neural activity. J. Neurosci. 6:1338–1348.

HAVERKAMP, L. J. AND R. W. OPPENHEIM. 1986. Behavioral development in the absence of neural activity: Effects of chronic immobilization on amphibian embryos. J. Neurosci. 6:1332–1337.

HAWRYSHYN, C. W. 1992. Polarization vision in fish. Am. Sci. 80:164–175.

HAY, D. A. 1976. The behavioral phenotype and mating behavior of two inbred strains of *Drosophila melanogaster* Behav. Gen. 6:161–170.

HAY, D. E. AND J. D. MCPHAIL. 1975. Mate selection in threespine sticklebacks (*Gasterosteus*). Can. J. Zool. 53:441–450.

HAYASHI, K. 1985. Alternative reproductive strategies in the water strider *Gerris elongatus* (Heteroptera, Gerridae). Behav. Ecol. Sociobiol. 16:301–306.

HAYES, K. J. AND C. HAYES. 1951. The intellectual development of a home-raised chimpanzee. Proc. Am. Phil. Soc. 95:105

HEDRICK, A. V. 1986. Female preference for male calling bout duration in a field cricket. Behav. Ecol. Soiobiol. 19:73–77.

HEDRICK, A. V. AND E. J. TEMELES. 1989. The evolution of sexual dimorphism in animals: Hypotheses and tests. Trends Ecol. Evol. 4:136–138.

HEILIGENBERG, W. 1977. *Principles of Electrolocation and Jamming Avoidance in Electric Fish.* Springer-Verlag. Berlin. 85 pp.

HEINRICH, B. 1979. Foraging strategies of caterpillars. Leaf damage and possible predator avoidance strategies. Oecologia 42:325–337.

HEINRICH, B. AND S. L. COLLINS. 1983. Caterpillar leaf damage, and the game of hide-and-seek with birds. Ecology 64:592–602.

HEINROTH, O. 1910. Beiträge zur Bwlogie, insbiesonder Psychologie und Ethologie der Anatiden, Verhand, d. as V. Intern. Ornithol. Kongr. Berlin, pp. 589–702.

HEINSOHN, R. G. 1991. Kidnapping and reciprocity in cooperatively breeding white-winged choughs. Anim. Behav. 41:1097–1100.

HEINSOHN, R. G., A. COCKBURN, AND R. A MULDER. 1990. Avian cooperative breeding: Old hypotheses and new directions. Trends Ecol. and Evol. 5:403–407.

HELBIG, A. J. 1991. Inheritance of migratory direction in a bird species: A cross-breeding experiment with SE- and SW-migrating blackcaps (*Sylvia atricapilla*). Behav. Ecol. Sociobiol. 28:9–12.

HELBIG, A. J., P. BERTHOLD, AND W. WILTSCHKO. 1989. Migratory orientation of blackcaps (*Sylvia atricapilla*): Population specific shifts of direction during the autumn. Ethology 82:307–315.

HENNESSY, D. F. 1986. On the deadly risk of predator harassment. Ethology 72:72–74.

HERRNKIND, W. F. 1972. Orientation in shore-living arthropods. In: H. E. Winn and B. L. Olla, eds., *Behavior of Marine Animals*, vol. 1. Plenum. New York, pp. 1–59.

HERRNKIND, W. P. AND P. KANCIRUK, 1978. Mass migration of the spiny lobster, *Panulirus argus* (Crustacea: Palinuridae): Synopsis and orientation. In: K. Schmidt-Koenig and W. T. Keeton, eds., *Animal Migration, Navigation, and Homing.* Springer-Verlag. New York, pp. 430–439.

HERRNSTEIN, R. J., D. H. LOVELAND, AND C. CABLE. 1976. Natural concepts in pigeons. J. Exp. Psych. An. Behav. Proc. 2:285–302.

HERTER, K. 1962. *Der Temperatursinn der Tiere.* Ziensen Verlag. Wittenbery, Germany.

HEWS, D. K. 1988. Alarm response in larval western toads, *Bufo boreas*: Release of larval chemicals by a natural predator and its effect on predator capture efficiency. Anim. Behav. 36:125–133.

HILGARD, E. R. AND R. C. ATKINSON. 1967. *Introduction to Psychology*, 4th ed. Harcourt, Brace and World. New York. 686 pp.

HILGARD, E. R., 1956. *Theories of Learning.* Appleton-Century-Crofts. New York.

HINDE, R. 1981. Animal signals: Ethological and games theory approaches are not incompatible. Anim. Behav. 29:535–542.

HINDE, R. A. 1956. The biological significance of territories in birds. Ibis 98:340–369.

HINDE, R. A. 1970. *Animal Behavior: A Synthesis of Ethology and Comparative Psychology*, 2nd ed. McGraw-Hill. New York. 876 pp.

HINDE, R. A. 1982. *Ethology.* Oxford University Press, Oxford. 320 pp.

HINGSTON, R. W. G. 1927a. Protective devices in spiders' snares, with a description of seven new species of orb-weaving spiders. Proc. Zool. Soc. London 1927:259–293.

HINGSTON, R. W. G. 1927b. Field observations on

spider mimics. Proc. Zool. Soc. London 1927:841–859.

HOAGE, R. J. AND L. GOLDMAN, EDS. 1986. *Animal Intelligence, Insights into the Animal Mind.* Smithsonian Institution Press. Washington DC. 207 pp.

HOCKING, B. 1964. Fire melanism in some African grasshoppers. Evolution 18:332–335.

HODOS, W. AND C. B. G. CAMPBELL. 1969. Scala naturae: Why there is no theory in comparative psychology. Psychol. Rev. 76:337–350.

HOFFMANN, A. A. AND Z. CACOYIANNI. 1990. Territoriality in *Drosophila melanogaster* as a conditional strategy. Anim. Behav. 40:526–537.

HOFFMANN, K. 1954. Versuche zu der im Richtungsfinden der Vögel enthaltenen Zeitschätzung. Z. Tierpsychol. 11:453–475.

HOFFMANN, K. 1976. The adaptive significance of biological rhythms corresponding to geophysical cycles. In: J. W. Hastings and H.-G. Schweiger, eds., *The Molecular Basis of Circadian Rhythms*. Abakon Verlagsgesellschaft. Berlin, pp. 63–75.

HOGG, J. T. 1988. Copulatory tactics in relation to sperm competition in Rocky Mountain bighorn sheep. Behav. Ecol. Sociobiol. 22:49–59.

HOGLUND, L. B. AND M. ASTRAND. 1973. Preferences among juvenile char (*Salvelinus alpinus* L.) to intraspecific odours and water current studied with the fluviarium technique. Inst. Freshwater Res. Drottningholm Report 53:21–30.

HOLDER, C. F. 1901. A curious means of defense. Sci. Am. 2 (Sept):186–187.

HÖLLDOBLER, B. 1976. Recruitment behavior. Home range orientation and territoriality in harvester ants, *Pogonomyrmes*. Behav. Ecol. Sociobiol. 1:3–44.

HÖLLDOBLER, B. AND E. O. WILSON. 1978. The multiple recruitment systems of the African weaver ant *Oecophylla longinoda* (Latreille) (Hymenoptera: Formicidae). Behav. Ecol. Sociobiol. 3:19–60.

HÖLLDOBLER, B. AND E. O. WILSON. 1990. *The Ants.* Belknap Press of Harvard University Press. Cambridge, MA. 732 pp.

HOLLEY, J. F. 1984. Adoption, parent–chick recognition and maladaptation in the herring gull *Larus argentatus*. Z. Tierpsychol. 64:9–14.

HOLLIS, K. L. 1984. The biological function of Pavlovian conditioning: The best defense is a good offense. J. Exp. Psych. An. Behav. Proc. 10:413–425.

HOLLIS, K. L. 1990. The role of Pavlovian conditioning in territorial aggression and reproduction. In: D. A. Dewsbury, ed., *Contemporary Issues in Comparative Psychology*. Sinauer Associates, Inc. Sunderland, MA. pp. 197–219.

HOLMES, W. 1940. The colour changes and colour patterns of *Sepia officinalis* L. Proc. Zool. Soc. London 110:17–35.

HOLMES, W. G. 1991. Predator risk affects foraging behavior of pikas: Observational and experimental evidence. Anim. Behav. 42:111–119.

HOLMES, W. G. AND P. W. SHERMAN. 1982. The ontogeny of kin recognition in two species of ground squirrels. Am. Zool. 22:491–517.

HOLMES, W. G. AND P. W. SHERMAN. 1983. Kin recognition in animals. Am. Sci. 71:46–55.

HONEYCUTT, R. L. 1992. Naked mole-rats. Am. Sci. 80:43–53.

HOOGLAND, J. L. 1982. Prairie dogs avoid extreme inbreeding. Science 215:1639–1641.

HOOGLAND, J. L. AND P. W. SHERMAN. 1976. Advantages and disadvantages of Bank swallow (*Riparia riparia*) coloniality. Ecol. Mongr. 46:33–58.

HOPKINS, C. D. 1972. Sex differences in electric signaling in an electric fish. Science 176:1035–1037.

HOPKINS, C. D. 1974. Electric communication in fish. Am. Sci. 62:426–437.

HOPKINS, C. D. 1977. Electric communciation. In: T. A. Sebeok, ed., *How Animals Communicate*. Indiana University Press. Bloomington, pp. 263–289.

HOPKINS, C. D. 1980. Evolution of electric communication channels of mormyrids. Behav. Ecol. Sociobiol. 7:1–13.

HOPKINS, C. D. 1986a. Temporal structure of non-propagated electric communication signals. Brain Behav. Evol. 28:43–59.

HOPKINS, C. D. 1986b. Behavior of Moryridae. In: T. H. Bullock and W. Heiligenberg, eds., *Electroreception*. John Wiley & Sons. New York, pp. 527–576.

HOPKINS, C. D. 1988. Neuroethology of electric communication. Ann. Rev. Neurosci. 11:497–535.

HOPKINS, C. D. AND A. H. BASS. 1981. Temporal coding of species recognition signals in an electric fish. Science 212:855–857.

HORSMANN, U., H. G. HEINZEL, AND G. WENDLER. 1983. The phasic influence of self-generated air current modulations on the locust flight motor. J. Comp. Physiol. 150:427–438.

HOTTA, Y. AND S. BENZER. 1972. Mapping of behavior in *Drosophila* mosaics. Nature (London) 240:527–535.

HOUCK, L. D. 1988. The effect of body size on male courtship success in a plethodontid salamander. Anim. Behav. 36:837–842.

HOUCK, L. D. AND N. L. REAGAN. 1990. Male courtship pheromones increase female receptivity in a plethodontid salamander. Anim. Behav. 39:729–734.

HOUSTON, A. I. AND J. M. MCNAMARA. 1988. Fighting for food: A dynamic version of the hawk–dove game. Evol. Ecol. 2:51–64.

HOWARD, E. 1974. Hormonal effects on the growth and DNA content of the developing brain. In: W. Himwich, ed., *Biochemistry of the Developing Brain*, vol. 2. Marcel Dekker. New York, pp. 1–68.

HOWARD, W. E. 1960. Innate and environmental dispersal of individual vertebrates. Am. Mid. Nat. 63:152–161.

HOWLETT, R. J. AND M. E. N. MAJERUS. 1987. The understanding of industrial melanism in the peppered moth (*Biston betularia*) (Lepidoptera: Geometridae). Biol. J. Linn. Soc. 30:31–44.

HOY, R. R. 1978. Acoustic communication in crickets. A model system for the study of feature detection. Fed. Proc. 37:2316–2323.

HUCK, U. W., J. B. LABOV, AND R. D. LISK. 1986. Food-restricting young hamsters (*Mesocricetus auratus*) affects sex ratio and growth of subsequent offspring. Biol. Reprod. 35:592–598.

HUDSPETH, A. J. 1983. The hair cells of the inner ear. Sci. Am. 248 (Jan):54–54.

HUNSAKER, D. AND T. C. HAHN. 1965. Vocalization of the South American tapir, *Tapirus terrestris*. Anim. Behav. 13:69–78.

HUNTER, M. L. AND J. R. KREBS. 1979. Geographical variation in the song of the Great tit *Parus major* in relation to ecological factors. J. Anim. Ecol. 48:759–785.

HUNTINGFORD, F. A. 1986. Development of behaviour in fish. In: T. J. Pitcher, ed., *The Behavior of Teleost Fishes*. Johns Hopkins University Press. Baltimore, pp. 47–68.

HUXLEY, J. 1914. The courtship habits of the Great crested grebe *Podiceps cristatus*. Proc. Zool. Soc. London 2:491–562.

HUXLEY, J. 1923. Courtship activities in the Redthroated Diver *Colymbus stellatus pontopp*; together with a discussion on the evolution of courtship in birds. J. Linn. Soc. London 25:253–292.

IMMELMANN, K. 1963. Drought adaptations in Australian desert birds. Proc. 13th Int. Ornithol. Congr., 1962, pp. 649–657.

IMMELMANN, K. 1969. Über den Einfluss frühkinlicher Erfahrungen auf die geschlechtliche Objektfixierung bei Estrildiden. Z. Tierpsychol. 26:677–691.

IMMELMANN, K. 1972. The influence of early experience upon the development of social behavior in estrildine finches. Proc. 15th Int. Ornith. Congr., The Hague, 1970, pp. 316–338.

IMMELMANN, K. 1980. *Introduction to Ethology*. Plenum Press. New York, 230 pp.

IMMELMANN, K. AND S. J. SUOMI. 1981. Sensitive phases in development. In: K. Immelmann, G. W. Barlow, L. Petrinovich, and M. Main, eds., *Behavioral Development*. Cambridge University Press. Cambridge, pp. 395–431.

IMS, R. A. 1987a. Male spacing systems in microtine rodents. Am. Nat. 130:475–484.

IMS, R. A. 1987b. Responses in spatial organization and behaviour to manipulations of the food resource in the vole *Clethrionomys rufocanus*. J. Anim. Ecol. 56:585–596.

IMS, R. A. 1988. Spatial clumping of sexually receptive females induces space sharing among male voles. Nature 335:541–543.

INOUYE, S. T. AND H. KAWAMURA. 1979. Persistence of circadian rhythmicity in a mammalian hypothalamic "island" containing the suprachiasmatic nucleus. Proc. Nat. Acad. Sci. USA 76:5962–5966.

IOALÈ, P. 1983. Effects of anesthesia on the nasal mucosae on the homing behavior of pigeons. Z. Tierpsychol. 61:102–110.

IOALÈ, P., M. NOZZOLINI, AND F. PAPI. 1990. Homing pigeons do extract directional information from olfactory stimuli. Behav. Ecol. Sociobiol. 26:301–305.

ITO, Y. 1989. The evolutionary biology of sterile soldiers in aphids. Trends Ecol. Evol. 4:69–73.

JACKSON, F. R., T. A. BARGIELLO, S.-H. YUN, AND M. W. YOUNG. 1986. Product of *per* locus in *Drosophila* shares homology with proteoglycans. Nature 320:185–188.

JACKSON, R. R. AND R. S. WILCOX. 1990. Aggressive mimicry, prey-specific predatory behaviour and predator-recognition in the predator–prey interactions of *Portia fimbriata* and *Euryattus* sp., jumping spiders from Queensland. Behav. Ecol. Sociobiol. 26:111–119.

JAKOB, E. M. 1991. Costs and benefits of group living for pholcid spiderlings: Losing food, saving silk. Anim. Behav. 41:711–722.

JAMIESON, I. G. 1986. The functional approach to behavior: Is it useful? Am. Nat. 127:195–208.

JAMIESON, I. G. 1989. Behavioral heterochrony and the evolution of birds' helping at the nest: An unselected consequence of communal breeding? Am. Nat. 133:394–406.

JAMIESON, I. G. 1991. The unselected hypothesis for the evolution of helping behavior: Too much or too little emphasis on natural selection? Am. Nat. 138:271–282.

JAMIESON, I. G. AND J. L. CRAIG. 1987. Critique of helping behavior in birds: A departure from functional explanations. In: P. P. G. Bateson and P. Klopfer, eds., *Perspectives in Ethology*, vol. 7. Plenum. New York, pp. 79–98.

JAN, Y., L. JAN, AND M. DENNIS. 1977. Two mutations of synaptic transmission in *Drosophila*. Proc. Roy. Soc. London 198:87–108.

JANETOS, A. C. 1980. Strategies of female mate choice: A theoretical analysis. Behav. Ecol. Sociobiol. 7:107–112.

JANUS, C. 1988. The development of responses to naturally occurring odors in spiny mice *Acomys cahirinus*. Anim. Behav. 36:1400–1406.

JAROSZ, S. J., T. J. KUEHL, AND W. R. DUKELOW. 1977. Vaginal cytology, induced ovulation, and gestation in the squirrel monkey (*Saimiri sciureus*). Biol. Reprod. 16:97–103.

JARVIS, J. U. M. 1981. Eusociality in a mammal: Cooperative breeding in naked mole-rat colonies. Science 212:571–573.

JAYNES, J. 1969. The historical origins of "ethology" and "comparative psychology." Anim. Behav. 17:601–606.

JEFFREYS, A. J., V. WILSON, AND S. L. THEIN. 1985. Hypervariable 'minisatellite' regions in human DNA. Nature 314:67–73.

JEGLA, T. C. AND T. L. POULSON. 1969. Circannian rhythms. I. Reproduction in the cave crayfish *Orconectes pellucidus inermis*. Comp. Biochem. Physiol. 33:347–355.

JENNI, D. A. AND B. J. BETTS. 1978. Sex differences in nest construction, incubation, and parental behavior in the polyandrous American Jacana (*Jacana spinosa*). Anim. Behav. 26:207–218.

JENNI, D. A. AND G. COLLIER. 1972. Polyandry in the American Jacana (*Jacana spinosa*). Auk 89:743–765.

JENNINGS, H. S. 1906. *Behavior of the Lower Organisms*. Columbia University Press, New York. 366 pp.

JOELS, M. AND E. R. DE KLOET. 1989. Effects of glucocorticoids and norepinephrine on the excitability in the hippocampus. Science 245:1502–1505.

JOHNSGARD, P. A. 1967. *Animal Behavior*. William C. Brown, Dubuque.

JOHNSON, D. F. AND G. COLLIER. 1989. Patch choice and meal size of foraging rats as a function of the profitability of food. Anim. Behav. 38:285–297.

JOHNSON, D. L. AND A. M. WENNER. 1966. A relationship between conditioning and communication in honey bees. Anim. Behav. 14:261–265.

JOHNSON, P. B. AND A. D. HASLER. 1980. The use of chemical cues in the upstream migration of coho salmon, *Oncorhynchus kisutch*. Waldbaum J. Fish Biol. 17:67–73.

JOHNSTON, T. D. 1988. Developmental explanation and the ontogeny of birdsong: Nature/nurture redux. Behav. Brain Sci. 11:617–663.

JONES, F. R. H. 1955. Photo-kinesis in the ammocoete larva of the brook lamprey. J. Exp. Biol. 34:492–503.

JOSLIN, P. W. P. 1966. Summer activities of two timber wolf (*Canis lupus*) packs in Algonquin Park. MS thesis, University of Toronto. 99 pp.

JOULE, J. AND G. N. CAMERON. 1975. Species removal studies. I. Dispersal strategies of sympatric *Sigmodon hispidus* and *Reithrodontomys fulvescens* populations. J. Mamm. 56:378–396.

JUDD, W. W. 1955. Observations on the blue-tailed skink, *Eumeces fasciatus*, captured in Rondeau Park, Ontario and kept in captivity over winter. Copeia 1955:135–136.

KALAT, J. W. 1983. Evolutionary thinking in the history of the comparative psychology of learning. Neurosci. and Biobehav. Rev. 7:309–314.

KALMIJN, A. J. 1984. Theory of electromagnetic orientation: a further analysis. In: L. Bolis, R. D. Keynes, and S. H. P, Maddrell, eds. *Comparative Physiology of Sensory Systems*. Cambridge University Press. Cambridge, pp. 525–560.

KALMIJN, A. J. 1966. Electro-perception in sharks and rays. Nature (London) 212:1232–1233.

KALMIJN, A. J. 1971. The electric sense of sharks and rays. J. Exp. Biol. 55:371–383.

KALMIJN, A. J. 1978. Experimental evidence of geomagnetic orientation in elasmobranch fishes. In: K. Schmidt-Koenig and W. T. Keeton, eds., *Animal Migration, Navigation, and Homing*. Springer-Verlag. New York, pp. 347–353.

KALMUS, H. AND C. R. RIBBANDS. 1952. The origin of the odours by which honeybees distinguish their companions. Proc. Roy. Soc. B. 140:50–59.

KAMIL, A. C. 1983. Optimal foraging theory and the psychology of learning. Am. Zool. 23:291–302.

KAMIL, A. C. AND J. E. MAULDIN. 1988. A comparative-ecological approach to the study of learning. In: R. C. Bolles and M. D. Beecher, eds., *Evolution and Learning.* Lawrence Erlbaum. Hillsdale, NJ, pp. 117–133.

KAMIL, A. C. AND S. I. YOERG, 1982. Learning and foraging behavior. In: P. P. G. Bateson and P. H. Klopfer, eds., *Perspectives on Ethology*, vol. 5. Plenum. New York, pp. 325–364.

KANDEL, E. 1976. *Cellular basis of behavior. An introduction to behavioral neurobiology.* W. H. Freeman & Co., San Francisco. 727 pp.

KANKEL, D. R. AND HALL, J. C., 1976. Fate mapping of nervous system and other internal tissues in genetic mosaics of *Drosophila melanogaster.* Dev. Biol. 48:1–24.

KARAKASHIAN, M. W. AND H.-G. SCHWEIGER, 1976. Circadian properties of the rhythmic system in individual nucleated and enucleated cells of *Acetabularia mediterreanea.* Exp. Cell Res. 97:366–377.

KARAKASHIAN, S. J., M. GYGER, AND P. MARLER. 1988. Audience effects on alarm calling in chickens (*Gallus gallus*). J. Comp. Psychol. 102:129–135.

KASAL, C., M. MENAKER, AND R. PEREZ-POLO. 1979. Circadian clock in culture: N. acetyltransferase activity of chick pineal glands oscillates *in vitro.* Science 203:656.

KATZ, Y. B. 1985. Sunset and the orientation of European robins (*Erithacus rubecula*). Anim. Behav. 33.825–828.

KAWAI, M. 1965. Newly acquired pre-cultured behavior of the natural troop of Japanese monkeys on Koshima Islet. Primates 6:1–30.

KAWAMURA, S. 1959. Sub-culture propagation among Japanese macaques. Primates 2:43–60.

KAWATA, M. 1985. Mating system and reproductive success in a spring population of the red-backed vole, *Clethrionomys rufocanus* bedfordiae. Oikos 45:181–190.

KEENLEYSIDE, M. H. A. 1978. Parental care behavior in fishes and birds. In: E. S. Reese and F. Lighter, eds., *Contrasts in Behavior.* John Wiley & Sons. New York, pp. 1–19.

KEENLEYSIDE, M. H. A. 1979. Diversity and adaptation in fish behavior. Springer-Verlag. New York. 209 pp.

KEETON, W. T. 1971. Magnets interfere with pigeon homing. Proc. Nat. Acad. Sci. 68:102–106.

KEETON, W. T., M. L. KREITHEN, AND K. L. HERMAYER. 1977. Orientation by pigeons deprived of olfaction by nasal tubes. J. Comp. Physiol. 114:289–299.

KEETON, W. T., T. S. LARKIN, AND D. M. WINDSOR. 1974. Normal fluctuations in the earth's magnetic field influence pigeon orientation. J. Comp. Physiol. 95:95–103.

KELLER, F. S. 1941. Light aversion in the white rat. Psychol. Rec. 4:235–250.

KELLEY, D. B. 1988. Sexually dimorphic behaviors. Ann. Rev. Neurosci. 11:225–251.

KELLEY, D. B. AND D. L. GORLICK. 1990. Sexual selection and the nervous system. BioScience 40(Apr):275–283.

KEMP, A. C. 1971. Some observations on the sealed-in nesting method of hornbills (Family: Bucerotidae). Ostrich (Suppl.) 8:149-155.

KEMP, A. C. 1976. A study of the ecology, behaviour, and systematics of *Tockus* hornbills (Aves: Bucerotidae). Transv. Mus. Mem. no. 20, 125 pp.

KERR, W. E., R. ZUCCHI, J. T. NAKAKAIRA, AND J. E. BUTOLO. 1962. Reproduction in the social bees. J. NY Entomol. Soc. 70:265–270.

KESSEL, E. L. 1955. Mating activities of balloon flies. Syst. Zool. 4:97–104.

KETTLEWELL, H. B. D. 1955. Selection experiments on industrial melanism in the Lepidoptera. Heredity 9:323–343.

KETTLEWELL, H. B. D. 1956. Further selection experiments on industrial melanism in the Lepidoptera. Heredity 10:287–301.

KHANNA, S. M. AND D. G. B. LEONARD. 1982. Basilar membrane tuning in the cat cochlea. Science 215:305–306.

KIEPENHEUER, J. 1978. A repetition of the deflector loft experiment. Behav. Ecol. Sociobiol. 3:393–395.

KIEPENHEUER, J. 1979. Pigeon homing: Deprivation of olfactory information does not affect the deflector effect. Behav. Ecol. Sociobiol. 6:11–22.

KIEPENHEUER, J. 1982. The effect of magnetic anomalies on the homing behavior of pigeons. In: F. Papi and H. G. Walraff, eds., *Avian Navigation*. Springer-Verlag. Berlin, pp. 120–128.

KIEPENHEUER, J. 1986. Are site-specific airborne stimuli relevant for pigeon navigation only when matched by other release-site information? Naturwiss. 73:42–43.

KILTIE, R. A. 1988. Countershading: Universally deceptive or deceptively universal? Trends Ecol. and Evol. 3:21–23.

KILTIE, R. A. 1989. Wildfire and the evolution of dorsal melanism in fox squirrels, *Sciurus niger*. J. Mamm. 70:726–739.

KIMBLE, G. A. 1961. *Hilgard and Marguis' Conditioning and Learning*. Appleton-Century-Crofts. New York.

KING, A. P. AND M. J. WEST. 1983. Epigenesis of cowbird song—a joint endeavour of males and females. Nature 305:704–706.

KINNE, O. 1975. Migratory cycles. In: O. Kinne, ed., *Marine Ecology*, vol. 2, part 2 John Wiley & Sons. New York, pp. 829–844.

KIRKPATRICK, M. 1982. Sexual selection and the evolution of female choice. Evolution 36:1–12.

KIRKPATRICK, M. AND C. D. JENKINS. 1989. Genetic segregation and the maintenance of sexual reproduction. Nature 339:300–301.

KIRSCHVINK, J. L. AND J. L. GOULD. 1981. Biogenic magnetite as a basis for magnetic field detection in animals. BioSystems 13:181–201.

KIRSCHVINK, J. L., A. E. DIZON, AND J. A WESTPHAL. 1986. Evidence from strandings for geomagnetic sensitivity in cetaceans. J. Exp. Biol. 120:1–24.

KIRSCHVINK, J. L., M. M. WALKER, S.–B. CHANGE, A. E. DIZON, AND K. A. PETERSON. 1985. Chains of single–domain magnetite particles in chinook salmon, *Oncorhynchus tshawytscha*. J. Comp. Physiol. A 157:375–381.

KLAPOW, L. A. 1972. Natural and artificial rephasing of a tidal rhythm. J. Comp. Physiol. 79:233–258.

KLEIMAN, D. G. 1977. Monogamy in mammals. Quart. Rev. Biol. 52:39–69.

KLEIN, M. AND E. F. KANDEL. 1978. Presynaptic modification of voltage dependent Ca^{2+} current: Mechanism for behavioral sensitization in *Aplysia californica*. Proc. Nat. Acad. Sci. USA. 75:3512–3516.

KLINOWSKA, M. 1985. Cetacean live stranding sites relate to the geomagnetic topography. Aquatic Mammals 11:27–32.

KLINOWSKA, M. 1986. Cetacean live stranding sites relate to geomagnetic disturbances. Aquatic Mammals 11:109–119.

KLOPFER, P. 1988. Metaphors for development: How important are experiences early in life? Dev. Psychobiol. 21:671–678.

KLOPFER, P. H. 1971. Mother love: What turns it on? Am. Sci. 59:404–407.

KLUG, W. S. AND M. R. CUMMINGS. 1983. *Concepts of Genetics*. Charles E. Merrill Publishing Company, Columbus, OH. 614 pp.

KNOWLTON, N. 1979. Reproductive synchrony, parental investment, and the evolutionary dynamics of sexual selection. Anim. Behav. 27:1022–1033.

KNUDSEN, E. I. 1981. The hearing of the barn owl. Sci. Am. 245 (Dec):113–125.

KNUDSEN, E. I. 1982. Auditory and visual maps of space in the optic tectum of the owl. J. Neurosci. 2:1177–1194.

KNUDSEN, E. I. AND M. KONISHI. 1978. A neural

map of auditory space in the owl. Science 200:795–797.

KOENIG, W. D. AND P. B. STACEY. 1990. Acorn woodpeckers: Group-living and food storage under contrasting ecological conditions. In: P. B. Stacey and W. D. Koenig, eds., *Cooperative Breeding in Birds*. Cambridge University Press. Cambridge, pp. 415–453.

KÖHLER, W. 1927. *The Mentality of Apes*. Harcourt, Brace, New York. 336 pp.

KOLATA, G. 1984. Studying leaning in the womb. Science 225:302–303.

KOLTERMANN, R. 1971. 24-std-Periodik in der Langzeiterinneryng an Duft-und Farbsignale ber der Honigbiene. Z. vergl. Physiol. 75:49–68.

KONISHI, M. 1965. The role of auditory feedback in the control of vocalization in the white-crowned sparrow. Z. Tierpsychol. 22:770–783.

KONISHI, M. 1973. Locatable and non-locatable acoustic signals for barn owls. Am. Nat. 107:775–785.

KONISHI, M. 1985. Birdsong: From behavior to neuron. Ann. Rev. Neurosci. 8:125–170.

KONOPKA, R. J. AND S. BENZER. 1971. Clock mutants of *Drosophila melanogaster*. Proc. Nat. Acad. Sci. USA 68:2112–2116.

KONOPKA, R. J. AND S. WELLS. 1980. *Drosophila* clock mutations affect the morphology of a brain neurosecretory cell group. J. Neurobiol. 11:411–415.

KOOB, G., M. LeMOAL, AND F. E. BLOOM. 1984. The role of endorphins in neurobiology, behavior, and psychiatric disorders. In: C. B. Nemeroff and A. J. Dunn, eds., *Peptides, Hormones, and Behavior*. Spectrum Publications. New York, pp. 349–383.

KOWALSKI, U., R. WILTSCHKO, AND E. FULLER. 1988. Normal fluctuation of the geomagnetic field may affect initial orientation in pigeons. J. Comp. Physiol. 163:593–600.

KRAMER, B. 1990. Sexual signals in electric fishes. Trends Ecol. Evol. 5:247–250.

KRAMER, D. L. AND D. M. WEARY, 1991. Exploration versus exploitation: A field study of time allocation to environmental tracking by foraging chipmunks. Anim. Behav. 41:443–449.

KRAMER, G. 1949. Über Richtungstendenzen bei der nachtlichen Zugenruhe gekafigter Vögel. In: E. Mayr and E. Shuz, eds., *Ornithologie als biologische Wissenschaft*. Carl Winter. Heidelberg, pp. 269–283.

KRAMER, G. 1950. Weitere Analyse der Faktoren, welche die Zugaktivität des gekäfigten Vogels orientieren. Naturwiss. 37:377–378.

KRAMER, G. 1951. Eine neue Methode zur Erforschung der Zugorientierung und die bisher damit erzielten Ergebnisse. Proc. 10th Int. Ornithol. Congr., pp. 269–280.

KRAUSE, K., G. CREMER-BARTELS, AND G. MITOSKAS. 1985. Effects of low magnetic field on human and avian retina. In: G. M. Brown and S. D. Wainwright, eds., *The Pineal Gland: Endocrine Aspects. Advances in the Biosciences* vol. 53. Pergamon Press, New York. pp. 209–215.

KREBS, J. 1985. Sociobiology ten years on. New Sci. Vol. 1476:40–43.

KREBS, J. R. 1978. Optimal foraging: Decision rules for predators. In: J. R. Krebs and N. B. Davies, eds., *Behavioural Ecology, An Evolutionary Approach*. Sinauer Associates, Inc. Sunderland, MA.

KREBS, J. R. AND M. I. AVERY. 1984. Chick growth and prey quality in the European bee-eater (*Merops apiaster*). Oecologia 64:363–368.

KREBS, J. R. AND N. B. DAVIES. 1978. *An Introduction to Behavioural Ecology*. Sinauer Associates Inc. Sunderland, MA. 292 pp.

KREBS, J. R. AND R. DAWKINS. 1984. Animal signals: Mind-reading and manipulation. In: *Behavioural Ecology—An Evolutionary Approach*. Sinauer Associates. Sunderland, MA, pp. 380–402.

KREBS J. R., J. T. ERICHSEN, M. I. WEBBER, AND E. L. CHARNOV. 1977. Optimal prey selection in the Great Tit (*Parus major*). Anim. Behav. 25:30–38.

KREITHEN, M. L. AND D. B. QUINE. 1979. Infrasound detection by the homing pigeon: A behavioral audiogram. J. Comp. Physiol. 129:1–4.

KREITHEN, M. L. AND W. T. KEETON. 1974. Detection of polarized light by the homing pigeon, *Columba livia*. J. Comp. Psychol. 89:83–92.

KREULEN, D. 1975. Wildebeest habitat selection on the Serengeti plains, Tanzania, in relation to calcium and lactation: A preliminary report. East Afr. Wildl. J. 13:297–304.

KROODSMA, D. E. 1974. Song learning, dialects, and dispersal in Bewick's wren. Z. Tierpsychol. 66:189–226.

KROODSMA, D. E. AND B. E. BYERS. 1991. The function(s) of bird song. Am. Zool. 31:318–328.

KROODSMA, D. E. AND R. PICKERT. 1980. Environmentally dependent sensitive periods for avian vocal learning. Nature 288:477–479.

KROODSMA, D. E. AND M. KONISHI. 1991. A suboseine bird (eastern phoebè, *Sayornis phoebe*) develops normal song without auditory feedback. Anim. Behav. 42:477–487.

KUMMER, H., W. GOTZ, AND W. ANGST. 1974. Triadic differentiation: An inhibitory process protecting pair bonds in baboons. Behaviour 49:62–87.

KUNG, C., S.-Y. CHANG, Y. SATOW, J. VAN HOUTEN, AND H. HANSMA. 1975. Genetic dissection of behavior in *Paramecium*. Science 188:898–904.

KUTTERBACH, D. A., C. WALCOTT, R. J. REEDER, AND R. B. FRANKEL. 1982. Iron-containing cells in the honey bee (*Apis mellifera*). Science 218:695–697.

KYRIACOU, C. P. AND J. C. HALL. 1980. Circadian rhythm mutations in *Drosophila melanogaster* affect short term fluctuations in the male's courtship song. Proc. Nat. Acad. Sci. USA 77:6729–6733.

LABOV, J. B. 1981. Pregnancy blocking in rodents: Adaptive advantages for females. Am. Nat. 118:361–371.

LACEY, E. A. AND P. W. SHERMAN. 1991. Social organization of naked mole-rat colonies: Evidence of divisions of labor. In: P. W. Sherman, J. U. M. Jarvis, and R. D. Alexander, eds., *The Biology of the Naked Mole-Rat*. Princeton University Press. Princeton, NJ, pp. 275–336.

LACK, D. 1939. The display of the black cock. Brit. Birds 32:290-303.

LACK, D. 1943. *The Life of the Robin*. Penguin Books, London. 239 pp.

LACK, D. 1968. Bird migration and natural selection. Oikos. 19:1–9.

LACK, D. 1968. Ecological adaptations for breeding in birds. Methuen. London. 409 pp.

LACY, R. C. AND P. W. SHERMAN. 1983. Kin recognition by phenotype matching. Am. Nat. 121:489–512.

LAIDLAW, H. H., JR. AND R. E. PAGE, JR. 1984. Polyandry in honey bees (*Apis mellifera* L.): Sperm utilization and intracolony genetic relationships. Genetics 108:985–997.

LALL, A. B., H. H. SELIGER, W. H. BIGGLEY, AND J. E. LLOYD. 1980. The ecology of colors of firefly bioluminescence. Science 210:560–562.

LAMBERT, S. AND G. M. FERGUSON. 1985. Blood ejection frequency by *Phrynosoma cornutum* (Iguanidae). Southwestern Nat. 30:616–617.

LANDE, R. 1981. Models of speciation by sexual selection on polygenic traits. Proc. Nat. Acad. Sci. USA 78:3721–3725.

LANG, H. J. 1964. Über lunarperiodische Schwankungen der Farbemfindlichkeit beim Guppy (*Lebistes reticulatus*). Verh. dt. zool. Ges. 58:379–386.

LANG, H. J. 1967. Über das Lichtrückenverhalten des Guppy (*Lebistes reticulatus*) in farbigen und farblosen Lichtern. Z. vergl. Physiol. 56:296–340.

LANYON, W. E. 1963. Experiments on species discrimination in (*Myiarchus*) flycatchers. Am. Mus. Novit. 2126:1–16.

LARKIN, T. AND W. T. KEETON. 1978. An apparent lunar rhythm in the day-to-day variations in initial bearings of homing pigeons. In: K. Schmidt-Koenig and W. T. Keeton, eds., *Animal Migration, Navigation, and Homing*. Springer-Verlag. New York, pp. 92–106.

LASHLEY, K. 1950. In search of the engram. Soc. Exp. Biol. Symp. IV:454–482.

LAWRENCE, P. A. 1981. A general cell marker for clonal analysis of *Drosophila* development. J. Embryo. Exper. Morphol. 64:321–332.

LEAKE, L. D. 1986. Leech Retzius cells and 5-hydroxytryptamine. Comp. Biochem. Physiol. 83C:229–239.

LEASK, M. J. M. 1977a. A physico-chemical mechanism for magnetic field detection by migratory birds and homing pigeons. Nature (London) 267:144–146.

LEASK, M. J. M. 1977b. Primitive models of magnetoreception. In: K. Schmidt-Koenig and W. T. Keeton eds., *Animal Migration, Navigation and Homing*. Springer-Verlag. New York. pp. 318–322.

LEBOEUF, B. J. 1967. Interindividual associations in dogs. Behaviour 29:268–295.

LEBOEUF, B. J. 1974. Male–male competition and reproductive success in elephant seals. Am. Zool. 14:163–176.

LEDNOR, A. J. 1982. Magnetic navigation in pigeons: Possibilities and problems. In: F. Papi and H. G. Walraff, eds., *Avian Navigation*. Springer-Verlag. Berlin, pp. 109–119.

LEE, S. AND L. M. BOOT. 1956. Spontaneous pseudopregnancy in mice. Acta Physiol. Pharmacol. Neer. 5:213–215.

LEHMAN, M. N., R. SILVER, W. R. GLADSTONE, R. M. KAHN, M. GIBSON, AND E. BITTMAN. 1987. Circadian rhythmicity restored by neural transplant. Immunocytochemical characterization

of the graft and its integration with the host brain. J. Neurosci. 7:1626–1639.

LEHRMAN, D. S. 1953. A critique of Konrad Lorenz's theory of instinctive behavior. Quart. Rev. Biol. 28:337–363.

LEHRMAN, D. S. 1970. Semantic and conceptual issues in the nature–nurture problem. In: L. R. Aronson, E. Tobach, D. S. Lehrman, and J. S. Rosenblatt, eds., *Development and Evolution of Behavior*; essays in honor of T. C. Schneirla. W. H. Freeman and Company, San Francisco, pp. 17–52.

LEIBRECHT, B. C. AND H. R. ASKEW. 1980. Habituation from a comparative perspective. In: M. R. Denny, ed., *Comparative Psychology: An Evolutionary Analysis of Animal Behavior*. John Wiley and Sons. New York, pp. 208–229.

LEIGHTON, M. 1986. Hornbill social dispersion: Variations on a monogamous theme. In: D. I. Rubenstein and R. W. Wrangham, eds., *Ecological Aspects of Social Evolution*. Princeton University Press. Princeton, NJ, pp. 108–130.

LEMON, R. E. 1966. Geographic variation in the song of cardinals. Can. J. Zool. 44:413–428.

LENNEBERG, E. H. 1967. *Biological Foundations of Language*. John Wiley & Sons. New York. 489 pp.

LENT, C. M. AND W. H. WATSON III. 1989. Introduction to the symposium: Behavioral neuromodulators: Cellular, comparative and evolutionary patterns. Am. Zool. 29:1211–1212.

LENT, C. M., M. H. DICKINSON, AND C. G. MARSHALL. 1989. Serotonin and leech feeding behavior: Obligatory neuromodulation. Am. Zool. 29:1241–1254.

LEUCHT, T. 1984. Responses to light under varying magnetic conditions in the honey bee, *Apis mellifica*. J. Comp. Physiol. A 154:865–870.

LEWIS, E. R. AND P. M. NARINS, 1985. Do frogs communicate with seismic signals? Science 227:187–189.

LIBERG, O. AND T. VON SCHANTZ. 1985. Sex-biased

philopatry and dispersal in birds and mammals: The Oedipus hypothesis. Am. Nat. 126:129–135.

LICHT, P. 1984. Reptiles. In: G. E. Lamming, ed., *Marshall's Physiology of Reproduction*; vol. 1, Reproductive cycles of vertebrates. Churchill Livingstone. Edinburgh, pp. 206–282.

LICKLITER, R. AND G. GOTTLIEB. 1988. Social specificity: Interaction with own species is necessary to foster species-specific maternal preferences in ducklings. Dev. Psychobiol. 21:311–321.

LIGON, J. D. AND P. B. STACEY. 1991. The origin and maintenance of helping behavior in birds. Am. Nat. 133: 254–258.

LIN, N. AND C. D. MICHENER. 1972. Evolution of sociality in insects. Quart. Rev. Biol. 47:131–159.

LINCOLN, F. C. 1950. Migration of birds. U.S. Fish and Wild. Serv. Washington D.C. Circ. 16:1–102.

LINDAUER, M. AND H. MARTIN. 1968. Die Schwereorientierung der Bienen unter dem Einfluss des Erdmagnetfelds. Z. vergl Physiol. 60:219–243.

LINDAUER, M. AND H. MARTIN. 1972. Magnetic effects on dancing bees. In: S. R. Galler, K. Schmidt-Koenig, G. J. Jacobs, and R. E. Belleville, eds., *Animal Migration and Navigation*. NASA Spec. Publ. NASA SP-262, pp 559–567.

LINDEN, E. 1992. A curious kinship: Apes and humans. National Geographic 181(3):2–45.

LINDSTROM, A., D. HASSELQUIST, S. BENSCH, AND M. GRAHN, 1990. Asymmetric contests over resources for survival and migration: A field experiment with bluethroats. Anim. Behav. 40:453–461.

LISSMANN, H. W. 1963. Electric location by fishes. Sci. Am. 208 (Mar):50–59.

LITTE, M. 1977. Behavioral ecology of the social wasp, *Mischocyttarus mexicanus*. Behav. Ecol. Sociobiol. 2:229–246.

LITTLEJOHN, M. J. 1977. Long-range acoustic communication in anurans: An integrated evolutionary approach. In: D. H. Taylor and S. I. Guttman, eds., *The Reproductive Biology of Amphibians*. Plenum Press. New York, pp. 263–294.

LITTLEJOHN, M. J. AND J. J. LOFTUS-HILLS. 1968. An experimental evaluation of premating isolation in the *Hyla ewingi* complex (Anura: Hylidae). Evolution 22:659–663.

LIU, X., L. LORENZ, Q. YU, J. C. HALL, AND M. ROSBASH. 1988. Spatial and temporal expression of the *period* gene in *Drosophila melanogaster*. Genes and Development 2:228–238.

LIVINGSTONE, M. S. 1981. Two mutations in *Drosophila* affect synthesis of octopamine, dopamine, and serotonin by altering the activities of two different amino acid carboxylases. Neurosci. Abst. 7:351.

LLOYD, J. E. 1968. Illumination, another function of firefly flashes? Entomological News 79:265–268.

LOHMANN, K. J. 1991. Magnetic orientation by hatchling loggerhead sea turtles. J. Exp. Biol. 155:37–49.

LOHMANN, K. J. 1992. How sea turtles navigate. Sci. Am. 266 (Jan):100–106.

LOHMANN, K. J. AND A. O. D. WILLOWS. 1987. Lunar-modulated geomagnetic orientation by a marine mollusk. Science 235:331–334.

LORENZ, K. 1935. Der Kumpan in der Umwelt des Vogels. J. Ornithol. 83:137–413.

LORENZ, K. 1937. Über die Bildung des Instinkbegriffes. Naturwiss. 25:289–331.

LORENZ, K. 1950. The comparative method in studying innate behavior patterns. In: Physiological mechanisms in animal behavior. Symp. Soc. Exp. Biol. IV:221–268. Cambridge University Press, Cambridge.

LORENZ, K. 1952. *King Solomon's Ring*. Thomas Y. Crowell Co. New York.

LORENZ, K. 1958. The evolution of behavior. Sci. Am. 199 (Dec):67–78.

LORENZ, K. 1972. Comparative studies on the behavior of Anatinae. In: P. H. Klopfer and J. P. Hailman, eds., *Function and Evolution of Behavior: An Historical Sample from the Pens of Ethologists*. Addison-Wesley Publishing Co. Reading, MA, pp. 231–258.

LORENZ, K. 1981. *The Foundations of Ethology.* Springer-Verlag, New York. 380 pp.

LORENZ, K. AND N. TINBERGEN. 1938. Taxis and Instinkhandlung in der Eirollbewegung der Graugans. Z. Tierpsychol. 2:1–29.

LOVEGROVE, B. G. 1991. The evolution of eusociality in molerats (Bathyergidae): A question of risks, numbers and costs. Behav. Ecol. Sociobiol. 28:37–45.

LOVEGROVE, B. G. AND C. WISSEL. 1988. Sociality in mole-rats: Metabolic scaling and the role of risk sensitivity. Oecologia (Berlin) 74:600–606.

LOW, B. S. 1990. Marriage systems and pathogen stress in human societies. Am. Zool. 30:325–339.

LUCIA, C. M. AND D. R. OSBORNE. 1983. Sunset as an orientation cue in white-throated sparrows. Ohio J. Sci. 83:185–188.

MACNAIR, M. R. AND G. A. PARKER. 1978. Models of parent-offspring conflict. II. Promiscuity. Anim. Behav. 26:111–122.

MACNAIR, M. R. AND G. A. PARKER. 1979. Models of parent–offspring conflict. III. Intrabrood conflict. Anim. Behav. 27:1202–1209.

MACPHAIL, E. M. 1987. The comparative psychology of intelligence. Behav. Brain Sci. 10:645–695.

MACROBERTS, M. H. AND B. R. MACROBERTS. 1976. Social organization and behavior of the Acorn Woodpecker in central coastal California. Ornith. Mongr. 21:1–115.

MADDOCK, L. 1979. The "migration" and grazing succession. In: A. R. E. Sinclair and M. Norton-Griffiths, eds., *Serengeti: Dynamics of an Ecosystem*. University of Chicago Press. Chicago, pp. 104–129.

MADISON, D. M., A. T. SCHOLZ, J. C. COOPER, R. M. HORRAL, AND A. D. HASLER. 1973. Olfactory hypotheses and salmon migration: A synopsis of recent findings. Fish. Res. Bd. Can. Tech. Report 414. 35 pp.

MAETERLINCK, M. 1901. *The Life of the Bee*. Dodd Mead. New York. 427 pp.

MALCOLM, S. B. 1987. Monarch butterfly migration in North America: Controversy and conservation. Trends Ecol. Evol. 2:135–139.

MALLOT, R. W. AND J. W. SIDDALL. 1972. Acquisition of the people concept in pigeons. Psychol. Rep. 31:3–13.

MANNING, A. 1979. *An Introduction to Animal Behavior*, 3rd ed. Addison-Wesley Publishing Co. Reading, MA. 329 pp.

MANZURE, M. AND E. FUENTES. 1979. Polygyny and agonistic behavior in the tree-dwelling lizard, *Liolaemus tenuis*. Behav. Ecol. Sociobiol. 6:23–28.

MARKL, H. 1985. Manipulation, modulation, information, cognition: Some of the riddles of communication. In: B. Hölldobler and M. Lindauer, eds., *Experimental Behavioral Ecology and Sociobiology*. Sinauer Asscociates, Inc. Sunderland, MA. pp. 163–194.

MARKOW, T. A., M. QUAID, AND S. KERR. 1978. Male mating experience and competitive courtship success in *Drosophila melanogaster*. Nature 276:821–822.

MARLER, P. 1952. Variation in the song of the chaffinch (*Fringilla coelebs*). Ibis 94:458–472.

MARLER, P. 1955. Characteristics of some animal calls. Nature 176:6–7.

MARLER, P. 1957. Specific distinctiveness in the communication signals of birds. Behaviour 11:13–39.

MARLER, P. 1959. Developments in the study of animal communication. In: P. R. Bell, ed., *Darwin's Biological Work*. Cambridge University Press. New York. pp. 150–206.

MARLER, P. 1965. Communication in monkeys and apes. In: I. DeVore, ed., *Primate Behavior: Field Studies of Monkeys and Apes*. Holt, Rinehart and Winston. New York. 654 pp.

MARLER, P. 1967. Animal communication signals. Science 157:769–774.

MARLER, P. 1970. A comparative approach to vocal learning: Song development in white-crowned sparrows. J. Comp. Physiol. Psychol. Monogr. 71:1–25.

MARLER, P. 1973. A comparison of vocalizations of red-tailed monkeys and blue monkeys *Cercopithecus ascanius* and *C. mitis*, in Uganda. Z. Tierpsychol. 33:223–247.

MARLER, P. 1987. Sensitive periods and the roles of specific and general sensory stimulation in birdsong learning. In: J. R. Rauschecker and P. Marler, eds., *Imprinting and Cortical plasticity: Comparative Aspects of Sensitive Periods*. John Wiley and Sons, Inc., New York, pp. 99–135.

MARLER, P. AND M. TAMURA. 1962. Song "dialects" in three populations of white-crowned sparrows. Condor 64:368–377.

MARLER, P. AND M. TAMURA. 1964. Culturally transmitted patterns of vocal behavior in sparrows. Science 146:1483–1486.

MARLER, P. AND W. J. HAMILTON III. 1966. *Mechanisms of Animal Behavior*. John Wiley & Sons. New York. 771 pp.

MARLER, P., S. KARAKASHIAN, AND M. GYGER. 1991. Do animals have the option of withholding signals when communication is inappropriate? The audience effect. In: C. A. Ristau, ed., *Cognitive Ethology, the Minds of Other Animals*. Lawrence Erlbaum. Hillsdale, NJ, pp. 187–208.

MARTAN, J. AND B. A. SHEPHERD. 1976. The role of the copulatory plug in reproduction of the guinea pig. J. Exp. Zool. 196:79–84.

MARTEN, K. AND P. MARLER. 1977. Sound transmission and its significance for animal vocalization. I. Temperate habitats. Behav. Ecol. Sociobiol. 2:271–290.

MARTEN, K., D. QUINE, AND P. MARLER. 1977. Sound transmission and its significance for animal vocalization. II. Tropical forest habitats. Behav. Ecol. Sociobiol. 2:291–302.

MARTIN, G. R. AND W. R. A. MUNTZ. 1978. Spectral sensitivity of the red and yellow oil droplet fields of the pigeon (*Columba livia*). Nature 274:620–621.

MARTIN H. AND M. LINDAUER. 1977. Der Einfluss der Erdmagnetfelds und Schwereorientierung der Honigbiene, J. Comp. Physiol. 122:145–187.

MARTIN, P. AND T. M. CARO. 1985. On the function of play and its role in behavioral development. Adv. Study Behav. 15:59–103.

MARX, J. 1980. Ape-language controversy flares up. Science 207:1330–1333.

MASCHWITZ, V. 1966. Alarm substances and alarm behavior in social insects. Vitamins and Hormones 24:267–290.

MASKELL, M. D., T. PARKIN, AND E. VERSPOOR. 1977. Apostatic selection by sticklebacks upon a dimorphic prey. Heredity, London 39:83–89.

MASSEY, A. 1988. Sexual interactions in red-spotted newt populations. Anim. Behav. 36:205–210.

MAST, S. O. 1938. Factors involved in the process of orientation of lower organisms in light. Biol. Rev. 13:186–224.

MATHER, J. G. AND R. R. BAKER. 1981. Magnetic sense of direction in woodmice for route-based navigation. Nature 291:152–155.

MATTHEWS, G. V. T. 1951. An experimental investigation of navigation in homing pigeons. J. Exp. Biol. 28:508–536.

MATTHEWS, L. H. 1971. *The Life of Mammals*, vol. 2. Universe Books. New York.

MAXSON, S. J. AND L. W. ORING. 1980. Breeding season time and energy budgets of the polyandrous spotted sandpiper. Behaviour 74:200–263.

MAYNARD SMITH, J. 1965. The evolution of alarm calls. Am. Nat. 99:59–63.

MAYNARD SMITH, J. 1971. What use is sex? J. Theor. Biol. 30:319–335.

MAYNARD SMITH, J. 1972. *On Evolution*. Edinburgh University Press. Edinburgh. 125 pp.

MAYNARD SMITH, J. 1974. The theory of games and the evolution of animal conflicts. J. Theoret. Biol. 47:209–221.

MAYNARD SMITH, J. 1976. A short-term advantage for sex and recombination through sib-competition. J. Theoret. Biol. 63:245–258.

MAYNARD SMITH, J. 1976. Evolution and the theory of games. Am. Sci. 64:41–45.

MAYNARD SMITH, J. 1977. Parental investment: A prospective analysis. Anim. Behav. 25:1–9.

MAYNARD SMITH, J. 1979. Game theory and the evolution of behavior. Proc. Roy. Soc. London B. 205:475–488.

MAYNARD SMITH, J. 1982. *Evolution and the Theory of Games*. Cambridge University Press. Cambridge. 224 pp.

MAYNARD SMITH, J. 1985. Sexual selection, handicaps and true fitness. J. Theoret. Biol. 115:1–8.

MAYNARD SMITH, J. 1991. Theories of sexual selection. Trends Ecol. Evol. 6:146–151.

MAYNARD SMITH, J. AND G. A. PARKER. 1976. The logic of asymmetric contests. Anim. Behav. 24:159–175.

MAYNARD SMITH, J. AND G. R. PRICE. 1973. The logic of animal conflict. Nature 246:15–18.

MAYR, E. 1983. How to carry out the adaptationist program? Am. Nat. 121:324–334.

MCCANN, T. S. 1981. Aggression and sexual activity of male southern elephant seals. J. Zool. London 195:295–310.

MCCLINTOCK, M. K. 1983. Pheromonal regulation of the ovarian cycle: Enhancement, suppression, and synchrony. In: J. G. Vandenbergh, ed., *Pheromones and Reproduction in Mammals*. Academic Press. New York, pp. 113–149.

MCCRACKEN, G. F. 1984. Communal nursing in Mexican free-tailed bat maternity colonies. Science 223:1090–1091.

MCEWEN, B. S. 1976. Interactions between hormones and nerve tissue. Sci. Am. 235 (July):48–58.

MCFALL-NGAI. M. J. 1990. Crypsis in the pelagic environment. Am. Zool. 30:175–188.

MCFARLAND, D. (ED.). 1981. *The Oxford Companion to Animal Behavior*. Oxford University Press.

MCGOWAN, K. J. AND G. E. WOOLFENDEN. 1989. A sentinel system in the Florida scrub jay. Anim. Behav. 37:1000–1006.

MCGUIRE, B. AND M. NOVAK. 1984. A comparison of maternal behavior in the meadow vole (*Microtus pennsylvanicus*), prairie vole (*M. ochrogaster*), and pine vole (*M. pinetorum*). Anim. Behav. 32:1132–1141.

MCGUIRE, B. 1988. The effects of cross-fostering on parental behavior of meadow voles (*Microtus pennsylvanicus*). J. Mamm. 69:332–341.

MCKAYE, K. R. 1977. Competition for breeding sites between the cichlid fishes of Lake Jiloa, Nicaragua. Ecology 58:291–302.

MCKAYE, K. R. 1981. Natural selection and the evolution of interspecific brood care in fishes. In: R. D. Alexander and D. W. Tinkle, eds., *Natural Selection and Social Behavior*. Chiron Press. New York, pp. 173–183.

MCKAYE, K. R. AND N. M. MCKAYE. 1977. Communal care and kidnapping of young by parental cichlids. Evolution 31:674–681.

MCLAREN, A. AND D. MICHIE. 1960. Control of prenatal growth in mammals. Nature 187:363–365.

MCMILLAN, J. P., H. C. KEATTS, AND M. MENAKER. 1975. On the role of the eyes and brain photoreceptors in the sparrow: Entrainment to light cycles. J. Comp. Physiol. 102:251–256.

MCNAMARA, J. M., S. MERMAD, AND A. HOUSTON.

1991. A model for risk-sensitive foraging for a reproducing animal. Anim. Behav. 41:787–792.

McShea, W. J. 1989. Reproductive synchrony and home range size in a territorial microtine. Oikos 56:182–186.

Mech, L. D. 1966. *The Wolf: The ecology and behavior of an endangered species*. Natural History Press, Garden City, N.Y. 384 pp.

Mech, L. D. 1977. Population trend and winter deer consumption in a Minnesota wolf pack. In: J. Phillips and C. Jonkel, eds., *Proc. 1975 Predator Symposium*. University of Montana Press. Missoula, pp. 55–83.

Mellgren, R. L. (ed.). 1983. *Animal Cognition and Behavior*. North Holland Publishing Co. Amsterdam. 513 pp.

Meltzoff, A. N. 1988 The human infant as *Homo Imitans*. In: T. R. Zentall, B. G. Galef Jr., eds. *Social Learning: Psychological and Biological Perspectives*. Lawrence Erlbaum. Hillsdale, N.J. pp. 319–341.

Menaker, M. 1968. Light reception by the extraretinal receptors in the brain of the sparrow. Proc. 76th Ann. Conv. Am. Psychol. Assoc. pp. 299–300.

Mendoza, S. P. and W. A. Mason. 1989. Behavioral and endocrine consequences of heterosexual pair formation in squirrel monkeys. Physiol. Behav. 46:597–603.

Mendoza, S. P., C. L. Coe, E. L. Lowe, and S. Levine. 1979. The physiological response to group formation in adult male squirrel monkeys. Psychoneuroendocrinol. 3:221–229.

Menzel, R. 1989. Bee-havior and the neural systems and behavior course. In: T. J. Carew and D. B. Kelley, eds., *Perspectives in Neural Systems and Behavior*, MBL Lectures in Biology, vol. 10. Alan Liss, Inc. New York, pp. 249–266.

Menzel, R., L. Chittka, S. Eichmüller, S. Geiger, K. Pietsch, and P. Knoll. 1990. Dominance of celestial cues over landmarks disproves map-like orientation in honey bees. Z. Naturforsch. 45c:723–726.

Mergenhagen, D. 1984. Genetic characterization of a *Chlamydomonas* clock mutant. Eur. J. Cell Biol. 33:13–18.

Mergenhagen, D. and H.-G. Schweiger. 1973. Recording the oxygen production of a single *Acetabularia* cell for a prolonged period. Exp. Cell Res. 81:360–364.

Mergenhagen, D. and H.-G. Schweiger. 1975. Circadian rhythm of oxygen evolution in cell fragments of *Acetabularia*. Exp. Cell Res. 92:127–130.

Merkel, F. W. 1971. Orientation behavior of birds in Kramer cages under different physical cues. Ann. N.Y. Acad. Sci. 188:283–294.

Merkel, F. W. and H. G. Fromme. 1958. Untersuchungen über das Orientierungsvermögen nachtlich ziehender Rotkehlchen (*Erithacus rubecula*). Naturwiss. 45:499–500.

Merkel, F. W. and W. Wiltschko. 1965. Magnetismus und Richtungsfinden zugenruhiger Rotkehlchen (*Erithacus rubecula*). Vogelwarte 22:168–173.

Metzgar, L. H. 1967. An experimental comparison of screech owl predation on resident and transient white-footed mice (*Peromyscus leucopus*). J. Mamm. 48:387–391.

Meyer, D. L., W. Heiligenberg, and T. H. Bullock. 1976. The ventral substrate response. A new postural control mechanism in fishes. J. Comp. Physiol. A 109:59–68.

Meyerriecks, A. J. 1960. Comparative breeding behavior of four species of N. American herons. Nuttall Ornithol. Club Publications 2:1–158.

Michelsen, A., B. B. Andersen, W. H. Kirchner, and M. Lindauer. 1989. Honeybees can be reruited by a mechanical model of a dancing bee. Naturwiss. 76:277–280.

Michod, R. E. and B. R. Levin. 1988. *The Evolution of Sex: An Examination of Current Ideas*.

Sinauer Associates, Inc. Sunderland, MA. 342 pp.

MILINSKI, M. AND T. C. M. BAKKER. 1990. Female sticklebacks use male coloration in mate choice and hence avoid parasitized males. Nature 344:330–333.

MILLER, D. B. 1980. Maternal vocal control of behavioral inhibition in mallard ducklings (*Anas platyrhynchos*). J. Comp. Physiol. Psychol. 94:606–623.

MILLER, D. B. AND G. GOTTLIEB. 1978. Maternal vocalizations of mallard ducks (*Anas platyrhynchos*). Anim. Behav. 26:1178–1194.

MILLER, R. C. 1922. The significance of the gregarious habit. Ecology 3:122–126.

MINEKA, S. AND M. COOK. 1988. Social learning and the acquisition of snake fear in monkeys. In: T. R. Zentall and B. G. Galef, Jr., eds., *Social Learning: Psychological and Biological Perspectives*. Lawrence Erlbaum. Hillsdale, NJ, pp. 51–73.

MINTZER, A. 1982. Nestmate recognition and incompatibility between colonies of the Acaciaant (*Pseudomyrmex ferruginea*). Behav. Ecol. Sociobiol. 10:165–168.

MOCK, D. W. 1983. On the study of avian mating systems. In: A. H. Brush and G. A. Clark, Jr., eds., *Perspectives in Ornithology*. Cambridge University Press. Cambridge, pp. 55–84.

MOEHLMAN, P. 1986. Ecology of cooperation in canids. In: D. I. Rubenstein and R. W. Wrangham, eds., *Ecological Aspects of Social Evolution*. Princeton University Press. Princeton, NJ, pp. 64–86.

MOEHLMAN, P. D. 1979. Jackal helpers and pup survival. Nature (London) 277:382–383.

MÖGLICH, M., V. MASCHWITZ, AND B. HÖLLDOBLER. 1974. Tandem calling: A new kind of signal in ant communication. Science 186:1046–1047.

MOHL, B. 1985. The role of proprioception in locust flight control. II. Information signalled by forewing stretch receptors during flight. J. Comp. Physiol. A 156:103–116.

MOISEFF, A. AND M. KONISHI. 1981. Neuronal and behavioral sensitivity to binaural time differences in the owl. J. Neurosci. 1:40–48.

MOISEFF, A., G. S. POLLACK, AND R. R. HOY. 1978. Steering responses of flying crickets to sound and ultrasound: Mate attraction and predator avoidance. Proc. Nat. Acad. Sci. USA 74:4025–4056.

MØLLER, A. P. 1988. False alarm calls as a means of resource usurpation in the great tit *Parus major*. Ethology 79:25–30.

MØLLER, A. P. 1991. Parasite load reduces song output in a passerine bird. Anim. Behav. 41:723–730.

MØLLER, A. P. AND T. R. BIRKHEAD. 1989. Copulation behaviour in mammals: Evidence that sperm competition is widespread. Biol. J. Linn. Soc. 38:119–131.

MOLLER, P. 1976. Electric signals and schooling behavior in a weakly electric fish, *Marcusenius cyprinoides* L. (mormyriformes). Science 193:697–699.

MOMENT, G. B. 1962. Reflexive selection: A possible answer to an old puzzle. Science 136:262–263.

MONTGOMERY, J. C. 1984. Noise cancellation in the eletrosensory system of the thornback ray: common mode rejection of input produced by the animal's own ventilatory movement. J. Comp. Physiol. A 155:103–111.

MOORE, A. J. AND P. J. MOORE. 1988. Female strategy during mate choice: Threshold assessment. Evolution 42:387–391.

MOORE, B. R. 1980. Is the homing pigeon's map geomagnetic? Nature 285:69–70.

MOORE, B. R. 1988. Magnetic fields and orientation in homing pigeons. Experiments of the late W. T. Keeton. Proc. Nat. Acad. Sci. 85:4907–4909.

MOORE, F. E. AND P. A. SIMM. 1986. Risk-sensitive

foraging by a migratory bird (*Dendroica coronata*). Experientia 42:1054–1056.

MOORE, F. R. 1978. Sunset and the orientation of a nocturnally migrant bird. Nature. (London) 274:154–156.

MOORE, F. R. 1980. Solar cues in the orientation of the savannah sparrow (*Passerculus sandwichensis*). Anim. Behav. 33:657–663.

MOORE, F. R. 1986. Sunrise, skylight polarization, and early morning orientation of night-migrating warblers. Condor 88:493–498.

MOORE, F. R. 1987a. Sunset and the orientation behavior of migratory birds. Biol. Rev. 62:65–86.

MOORE, F. R. 1987b. Moonlight and the migratory orientation of savannah sparrows (*Passerculus sandwichensis*). Ethology 75:155–167.

MOORE, R. Y. AND V. B. EICHLER, 1972. Loss of a circadian adrenal corticosterone rhythm following suprachiasmatic lesion in the rat. Brain Res. 42:201–206.

MOORE, R. Y. 1979. The retinohypothalamic tract, suprachiasmatic hypothalamic nucleus and central neural mechanisms of circadian rhythm regulation. In: M. Suda, O. Hayaishi, and H. Nakagawa, eds., *Biological Rhythms and Their Central Mechanisms*. Elsevier/North Holland Biomedical Press. Amsterdam, pp. 343–354.

MOORE-EDE M. AND F. SULZMAN. 1977. The physiological basis of circadian timekeeping in primates. Fed. Proc. 20:17–25.

MORAN, G. 1984. Vigilance behaviour and alarm calls in a captive group of meerkats, *Suricata suricatta*. Z. Tierpsychol. 65:228–240.

MORGAN, C. L. 1894. *An Introduction to Comparative Psychology.* Scribner, New York.

MORRIS, D. 1954. The reproductive behavior of the Zebra finch *Poephilia guttata* with special reference to pseudofemale behavior and displacement activities. Behaviour 6:271–322.

MORRIS, D. 1956. The feather postures of birds and the problem of the origin of social signals. Behaviour 9:75–113.

MORRIS, D. 1957. Typical intensity and its relation to the problem of ritualization. Behaviour 11:1–12.

MORRIS, D. 1958. The reproductive behavior of the ten-spined stickleback (*Pygosteus pungitius* L.). Behaviour Suppl. 6:1–154.

MORTON, E. S. 1975. Ecological sources of selection in avian sounds. Am. Nat. 109:17–34.

MOUM, S. E. AND R. L. BAKER. 1990. Colour change and substrate selection in larval *Ischnura verticalis* (Coenagrionidae: Odonata). Can. J. Zool. 68:221–224.

MOWRER, O. H. 1940. An experimental analogue of "regression" with incidental observations on "reaction-formation." J. Abn. Soc. Psychol. 35:56–87.

MOYNIHAN, M. 1956. Notes on the behavior of some North American gulls. I. Aerial hostile behavior. Behaviour 10:126–178.

MOYNIHAN, M. 1958. Notes on the behavior of some North American gulls. II. Nonaerial hostile behavior of adults. Behaviour 12:95–182.

MOYNIHAN, M. 1967. Comparative aspects of communication in New World Primates. In: D. Morris, ed., *Primate Ethology*. Weidenfeld and Nicholson. London. pp. 236–266.

MOYNIHAN, M. 1970. Some behavior patterns of platyrrhine monkeys. II. *Saguinus geoffroyi* and some other tamarins. Smithsonian Contributions in Zoology 28:1–77.

MUGGLETON, J. 1979. Non-random mating in wild populations of polymorphic *Adalia bipunctata*. Heredity 42:57–65.

MULLEN, R. 1977a. Site of *pcd* gene action and Purkinje cell mosaicism in cerebella of chimeric mice. Nature 270:245–247.

MULLEN, R. 1977b. Genetic dissection of the CNS with mutant–normal mouse and rat chimeras. In: W. M. Cowan and J. A. Ferrendelli, eds.,

Society for Neuroscience Symposia 2:47–65. Soc. Neuroscience, Bethesda, MD.

MÜLLER-SCHWARZE, D. 1971. Pheromones in black-tailed deer (*Odocoileus hemionus colombianus*). Anim. Behav. 19:141–152.

MULLER, H. J. 1932. Some genetic aspects of sex. Am. Nat. 66:118–138.

MUNN, N. L. 1950. *Handbook of Psychological Research on the Rat.* Houghton Mifflin. Boston. 598 pp.

MURTON, R. K., N. J. WESTWOOD, AND R. J. P. THEARLE. 1973. Polymorphism and the evolution of a continuous breeding season in the pigeon, *Columba livia.* J. Reprod. and Fertility Suppl. 19:563–577.

MYERS, C. W. AND J. W. DALY. 1983. Dart-poison frogs. Sci. Am. 248 (Feb):120–133.

NAKASHIMA, H. 1986. Phase shifting of the circadian conidiation rhythm in *Neurospora crassa* by calmodulin antagonists. J. Biol. Rhythms 1:163–169.

NARINS, P. M., O. J. REICHMAN, J. U. M. JARVIS, AND E. R. LEWIS. 1992. Seismic signal transmission between burrows of the Cape mole-rat, *Georychus capensis.* J. Comp. Physiol. A 170:13–21.

NAUMANN, K., M. L. WINSTON, K. N. SLESSOR, G. D. PRESTWICH, AND F. X. WEBSTER. 1991. Production and transmission of honey bee queen (*Apis mellifera* L.) mandibular gland pheromone. Behav. Ecol. Sociobiol. 29:321–332.

NEILL, S. R. ST. J. AND J. M. CULLEN. 1974. Experiments on whether schooling by their prey affects the hunting behaviour of cephalopods and fish predators. J. Zool. London 172:549–569.

NELSON, J. B. 1967. Coloniality and cliff-nesting in the gannet. Ardea 55:60–90.

NEUMANN, D. 1976. Entrainment of semi-lunar rhythm. In: P. Decoursey, ed., *Biological Rhythms in the Marine Environment.* University of South Carolina Press. Columbia, pp. 115–127.

NEUWILLER, G. 1984. Foraging, echolocation and audition in bats. Naturwiss. 71:446–455.

NEUWILLER, G. 1990. Auditory adaptations for prey capture in echolocating bats. Physiol. Rev. 70:615–641.

NEW, J. G. AND D. BODZNICK. 1990. Medullary electrosensory processing in the little skate. II. Suppression of self-generated electrosensory interference during respiration. J. Comp. Physiol. A 167:295–307.

NEWCOMBE, C. AND G. HARTMAN. 1973. Some chemical signals in the spawning behavior of Rainbow trout (*S. gairdneri*). J. Fish. Res. Bd. Can. 30:995–997.

NEWMAN, E. A. AND P. H. HARTLINE. 1982. The infrared "vision" of snakes. Sci. Am. 246 (Mar):116–124.

NICOLAI, J. 1964. Der Brutparasitismus der Viduinae als ethologisches problem. Z. Tierpsychol. 21:9–204.

NOBLE, G. K. 1936. Courtship and sexual selection in the flicker (*Colaptes auratus luteus*). Auk 53:269–282.

NOBLE, G. K. 1939. The role of dominance in the social life of birds. Auk 56:263–273.

NOBLE, G. K. AND E. J. FARRIS. 1929. The method of sex recognition in the wood frog, *Rana sylvatica* Le Conte. Am. Mus. Novit. 363:1–17.

NOLAN, V. 1978. The ecology and behavior of the prairie warbler *Dendroica discolor.* Ornith. Monogr. 26, 595 pp.

NORDEEN, E. J. AND K. W. NORDEEN. 1990. Neurogenesis and sensitive periods in avian song learning. Trends in Neurosci. 13:31–36.

NORDEEN, K. W., E. J. NORDEEN, AND A. P. ARNOLD. 1986. Estrogen establishes sex differences in androgen accumulation in zebra finch brain. J. Neurosci. 6:734–738.

NORDENG, H. 1971. Is the local orientation of anadromous fishes determined by pheromones? Nature 233:411–413.

NORDENG, H. 1977. A pheromone hypothesis for

homeward migration in anadromous salmonids. Oikos 28:155–159.

NORRIS, K. S. AND C. H. LOWE. 1964. An analysis of background color-matching in amphibians and reptiles. Ecology 45:565–580.

NOTTEBOHM, F. 1969. The song of the chingolo (*Zonotrichia capensis*) in Argentina: Description and evaluation of a system of dialects. Condor 71:299–315.

NOTTEBOHM, F. 1970. Ontogeny of bird song. Science 167:950–956.

NOTTEBOHM, F. 1972. The origins of vocal learning. Am. Nat. 106:116–140.

NOTTEBOHM, F. 1975. Continental patterns of song variability in *Zonotrichia capensis*: Some possible ecological correlates. Am. Nat. 109:605–624.

NOTTEBOHM, F. AND J. P. ARNOLD. 1976. Sexual dimorphism in vocal control areas of the songbird brain. Science 194:211–213.

NOWAK, E. AND P. BERTHOLD. 1991. Satellite tracking: A new method in orientation research. In: P. Berthold, ed., *Orientation in Birds*. Birkhäuser. Basel, Switzerland, pp. 307–322.

OCLESE, J., S. REUSS, AND L. VOLLRATH. 1985. Evidence for the involvement of the visual system in mediating magnetic field effects on pineal melatonin synthesis in the rat. Brain Res. 333:382–384.

OEHMKE, M. G. 1973. Lunar periodicity in the flight activity of honey bees. J. Interdiscipl. Cycle Res. 4:319–335.

O'HARA, R. K. AND A. R. BLAUSTEIN. 1981. An investigation of sibling recognition in *Rana cascadae* tadpoles. Anim. Behav. 29:1121–1126.

O'LEARY, D. P., J. VILCHES-TROYA, D. F. DUNN, AND A. CAMPOZ–MUNOZ. 1981. Magnets in guitarfish vestibular receptors. Experientia 37:86–88.

ORCHINIK, M., P. LICHT, AND D. CREWS. 1988. Plasma steroid concentrations change in response to sexual behavior in *Bufo marinus*. Horm. Behav. 22:338–350.

ORIANS, G. H. 1969. On the evolution of mating systems of birds and mammals. Am. Nat. 103:589–603.

ORIANS, G. H. AND G. M. CHRISTMAN. 1968. A comparative study of red-winged, tricolored and yellow-headed blackbirds. University of California Publications. Zoology 84:81 pp.

ORING, L. W. AND D. B. LANK. 1986. Polyandry in spotted sandpipers: The impact of environment and experience. In: D. I. Rubenstein and R. W. Wrangham, eds., *Ecological Aspects of Social Evolution*. Princeton University Press. Princeton, NJ, pp. 21–42.

O'SHEA, M. AND P. D. EVANS. 1979. Potentiation of neuromuscular transmission by an octopaminergic neurone in the locust. J. Exp. Biol. 79:169–190.

OSTFELD, R. S. 1985. Limiting resources and territoriality in microtine rodents. Am. Nat. 126:1–15.

OSTFELD, R. S. 1990. The ecology of territoriality in small mammals. Trends Ecol. Evol. 5:411–415.

OWEN, D. 1980. *Camouflage and Mimicry*. University of Chicago Press. Chicago. 158 pp.

PACKER C., L. HERBST, A. E. PUSEY, J. D. BYGOTT, J. P. HANBY, S. J. CAIRNS, AND M. B. MULDER. 1988. Reproductive success of lions. In: T. H. Clutton-Brock, ed., *Reproductive Success, Studies of Individual Variation in Contrasting Breeding Systems*. University of Chicago Press. Chicago, pp. 363–383.

PACKER, C. 1977. Reciprocal altruism in *Papio anubis*. Nature (London) 265:441–443.

PACKER, C. 1986. The ecology of felid sociality. In: D. J. Rubenstein and R. W. Wrangham, eds., *Ecological Aspects of Social Evolution*. Princeton University Press. Princeton, NJ. pp. 429–451.

PACKER, C. AND A. E. PUSEY. 1982. Cooperation and competition within coalitions of male lions: Kin selection or game theory? Nature 296:740–742.

PACKER, C. AND A. E. PUSEY. 1987. Intrasexual cooperation and the sex ratio in African lions. Am. Nat. 130:636–642.

PACKER, C., D. SCHEEL, AND A. E. PUSEY. 1990. Why lions form groups: Food is not enough. Am. Nat. 136:1–19.

PAGE, R. E., JR. AND R. A. METCALF. 1982. Multiple mating, sperm utilization, and social evolution. Am. Nat. 119:263–281.

PAGE, T. L. 1983. Regeneration of rhythms in lobectomized roaches. J. Comp. Physiol. 152:231–240.

PAGE, T. L. 1985. Circadian organization in cockroaches: Effects of temperature cycles on locomotor activity. J. Insect Physiol. 31:235–242.

PALMER, J. D. 1966. How a bird tells the time of day. Nat. Hist. 75:48–53.

PALMER, J. D. 1990. The rhythmic lives of crabs. BioScience 40:352–357.

PANTELOURIS, E. M. 1965. *The common liver fluke, Fasciola hepatica L.* Pergamon Press, New York.

PAPI, F. 1986. Pigeon navigation: Solved problems and open questions. Monit. Zool. Ital. (N. S.) 20:471–517.

PAPI, F. AND L. PARDI. 1963. On the orientation of sandhoppers (Amphipoda, Talitridae). Biol. Bull. 124:97–105.

PAPI, F., L. FIORE, V. FIASCHI, AND S. BENVENUTI. 1971. The influence of olfactory nerve section on the homing capacity of carrier pigeons. Monit. Zool. Ital. (N.S.) 5:265–267.

PAPI, F., L. FIORE, V. FIASCHI, AND S. BENVENUTI. 1972. Olfaction and homing in pigeons. Monit. Zool. Ital. (N.S.) 6:85–95.

PAPI, F., W. T. KEETON, A. I. BROWN, AND S. BENVENUTI. 1978. Do American and Italian pigeons rely on different homing mechanisms? J. Comp. Physiol. 128:303–317.

PARDI, L. 1954. Über die orientierung von *Tylos latreillii* Aud. and Sav. (Isopoda terrestria). Z. Tierpsychol. 11:175–181.

PARKER, G. A. 1970. Sperm competition and its evolutionary consequences in the insects. Biol. Rev. 45:525–567.

PARKER, G. A. 1974. Assessment strategy and the evolution of fighting behavior. J. Theoret. Biol. 47:223–243.

PARKER, G. A. 1984. Sperm competition and the evolution of animal mating strategies. In: R. L. Smith, ed., *Sperm Competition and the Evolution of Animal Mating Systems.* Academic Press. Orlando, FL, pp. 2–60.

PARKER, G. A. 1985. Models of parent–offspring conflict. V. Effects of the behaviour of two parents. Anim. Behav. 33:519–533.

PARKER, G. A. AND M. R. MACNAIR. 1978. Models of parent–offspring conflict. I. Monogamy. Anim. Behav. 26:97–111.

PARKER, G. A. AND M. R. MACNAIR. 1979. Models of parent–offspring conflict. IV. Suppression: Evolutionary retaliation of the parent. Anim. Behav. 27:1210–1235.

PARKER, G. A., R. R. BAKER, AND V. G. F. SMITH. 1972. The origin and evolution of gamete dimorphism and the male-female phenomenon. J. Theoret. Biol. 36:529–553.

PARKES, A. S. AND BRUCE, H. M. 1961. Olfactory stimuli in mammalian reproduction. Science 134:1049–1054.

PARRISH, J. K. 1989. Re-examining the selfish herd: Are central fish safer? Anim. Behav. 38:1048–1053.

PARTRIDGE, L. 1980. Mate choice increases a component of offspring fitness in fruit flies. Nature 283:290–291.

PATTERSON, F. 1978. The gestures of a gorilla: Sign language acquisition in another pongid species. Brain and Language 5:72–97.

PAVLOV, I. 1927. *The Conditioned Reflex*, G. V. Anrep, trans. Oxford University Press. London.

PAYNE, R. B. 1962. How the barn owl locates its prey by hearing. Living Bird 1:151–159.

PAYNE, R. B. 1977. The ecology of brood parasitism in birds. Ann. Rev. Ecol. Syst. 8:1–28.

PAYNE, R. B. 1981. Population structure and social behavior: Models for testing the ecological significance of song dialects in birds. In: R. D. Alexander and D. W. Tinkle, eds., *Natural Selection and Social Behavior*. Chiron Press. New York, pp. 108–120.

PAYNE, R. B. AND K. PAYNE. 1977. Social organization and mating success in local song populations of village indigobirds (*Vidua chalybeata*). Z. Tierpsychol. 45:113–173.

PAYNE, R. S. 1972. The song of the whale. In: *The Marvels of Animal Behavior*. National Geographic Society. Washington, DC. pp. 144–167.

PAYNE, R. S. AND D. WEBB. 1971. Orientation by means of long range acoustic signaling in baleen whales. Ann. N.Y. Acad. Sci. 188:110–142.

PEARSON, K. G. 1987. Central pattern generation: A concept under scrutiny. In H. McLennan, J. R. Ledsome, C. H. S. McIntosh, and D. R. Jones, eds., *Advances in Physiological Research*. Plenum Press. New York, pp. 167–185.

PEARSON, K. G., D. N. REYE, AND R. M. ROBERTSON. 1983. Phase-dependent influences of wing stretch receptors on flight rhythm in the locust. J. Neurophysiol. 49:1168–1181.

PEEKE, H. V. S., M. J. HERZ, AND J. E. GALLAGHER. 1971. Changes in aggressive interaction in aggressive territorial Convict Cichlids (*Cichlasoma nigrofasciatum*). A study of habituation. Behaviour 40:43–54.

PELKWIJK, J. J. TER AND N. TINBERGEN. 1937. Eine reizbiologische Analyse einiger Verhaltensweisen von *Gasterosteus aculeatus* L. Z. Tierpsychol. 1:193–200.

PENGELLEY, E. T. AND S. J. ASMUNDSON. 1970. The effect of light on the free-running circannual rhythm of the golden-mantled ground squirrel, *Citellus lateralis*. Comp. Biochem. Physiol. 30:177–183.

PENGELLEY, E. T. AND S. J. ASMUNDSON. 1971. Annual biological clocks. Sci. Am. 224 (Apr):72–79.

PEPPERBERG, I. M. 1987a. Evidence for conceptual quantitative abilities in the African Grey Parrot: Labeling of cardinal sets. Ethology 75:37–61.

PEPPERBERG, I. M. 1987b. Acquisition of same/different concept by an African Grey parrot (*Psittacus erithacus*): Learning with respect to categories of color, shape and material. Anim. Learn. & Behav. 15:423–432.

PEPPERBERG, I. M. 1991. A communicative approach to animal cognition: A study of conceptual abilities of an African Grey Parrot. In: C. A. Ristau, ed., *Cognitive Ethology, the Minds of Other Animals*. Lawrence Erlbaum. Hillsdale, NJ, pp. 153–186.

PERDECK, A. C. 1967. Orientation of starlings after displacement to Spain. Ardea 51:91–104.

PEREIRA, M. E. AND M. K. IZARD. 1989. Lactation and care for unrelated infants in forest-living ringtailed lemurs. Am. J. Primatol. 18:101–108.

PETERSON, R. O. 1977. *Wolf Ecology and Prey Relationships on Isle Royale*. National Park Service. Scientific Monograph Series no. 11. U.S. Government Printing Office. Washington, DC. 210 pp.

PETRIE, M., T. HALLIDAY, AND C. SANDERS. 1991. Peahens prefer peacocks with elaborate trains. Anim. Behav. 41:323–331.

PETRINOVICH, L. 1988. Individual stability, local variability and the cultural transmission of song in white-crowned sparrows (*Zonotrichia leucophrys nuttalli*). Behaviour 107:208–240.

PETRINOVICH, L. AND L. F. BAPTISTA. 1984. Song dialects, mate selection and breeding success in white-crowned sparrows. Anim. Behav. 32:1078–1088.

PFEFFER, P. 1967. Le moufflon de Corse (*Ovis ammon musimom* Schreber 1782) position systématique, écologie et éthologie canparées. Mammalia 31 supplement 262 pp.

PHILLIPS, J. B. 1986. Two magnetoreception pathways in a migratory salamander. Science 233:765–767.

PHILLIPS, J. B. 1987. Specialized visual receptors respond to magnetic field alignment in the blowfly (*Calliphora vicina*). Soc. for Neurosci. Abstr. 13:397.

PHILLIPS, J. B. AND J. A. WALDVOGEL. 1982. Reflected light cues generate the short-term deflector loft effect. In: F. Papi and H. G. Walraff, eds., *Avian Navigation*. Springer-Verlag. Berlin, pp. 190–202.

PHILLIPS, J. B. AND J. A. WALDVOGEL. 1988. Celestial polarized light patterns as a calibration reference for sun compass of homing pigeons. J. Theoret. Biol. 131:55–67.

PHOENIX, C., R. GOY, A. GERALL, AND W. YOUNG. 1959. Organizing action of prenatally administered testosterone proprionate on the tissues mediating mating behavior in the female guinea pig. Endocrinol. 65:369–382.

PICKARD, G. E. AND F. W. TUREK. 1982. Splitting of the circadian rhythm of activity is abolished by unilateral lesions of the suprachiasmatic nuclei. Science 215:119–1121.

PIETRAS, R. J. AND D. G. MOULTON. 1974. Hormonal influences on odor detection in rats: Changes associated with the estrous cycle, pseudopregnancy, ovariectomy, and administration of testosterone proprionate. Physiol. Behav. 12:475–491.

PIETREWICZ, A. T. AND A. C. KAMIL. 1979. Search image formation in the Blue Jay (*Cyanocitta cristata*). Science 204:1332–1333.

PIETREWICZ, A. T. AND A. C. KAMIL. 1981. Search images and the detection of cryptic prey: An operant approach. In: A. C. Kamil and T. D. Sargent, eds., *Foraging Behavior, Ecological, Ethological and Psychological Approaches*. Garland STPM Press. New York, pp. 311–331.

PIETSCH, T. W. AND D. B. GROBECKER. 1978. The complete angler: Aggressive mimicry in the antennariid angler fish. Science 201:369–370.

PLOMIN, R., J. C. DeFRIES, AND G. E. McCLEARN. 1980. *Behavioral Genetics—A Primer*. W. H. Freeman & Co., San Francisco. 417 pp.

PLUMMER, M. R. AND J. M. CAMHI. 1981. Discrimination of sensory signals from noise in the escape system of the cockroach: The role of wind acceleration. J. Comp. Physiol. 142:347–357.

POLIS, G. A. 1990. Introduction. In: G. A. Polis, ed., *The Biology of Scorpions*. Stanford University Press. Stanford, pp. 1–8.

POOLE, T. B. 1966. Aggressive play of polecats. Symp. Zool. Soc. London 18:23–24.

POPOV, A.V. AND V. F. SHUVALOV. 1977. Phonotactic behavior of crickets. J. Comp. Physiol. A 119:111–126.

POPP, J. W. 1987. Resource value and dominance among American goldfinches. Bird Behav. 7:73–77.

PORTER, R. H., V. J. TEPPER, AND D. M. WHITE. 1981. Experimental influences on the development of huddling preferences and "sibling" recognition in spiny mice. Dev. Psychobiol. 14:375–382.

POTTS, W. K., C. J. MANNING, AND E. K. WAKELAND. 1991. Mating patterns in seminatural populations of mice influenced by MHC genotype. Nature 352:619–621.

POUGH, F. H. 1976. Multiple cryptic effects of cross-banded and ringed patterns of snakes. Copeia 1976:834–836.

POUGH, F. H. 1988. Mimicry and related phenomena. In: C. Gans, and R. B. Huey, eds., *Biology of the Reptilia*, vol. 16. Alan Liss, Inc. New York, pp. 153–234.

POWER, H. W. 1975. Mountain bluebirds: Experimental evidence against altruism. Science 189:142–143.

POWER, H. W. 1976. A response to Emlen's criticism. Science 191:809–810.

PREMACK, D. 1976. *Intelligence in Ape and Man*. Lawrence Erlbaum. Hillsdale, NJ. 370 pp.

PREMACK, D. 1978. On the abstractness of human concepts: Why it would be difficult to talk to a pigeon. In: S. H. Hulse, H. Fowler, and W. K. Honig, eds., *Cognitive Processes in Animal Behavior*. Lawrence Erlbaum. Hillsdale, NJ, pp. 423–451.

PROHAZKA, D., M. A. NOVAK, AND J. S. MEYER. 1986. Divergent effects of early hydrocortisone treatment on behavioral and brain development in meadow and pine voles. Dev. Psychobiol. 19:521–535.

PROMPTOFF, A. N. 1930. Die geographische variabilitat des Buchfinkenschlags (*Fringilla coelebs* L.). Biol. Zentralble. 50:478–503.

PYKE, G. 1979. The economics of territory size and time budget in the golden-winged sunbird. Am. Nat. 114:131–145.

PYKE, G. H., H. R. PULLIAM, AND E. L. CHARNOV. 1977. Optimal foraging: A selective review of theory and tests. Quart. Rev. Biol. 52:137–154.

PYLE, D. W. AND M. H. GROMKO. 1978. Repeated mating by female *Drosophila melanogaster*: The adaptive importance. Experientia 34:449–450.

QUINLAN, R. J. AND J. M. CHERRETT. 1977. The role of substrate preparation in the symbiosis between the leaf-cutting ant *Acromyrmex octospinosus* (Reich) and its food fungus. Ecolog. Entomol. 2:161–170.

QUINN, T. P. 1980. Evidence for celestial and magnetic compass orientation in lake migrating sockeye salmon fry. J Comp. Physiol. A 147:547–552.

QUINN, T. P. 1982. Intra-specific differences in sockeye salmon by compass orientation mechanisms. Proc. Salmon and Trout Migratory Behavior Symp. Seattle, pp. 79–85.

QUINN, T. P. AND A. H. DITTMAN. 1990. Pacific salmon migrations and homing: Mechanisms and adaptive significance. Trends Ecol. Evol. 5:174–177.

QUINN, T. P. AND C. A. BUSAK. 1985. Chemosensory recognition of siblings in juvenile coho salmon (*Oncorhynchus kisutch*). Anim. Behav. 33:51–56.

QUINN, T. P. AND E. L. BRANNON. 1982. The use of celestial and magnetic cues by orienting sockeye salmon smolts. J. Comp. Physiol. 147:547–552.

QUINN, T. P. AND G. M. TOLSON. 1986. Evidence of chemically mediated population recognition in coho salmon (*Oncorhynchus kisutch*). Can. J. Zool. 64:84–87.

QUINN, T. P. AND T. J. HARA. 1986. Sibling recognition and olfactory sensitivity in juvenile coho salmon (*Oncorhynchus kisutch*). Can. J. Zool. 64:921–925.

QUINN, W. G., W. A. HARRIS, AND S. BENZER. 1974. Conditioned behavior in *Drosophila melanogaster*. Proc. Nat. Acad. Sci. USA 71:708–712.

RAISMAN, G. AND K. BROWN-GRANT. 1977. The "suprachiasmatic syndrome": Endocrine and behavioral abnormalities following lesions in the suprachiasmatic nuclei in the female rat. Proc. Roy. Soc. London B. 198:297–314.

RALPH, M. R. AND M. MENAKER. 1988. A mutation in the circadian system in the Golden Hamster. Science 241:1225–1227.

RALPH, M. R., F. C. DAVIS, AND M. MENAKER. 1988. Suprachiasmatic nucleus transplantation restores donor specific circadian rhythms to arryhthmic hosts. Soc. for Neurosci. Abstr. 14:462.

RAMIREZ, J. M. AND K. G. PEARSON. 1990. Chemical deafferentation of the locust flight system by phentolamine. J. Comp. Physiol. A 167:485–494.

RANDALL, J. A. 1991. Mating strategies of a nocturnal, desert rodent (*Dipodomys spectabilis*). Behav. Ecol. Sociobiol. 28:215–220.

RASA, O. A. E. 1986. Coordinated vigilance in dwarf mongoose family groups: The 'Watchman's Song' hypothesis and the costs of guarding. Ethology 71:340–344.

RAYBOURN, M. S. 1983. The effects of direct-current magnetic fields on turtle retinas in vitro. Science 230:715–717.

RAYNAUD, A. AND C. PIEAU. 1985. Embryonic development of the genital system. In: C. Gans and F. Billet, eds., *Biology of the Reptilia*. John Wiley & Sons. New York, pp. 149–300.

RAYNOR, L. S. AND G. W. UETZ. 1990. Trade-offs in foraging success and predation risk with spatial position in colonial spiders. Behav. Ecol. Sociobiol. 27:77–85.

REAL, L. A. 1981. Uncertainty and pollinator–plant interactions: The foraging behavior of bees and wasps on artificial flowers. Ecology 62:20–26.

REAL, L. A., J. OTT, AND E. SILVERFINE. 1982. Optimal foraging: Some simple stochastic models. Behav. Ecol. Sociobiol. 10:251–263.

REES, H. D. AND H. E. GRAY. 1984. Glucocorticoids and mineralocorticoids: Actions on brain and behavior. In: C. B. Nemeroff and A. J. Dunn, eds., *Peptides, Hormones, and Behavior*. Spectrum Publications. New York, pp. 579–644.

REEVE, H. K., D. F. WESTNEAT, W. A. NOON, P. W. SHERMAN, AND C. F. AQUADRO. 1990. DNA "fingerprinting" reveals high levels of inbreeding in colonies of the eusocial naked mole-rat. Proc. Nat. Acad. Sci. USA 87:2496–2500.

REICHERT, H., C. H. F. ROWELL, AND C. GRISS. 1985. Course correction circuitry translates feature detection into behavioral action in locusts. Nature 315:142–144.

REILLY, L. M. 1987. Measurements of inbreeding and average relatedness in a termite population. Am. Nat. 130:339–349.

REIMCHEN, T. E. 1989. Shell color ontogeny and tubeworm mimicry in a marine gastropod *Littorina mariae*. Biol. J. Linn. Soc. 36:97–109.

REITER, R. J. 1980. The pineal and its hormones in the control of reproduction in mammals. Endocrine Rev. 1:109–131.

RENNER, M. 1959. Time sense in bees. Nat. Hist. 68:434–440.

RENSCH, B. AND G. DUCKER. 1959. Die Spiele von Mungo und Ichneumon. Behavior 14:185–213.

RESCORLA, R. A. 1967. Pavlovian conditioning and its proper control procedures. Psychol. Rev. 74:71–80.

RESCORLA, R. A. 1988. Behavioral studies of Pavlovian conditioning. Ann. Rev. Neurosci. 11:329–352.

REUSS S. AND J. OLCESE. 1986. Magnetic field effects on the rat pineal gland: role of retinal activation by light. Neurosci. Lett. 64:97–101.

REUSS, S., P. SEMM, AND L. VOLLRATH. 1983. Different types of magnetically sensitive cells in the rat pineal gland. Neurosci. Lett. 40:23–26.

REYE, D. N. AND K. G. PEARSON. 1988. Entrainment of the locust central flight oscillator by wing stretch receptor stimulation. J. Comp. Physiol. A 162:77–89.

REYER, H.-U. 1980. Flexible helper structure as an ecological adaptation in the red Kingfisher (*Ceryle rudis rudis L.*). Behav. Ecol. Sociobiol. 6:219–227.

REYER, H.-U. 1984. Investment and relatedness: A cost/benefit analysis of breeding and helping in the pied kingfisher (*Ceryle rudis*). Anim. Behav. 32:1163–1178.

REYER, H.-U. 1986. Breeder–helper interactions in the pied kingfisher reflect costs and benefits of cooperative breeding. Behaviour 96:277–303.

RHEINLAENDER, J. H., C. GERHARDT, D. D. YAGER, AND R. R. CAPRANICA. 1979. Accuracy of phonotaxis by the green treefrog (*Hyla cinerea*). J. Comp. Physiol. A 133:247–255.

RIBBINK, A. J. 1977. Cuckoo among Lake Malawi cichlid fish. Nature 267:243–244.

RICCIUTTI, E. 1978. Night of the grunting fish. Audubon 80(4):92–97.

RICHTER, C. P. 1975. II. Biological rhythm. Astronomical references in biological rhythms. In: J. T. Fraser and N. Lawrence, eds., *The Study of Time.* Springer. New York, pp. 39–53.

RIDLEY, M. 1978. Paternal care. Anim. Behav. 26:904–932.

RIECHERT, S. E. 1979. Games spiders play. II. Resource assessment strategies. Behav. Ecol. Sociobiol. 6:121–128.

RIECHERT, S. E. 1981. The consequences of being territorial: Spiders, a case study. Am. Nat. 117:871–892.

RIECHERT, S. E. 1982. Spider interaction strategies: Communication vs. coercion. In: P. N. Witt and J. Rovner, eds., *Spider Communication: Mechanisms and Ecological Significance.* Princeton University Press, Princeton, NJ, pp 281–315.

RIECHERT, S. E. 1984. Games spiders play. III: Cues underlying context-associated changes in agonistic behavior. Anim. Behav. 32:1–15.

RIECHERT, S. E. 1986a. Between population variation in spider territorial behavior: Hybrid-pure population line comparisons. In: M. D. Huettel, ed., *Evolutionary Genetics of Invertebrate Behavior: Progress and Prospects.* Plenum Press, New York, pp. 33–42.

RIECHERT, S. E. 1986b. Spider fights as a test of evolutionary game theory. Am. Sci. 74:604–610.

RIECHERT, S. E. AND C. R. TRACY. 1975. Thermal balance and prey availability: Bases for a model relating web-site characteristics to spider reproductive success. Ecology 56:265–285.

RIEDMAN, M. L. 1982. The evolution of alloparental care and adoption in mammals and birds. Quart. Rev. Biol. 57:405–435.

RISSMAN, E. F. 1990. The musk shrew, *Suncus murinus*, a unique animal model for the study of female behavioral endocrinology. J. Exp. Zool. Suppl. 4:207–209.

RISTAU, C. A., ED. 1991. *Cognitive Ethology, the Minds of other Animals.* Lawrence Erlbaum. Hillsdale, NJ. 332 pp.

RITZMANN, R. E. AND POLLACK, A. J. 1986. Identification of thoracic interneurons that mediate giant interneuron-to-motor pathways in the cockroach. J. Comp. Physiol. A 159:639–654.

ROBBINS, R. K. 1981. The "false head" hypothesis: Predation and wing pattern variation of Lycaenid butterflies. Am. Nat. 118:770–775.

ROBERTS, J. L. 1984. The biosynthesis of peptide hormones. In: C. B. Nemeroff and A. J. Dunn, eds., *Peptides, Hormones, and Behavior.* Spectrum Publications. New York, pp. 99–118.

ROBERTS, S. 1965. Photoreception and entrainment of cockroach activity rhythms. Science 148:958–959.

ROBERTS, S. 1974. Circadian rhythms in cockroaches. Effects of optic lobe lesions. J. Comp. Physiol. 88:21–30.

ROBERTSON, D. R. 1972. Social control of sex reversal in a coral reef fish. Science 177:1007–1009.

ROBERTSON, L. AND J. TAKAHASHI. 1988a. Circadian clock in cell culture. I. Oscillation of melatonin release from dissociated chick pineal cells in flow-through microcarrier culture. J. Neurosci. 8:12–21.

ROBERTSON, L. AND J. TAKAHASHI. 1988b. Circadian clock in cell culture. II. In vitro photic entrainment of melatonin oscillation from dissociated chick pineal cells. J. Neurosci. 8:22–30.

ROBERTSON, R. J. AND B. J. STUTCHBURY. 1988. Experimental evidence for sexually selected infanticide in tree swallows. Anim. Behav. 36:749–753.

ROBERTSON, R. M. AND K. G. PEARSON. 1983. Interneurons in the flight system of the locust:

Distribution, connections, and resetting properties. J. Comp. Neurol. 215:33–50.

ROBERTSON, R. M. AND K. G. PEARSON. 1984. Interneuronal organization in the flight system of the locust. J. Insect Physiol. 30:95–101.

ROBERTSON, R. M. AND K. G. PEARSON. 1985. Neural circuits in the flight system of the locust. J. Neurophysiol. 53:110–128.

ROBINSON, G. E. 1986. The dance language of the honey bee: The controversy and its resolution. Am. Bee J. 126:184–189.

ROBINSON, G. E. 1987. Regulation of honey bee age polyethism by juvenile hormone. Behav. Ecol. Sociobiol. 20:329–338.

ROBINSON, G. E., A. STRAMBI, C. STRAMBI, Z. L. PAULINO-SIMOES, S. O. TOZETO, AND J. M. NEGRAES BARBOSA. 1987. Juvenile hormone titers in European and Africanized honey bees in Brazil. Gen. Comp. Endocrinol. 66:457–459.

ROBINSON, G. E., R. E. PAGE, C. STRAMBI, AND A. STRAMBI. 1989. Hormonal and genetic control of behavioral integration in honey bee colonies. Science 246:109–112.

ROEDER, K. D. 1963 (and revised edition, 1967). *Nerve Cells and Insect Behavior*. Harvard University Press. Cambridge, MA.

ROMANES, G. J. 1882. *Animal Intelligence*. D. Appleton and Company, New York. 520 pp.

ROMANES, G. J. 1884. *Mental Evolution in Animals*. Keegan, Paul, Trench and Company, London. 411 pp.

ROMANES, G. J. 1889. *Mental Evolution in Man*. D. Appleton and Company, New York. 452 pp.

ROSBASH, M. AND J. C. HALL. 1985. Biological clocks in *Drosophila,* finding the molecules that make them tick. Cell 43:3–4.

ROSBASH, M. AND J. C. HALL. 1989. The molecular biology of circadian rhythms. Neuron 3:387–398.

ROSEN, R. A. AND D. C. HALES. 1981. Feeding of paddlefish, *Polyodon spathula*. Copeia 1981:441–455.

ROSENBLATT, J. S., A. D. MAYER, AND A. L. GIORDANO. 1988. Hormonal basis during pregnancy for the onset of maternal behavior in the rat. Psychoneuroendocrinol. 13:29–46.

ROSIN, R. 1978. The honey bee "language" controversy. J. Theoret. Biol. 72:589–602.

ROSIN, R. 1980a. Paradoxes of the honey-bee "dance language" hypothesis. J. Theoret. Biol. 84:775–800.

ROSIN, R. 1980b. The honey-bee "dance language" hypothesis and the foundations of biology and behavior. J. Theoret. Biol. 87:457–481.

ROSIN, R. 1984. Further analysis of the honey bee "dance language" controversy. I. Presumed proofs for the "dance language" hypothesis by Soviet scientists. J. Theoret. Biol. 107:417–442.

ROSS, K. G. 1986. Kin selection and the problem of sperm utilization in social insects. Nature 323:798–800.

ROSS, R. M. 1990. The evolution of sex-change mechanisms in fishes. Environ. Biol. Fishes 29:81–93.

ROTHENBULER, W. C. 1964a. Behaviour genetics of nest cleaning in honeybees I. Responses of four in-bred lines to disease-killed brood. Anim. Behav. 12:578–583.

ROTHENBULER, W. C. 1964b. Behaviour genetics of nest cleaning in honey-bees IV. Responses of F-1 and backcross generations to disease-killed brood. Am. Zool. 4:111–123.

ROTHSTEIN, S. I. AND R. C. FLEISCHER. 1987. Vocal dialects and their possible relation to honest status signaling in the brown-headed cowbird. Condor 89:1–23.

ROWELL, C. H. F. 1989. Descending interneurones of the locust reporting deviation from flight course: What is their role in steering? J. Exp. Biol. 146:177–194.

ROWLAND, W. J. AND P. SEVENSTER. 1985. Sign stimuli in the threespine stickleback (*Gasterosteus aculeatus*): A re-examination and extension of some classic experiments. Behaviour 93:241–257.

ROWLEY, I. C. R. 1981. The communal way of life in the Splendid Wren, *Malurus splendens*. Z. Tierpsychol. 55:228–267.

ROWLEY, I. C. R. AND E. M. RUSSELL. 1990. Splendid Fairy-wrens: Demonstrating the importance of longevity. In: P. B. Stacey and W. D. Koenig, eds., *Cooperative Breeding in Birds*. Cambridge University Press, Cambridge, England. pp. 3–30.

RUBENSTEIN, D. I. 1981. Population density, resource patterning and territoriality in the Everglades pygmy sunfish. Anim. Behav. 29:155–172.

RUDOLPH, K., A. WIRZ–JUSTICE, K. KRAUCHI, AND H. FEER. 1988. Static magnetic fields decrease nocturnal pineal cAMP in the rat. Brain Res. 446:159–160.

RUMBAUGH, D. M. 1977. The emergence and the state of ape language research. In: G.H. Bourne, ed., *Progress in Ape Research*. Academic Press. New York, pp. 75–83.

RUMBAUGH, D. M., T. V. GILL, AND E. C. VON GLASERFELD. 1973. Reading and sentence completion by a chimpanzee. Science 182:731–733.

RUSAK, B. AND G. GROSS. 1982. Suprachiasmatic stimulation phase shifts rodent circadian rhythms. Science 215:1407–1409.

RUSHFORTH, N. B. 1965. Behavioral studies of the coelenterate *Hydra pirardi* Brien. Anim. Behav. Suppl. 1:30–42.

RUSSELL, E. M. AND I. C. R. ROWLEY. 1988. Helper contributions to reproductive success in the splendid fairy-wren *Malurus splendens*. Behav. Ecol. Sociobiol. 22:131–140.

RYAN, M. J., J. H. FOX, W. WILZYNSKI, AND A. S. RAND. 1990. Sexual selection for sexual exploitation in the frog *Physalaemus pustulosu*. Nature (London) 343:66–67.

SADAUSKAS, K. K. AND SHURANOVA ZH.P. 1984. Effects of pulsed magnetic field on electrical activity of crayfish neurons. Biofiska 29:681–683.

SANTSCHI, F. 1911. Le mechanisme d'orientation chez les fourmis. Rev. Suisse Zool. 19:117–134.

SARGENT, M. L., I. E. ASHKENAZI, E. M. BRADBURG, V. G. BRUCE, C. F. EHRET, J. F. FELDMAN, M. W. KARAKASHIAN, R. J. KONOPKA, D. MERGENHAGEN, G. A. SCHÜTZ, H.-G. SCHWEIGER, AND T. E. A. VANDEN DRIESSCHE. 1976. The role of genes and their expression. In: J. W. Hastings and H.-G. Schweiger, eds., *The Molecular Basis of Circadian Rhythms*. Abakon Verlagsgesellschaft. Berlin, pp. 295–310.

SARGENT, T. D. 1962. A study of homing in the bank swallow (*Riparia riparia*). Auk 79:234–246.

SARGENT, T. D. 1968. Cryptic moths: Effects on background selection of painting the circumocular scales. Science 159:100–101.

SARGENT, T. D. 1969. Behavioural adaptations of cryptic moths. III: Resting attitudes of two bark-like species, *Melanolophia canadaria* and *Catocala ultronia*. Anim. Behav. 17:670–672.

SARGENT, T. D. 1976. *Legion of Night: The Underwing Moths*. University of Massachusetts Press. Amherst. 222 pp.

SASSOON, D. AND D. KELLEY. 1986. The sexually dimorphic larynx of *Xenopus laevis*: Development and androgen regulation. Am. J. Anat. 177:457–472.

SAUER, E. G. F. 1957. Die Sterenorientierung nächtlich ziehender Grasmücken (*Sylvia atricapilla, borin and curruca*). Z. Tierpsychol. 14:29–70.

SAUER, E. G. F. 1961. Further studies on the stellar orientation of nocturnally migrating birds. Psychol. Forschung 26:224–244.

SAUER, E. G. F. 1963. Migration habits of golden plovers. Proc. 13th. Inter. Ornithol. Congr. 454–467.

SAUER, E. G. F. AND E. M. SAUER. 1960. Star navigation in nocturnal migrating birds. The 1958 planetarium experiments. Cold Spring Harbor Symp. Quant. Biol. 25:463–473.

SAUNDERS, D. S. 1977. *An Introduction to Biological Rhythms.* Halstead Press, a division of John Wiley & Sons. New York. 170 pp.

SAVAGE-RUMBAUGH, E. S. 1986. *Ape Language from Conditioned Response to Symbol.* Columbia University Press. New York. 433 pp.

SAVAGE-RUMBAUGH, E. S., D. M. RUMBAUGH, AND S. BOYSEN. 1978a. Symbolic communication between two chimpanzees (*Pan troglodytes*). Science 201:641–644.

SAVAGE-RUMBAUGH, E. S., D. M. RUMBAUGH, AND S. BOYSEN. 1978b. Linguistically mediated tool use and exchange by chimpanzees *Pan troglodytes.* Behav. Brain Sci. 1:539–554.

SAVAGE-RUMBAUGH, E. S., D. M. RUMBAUGH, AND S. BOYSEN. 1980. Do apes use language? Am. Sci. 68:49–61.

SAWAKI, Y., I. NIHONMATSU, AND H. KAWAMURA. 1984. Transplantation of the neonatal suprachiasmatic nuclei into rats with complete bilateral suprachiasmatic lesions. Neurosci. Res. 1:67–72.

SCHALLER, G. 1972. The sociable kingdom. In: *The Marvels of Animal Behavior.* National Geographical Society. Washington, DC, pp. 66–87.

SCHALLER, G. B. 1963. *The Mountain Gorilla.* University of Chicago Press. Chicago. 431 pp.

SCHALLER, G. B. 1972. *The Serengeti Lion.* University of Chicago Press. Chicago. 480 pp.

SCHAPIRO, S., M. SALAS, AND K. VUKOVICH. 1970. Hormonal effects on ontogeny of swimming ability in the rat: Assessment of central nervous system development. Science 168:147–151.

SCHEICH, H., G. LANGER, C. TIDEMAN, R. B. COLES, AND A. GUPPY. 1986. Electroreception and electrolocation in platypus. Nature 319:401–402.

SCHELLER R. H. AND R. AXEL. 1984. How genes control an innate behavior. Sci. Am. 250 (Mar):54–62.

SCHERMULY, L. AND R. KLINKE. 1990. Infrasound sensitive neurones in the pigeon cochlear ganglion. J. Comp. Physiol. A 166:355–363.

SCHLEIDT, W. 1961a. Über die Auslösung des Kollern beim Truthahn (*Meleagris galopavo*) Z. Tierpsychol. 11:417–435.

SCHLEIDT, W. 1961b. Reaktionen von Truthühnern auf fliegende Raubvögel und Veruche zur Analyse ihrer AAM's. Z. Tierpsychol. 18:534–560.

SCHLENOFF, D. H. 1985. The startle responses of blue jays to *Catocala* (Lepidoptera: Noctuidae) prey models. Anim. Behav. 33:1057–1067.

SCHMIDT-KOENIG, K. 1987. Bird navigation: Has olfactory orientation solved the problem? Quart. Rev. Biol. 62:31–47.

SCHMIDT-KOENIG, K. AND H. J. SCHLICHTE. 1972. Homing in pigeons with impaired vision. Proc. Nat. Acad. Sci. USA 69:2446–2447.

SCHNEIDER, D. 1966. Chemical sense communication in insects. Symp. Soc. Exp. Biol. 20:273–297.

SCHNEIDER, D. 1974. The sex-attractant receptor of moths. Sci. Am. 231 (Jul):28–35.

SCHNEIRLA, T. C. 1971. *Army Ants: A Study in Social Organization.* W. H. Freeman and Co. San Francisco. 349 pp.

SCHNEIRLA, T. C. AND J. S. ROSENBLATT. 1961. Behavioral organization and genesis of the social bond in insects and mammals. Amer. J. Orthopsychiat. 31:223–253.

SCHNEIRLA, T. C., J. S. ROSENBLATT, AND E. TOBACH. 1963. Maternal behavior in the cat. In: H. L. Rheingold, ed., *Maternal Behavior in Mammals.* John Wiley & Sons. New York, pp. 122–168.

SCHOENER, T. W. 1968. Sizes of feeding territories among birds. Ecology 49:123–141.

SCHOENER, T. W. 1983. Simple models of optimal feeding-territory size: A reconciliation. Am. Nat. 121:608–629.

SCHOENER, T. W. 1987. A brief history of optimal foraging ecology. In: A. C. Kamil, J. R. Krebs, and H. R. Pulliam, eds., *Foraging Behavior*. Plenum Press. New York, pp. 5–67.

SCHOLZ, A. T., R. M. HORRAL, J. C. COOPER, A. D. HASLER, D. M. MADISON, J. J. POFF, AND R. DALY. 1975. Artificial imprinting of salmon and trout in Lake Michigan. Wisconsin Dept. Nat. Res. Fishery Management Report 80. 45 pp.

SCHOLZ, A. T., R. M. HORRALL, J. C. COOPER, AND A. D. HASLER. 1976. Imprinting to chemical cues: The basis for home stream selection in salmon. Science 192:1247–1249.

SCHÖNE, H. 1984. *Spatial Orientation*. Princeton University Press. Princeton, NJ. 347 pp.

SCHÖNE, H. AND H. SCHÖNE. 1963. Balz und andere Verhaltensweisen der Mangrove-krabbe *Goniopsis cruentata* Ltr. und das eulitoralen Brachyuren. Z. Tierpsychol. 20:641–656.

SCHULLER, G., K. BEUTER, AND R. RUBSAMEN. 1975. Dynamic properties of the compensation system for Doppler shifts in the bat, *Rhinolophus ferrumequinum*. J. Comp. Physiol. 97:113–125.

SCHULTEN, K. 1982. Magnetic field effects in chemistry and biology. Adv, Solid-State Phys. 22:61–83.

SCHULTEN, K. AND A. WINDMUTH. 1986. Model for a physiological magnetic compass. In: M. G. Kiepenheuer and J. Boccasra, eds., *Biophysical Effects of Steady Magnetic Fields*. Springer-Verlag. New York, pp. 96–106.

SCHUTZ, VON F. 1965. Sexuelle Prägung bei Anatiden. Z. Tierpsychol. 22:50–103.

SCHÜZ, E. 1949. Die Spät-Auflassung ostpreussischer Jungstörche in West-Deutschland durch die Vogelwarte Rossiten 1933. Vogelwarte 15:63–78.

SCHWAB, R. G. 1971. Circannian testicular periodicity in the European starling in the absence of photoperiodic change. In: M. Menaker, ed., *Biochronometry*. National Academy of Science. Washington, DC, pp. 428–447.

SCHWAGMEYER, P. L. 1979. The Bruce effect: An evaluation of male/female advantages. Am. Nat. 114:932–938.

SCHWAGMEYER, P. L. AND G. A. PARKER. 1990. Male mate choice as predicted by sperm competition in thirteen-lined ground squirrels. Nature 348:62–64.

SCHWIND, R. 1983. Zonation of the optical environment and zonation in the rhabdom structure within the eye of the back swimmer, *Notonecta glauca*. Cell Tissue Res. 232:53–62.

SCHWIND, R. 1991. Polarization vision in water insects and insects living on a moist substrate. J. Comp. Physiol. A 169:531–540.

SCHWIPPERT, W. W., T. W. BENEKE, AND J. -P. EWERT. 1990. Response of medullary neurons to visual stimuli in the common toad. II. An intracellular recording and cobalt-lysine labeling study. J. Comp. Physiol. A 167:509–520.

SCOTT, J. P. 1962. Critical periods in behavioral development. Science 138:949–958.

SCOTT, J. P. AND J. L. FULLER. 1965. *Genetics and the Social Behavior of the Dog*. University of Chicago Press, Chicago. 281 pp.

SEARCY, W. A. 1979. Female choice of mates: A general model for birds and its application to red-winged blackbirds (*Agelaius phoeniceus*). Am. Nat. 114:77–100.

SEARCY, W. A., D. ERIKSSON, AND A. LUNDBERG. 1991. Deceptive behavior in pied flycatchers. Behav. Ecol. Sociobiol. 29:167–175.

SEARLE, L. V. 1949. The organization of hereditary maze-brightness and maze-dullness. Gen. Psychol. Monogr. 39:279–335.

SEIBT, U. AND W. WICKLER. 1987. Gerontophagy versus cannibalism in the social spiders *Stegodyphus mimosarum* Pavesi and *Stegody-*

phus dumicola Pocock. Anim. Behav. 35:1903–1905.

SEITZ, A. 1940. Die Paarbildung bei einigen Cichliden. Z. Tierpsychol. 4:40–84.

SELANDER, R. K. 1966. Sexual dimorphism and differential niche utilization in birds. Condor 68:113–151.

SELIGMAN, M. E. P. 1970. On the generality of the laws of learning. Psychol. Rev. 77:406–418.

SELSET, R. AND K. B. DOVING. 1980. Behavior of mature anadromous char (*Salmo alpinus* L.) towards odorants produced by smolts of their own population. Acta Physiologica Scand. 108:113–122.

SEMM, P. AND C. DEMAINE. 1986. Neurophysiological properties of magnetic cells in the pigeon's visual system. J. Comp Physiol. A 156:619–625.

SEMM, P. AND R. C. BEASON. 1990. Responses to small magnetic variations by the trigeminal system of the bobolink. Brain Res. Bull. 25:735–740.

SEMM, P., D. NOHR, C. DEMAINE, AND W. WILTSCHKO. 1984. Neural basis of the magnetic compass: Interactions of visual, magnetic and vestibular inputs in the pigeon's brain. J. Comp. Physiol. A 155:283–288.

SEMM, P., T. SCHNEIDER, AND L. VOLLRATH. 1980. Effects of an earth-strength magnetic field on electrical activity of pineal cells. Nature 288:607–609.

SERVENTY, D. L. 1971. Biology of desert birds. In: D. S. Farner, J. R. King, and K. C. Parkes, eds., *Avian Biology*, vol. 1. Academic Press. New York, pp. 287–339.

SEYFARTH, R. M. 1976. Social relationships among adult female baboons. Anim. Behav. 24:917–938.

SEYFARTH, R. M. 1977. A model of social grooming among adult female monkeys. J. Theoret. Biol. 65:671–698.

SEYFARTH, R. M., D. L. CHENEY, AND P. MARLER. 1980. Monkey responses to three different alarm calls: Evidence of predator classification and semantic classification. Science 210:801–803.

SHALTER, M. D. 1978. Localization of passerine seet and mobbing calls by goshawks and pygmy owls. Z. Tierpsychol. 46:260–267.

SHALTER, M. D. AND W. M. SCHLEIDT. 1977. The ability of barn owls, *Tyto alba*, to discriminate and localize avian alarm calls. Ibis 119:22–27.

SHERMAN, P. W. 1977. Nepotism and the evolution of alarm calls. Science 197:1226–1253.

SHERMAN, P. W. 1980a. The limits of ground squirrel nepotism. In: G. W. Barlow and J. Silverberg, eds., *Sociobiology: Beyond Nature/Nurture?* Westview Press. Boulder, CO, pp. 505–544.

SHERMAN, P. W. 1980b. The meaning of nepotism. Am. Nat. 116:604–606.

SHERMAN, P. W. 1981. Kin recognition in animals. Behav. Ecol. Sociobiol. 8:251–259.

SHERMAN, P. W. 1985. Alarm calls of Belding's ground squirrels to aerial predators: Nepotism or self-preservation? Behav. Ecol. Sociobiol. 17:313–323.

SHERMAN, P. W. 1988. The levels of analysis. Anim. Behav. 36:616–619.

SHERMAN, P. W., J. U. M. JARVIS, AND S. H. BRAUDE, 1992. Naked mole rats. Sci. Am. 267 (Aug):72–78.

SHIELDS, W. M. 1982. *Philopatry, Inbreeding, and the Evolution of Sex*. State University of New York Press. Albany.

SHIELDS, W. M. 1984. Barn swallow mobbing: Self-defence, collateral kin defence, group defence, or parental care? Anim. Behav. 32:132–148.

SHIELDS, W. M. 1987. Dispersal and mating systems: Investigating their causal connections. In: B. D. Chepko-Sade and Z. T. Halpin, eds., *Mammalian Dispersal Patterns*. The University of Chicago Press. Chicago, pp. 3–24.

SHIOTSUKA, R., J. JOVONOVICH, AND J. A. JOVONOVICH. 1974. In vitro data on drug sensitivity: Circadian and ultradian corticosterone rhythms in adrenal organ cultures. In: J. Aschoff, F. Ceresa and F. Halberg, eds., *Chronobiological Aspects of Endocrinology.* Schattauer-verlag. Stuttgart, pp. 255–267.

SIDMAN, R. L., M. C. GREEN, AND S. H. APPEL. 1965. *Catalog of the Neurological Mutants of the Mouse.* Harvard University Press, Cambridge, MA. 82 pp.

SIEGEL, R. G. AND W. K. HONIG. 1970. Pigeon concept formation: Successive and simultaneous acquisition. J. Exp. Analysis Behav. 13:385–390.

SIGG, H. AND J. FALETT. 1985. Experiments on respect of possession and property in hamadryas baboons (*Papio hamadryas*). Anim. Behav. 33:978–984.

SILBERGLIED, R. E., J. G. SHEPHERD, AND J. L. DICKINSON. 1984. Eunuchs: The role of apyrene sperm in Lepidoptera. Am. Nat. 123:255–265.

SILK, J. B., C. B. CLARK-WHEATLEY, P. S. RODMAN, AND A. SAMUELS. 1981. Differential reproductive success and facultative adjustment of sex ratios among captive female bonnet macaques (*Macaca radiata*). Anim. Behav. 29:1106–1120.

SILVER, R., P. WITKOVSKY, P. HORVATH, V. ALONES, C. J. BARNSTABLE, AND M. N. LEHMAN. 1988. Coexpression of opsin- and VIP-like–immunoreactivity in CSF-containing neurons of the avian brain. Cell Tissue Res. 253:189–198.

SIMMONS, J. A. 1973. The resolution of target range by echolocating bats. J. Acoust. Soc. Am. 54:157–173.

SIMMONS, J. A. 1974. The response of the Doppler shift echolocation system in the bat *Rhinolophus ferrumequinum.* J Acoust. Soc. Am. 56:672–682.

SIMMONS, J. A. 1979. The perception of phase information in bat sonar. Science 204:1336–1338.

SIMMONS, J. A., B. M. FENTON, AND M. J. O'FARRELL. 1979. Echolocation and pursuit of prey by bats. Science 203:16–21.

SIMMONS, J. A., D. J. HOWELL, AND N. SUGA. 1975. Information content of bat sonar echoes. Am. Sci. 63:204–215.

SIMMONS, L. W. 1988. Male size, mating potential and lifetime reproductive success in the field cricket, *Gryllus bimaculatus* (De Geer). Anim. Behav. 36:372–379.

SIMPSON, A. E. AND M. J. A. SIMPSON. 1985. Short-term consequences of different breeding histories for captive rhesus macaque mothers and young. Behav. Ecol. Sociobiol. 18:83–89.

SIMPSON, S. M. AND B. K. FOLLETT. 1981. Pineal and hypothalamic pacemakers: Their role in regulating circadian rhythmicity in Japanese quail. J. Comp. Physiol. 144:381–389.

SINCLAIR, S. 1985. *How Animals See—Other Visions of Our World.* Facts on File Publications. New York.

SIWICKI, K. K., C. EASTMAN, G. PETERSEN, M. ROSBASH, AND J. C. HALL. 1988. Antibodies to the *period* gene product of *Drosophila* reveal diverse tissue distribution and rhythmic changes in the visual system. Neuron 1:141–159.

SKINNER, B. F. 1953. *Science and Human Behavior.* Free Press Paperback, Macmillan. New York. 461 pp.

SKIPPER, R. AND G. SKIPPER. 1957. Those British-bred pompadours—the story completed. Water Life and Aquarium World 12(2):63–64.

SKUTCH, A. F. 1935. Helpers at the nest. Auk 52:257–273.

SKUTCH, A. F. 1961. Helpers among birds. Condor 63:198–226.

SKUTCH, A. F. 1987. *Helpers at Birds' Nests.* University of Iowa Press. Iowa City. 298 pp.

SLATER, P. J. B., L. A. EALES, AND N. S. CLAYTON. 1988. Song learning in zebra finches (*Taeniopygia guttata*): Progress and prospects. In: J. S. Rosenblatt, C. Beer, M. Busnel, and P. J. B. Slater, eds., *Advances in the Study of Behavior*, vol. 18. Academic Press. New York, pp. 1–34.

SMITH, J. N. M. 1974. The food searching behavior of two European thrushes. II. The adaptiveness of the search patterns. Behaviour 49:1–61.

SMITH, J. N. M. AND H. P. A. SWEATMAN. 1974. Food searching behavior of titmice in patchy environments. Ecology 55:1216–1232.

SMITH, R. J. F. 1982. The adaptive significance of the alarm substance—fright reaction system. In: T. J. Hara, ed., *Chemoreception in Fishes*. Elsevier Scientific Publishing Co. New York, pp. 327–342.

SMITH, R. J. F. 1989. The response of *Asterropteryx semipunctatus* and *Gnatholepis anjerensis* (Pisces, Gobiidae) to chemical stimuli from injured conspecifics, an alarm response in gobies. Ethology 81:279–290.

SMITH, R. L. 1979. Repeated copulation and sperm precedence: Paternity assurance for a male brooding water bug. Science 205:1029–1031.

SMITH, W. J. 1969. Messages of vertebrate communication. Science 165:145–150.

SMITH, W. J. 1977. *The Behavior of Communicating, An Ethological Approach*. Harvard University Press. Cambridge, MA. 545 pp.

SMITH, W. J. 1991. Animal communication and the study of cognition. In: C. A. Ristau, ed., *Cognitive Ethology, the Minds of Other Animals*. Lawrence Erlbaum. Hillsdale, NJ, pp. 209–230.

SMOTHERMAN, W. P. 1982. Odor aversion learning by the rat fetus. Physiol. Behav. 29:769–771.

SNOW, B. K. 1974. Lek behavior and breeding of Guy's hermit hummingbird *Phaethornis guy*. Ibis 116:278–297.

SNOW, D. W. 1963. The evolution of manakin displays. Proc. 13th Int. Ornithol. Congr. 1:553–561.

SNYDER, R. L. AND C. D. CHENEY. 1975. Homing performance of anosmic pigeons. Bull. Pschonom. Soc. 6:592–594.

SOLLARS, P. J. AND D. P. KIMBLE. 1988. Cross species transplantation of fetal hypothalamic tissue restores circadian locomotor rhythm to SCN-lesioned hosts. Soc. for Neurosci. Abstr. 14:49.

SOLOMON, D. J. 1973. Evidence for pheromone-influenced homing by migrating Atlantic salmon, *Salmo salar* (L.). Nature (London) 244:231–232.

SONG, P.-S. AND K. L. POFF. 1989. Photomovement. In: *The Science of Photobiology*. Plenum Press. New York, pp. 305–346.

SOTTHIBANDHU, S. AND R. BAKER. 1979. Celestial orientation by the large yellowing moth, *Noctua pronuba* L. Anim. Behav. 27:786–800.

SOUTHERN, H. N. 1954. Mimicry in cuckoos' eggs. In: J. Hurley, A. C. Hardy, E. B. Ford, eds. *Evolution as a Process*. George Allen and Unwin, London, pp. 219–232.

SOUTHERN, W. E. 1978. Orientation responses of ring-billed gull chicks. In: K. Schmidt-Koenig and W. T. Keeton, eds., *Animal Migration, Navigation and Homing*. Springer-Verlag. Berlin, pp. 311–317.

SPENCER, H. 1855. *Principles of Psychology*. D. Appleton and Company, New York.

SPRINGER, S. 1948. Oviphagous embryos of the sand shark, *Carcharias taurus*. Copeia 3:153–157.

SPROTT, R. L. AND J. STAATS. 1975. Behavioral studies using genetically defined mice—A bibliography. Behav. Gen. 5:27–82.

SPROTT, R. L. AND J. STAATS. 1978. Behavioral studies using genetically defined mice—A bibliography (July 1973– July 1976). Behav. Gen. 8:183–206.

SPROTT, R. L. AND J. STAATS. 1979. Behavioral studies using genetically defined mice—A bibliography (July 1976–August 1978). Behav. Gen. 9:87–102.

STACEY, N. E. 1987. Role of hormones and pheromones in fish reproductive behavior. In: D. Crews, ed., *Psychobiology of Reproductive Be-*

havior. Prentice Hall. Englewood Cliffs, NJ, pp. 28–60.

STACEY, P. B. 1979. Habitat saturation and communal breeding in the Acorn woodpecker. Anim. Behav. 27:1153–1166.

STACEY, P. B. AND C. E. BOCK. 1978. Social plasticity in the Acorn Woodpecker. Science 202:1298–1300.

STACEY, P. B. AND J. D. LIGON. 1987. Territory quality and dispersal options in the acorn woodpecker, and a challenge to the habitat saturation model of cooperative breeding. Am. Nat. 120:654–676.

STALLCUP, J. A. AND G. E. WOOLFENDEN. 1978. Family status and contributions to breeding by Florida Scrub jays. Anim. Behav. 26:1144–1156.

STAMPS, J. A. 1991. Why evolutionary issues are reviving interest in proximate behavioral mechanisms. Am. Zool. 31:338–348.

STEHLE, J., S. REUSS, H. SCHRÖEDER, M. HENSCHEL, AND L. VOLLRATH. 1988. Magnetic effects on pineal n-acetyltransferase activity and melatonin content in the gerbil—role of pigmentation and sex. Physiol. Behav. 44:91–94.

STEHN, R. A. AND M. E. RICHMOND. 1975. Male induced pregnancy termination in the prairie vole, *Microtus ochrogaster*. Science 187:1210–1211.

STEINACH, E. 1940. *Sex and Life*. The Viking Press, New York. 252 pp.

STENT, G. S. 1981. Strength and weakness of the genetic approach to the development of the nervous system. Ann. Rev. Neurosci. 4:163–194.

STEPHENS, D. W. 1987. On economically tracking a variable environment. Theoret. Pop. Biol. 32:15–25.

STEPHENS, D. W. AND E. L. CHARNOV. 1982. Optimal foraging: some simple stochastic models. Behav. Ecol. Sociobiol. 10:251–263.

STEPHENS, D. W. AND J. R. KREBS. 1986. *Foraging Theory*. Princeton University Press. Princeton, NJ.

STEPHENS D. W. AND S. R. PATON. 1986. How constant is the constant of risk-aversion? Anim. Behav. 34:1659–1667.

STETSON, M. H. AND M. WATSON-WHITMYRE. 1976. Nucleus suprachiasmaticus. The biological clock in the hamster? Science 191:197–199.

STEVENSON, P. A. AND W. KUTSCH. 1987. A reconsideration of the central pattern generator concept for locust flight. J. Comp. Physiol. A 161:115–129.

STRUHSAKER, T. T. 1967. Auditory communication among vervet monkeys (*Cercopithecus aethiops*). In: S. A. Altmann, ed., *Social Communication Among Primates*. University of Chicago Press. Chicago, pp. 281–324.

STUART, A. 1963a. Origin of the trail in the termites *Naustitermes corniger* (Motulsky) and *Zootermopsis nevadensis* (Hagan) Isoptera. Physiol. Zool. 36:69–84.

STUART, A. 1963b. Studies on the communication of alarm in the termite, *Zootermopsis nevadensis* (Hagan) Isoptera. Physiol. Zool. 36:85–96.

STUART, A. 1967. Alarm, defense, and construction behavior relationships in termites (Isoptera). Science 156:1123–1125.

SUGA, N. 1990. Biosonar and neural computation in bats. Sci. Am. 262 (June):60–68.

SUGIYAMA, Y. 1965. Behavioral development and social structure in two troops of hanuman langurs (*Presbytis entellus*). Primates 6:213–247.

SUGIYAMA, Y. 1984. Proximate factors of infanticide among langurs at Dharwar: A reply to Boggess. In: G. Hausfater and S. B. Hrdy, eds., *Infanticide: Comparative and Evolutionary Perspectives*. Aldine. New York, pp. 311–314.

SVARE, B. 1988a. Some trends in the responses studied and the species employed by behavioral endocrinologists. Horm. Behav. 22:139–142.

SVARE, B. 1988b. Genotype modulates the aggression-promoting quality of progesterone in pregnant mice. Horm. Behav. 22:90–99.

SWEENEY, B. M. 1976. Evidence that membranes are components of circadian oscillators. In: J. W. Hastings and H.-G. Schweiger, eds., *Molecular Basis of Circadian Rhythms*. Abakon Verlagsgesellschaft. Berlin, pp. 267–281.

SWEENEY, B. M. AND F. T. HAXO. 1961. Persistence of a photosynthetic rhythm in enucleated *Acetabularia*. Science 134:1361–1363.

SYMINGTON, M. McFARLAND. 1987. Sex ratio and maternal rank in wild spider monkeys: When daughters disperse. Behav. Ecol. Sociobiol. 20:421–425.

TABORSKY, M. AND D. LIMBERGER. 1981. Helpers in fish. Behav. Ecol. Sociobiol. 8:143–145.

TAKAHASHI, J. S. AND M. MENAKER. 1979. Brain mechanisms in avian circadian systems. In: M. Suda, O. Hayaishi, and H. Nakagawa, eds. *Biological Rhythms and Their Central Mechanisms*. Elsevier/North Holland. Amsterdam, pp. 95–109.

TAKAHASHI, J. S., H. HAMM, AND M. MENAKER. 1980. Circadian rhythms of melatonin release from individual superfused chicken pineal glands *in vitro*. Proc. Nat. Acad. Sci. USA 77:2319–2322.

TAMURA, N. 1989. Snake-directed mobbing by the Formosan squirrel *Callosciurus erythraeus thaiwanensis*. Behav. Ecol. Sociobiol. 24:175–180.

TANOUYE, M. A., C. A. FERRUS, AND S. C. FUJITA. 1981. Abnormal action potentials associated with the *Shaker* complex locus in *Drosophila*. Proc. Nat. Acad. Sci. USA 78:6548–6552.

TARBURTON, M. K. AND E. O. MINOT. 1987. A novel strategy of reproduction in birds. Anim. Behav. 35:1898–1899.

TAUTZ, J. 1979. Reception of particle oscillation in a medium—An unorthodox sensory capacity. Naturwiss. 66:452–461.

TAUTZ, J. AND H. MARKL. 1978. Caterpillars detect flying wasps by hairs sensitive to airborne vibration. Behav. Ecol. Sociobiol. 4:101–110.

TAYLOR, D. H. AND K. ADLER. 1973. Spatial orientation by salamanders using plane polarized light. Science 181:285–287.

TEMELES, E. J. 1989. Effect of prey consumption on foraging activity of northern harriers. Auk 106:353–357.

TEPPERMAN, J. 1980. *Metabolic and Endocrine Physiology*, 4th ed. Year Book Medical Publishers, Chicago.

TERBORGH, J. AND A. W. GOLDIZEN. 1985. On the mating system of the cooperatively breeding saddle-backed tamarin (*Saguinus fuscicollis*). Behav. Ecol. Sociobiol. 16:293–299.

TERKEL, J. 1972. A chronic cross-transfusion technique in freely behaving rats by use of a single heart catheter. J. Appl. Physiol. 33:519–522.

TERKEL, J. AND J. S. ROSENBLATT. 1972. Humoral factors underlying maternal behavior at parturition: Cross-transfusion between freely moving rats. J. Comp. Physiol. Psychol. 80:365–371.

TERMAN, M. AND J. S. TERMAN. 1970. Circadian rhythm of brain self-stimulation behavior. Science 168:1242–1244.

TERNES, J., D. FARNER, AND M. DEAVORS. 1967. A free-running circadian operant behavior rhythm in the chimpanzee. U.S..Air Force Tech. Doc. Rep. ARL-TR 67-13, 1–18.

TERRACE, H. S., L. A. PETITTO, R. J. SANDERS, AND T. G. BEVER. 1979. Can an ape create a sentence? Science 206:891–900.

TERRILL, S. B. 1991. Evolutionary aspects of orientation and migration in birds. In: P. Berthold, ed., *Orientation in Birds*. Birkhäuser Verlag. Basel, Switzerland, pp. 180–201.

THEODORAKIS, C. W. 1989. Size segregation and the effects of oddity on predation risk in minnow schools. Anim. Behav. 38:496–502.

THIELCKE, I. G. 1965. Gesangsgeographische Variation des Gartenbaum läufers (*Certhia brachydactyla*) in Hinblick auf das Artbildungsproblem. Z. Tierpsychol. 22:542–566.

THORNDIKE, E. L. 1898. Animal intelligence: An experimental study of the associative process in animals. Psych. Monogr. 2(8).

THORNDIKE, E. L. 1911. *Animal Intelligence: Experimental Studies*. Macmillan, New York. 109 pp.

THORNHILL, R. 1980. Mate choice in *Hylobittacus apicalis* (Insecta: Mecoptera) and its relation to some models of female choice. Evolution 34:519–538.

THORNHILL, R. 1981. *Panorpa* (Mecoptera: Panorpidae) scorpionflies: Systems for understanding resource-defense polygyny and alternative male reproductive efforts. Ann. Rev. Ecol. Syst. 12:355–386.

THORNHILL, R. AND J. ALCOCK. 1983. The evolution of insect mating systems. Harvard University Press. Cambridge, MA. 547 pp.

THORPE, W. H. 1963. *Learning and Instinct in Animals*, 2nd ed. Methuen. London.

THORPE, W. H. 1972. Duetting and antiphonal song in birds. Its extent and significance. Behaviour Suppl. 18:1–197.

THORPE, W. H. AND F. G. W. JONES. 1937. Olfactory conditioning in a parasitic insect and its relation to the problem of host selection. Proc. Roy. Soc., B, 124:56–81.

TINBERGEN, L. 1960. The natural color of insects in pinewoods. I. Factors influencing the intensity of predation by songbirds. Archives Néerlandaises de Zoologie. 13:265–343.

TINBERGEN, N. 1942. An objectivistic study of the innate behaviour of animals. Biblio. Biotheor. 1:39–98.

TINBERGEN, N. 1948. Social releasers and the experimental method required for their study. Wilson Bull. 60:5–52.

TINBERGEN, N. 1951. *The Study of Instinct*. Oxford Clarendon Press, London. 228 pp.

TINBERGEN, N. 1952a. The curious behavior of the stickleback. Sci. Am. 187 (Dec):22–26.

TINBERGEN, N. 1952b. Derived activities: Their causation, biological significance, origin and emancipation during evolution. Quart. Rev. Biol. 27:1–32.

TINBERGEN, N. 1960. *The Herring Gull's World*. Doubleday, Garden City, New York. 255 pp.

TINBERGEN, N. 1963. On aims and methods of ethology. Z. Tierpsychol. 20:410–433.

TINBERGEN, N. 1965. Behavior and natural selection. In: A. J. Moore, ed., *Ideas in Evolution and Behavior*, vol. 6. Proc. XVIth Int. Congr. Zool. Natural History Press, New York. 519–542.

TINBERGEN, N. AND A. C. PERDECK. 1950. On the stimulus situation releasing the begging response in the newly hatched herring gull chick (*Larus argentatus argentatus* Pont.) Behaviour 3:1–38.

TINBERGEN, N. AND W. KRUYT. 1938. Über die Oreintierung des Bienenwolfes (*Philanthus triangulum* Fabr.) III. Die Bevorzugung bestimmter Wegmarken. Z. vergl. Physiol. 25:292–334.

TINBERGEN, N., G. J. BROEKHUYSEN, F. FEEKES, J. C. W. HOUGHTON, H. KRUUK, AND E. SZULC. 1962. Egg-shell removal by the black-headed gull, *Larus ridibundus* L., a behaviour component of camouflage. Behaviour 19:74–118.

TODD, J. H. 1971. The chemical language of fish. Sci. Am. 224 (May):98–108.

TODD, J. H., J. ATEMA, AND J. E. BARDACH. 1967. Chemical communication in social behavior of a fish, the yellow bullhead (*Ictalurus natalis*). Science 158:672–673.

TOLMAN, E. C. 1924. The inheritance of maze-learning ability in rats. J. Comp. Psychol. 4:1–18.

TOLMAN, E. C. 1948. Cognitive maps in rats and men. Psychol. Rev. 55:189–208.

TOLMAN, E. C. AND C. H. HOLZIK. 1930. Introduction and removal of reward and maze performance in rats. Univ. of Calif. Publ. Psychol. 4:257–275.

TONGIORGI, P. 1970. Evidence of a moon orientation in the wolf spider, *Arctosa variana* C. L. Koch (araneae, Lycosidae). Bull. Mus. Nat. Hist. Natur. 41:243–249.

TOWNE, W. F. AND J. L. GOULD. 1985. Magnetic field sensitivity in honeybees. In: J. H. Kirschvink, D. S. Jones, and B. J. MacFadden, eds., *Magnetite, Biomineralization and Magnetoreception in Organisms—A New Biomagnetism*. Plenum Press. New York, pp. 385–406.

TOWNE, W. F. AND J. L. GOULD. 1988. The spatial precision of the honeybees' dance communication. J. Insect Behav. 1:129–155.

TOWNE, W. F. AND W. H. KIRCHNER. 1989. Hearing in honey bees: Detection of air-particle oscillations. Science 244:686–688.

TRANIELLO, J. F. A. AND S. N. BESHERS. 1991. Maximization of foraging efficiency and resource defense by group retrieval in the ant *Formica schaufussi*. Behav. Ecol. Sociobiol. 29:283–289.

TREISMAN, M. AND R. DAWKINS. 1976. The "cost of meiosis": Is there any? J. Theoret. Biol. 63:479–484.

TRIVERS, R. L. 1971. The evolution of reciprocal altruism. Quart. Rev. Biol. 46:35–57.

TRIVERS, R. L. 1972. Parental investment and sexual selection. In: B. Campbell, ed., *Sexual Selection and the Descent of Man, 1871–1971*. Aldine. Chicago, pp. 136–179.

TRIVERS, R. L. 1974. Parent–offspring conflict. Am. Zool. 14:249–264.

TRIVERS, R. L. AND D. E. WILLARD. 1973. Natural selection of parental ability to vary the sex ratio of offspring. Science 179:90–92.

TRIVERS, R. L. AND H. HARE. 1976. Haplodiploidy and the evolution of the social insects. Science 191:249–263.

TRUMAN, J. W. 1971. Circadian rhythms and phys-

iology with special reference to neuroendocrine processes in insects. In: *Circadian Rhymicity*. Center for Agricultural Publishing and Documentation Wageningen. pp. 111–135.

TRUMAN, J. W. 1972. Physiology of insect rhythms. II. The silk moth brain as the location of the biological clock controlling eclosion. J. Comp. Physiol. 81:99–114.

TRUMAN, J. W. AND L. M. RIDDIFORD. 1970. Neuroendocrine control of ecdysis in silkmoths. Science 167:1624–1626.

TRUMLER, E. 1956. Das "Rossigkeitsgesicht" und ahnliches Ausdrucksverhalten bei Einhufern Z. Tierpsychol. 16:478–488.

TRYON, R. C. 1940. Studies in individual differences in maze ability VII. The specific components of maze ability and a general theory of psychological components. J. Comp. Physiol. Psychol. 30:283–335.

TURCHIN, P. AND P. KAREIVA. 1989. Aggregation in *Aphis varians*: An effective strategy for reducing predation risk. Ecology 70:1008–1016.

TUREK, F. W., J. P. MCMILLAN, AND M. MENAKER. 1976. Melatonin: Effects on the circadian locomotor rhythm of sparrows. Science 194:1441–1443.

TYCHSEN, P. H. AND B. S. FLETCHER. 1971. Studies on the rhythm of mating in the Queensland Fruit fly, *Dacus tryoni*. J. Insect Physiol. 17:2139–2156.

UNDERWOOD, H. AND T. SIOPES. 1984. Circadian organization in quail. J. Exp. Zool. 232:557–566.

UNDERWOOD, H., R. KEITH BARRETT, AND T. SIOPES. 1990a. The quail's eye: A biological clock. J. Biol. Rhythms 5:257–265.

UNDERWOOD, H., R. KEITH BARRETT, AND T. SIOPES. 1990b. Melatonin does not link the eyes to the rest of the circadian system in quail: A neural pathway is involved. J. Biol. Rhythms 5:349–361.

URQUHART, F. A. 1987. *The Monarch Butterfly:*

International Traveler. Nelson-Hall. Chicago. 232 pp.

VAN DEN ASSEM, J. AND J. VAN DER MOLEN. 1969. The waning of the aggressive response in the three-spined stickleback upon constant exposure to a conspecific. I. A preliminary analysis of the phenomenon. Behaviour 34:286–324.

VAN LAWICK-GOODALL, J. 1968. Behavior of free-living chimpanzees of the Gombe Stream area. Anim. Behav. Monogr. 1:165–311.

VAN LAWICK-GOODALL, J. 1971. *In the Shadow of Man*. Dell. New York. 304 pp.

VAN RHINJN, J. G. AND R. VODEGEL. 1980. Being honest about one's intentions: An evolutionarily stable strategy for animal conflicts. J. Theoret. Biol. 85:623–641.

VAN RIPER, W. AND E. R. KALMBACH. 1952. Homing not hindered by wing magnets. Science 115:577–578.

VAN TETS, G. F. 1965. A comparative study of some social communication patterns in the Pelecaniformes. Am. Ornithol. Union Ornithological Monogr. 2:1–88.

VAN VALEN, L. 1973. A new evolutionary law. Evol. Theory 1:1–30.

VANDER WALL, S. B. 1982. An experimental analysis of cache recovery in Clark's nutcracker. Anim. Behav. 30:84–94.

VANDER WALL, S. B. AND H. E. HUTCHINS. 1983. Dependence of Clark's nutcracker, *Nucifraga columbiana*, on conifer seeds during the postfledgling period. Can. Field Nat. 97:208–214.

VANDER WALL, S. B. AND R. P. BALDA. 1977. Coadaptations of the Clark's Nutcracker and the Pinyon pine for efficient seed harvest and dispersal. Ecol. Monogr. 47:89–111.

VANĚČĚK, J., A. PAVLIK, AND H. ILLNEROVÁ. 1987. Hypothalamic melatonin receptor sites revealed by autoradiography. Brain Res. 435:359–362.

VARGO, E. L. AND L. PASSERA. 1991. Pheromonal and behavioral queen control over the production of gynes in the Argentine ant *Iridomyrmex humilis* (Mayr). Behav. Ecol. Sociobiol. 28:161–169.

VAUGHAN, T. A. 1986. *Mammalogy*. Saunders College Publishing. Philadelphia. 576 pp.

VEPSALAINEN, K. AND M. NUMMELIN. 1985. Male territoriality in the waterstrider *Limnoporus rufoscutellatus*. Ann. Zool. Fenn. 22:441–448.

VERNER, J. AND M. F. WILLSON. 1966. The influence of habitats on mating systems of North American passerine birds. Ecology 47:143–147.

VERRELL, P. A. 1988. The chemistry of sexual persuasion. New Sci. 118:40–43.

VERWEY, J. 1930. Die paarungsbiologie des Fischreihers. Zool. Jahrbücher, Abteilungen Physiol. 48:1–120.

VILLEE, C. A., E. P. SOLOMON, & P. W. DAVIS. 1985. *Biology*. Saunders College Publishing, Philadelphia, PA.

VINCE, M. 1969. Embryonic communication, respiration, and the synchronization of hatching. In: R. A. Hinde, ed., *Bird Vocalizations:Their Relations to Current Problems in Biology and Psychology*. Cambridge University Press. Cambridge, England pp. 233–260.

VINCENT, L. E. AND M. BEKOFF. 1978. Quantitative analysis of the ontogeny of predatory behavior in coyotes, *Canis latrans*. Anim. Behav. 26:225–231.

VISALBERGHI, E. AND E. ALLEVA. 1979. Magnetic influences on pigeon homing. Biol. Bull. 156:246–256.

VOM SAAL, F. S. 1981. Variation in phenotype due to random intrauterine positioning of male and female fetuses in rodents. J. Reprod. Fert. 62:633–650.

VOM SAAL, F. S. 1983. Variation in infanticide and parental behavior in male mice due to prior intrauterine proximity to female fetuses:

Elimination by prenatal stress. Physiol. Behav. 30:675–681.

VOM SAAL, F. S. AND F. H. BRONSON. 1978. In utero proximity of female mouse fetuses to males: Effect on reproductive performance during later life. Biol. Reprod. 19:842–853.

VOM SAAL, F. S. AND F. H. BRONSON. 1980a. Sexual characteristics of adult female mice are correlated with their blood testosterone levels during prenatal development. Science 208:597–599.

VOM SAAL, F. S. AND F. H. BRONSON. 1980b. Variation in length of the estrous cycle in mice due to former intrauterine proximity to male fetuses. Biol. Reprod. 22:777–780.

VON BEKESY, G. 1956. Current status of theories of hearing. Science 123:779–783.

VON FRISCH, K. 1950. Die Sonne als Kompass im Leben der Bienen. Experientia 6:210–221.

VON FRISCH. K. 1967. *The Dance Language and Orientation of Bees*. Harvard University Press. Cambridge, MA.

VON FRISCH, K. 1971. *Bees: Their Vision, Chemical Senses and Language*, rev. ed. Cornell University Press. Ithaca, NY. 157 pp.

VOS, G. J. DE. 1979. Adaptedness of arena behaviour in black grouse (*Tetrao tetrix*) and other grouse species (Tetraonidae). Behaviour 68:277–314.

VOSS, R. 1979. Male accessory glands and the evolution of copulatory plugs in rodents. Occ. Pap. Mus. Zool. Univ. Mich. 968:1–27.

WAAGE, J. K. 1979. Dual function of the damselfly penis: Sperm removal and transfer. Science 203:916–918.

WAAS, J. R. 1990 An analysis of communication during the aggressive interactions of the Little Blue Penguin (*Eudyptula minor*) In: L. S. Davis and J. T. Darby, eds. *Penguin Biology*. Academic Press. New York, pp. 345–376.

WAAS, J. R. 1991. The risks and benefits of signaling aggressive motivation: A study of cave-dwelling little blue penguins. Behav. Ecol. Sociobiol. 29:139–146.

WAGNER, G. 1976. Das Orientierungverhalten von Brieftauben im erdmagnetisch gestörten Gebiete des Chasseral. Rev. Suisse Zool. 83:883–890.

WAKE, M. 1977. Fetal maintenance and its evolutionary significance in the Amphibia: Gymnophiona. J. Herp. 11:379–386.

WAKE, M. 1980. Fetal tooth development and adult replacement in *Dermophis mexicanus* (Amphibia: Gymnophiona). J. Morph. 166:203–216.

WALCOTT, C. 1977. Magnetic fields and the orientation of homing pigeons under sun. J. Exp. Biol. 70:105–123.

WALCOTT, C. 1978. Anomalies in the earth's magnetic field increase the scatter of pigeon vanishing bearings. In: K. Schmidt-Koenig and W. T. Keeton, eds., *Animal Migration, Navigation and Homing*. Springer-Verlag. Berlin, pp. 143–151.

WALCOTT, C. 1980. Homing-pigeons vanishing directions at magnetic anomalies are not altered by magnets. J. Exp. Biol. 86:349–352.

WALCOTT, C. 1982. Is there evidence for a magnetic map in homing pigeons? In: F. Papi and H. G. Walraff, eds., *Avian Navigation*. Springer-Verlag. Berlin, pp. 99–108.

WALCOTT, C. 1991. Magnetic maps in pigeons. In: P. Berthold, ed., *Orientation in Birds*. Birkhauser Verlag. Basel, Switzerland, pp. 38–51.

WALCOTT, C. AND R. P. GREEN. 1974. Orientation of homing pigeons altered by a change in the direction of an applied magnetic field. Science 184:180.

WALCOTT, C., J. L. GOULD, AND J. L. KIRSCHVINK. 1979. Pigeons have magnets. Science 205:1027–1029.

WALD, G. 1968. Molecular basis of visual excitation. Science 162:230–239.

WALDMAN, B. 1987. Mechanisms of kin recognition. J. Theoret. Biol. 128:159–185.

WALDMAN, B., P. C. FRUMHOFF, AND P. W. SHERMAN. 1988. Problems of kin recognition. Trends Ecol. Evol. 3:8–13.

WALDVOGEL, J. A. 1989. Olfactory orientation by birds. In: D. M. Power, ed., *Current Ornithology*, vol. 6. Plenum Press. New York, pp. 269–321.

WALDVOGEL, J. A. AND J. B. PHILLIPS. 1982. New experiments involving permanent-resident deflector loft birds. In: F. Papi and H. G. Walraff, eds., *Avian Navigation*. Springer-Verlag. Berlin, pp. 179–189.

WALDVOGEL, J. A., J. B. PHILLIPS, AND A. I. BROWN. 1988. Changes in short-term deflector loft effect are linked to the sun compass of homing pigeons. Anim. Behav. 36:150–158.

WALDVOGEL, J. A., S. BENVENUTI, W. T. KEETON, AND F. PAPI. 1978. Homing pigeon orientation influenced by deflected winds at home loft? J. Comp. Physiol. 128:297–301.

WALKER, M. M. AND M. E. BITTERMAN. 1985. Conditioned responding to magnetic fields by honeybees. J. Comp. Physiol. A 157:67–71.

WALKER, M. M., J. L. KIRSCHVINK, S.-B. R. CHANG, AND A. E. DIZON. 1984. A candidate magnetoreceptor organ in yellowfin tuna, *Thunnus albacares*. Science 224:751–753.

WALRAFF, H. G. 1980. Olfaction and homing in pigeons: Nerve-section experiments, critique, hypotheses. J. Comp. Physiol. 139:209–224.

WALRAFF, H. G. 1981. The olfactory component of pigeon navigation: Steps of analysis. J. Comp. Physiol. 143:411–422.

WALTER, H. 1979. *Eleanora's Falcon: Adaptations to Prey and Habitat in a Social Raptor*. University of Chicago Press. Chicago.

WANG, X. AND M. NOVAK. 1992. Influence of the social environment on parental behavior and pup development of meadow voles (*Microtus pennsylvanicus*) and prairie voles (*M. ochrogaster*). J. Comp. Psychol. 106:163–171.

WARING, G. H. 1969. The blow sound of pronghorn (*Antilocapra americana*). J. Mamm. 50:647–648.

WARNER, R. R. 1978. The evolution of hermaphroditism and unisexuality in aquatic and terrestrial vertebrates. In: E. S. Reese and F. J. Lighter, eds., *Contrasts in Behavior*, vol. 3. John Wiley & Sons. New York, pp. 77–101.

WARNER, R. R. 1984. Mating behavior and hermaphroditism in coral reef fishes. Am. Sci. 72:128–136.

WASER, P. M. AND W. T. JONES. 1983. Natal philopatry among solitary mammals. Quart. Rev. Biol. 58:355–390.

WATERMAN, T. H. 1961. The *Physiology of Crustacea*, vol. 2. Academic Press. New York.

WATERMAN, T. H. 1975. The optics of polarization sensitivity. In: A. W. Snyder and R. Menzel, eds. *Photoreceptor Optics*. Springer-Verlag. New York, pp. 239–371.

WATERMAN, T. H. 1989. *Animal Navigation*. W. H. Freeman and Co. New York. 243 pp.

WATSON, J. B. 1930. *Behaviorism*. W. W. Norton & Company, Inc., New York.

WATTS, C. R. AND A. W. STOKES. 1971. The social order of turkeys. Sci. Am. 224 (June):112–118.

WEATHERHEAD, P. J. AND R. D. MONTGOMERIE. 1991. Good news and bad news about DNA fingerprinting. Trends Ecol. Evol. 6:173–174.

WEBER, N. A. 1972. The attines: The fungus-culturing ants. Am. Sci. 60:448–456.

WEEKS, J. C. AND J. W. TRUMAN. 1984. Neural organization of peptide-activated ecdysis behaviors during the retention of the proleg motor pattern despite loss of the prolegs at pupation. J. Comp. Physiol. A 155:423–433.

WEEKS, J. C. AND R. B. LEVINE. 1990. Postembryonic neuronal plasticity and its hormonal control during insect metamorphosis. Ann. Rev. Neurosci. 13:183–194.

WEEKS, J. C., G. A. JACOBS, AND C. I. MILES. 1989. Hormonally mediated modifications of neuronal structure, synaptic connectivity, and behavior during metamorphosis of the tobacco hornworm, *Manduca sexta*. Am. Zool. 29:1331–1344.

WEHNER, R. 1981. Spatial vision in arthropods. In: H. Autrum, ed., *Handbook of Sensory Physiology*, vol. VII/6C. Spring-Verlag. Berlin. pp. 287–417.

WEHNER, R. 1983. Celestial and terrestrial navigation: Human strategies–insect strategies. In: F. Huber and H. Markl, eds., *Neuroethologie and Behavioral Physiology*. Springer-Verlag. Berlin, pp. 336–381.

WEHNER, R. 1987. Matched filters—neural models of the external world. J. Comp. Physiol. 161:511–531.

WEHNER, R. AND F. RABER. 1979. Visual memory in desert ants *Cataglyphis bicolor* (Hymenoptera: Formididae). Experientia 35:1569–1571.

WEHNER, R. AND I. FLATT, 1972. The visual orientation of desert ants, *Cataglyphis bicolor,* by means of terrestrial cues. In: R. Wehner, ed., *Information Processing in Visual Systems of Arthropods*. Springer, Berlin. pp. 295–302.

WEHNER, R. AND M. V. SRINIVASAN. 1981. Searching behavior of desert ants, genus *Cataglyphis* (Formicidae. Hymenoptera). J. Comp. Physiol. 142:315–338.

WEIGMANN, C. AND J. LAMPRECHT. 1991. Intraspecific nest parasitism in bar-headed geese, *Anser indicus*. Anim. Behav. 41:677–688.

WEINTRAUB, A., R. BOCKMAN, AND D. KELLEY 1985. Prostaglandin E2 induces sexual receptivity in female *Xenopus laevis*. Horm. Behav. 19:386–399.

WEISMANN, A. 1889. *Essays upon Heredity and Kindred Biological Problems*, trans. by E. B. Poulton, S. Schonland, and A. E. Shipley. Clarendon Press. Oxford.

WELKER, H. A., P. SEMM, R. P. WILLIG, J. C. COMMENTZ, W. WILTSCHKO, AND L. VOLLRATH. 1983. Effects of an artificial magnetic field on serotonin N-acetyltransferase activity and mela-

tonin content of the rat pineal gland. Exp. Brain Res. 50:426–432.

WELLS, K. D. 1977. The courtship of frogs. In: D. H. Taylor and S. I. Guttman, eds., *The Reproductive Biology of Amphibians*. Plenum Press. New York, pp. 233–262.

WELTY, J. C. 1975. *The Life of Birds*. Second ed. Saunders, Philadelphia.

WENNER, A. M. 1964. Sound communication in honeybees. Sci. Am. 210 (Apr):116–125.

WENNER, A. M. 1971. *The Bee Language Controversy—An Experience in Science*. Educational Programs Improvement Corp., P.O. Box 3406. Boulder, CO. 109 pp.

WENNER, A. M. AND D. L. JOHNSON. 1966. Simple conditioning in honeybees. Anim. Behav. 14:149–155.

WENNER, A. M. AND P. H. WELLS. 1987. The honeybee dance language controversy: The search for "truth" vs. the search for useful information. Am. Bee J. 127:130–131.

WENNER, A. M. AND P. H. WELLS. 1990. *Anatomy of a Controversy—The Question of a "language" among Bees*. Columbia University Press. New York. 399 pp.

WERREN, J. H., M. R. GROSS, AND R. SHINE. 1980. Paternity and the evolution of male parental care. J. Theoret. Biol. 82:619–631.

WEST EBERHARD, M. J. 1975. The evolution of social behavior by kin selection. Quart. Rev. Biol. 50:1–33.

WEST, M. J. AND A. P. KING. 1988. Female visual displays affect the development of male song in the cowbird. Nature 334:244–246.

WEST, M. J., A. P. KING, AND D. H. EASTZER. 1981. The cowbird: Reflections on development from an unlikely source. Am. Sci. 69:56–66.

WESTBY G. W. M. AND K. J. PARTRIDGE. 1986. Human homing: Still no evidence despite geomagnetic controls. J. Exp. Biol. 120: 325–331.

WETZEL, D. AND D. KELLEY. 1983. Androgen and gonadotropin control of the mate calls of male South African clawed frogs, *Xenopus laevis*. Horm. Behav. 17:388–404.

WHITE, C. S., D. M. LAMBERT, C. D. MILLAR, P. M. STEVENS. 1991. Is helping behavior a consequence of natural selection? Am. Nat. 133:246–253.

WHITTEN, W. F., F. H. BRONSON, AND J. A. GREENSTEIN. 1968. Estrus inducing pheromone of male mice: Transport by movement of air. Science 161:584–585.

WICKLER, W. 1972. *Mimicry*. World University Library, McGraw-Hill. New York. 255 pp.

WICKLER, W. AND U. SEIBT. 1983. Monogamy: An ambiguous concept. In: P. Bateson ed., *Mate Choice*. Cambridge University Press. Cambridge, pp. 33–52.

WICKSTEN, M. K. 1980. Decorator crabs. Sci. Am. 242 (Feb):146–154.

WIGGINS, D. A. AND R. D. MORRIS. 1986. Criteria for female choice of mates: Courtship feeding and paternal care in the common tern. Am. Nat. 128:126–129.

WILCOX, R. S. 1972. Communication by surface waves: Mating behavior of a water strider (Gerridae). J. Comp. Physiol. 80:255–266.

WILCOX, R. S. 1979. Sex discrimination in *Gerris remigis*: Role of a water surface wave signal. Science 206:1325–1327.

WILCOX, R. S. AND J. R. SPENCE. 1986. The mating system of two hybridizing species of water striders (Gerridae). I. Ripple signal functions. Behav. Ecol. Sociobiol. 19:79–85.

WILCOX R. S. AND T. RUCKDESCHEL. 1982. Food threshold territoriality in a water strider (*Gerris remigis*). Behav. Ecol. Sociobiol. 11:85–90.

WILEY, R. H. 1974. Evolution of social organization and life history patterns among grouse (Aves: Tetraonidae). Quart. Rev. Biol. 49:209–227.

WILEY, R. H. AND D. G. RICHARDS. 1978. Physical constraints on acoustic communication in the atmosphere: Implications for the evolution of animal vocalizations. Behav. Ecol. Sociobiol. 3:69–94.

WILKINSON, G. S. 1984. Reciprocal food sharing in the vampire bat. Nature 308:181–184.

WILKINSON, G. S. 1990. Food sharing in vampire bats. Sci. Am. 262 (Feb):76–82.

WILLIAMS, C. B. 1965. *Insect Migration*. Collins. London. 237 pp.

WILLIAMS, G. C. 1966. *Adaptation and Natural Selection*. Princeton University Press. Princeton, NJ. 307 pp.

WILLIAMS, G. C. 1966. Natural selection, the costs of reproduction, and a refinement of Lack's principle. Am. Nat. 100:687–690.

WILLIAMS, G. C. 1975. *Sex and Evolution*. Princeton University Press. Princeton, NJ. 200 pp.

WILLIAMS, H. 1990. Models for song learning in the zebra finch: Fathers or others? Anim. Behav. 39:745–757.

WILLIAMS, T. C. AND J. M. WILLIAMS. 1978. An oceanic mass migration of land birds. Sci. Am. 239(Oct):166–176.

WILSON, D. M. 1961. The central nervous control of flight in a locust. J. Exp. Biol. 38:471–490.

WILSON, D. M. AND T. WEIS-FOGH. 1962. Patterned activity of co-ordinated motor units studied in flying locusts. J. Exp. Biol. 40:643–667.

WILSON, E. O. 1963. Pheromones. Sci. Am. 208 (May):100–114.

WILSON, E. O. 1965. Chemical communication in the social insects. Science 149:1064–1071.

WILSON, E. O. 1968. Chemical systems. In: T. A. Sebeok, ed., *Animal Communication: Techniques of Study and Results of Research*. Indiana University Press. Bloomington, pp. 75–102.

WILSON, E. O. 1971. *The Insect Societies*. Belknap/Harvard University Press, Cambridge, Massachusetts. 548 pp.

Wilson, E. O. 1972. Animal Communication. Sci. Am. 227 (Sept.):52–60.

WILSON, E. O. 1975. *Sociobiology*. Belknap Press

of Harvard University Press. Cambridge, MA. 697 pp.

WILSON, E. O. 1980. *Sociobiology, the Abridged Edition*. Belknap Press of Harvard University Press. Cambridge, MA. 366 pp.

WILSON, E. O. AND W. H. BOSSERT. 1963. Chemical communication among animals. Recent Prog. Hormone Res. 19:673–716.

WILTSCHKO, R. 1983. The ontogeny of orientation in young pigeons. Comp. Biochem. Physiol. A 76:701–708.

WILTSCHKO, R. AND W. WILTSCHKO. 1981. The development of sun compass orientation in young homing pigeons. Behav. Ecol. Sociobiol. 9:135–141.

WILTSCHKO R. AND W. WILTSCHKO. 1987. Pigeon homing: Olfactory experiments with inexperienced birds. Naturwiss. 74:94–95.

WILTSCHKO, R. AND W. WILTSCHKO. 1989. Pigeon homing: Olfactory orientation—a paradox. Behav. Ecol. Sociobiol. 24:163–173.

WILTSCHKO, W. 1968. Über den Einfluss statischer Magnetfelder auf die Zugorientierung der Rotkehlchen (*Erithacus rubecula*). Z. Tierpsychol. 25:537–558.

WILTSCHKO, W. 1978. Further analysis of the magnetic compass of migratory birds. In: K. Schmidt-Koenig and W. T. Keeton, eds., *Animal Migration, Navigation and Homing*. Springer-Verlag. Berlin, pp. 302–310.

WILTSCHKO, W. 1982. The migratory orientation of Garden Warblers *Sylvia borin*. In: F. Papi and H.-G. Walraff, eds., *Avian Navigation*. Springer-Verlag. Berlin, pp. 50–58.

WILTSCHKO W. AND E. GWINNER. 1974. Evidence for an innate magnetic compass in Garden Warblers. Naturwiss. 61:406.

WILTSCHKO, W. AND R. WILTSCHKO. 1972. Magnetic compass of European Robins. Science 176:62–64.

WILTSCHKO, W. AND R. WILTSCHKO. 1981. Disorientation of inexperienced young pigeons af-

ter transportation in total darkness. Nature (London) 291:433–434.

WILTSCHKO, W. AND R. WILTSCHKO. 1988a. Magnetic orientation in birds. In: R. F. Johnston, ed., *Current Ornithology*, vol. 5. Plenum Press. New York, pp. 67–121.

WILTSCHKO, W. AND R. WILTSCHKO. 1988b. Magnetic vs. celestial orientation in migrating birds. Trends Ecol. Evol. 3:13–15.

WILTSCHKO, W., P. DAUM, A. FERGENBAUER-KIMMEL, AND R. WILTSCHKO. 1987. The development of the star compass in Garden Warblers, *Sylvia borin*. Ethology 74:285–295.

WILTSCHKO, W., R. WILTSCHKO, AND C. WALCOTT. 1987. Pigeon homing: Different effects of olfactory deprivation in different countries. Behav. Ecol. Sociobiol. 21:333–342.

WILTSCHKO, W., R. WILTSCHKO, AND M. JAHNEL. 1987. The orientation behavior of anosmic pigeons in Frankfurt a.M. Germany. Anim. Behav. 35:1324–1333.

WILTSCHKO, W., R. WILTSCHKO, AND W. T. KEETON. 1976. Effects of a "permanent" clock-shift on the orientation of young homing pigeons. Behav. Ecol. Sociobiol. 1:229–243.

WILTSCHKO, W., R. WILTSCHKO, W. T. KEETON, AND R. MADDEN. 1983. Growing up in an altered magnetic field affects the initial orientation of young homing pigeons. Behav. Ecol. Sociobiol. 12:135–142.

WINGFIELD, J. C. 1984a. Environmental and endocrine control of reproduction in the song sparrow, *Melospiza melodia*. I. Temporal organization of the breeding cycle. Gen. Comp. Endocrinol. 56:406–416.

WINGFIELD, J. C. 1984b. Environmental and endocrine control of reproduction in the song sparrow, *Melospiza melodia*. II. Agonistic interactions as environmental information stimulating secretion of testosterone. Gen. Comp. Endocrinol. 56:417–424.

WINGFIELD, J. C. AND M. C. MOORE. 1987. Hormonal, social, and environmental factors in the reproductive biology of free-living male birds. In: D. Crews, ed., *Psychobiology of Reproductive Behavior*. Prentice Hall. Englewood Cliffs, NJ, pp. 148–175.

WINSTON, M. L. AND K. N. SLESSOR. 1992. The essence of royalty: Honey bee queen pheromone. Am. Sci. 80:374–385.

WISBY, W. J. AND A. D. HASLER. 1954. The effect of olfactory occlusion on migrating silver salmon (*O. kisutch*). J. Fish. Res. Bd. Can. 11:472–478.

WITTENBERGER, J. F. 1978. The evolution of mating systems in grouse. Condor 80:126–137.

WITTENBERGER, J. F. 1979. The evolution of mating systems in birds and mammals. In: P. Marler and J. G. Vandenbergh, eds., *Handbook of Behavioral Neurobiology*, vol. 3, Social behavior and communication. Plenum Press. New York, pp. 271–349.

WITTENBERGER, J. F. 1981. *Animal Social Behavior*. Duxbury Press. Boston. 722 pp.

WITTENBERGER, J. F. AND R. L. TILSON. 1980. The evolution of monogamy: Hypotheses and evidence. Ann. Rev. Ecol. Syst. 11:197–232.

WOLF, L. L. 1975. "Prostitution" behavior in a tropical hummingbird. Condor 77:140–144.

WOLFSON, A. 1948. Bird migration and the concept of continental drift. Science 108:23–30.

WOOD-GUSH, D. G. M. 1972. Strain differences in response to sub-optimal stimuli in the fowl. Anim. Behav. 20:72–76.

WOOLBRIGHT, L. L., E. J. GREENE, AND G. C. RAPP. 1990. Density-dependent mate searching strategies of male wood frogs. Anim. Behav. 40:135–142.

WOOLFENDEN, G. E. 1975. Florida scrub jay helpers at the nest. Auk 92:1–15.

WOOLFENDEN, G. E. AND J. W. FITZPATRICK. 1978. The inheritance of territory in group breeding birds. BioScience 28:104–108.

WOOLFENDEN, G. E. AND J. W. FITZPATRICK. 1984. *The Florida Scrub Jay, Demography of a Co-operative-Breeding Bird*. Princeton University Press. Princeton, NJ. 406 pp.

WOOLFENDEN, G. E. AND J. W. FITZPATRICK. 1990. Florida scrub jays: A synopsis after 18 years of study. In: P. B. Stacey and W. D. Koenig, eds., *Cooperative Breeding in Birds*. Cambridge University Press. Cambridge, pp. 241–266.

WOOLUM, J. C. 1991. A re-examination of the role of the nucleus in generating the circadian rhythm in *Acetabularia*. J. Biol. Rhythms 6:129–136.

WOURMS, M. K. AND F. E. WASSERMAN. 1985. Butterfly wing markings are more advantageous during handling than during the initial strike of an avian predator. Evolution 39:845–851.

WOYKE, J. 1964. Causes of repeated mating flights by queen honeybees. J. Apic. Res. 3:17–23.

WRANGHAM, R. D. 1980. Female choice of least costly males; a possible factor in the evolution of leks. Z. Tierpsychol. 54:357–367.

WRIGHT, W. G. 1988. Sex change in the mollusca. Trends Ecol. Evol. 3:137–140.

WU, J., J. M. WHITTIER, AND D. CREWS. 1985. Role of progesterone in the control of female sexual receptivity in *Anolis carolinensis*. Gen. Comp. Endocrinol. 58:402–406.

WYERS, E. J., H. V. S. PEEKE, AND M. J. HERZ. 1973. Behavioral habituation in invertebrates. In: H. V. S. Peeke and M. J. Herz, eds., *Habituation*, vol. 1, Behavioral studies. Academic Press. New York. pp. 1–57

WYLLIE, I. 1981. *The Cuckoo*. Batsford. London. 176 pp.

WYSOCKI, C. J., G. WHITNEY, AND D. TUCKER. 1977. Specific anosmia in the laboratory mouse. Behav. Gen. 7:171–188.

YAMAZAKI K., A. E. BOYSE, V. MIKE, H. T. THALER, B. J. MATHIESON, J. ABBOTT, J. BOYSE, Z. A. ZAYAS, AND L. THOMAS. 1976. Control of mating preferences in mice by genes in the major histocompatibility complex. J. Exp. Med. 144:1324–1335.

YAMAZAKI, K., M. YAMAGUCHI, E. A. BOYSE, AND L. THOMAS. 1980. The major histocompatibility complex as a source of odors imparting individuality among mice In: D. Müller-Schwartze and R.M. Silverstcin, eds., *Chemical Signals*. Plenum Press. New York, pp. 267–273.

YAMAZAKI, K., M. YAMAGUCHI, P. W. ANDREWS, B. PEAKE, AND E. A. BOYSE. 1978. Mating preferences of F$_2$ segregants of crosses between MHC-congenic mouse strains. Immunogenetics 6:253–259.

YASUKAWA, K. 1981. Male quality and female choice of mate in the red-winged blackbird (*Agelaius phoeniceus*). Ecology 62:922–929.

YASUKAWA, K., E. I. BICK, D. W. WAGMAN, AND P. MARLER. 1982. Playback and speaker replacement experiments on song-based neighbor, stranger, and self-discrimination in male Red-winged blackbirds. Behav. Ecol. Sociobiol. 10:211–215.

YDENBERG, R. C. 1984. The conflict between feeding and territorial defcnse in the great tit. Behav. Ecol. Sociobiol. 15:103–108.

YDENBERG, R. C. AND J. R. KREBS. 1987. The trade-off between territorial defense and foraging in the great tit (*Parus major*). Am. Zool. 27:337–346.

YEAGLEY, H. L. 1947. A preliminary study of a physical basis of bird navigation. J. Appl. Phys. 18:1035–1063.

YEAGLEY, H. L. 1951. A preliminary study of a physical basis of bird navigation. II. J. Appl. Phys. 22:746–760.

YLONEN, H., T. KOJOLA, AND J. VIITALA. 1988. Changing female spacing behavior and demography in an enclosed population of *Clethrionomys glareolus*. Holarct. Ecol. 11:286–292.

YNTEMA, C. L. 1981. Characteristics of gonads and oviducts in hatchlings and young of *Chelydra serpentina* resulting from three incubation temperatures. J. Morph. 167:297–304.

YORKE, E. 1981. Sensitivity of pigeons to small magnetic field variations. J. Theoret. Biol. 89:533–537.

YOUNG, D. 1989. *Nerve Cells and Behaviour.* Cambridge University Press. Cambridge. 236 pp.

YOUTHED, G. J. AND R. C. MORAN. 1969. The lunar day activity rhythm of myrmeleontid larvae. J. Insect Physiol. 15:1259–1271.

ZACH, R. 1978. Selection and dropping of whelks by Northwestern crows. Behaviour 67:134–148.

ZACH, R. 1979. Shell dropping: Decision making and optimal foraging in Northwestern crows. Behaviour 68:106–117.

ZAGOTTA, W. N., S. GERMERAAD, S. S. GARBER, T. HOSHI, AND R. W. ALDRICH. 1989. Properties of ShB A-type potassium channels expressed in *shaker* mutant *Drosophila* by germline transformation. Neuron 3:773–782.

ZAHAVI, A. 1974. Communal nesting by the Arabian babbler: A case of individual selection. Ibis 116:84–87.

ZAHAVI, A. 1975. Mate selection—a selection for a handicap. J. Theoret. Biol. 53:205–214.

ZAHAVI, A. 1977. Reliability in communication systems and the evolution of altruism. In: B. Stonehouse and M. C. Perrins, eds., *Evolutionary Ecology*. Macmillan. London, pp. 253–259.

ZAHAVI, A. 1980. Ritualization and the evolution of movement signals. Behaviour 72:77–81.

ZIMMERMAN, N. H. AND M. MENAKER. 1979. The pineal gland: A pacemaker within the circadian system of the house sparrow. Proc. Nat. Acad. Sci. USA 76:999–1003.

ZINK, R. M. AND G. F. BARROWCLOUGH. 1984. Allozymes and song dialects: A reassessment. Evolution 38:444–448.

ZIPPELIUS, H. 1972. Die Karawanenbildung bei Feld-und Hausspitzmaus. Z. Tierpsychol. 30:305–320.

ZOEGER, J., J. DUNN, AND M. FULLER. 1981. Magnetic material in the head of a common Pacific dolphin. Science 213:892–894.

ZUSNE L. AND B. ALLEN. 1981. Magnetic sense in humans? Percept. Mot. Skills 52:910.

ZWISLOCKI, J. J. 1981. Sound analysis in the ear: A history of discoveries. Am. Sci. 69:184–192.

Photo Credits

11.28: Toni Angermayer/Photo Researchers. Figure 11.31: Tom McHugh/Photo Researchers. **Chapter 12** Figure 12.1: Ross Hutchins/Photo Researchers. Figure 12.2: Francois Gohier/Ardea London. Figure 12.3: Catherine Craig/Sinauer Associates. Figure 12.4: Courtesy Dr. Robert Jackson, University of Canterbury, New Zealand. Figure 12.5: Ira Kirschenbaum/Stock, Boston. Figure 12.6: R. C. Hermes/Photo Researchers. Figure 12.7a): Tom McHugh/Photo Researchers. Figure 12.7b): Z. Leszczynski/Animals Animals. Figure 12.8a): Joe McDonald/Animals Animals. Figure 12.8b): *Scientific American*, May, 1973, R. Igor Gamow and John F. Harris. Figures 12.10 and 12.11: Tom McHugh/Photo Researchers. Figure 12.13a): Courtesy Barth Falkenberg, School of Biological Sciences, University of Nebraska, Lincoln. Figure 12.13b)(left): Courtesy Al Kamil, School of Biological Sciences, University of Nebraska. Figure 12.13b) (right): H. J. Vermes/Courtesy Ted Sargent. Figure 12.16: Kent & Donna Dannen/Photo Researchers. Figure 12.18: J. B. & S. Bottomley/Ardea London. Figure 12.19: Breck Kent/Animals Animals. Figure 12.20: Z. Leszczynski/Animals Animals. Figure 12.21: Runk Schoenberger/Grant Heilman Photography. Figure 12.22: Ray Richardson/Animals Animals. **Chapter 13** Figure 13.1: L. P. Brower, Dept. of Zoology, University of Florida at Gainesville. Figure 13.2: Robert and Linda Mitchell. Figure 13.8: Courtesy Dr. H. B. Kettlewell. Figure 13.9: H. J. Vermes/Courtesy Ted Sargent. Figure 13.11: Courtesy F. Harvey Pough. Figure 13.17: Courtesy T. E. Reimche, Dept. of Biology, University of Victoria. Figure 13.19: Leonard Lee Rue, from National Audubon Society/Photo Researchers. Figure 13.21: Anup & Manoj Sham/Animals Animals. Figure 13.24: Courtesy Dr. Edmund D. Brodie, Jr. , Dept. of Biology, University of Texas at Arlington. Figure 13.25: Courtesy Thomas Eisner & Daniel Aneshansley/Cornell University. **Chapter 14** Figure 14.1: R. M. Meadows/Peter Arnold. Figure 14.3 a): David Scharf. Figure 14.5 a): Walter Dawn/National Audubon Society/Photo Researchers. Figure 14.5 b): Courtesy Scripps Institute of Oceanography. Figure 14.8a): Wendell Metzen/Bruce Coleman. Figure 14.8 b): Dr. E. R. Degginger. Figure 14.9: Al Grotell. Figure 14.10: Omikron/Photo Researchers. Figure 14.11: Gordon S. Smith/Photo Researchers. Figure 14.12: Michael Fogden/DRK Photo. Figure 14.13: Howard Earl Uible/Photo Researchers. Figure 14.15: Courtesy Darryl T. Gwynne, Biology Dept., University of Toronto. Figure 14.16: H. Carl Germart. Figure 14.17: David Woodfall/NHPA. Figure 14.19 a): Hans Pfletschinger/Peter Arnold Figure 14.19 b): Jonathan Waage, Division of Biology & Medicine, Brown University. Figure 14.20: Irene Vandermolen/Photo Researchers. Figure 14.21: Clark & Handley/National Audubon Society/Photo Researchers. Figure 14.22: Courtesy Dr. Mark A. Bellis. Figure 14.23: S. J. Arnold, Dept. of Biology, University of Chicago. Figure 14.24: Richard Matthews/Planet Earth Pictures. Figure 14.25: Tom McHugh/Photo Researchers. Figure 14.27: Courtesy Stephen Goodenough. Figure 14.27: Eric Hosking. Figure 14.28: John Cancalosi/DRK Photo. **Chapter 15** Figure 15.1a) and b): Courtesy Marva Lee Wake, Dept. of Integrative Biology, University of California, Berkeley. Figure 15.2: Gordon S. Smith,National Audubon Society/Photo Researchers. Figure 15.7: Aaron Norman. Figure 15.8: Courtesy Robert L. Smith, Dept. of Entomology, University of Arizona. Figure 15.9: Courtesy Terrance Mace, University of Puget Sound Figure

15.10: Eric & David Hosking. Figure 15.11: Stephen Dalton/Photo Researchers. Figure 15.12: Al Lowry/Photo Researchers. Figure 15.13: C. Behnke/Animals Animals. Figure 15.15: Verna R. Johnston/Photo Researchers. Figure 15.17: David Overcash/Bruce Coleman. Figure 15.18: P. Ward/Bruce Coleman. Figure 15.19: Tom McHugh/Photo Researchers. **Part Three Opener**: S. Dalton/Animals Animals. **Chapter 16** Figure 16.1: Courtesy Dr. Simon Pollard, Dept. of Entomology, University of Alberta, Edmonton. Figure 16.3: Dr. Gary McCracken, Dept. of Zoology, University of Tennessee, Knoxville. Figure 16.6: Courtesy W. E. Bemis. Figure 16.11: Courtesy Dr. Betty McGuire, University of Massachusetts, Amherst. **Chapter 17** Figure 17.1: Pierre Berger/Photo Researchers. Figure 17.2: Anthony Bannister/NHPA. Figure 17.3b): Roy L. Caldwell, Dept. of Integrative Biology, University of California. Figure 17.6: Courtesy Joseph Waas, *Penguin Biology* by Lloyd Davis & John Darby, Academic Press, Inc., H.B.J., San Diego, 1990. Figure 17.9 a): Townsend P. Dickinson/Comstock. Figure 17.9b): Courtesy Dr. Frans B. M. de Waal, Yerkes Primate Center. Figure 17.10: Courtesy John D. Palmer. Figure 17.14: Jen and Des Bartlett/Photo Researchers. Figure 17.15: Dwight R. Kuhn. Figure 17.16: Toni Angermayer/Photo Researchers. Figure 17.18: Courtesy Stephen Goodenough. Figure 17.19: Runk/Schoenberger/Grant Heilman Photography. Figure 17.21: Courtesy Dr. Dietrich Schneider, Max-Planck Institute. Figure 17.22: Stimson Wilcox, SUNY, Binghamton. Figure 17.24: Arthur Ambler/National Audubon Society/Photo Researchers. Figure 17.25: Michael Moblick. Figure 17.26: Raymond A. Mendez/Animals Animals. Figure 17.31: Susan Kuklin/Photo Researchers. Figures 17.34 and 17.35: Courtesy Georgia State University & Yerkes/Emory. **Chapter 18** Figure 18.2: R. M. Meadows/Peter Arnold. Figure 18.3: S. Dalton/Animals Animals. Figure 18.5: Pierre Boulat,Paris. Figure 18.7 a): Courtesy Dr. Lynne D. Houck, Dept. of Ecology and Evolution, University of Chicago . Figure 18.8: Stephen Dalton/Photo Researchers. Figure 18.9: Walther Rohdich/Frank Lane Picture Agency. Figure 18.10: Eugene Gordon/Photo Researchers. Figure 18.11: Daniel Cox/Natural Selection. Figure 18.13: Photo Researchers. Figure 18.14: Gene Wolfsheimer/National Audubon Society/Photo Researchers. Figure 18.16: Fritz Siedel/Frank Lane Picture Agency. Figure 18.27: Mark Moffett/Minden Pictures, Inc. Figure 18.30: Art Gingert/Comstock. Figure 18.31: Roy P. Fontaine/Photo Researchers. **Chapter 19** Figure 19.1: Milton H. Tierney, Jr./Visuals Unlimited. Figure 19.2: From *Gerhard marcuse cichlid of Lake Malawi* by H. Axelrod & Warren Burgess, 1988, 12th ed., TFH Publishing, Neptune City, NJ. Figure 19.5: Kenneth W. Fink/Bruce Coleman. Figure 19.9: Richard R. Hansen/Photo Researchers. Figure 19.12: Courtesy Patricia D. Moehlman/Wildlife Conservation International. Figure 19.15: Tom J. Ulrich/Visuals Unlimited. Figure 19.17: M. K. & I. M. Morcombe/NHPA. Figure 19.18: Nigel Dennis/NHPA. Figure 19.21: Leonard Lee Rue III/Animals Animals. Figure 19.22: Ian Cleghorn/Photo Researchers. Figure 19.24: Leanne Nash. Figure 19.25: R. Williamson/Visuals Unlimited. Figure 19.28: Scott Camazine, Neurobiology and Behavior, Cornell University. Figure 19.29: Raymond A. Mendez/Animals Animals. **Appendix** Figure 2: Courtesy Professor Stephen Mann, School of Chemistry, University of Bath.

Author Index

Subject Index

Species Index

Sylvia:
  atricapilla, 326, See also, Black-cap
  borin, 326, 362 See also, Warbler, garden
Symphysodon discus, 500–501, See also, Cichlid, discus

Tachycineta bicolor, 482–483, See also, Swallow, tree
Tadarida brasiliensis, 334, 375, 529, See also, Bat, Mexican free-tailed
Taeniopygia guttata, 222, 230, 248, 250, 266, See also, Finch, zebra
Talitrus saltator, 350, See also, Beach hopper
Talorchestia: See also, Beach hopper
  latreillii, 352
  martensii, 350–351
Tamarin, 578–579
  rufous-naped, 565
  saddle-backed, 651
Tapir, South American, 618
Tapirus terrestris, 618, See also, Tapir, South American
Taricha torosa, 278, See also, Newt, California
Tayassu tajacu, 496, See also, Peccary, collared
Teleogryllus, 63, See also, Cricket
  commodus, 62
  oceanicus, 62
Termite, 204, 518, 619, 667, 673, 681
Tern:
  Arctic, 330, **336**
  common, 610
Tetragnatha foliferens, 422
Thamnophis sirtalis parietalis 228–230, See also, Garter snake, red-sided
Thecla togarna, 431
Thrip, 667
Thryomanes bewickii, 139, See also, Wren, Bewick's
Thunnus albacares, 686, See also, Tuna, yellowfin
Tiger, 60
Timalidae, 629, *also,* See Babbler
Tit, great, 400, 409, 538–539, 583–584, 614
Titmouse, 396
  plain, 602
Toad, 113, 182–183, 185–188, 439–440, 442, 570, 598, 607, 613, 617
  American, 599
  common, 185–188
  marine, 237
  natterjack, 474

Treecreeper, short-toed, 602
Treefrog, 599
Trichogaster trichopterus, 131, See also, Gourami, blue
Tritonia, 358, See also, Snail, marine
Triturus vulgaris, 552, See also, Newt, smooth
Trout, rainbow, 407, 421–422
Tuna, 375
  yellowfin, 686
Turdoides squamiceps, 653, See also, Babbler, Arabian
Turkey, 128, 634, 662–663, 680
Turtle, 329, 382, 489, 687
  alligator snapping, 382
  green, 328, 330
  loggerhead, **333,** 358–359
  musk, 427
  painted, 230
  sea, 333, 359
  snapping, 459, 460
Tylos: See also, Isopod
  latreillii, 347–348
  punctatus, 347–348
Tyto alba, 170–172, 645–646, See also, Owl, barn

Uca: See also, Crab, fiddler
  annulipes, 598
  beebei, 566
  pugilator, 598
  pugnax, 289
  rhizophorae, 598

Vespula, 427, See also, Wasp
Vireo, Red-eyed, 599
Vole, 58, 488, 539–544
  bank, 541–542
  gray-sided, 540–544
  meadow 58, 232–233, 543
  pine, 232–233, 544
  prairie, 58, 504, 542, 544
  red-backed, 542

Walrus, 336
Warbler, 335, 348
  garden, 326, 350, 362
  prairie, 510
  willow, 296
Wasp, 166, 181, 198, 204, 427, 429, 600, 671, 673, 680
  Mischocyttarus mexicanus, 673, 674
  paper, 674
  digger, 108–110, 135–136, 340
  parasitic, 93

Water bug, giant, 442, 502–503
Water moccasin, 385
Water scavenger, 383
Water strider, 105, 574–575
Waxbills, 629
Weasel, long-tailed, 647
Weaver bird, 101, 102
Whale, 15, 214, 360, 382
  baleen, 380
  finback, 581
  gray, 330, 336
  humpback, 336, 382, 382
  toothed, 375
Whelk, 397–399, 409
White eyes, 629
Widow bird, long-tailed paradise, 21-**22**
Wildebeast, 330, 334, 380, 532
Wolf, 428, 548, 601–602
Woodmice, European 360
Woodpecker, 567
  acorn, 654, 655–656, 658
  flicker, 19
Worm, 295, 400, 409, 458–459, 489
  Samoan palolo, 293, 295
Wrasse, Pacific cleaner, 384–386, 461–462
Wren:
  Bewick's, 139
  marsh, 252
  splendid fairy-, 656

Xanthocephalus xanthocephalus, 566, See also, Blackbird, yellow-headed
Xenopus laevis, 223–224, 278 See also, Frog, clawed
Xysticus cristatus, 608, See also, Spider, crab

Yellowlegs, lesser, 330

Zebra, 413, 557
Zonotrichia:
  albicolis, 336, 348, See also, Sparrow, white-throated
  capensis, 585, See also, Sparrow, South American rufous-collared
  leucophrys, 250, 312, 602, See also, Sparrow, white-crowned
Zootermopsis nevadensis, 619, See also, Termite
Zosteropidae, 629, See also, White eyes

Common and Scientific Names for Selected Species Discussed in the Text

Alligator, American	*Alligator mississippiensis*
Anemonefish, Nosestripe	*Amphiprion akallopisos*
Anole, Green	*Anolis carolinensis*
Ant Lion	*Myrmeleon obscurus*
Ant, Acacia	*Pseudomyrmex ferruginea*
Ant, African Weaver	*Oecophylla longinoda*
Ant, Desert	*Cataglyphis bicolor*
Ant, Fire	*Solenopsis saevissma*
Ant, Slave-making	*Harpagoxenus sublaevis*
Ant, Slave-making	*Leptothorax duloticus*
Axolotl, Mexican	*Ambystoma mexicanum*
Babbler, Arabian	*Turdoides squamiceps*
Babbler, Gray-crowned	*Pomatostomus temporalis*
Baboon, Hamadryas	*Papio hamadryas*
Baboon, Olive	*Papio anubis*
Backswimmer	*Notonecta glauca*
Bat, Big Brown	*Eptesicus fuscus*
Bat, Fishing	*Noctilio leporinus*
Bat, Hammer-headed	*Hypsignathus monstrosus*
Bat, Mexican Free-tailed	*Tadarida brasiliensis*
Bat, Vampire	*Desmodus rotundus*
Beau Gregory	*Eupomacentris leucostictus*
Bee, Sweat	*Lasioglossum zephrum*
Bee-eater, European	*Merops apiaster*
Bee-eater, White Fronted	*Merops bullockoides*
Beetle, Rove	*Leistotrophus versicolor*
Bird of Paradise	*Lophorina superba*
Blackbird, Red-winged	*Agelaius phoeniceus*
Blackbird, Yellow-headed	*Xanthocephalus xanthocephalus*
Blackcap	*Sylvia atricapilla*
Blowfly	*Calliphora vicina*
Bluebird, Eastern	*Sialia sialis*
Bluebird, Mountain	*Sialia currucoides*
Bluethroat	*Luscinia svecica*
Bobolink	*Dolichonyx oryzivorus*
Booby, Blue-footed	*Sula nebouxii*
Bowerbird, Satin	*Ptilonorhyncus violaceus*
Bug, Assassin	*Platymeris rhadamantus*
Bunting, Indigo	*Passerina cyanea*
Butterfly, Monarch	*Danaus plexippus*
Butterfly, Speckled Wood	*Pararge aegeria*
Butterfly, White (or Cabbage)	*Peiris rapae*
Caiman	*Caiman crocodilus*

Canary	*Serinus canarius*
Cat, Domestic	*Felis catus*
Catfish, Yellow Bullhead	*Ictalurus natalis*
Cheetah	*Acinonyx jubatus*
Chicken	*Gallus domesticus*
Chimpanzee	*Pan troglodytes*
Chimpanzee, Pygmy	*Pan paniscus*
Chough, White-winged	*Corcorax melanorhamphos*
Cichlid, Convict	*Cichlasoma nigrofasciatum*
Clamworm	*Nereis pelagica*
Cockatoo, Sulfur-crested	*Kakatoe galerita*
Cockroach, American	*Periplaneta americana*
Cowbird, Brown-headed	*Molothrus ater*
Coyote	*Canis latrans*
Crab, Fiddler	*Uca annulipes*
Crab, Fiddler	*Uca pugilator*
Crab, Fiddler	*Uca pugnax*
Crab, Fiddler	*Uca rhizophorae*
Crab, Spider	*Inachus phalangium*
Cricket, Field	*Gryllus bimaculatus*
Cricket, Field	*Gryllus integer*
Crow, Carrion	*Corvus corone*
Crow, Northwestern	*Corvus caurinus*
Cuckoo, European	*Cuculus canorus*
Curlew, Bristle-thighed	*Numenius tahitiensis*
Cuttlefish	*Sepia offinalis*
Damselfish, Humbug	*Dascyllus aruanus*
Deer, Black-tailed	*Odocoileus hemionus columbianus*
Deer, Red	*Cervus elaphus*
Deer, Sitka	*Cervus nippon*
Dickcissel	*Spiza americana*
Discus	*Symphysodon discus*
Dog	*Canis familiaris*
Dogwhelk	*Nucella lapella*
Duck, Mallard (and Peking)	*Anas platyrhynchos*
Duck-billed Platypus	*Ornithorhynchus anatinus*
Eel, American	*Anguilla anguilla*
Fighting Fish, Siamese	*Betta splendens*
Finch, Bengalese	*Lonchura striata*
Finch, Bicheno	*Poephila bichenovi*
Finch, Cutthroat	*Amadina fasciata*
Finch, Zebra	*Taeniopygia guttata*
Flicker, Common	*Colaptes auratus*
Fly, Fruit	*Drosophila melanogaster*
Fly, Fruit	*Drosophila pseudoobscura*
Fly, House	*Musca domestica*
Fly, Seaweed	*Coelopa frigida*
Flycatcher, Pied	*Ficedula hypoleuca*
Frog, California Yellow-legged	*Rana muscosa*
Frog, Cascades	*Rana cascadae*
Frog, Clawed	*Xenopus laevis*
Frog, Cricket	*Acris crepitans*
Frog, Southern Cricket	*Acris gryllus*
Frog, White-lipped	*Leptodactylus albilabris*
Frog, Wood	*Rana sylvatica*
Gannet	*Sulsa bassana*
Gazelle, Grant's	*Gazella granti*
Gazelle, Thomson's	*Gazella thomsoni*
Gecko, Leopard	*Eublepharis macularius*
Goat	*Capra hircus*

Goose, Bar-headed	*Anser indicus*
Goose, Greylag	*Anser anser*
Gorilla	*Gorilla gorilla*
Goshawk	*Accipter gentilis*
Gourami, Blue	*Trichogaster trichopterus*
Great tit	*Parus major*
Grebe, Great Crested	*Podiceps cristatus*
Grosbeak, Black-headed	*Pheucticus melanocephalus*
Ground Squirrel, Belding's	*Spermophilus beldingi*
Ground Squirrel, Golden-mantled	*Citellus lateralis*
Ground Squirrel, Round-tailed	*Spermophilus tereticaudus*
Ground Squirrel, Thirteen-lined	*Spermophilus tridecemlineatus*
Grunion	*Leuresthes tenuis*
Gull, Blackheaded	*Larus ridibundus*
Gull, Galapagos swallow tailed	*Larus furcatus*
Gull, Herring	*Larus argentatus*
Gull, Laughing	*Larus atricilla*
Gull, Ring-billed	*Larus delawarensis*
Gull, Swallow-tailed	*Larus furcatus*
Guppy	*Poecilia reticulata*
Hamster, Golden	*Mesocricetus auratus*
Harrier, Northern	*Circus cyaneus*
Hawk, Harris'	*Parabuteo unicinctus*
Hawkmoth	*Laothoe populi*
Heron, Gray	*Ardea cinerea*
Honeybee	*Apis mellifera*
Horned Toad	*Phrynosoma cornutum*
Hornworm, Tobacco	*Manduca sexta*
Hummingbird, Rufous	*Selasphorus rufus*
Ibis, Bald	*Geronticus eremitta*
Jacana, Northern	*Jacana spinosa*
Jackal, Black(or Silver)backed	*Canis mesomelas*
Jay, Blue	*Cyanocitta cristata*
Jay, Canada	*Perisoreus canadensis*
Jay, Pinyon	*Gymnorhinus cyanocephalus*
Jay, Scrub	*Aphelocoma coerulescens*
Junco, Dark-eyed	*Junco hyemalis*
Kangaroo rat, Banner-tailed	*Dipodomys spectabilis*
Killdeer	*Charadrius vociferus*
Kingfisher, Pied	*Ceryle rudis*
Kittiwake	*Rissa tridactyla*
Knife fish, African	*Gymnarchis niloticus*
Knife fish, South American	*Eigenmannia virescens*
Ladybird beetle, Two spot	*Adalia bipunctata*
Lamprey, Brook	*Lampetra planeri*
Langurs, Hanuman	*Presbytis entellus*
Leech, European	*Hirudo medicinalis*
Lion	*Panthera leo*
Liver Fluke	*Fasciola hepatica*
Lobster, Spiny	*Panulirus argus*
Locust, Desert	*Schistocerca gregaria*
Locust, Migratory	*Locusta migratoria*
Lovebird, Masked	*Agapornis personata*
Macaque, Stumptail	*Macaca arctoides*
Manakin, Long-tailed	*Chiroxiphia linearis*
Mantis, Praying	*Mantis religiosa*
Marmot, Yellow-bellied	*Marmota flaviventris*
Meerkat	*Suricata suricatta*
Mole-rat, Cape	*Georychus capensis*
Mole-rat, Damara	*Cryptomys damarensis*

Mole-rat, Naked	*Heterocephalus glaber*
Mongoose, Dwarf	*Helogale parvula*
Mongoose, Dwarf	*Helogale undulata rufula*
Monkey, Rhesus	*Macaca mulatta*
Monkey, Spider	*Ateles paniscus*
Monkey, Squirrel	*Saimiri sciureus*
Monkey, Vervet	*Cercopithecus aethiops*
Moth, Peppered	*Biston betularia*
Moth, Silk	*Bombyx mori*
Moth, Underwing	*Catocala cerogama*
Moth, Underwing	*Catocala retecta*
Moth, Yellow Underwing	*Noctua pronuba*
Mouflon	*Ovis ammon*
Mouse, Deer	*Peromyscus maniculatus*
Mouse, House	*Mus musculus*
Mouse, Spiny	*Acomys cahirinus*
Mouse, White-footed	*Peromyscus leucopus*
Newt, California	*Taricha torosa*
Newt, Eastern Red Spotted	*Notophthalmus viridescens*
Newt, Smooth	*Triturus vulgaris*
Nutcracker, Clark's	*Nucifraga columbiana*
Octopus	*Octopus cyanea*
Opossum, South American	*Didelphis marsupialis*
Opossum, Virginia	*Didelphis virginiana*
Orangutan	*Pongo pygmaeus*
Oriole, Black-backed	*Icterus balbula*
Owl, Barn	*Tyto alba*
Owl, Pygmy	*Glaucidium perlatum; G. brasilianum*
Owl, Screech	*Otus asio*
Oystercatcher	*Haematopus ostralegus*
Paddlefish, American	*Polyodon spathula*
Parrot, African Grey	*Psittacus erithacus*
Peccary, Collared	*Tayassu tajacu*
Penguin, Little Blue	*Eudyptula minor*
Perch, Yellow	*Perca fluviatilis*
Pigeon (or Rock Dove)	*Columba livia*
Pika, Collared	*Ochotona collaris*
Pike, Northern	*Esox lucius*
Plaice	*Pleuronectes platessa*
Pronghorn	*Antilocapra americana*
Quail, Bobwhite	*Colinus virginianus*
Quail, Japanese	*Coturnix coturnix japonica*
Rat, Brown (or Roof)	*Rattus rattus*
Rat, Norway	*Rattus norvegicus*
Redshank	*Tringa totanus*
Robin, European	*Erithacus rubecula*
Salamander, Mountain Dusky	*Desmognathus ochrophaeus*
Salmon, Atlantic	*Salmo salar*
Salmon, Chinook	*Oncorhynchus ishawytscha*
Salmon, Coho	*Oncorhynchus kisutch*
Salmon, Sockeye	*Oncorhynchus nerka*
Sandpiper, Spotted	*Actitis macularia*
Scorpion, Sand	*Paruoctonus mesaensis*
Sculpin, Mottled	*Cottus bairdi*
Sea hare, California	*Aplysia californica*
Seal, Northern Elephant	*Mirounga angustirostis*
Seal, Southern Elephant	*Mirounga leonina*
Shark, Sand	*Carcharias taurus*
Shearwater, Greater	*Puffinus gravis*
Sheep, Rocky Mountain Bighorn	*Ovis canadensis*

Shrew, Musk	*Suncus murinus*
Shrimp, Grass	*Palaemonetes vulgaris*
Silverside	*Menidia menidia*
Skunk, Spotted	*Spilogale putorius*
Skunk, Striped	*Mephitis mephitis*
Snail, Banded	*Cepaea nemoralis*
Snake, Hognose	*Heterodon platirhinos*
Snake, Northern Water	*Natrix sipedon*
Snake, Red-sided Garter	*Thamnophis sirtalis parietalis*
Sparrow, House	*Passer domesticus*
Sparrow, Java	*Padda oryzivora*
Sparrow, Rufous-collared	*Zonotrichia capensis*
Sparrow, Savannah	*Passerculus sandwichensis*
Sparrow, Song	*Melospiza melodia*
Sparrow, Tree	*Spizella arborea*
Sparrow, White-crowned	*Zonotrichia leucophrys*
Sparrow, White-throated	*Zonotrichia albicolis*
Spider, Bowl and Doily	*Frontinella pyramitela*
Spider, Crab	*Xysticus cristatus*
Spider, Funnel Web Building	*Agelenopsis aperta*
Spider, Jumping	*Portia fimbriata*
Spider, Wolf	*Arctosa variana*
Squirrel, Flying	*Glaucomys volans*
Squirrel, Fox	*Sciurus niger*
Starling, European	*Sturnus vulgaris*
Stickleback, Three-spined	*Gasterosteus aculeatus*
Sunbird, Golden-winged	*Necterinia reichenowi*
Sunfish, Bluegill	*Lepomis macrochirus*
Swallow, Bank	*Riparia riparia*
Swallow, Cliff	*Hirundo pyrrhonota*
Swallow, Tree	*Tachycineta bicolor*
Swiftlets, White-rumped	*Aerodramus spodiopygius*
Tamarin, Rufous-naped	*Saguinus geoffroyi*
Tamarin, Saddle-backed	*Saguinus fuscicollis*
Tapir, South American	*Tapirus terrestris*
Tern, Arctic	*Sterna paradisaea*
Tern, Common	*Sterna hirundo*
Toad, American	*Bufo americanus*
Toad, Common	*Bufo bufo*
Toad, Marine	*Bufo marinus*
Toad, Natterjack	*Bufo calamita*
Treefrog, Gray	*Hyla versicolor*
Trout, Rainbow	*Salmo gairdneri*
Tuna, Yellowfin	*Thunnus albacares*
Turkey, Wild	*Meleagris gallopavo*
Turtle, Green	*Chelonia mydas*
Turtle, Loggerhead	*Caretta caretta*
Turtle, Musk	*Sternotherus odoratus*
Turtle, Painted	*Chrysemys picta*
Turtle, Snapping	*Chelydra serpentina*
Vole, Bank	*Clethrionomys glareolus*
Vole, Gray-sided	*Clethrionomys rufocanus*
Vole, Meadow	*Microtus pennsylvanicus*
Vole, Pine	*Microtus pinetorum*
Vole, Prairie	*Microtus ochrogaster*
Warbler, Garden	*Sylvia borin*
Warbler, Prairie	*Dendroica discolor*
Warbler, Willow	*Phylloscopus trochilus*
Wasp, Digger	*Philanthus triangulum*
Wasp, Digger	*Sphex ichneumoneus*

Wasp, Paper	*Polistes metricus*
Water Bug, Giant	*Abedus herberti*
Water Bug, Giant	*Lethocerus americanus*
Water Scavenger	*Hydrochara obtusata*
Whale, Gray	*Eschrichtius robustus*
Whale, Humpback	*Megaptera novae-angliae*
Widow Bird, Long-tailed paradise	*Steganura paradisaea*
Wild Dog, African	*Lycaon pictus*
Woodmouse, European	*Apodemus sylvaticus*
Woodpecker, Acorn	*Melanerpes formicivorus*
Wrasse, Pacific cleaner	*Labroides dimidiatus*
Wren, Bewick's	*Thryomanes bewickii*
Wren, Marsh	*Cistothorus palustris*
Wren, Splendid fairy	*Malurus splendens*